Metals: Metallurgy and Applications

Metallurgy Training and Self-Study

Metals: Metallurgy and Applications

Metallurgy Training and Self-Study

RON SCOTT

FREng., BSc., C.Eng., FIMMM

Typeset in Minion

Editing, design, typesetting and publishing by UK Book Publishing
www.ukbookpublishing.com

ISBN: 978-1-913179-26-7

Contents

Detailed Contents

CHAPTER 7

Corrosion 203

CHAPTER 11

Prediction of Micro-Structural Evolution in Steel   442

Preface

Throughout history there has always been a common theme within the business environment that it is important to understand that the commercial aspects of business can represent a serious challenge and that there are always several companies competing for the same contract, creating a "win or lose" situation. To succeed as a business, or as a community or as a nation, there is a need to encourage creativity, team spirit and to nurture the difficult to define need for "the ability to generate innovative ideas". Equally important, companies must also be backed by a competent and motivated workforce. Competency can only be achieved by appropriate training and by allowing time and providing encouragement for self-study and therefore, it is important for a business to create a working environment where competency can develop.

Businesses compete in an economic climate where only the best prosper, almost equating to a Darwinian "survival of the fittest". The impact of the notion of "the survival of the fittest" in the commercial environment was uniquely described by Nichols and Pye [Pye, A. (1992), "Getting your process right." Engineering, September 1992, p14].

> "Two industry executives were in the African bush, discussing the relative performance of their companies, when a ravenous lion appeared on the horizon. "Run for it," said one executive. The other took a pair of running shoes from his rucksack and bent down to put them on. "What are you doing?" said the first executive. "You can't out-run the lion." "I don't need to out-run the lion," said the kneeling businessman, "I need to out-run you."

There are many examples where poorly engineered or poorly manufactured products have resulted in failure and expensive settlements. It is the responsibility of a company to ensure that their products do not fail. Failures often raise issues of competency of the people involved and necessitates acknowledgement of the often specialist "knowledge" requirements and training needs that have to be met to ensure that the company can deliver quality products and services. Engineering judgement is an important factor in the design and manufacture of reliable parts and products. It requires engineering staff with the ability to acquire and appraise data to enable clear decision making, to access and exploit information technology and to participate in continuing professional development to broaden and enhance their expertise.

This document has been prepared to assist in the provision of metallurgical training which is essential in all metal-related businesses. The importance of metallurgical training has also been identified by the 2014 European roadmap for Metallurgy.

This is not a comprehensive documentation of metallurgical theory and practice, but a simple and basic introduction to the salient aspects which are essential to achieve a level of awareness and understanding in metallurgy to assist with the control of the quality of product, solving problems and preventing errors. The approach covers insight into the timeline of key concepts and discoveries that guided the science, the basic methods to control properties and the important effect that microstructure has on the final product's mechanical properties together with a review of mechanical testing, heat treatment, fracture, welding and corrosion.

In the UK it is no longer business as usual. The current political climate has forced Brexit "for better or for worse". However, we all remember the UK success at the 2016 Olympics which demonstrated the effect generous lottery funding used to finance training can have on success. There is hope that if similar priority can be given to the next generation of engineers and technicians and funding can be provided for technical training, there will be the ability and tenacity in today's generation of young engineers to gain top ranking in the world and to provide the creativity, team spirit and "innovation" that the country will need.

Ron Scott April 2019

CHAPTER 1

Background to Key Events and Developments

1.1 INTRODUCTION

Steel has been available as a high tonnage engineering material for over 160 years; since the introduction of the Bessemer process around the mid-1850s, the Open-Hearth Furnace (Siemens-Martin) in 1863 and the Electric Arc Furnace in 1878 (Siemens patent). During the intervening period of time the unique advantages of steel compared to other competing materials resulted in steel achieving "fitness for service" for several important applications where iron and steel has been established as the material of choice. The unique advantages of iron and steel include:

- The availability of high-quality iron ores and the ancient discovery of the ability to reduce the iron ore to iron using charcoal initially then coke and limestone in a blast furnace from 1750 onwards
- The diversity in methods of manufacture (Cast, forged, powder metallurgy, additive manufacture, rolled long and flat products, fabricated by welding, machined, heat treated and surface engineered)
- The range of strengths and toughness that can be achieved (strengths from 150MPa to 2000MPa) and the ability to change the properties by heat treatment (the allotropic transformation)
- It's a cost-effective material in many applications and can be recycled.

The main applications include products which form the basic infrastructure of the modern world:

- Military – armour plate, projectiles, missile bodies, gun barrels
- Shipbuilding, full rail infrastructure, aircraft landing gear

- Automotive vehicles, earthmoving vehicles, implements of agriculture
- Structural applications, bridges, buildings, concrete reinforcement
- Industrial process plant for oil and gas pipelines and storage tanks.

During the past 50 years the steel industry has gone through a period of change. This has involved a substantial downsize in the UK steel industry with a considerable increase in capacity in the Far East. China achieved an output of 808 million tonnes of steel in 2016, which is almost equal to the total world steel output in the year 2000 of 840 million tonnes. Fifty years ago, the UK steel industry was led by a major enterprise known as the British Steel Corporation which employed over 320,000 people and produced over 25 million tonnes of steel per year and was ranked 2nd in Europe. The British Steel Corporation was formed in 1967 by nationalisation of the largest 12 private companies. In 2016 the UK steel output was 8 million tonnes. The industry employed 32,000 people and the UK was ranked 7th in Europe.

My career began in the steel industry around 1970 in North West County Durham near where I was born. I was employed in the nationalised works of the company "Consett Iron Company" (CIC). CIC began in the 1840s using local ores and coal and was closed by the British Steel Corporation in 1980 with complete demolition. Throughout its corporate history CIC was one of the more enlightened steel manufacturers. In the 1880s it made the correct choice of open hearth and basic Bessemer process adding up to 10 open hearth furnaces and making substantial profits manufacturing plate steel for shipbuilding (see Table 1.1).

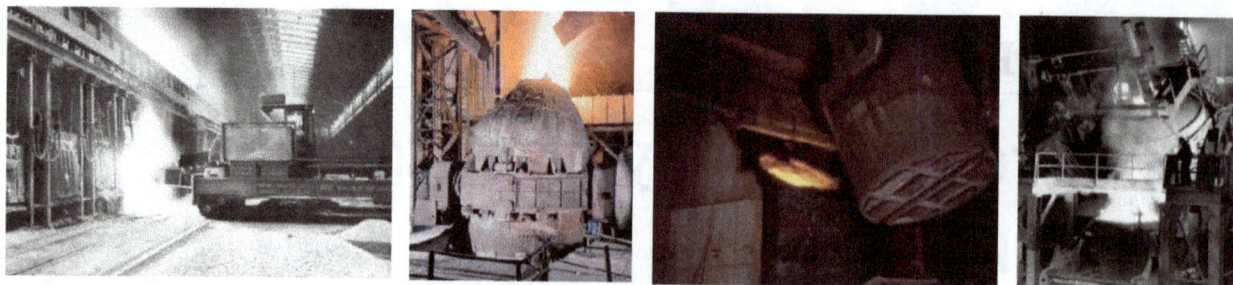

Steelmaking: LHS = The older – Obsolete, Open Hearth Siemens-Martin and Centre=Bessemer, RHS = The newer – BOS and Electric

Year	Net Profit £'000s	Profit on Capital %
1865	39	12
1870	102	24.5
1875	215	33.7
1880	104	15.6
1885	60	8.6
1890	366	38.6
1895	115	8.6
1900	673	38.7
1905	245	13.6
1910	221	12.4

Table 1.1 *Profits of Consett Iron Company in the late 19th and early 20th century, RHS = Thomas converter (Basic Bessemer invented by Sidney Gilchrist Thomas). World wide there are several converters erected as monuments to the early age of steel. This vessel is at Phoenix Lake, Dortmund, Germany. Another can be seen at Kawasaki City Museum, which is located in Todoroki Ryokuchi Park*

Source – Enterprise and Technology, Ulrich Wenenroth, 1994 ISBN 978-0-521-38425-4).

In the 1960s when the economics of steel making technology changed again, CIC moved into BOS (Basic Oxygen Steelmaking) with a cycle time of 45 mins compared to 8 hrs for an open-hearth furnace. If it had added an Electric furnace, Vacuum Degas and Continuous Casting it would have survived. One technical book that survived the shutdown was prepared by the CIC R&D director (Dr T F Pearson) during the 1960s and outlines the development of steels for low temperature service which also describes the steel development work that he carried out in CIC.[1]

Working in the Steel Industry provided a unique practical introduction to iron and steel technology and provided many memories. One amusing tale related to the Consett Iron Company was:

a technical person had reviewed a report and had put a comment "spherical objects" against one paragraph that he could not agree with. The report was then passed to the managing director, Mr George. The report was returned with the comments – who is Mr Spherical and what does he object to!

In 1979 I was awarded a travel fellowship by the Metals Society which allowed me to visit 10 European Steel Works to study the manufacture of high carbon steel. All the works I visited were financially stable and showed no indications of lack of business or threats of closure. One works was still operating a steam driven blooming mill. I returned to the UK to learn that I was to be made redundant due to the closure of Consett Works.

There are several nostalgic videos of Consett Iron Company showing the works and the community. The videos are all positive and none predicted the future plant closure and the resultant 35% unemployment in the area.[2, 3] When I worked in Korea in 1979, at POSCO Pohang, a Korean director used the closure of Consett Works as part of a motivational talk to his staff to encourage them to work hard or face a similar future. (The lack of hardworking people was not the reason for the downfall of BSC!) When Consett Steel Works closed I left the steel industry and decided to stay in the area and established Metaltech Ltd in Consett. The main business areas were heat treatment, metallurgical consultancy and engineering failures.

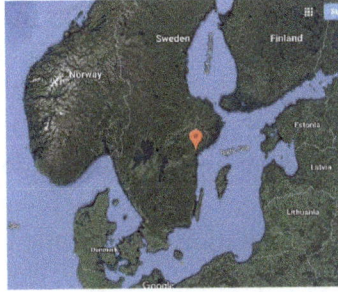

◄ *Steel plants where I have worked: Oxelosund, Sweden; Consett, England; POSCO Pohang, Korea*

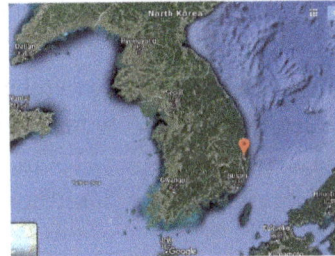

▼ **Figure 1.1** *Consett Works, Top LHS = View of the BOS steel plant Top RHS and centre = Blast furnaces ironmaking. Lower LHS = General Offices Centre = Sinter plant, Lower RHS = power station*

1.1.1 The Decline of The UK Steel Industry

Figures 1.2a and 1.2b show some data relating to the decline of the UK Steel Industry from a recent briefing paper for the House of Commons.

What was the root cause for the decimation of the UK steel Industry? The reasons were partly due to the market forces but also an inheritance of the bad decisions made by government in the 1970s. In government ownership and based on a parliamentary debate it was decided to invest £3 billion to expand the industry from 20Mt per year to 30Mt per year. After spending the investment money, the market share had dropped to 12Mt per year, partly due to imports which were transported into the UK at a cost of £25 per tonne. With the burden of the interest repayment, the steel industry went into a downward spiral!

In addition, the expansion plan included the building of a large blast furnace in Teesside to produce 10,000 tonnes of iron per day. To create an order book for Teesside to allow the large blast furnace to operate there were several compulsory closures.

(This included Consett and Ravenscraig.) The cash crises fragmented the steel industry and even after the massive debt write-off by the British government followed by privatisation, it still struggled and continued downhill. Relieved of debt it was re-privatised in 1988 as the British Steel Corporation. It subsequently became Corus after a merger with a Dutch rival, Hoogovens, and in 1999, Corus was purchased by India's Tata Steel in 2006.

The more recent events include the closure of Iron and Steelmaking on Teesside (2015); the sale of Scunthorpe Works by Tata to Greybull Capital, the investment office of the Meyohas and Perlhagen families, for a nominal sum of £1 (2016) who revived the name "The British Steel Corporation". Unfortunately, this turned out to be only a temporary solution and on the, 21st May 2019 British Steel announced that it would enter into administration. After being run by the administrator for over 5 months, on the 11th November 2019 it was announced that British Steel would be sold to the Chinese firm Jingye, for a sum of £50m. This deal would prevent the closure of the Scunthorpe site and would safeguard 4,000 jobs.

The rise and fall of British Steel

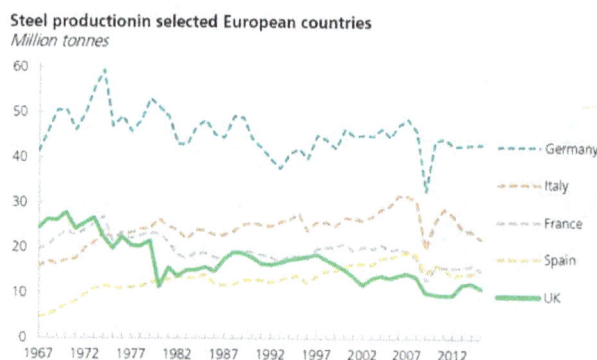

Politicians are active

Figure 1.2a LHS = Steel output of European Countries RHS = Briefing paper on UK steel industry[4]

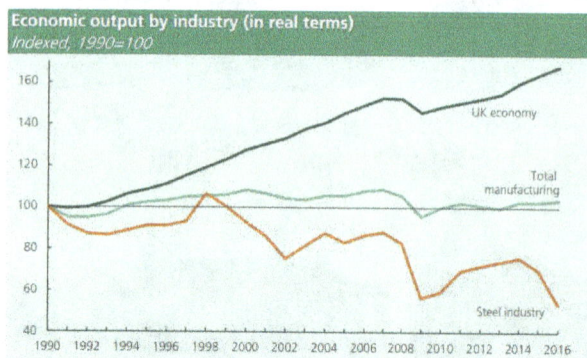

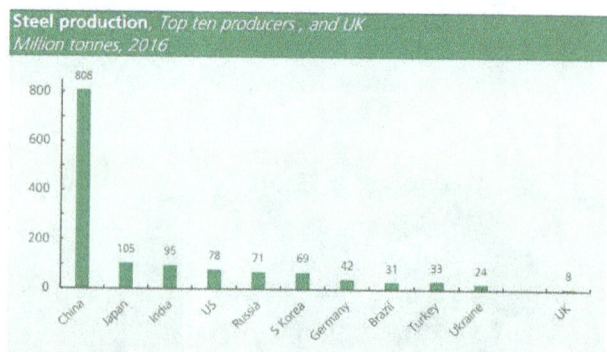

Figure 1.2b LHS = Steel industry output, RHS = Top ten producers and the UK 2016[4]

Jingye promised to invest around £1.2bn in British Steel over the next decade.

The Liberty House Group acquired the Dalzell Plate Mill (2016), Hartlepool Pipe Mill (2017), and the Rotherham and Stocksbridge Speciality Steels including arc steel making and rolling mills for £100M (2017). The remains of Tata, which included Port Talbot in the UK, tried to merge with Thyssenkrupp (2019) but the European Commission prohibited the proposed transaction. In October 2019 Tata announced that it would make cost savings of over £900m by removing 2500 jobs in its steelplants in Europe and planned to keep Port Talbot open. In the same month ThyssenKrupp filed a complaint with an EU court against the European Commission for blocking its plan to merge with Tata Steel.

The concept of the nationalisation adopted by government in the 1970s was based on the premise that private companies could not find the substantial sources of finance that were essential to develop a world class steel industry. This was long forgotten during the era of privatisation in the eighties. One consequence of the shrinkage of the UK metals industry has been that the UK's technical compentancy in metallurgy has shrunk and we are left reading about steel innovations associated with steel made and used abroad. The USA still regard their steel industry as a strategic defence industry and carried out a study to compare the defence potential of the remains of the USA steel industry after the influx of foreign ownership compared with China's burgeoning steel empire.[5]

Associated with the downward movement in steel activity in the UK (320,000 employed 1967, 32,000 in 2016) and the decline of other activities, there was a concurrent reduction in the skills and knowledge of steels and other materials. Currently the service sector dominates economic activity, in the UK, and accounts for 84% of the employment in the UK. The rise of this sector has been at the expense of jobs in manufacturing.

During the same period of time the number of university courses for metallurgy decreased and the understanding of many aspects of metals technology has been lost and some aspects of the skills and knowledge are now probably irrecoverable.

An excellent historical perspective is given in the book "British Iron and Steel AD1800-2000 and Beyond" (IOM 2001 ISBN 186125119X)[6]. In particular, pages 194 and 195 in the paper by Terry Gladman and Brian Pickering prepared around 2001 with their final comment,

"The steel industry and steel itself has been rather denigrated in the last twenty years or so, and this is nothing new. But it does lead to some loss of confidence and people involved in the industry must not lose faith in the viability of this pre-eminent engineering material. The steel industry has risen to challenges in the past and no doubt will do so again successfully. But we must not delude ourselves in thinking the changes required, indeed which are essential, will be easy."

1.1.2 The Sudden Decline of The UK Aluminium Industry

Within a period of three years from 2009 to 2012 the UK closed 87 percent of its primary aluminium capacity dropping from 372,000tonnes per year to a remaining capacity of 47,000 tonnes per year. Caught in a similar financial problem to the steel industry where the cost of production exceeded the selling price, the businesses that carried out primary aluminium production suffered a sudden decline in profitability. Some of the significant additional costs were the high energy costs, largely due to UK climate policies and high gas prices, which in theory were under governmental control.

The high energy costs resulted in the closure of two major aluminium smelters. In 2009 the Anglesey smelter, owned by Anglesey Aluminium (a joint venture between Rio Tinto and Kaiser Aluminium) with a 145,000 tonne per year capacity ceased operation and closed. This was followed in 2012, by the closure of the 180,000 tonne per year Lynemouth smelter, owned by Rio Tinto, which was a major financial blow for Northumberland.[7,8]

The only remaining smelter in the UK is the 47,000 tonne per year Lochaber plant currently owned by the GFG alliance the company founded by Indian businessman P.K. Gupta. The electrical energy for the plant is supplied by the 87.75 MW hydro-electric power station which is the largest in the UK.

During the period of decline of primary aluminium output in the UK, the output in China has significantly increased. China now makes over half the world's aluminium. The Chinese government's financial support for China's aluminium industry is *"on an order of magnitude larger"* than in other countries according to Alcoa chief executive officer Roy Harvey in 2019.

In similar circumstances to the UK, Chinese aluminium was very competitively priced in the USA causing aluminium smelters to be uncompetitive and suffer reduced market share. The number of aluminium smelters in the United States was reduced. Eight smelters either shut down or scaled back operations since 2015, and about 3,500 jobs in the aluminium industry disappeared in the subsequent 18 months.

In the USA a national security issue developed. High purity aluminium is used in the manufacture of Boeing's F-18 and Lockheed Martin's F-35, as well as armoured vehicles. The closure of smelters meant that the USA had one remaining domestic smelter of high purity aluminium. This was Century Aluminum's Hawesville, Ky. Plant, which was also under price pressure and decreasing market share.

This represented a cause for concern regarding national security and was one of the reasons for the USA import duty on Aluminium imports imposed in 2019.

Allowing the UK aluminium industry to suddenly decline is strange, because the metal is regarded as environmentally useful. Aluminium's combination of low density, strength and corrosion resistance make it ideal for many applications including military and civil. One important application is the weight reduction of vehicles to allow improved efficiency and lower emissions. In addition, aluminium is 100% recyclable where the re-melting of re-cycled scrap requires only 5% of the energy needed to produce the metal in the first place.

1.1.3 The decline in UK's productivity and high electricity costs

An important aspect of the decline in the UK's competitiveness has been the decline in productivity against other countries. The decline in productivity is shown in Figure 1.3 Upper. Growth has been relatively slow since the financial crisis (banking crisis) around 2008. The Office for National Statistics has reported that the productivity in the fourth quarter of 2018 was 18.3% below its pre-downturn trend.

After 2009, UK electricity prices remained higher than in France and Germany, and continued to rise over 2010-15. (Figure 1.3 Lower) In France, industrial electricity prices stabilised over 2009-13, and fell over 2013-15, while in Germany electricity prices fell over 2009-15, arriving in 2015 at the lowest of prices in the three countries. Both low productivity and high electricity costs will have a negative effect on all future manufacturing in the UK.[8]

Source: Financial Times

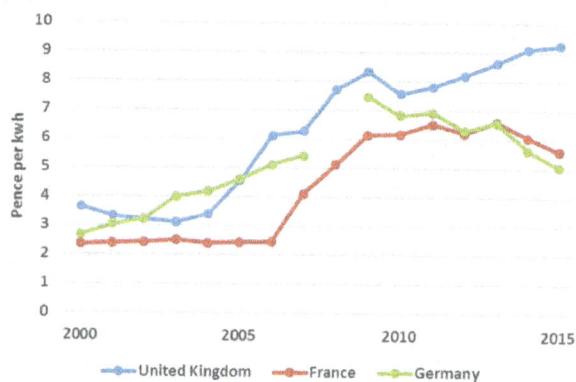

Source: BEIS.

Figure 1.3 Upper = UK's productivity lags behind the OECD average (Organization for Economic Cooperation and Development (OECD) is a unique forum of the governments of 34 democracies with market economies) Lower = Average industrial electricity prices excl. taxes in the UK, France and Germany, 2000-2015[8]

1.2 THE CHINESE PLAN TO TAKE A DOMINANT ROLE IN ALL METAL SUPPLY

This important strategy can be found in reports prepared by China Metallurgical Industry Planning and Research Institute (www.MPI1972.com). China has taken a dominant role in the supply of many metals, number one in titanium and aluminium[9], number three in copper, number one in zinc, number one in tungsten[9] and China has become the world's dominant producer of REEs, producing 90% of the global REEs output.[10] Between the years 2000 and 2015 the Chinese steel output increased by 14% every year. Between 1800 and 1890 the UK made 50% of the world steel. From 2015 China makes 50% of the total world output as shown in Figure 1.4a. The impressive marketing can be seen in their "Global/Chinese Steel Demand Forecast and Competitiveness of Chinese

TRANSFER OF THE WORLD STEEL INDUSTRY

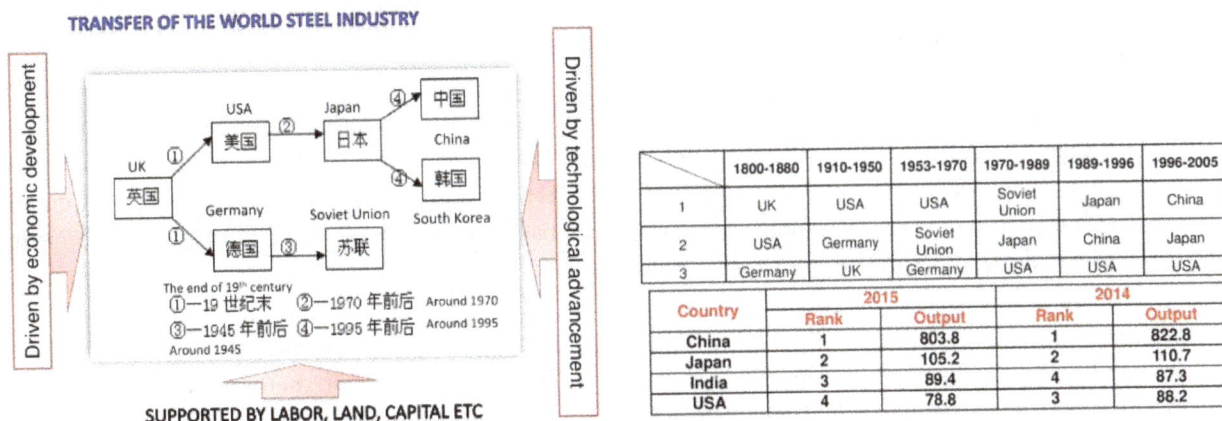

	1800-1880	1910-1950	1953-1970	1970-1989	1989-1996	1996-2005
1	UK	USA	USA	Soviet Union	Japan	China
2	USA	Germany	Soviet Union	Japan	China	Japan
3	Germany	UK	Germany	USA	USA	USA

Country	2015		2014	
	Rank	Output	Rank	Output
China	1	803.8	1	822.8
Japan	2	105.2	2	110.7
India	3	89.4	4	87.3
USA	4	78.8	3	88.2

Figure 1.4a *The changing history of the top three steel producers. Between 1800 and 1890 the UK made 50% of the world steel. From 2015 China makes 50% of the total world output. From https://slideplayer.com/slide/12716709/*

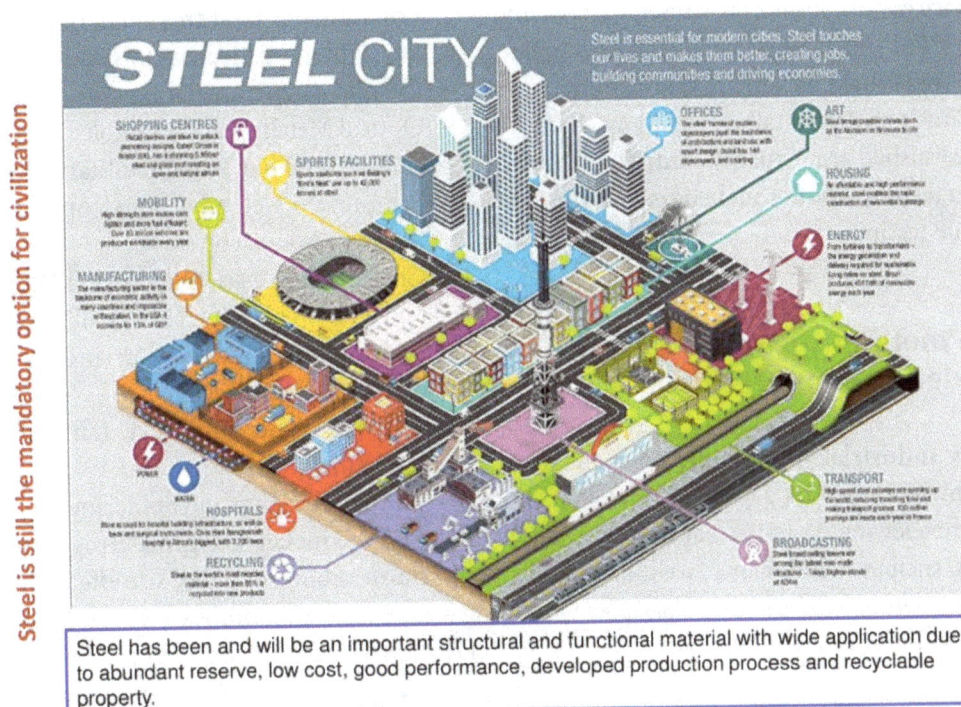

Steel has been and will be an important structural and functional material with wide application due to abundant reserve, low cost, good performance, developed production process and recyclable property.

Figure 1.4b *Bottom=Impressive marketing Image of the uses of steel in society used in a Global/Chinese Steel Demand Forecast. From https://slideplayer.com/slide/12716709/*

Steel Plate at https://slideplayer.com/slide/12716709/ Figure1.4b shows an impressive image used in this marketing report.

China's substantial income has allowed investment into foreign high-tech businesses which, according to the US, is "acquiring the crown jewels by stealth". In addition to its planned financial strategy into metals, China has made significant defence spending and now has the second largest spending after the US. A major concern if it is intended to support their interests in the South China Sea and Taiwan.[11]

1.3 THE OBJECTIVES OF THIS TRAINING MANUAL

The purpose of this training document is to provide the essential knowledge needed to gain a working understanding of the metallurgical requirements currently required to support industry. An additional reason for the training guide has been to prepare a reference document that includes some key diagrams and important items of information.

In the last 20 years metallurgical knowledge has grown ever more complex despite the better "understanding" now available and the easy access to knowledge on the internet. Training the future workforce represents a major task and needs to be combined with ways to ensure that each person is committed to continuous professional development. Within the engineering profession, especially in the areas of design, quality assurance, production engineering, welding, casting, NDT, customer complaints and the associated examination of failures, the need for metallurgical knowledge has now become an essential requirement.

This can be demonstrated by consideration of the metallurgical principles involved with the carburizing of an automotive gear. The heat treatment must be reliably controlled and production parts adequately tested to ensure that the integrity of the gears meet the required service conditions for the subsequent twenty or more years. For this to be achieved involves material selection (hardenability), method of manufacture (forging casting, powder metallurgy) and mechanical testing of the finished gear (quality control and product validation testing – tooth root bending fatigue and contact fatigue testing).

1.3.1 The metallurgical knowledge needed to support industry

Within many industrial sectors there is a need to have an appreciation of the aspects of metals technology that are relevant to a particular industry or job requirements, such as maintenance, welding, quality, design or management. Some basic examples where metallurgical knowledge is essential are:

- The design and validation testing of parts, components and structures
- The examination of failed parts and identification of potential failure modes
- The review of codes, standards and specifications
- Quality assurance and quality control in the materials sectors of industry (FMEA, 8D, Control Plans)
- Understanding test certification (chemical analysis, mechanical properties)
- The optimization of cost, quality and delivery, value analysis and lean manufacture
- Examination of failed parts (Parts Performance Analysis PPA)

Some of the essential knowledge requirements would include the following:

1. Structure of metals, The Periodic Table of Elements, Properties, Processing
2. Crystallographic structure (BCC, FCC and HCP) grain structure, dislocations, lattice defects
3. Metal composition and alloy type – Cast irons, steels, stainless steels, nickel base alloys, cobalt base alloys, zinc, copper, aluminium, titanium, tin and lead bearing alloys
4. Diffusion and heat flow, convection, conduction and radiation, heat transfer coefficient for quench
5. Temperature measurements, thermocouples type N, type R, Infrared, Thermal Imaging Camera
6. Equilibrium diagrams-Isomorphous, Eutectic, Peritectic, Phase rule, Gibbs free energy
7. Thermal analysis, Dilatometry, Phase transformations and the Heat treatment of metals
8. Strengthening Mechanisms and Phases present in the microstructure, Grain size
9. Optical Metallography, sample selection, preparation and etching, Plus Quantitative Metallography
10. Scanning Electron Microscope, Transmission Electron Microscope
11. Chemical analysis, OES, XRF, EDX, Leco fusion and XRD to identify the crystalline phases present
12. Melting and solidification and its relevance to casting and welding
13. Alloying solid solubility (substitutional and interstitial)
14. Welding, Brazing and Soldering
15. Gases in metals-Oxygen, Nitrogen and Hydrogen
16. Hardenability and the "size effect" during heating and cooling, Jominy, Grossman Factors
17. Non-metallic inclusions, oxides and sulphides and inclusion modification with Calcium and REM
18. Hot working, cold working, recovery, recrystallization and grain growth
19. Solution treatment and Precipitation hardening
20. Metal Matrix Composites, Bearing alloys, Polymers, Ceramics, Elastomers, Glasses, Composites.
21. Powder metallurgy, manufacture of powders, compaction, sintering and heat treatment
22. Additive Manufacture (Rapid Prototyping) (3D Printing)
23. Metal forging processes, cold forging, roll forging, hand forging, upset, closed and open die forging
24. Machinability of metals, Free cutting steels either high sulphur or lead based
25. Mechanical metallurgy, Load paths, Services stresses and Residual stress
26. Mechanical properties. Tensile, Impact, hardness, fracture toughness, fatigue, creep, stress rupture

27. Corrosion and protection against corrosion, Pourbaix Diagram, Galvanic or Electrochemical Series, Current Potential Diagram
28. Surface engineering for wear and corrosion, PVD, CVD, nitriding, carburizing, plating processes
29. Failure modes, fracture, fatigue, creep, corrosion, wear and a knowledge of the history of failures
30. Computational metallurgy (Ashby diagrams, Solidification software for cast metals (Magma)), Phase diagram construction and corrosion studies (Thermo-Calc & Dictra[12] and Factsage[13], and heat treatment simulation (Sysweld, QT Steel)). In addition, information on mechanical property and corrosion data, plus the interpretation and understanding of Codes and Standards where the skill of "technical judgement" is essential.

1.3.2 When things go wrong

When failures occur, there is a need to establish the cause of the failure, to prevent the re-occurrence of further failures. For example, Figure 1.5 shows three failed bolts each requiring a different cure. To carry out a failure investigation a procedure for a failure analysis would be essential. Some initial suggestions for inclusion in such a procedure would be:

- To establish a comprehensive description of the failure including past history
- To document the events and service conditions especially at the time of failure
- To develop an understanding of the design, manufacture and how components or parts operate
- To understand and appreciate the mechanical "**load path**" and types of loads and whether the applied loads were static, dynamic or cyclic
- To establish the potential failure mode and the main contributing factors
- To prepare the conclusions and make recommendation to avoid further failures

API RP 571, although limited to refineries, provides an excellent review of failure mechanisms together with photographs and is a recommended source for guidance on the failure of metals in service.[13]

1.3.3 Material Selection and the Design Process

Material selection is a fundamental part of any design process. For many products there will be an established history of the materials used, and where the performance has been without problems, design "carry through" is often adopted and future parts are made using similar materials. This design practice has played an important part in the continuous improvement in the reliability of parts in the automotive industry.

However, where product improvements are required such as the use of cheaper materials to give greater product competitiveness, or the selection of a lighter material to increase fuel economy and reduce emissions – which is currently very important in the automotive manufacturing sector – then a quantitative method of material selection has to be applied at the design stage. In addition, the design process must avoid the selection of inappropriate or poor-quality materials, which could result in product failure or poor performance. A study by Noori Brifcani and others reviews the methods available.[14]

There are probably over 160,000 materials available, which gives an insight into the need for tools and assistance in the task of material selection. Material engineers have grouped materials into categories: Metals & Alloys, Polymers, Ceramics, Elastomers, Glasses, Composites.

The materials inside each group have a similarity of properties. Each material is defined by its properties which are usually measured by test work carried out in accordance to standards (BS, DIN, ISO or ASTM). The properties can be grouped into Mechanical Properties, Physical Properties, Thermal

Figure 1.5 Two fatigue fractures and a hydrogen embrittlement failure

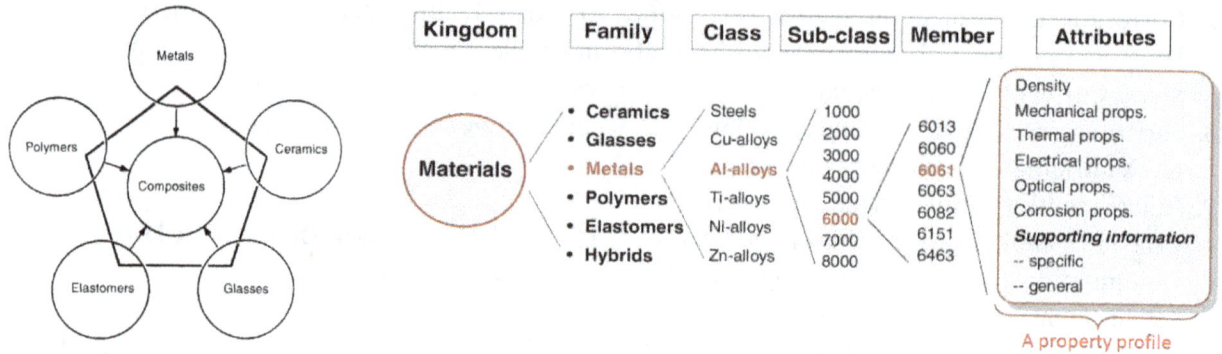

Figure 1.6 *LHS = Materials, RHS = The kingdom of materials divided into families, classes, sub classes and members and each member can be characterised by a set of attributes which are known as "a property profile"*[15]

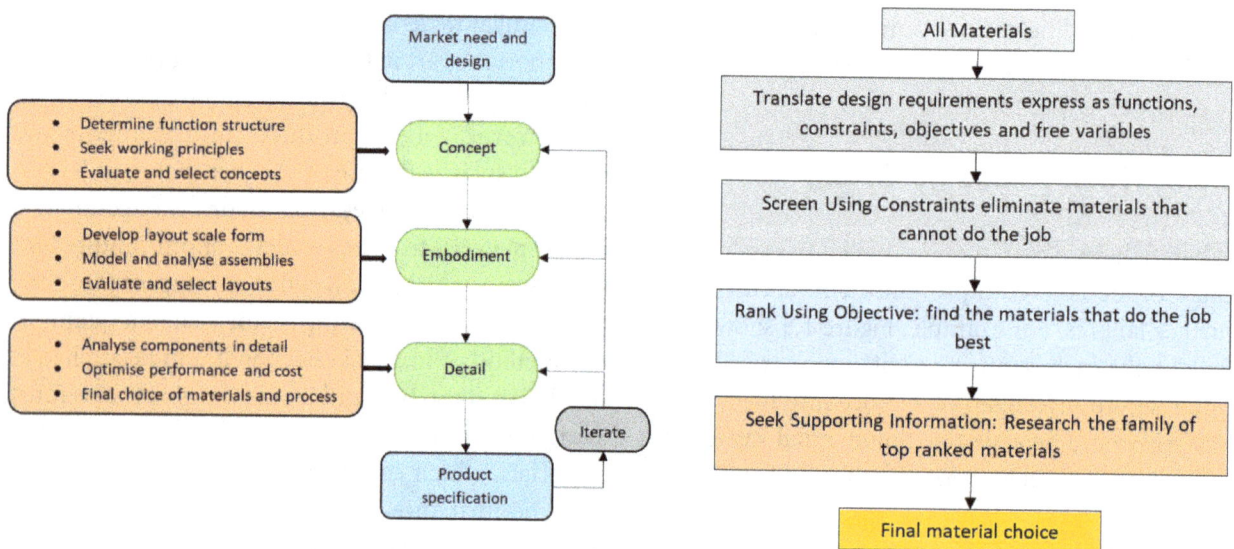

Figure 1.7 *LHS = The stages in a design process, RHS Material selection process (adapted from 15)*

Properties and others, such as Corrosion resistance, Weldability and Cost. Some material properties can be given a quantitative value, such as the Strength (MPa) and Toughness (Charpy Energy J). For other material properties, such as corrosion and wear resistance, machinability, and weldability, numerical values are difficult to apply and materials are usually described in a linguistic fashion and rated as Very Good, Good, Fair and Poor to allow evaluation. Figure 1.6 shows the kingdom, family and class relationships which tries to apply some logic to differentiate between various grades. The choice of material available is often constrained by commercial availability of materials and for most applications the commercial aspects can limit the material selection process both in terms of the optimum material and methods of manufacture.

The selection of a material should be carried out in parallel with the initial design and product development, as the material selected will have individual properties that influence the manufacturing process such as welding, casting and forging and therefore how it should be designed. Professor Ashby's approach to design and material selection are shown in the diagrams in Figure 1.7.

Design begins with a "market" requirement. It is essential to document the needs of the product precisely. A typical design procedure as shown in Figure 1.7 LHS would be followed to create a detailed PDS (product design specification).

There are several sources of materials information for design:

www.azom.com AZoM is the leading online publication for the Materials Science community. Our MissionAZoM educates and informs a worldwide audience of researchers, engineers and scientists with the latest industry news, information and insights from the Materials Science industry

www.matweb.com MatWeb's searchable database of material properties

www.makeitfrom.com Material Properties Database

www.citrination.com AI-Powered Materials Data Platform

www.matmatch.com Discover, evaluate and source the best materials for your projects.

Professor Ashby developed the chart concept for material selection in the 1990s and examples are shown in the following Figures 1.8 to 1.9. The charts represent a comprehensive resource of material data and logical rules to find the optimum material for a specific set of requirements. They are also now supplemented by Tables on the methods of manufacture[16]. Two sets of data are shown fracture toughness vs yield strength in Figure 1.8 and approximate cost of raw materials

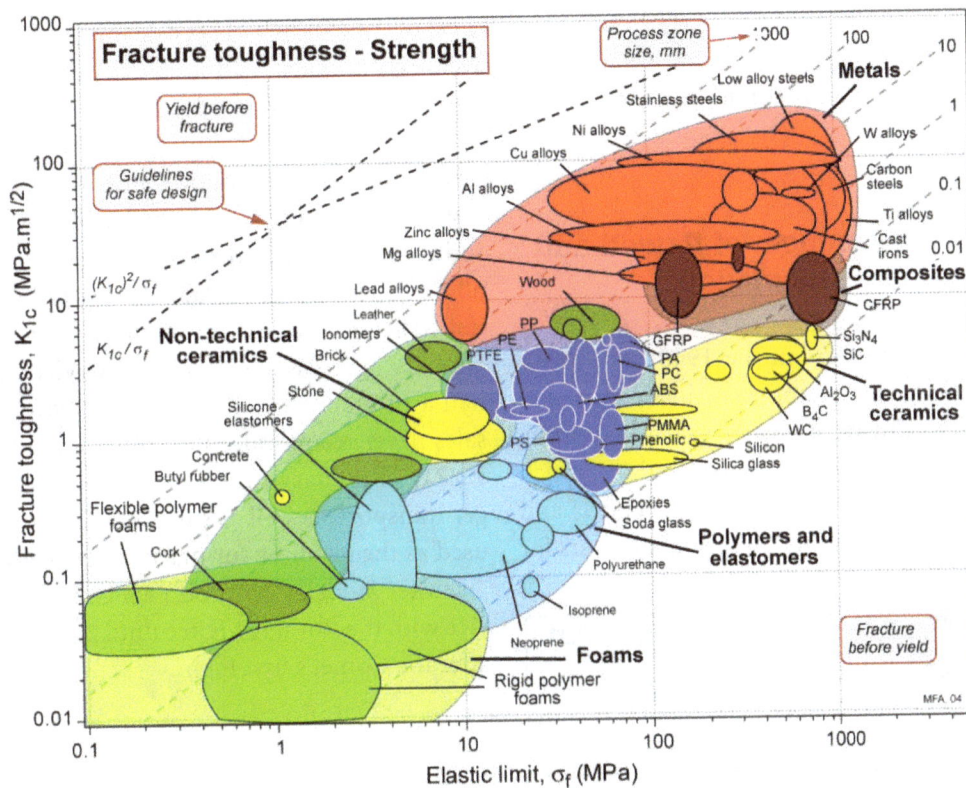

Figure 1.8 *Fracture toughness vs yield strength (16)*

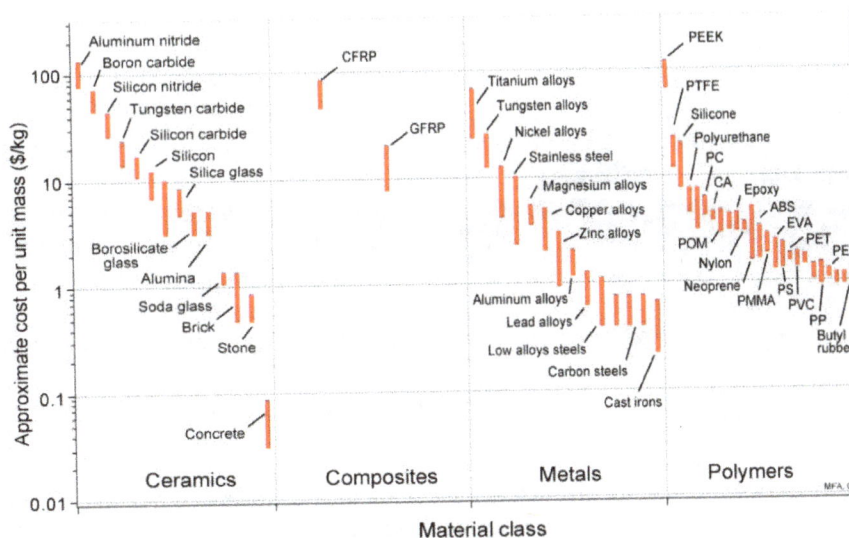

Figure 1.9 *Approximate price of material (16)*

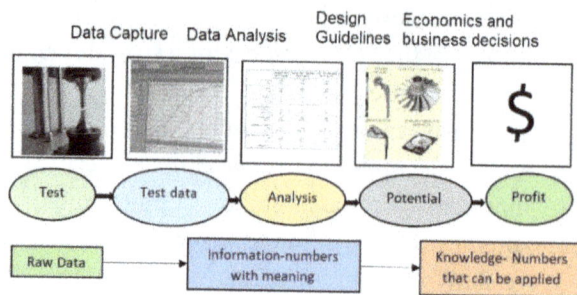

Figure 1.10 Materials information needed for design (adapted from 15)

in Figure 1.9. Figure 1.10 shows the information and knowledge requirements needed to justify the use of new materials.

1.3.4 The consequences of flawed and inadequate knowledge

With some major projects in the 1940s and 1950s material knowledge was found to be inadequate and flawed especially during the early years of welding (Liberty ships and Kings Bridge in Australia[17], which was the subject of a Royal Commission report) and aviation (Comet) that resulted in several disasters for

Date of Comet Crash	Where and How	Fatalities
October 1952	Rome Failed to become airborne	None
March 1953	Karachi Failed to become airborne	11
May 1953	Calcutta 6 mins after take-off	43
January 1954	Ciampino 20 mins after take-off. Mid-air break-up	35
April 1954	Rome to Cairo, Mid-air break-up	21

Table 1.2 The crash history of the Comet

which the underlying root cause was not known or not identified at the time. The Table below shows the crash history of the Comet. After the April 1954 incident the whole fleet was grounded.

An important outcome of the significant early failures, was the provision of funding that led to the understanding of the many important metallurgical mechanisms that result in metal failures: fatigue, fracture, creep and stress rupture, corrosion. It also resulted in improved mechanical characterisation of metals. Two key failure events that led to the metallurgical understanding of "brittle fracture" and "fatigue" were events linked with sea and air travel –the Liberty ships and Comet. Figures 1.11 and 1.12

Design and manufacturing flaws, such as improper riveting and stress around some of the square windows, were rectified. The worldwide aerospace industry benefited from the improved knowledge, allowing rival manufacturers to avoid the structural deficiencies that had led to failure. The redesigned Comet 4 series was returned to service in 1958 and had a productive career of over 30 years. However, sales never recovered. The Comet was adapted for a variety of military roles such as medical and passenger transport, as well as surveillance. The Comet was used as the airframe for maritime patrol variant, the "Hawker Siddeley Nimrod", which remained in service with the Royal Air Force until 2011, over 60 years after the Comet's first flight.

Ship Failures and loss of life

- Titanic (14 April 1912) Weak rivets and brittle steel hull
- Liberty Ships Welded construction, MMA, 250 ships fractured, 19 broke into two pieces
- The understanding of fracture mechanics begins as the reasons for failure are investigated, others were stress concentration, weld imperfections, residual stresses and susceptible steel
- Derbyshire(10th September 1980). Ship failures still happen

Figure 1.11 The Liberty ships were built to carry strategic supplies from the USA to the UK during WW2. Several of the ships experienced brittle fracture. The subsequent research to find the cause of the fracture led to a better understanding of brittle fracture especially of welded structures (17). The problems were never completely solved and there were subsequent failures with loss of life such as The Derbyshire 10 Sept 1980 (18 and 19).

Did real life imitate a movie?

- Nevil Shutes book "No Highway" (1948) and movie (1951) predicted fatigue failure of plane and the events of denial
- De Havilland Comet 1 first commercial aircraft with jet engine. First flight 1952. Operated at 40,000 feet with cabin pressure at 8000 feet
- Two catastrophic accidents 1954. All dead
- RAE carried out pressure tests and examined failure. Confirmed structural failure by fatigue

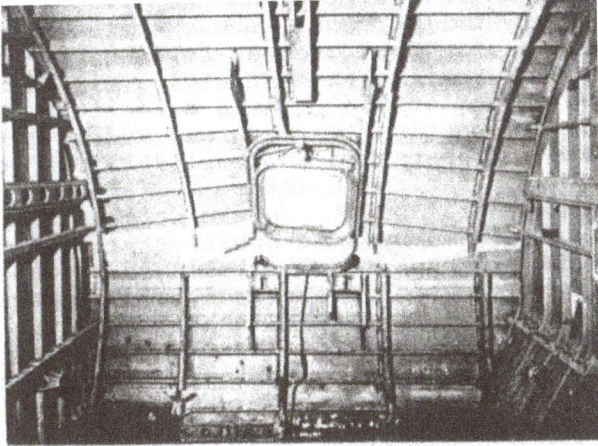

Figure 1.12 Comet – Fatigue fracture. The film "No Highway in the Sky" was shown in 1951 staring James Stewart and represents one of the more unfortunate examples of life imitating the movies. It appears to be a startling premonition of the real events that destroyed the Comet aircraft three years later. The Comet was the first commercial aircraft to use jet engines, and was designed and manufactured by the de Havilland Aircraft Co. Ltd. Nevil Shute published his novel, No Highway, in 1948. Ironically, he began his career as an engineer with the de Havilland Aircraft Co. Ltd. The novel No Highway focussed upon the idea of metal fatigue on the plane structure. It told the story of a scientist who was convinced of the imminent failure, and fights to ground a new fleet of jet planes in which he perceives a fatal design flaw. The actual Comet failure caused severe commercial damage to the UK aerospace industry and left the way for Boeing and Douglas to dominate the market[20].

1.3.5 Avoiding Failures: The balance between "service stresses" and "mechanical properties"

- Brittle fracture has caused major failures and therefore appropriate level of "toughness" is essential.
- Fatigue is an insidious form of failure and therefore sufficient "endurance" is essential.
- To prevent engineering failures appropriate mechanical property data of the actual material supply needs to be measured and used in the design process depending on the expected failure modes.

There are some essential tasks:

- It is important to establish the expected service loads, particularly the "load path" by using either prototype test work and the use of strain gauges, or by theoretical calculation often assisted by FEM (Finite Element Method)
- To use mechanical property data effectively it is important to understand how mechanical properties are measured and what they represent, so that the limitations and significance are understood. Resistance to failure by plastic deformation is quantified by yield strength; the resistance of a cracked part to fracture is measured by fracture toughness. The resistance to fatigue failure is given by a stress vs life curves and for creep strain vs life (at constant load and temperature).
- Based on a comparison of the service loads and the mechanical properties it is a simple matter to ensure that the essential material properties are the greater; preferably with a safety margin to allow for statistical scatter in data.

axial loading

torsion

bending

Service load types

Estimation of service loads

Mechanical property measurement

FORCE MEASUREMENT

TEST SPECIMEN

FIXED HEAD

GRIPS FOR HOLDING SPECIMEN FIRMLY

STRAIN GAUGE

GAUGE MARKS

MOVABLE HEAD

CONSTANT RATE OF MOTION

THICKNESS = t

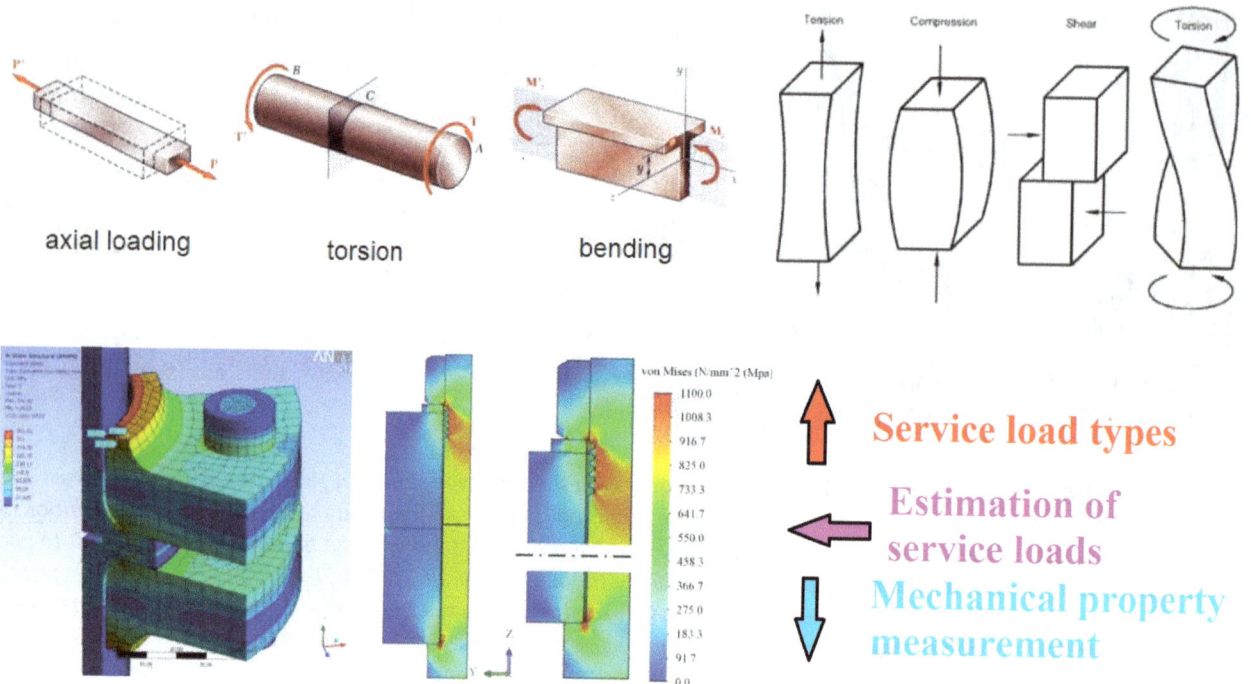

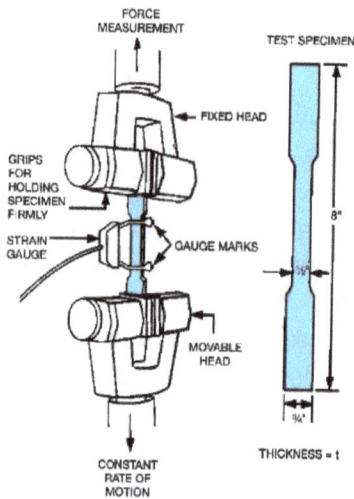

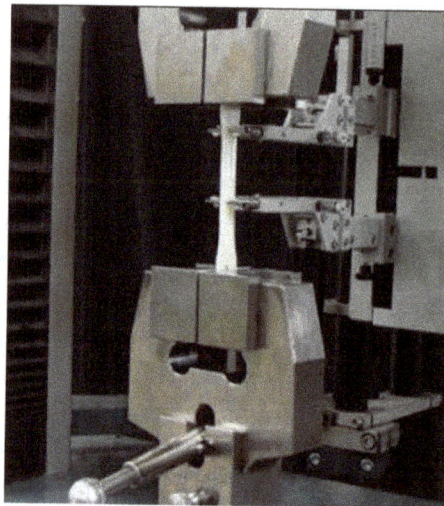

Ductile and brittle fractures

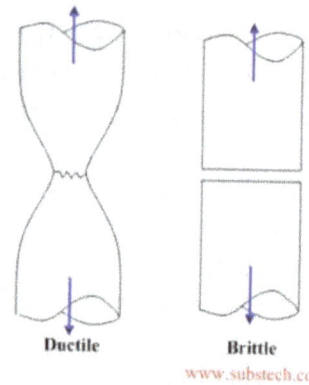

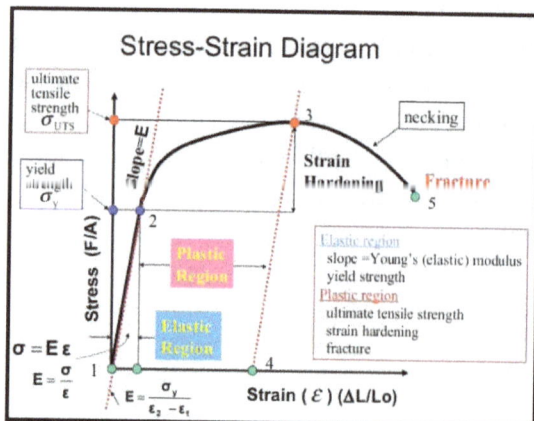

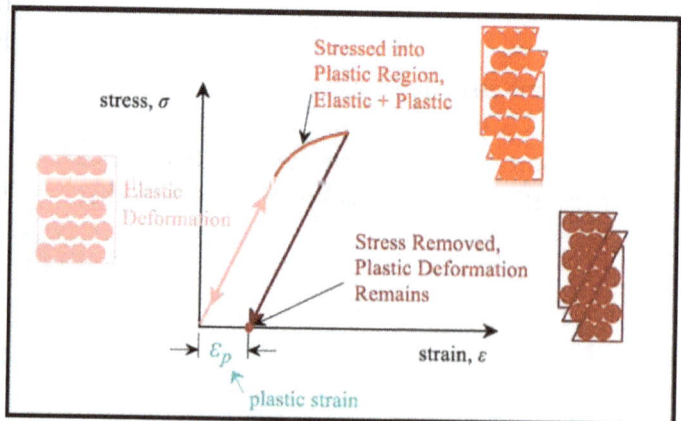

Ductile

Brittle

www.substech.com

Stress-Strain Diagram

ultimate tensile strength σ_{UTS}

yield strength σ_y

necking

Strain Hardening

Fracture

Plastic Region

Elastic Region

Elastic region
slope = Young's (elastic) modulus
yield strength
Plastic region
ultimate tensile strength
strain hardening
fracture

$\sigma = E\varepsilon$

$E = \dfrac{\sigma}{\varepsilon}$

$E = \dfrac{\sigma_y}{\varepsilon_2 - \varepsilon_1}$

Strain (ε) ($\Delta L/Lo$)

Stress (F/A)

stress, σ

Stressed into Plastic Region, Elastic + Plastic

Elastic Deformation

Stress Removed, Plastic Deformation Remains

ε_p

strain, ε

plastic strain

Figure 1.13 Top 4 diagrams relate to service loads and the "load path". The lower five images to tensile testing. Lower RHS = The pink balls demonstrate atom movement during elastic loading and the red balls the permanent movement of atoms during plastic loading. Plastic deformation is the onset of the movement of small imperfections in the crystal structure known as "dislocations" (see section 2.3.4.2. for explanation of dislocations))

Modern Vickers

Rockwell

Brinell

Old Vickers

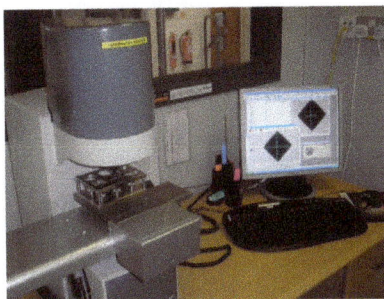   

Brinell Hardness Test

Vickers Hardness Test

Statistics

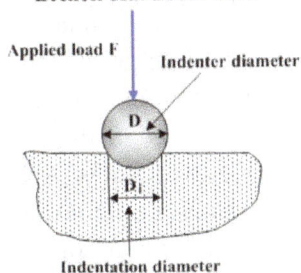 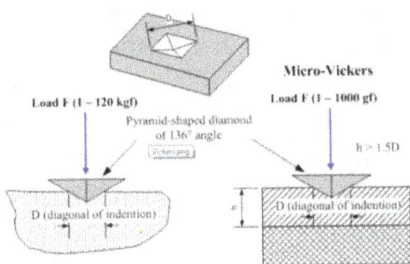 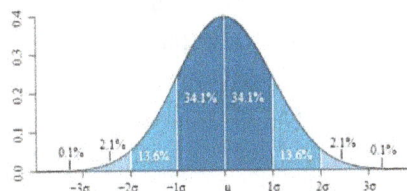

Rockwell C Hardness Test

Rockwell B Hardness Test

Rockwell Superficial Hardness Test

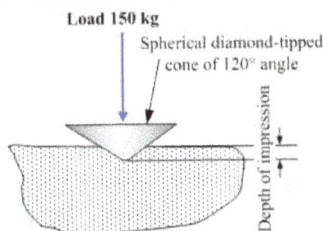 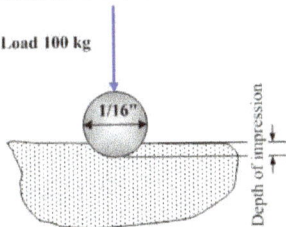 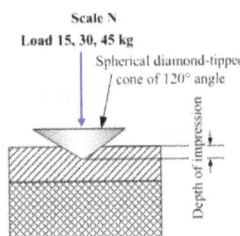 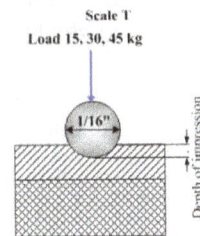

Pendulum impact test machine

Charpy Test

 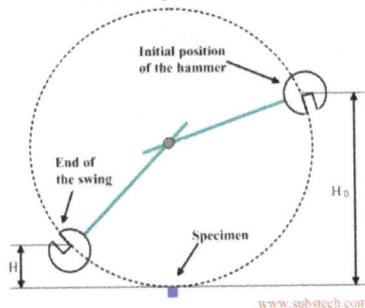 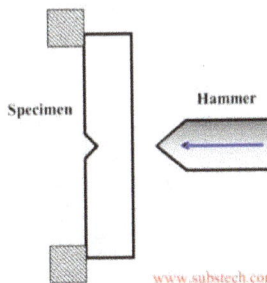

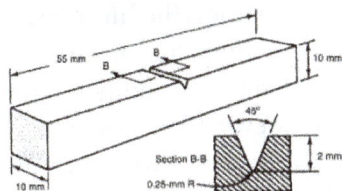 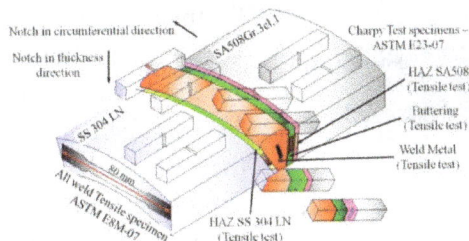 

Figure 1.14 *Top 10 show hardness modern Vickers, Rockwell, Brinell. The mid 5 show common hardness test. The statistics reference – There is scatter in both loads on structures and the strength of materials, Weibull, log normal, normal. Lower diagrams show the basic method of toughness measurement.*

1.4 FROM PROBLEMS TO SOLUTIONS IN 60 YEARS (1954 TO 2013). GUIDELINES TO PREVENT BRITTLE FRACTURE, FATIGUE, CREEP AND CORROSION FAILURE MODES

Failures always have consequences. This has been the case since civilisation began.

One of the earliest known principles of construction law can be found in the Code of Hammurabi. Hammurabi was the sixth king of Babylon and ruled from 1792 BC to 1750 BC. The Code of Hammurabi contained 282 laws inscribed on twelve stone tablets which were placed in public view. Hammurabi's Code was one of the earliest written codes of law in recorded history. Several of the laws pertained to the built environment and demonstrate their views on failures.

> Law 229. If a builder builds a house for someone, and does not construct it properly, and the house which he built falls in and kills its owner, then that builder shall be put to death.
>
> Law 230. If it kills the son of the owner, the son of that builder shall be put to death.

In the Roman era structural validation methods were deemed important. The Romans are believed to have had a unique method to carry out integrity and durability testing of bridges. The design engineer was placed beneath the bridge followed by chariots using the bridge.[22]

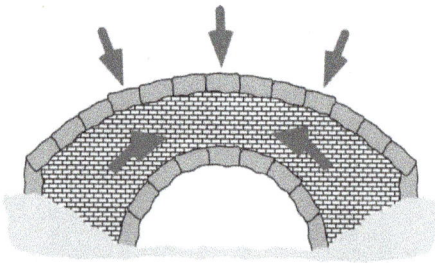

Financial Damage $42-84BILLION!?

POST-SPILL CLEAN-UP	US GOVT CRIMINAL SUIT	GULF ECO RESTORATION	SETTLEMENT OF CLASS ACTION SUIT – GULF RESIDENTS & BUSINESSES	EPA CIVIL SUIT (CLEAN WATER ACT)	GULF STATE LAWSUITS
DONE	SETTLED	SETTLED	COURT SETTLEMENT BUT TOTAL STILL INCREASING	IN PROCESS; 9/4/2014 RULED AS "GROSS NEGLIGENCE"	FILED; not yet litigated
$14B	$4B	$2B	$12B +	Est $18 to $18B	UP TO $34B

Figure 1.15 Upper = Roman Arch bridge aimed for compressive stresses. Lower = Cost for Macondo

The modern-day approach is less brutal but can result in substantial costs: as experienced by BP due to errors in the Macondo [Deepwater Horizon in the Mexican Gulf] project failure (@ 2018 $65 Billion).[23]

1.4.1 Major Engineering Failures and Lessons Learnt

There are several major texts covering the failure of engineering components which provide useful insight into the methods and procedures involved in failure investigations and the prevention of failures[24-33]. In addition, the CD ROM compiled by the American Society of Metals (ASM) provided one of the first comprehensive technical databases on failures (ASM CD ROM,1997.)[34] However, perhaps one of the most memorable illustrations of the importance of component failure is given by Harrison[35], when relating a story from the Welding Institute.

> "A short story concerning Dr Geoff Egan, who is well known in the fracture field illustrates the type of problem that may confront those at the forefront of fracture research. Dr Egan was flying back to Europe from a conference in the USA. He awoke as the plane crossed the coast of France and saw a group of people looking out of one of the windows. The Captain approached him and said 'Dr Egan, I understand that you are an expert on fatigue and fracture, I would like to ask your advice about a problem we have, could you come and look out of this window'. From the window a crack about 1m long could be seen in the top of the wing. Dr Egan said, 'I can give you two pieces of advice: firstly, get this plane on the ground as fast as you can; and secondly slow down!' It is to be hoped that advice from experts in fracture is generally less conflicting. The crack in question grew another 75mm during the flight."

Fortunately, Dr Egan did not witness the "in-service" failure of the aircraft wing, and hopefully none of us are put in the position of testing our ability to give "less conflicting" advice. However, there is no doubt that we would certainly come away from such an experience with a much clearer understanding of the importance of avoiding failure in engineering components.

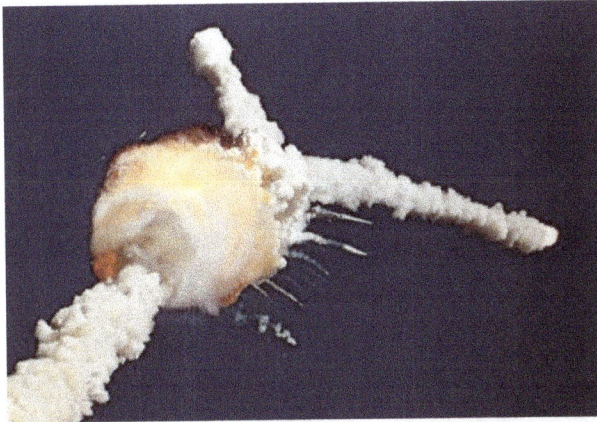

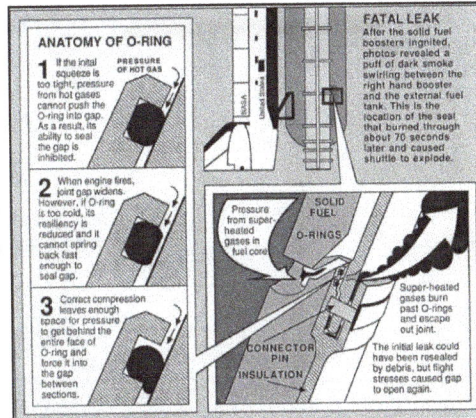

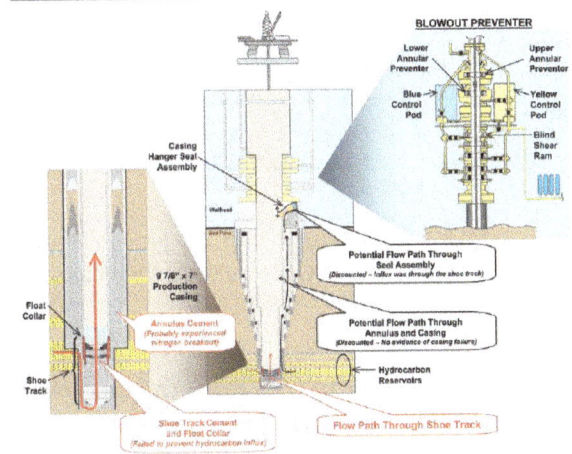

Figure 1.16 Some examples of Major engineering failures: Challenger (36 and 37), Bhopal (38) and Deepwater Horizon (39)

1.4.2 From Safety Factors to PD 6493 and then to BS 7910- 2013 (Damage Tolerance)

During the industrial revolution and for a long time afterwards the basic approach to ensuring that engineering parts were "fit for service" was the use of factors of safety or a margin of safety gained from service experience. This was a simple approach, and it tried to guarantee that the strength of a part was greater than the expected service loads. Table 1.3 shows some historical factors of safety that have been used for various parts and materials. Some statistical considerations associated with safety factors often used to illustrate "mechanical reliability" is shown in Figure 1.17 and 1.18.

A major weakness of this type of approach was that it only considered strength and neglected toughness and other failure modes. The safety factor can cover for potentially unknown-known factors (factor of ignorance) which eventually erode the margin of safety and lead to potential failures as shown in Figure

▶ **Table 1.3** *Factor of safety for various application (40)*

Class of service	a	b	c	d	Safety factor F= a.b.c.d
Boilers	2	1	1	2.25-3	4.5-6
Piston and con rods for double acting engines	1.5-2	3	2	1.5	13.5-18
Shaft with bandwheel, flywheel or armature	1.5-2	3	1	1.5	6.75-9
Mill shafting	2	3	2	2	24
Steel work in building	2	1	1	2	4
Steel work in bridges	2	1	1	2.5	5
Cast iron wheel rims	2	1	1	10	20
Steel wheel rims	2	1	1	4	8
Material	a	b	c	d	Safety Factor
Cast iron and other castings	2	1	1	2	4
Wrought iron or mild steel	2	1	1	1.5	3
Oil-tempered or nickel steel	1.5	1	1	1.5	2.25
Hardened steel	1.5	1	1	2	3
Bronze and brass, rolled or forged	2	1	1	1.5	3

a = the UTS/YS (or 0.2%PS) ratio

b = 1 for dead load, 2 for load cycling 0 to +x

c = 1 for gradually applied, 2 for sudden applied load

d = a factor of ignorance to provide against unknown conditions; varies between 1.5 and 3 but may be as high as 10

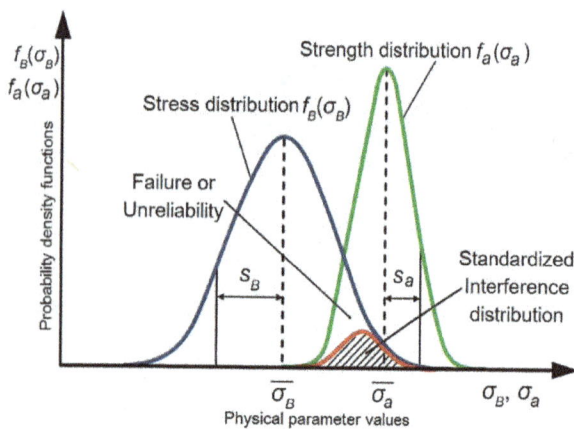

Figure 1.17 *Importance of statistical considerations*

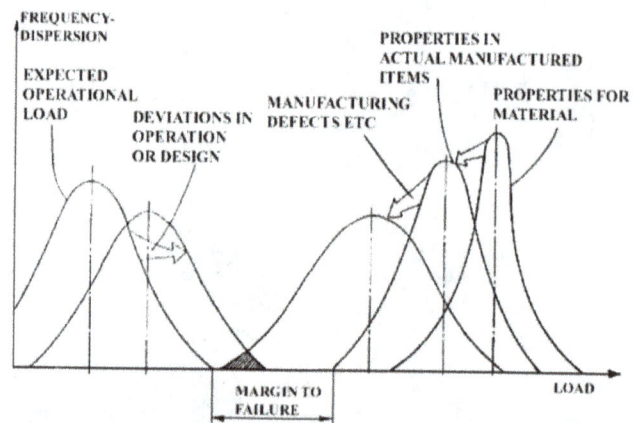

Figure 1.18 *Effect of manufacturing defects and high loads on safety margin*

1.17 and 1.18. **It could not cover for the unknown-unknowns, nor can modern design methods!**

The practical experience in the field of failures and new knowledge on fracture toughness, fatigue and welding led to the development of guidance documents and technical specification and codes that could assist the design engineer to avoid failures. The initial approach focussed on the two main problems: fracture and fatigue. In1980, PD 6493 (*PD=published document, which was intended as a guidance document*) was released which outlined a fracture/fatigue assessment procedure. It was published at the height of the North Sea Oil construction boom and gave essential guidance on how to handle weld imperfections. This was important since in some cases it was observed that a repair procedure could create more damage to the structure than the initial defect that it tried to repair.

1.4.2.1 Understanding the failure modes and the success of PD 6493

To use and appreciate PD 6493 or BS 7910 requires knowledge of fracture mechanics, fatigue, creep and corrosion. Figures 1.19 to 1.23 provide some visual images and an introduction of the important aspects of fracture toughness, fatigue, creep and corrosion.

Brittle Fracture-The Fracture Mechanics approach was the first method that involved the service stress, the potential "flaw" or defect in the part, component or structure and a measure of the toughness. (Figure 1.19 top RHS). The other parts of Figure 1.19 show important aspects relating to fracture control.

The need to use fracture mechanics was the realization that all structures contain defects and imperfections that act as cracks and locally amplify

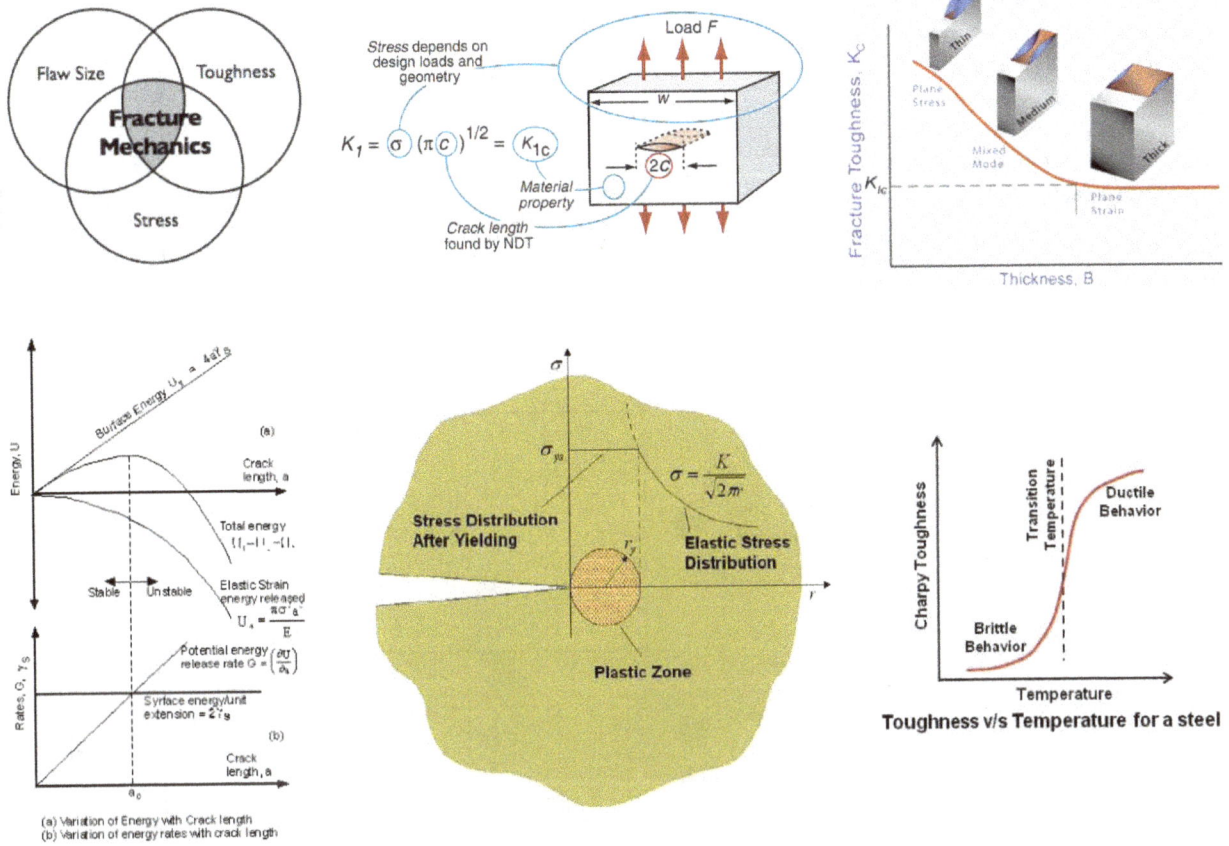

Figure 1.19 *Top=Fracture toughness Venn, Centre= Key equation, RHS = lack of "similitude" with toughness thicker samples lower toughness). Lower LHS = Energy approach Griffiths. Centre = Stress approach Irwin. RHS = Charpy ITT*

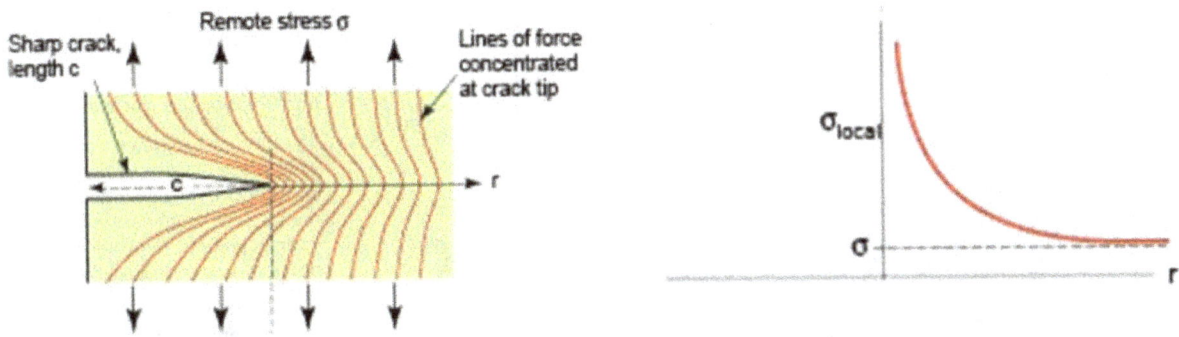

Figure 1.20 *Lines of force and local stress variation from a body with sharp crack. The stress gradient at a crack tip is defined in fracture toughness terms by the K (SIF) and determined by stress analysis. (41)*

stress at crack tips (Fig 1.20 and 1.21). **K** the **stress intensity factor** is used to represent the stress field ahead of the tip. The **stress intensity factor** can be viewed as a **"scaling factor"** that characterises the severity of the crack situation as affected by crack size, stress, and geometry. A given material can resist a crack without brittle fracture occurring as long as this **K** is below a critical value **Kc**, which is a property of the material called the fracture toughness. Values of **Kc** vary widely for different materials and are affected by temperature and loading rate, and also by the thickness of the part under consideration. (For

fracture K = **stress intensity factor** > Kc = material fracture toughness.)

Figure 1.21 compares the traditional approach where the design was based on yield strength with the fracture mechanics approach based on toughness, flaw size and service stress. Figure 1.22 shows an approximate way to determine the "critical defect size" (the maximum allowable defect for a particular service stress yield strength and fracture toughness) before brittle fracture would occur, assuming the material behaves as a linear elastic material. Figure 1.23 show the origin of potential defects

Figure 1.21 Fracture Mechanics
Approach – Material Fracture
Toughness (KIC) Divided by
the Stress Intensity Factor in a
Service Part (K) instead of Yield
Strength Divided by the Stress
in a Service Part

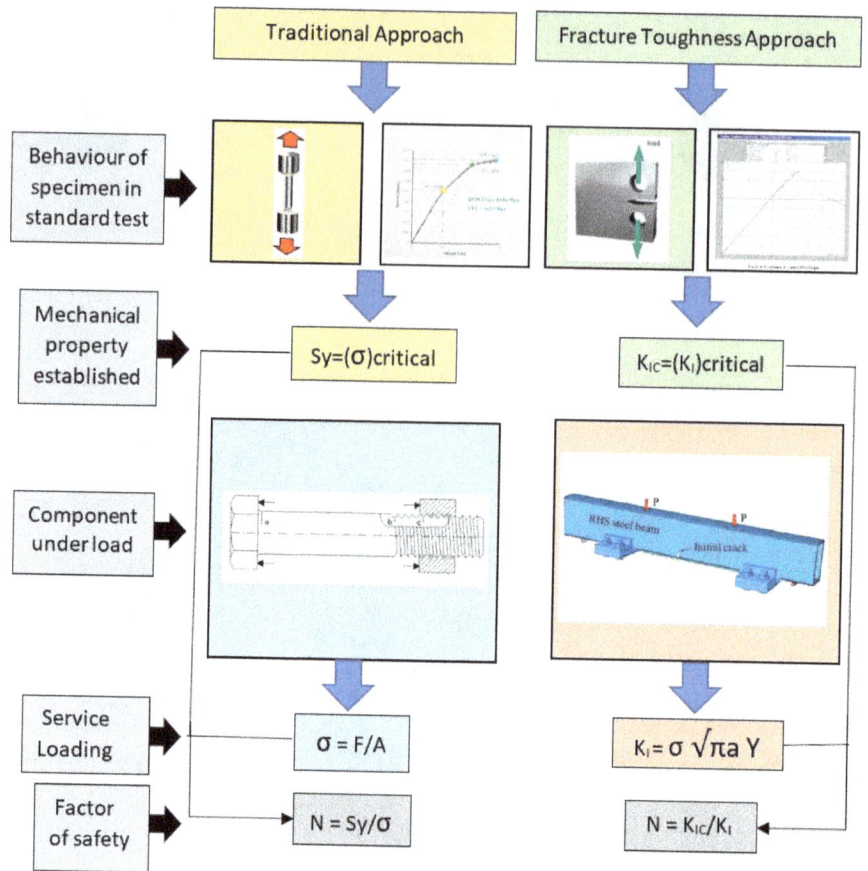

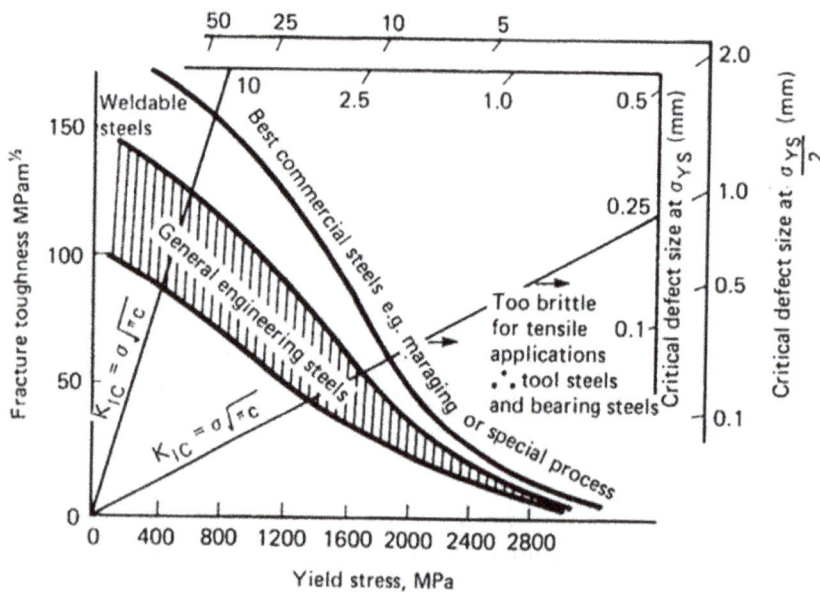

Ratio analysis diagram for quenched and tempered steels.

1. The "X" axis represents the yield strength of the steel

2. The "Y" axis represents the fracture toughness of the steel

3. A vertical line is drawn from the yield strength value of the steel and a horizontal line drawn from the fracture toughness value. The intersection of the two lines is noted

4. A line is drawn from the zero point on the graph through the intersection of the yield strength and fracture toughness lines and extrapolated to the outer two scales which shows the critical defect size

5. The inner scale represents service loads at the yield strength of the steel and the outer scale represents service loads at half of the steel's yield strength

Figure 1.22 Diagram that allows the calculation of an approximate critical defect size, which can be used to indicate how sensitive the part or component is to the expected surface imperfections

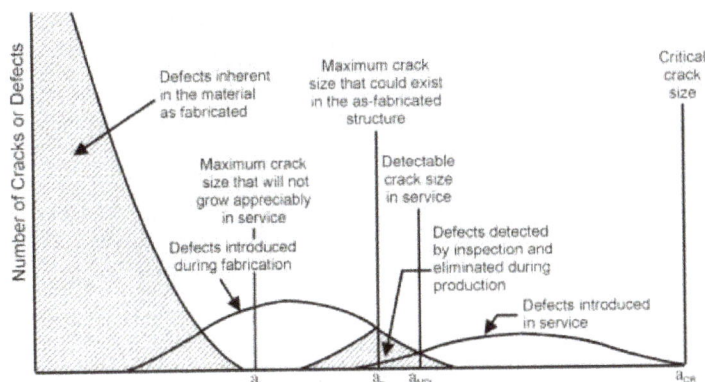

Figure 1.23 *Origins of defects in parts.
From http://www.afgrow.net/applications/
DTDHandbook/Sections/page3_0.aspx*

Fatigue- The physical damage caused by repeated cyclic loads on a material is different from a simple static load. The failure occurs by a brittle fracture mode regardless of the ductility or toughness of the material and the fatigue failure occurs at nominal stress levels below static uni-axial elastic limit. The main problem with fatigue failure is that it is an insidious form of failure and can be sudden and catastrophic

It has been noted at Institute of Mechanical Engineer Conferences that the physics of fatigue has been well known for over 100 years however, application of this knowledge by engineers still poses challenges. Figure 1.24 shows images of the traditional S-N approach and the "Crack Growth" method. Figure 1.25 shows the development of slip and the initiation intrusions and extrusions and striation on a fracture surface.

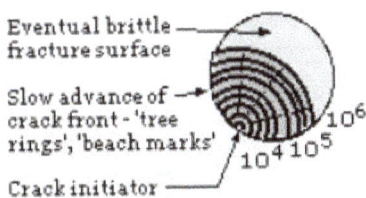

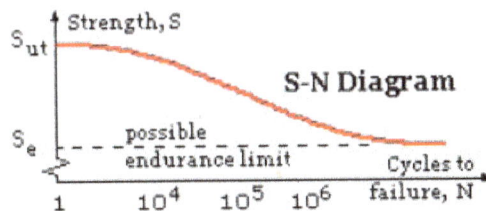

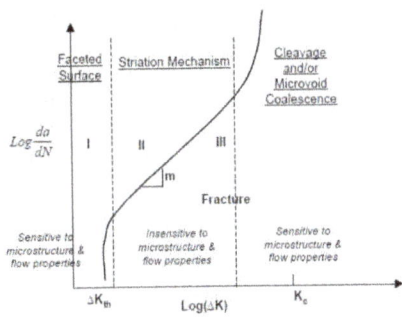

$$\log\left(\frac{da}{dN}\right) = m\log(\Delta K) + \log C$$

Taking out the logs gives:

$$\frac{da}{dN} = C\Delta K^m$$

◀ *Figure 1.24* *Aspects of Fatigue
Top LHS = Beach marks, RHS = S/N
curve. Lower LHS crack growth, RHS
Paris law*

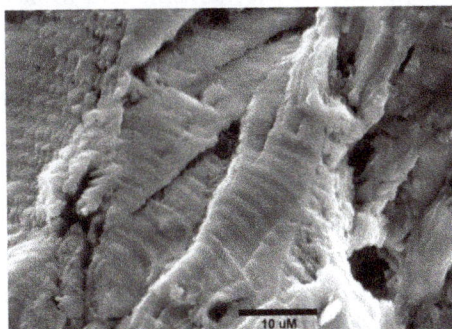

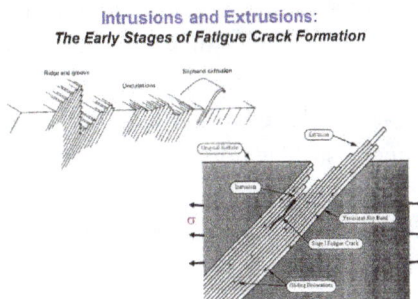

Figure 1.25 *Fatigue – LHS + Striations which requires magnification of X1000 to X20K to see. RHS = Surface fatigue initiation intrusions and extrusions are not always visible.*

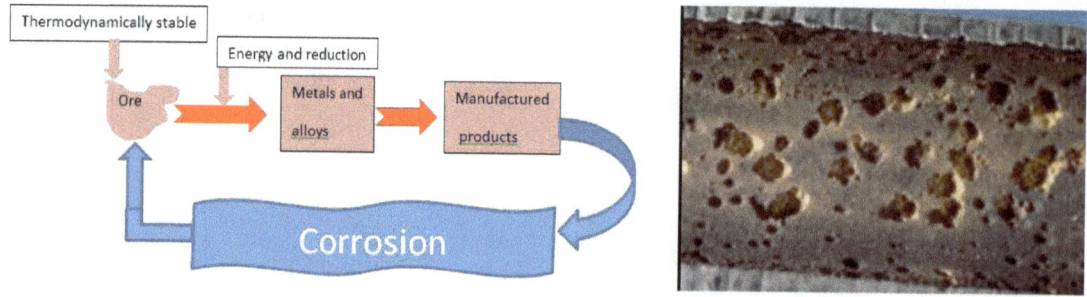

Figure 1.26 *LHS = Corrosion process reverse of metal extraction, RHS = Pitting corrosion (typical where a protective layer is formed but breaks in specific areas (example stainless and chromium oxide layer*

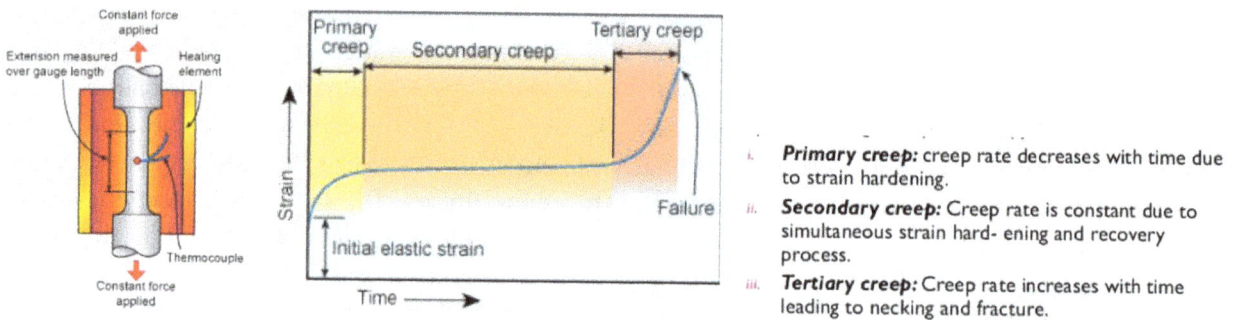

i. **Primary creep:** creep rate decreases with time due to strain hardening.

ii. **Secondary creep:** Creep rate is constant due to simultaneous strain hard- ening and recovery process.

iii. **Tertiary creep:** Creep rate increases with time leading to necking and fracture.

Creep

- Metals/alloys do not exhibit time dependent deformation under normal service condition.
- Metals subjected to a constant load at elevated temperatures will undergo a time dependent increase in length.

At what temperature the material will creep?

- Different metals have different melting temperatures. e.g. Pb 327°C, W 3407°C.
- Material will creep when the temperature will be > 0.5Tm (Tm = absolute melting temperature).

Metal	Melting temp.		0.5xMelting Temp	
Lead	327°C	600°K	327°K	27°C
W	3407	3680	1840	1567

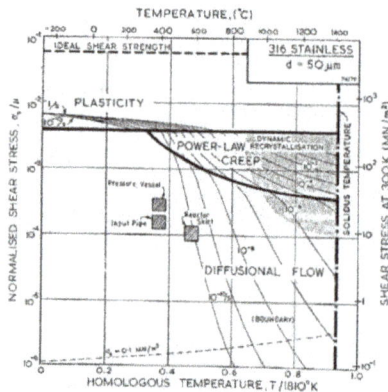

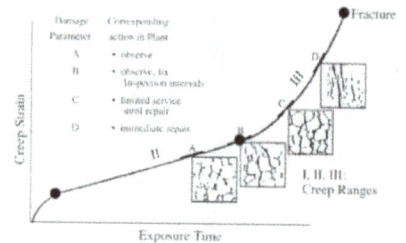

Figure 1. Neubauer's classification of creep damage from observation of replicas and consequent action to be taken.

Figure 1.27 *Aspects of Creep. Top = Creep testing and results Bottom LHS = Background Centre= Creep mechanisms can be summarised in a deformation mechanism map. RHS = Microstructural creep damage*

Corrosion is a natural process that converts a metal or alloy into a more chemically-stable form such as oxide, hydroxide, or sulphide. It is the gradual destruction of metals by chemical and/or electrochemical reaction with their environment. The oxide that forms can either be tenaciously attached to the metal which, would prevent any further corrosion such as the oxide formed on aluminium, titanium and stainless steel or the oxide can be easily removed to allow further corrosion of the base metal such as unprotected mild steel. Figure 1.26 LHS shows the journey of a metal from a mined ore, to a metal and then back to an oxide due to corrosion. The RHS shows pitting due to breaks in a protective layer of oxide.

Creep is the tendency of a solid material to move slowly or deform permanently under the influence of a constant mechanical stress. It can occur as a result of long-term exposure to high levels of stress that are below the yield strength of the material. To control the amount of creep there is a need to understand creep and its mechanisms and to have data on the mathematical relationships between the creep variables (time, stress and temperature) to allow material selection and operating condition to be used which avoid creep failures. Figure 1.27 shows hot tensile testing (top LHS), the classical three stages of creep (top centre and top RHS). The lower diagrams show a deformation mechanism map and on the RHS the type of microstructural creep damage that occurs in service and needs to be evaluated by onsite metallography, taking replicas or the removal or samples.

1.4.2.2 The improvement of PD 6493 1980 and the creation of BS 7910

Since its first publication in 1980, PD 6493 has undergone continuous improvement, changing to BS 7910 in 1999 and adopting some methods from other established procedures such as the UK power industry's fracture assessment procedure R6 code (in particular the Failure Assessment Diagram approach), the creep assessment procedure PD 6539 and the gas transmission industry's approach to assessment of locally thinned areas in pipelines.

TWI (The Welding Institute) has played a major role in the development and application of BS 7910

and the use of the standard for determining "fitness for service". There are several comprehensive presentations and published papers on many aspects of BS7910 covering the history, development and applications by Professor Isabel Hadley. TWI can provide training in the use and application of BS 7910 and have a software known as crackWISE to assist in "fitness for service" considerations.[42]

The FITNET fracture assessment methods were technically regarded as improvements over the initial BS 7910 methods; for example, weld strength mismatch could be analysed by using FITNET. Corrosion assessment methods in FITNET were also more comprehensive than those of BS 7910. In view of

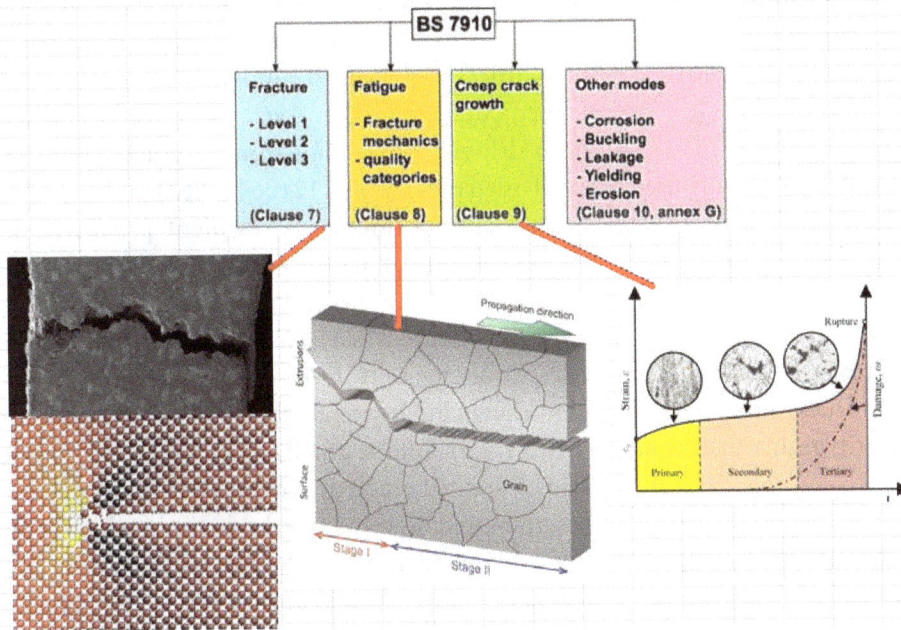

◀ *Figure 1.28* Failure modes covered in BS 7910. BS 7910 provides design guidance.

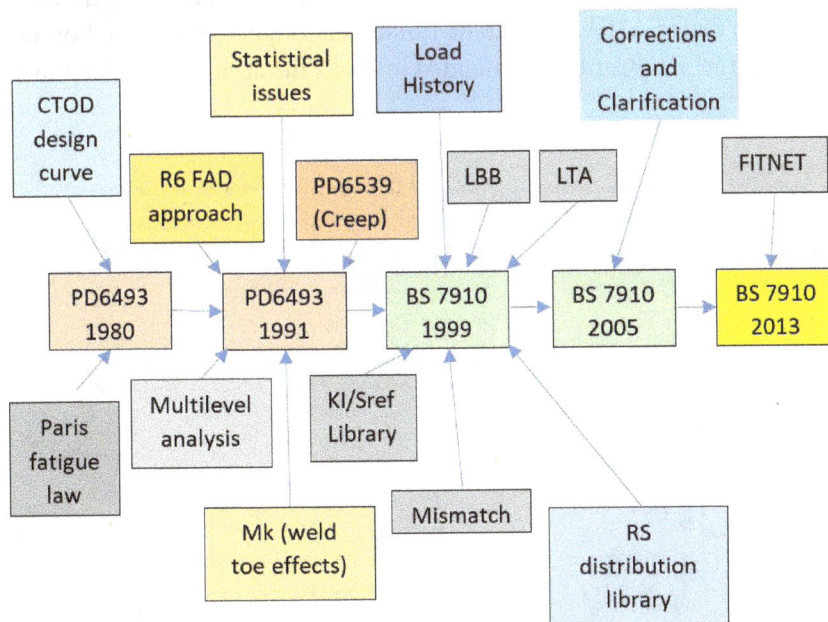

◀ *Figure 1.29* Evolution of BS 7910 From (42)

the opportunity to improve, the BS 7910 committee incorporated many elements of the FITNET procedure into the 2013 edition of BS 7910.

Through these various stages of development and by including the knowledge gained through the SINTAP and FITNET projects, the PD eventually became the standard BS 7910-2013 which provides essential guidance to avoid the main failure modes in service. The BS 7910-2013 represents a very comprehensive set of rules to assist design and manufacture.[42]

FITNET was a European "fitness for service" project that ran between 2002 and 2006. (http://eurofitnet.org/fitnet/FITNETFinalTechnRepwithProjManagement30Jan07.pdf) It led to an improved knowledge base by incorporating some assessment methods from SINTAP.

SINTAP was an earlier European research project (http://www.eurofitnet.org/sintap_Procedure_version_1a.pdf)

- ECA: Engineering critical assessment, usually used to denote a fitness-for-service analysis of a cracked body
- LTA: Locally thinned area
- RS: Residual stress
- Mk. Stress concentration factor at weld toe
- FFS/FFP: Fitness-for-service/fitness-for-purpose, i.e. an analysis based on integrity rather than code compliance
- LBB: Leak before break
- CTOD: Crack tip opening displacement
- FAD: Failure assessment diagram
- LEFM: Linear elastic fracture mechanics

Figure 1.30 shows a failure assessment diagram (FAD). This approach was developed by the UK Central Electricity Generating Board (CEGB) as an approach to the assessment of imperfections that could lead to static ductile or brittle fracture. This was part of the program needed to demonstrate the integrity of nuclear pressure vessels and of large rotor forgings.

In the Failure Assessment Diagram (FAD):

- the vertical axes would be the ratio of the "fracture driving force" to the "fracture toughness"
- the horizontal axes would be the "applied load" to the "plastic collapse load".

Failure was predicted when either of these ratios exceeded unity. Interaction between brittle behaviour and plastic collapse was allowed for by a curve derived from a strip yield analysis. (Elastic analysis predicts singular stresses (infinite) at the crack-tip, which is unrealistic. In an effort to eliminate such an unrealistic prediction, Dugdale in 1960 introduced yielded or intact strip zones extending from the crack-tip. http://textofvideo.nptel.ac.in/112106065/lec30.pdf)

The Level 1 fracture assessment procedure in BS 7910 is similar to the initial PD 6493 but was renamed the 'Simplified Assessment' procedure. The Level 1 procedure was based on a conservative Failure Assessment Diagram (FAD).

As shown in Figure 1.31 the Level 1 FAD has Kr (or $\sqrt{\delta r}$), S r as axis of the graph, where Kr (or $\sqrt{\delta r}$) is the ratio of applied crack driving force (SIF) to fracture toughness and on the x axis the factor Lr which is the ratio of applied stress to flow strength (where flow strength is mean of yield and tensile strength, hence incorporating some plasticity).

Figure 1.32 shows the FITNET options for five different 'Routes' for fatigue assessment. Routes 1 to 3 are intended to assess the accumulated fatigue damage at

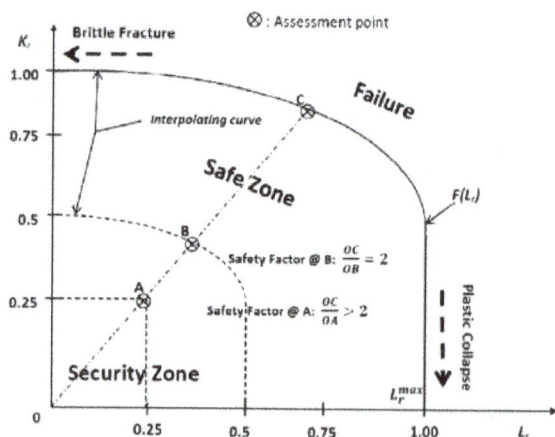

Figure 1.30 *R6 code FAD*

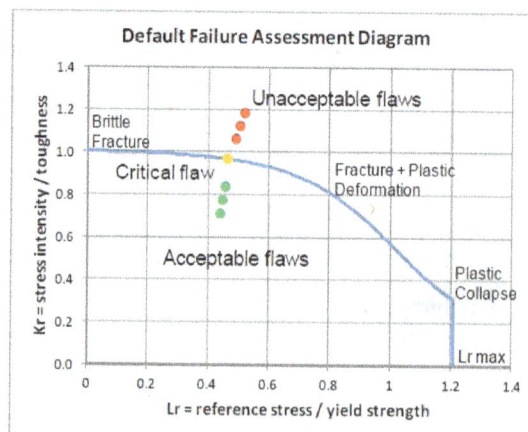

Figure 1.31 *The Level 1 FAD*

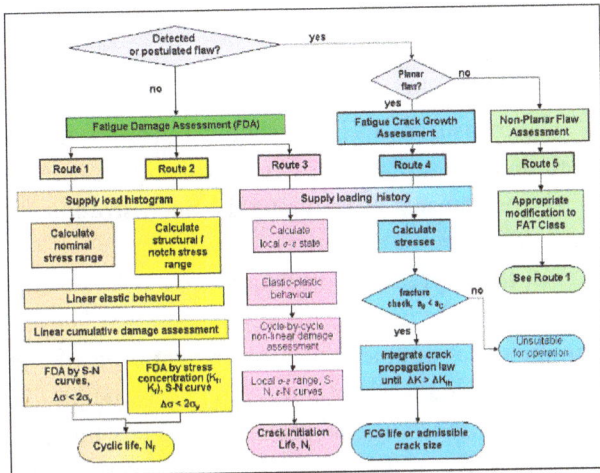

Figure 1.32 Summary of the five FITNET fatigue assessment Routes

critical locations in the absence of a pre-existing flaw. If real damage is present, then Routes 4 or 5 are used. Route 4 is the appropriate approach for assessing planar crack-like flaws, and uses a fatigue crack growth analysis to calculate the growth of a pre-existing flaw for the purpose of making a safe FFS assessment. Route 5 is used only to assess non-planar flaws, and uses an S-N curve (stress versus number of cycles) approach.

1.4.3 The Automotive Initiatives

Product deficiencies have been identified in many industries, particularly the automotive, and are often traced to a lack of competency in heat treatment, plating and coating for corrosion protection and cast metal quality. These were identified as key processes which were essential to provide the required mechanical integrity and durability to meet the "fitness for service" requirements of products supplied to the automotive industry. The automotive industry (AIAG –Automotive Industries Action Group)

began many "initiatives" to solve the weaknesses in the supply chain of automotive part manufacture and to encourage the spirit of continuous improvement. These audits make technical capability and proficiency an important factor.

AIAG offer know-how in the following CQI standards:

CQI-9 Heat Treat Assessment
CQI-17 Soldering System Assessment
CQI-11 Plating System Assessment
CQI-22 Cost of Poor-Quality Guide
CQI-12 Coating System Assessment
CQI-23 Moulding System Assessment
CQI-15 Welding System Assessment
CQI-27 Casting System Assessment

The first edition of CQI-9 which was a procedure to assess the process capability of a heat treatment company was released in March 2006. The 3rd addition introduced changes to SATs (System Accuracy Tests) and calibration frequency[43].

To specify, carry out and guarantee the quality of heat treatment requires a detailed knowledge of heat treatment technology. This requires an understanding of the effects of composition, microstructure and the possible ways in which specific metals can be heat treated. For steels the important aspects would be the chemical composition, the "mass effect" (thermal heating and hardenability aspects) of the parts and the control of quench technology. (These topics will be covered in more detail in Chapter 4.)

Failure to achieve the required quality in any process should be avoided since it leads to the additional costs for re-processing or the cost of damaged parts and valuable management time associated with the 8D and RCA process (root cause analysis) such as that shown in Figure 1.33. The old QA philosophy of "right first time" should still be the ambition. The main business asset required to achieve that objective would be competent, motivated and well-trained staff who have the required knowledge base and access to the appropriate information technology.

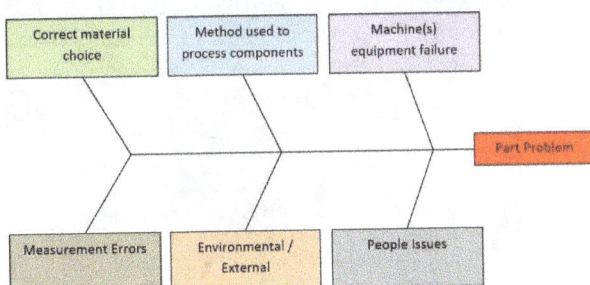

Figure 1.33 Ishikawa diagrams that show the causes of a specific event.

1.5 THE BACKGROUND TO METALLOGRAPHY, THERMAL ANALYSIS AND THE EQUILIBRIUM DIAGRAM

The fact that steel can be hardened has been known for thousands of years. However, an understanding of the exact scientific mechanism was delayed until the 1920s due to the lack of suitable methods to study the changes that occur during hardening.[45] In a similar way the understanding of the hardening mechanism in aluminium-copper alloys that were developed in the early 1900s[45] required the use of X-ray diffraction to establish the hardening mechanism.[46] The development of these techniques gave insight into many of the important aspects of metals technology such as composition, cooling rate, work hardening and strengthening mechanisms. Materials technology involves:

- Providing an explanation of how the various microstructures form and create the associated metallographic appearance
- The measurement of mechanical properties and understanding how the mechanical property values can be adjusted by changing the microstructure.

Part of the approach in this document will relate to the timeline evolution of the key technological developments that enabled a full understanding of the microstructural features and mechanical strength. In addition, important background to metals involves the knowledge of the timeline sequence of their discovery and also the relative occurrence on the planet which are shown on Figures 1.34 and 1.35 for the moon.

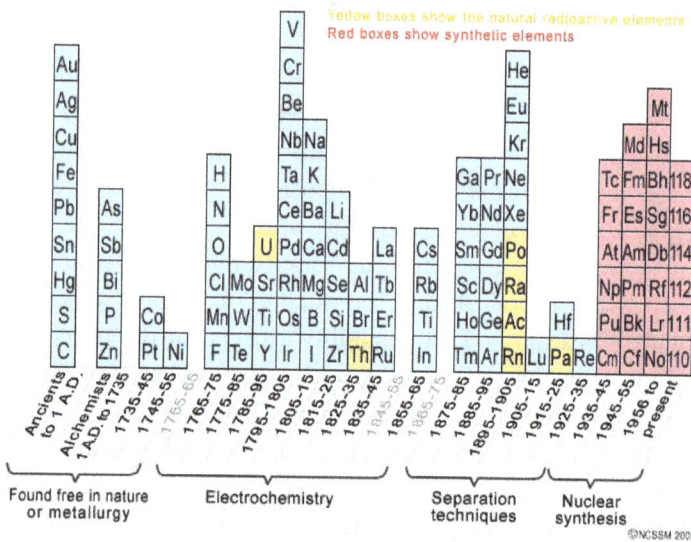

Figure 1.34 Year of discovery of metals RHS = Elements on Earth

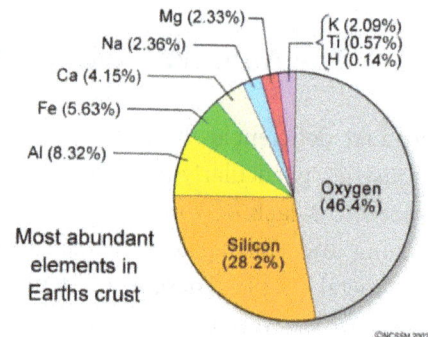

In 1972 Helium 3 found in moon soil samples. Identified as an ideal element for fusion. Very little on planet earth. Sufficient to financially justify mining. Fusion research still major R&D topic (Lockheed Martin Compact Fusion, ITER, CCFE JET) Plus rare earth elements

https://ktwop.com/2014/08/16/its-coming-but-dont-invest-just-yet-in-mining-helium-3-on-the-moon/

http://www.ccfe.ac.uk/assets/Documents/CPS16.49_May2016_low.pdf

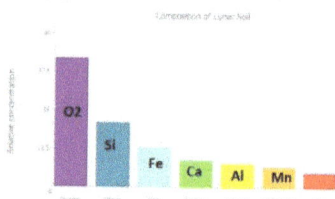

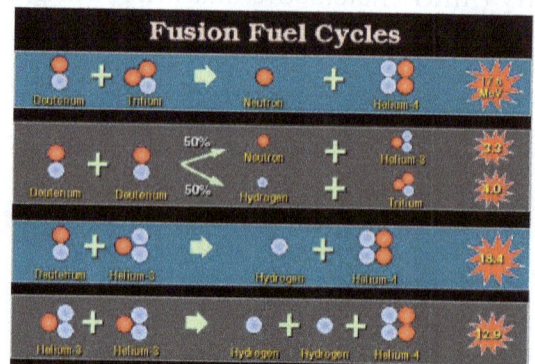

Figure 1.35 LHS = Elements found on the moon. RHS = The importance of Helium-3 to Nuclear Fusion (Clean nuclear without radiation problems – No neutron produced.

1.6 TIMELINE OF ASPECTS OF METALLURGY DEVELOPMENT

The initial development work that allowed a greater understanding of the metallurgy resulted from a combination of metallography and thermal analysis that resulted in the equilibrium diagram (1880 to 1920). Some of the key aspects of these technical discoveries are shown in the following list.

1.6.1 Optical metallography – Martens and Sorby 1860s [USSR claimed 1830]

Henry Clifton Sorby

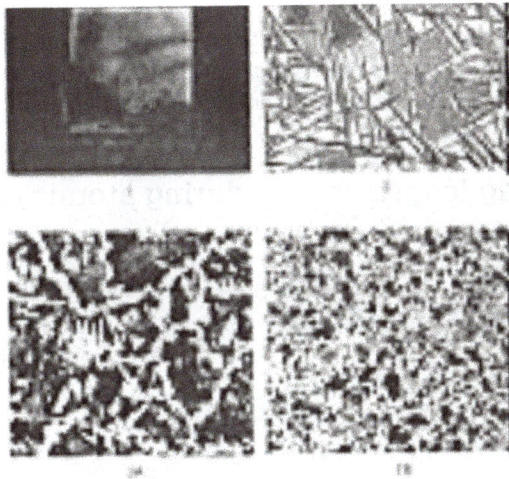

◄ *Figure 1.36 H C Sorby is credited with the first examination of metallic microstructure using an optical microscope Sorby in 1886 described a pearly compound in the steel microstructure. This phase eventually named pearlite after this first observation*

1.6.2 Platinum-Rhodium thermo-couple 1885 – Le Chatelier (Initial principle Seebeck 1820) used by Austin to produce equilibrium diagrams

Seebeck and the reverse (Peltier) Effects
~ millivolts/K for (Pb,Bi)Te

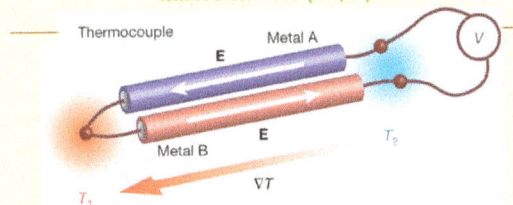

The *Seebeck effect* is the conversion of temperature differences directly into electricity.
Applications: Temperature measurement via thermocouples; thermoelectric power generators; thermoelectric refrigerators; recovering waste heat

Figure 1.37 *Le Chatelier was the first to use a rhodium-platinum alloy connected to platinum metal to produce a thermocouple-1885. Le Chatelier also recommended calibration of the thermocouple by the use of the fixed points of melting or boiling of pure substances. (Reference L. B. Hunt, The Early History of the Thermocouple Platinum Metals Rev., 1964, 8, (1), 23-28)*

1.6.3 Thermal analysis – Simple thermal analysis and differential thermal analysis DTA

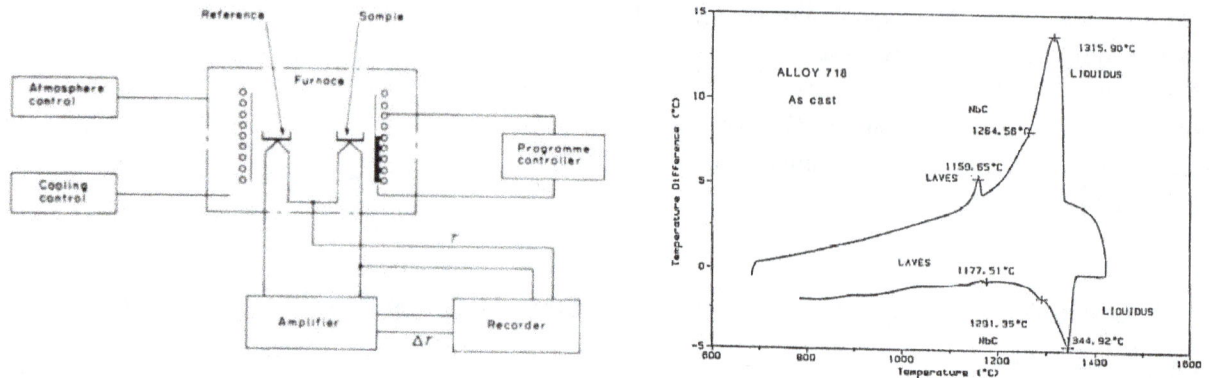

Figure 1.38 *Diagram of equipment and the output data of differential thermal analysis. The approach monitors the difference in temperature between a metal that undergoes no phase changes such as nickel or platinum and the specimen that is examined. The graph above shows the changes in 718 inconel against a platinum reference sample.*

1.6.4 Dilatometer—monitoring length change during atomic rearrangement

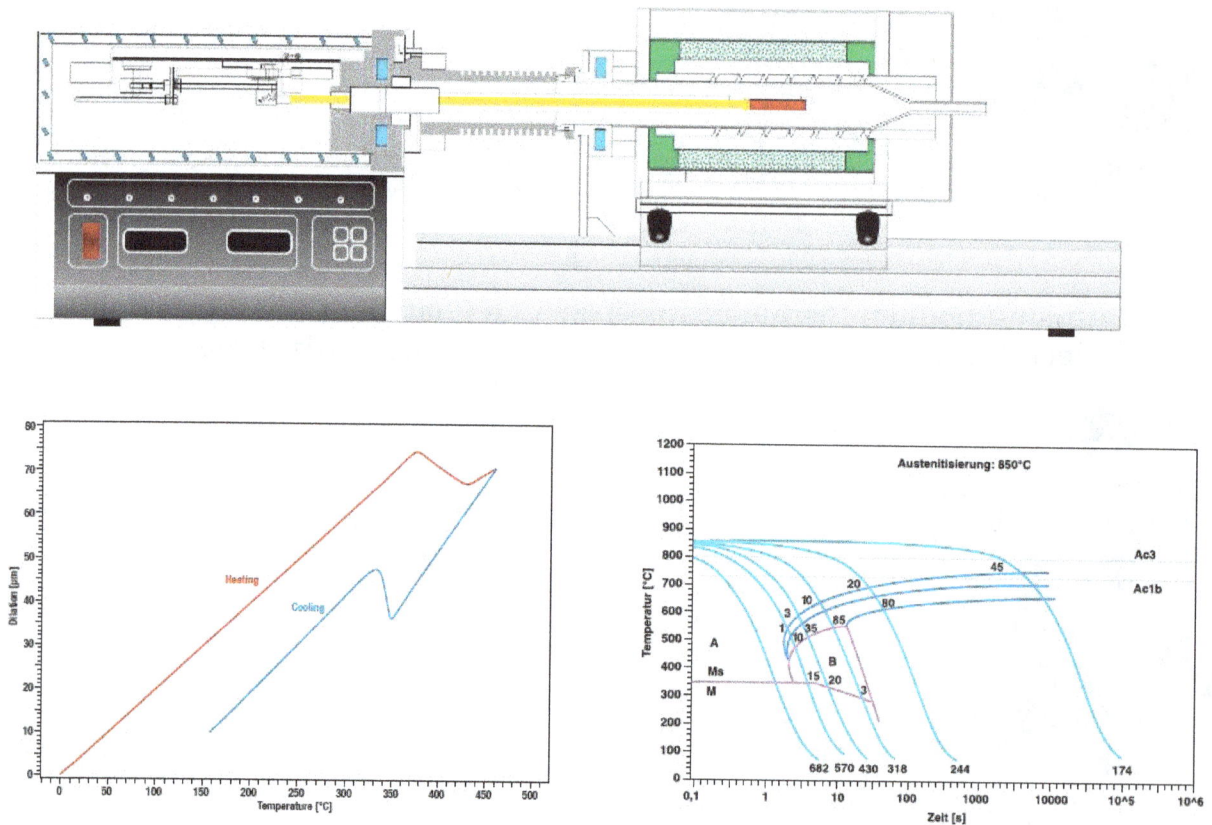

Figure 1.39 *A horizontal dilatometer showing a heating curve and a cooling curve with phase changes. On the RHS the cooling curve data that reveals the temperatures are rates of transformation are converted to a CCT diagram https:// www.linseis.com/wp-content/uploads/2018/06/L78_Rita_Quenching_Dilatometer_01.pdf*

1.6.5 Equilibrium diagrams – William Roberts-Austen developed the phase diagram for iron and carbon. 1898

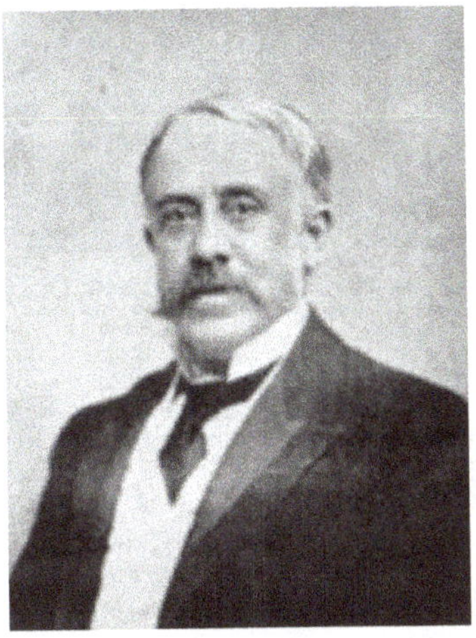

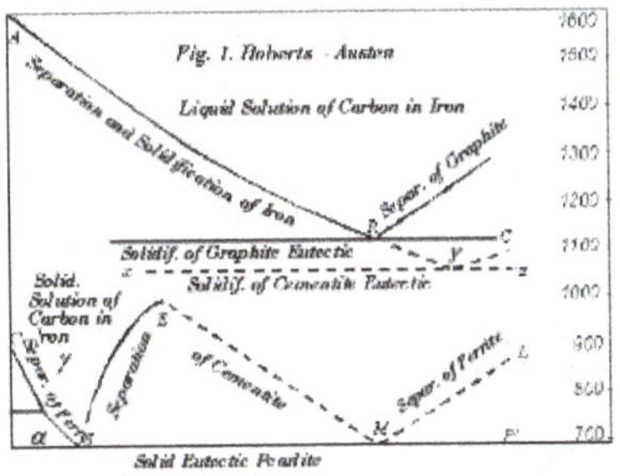

First phase diagram of iron - carbon by Roberts-Austen
Redrawn as close as possible to the (rather fuzzy) original with some minor omissions of (unreadable) letters

Figure 1.40 *William Chandler Roberts-Austen, a British metallurgist (1843–1902) was appointed to the Chair of Metallurgy at the Royal School of Mines (RSM) in1870 and started his researches into the structure and properties of steel. He published the first "T – x" diagrams or measured temperature (T) composition (x) diagrams. His first T – x diagram for the iron – carbon system appeared in 1897 He also carried out studies of diffusion in metals while in charge of the Mint*

1.6.6 Periodic classification of elements – Dmitri Mendeleev devised the Periodic Table of Elements.

I	II	III	IV	V	VI	VII	VIII		
H 1.01									
Li 6.94	Be 9.01	B 10.8	C 12.0	N 14.0	O 16.0	F 19.0			
Na 23.0	Mg 24.3	Al 27.0	Si 28.1	P 31.0	S 32.1	Cl 35.5			
K 39.1 Cu 63.5	Ca 40.1 Zn 65.4		Ti 47.9	V 50.9 As 74.9	Cr 52.0 Se 79.0	Mn 54.9 Br 79.9	Fe 55.9	Co 58.9	Ni 58.7
Rb 85.5 Ag 108	Sr 87.6 Cd 112	Y 88.9 In 115	Zr 91.2 Sn 119	Nb 92.9 Sb 122	Mo 95.9 Te 128	I 127	Ru 101	Rh 103	Pd 106
Ce 133 Au 197	Ba 137 Hg 201	La 139 Tl 204	Pb 207	Ta 181 Bi 209	W 184		Os 194	Ir 192	Pt 195
			Th 232		U 238				

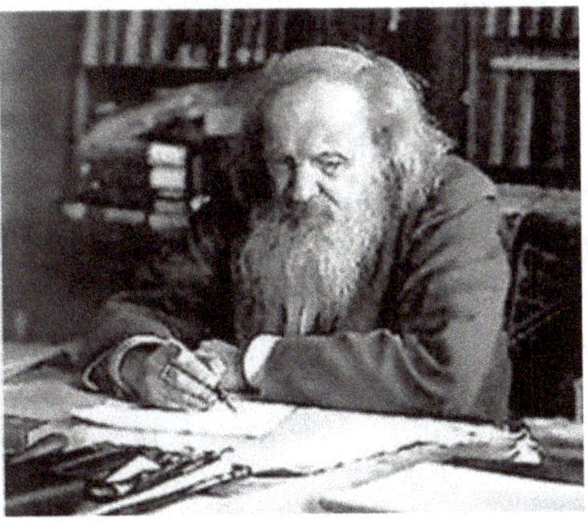

Figure 1.41 *Dmitri Mendeleev was born in Russia in 1834 and died in 1907. Mendeleev studied science at St. Petersburg and graduated in 1856. Mendeleev is best known for his work on the periodic table; arranging the 63 known elements into a Periodic Table based on atomic mass, which he published in Principles of Chemistry in 1869. His first Periodic Table was compiled on the basis of arranging the elements in ascending order of atomic weight and grouping them by similarity of properties.*

1.6.7 Gibbs free energy – J. Willard Gibbs published thermodynamics 1876

J W Gibbs (late 19th Century) combined 1st and 2nd Laws to express spontaneity of reactions in terms of measurable system parameters.

$$\Delta G = \Delta H - T\Delta S \tag{2}$$

ΔH: change in enthalpy (heat content)

T: absolute temp.

ΔG: change in (Gibbs) Free Energy. Units: J/mol

: a measure of the useful work system can perform

: must be – ve for spontaneous reaction

Gibbs Phase Rule   F = C – P + 1

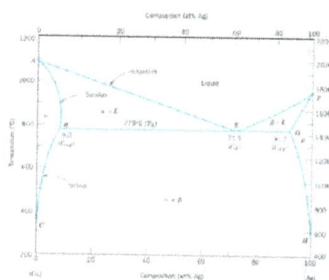

Apply to eutectic phase diagram

1 phase field: F = 2 – 1 + 1 = 2   Change T and C independently in phase field

2 phase field: F = 2 – 2 + 1 = 1   C depends on T – not independent

3 phase point: F = 2 – 3 +1 = 0   C and T defined only at one point
(Eutectic point)   (no degrees of freedom)

Major Step

A comment from a book by **ALEX FINDLAY, M.A., Ph.D., D.Sc. Lecturer on physical chemistry at the University of Birmingham, regarding the Phase Rule and its application in 1903 is a tribute to Gibb's mathematical brilliance:**

"Although we are indebted to the late Professor Willard Gibbs for the first enunciation of the Phase Rule, it was not till 1887 that its practical applicability to the study of Chemical Equilibria was made apparent. In that year Roozeboom disclosed the great generalization, which for upwards of ten years had remained hidden and unknown save to a very few, by stripping from it the garb of abstract Mathematics in which it had been clothed by its first discoverer. The Phase Rule was thus made generally accessible; and its adoption by Roozeboom as the basis of classification of the different cases of chemical equilibrium then known established its value, not only as a means of co-ordinating the large number of isolated cases of equilibrium and of giving a deeper insight into the relationships existing between the different systems, but also as a guide in the investigation of unknown systems".

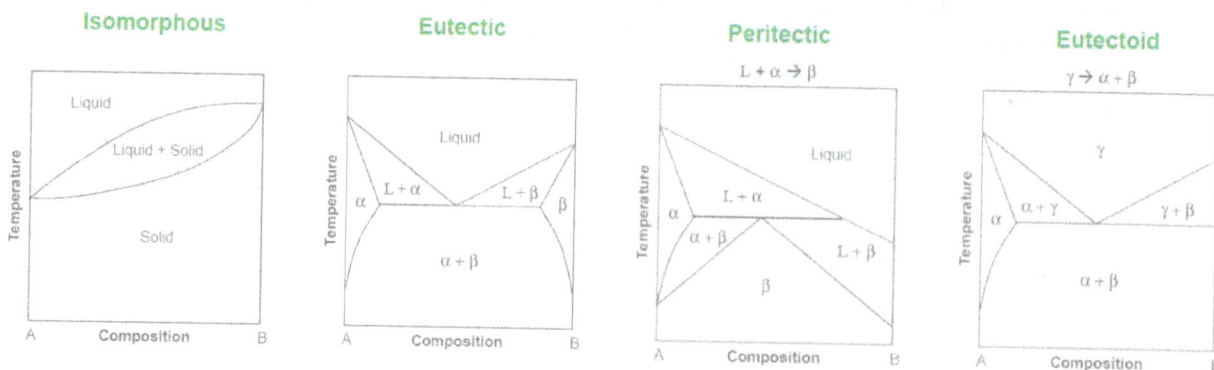

Figure 1.42 *Josiah Willard Gibbs (1839-1903) was an American mathematical physicist whose work in statistical mechanics laid the basis for the development of physical chemistry as a science. During the years from 1876 to 1878, he worked on the principles of thermodynamics, applying them to the complex processes involved in chemical reaction.*

1.6.8 XRD – Max von Laue discovered the diffraction of X-rays by crystals. 1912

1895 X-rays discovered by Roentgen

1914 First diffraction pattern of a crystal made by Knipping and von Laue

1915 Theory to determine crystal structure from diffraction pattern developed by Bragg.

1953 DNA structure solved by Watson and Crick

Now   Diffraction improved by computer technology; methods used to determine atomic structures and in medical applications

- **1912**: Max Theordor Felix von Laue (1879-1960) (U of Munich) thought that X-ray has a wavelength similar to interatomic distances in crystals
  Along with Walter Friedrich (research assistant) and Paul Knipping (PhD grad student), he did the first diffraction experiment on $CuSO_4$ crystal – Nobel Prize in Physics 1914.

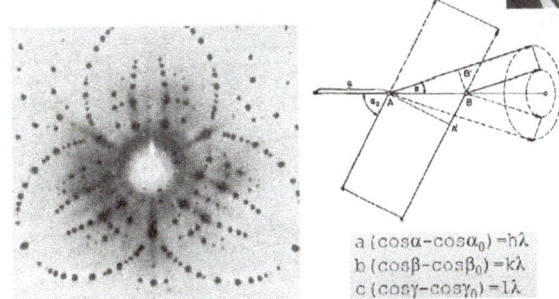

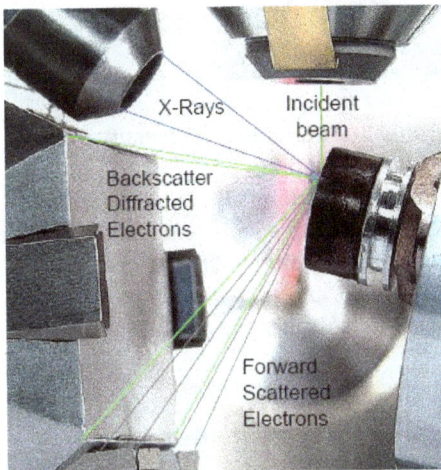

$$a\,(\cos\alpha - \cos\alpha_0) = h\lambda$$
$$b\,(\cos\beta - \cos\beta_0) = k\lambda$$
$$c\,(\cos\gamma - \cos\gamma_0) = l\lambda$$

Applications Example: Phase distribution & texture in laser welded duplex steel

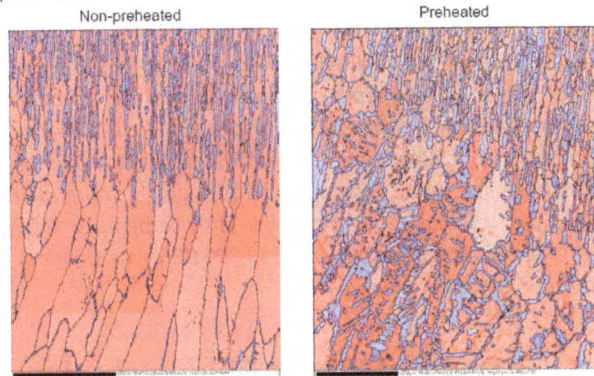

Non-preheated   Preheated

Phase maps (+ GBs): red = ferrite, blue = austenite   47/63

X-Rays   Incident beam

Backscatter Diffracted Electrons

Forward Scattered Electrons

Figure 1.43 *X-ray diffraction and the modern application in SEM-EBSD (electron back-scattered diffraction*

1.6.9 Gleeble – DSI (Dynamic Systems Inc)

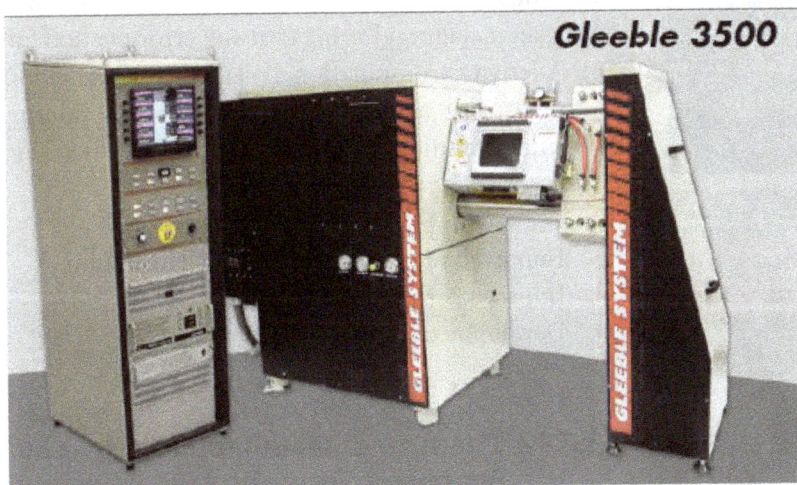

Gleeble 3500

Process Simulation

- Continuous casting
- Mushy zone processing
- Hot rolling
- Forging
- Extrusion
- Weld HAZ cycles
- Upset butt welding
- Diffusion bonding
- Continuous strip annealing
- Heat treating
- Quenching
- Powder metallurgy/sintering
- Synthesis (SHS)

Figure 1.44 *DSI was founded in 1957 by Dr. Warren Savage and others as a result of work being done at Rensselaer Polytechnic Institute (RPI) in welding research. The Gleeble creators wanted a machine that would allow them to simulate welding on a laboratory scale particularly the HAZ.*

1.6.10 Scanning Electron microscope

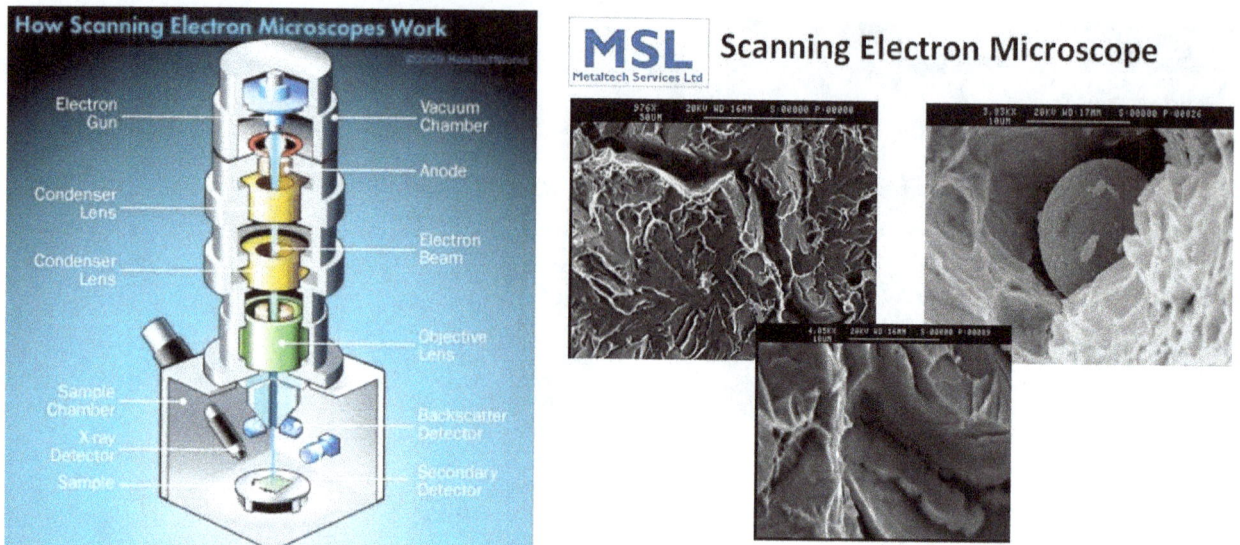

Figure 1.45 *In 1948 C. W. Oatley, then a lecturer in the Engineering Department of Cambridge University, became interested in conducting research in the field of electron optics and decided to re-investigate the SEM. Over the next 15 years a succession of his research students built 5 SEMs of increasingly improved performance culminating in the production of a commercial instrument by the Cambridge Instrument Company in 1965. One story told when the SEM was being considered as a commercial product, a group of Marketing experts were sent out to make an evaluation of the number of SEMs that could be sold. They came back with probably between 6 (six), and 10 (ten) would saturate the market. Today there are in excess of 50,000 SEMs worldwide. (In 1997) The Market research personnel obviously asked the wrong question.*

1.6.11 Sir Harshad Kumar Dharamshi Hansraj Bhadeshia, FRS, FREng, FNAE

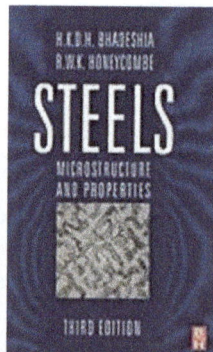

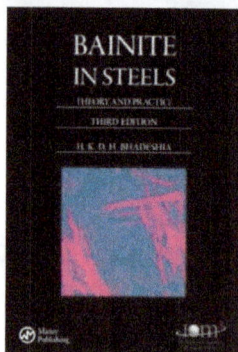

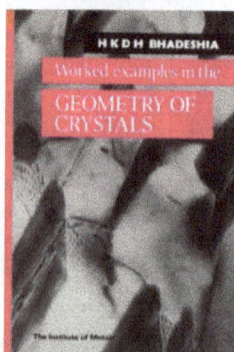

metallurgy and several books. Most of his work is open source and can be downloaded for personal study at the Cambridge website. He played a major role in the education of metallurgy of steel (various YouTube lectures) and in the moving away from "bucket metallurgy" using trial and error method for steel design to computational methods of steel design.

His major achievements have been in gaining an understanding and exploitation of bainite. This allowed:

• Improved rail steels used in the channel tunnel
• Improved armour plate
• Development of new bearing steels.

1.6.12 The Age of Nanotechnology. The Scanning Tunnelling Microscope and Atomic Force Microscope

Harry Bhadeshia obtained a PhD at Cambridge in 1981 and has been part of the academic staff at the University of Cambridge since then. He is the author of more than 500 published papers in the field of

Scanning Probe Microscopy (SPM)

In 1981, a new microscope became available which was powerful enough to allow a single atom to be viewed.

Typical STM schematic

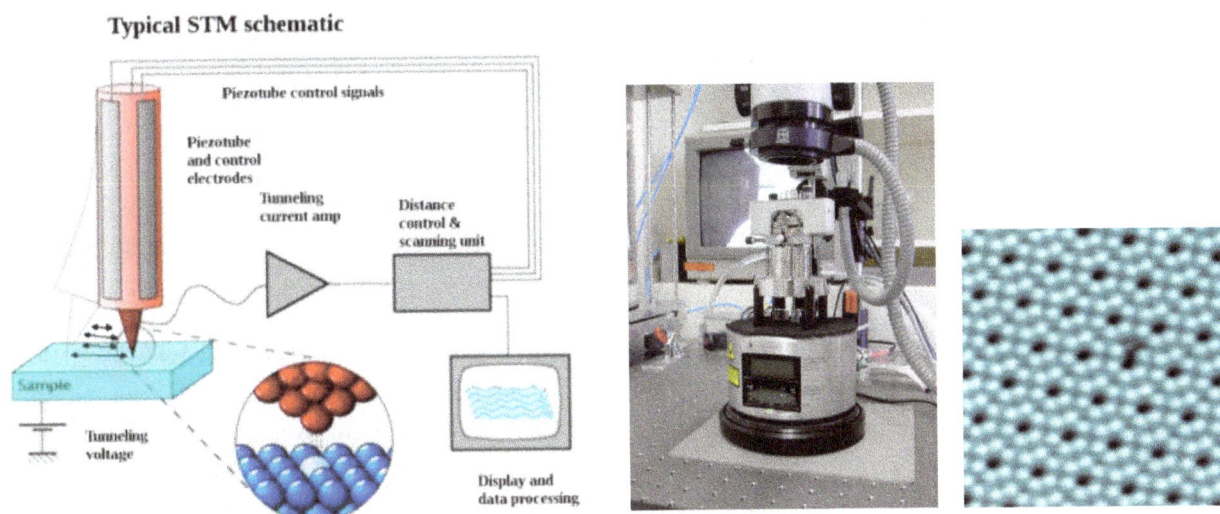

Figure 1.46 *Scanning Probe Microscopy (SPM). From file:///C:/Users/user/Downloads/SSP.228.AB_Scanning ElectrochemicalMicroscopy.pdf*

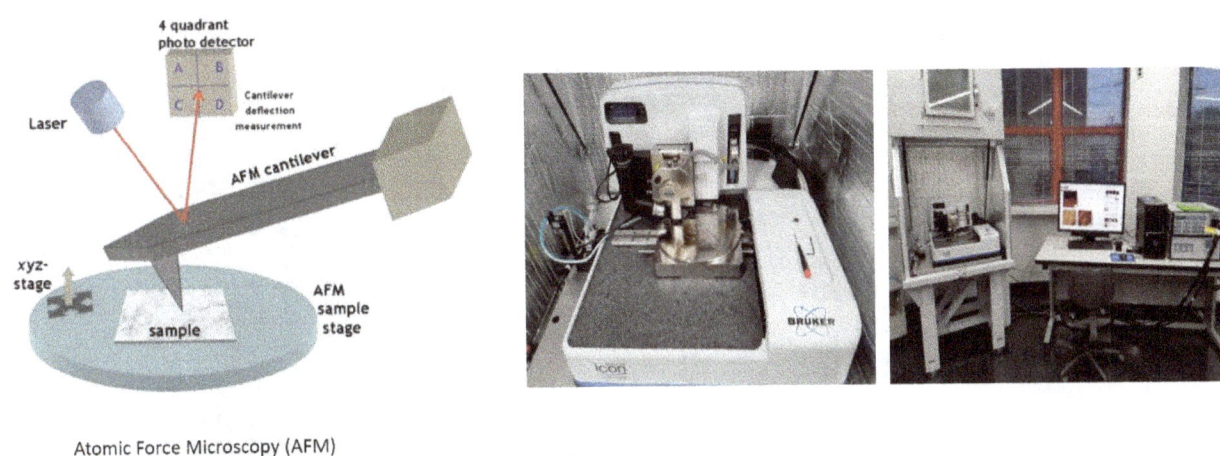

Atomic Force Microscopy (AFM)

Figure 1.47 *The Atomic Force Microscope (AFM)*

The device was the Scanning Tunnelling Microscope invented by G. Binning and H. Rohrer for which they received the Nobel Award. In the STM a sharp probe is scanned across a surface and the probe/sample interactions are monitored. To accomplish this a bias voltage is applied between probe and the surface, which causes electrons to travel between the probe and the surface. The current flow depends upon tip position relative to the surface, the applied voltage and local density state of the sample. STM uses the variation in output electric current to detect the height of individual atoms. In 1990 it was also discovered that the Scanning Tunnelling Microscope could also be used to pick up, move, and precisely place atoms, one at a time. This opens up the concept of manipulating materials atom by atom. It has been used in the development of catalysts, electronic and material science. There are now a range of similar microscopes under the name scanning probe microscope.

The Atomic Force Microscope (AFM) belongs to the family of the Scanning Probe Microscopy (SPM). The AFM senses interatomic forces that occur between a probe tip and a substrate. AFM made by Gerd Binnig and Cristoph Gerber in 1985. The AFM was developed to view non-conducting surfaces.

1.6.13 The Emergence of Computational Material Science

http://www.dierk-raabe.com/ (includes his book on computational material science and his website is one of the best sites with extensive technical content)

The material science was a rebranding of the term metallurgy in the early 1960s to try to broaden the awareness and include the knowledge of plastics, ceramics and composite materials. At that time many metallurgy departments were renamed "metallurgy and materials science". In the 1970s they dropped the term "metallurgy" altogether and became known as "departments of materials science and engineering". MSE attempted to unify metallurgy with polymer science, physical and inorganic chemistry, mineralogy, glass and ceramic technology and solid-state physics. Probably and impossible task.

In countries that have few raw materials, such as the UK, it is an essential to focus on how material science and engineering can contribute to gross national product (GNP) and the nation's "well-being". This typically requires some guidance, infrastructure and direction from the government normally given in the form of "road maps" outlining strategic industries and directing "seed funding" and grant aid.[47]

In the 1980s MSE acquired the concept of being characterised by the interaction between four variables: **structure, properties, performances and process**. This has commonly been visualized with the help of a tetrahedron indicating the interrelation of each of the "four elements" with each of the three others.[48]

To assist in the identification and development of new materials the development computational material science and Integrated Computational Materials Engineering have a new generation of tools.

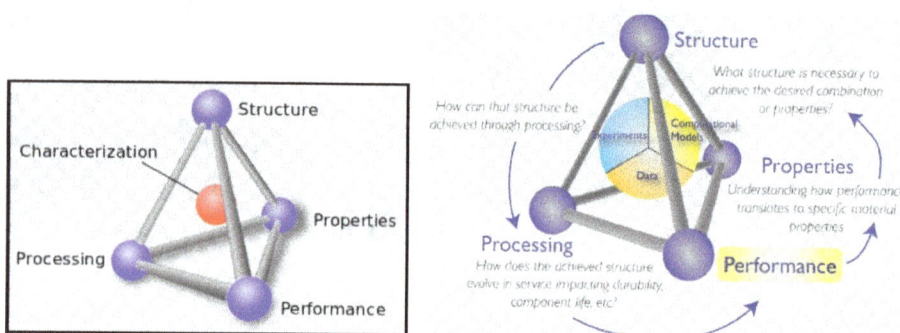

PROF. DR. DIERK RAABE

Currently he is Chief Executive of the Max-Planck Institut für Eisenforschung in Düsseldorf and Professor at RWTH Aachen University. His research interests are in microstructures, simulations and mechanical properties of metallic alloys. He wrote and edited several books His web site at http://www.dierk-raabe.com/ has world class leading edge content

Figure 1.48 *LHS = MSE, Centre= ICME, RHS = PROF. Dr. Dierk Raabe.*

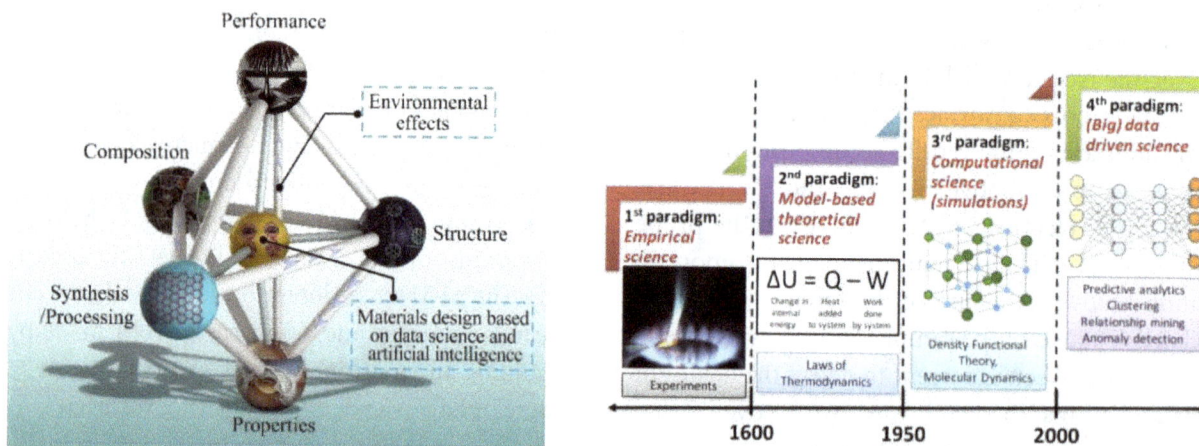

Figure 1.49 *LHS = The DIMSE hexahedron. Incorporating AI (Artificial Intelligence) into MSE from (49). RHS = Big Data From (50) (51)*

1.7 HARDENING MECHANISMS

It is now well established that mechanical properties are structure-sensitive and the actual strength and the toughness depends largely on size, grain size and shape and distribution of various micro-constituents present in the microstructure. (See Figure 1.50.) To improve yield strength the method would be to increase the intrinsic resistance to dislocation motion. (see section 2.3.4.2. for explanation of dislocations) However, there is a general trend that as the strength increases toughness deceases, so in most cases the optimum levels have to be found. The key metal strengthening mechanisms can be summarised as follows:

- Grain refinement (discovered by Hall and Petch)
- Strain or work hardening
- Solid solution hardening
- Precipitation hardening
- Martensitic transformation hardening.

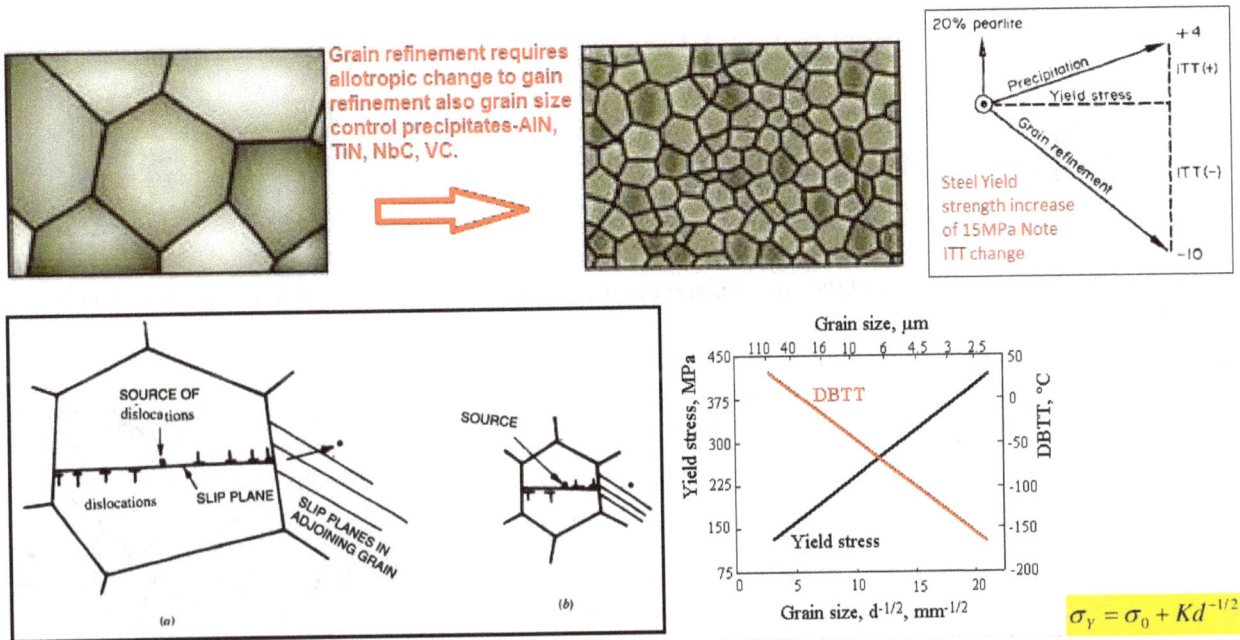

Grain refinement requires allotropic change to gain refinement also grain size control precipitates-AlN, TiN, NbC, VC.

Steel Yield strength increase of 15MPa Note ITT change

$$\sigma_Y = \sigma_0 + Kd^{-1/2}$$

Figure 1.50 *Grain refinement*

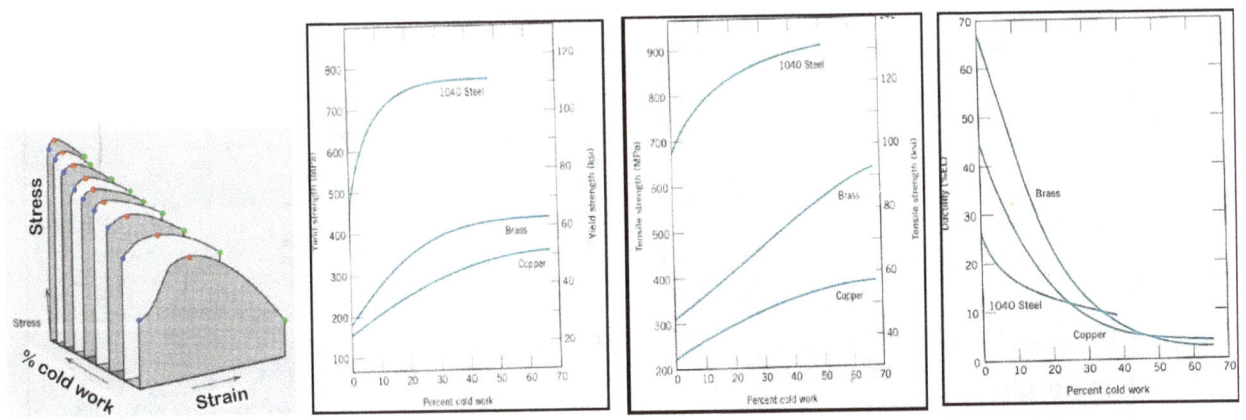

Figure 1.51 *Strain or work hardening*

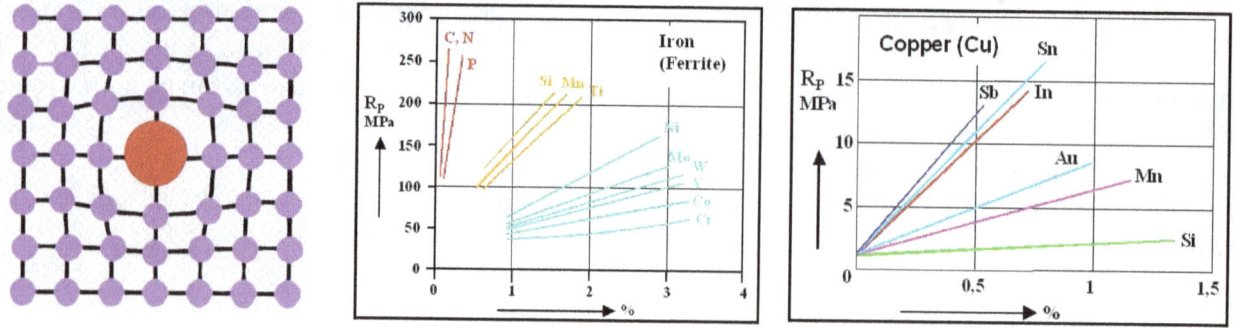

Figure 1.52 *Solid solution hardening*

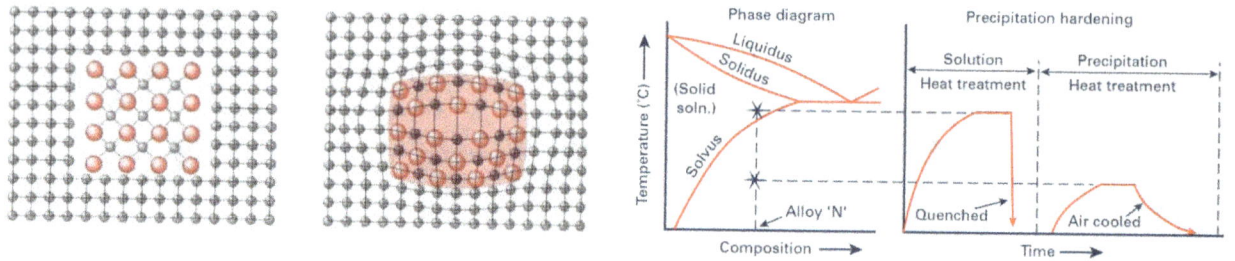

Figure 1.53 *Precipitation hardening LHS = non-coherent with no strengthening. RHS = coherent and strengthening (45 and 46)*

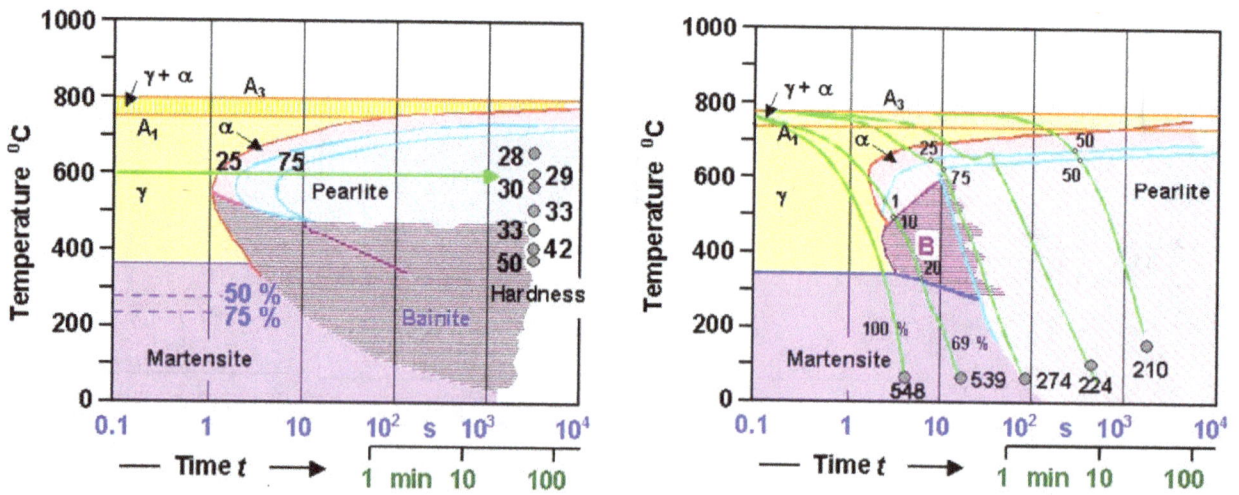

Figure 1.54 *Martensitic transformation hardening the target is change from austenite to tempered martensite. Top LHS = TTT with microstructures. Centre and RHS LOM and SEM images of bainite/martensite mixtures. Lower TTT (LHS) and CCT diagrams for a medium carbon steel*

1.7.1 The hardening mechanism in steel – Transformation Hardening – A brief explanation

To provide an understanding of the hardening of steel the following factors are important:

- Steel exists as a BCC (body centred cubic) atomic structure at room temperature. The BCC can only accommodate a very small amount of carbon. (carbon in solid solution in the iron matrix)
- At about 900°C (for pure iron the actual will depend upon carbon level) the atomic structure of pure iron changes to FCC (face centred cubic) which can accommodate up to 2% carbon.
- The change from BCC to FCC is known as the allotropic change.
- This change in atomic structure forms the basis of "transformation hardening"
- For example, the steel is heated to form the FCC atomic structure known as austenite (named after Roberts-Austen). The carbon is taken into solution in austenite.
- If the structure is then cooled rapidly the carbon is trapped and the transformation occurs by a shear mechanism (no diffusion) and forms martensite (named after Professor Adolf Martens) which is hard.
- The hardness of the martensite depends mainly on the carbon content of the steel. The type of graph used in most business organisations for predicting the martensite content based on the carbon content and hardness is shown in Figure 1.55 (From DIN 17020).
- If cooling occurs at a slower rate some carbon diffusion occurs to form a separate compound of iron carbide (Fe_3C) and the transformation creates bainite (named after Edgar Bain) and at even slower cooling rate allows complete separation of carbon as cementite bands in a ferrite/pearlite microstructure. (Named Pearlite-because it looked like mother of pearl in the microscope.) (See Fig 1.62 RHS and Figure 1.63)

The term heat treatment describes a process in which a part is intentionally subjected to a specific time-temperature sequence. The aim of heat treatment may be to change the mechanical properties of the part to allow further processing (Annealing) or to develop the mechanical properties suitable for the intended application (hardening and tempering).

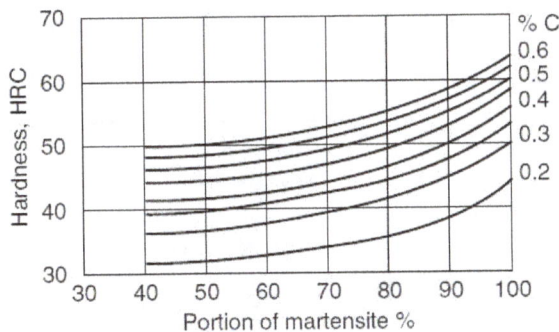

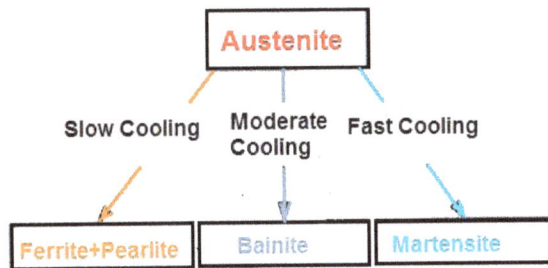

Figure 1.55 LHS = Relationship between as quenched hardness, carbon content and the percentage martensite in the microstructure. From DIN 17020. RHS = Effect of cooling rate on room temperature microstructure

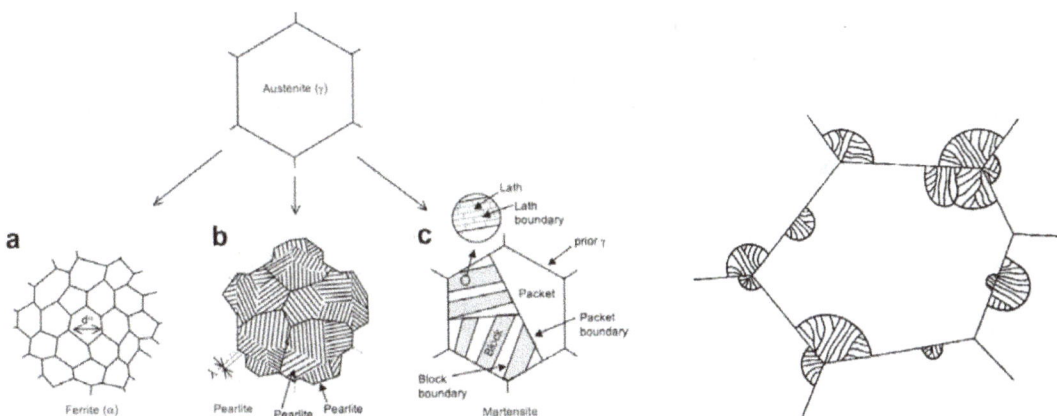

Figure 1.56 More detailed description of the formation of ferrite, pearlite and martensite

1.7.2 Hardening of copper and other alloys

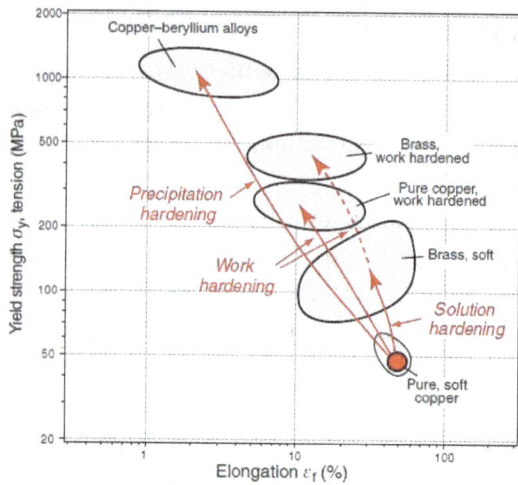

Figure 1.57 Strengthening mechanisms for copper alloys[52]

Alloy	Typical uses	Solution hardening	Precipitation hardening	Work hardening
Pure Al	Kitchen foil			✓✓✓
Pure Cu	Wire			✓✓✓
Cast Al, Mg	Automotive parts	✓✓✓	✓	
Bronze (Cu–Sn), Brass (Cu–Zn)	Marine components	✓✓✓	✓	✓✓
Non-heat-treatable wrought Al	Ships, cans, structures	✓✓✓		✓✓✓
Heat-treatable wrought Al	Aircraft, structures	✓	✓✓✓	✓
Low-carbon steels	Car bodies, structures, ships, cans	✓✓✓		✓✓✓
Low alloy steels	Automotive parts, tools	✓	✓✓✓	✓
Stainless steels	Pressure vessels	✓✓✓	✓	✓✓✓
Cast Ni alloys	Jet engine turbines	✓✓✓	✓✓✓	

Symbols: ✓✓✓ = Routinely used. ✓ = Sometimes used.

Figure 1.58 Metals and alloys typical uses and strengthening mechanisms[52]

1.8 MATERIALS OF INTEREST

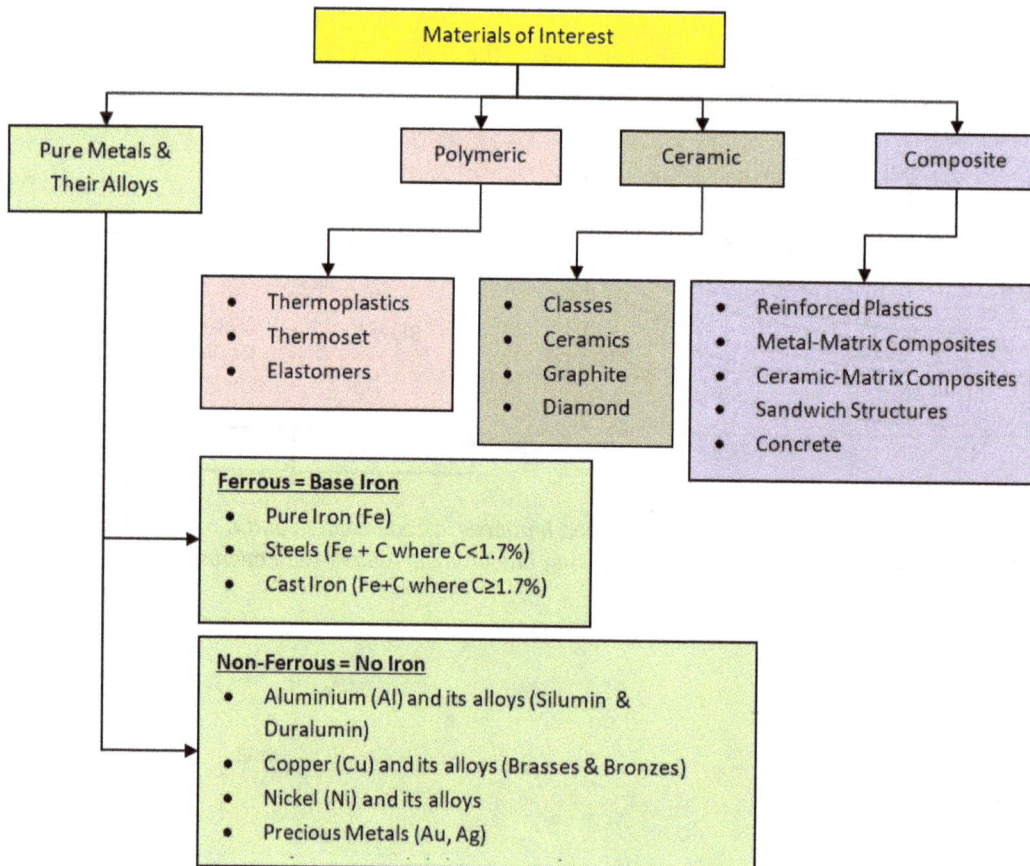

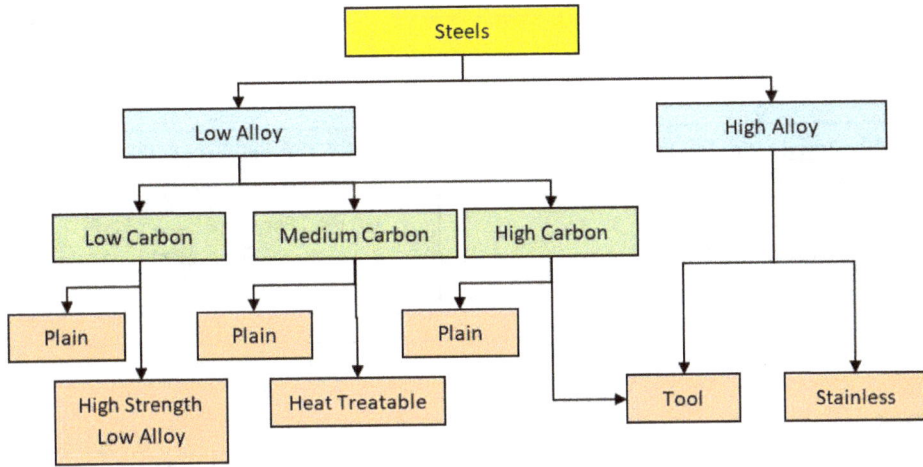

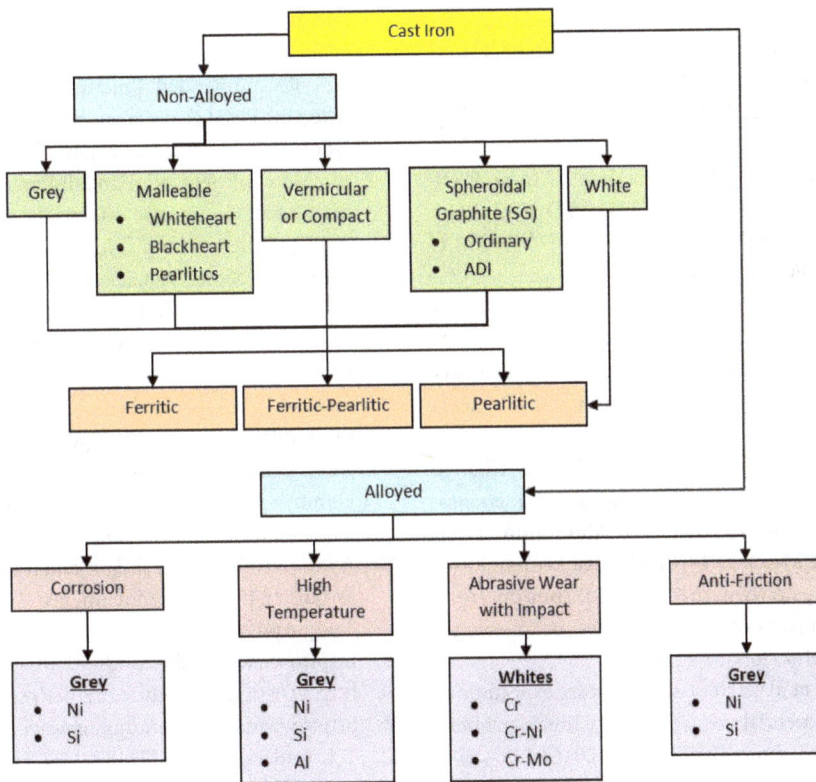

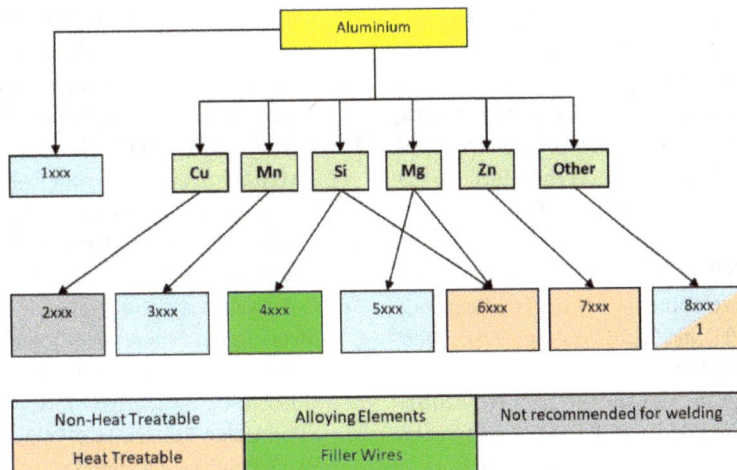

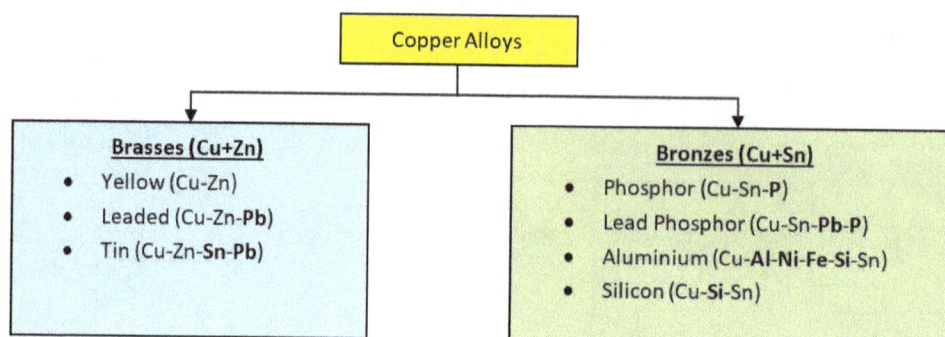

```
                        ┌─────────────────┐
                        │  Copper Alloys  │
                        └─────────────────┘
             ┌──────────────────┴──────────────────┐
```

Brasses (Cu+Zn)
- Yellow (Cu-Zn)
- Leaded (Cu-Zn-**Pb**)
- Tin (Cu-Zn-**Sn**-**Pb**)

Bronzes (Cu+Sn)
- Phosphor (Cu-Sn-**P**)
- Lead Phosphor (Cu-Sn-**Pb**-**P**)
- Aluminium (Cu-**Al**-**Ni**-**Fe**-**Si**-Sn)
- Silicon (Cu-**Si**-Sn)

REFERENCES

1. T F Pearson: Steels for low temperature service
2. https://player.bfi.org.uk/free/film/watch-consett-steel-1967-online,
3. https://www.youtube.com/watch?v=DB9hjSXYNh4, http://www.yorkshirefilmarchive.com/film/consett-steel-0
4. Chris Rhodes, UK steel industry: statistics and policy. Briefing paper, Number 07317, 2 January 2018
5. U.S. STEEL INDUSTRY ANALYSIS: IMPORTANCE OF DOMESTICALLY-PRODUCED STEEL TO OVERALL NATIONAL DEFENSE OBJECTIVES AND ECONOMIC AND MILITARY SECURITY January 2007
6. British Iron and Steel AD1800-2000 and Beyond". (IOM 2001 ISBN 186125119X)
7. The closure of the Lynemouth aluminium smelter: an analysis David Merlin-Jones April 2012, https://www.civitas.org.uk/content/files/aluminium2012.pdf
8. Report prepared for Committee on Climate Change Competitiveness impacts of carbon policies on UK energy-intensive industrial sectors to 2030, Aluminium Deep Dive, Final report March 2017 Cambridge Econometrics Cambridge, https://www.theccc.org.uk/publication/competitiveness-impacts-of-carbon-policies-on-uk-energy-intensive-industrial-sectors-to-2030-2/
9. Dimos Paraskevas et al Current status, future expectations and mitigation potential scenarios for China's primary aluminium industry. Procedia CIRP 48 (2016) 295 – 300, 300 https://ac.els-cdn.com/S2212827116301214/1-s2.0-S2212827116301214-main.pdf?_tid=a68f786d-077c-4c2c-ac0e-01e9f0d1a65b&acdnat=1549811849_ab55ee6b4589b08a071764a1cb62ccac and Tungsten 2017, https://minerals.usgs.gov/minerals/pubs/commodity/tungsten/mcs-2017-tungs.pdf and Hongzhang Chen, The Competitiveness Analysis of Tungsten Industry: A Case from China, International journal of management and economics invention, Volume 2, Issue 02, Pages-528-540, Feb-2016, file:///C:/Users/user/Downloads/295-Article%20Text-946-1-10-20180829.pdf
10. Xibo Wang et al, China's Rare Earths Production Forecasting and Sustainable Development Policy Implications, Sustainability 2017, 9, 1003, file:///C:/Users/user/Downloads/sustainability-09-01003.pdf
11. S.D. MUNI and VIVEK CHADHA Editors, ASIAN STRATEGIC REVIEW. INSTITUTE FOR DEFENCE STUDIES & ANALYSES NEW DELHI, PENTAGON PRESS, 2013.
12. J-O Andersson et al, THERMO-CALC & DICTRA, Computational Tools For Materials Science, Calphad, Vol. 26, No. 2, pp. 273-312, 2002 http://en.iric.imet-db.ru/PDF/35.pdf and C W Bale et al, FactSage thermochemical software and databases, 2010–2016, CALPHAD: Computer Coupling of Phase Diagrams and Thermochemistry 54 (2016) 35–53.
13. API Recommended Practice 571, Damage Mechanisms Affecting Fixed Equipment in the Refining Industries
14. Noori Brifcani et al, A Review of Cutting-edge Techniques for Material Selection, 2nd International Conference on Advanced Composite Materials and Technologies for Aerospace Applications, June 11-13, 2012, Wrexham, UK, http://collections.crest.ac.uk/3975/1/fulltext.pdf
15. Michael F Ashby, Material Selection in Mechanical Design, Fifth Edition Pub Butterworth 2017
16. http://www.grantadesign.com/download/pdf/teaching_resource_books/2-Materials-Charts-2010.pdf, http://www.diim.unict.it/users/fgiudice/pdfs/SM_2.1.pdf
17. Kings Bridge in Australia which was the subject of a Royal Commission report https://www.parliament.vic.gov.au/papers/govpub/VPARL1963-64No1.pdf
18. ASM. Introduction to fatigue and fracture 2012 http://www.asminternational.org/documents/10192/1849770/05361G_Sample.pdf
19. https://www.youtube.com/watch?v=9tN4xROtMjI
20. http://www.shipstructure.org/derby.shtml
21. http://www.aycyas.com/nohighwayinthesky.htm
22. T L Anderson, FRACTURE MECHANICS Fundamentals and Applications, Third Edition 2005 Pub Taylor and Francis.
23. http://ccrm.berkeley.edu/pdfs_papers/bea_pdfs/dhsgfinalreport-march2011-tag.pdf and https://uk.reuters.com/article/uk-bp-deepwaterhorizon/bp-deepwater-horizon-costs-balloon-to-65-billion-idUKKBN1F50O6
24. http://www.tech.plym.ac.uk/sme/FailureCases/List_Engineering_Successes_Failures.PDF
25. Reynolds,K.A. (1997) The Role of the Metallurgist in the Investigation of Component Failures and Vehicle Accidents. Journal of Inst. of Automative Eng. Assoc
26. Hutchings,F.R. and Unterweiser,P.M. (1981), Failure Analysis: The British Engine. Technical Reports, Am. Soc. Metals.
27. Colangelo,V.S. and Heiser,F.A. (1974), Analysis and Metallurgical Failures. Pub. John Wiley, New York.
28. Borer,R.B. and Peters,B.F. (1970), Why Metals Fail. Pub. Gordon and Breach, New York.
29. Engel,L. and Klingele,H. (1981) An Atlas of Metal Damage. Wolfe Publishing, London.

30. McCall,J.L. and French,P.M. (eds) (1978), Metallography in Failure Analysis. Plenum Press, New York.
31. Polushkin,E.P. (1956), Defects and Failures of Metals. Pub. Elsevier, Cat No 5511998.
32. Wulpi,D.J. (1985), Understanding How Components Fail. Pub. ASM. ISBN 0-87170-189-8.
33. ASM (1974), Source book in Failure Analysis. Pub. ASM.
34. ASM CD ROM (1997). CD ROM on Failures. Pub. ASM.
35. Harrison,J.D. (1980), The brittle fracture story PXIX. Engineering applications of fracture analysis. Proceedings of the first National Conference on Fracture held in Johannesburg, South Africa
36. http://www.cbsnews.com/pictures/challenger-shuttle-disaster/
37. http://pics-about-space.com/nasa-challenger-o-ring-failure?p=2
38. http://www.ndtv.com/photos/news/bhopal-gas-tragedy-verdict-and-after-7520 http://www.slideshare.net/philominide/bhopal-gas-tragedy-33689171
39. https://www.jsg.utexas.edu/news/2013/04/uncharted-waters-deepwater-horizon-oil-spill/ US Coast Guard
40. W. J. Jackson, FRACTURE TOUGHNESS IN RELATION TO STEEL CASTINGS DESIGN AND APPLICATION, https://www.sfsa.org/publications/misc/Fracture%20Toughness.pdf
41. Ashby, M. F., Shercliff, H., & Cebon, D. (2007). Materials: engineering, science, processing and design, Elsevier, Amsterdam, 167.
42. I Hanley http://www.twi-global.com/technical-knowledge/published-papers/bs-7910-history-and-future-developments-july-2009/
43. http://www.aiag.org/docs/default-source/Quality-/special-process-flyer_-may-2016.pdf?sfvrsn=0
44. R. E. Hackenberg, The historical development of phase transformations understanding in ferrous alloys Los Alamos National Laboratory, USA. Published by Woodhead Publishing Limited, 2012,
45. Olivier Hardouin Duparc, Alfred Wilm and the beginnings of Duralumin, Carl Hanser Verlag Munchen Z Metallkd 96 (2005) 4 page 398.
46. O.B.M. Hardouin Duparc, The Preston of the Guinier-Preston Zones. Guinier, METALLURGICAL AND MATERIALS TRANSACTIONS, VOLUME 41A, AUGUST 2010—1873.
47. https://www.ifm.eng.cam.ac.uk/uploads/Resources/Featherston__OSullivan_2014_-_A_review_of_international_public_sector_roadmaps-_advanced_materials_full_report.pdf
48. Callister https://abmpk.files.wordpress.com/2014/02/book_maretial-science-callister.pdf
49. Materials science and engineering: New vision in the era of artificial intelligence Tao Qiang a, Honghong Gao. https://arxiv.org/ftp/arxiv/papers/1804/1804.08293.pdf
50. https://aip.scitation.org/doi/10.1063/1.4946894
51. https://www.vttresearch.com/services/smart-industry/factory-of-the-future-(2)/vtt-propertune-for-optimal-material-design
52. Michael Ashby, Hugh Sherdiff, David Cebon, Materials engineering science processing and design BH 2007.

END OF CHAPTER 1 QUESTIONS

1. Draw a typical heat treatment cycle for a precipitation hardening alloy and give an explanation of how the hardening takes place.
2. Name an alloy that would be expected to allow precipitation hardening.
3. What element present on the moon could possibly justify the cost of extraction and return to Earth?
4. Who was responsible for the first application of metallography to iron and steel?
5. What hardening methods can be used to harden pure copper?
6. What hardening mechanisms can be used for cast nickel base alloys?
7. Explain what is meant by "transformation hardening" with reference to ferrous materials.
8. Explain the uses and importance of a dilatometer.
9. What happens inside the crystal structure when a metal is loaded beyond its yield point?
10. Name three methods used to measure hardness.
11. How would you measure the toughness of a high strength aero-space quality steel?
12. Name two major metal failures that led to major improvements in the understanding of why metals fail.
13. What specification would you use to validate a welded structure against brittle fracture?
14. Describe two method used to quantify fatigue properties so that the data can be used for design purposes.
15. If a tensile test was carried out on a ductile alloy what events would occur after the maximum load was reached?

16. The product design requirements for a strut in tensile requires a fracture toughness of 50MPa m$^{1/2}$ and a yield strength or 1000MPa. What materials are available to meet that requirement?

17. What test work would be required to establish the chosen material could meet these requirements?

18. A large counter-balance weight was required. What would be cheapest material that could be used for this purpose?

19. If the counterbalance was for a fork truck what material would represent the lowest cost?

20. Name four failure modes that should be considered during the process of material selection?

21. Name four alloy elements used in the manufacture of aluminium alloys?

22. Name three matrix microstructures that could be present in a white cast iron?

23. A steel stud with a yield strength of 1200MPa and a toughness of 50MPam$^{1/2}$ will be used in an application where it will be loaded to 600MPa. What will be the largest imperfection that can be allowed in the stud to avoid brittle fracture?

24. The fracture surface of a bolt has to be examined at a magnification of X1000 to assist with establishing the mode of failure. What facilities would be required to carry out this work?

25. Discuss the following statement and give some examples: Some metals and alloys cannot be heat treated in order to improve their mechanical properties.

CHAPTER 2

The structure of metal

2.1 INTRODUCTION

To understand how metal components and parts and structures behave in service, it is essential to have some knowledge of how metals are composed of atoms, which are combined together during solidification or phase transformation to form the microstructure of the metal or alloy.

The understanding of the basic building blocks that form a metals structure will allow insight and appreciation of the relationship between the micro-structure and the mechanical properties. This is an important subject area since the ability to predict and develop the specified mechanical properties of strength and toughness has always been a key objective in alloy development and metal manufacture.

There are 90 naturally occurring chemical elements on Earth; 18 elements are described as non-metals (e.g. noble gases and halogens) and 7 as half-metals. The remaining 65 chemical elements are metals, 60 of which are commercially available. Metals and alloys are essential for the well-being of industry and successful procurement and supply can be a major governmental issue. There are over 20 metals that are considered "critical" for Europe's industrial future.[1 to 4] In the USA stockpile requirements are given detailed consideration and the Secretary of Defence is responsible for preparing and submitting a report to Congress by January 15th of every other year. The report must include stockpile requirements and detail the key supply-demand assumptions used to arrive at its recommendations.[5]

The atomic structure considers the way in which the metal atoms consist of neutrons, protons and electrons. As the number of neutrons and protons increases, the atomic number increases as shown in the period table. The number in the top left-hand side of each box is the atomic number for that element.

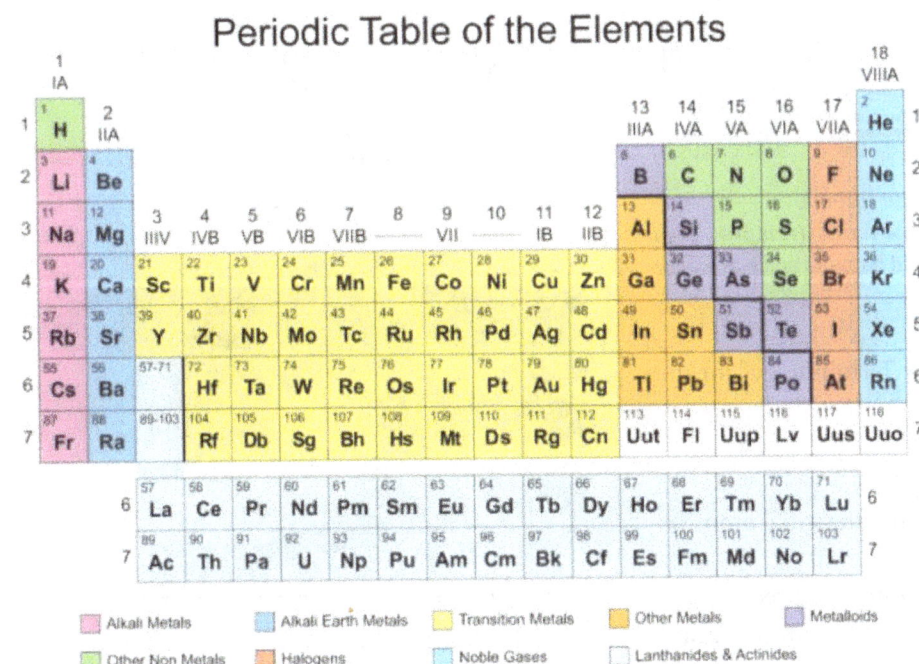

Figure 2.1 Periodic table

2.2 THE BUILDING OF THE CRYSTAL STRUCTURE AND THE DEVELOPMENT OF THE MICROSTRUCTURE

Metals are one of the most widely used engineering material. Some of their properties, e.g. elastic constants, can be directly related to the atomic structure of the metallic bonds between the atoms. The yield, flow, and fracture stress are controlled by microstructural features of metals, such as point defects, dislocations, grain boundaries, and second phase particles. The atomic structure primarily affects the chemical, physical, thermal, electrical, magnetic, and optical properties. Therefore, the levels of structure which are most relevant when considering engineering properties are:

- The crystal structure – how the atoms are linked together to form a crystal, and
- The microstructure – how the crystals are put together to form large pieces of metal which can be used in engineering applications.

Further consideration of the microstructure will be considered based on these two levels of structure.

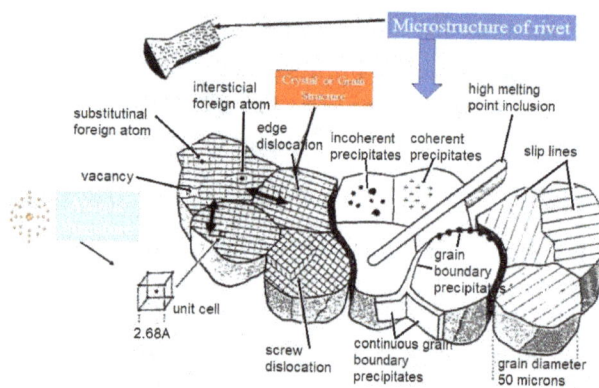

Figure 2.2 *The diagram shows the three levels—atomic, crystal and microstructure and new terms. (Adapted from -Lothar Engel et al An atlas of metal damage surface examination by SEM. Wolfe Science Books)*

Figure 2.2. gives some "New Terminology" for the Reader which, include:

- Grain Size, Grain boundary.
- Vacancy, Substitutional atom, Interstitial atom.
- Edge Dislocation, Screw dislocation.
- Incoherent precipitates, Coherent precipitates, Grain boundary precipitates, Continuous grain boundary precipitates.

- Non-Metallic Inclusion. Oxide Inclusions, Alumina (Al_2O_3), Silicate (SiO_2) Manganese Sulphide (MnS)
- Slip Lines.

It is helpful that Figure 2.2 shows these features as they are typically seen during the metallographic examination of a prepared specimen.

2.3 CRYSTAL STRUCTURE

Figure 2.2 shows the three levels of structure that combine to form the microstructure. The diagram also shows some features/defects that are commonly found in the microstructure. Many engineering materials, including almost all metals, consist of small crystals in which atoms are packed in regular, repeating, three-dimensional (3-D) patterns. When metal atoms come together in the solid-state, they generally arrange themselves into a particular regular 3-D pattern, or lattice. The structure remains intact since each atom readily donates some electrons to a common cloud, which then binds the structure together as shown in Figure 2.3.

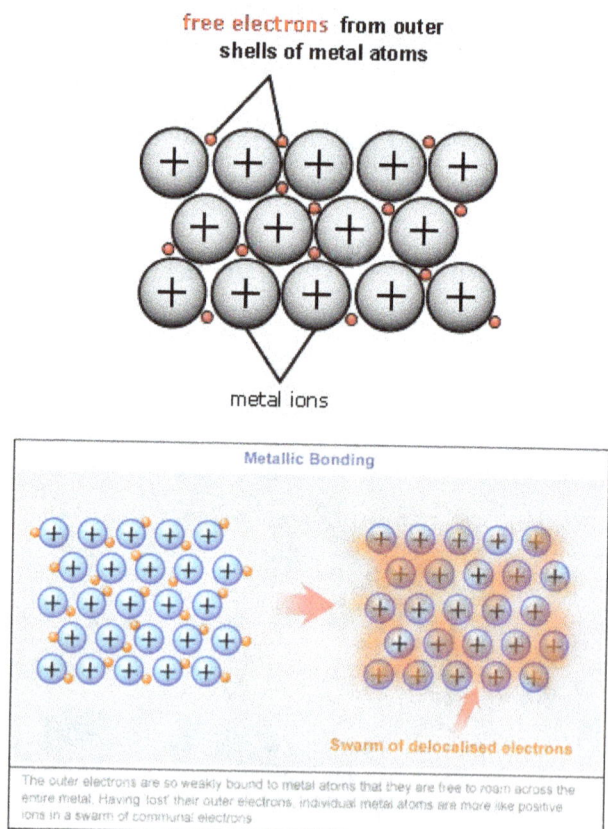

Figure 2.3 *Outer electrons detach and form a common cloud, which then binds the structure together*

The free electrons do not combine with any particular ion, rather remaining as a negatively charged cloud floating between the positive ions and acting rather like a glue holding the positive ions together. The presence of free electrons explains why metals are good conductors of electricity since they have free charge carriers which are easily moved through the solid when a voltage is applied. In contrast insulators have no free electrons, the atoms in the solid being bonded in a different manner.

When a larger collection of atoms as shown in Figure 2.4 are studied it can be seen that there is a repetition of a simple unit cell. This array of points (atom centre positions) is known as the space lattice or the Bravais lattice. In addition, each lattice can therefore be described by specifying the positions in a unit cell which is repeated. Auguste Bravais was born 23 August 1811, died 30 March 1863. He was a French physicist known for his work in crystallography. He is best remembered for Bravais lattice; his 1848 discovery that there are 14 unique lattices in three dimensional crystalline systems. All done without X-ray!

There are 14 different Bravais lattices which are divided into seven crystal systems. The lattices within one system share the same point group symmetry operations. Figure 2.5. shows this classification with the properties of each system regarding angles and lengths of their elementary cells.

The simplest space lattices are those based on a cube, i.e. simple cubic, body centred cubic and face centred cubic. The simple representation of the atom by a dot has some drawbacks, e.g. the crystal appears to be made up mainly of space; other representation has therefore been developed. The hard sphere model

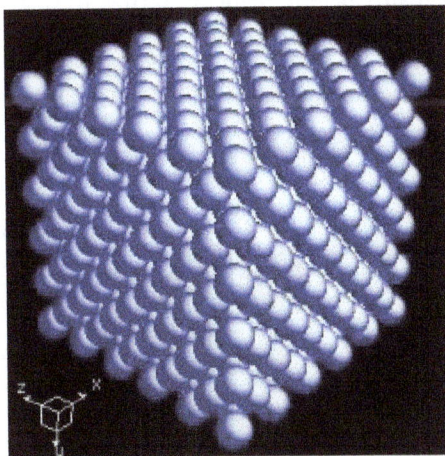

Figure 2.4 *Arrangement of atoms in a copper crystal (hard ball model)*

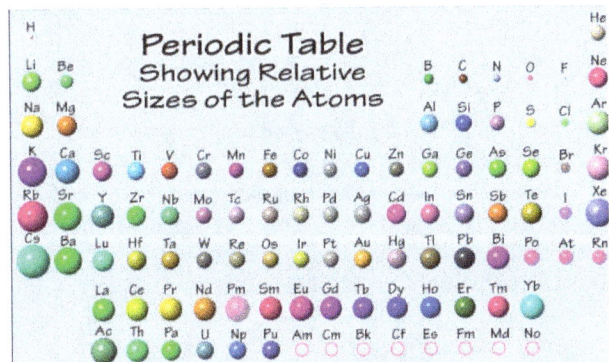

Figure 2.5 *Bravais lattices and relative size of atoms*

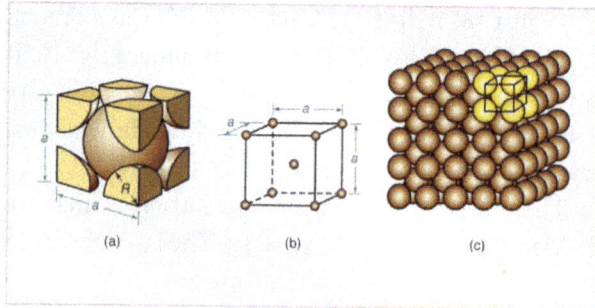

Body-Centered Cubic (BCC)

- Many metals such as iron, chromium, tungsten, molybdenum, and vanadium have the BCC crystal structure at room temperature.
- Metals with mixed bonding, such as iron, may have unit cells with less than the maximum packing factor.

Metal	Lattice constant a (nm)	Atomic radius R^* (nm)
Chromium	0.289	0.125
Iron	0.287	0.124
Molybdenum	0.315	0.136
Potassium	0.533	0.231
Sodium	0.429	0.186
Tantalum	0.330	0.143
Tungsten	0.316	0.137
Vanadium	0.304	0.132

*Calculated from lattice constants by using $R = \sqrt{3}a/4$.

Selected metals that have the BCC Crystal Structure at room temperature (20°C) and their Lattice Constants and Atomic Radii

Figure 2.6 BCC (Body centred cubic)

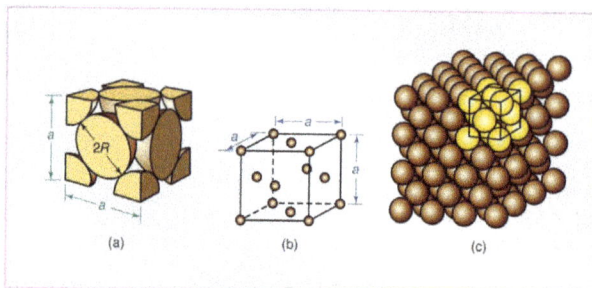

Face-Centered Cubic (FCC)

- The PF of 0.74 is for the closest packing possible of "spherical atoms."
- Many metals such as aluminum, copper, lead, nickel, and iron at elevated temperatures (912 to 1394°C) crystallize with the FCC crystal structure.
- Metals with only metallic bonding are packed as efficiently as possible.

Metal	Lattice constant a (nm)	Atomic radius R^* (nm)
Aluminum	0.405	0.143
Copper	0.3615	0.128
Gold	0.408	0.144
Lead	0.495	0.175
Nickel	0.352	0.125
Platinum	0.393	0.139
Silver	0.409	0.144

*Calculated from lattice constants by using $R = \sqrt{2}a/4$.

Selected metals that have the FCC Crystal Structure at room temperature (20°C) and their Lattice Constants and Atomic Radii

Figure 2.7 FCC (Face centred cubic)

has the atoms repeated as incompressible spheres, and the packing of these spheres is considered to represent the packing of the atoms. Some common structures are shown, using the hard sphere model, in Figure 2.6, 2.7 and 2.8. The simple cubic structure is not exhibited by any of the commercially important metals.

The body centred cubic structure (BCC) is shown by metals such as chromium (Cr), tungsten (W), and molybdenum (Mo).

The face-centred cubic structure (FCC) is exhibited by metals such as copper (Cu), nickel (Ni), aluminium (Al) and gold (Au).

Of the non-cubic systems, the only common structure with regard to metals is the hexagonal close-packed structure (hcp or cph), which is shown by metals such as zinc (Zn) and magnesium (Mg).

In order to build up a 3-D packing pattern it is often easier to begin by considering packing atoms two dimensionally in atomic planes and then to stack these planes on top of one another to give crystals.

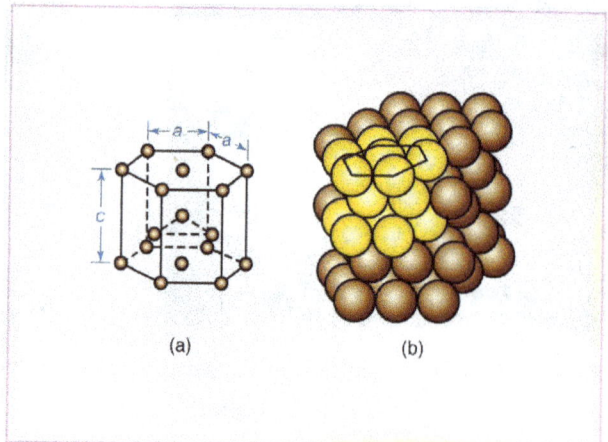

Hexagonal Close-Packed (HCP)

- Coordination number is 12.
- The ratio of the height c of the hexagonal prism of the HCP crystal structure to its basal side a is called the c/a ratio.
- The c/a ratio for an ideal HCP crystal structure consisting of uniform spheres packed as tightly together as possible is 1.633.

Metal	Lattice constants (nm)		Atomic radius R (nm)	c/a ratio	% deviation from ideality
	a	c			
Cadmium	0.2973	0.5618	0.149	1.890	+15.7
Zinc	0.2665	0.4947	0.133	1.856	+13.6
Ideal HCP				1.633	0
Magnesium	0.3209	0.5209	0.160	1.623	−0.66
Cobalt	0.2507	0.4069	0.125	1.623	−0.66
Zirconium	0.3231	0.5148	0.160	1.593	−2.45
Titanium	0.2950	0.4683	0.147	1.587	−2.81
Beryllium	0.2286	0.3584	0.113	1.568	−3.98

Selected metals that have the HCP Crystal Structure at room temperature (20°C) and their Lattice Constants, Atomic Radii and c/a ratios

Figure 2.8 HCP (Hexagonal close-packed)

An example of how we might pack atoms in a plane is the manner in which the red balls are set up on a snooker table. The reds are packed in a triangular arrangement so as to take up the least possible space on the table. This type of plane is thus termed a close-packed plane.

How can we add a further layer of atoms to convert the two-dimensional pattern into three dimensions?

The depressions where the balls/atoms meet provide perfect seats for the next layer of atoms to sit. By simply dropping atoms into alternate seats we can generate a second plane on top of the original plane. Then a third layer can be added and a fourth and so on until we have developed a reasonably sized piece of crystal with now a regularly repeating pattern of atoms in three dimensions. Figure 2.9 shows how the positioning of the third layer of balls can give an FCC crystal structure or an HCP crystal structure.

One major factor that can now be appreciated is an understanding of the elastic modulus of a material (Young's modulus). When a stress is applied to a material the stress attempts to move the atoms apart. The elastic modulus is a measure of stress needed to move the atoms apart as shown in Figure 2.10 shows the effect of forcing apart the atoms and the elastic modulus. Figure 2.11 shows the possibility of breaking bonds and creating "cracks" and the role of the "strain energy" introduced due to loads applied to the structure.

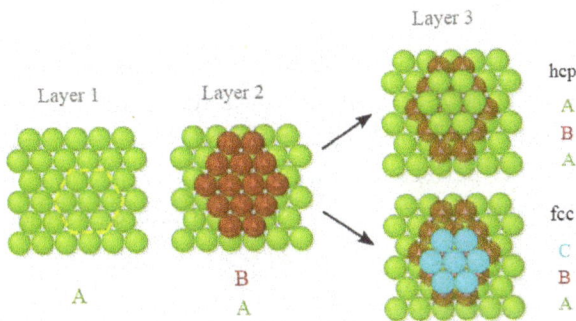

Figure 2.9 Stacking arrangements

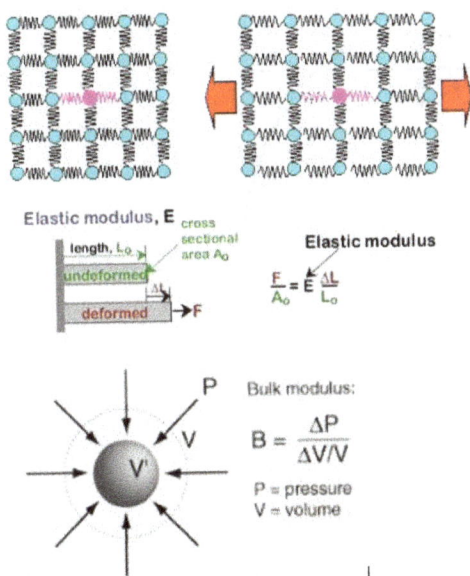

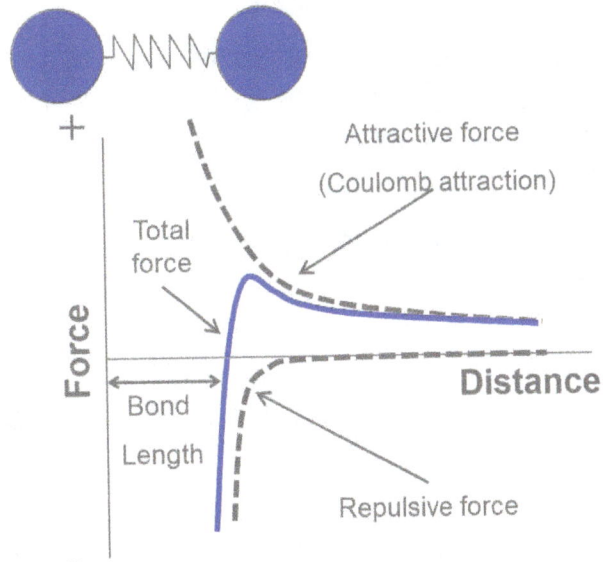

Figure 2.10 Elastic and Bulk moduli

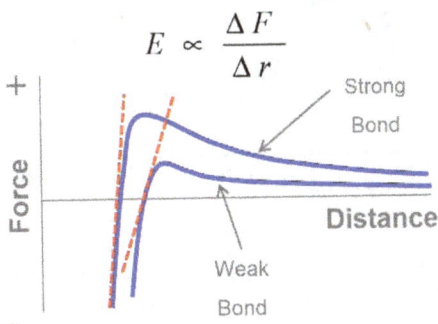

2.3.1 Surface energy and strain energy and fracture http://www.fracturemechanics.org/griffith.html

Figure 2.11 shows the possibility of breaking bonds and creating "cracks" and the role of the "strain energy" introduced due to loads applied to the structure and the "surface energy" associated with any crack surface.

The atom on the surface has fewer adjacent atoms and as a result it has higher energy of approximate Eo/2 (Eo= bond energy). The surface energy is calculated from this value multiplied by the number of atoms per unit area. Most materials have a surface energy of 0.5 to 10J/m². This is an important aspect of the fracture process. It was first adopted by Griffiths in his approach to the fracture of glass. The concept being to create a new fracture requires the provision of sufficient energy to create the new surface. This energy being provided from the strain energy.

One of Irwin's contributions to allow application of a these" fracture mechanics" concepts to metals was to include an addition term to account for the plasticity (energy consumed due to plasticity around the fracture) that occurs on either side of the fraction. Fracture energy =R = 2(γ+γp) where γp is the extrinsic work of fracture (plastic deformation).

2.3.2 Defects and imperfections in the crystal structure

The way in which metals form crystalline structures has been outlined, but only idealised structures with perfect order have been considered. In reality metals are never perfect, several types of imperfections are invariably present in the crystal structure and these

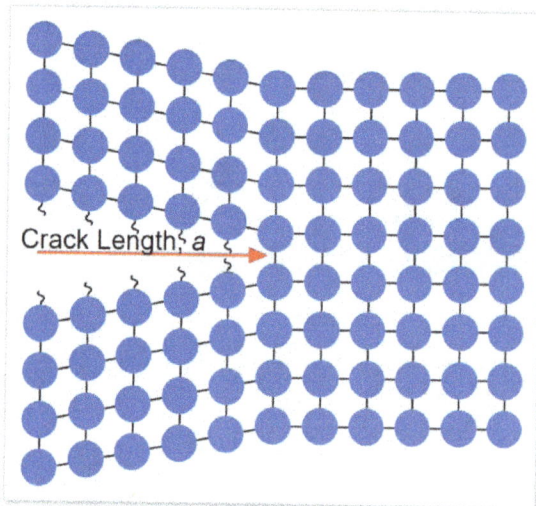

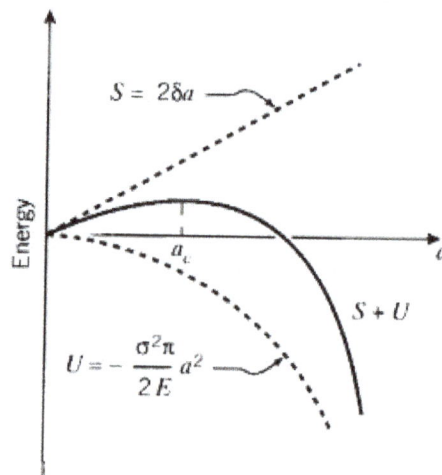

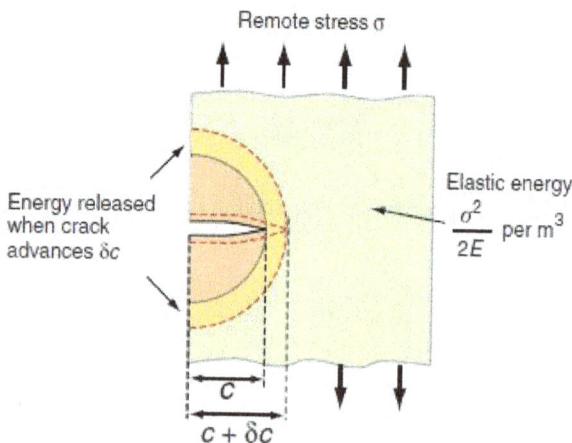

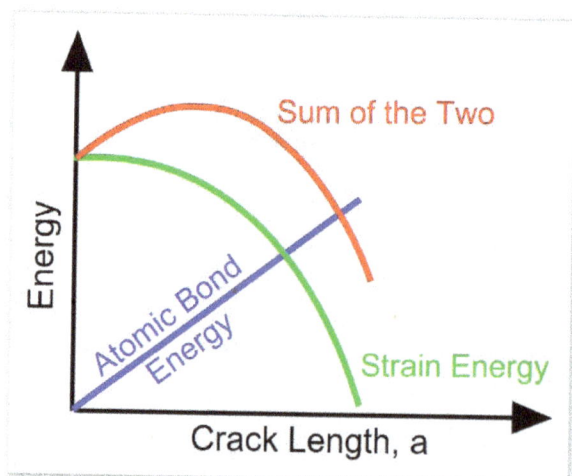

Figure 2.11 *Griffiths approach to fracture 4 diagrams depict the initial Griffith's concept of "fracture". The energy/work needed to break atomic bonds and the associated "Stain energy" that provides a driving force. Lower photos show some of the factors that LEFM and fatigue crack growth models do not consider*

imperfections play an important role in the development of the physical and mechanical properties of metals. (At a macro level we can have other imperfections associated with forging, rolling or casting, seams, laps, cracks – see Chapter 6, Figure 6.19 The possible origin of short cracks.)

The most common crystal imperfections, which are regarded as being the most important in affecting the properties of metals are: associated with (i) points, (ii) lines and (iii) planes.

How do these defects affect metal properties? As has already been mentioned the presence of defects in crystalline materials is often responsible for the characteristic properties associated with metals.

The diffusion of an atom is therefore dependent upon the presence of a vacancy on an adjacent site, and the rate of diffusion is therefore dependent upon two factors: how easily vacancies can form in the lattice, and how easy it is for an atom to move into a vacancy.

2.3.3 Point defects

These types of defect consist of disturbances in orderly array associated with a point in the Bravais lattice, the most common types being shown in Figure 2.12.

Point defects can influence physical properties such as density and electrical conductivity, but the most important consequence of point defects is that they provide a mechanism of mass transport in the solid state. This is illustrated in Figure 2.13.

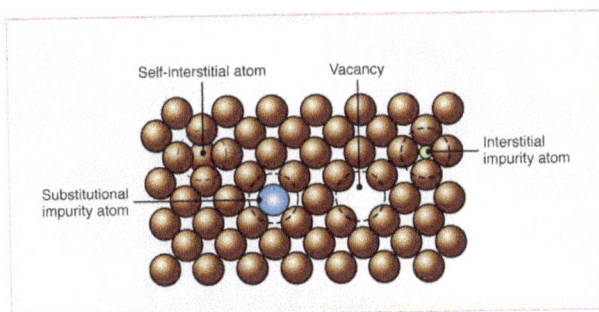

Figure 2.12 Point defects in the crystal structure

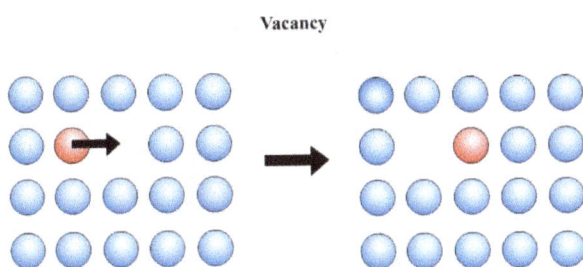

Figure 2.13 Diffusion controlled by vacancies

For more information http://www.doitpoms. ac.uk/tlplib/diffusion/diffusion_mechanism.php (DoITPoMS Cambridge University)

The diffusion of an atom is therefore dependent upon the presence of a vacancy on an adjacent site, and the rate of diffusion is therefore dependent upon two factors: how easily vacancies can form in the lattice, and how easy it is for an atom to move into a vacancy.

2.3.3.1 Diffusion

Diffusion is the process by which matter is transported from one part of a system to another as a result of random molecular motions. Transfer of heat by conduction is also due to random molecular motions, and there is an obvious analogy between the two processes. This was recognized by Fick (1855), who first put diffusion on a quantitative basis by adopting the mathematical equation of heat conduction derived some years earlier by Fourier (1822). The mathematical theory of diffusion in isotropic substances is therefore based on the hypothesis that the rate of transfer of diffusing substance through unit area of a section is proportional to the concentration gradient measured normal to the section.

2.3.3.2 Fick's First Law

By considering the flux of diffusing particles in one dimension (x direction) as illustrated in Figure 2.14. The particles can be atoms, molecules, or ions. Fick's first law for an isotropic medium can be written as:

$$J = -D\frac{\mathrm{d}c}{\mathrm{d}x}$$

Here J is the flux of particles (diffusion flux) and C their number density (concentration). The negative sign indicates opposite directions of diffusion flux and concentration gradient. Diffusion is a process which leads to an equalisation of concentration. The factor of proportionality, D, is denoted as the *diffusion coefficient* or as the *diffusivity* of the species considered.

Units: The diffusion flux is expressed in number of particles (or moles) traversing a unit area per unit time and the concentration in number of particles per unit volume. Thus, the diffusivity D has the dimension of *length^2 per time* and bears the units [cm^2 s-1] or [m^2 s-1].

$$J_x = -D \, \Delta C / \Delta x$$

Concentration *C*

ΔC

Δx

Flux J_x

Distance *x*

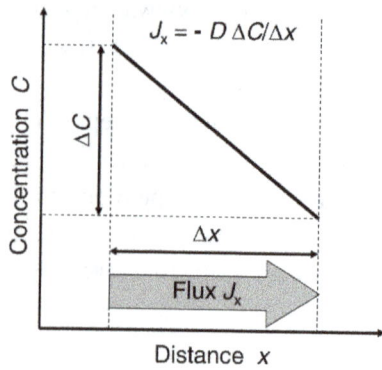

Carburising of gear involves a diffusion process

- The heat treatment preferred for heavily loaded gears to give the hardness and fatigue resistance –both tooth bending and contact fatigue resistance would be carburising
- A suitable carbon atmosphere inside the hot zone of a furnace allows carbon to diffuse into the surface. The atmosphere is measured by oxygen potential from an oxy-probe
- The time required to give the case depth can be determined by the mathematics of diffusion

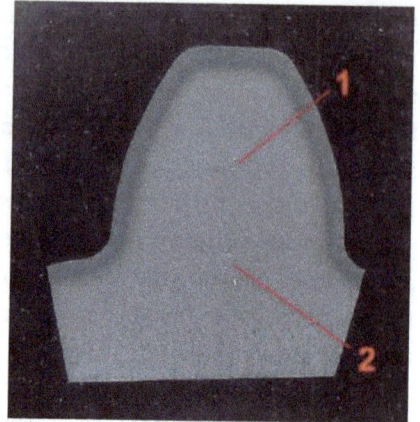

Figure 2.14 Flux of diffusing particles in one dimension and a carburised gear

Figure 2.14 Flux of diffusing particles in one dimension and a carburised gear

2.3.3.3 Fick's second law

$$\frac{dC}{dt} = D \frac{d^2 C}{dx^2}$$

The proportionality factor D is called diffusion coefficient and is usually measured in $cm^2 \, s^{-1}$. The first law equation above applies only if the concentration gradient -dc/dx does not change during the whole diffusion process (stationary case). Such is approximately the case, for carburising as gas diffuses through the part.

If, in the course of diffusion, concentration changes everywhere in the material (non-stationary case), Fick's second law of diffusion applies, which can be solved using an "erf" function.

Diffusion in solids is an important topic in the process metallurgy as well as the sciences of solid-state physics, physical chemistry, physical metallurgy, and materials science.

Diffusion processes allow many microstructural changes that occur during preparation, processing, and heat treatment of materials. Typical examples include solidification, nucleation of new phases, diffusive phase transformations, precipitation and dissolution of a second phase, homogenisation of alloys, recrystallisation, high-temperature creep, and thermal oxidation.

The practical application of the diffusion process is very diverse: the doping during the fabrication of microelectronic devices, solid electrolytes for batteries and fuel cells, surface hardening of steel by carburising or nitriding, joining of metals and ceramics by diffusion bonding, and sintering.[8]

2.3.4 Line defects and dislocation

In the 1930s the concept of dislocations in the metal microstructure was developed to explain why the actual strength of metals was significantly lower than predictions based on atomic theory.

2.3.4.1 Theoretical cohesive strength

This is the force necessary to break the atomic bonds. The evaluation of this value and the realisation that real materials failed at forces well below this value led to postulation and then the understanding of the role of dislocations.

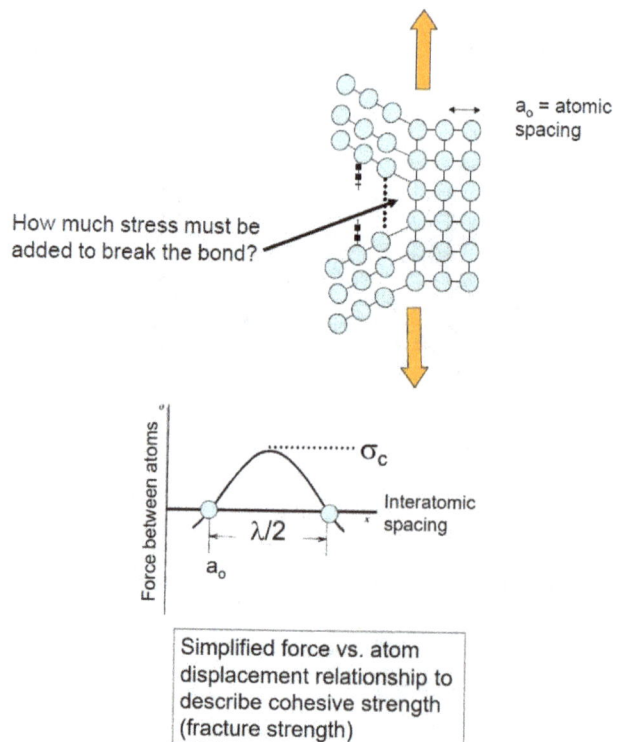

a_o = atomic spacing

How much stress must be added to break the bond?

Force between atoms

σ_c

Interatomic spacing

$\lambda/2$

a_o

Simplified force vs. atom displacement relationship to describe cohesive strength (fracture strength)

Figure 2.15 Bond breaking

$$\sigma_i \approx \sqrt{\frac{\gamma E}{a_0}} \approx \frac{E}{10}.$$

material	E (GPa)	UTS (GPa) predicted	UTS (GPa) experimental
glass	70	7	0.006
aluminum oxide	400	40	2
stainless steel	200	20	0.9
aluminum	70	7	0.5
nylon	70	7	0.05

Figure 2.16 *Theoretical cohesive strength*

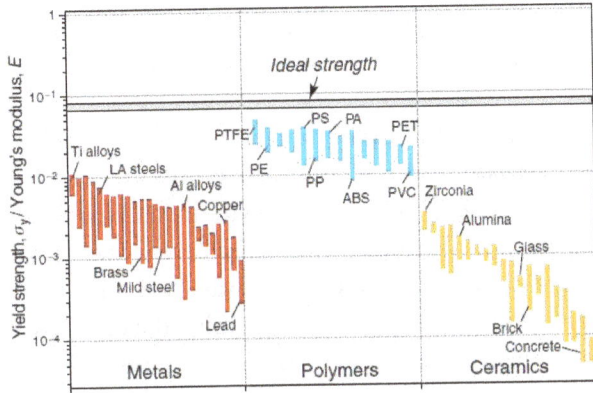

Figure 2.17 *Comparison of YS/Elastic modulus[2]*

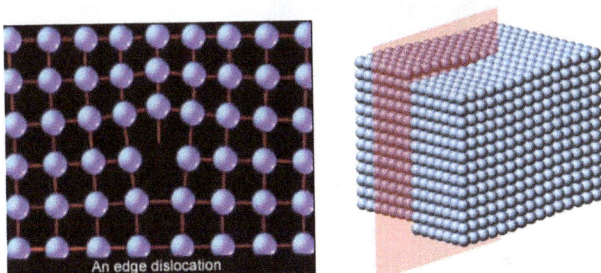

Figure 2.19 *A Screw dislocation*

2.3.4.2 Dislocations

The most common line defect is known as a dislocation and this corresponds to an interruption in the order of atoms along a line in the Bravais lattice. Two types of dislocation may occur:

- an edge dislocation – associated with an extra portion of a plane of atoms or half plane, which terminates within the crystal Figure 2.18
- a screw dislocation – which derives its name from the spiral path that is traced around the dislocation line by the atomic plane of atoms. Figure 2.19

Most dislocations found in metals are probably neither pure edge nor pure screw but exhibit components of both types, the so-called mixed dislocations, but these will not be considered in any further detail. For a detailed appreciation the reader is referred to the book shown in Figure 2.21.

The presence of dislocations in crystals allows metal to deform plastically (i.e. produce permanent deformation) at relatively low stress levels. The mechanism by which this occurs is shown in Figure 2.20 for an edge dislocation.

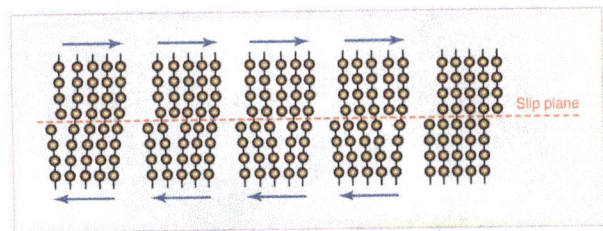

Figure 2.20 *Movement of an edge dislocation*

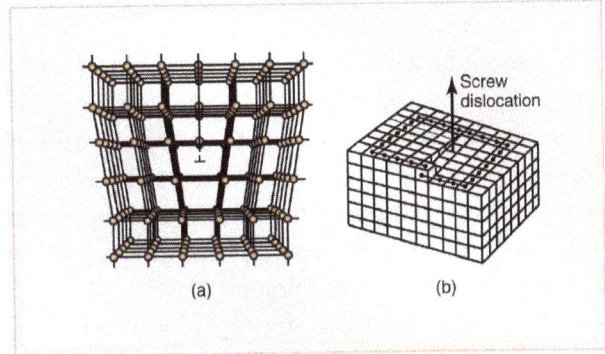

Figure 2.18 *An edge dislocation*

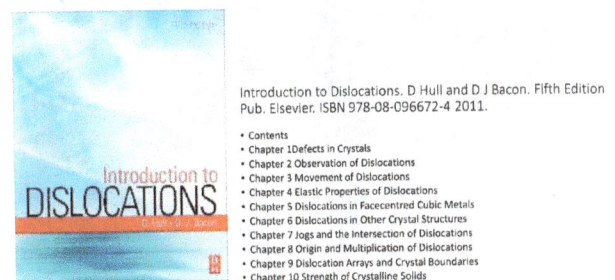

Figure 2.21 *Recommended book with comprehensive details on dislocation*

The concept of dislocations and their impact on strength and toughness was regarded as an elegant explanation by all concerned and gave a plausible explanation into the work hardening, deformation and fracture behaviour of metals.

Gordon (From: J E Gordon, Structures (or why things don't fall down) Penguin Books 1978) *made an interesting comparison when he compared the "clever" dislocation mechanisms in metals to the evolutionary developments in living biological material that occurs in nature.* The irony being that nature makes no use of metals and therefore could have taken no part in the creation or evolution of dislocation mechanisms in metals. It is almost tempting to enter the world of "science fiction" and to fantasise that the evolution of metals occurred somewhere else or in some other time period or in another part of the universe where metals were the preferred "biological" material. (Note – in the paragraph below "purposive" means having or done with a purpose.)

"The dislocation mechanism, which was originally postulated by Sir Geoffrey Taylor in 1934, has been the subject of intensive academic research over the last thirty years. It turns out to be an extraordinary subtle and complicated affair. What takes place inside so apparently simple a thing as a piece of

metal seems to be quite clever as many mechanisms in living biological tissues. Yet the funny thing is that this clever mechanism cannot possibly be purposive, if only because Nature has nothing, so to speak to gain from it, since she never makes any structural use of metals, which very seldom occur native in the metallic state anyway. However, this may be, dislocations in metals have been of enormous benefit to engineers and might have almost have been invented for their benefit, since they not only result in metals being tough but also enable them to be forged and worked and hardened."

The movement of dislocations is how metals achieve plastic strain. A single bond breaks and the "dislocation" progressively passes through the metal. Figure 2.22 Upper shows the appearance of slip lines in the microstructure, the first photograph was from the 1980's the second from the 1930's confirming the knowledge has been around for many years. The images shown in Figure 2.22 Lower display the slip appearing as extrusions on the outer surface of the metal. This was found to be possible source of fatigue crack initiation.

The more plastic strain that occurs the number/density of dislocations increases and they can bump together and lock-up and thus require a greater force to move them. Figure 2.23 shows a thin film transmission electron microscope image showing the build up of dislocations. This is referred to as work hardening.

Figure 2.24 uses a carpet anology pushing a raised part of the carpet along the floor. However, if there is an additional fold in the carpet when they meet the motion stops and they are pinned. Dislocations

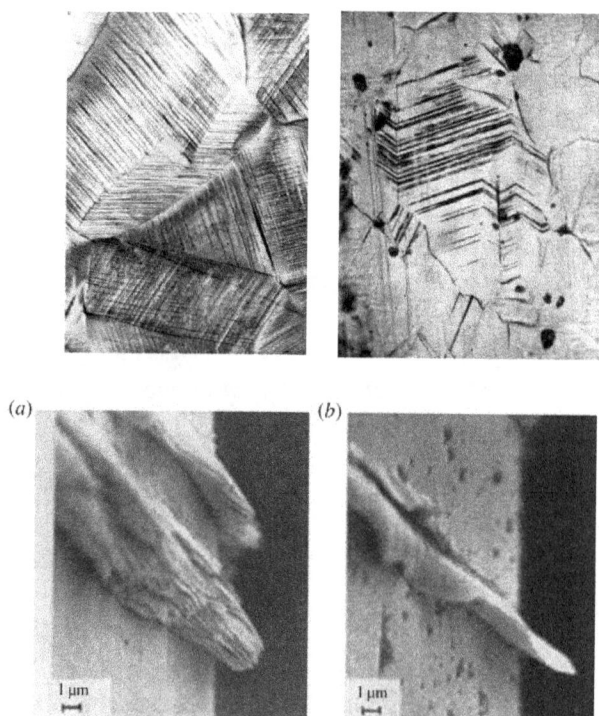

(a) *(b)*

Figure 2.22 *Upper = Micro with slip lines 1ˢᵗ photo = 1980s from Calister. 2ⁿᵈ Photo = 1930 from H F Moore Illinois University. Lower = 3ʳᵈ and 4ᵗʰ= Extrusions on copper due to slip*

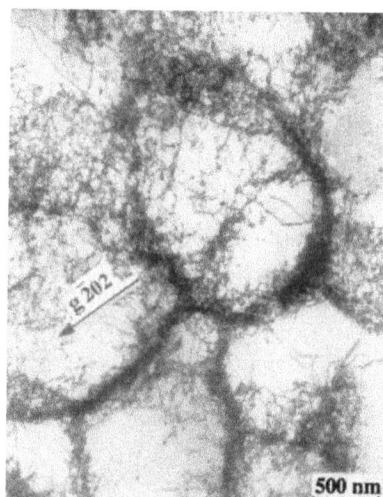

Figure 2.23 *A thin film sample viewed in a transmission electron microscope showing dislocations piling-up at grain boundaries*

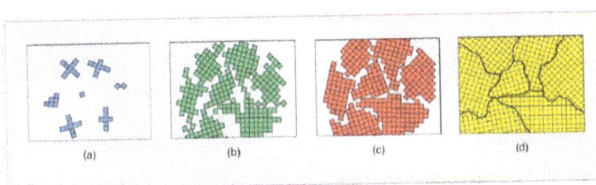

(a) Dislocation

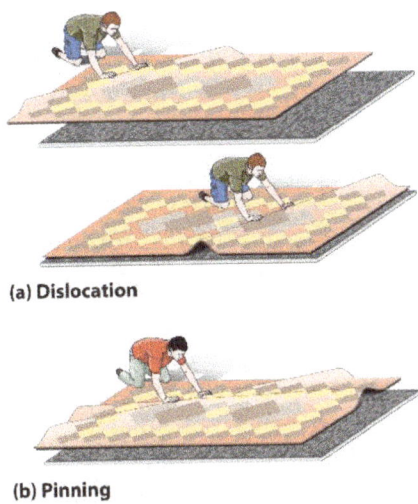

(b) Pinning

Figure 2.24 Carpet simulation of dislocation movement and pinning

behave in a similar fashion. Eventually the dislocation are "totally gridlocked" and cannot move and fracture initiates. (This is also a mechanism of fatigue crack initiation.) Thermal treatment allows atoms to move back to their preferred locations and the dislocation density drops (annealing treatment) and then ductility is restored and further mechanical deformation can be carried out.

2.3.5 Plane defects- grain boundaries

Plane defects consist of disturbances in "ordered state of atomic arrangement" associated with a plane in the "Bravais" lattice. The most common types of plane defects are the grain boundaries, as illustrated in Figure 2.25 These boundaries separate the grains or

crystals having different crystallographic orientations. In metals grain boundaries are created during solidification when crystals formed from different nuclei grow simultaneously and meet each other in the final stage of solidification.

The liquid state is composed of a random dispersion of atoms, that is there is no long or short-range order. Solidification involves the formation of solid crystalline material from the liquid. The first stage of the process is the formation of small regions of solid (nuclei) in the liquid. The second stage involves the growth of nuclei by the addition of atoms from the liquid onto the solid. The growth stage continues until crystals (or grains) come into contact with one another, at the grain boundaries. (See Figure 2.25 top LHS.) The final material can then be referred to as polycrystalline, and the grain boundaries are one of the main features of the microstructure of the metal or alloy. The top RHS shows the possible grain size modification that can happen the solid state due to temperature (recovery, recrystallisation and grain growth). The lower LHS diagram summarises the demarcation between cold working, warn working and hot working expressed as a percentage of the melting temperature of the metal. The image on the lower RHS shows the effect of mechanical working during hot deformation, which can "refine" the grain size depending upon the amount of deformation and the temperature. This is the basis of thermo-mechanically treatment during plate and strip rolling. (controlled rolling- best with a niobium addition which can delay the recrystallization)

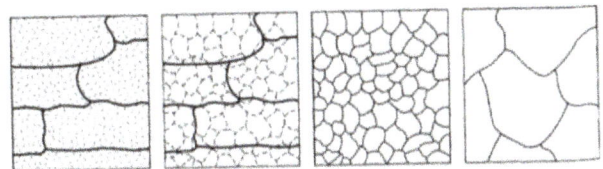

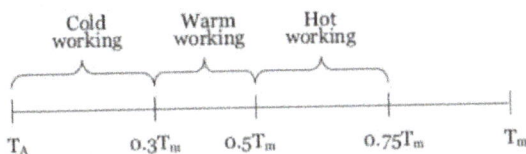

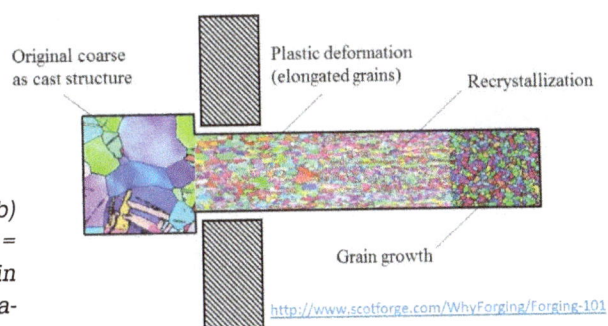

Figure 2.25 Top LHS = Solidification(a) Nucleation (b) Growth (c) Growth (d) Completely solid. Top RHS = Anneal- showing recovery, recrystallisation and grain growth. Lower = Classification of metal working temperatures, RHS = Closed die forging

2.3.5.1 The important effects of grain boundaries in the microstructure

Grain boundaries restrict the movement of dislocations, which can have a major beneficial effect on mechanical properties. This beneficial feature was discovered independently by O E Hall (Sheffield University 1951) and N J Petch (University of Leeds, work done between 1947 to 1949) and it is referred to as the Hall Petch equation.

The concept grain-boundary strengthening is based on the observation that grain boundaries impede dislocation movement and that the number of dislocations within a grain have an effect on how easily dislocations can traverse grain boundaries and travel from grain to grain.

Therefore, a change in the grain size can influence dislocation movement and the yield strength. There has been significant work on methods to refine the grain size in many metals. The grain size of steel can be refined by heat treatment to allow the solid-state transformation above the AC_3 (BCC structure to FCC structure). Some metals use cold deformation followed by an annealing heat treatment. Some methods involve the addition of grain refining elements or changing the rate of solidification or inoculation to alter the grain size. Figure 2.27 shows the effect of ferrite grain refinement in steel on the improvement of the ITT (Impact transition temperature) and yield strength.

Similar relationships with strength and toughness have been found with martensitic steels. The quantitative evaluation of a martensitic microstructure

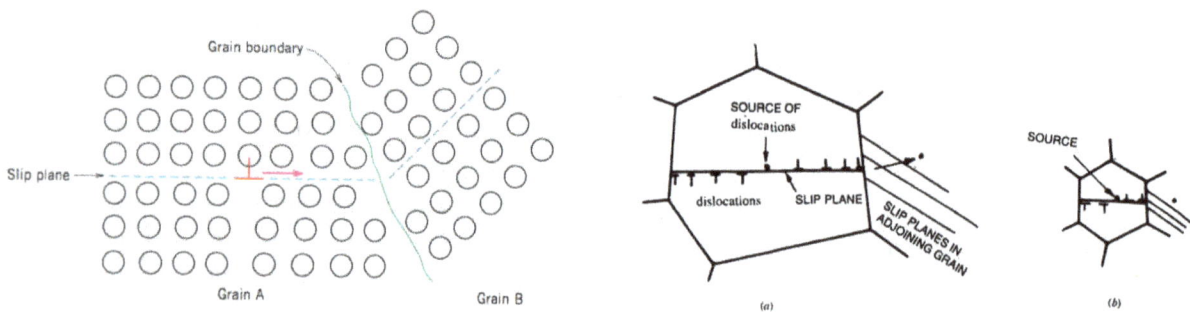

Figure 2.26 Grain boundary as a dislocation barrier and fine grain size creates biggest barrier.

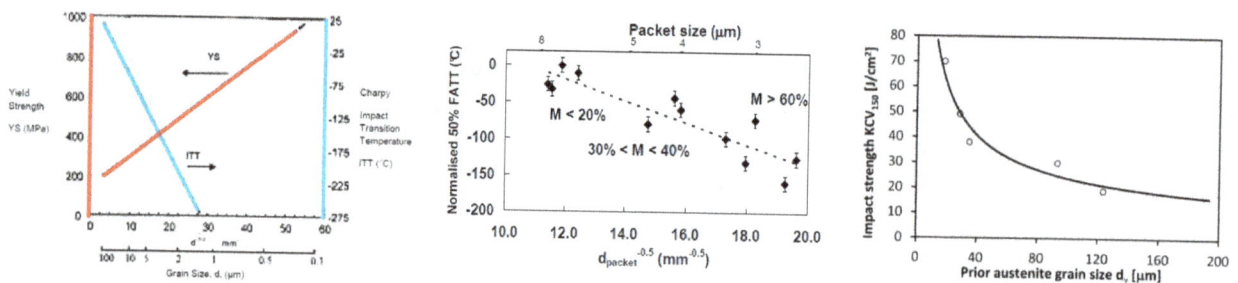

Figure 2.27 LHS = Effect of ferrite grain size (ferrite/pearlite steels). Centre = martensite "packet size" on the ITT (Impact transition temperature). RHS = PAGS vs Impact energy at -40°C for Hardox

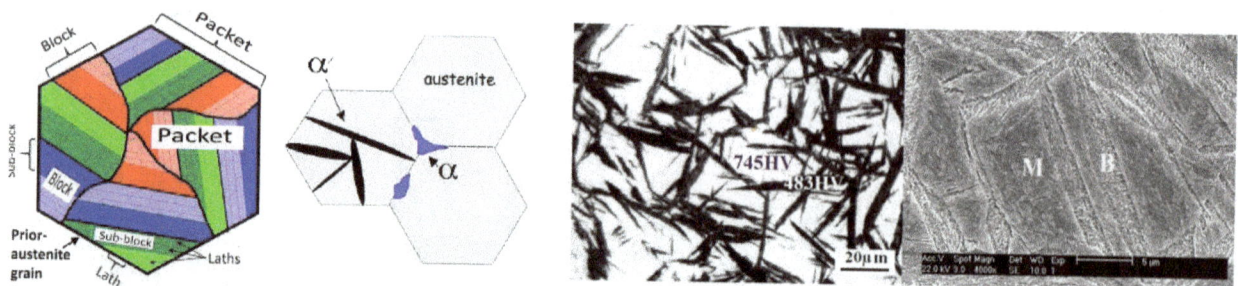

Figure 2.28 Left = Terms used for martensite, Centre = Martensite and bainite are contained within PAGS (prior austenite grain size), but the Allotriomorphic ferrite and pearlite can cross the PAGS so that PAGS cannot be seen in the microstructure of ferrite/perlite steel, Right = Martensite and retained austenite. RHS = Martensite bainite mixture (from Mat. Res. vol.21 no.1 São Carlos 2018 Epub Nov 27, 2017)

a)

b)

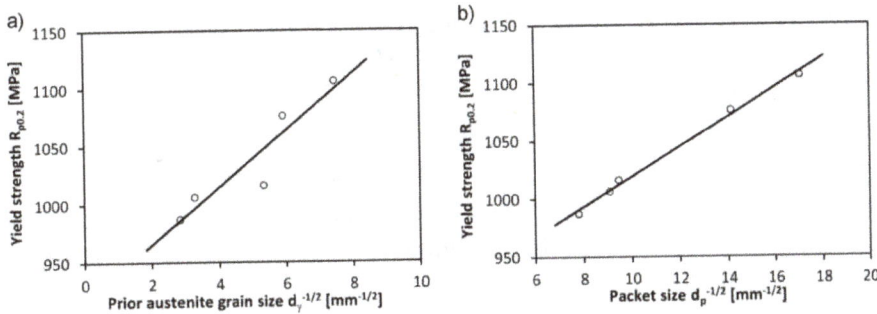

Figure 2.29 Effect of grain size and martensite packet size on yield strength

involves consideration of the prior austenite grains (PAG) each of which consist of packets, blocks and laths (Figure 2.28).

A packet is a region that consists of a number of parallel laths with the same habit planes. These laths can either be slightly mis-orientated or have very different orientations from each other.

A block is a series of laths with similar orientation. The use of orientation mapping by EBSD (Electron backscatter diffraction) has allowed a greater understanding of the martensitic microstructure and allowed the determination of block sizes that is not possible to determine by standard microscopy This understanding of block sizes has led some researchers to conclude that the block size, and not the packet or PAG (prior austenite grain) size (as previously thought), is the controlling size factor in the yield strength of a martensitic alloy.

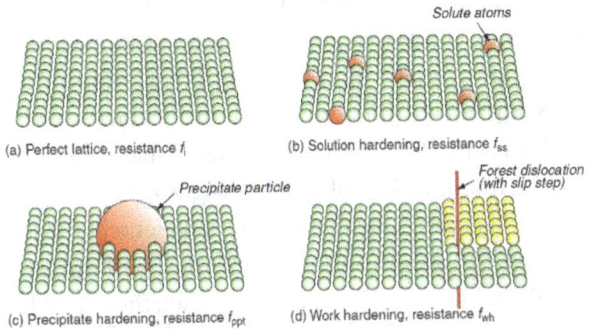

Figure 2.31 Dislocation's-eye view of the slip plane[7]

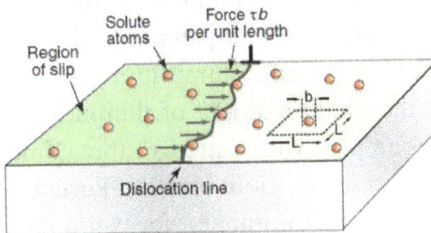

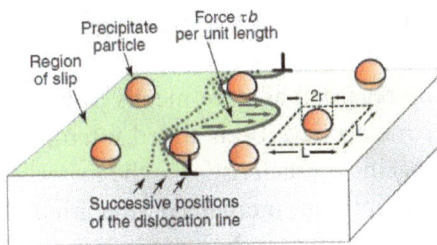

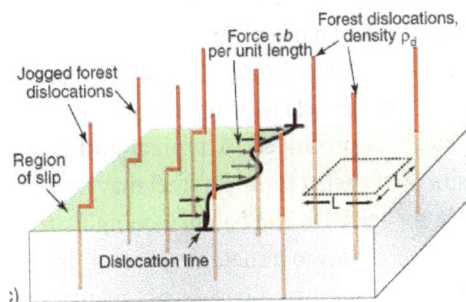

2.3.6 Summary of imperfections and dislocations

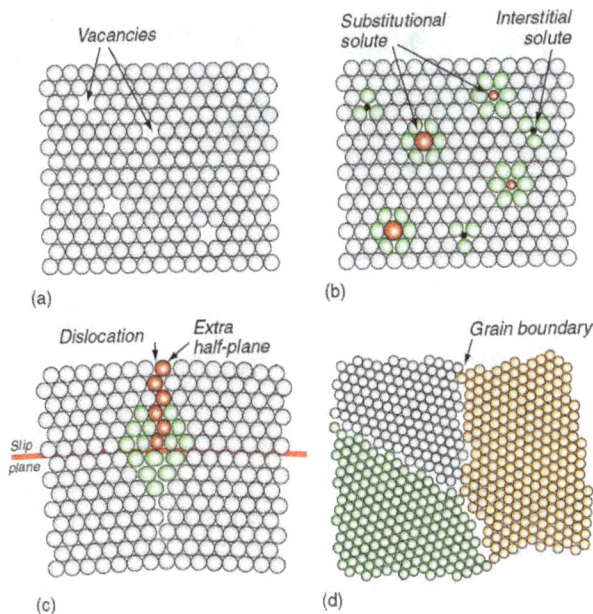

Figure 2.30 Summary of imperfections

Figure 2.32 Top= dislocation interactions[7]

2.4 ALLOYING ELEMENTS IN METALS

Most everyday metallic objects we see around us are usually made from alloys rather than pure metals. This is due to the fact that pure metals frequently do not possess the appropriate combination of properties required for a particular application.

Alloys, which are mixtures of two or more metal atoms, can be specifically designed to have required properties. They can have structures which are simple and basic, such as that for cartridge brass, an essentially binary alloy (made up of two metals) of 70% zinc and 30% copper, or they can be relatively complex such as those found in the nickel-based superalloys used for the manufacture of jet engine components. Such superalloys can have in excess of ten elements within their composition (Ni, Cr, Mo, V, AI, C and others).

When one metal is added to another, the alloy often takes on a new "identity" and mechanical properties are not predicted from the properties of the metals from which it was formed. A weak or soft metal may combine with another metal to produce a strong alloy with properties different from its parent metals.

For example, copper and aluminium are both fairly weak, but the addition of 10% of aluminium to copper produces an alloy with a strength three times that of pure copper (aluminium bronze). Pure iron is soft, and carbon in its commonest form, graphite, is mechanically weak, yet as little as 0.5% carbon in iron leads to the formation of a steel with high tensile properties.

Alloys can be regarded as pure metals to which atoms of other metals have been added intentionally, to impart specific characteristics to the material. Alloying is used to improve mechanical strength and other important properties such as corrosion resistance, hardenability, machineability etc.

For example, when chromium and nickel are added to steels, the corrosion resistance is significantly enhanced and the class of metals we know as the stainless steels is produced. A typical and common composition of such a stainless steel may be 18% chromium, 8% nickel, with the balance being iron.

The addition of atoms to a metal may result in the formation of a solid solution. Two terms relating to solid solutions that are important are: solvent and solute. The solvent is the metal present in the greatest amount, with the solute being that present in a lower percentage.

2.5 THE FORMATION AND STRUCTURE OF SOLID SOLUTIONS

A solid solution can exist over a range of compositions, unlike pure metals and chemical compounds, which have fixed compositions. At any composition within the allowable range the material is completely homogeneous. In general, the composition range over which the solid solution can exist is limited and, if a pure constituent is included in the range, then the solid solution is known as a primary solid solution. If the composition range does not contain a pure constituent then the solid solution is referred to as a secondary solid solution, or intermediate solid solution.

Two types of solid solution have been previously referred to in this document:

* substitutional solid solution
* interstitial solid solution.

2.5.1 Substitutional solid solution

If the atoms of the solvent metal and solute element are of similar sizes (not more than 15% difference), they form substitution solid solution, where part of the solvent atoms are substituted by atoms of the alloying element (see Figure 2.33).

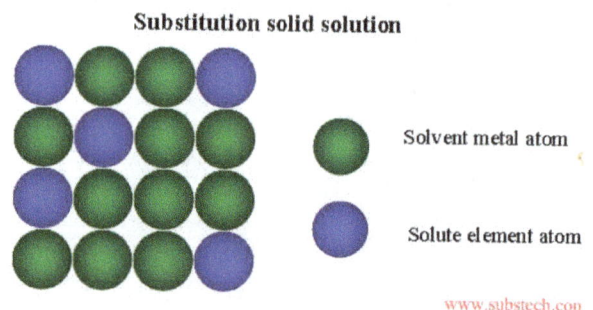

Solvent metal atom

Solute element atom

www.substech.con

Figure 2.33 Solid solution

2.5.2 Interstitial solid solution

If the atoms of the alloying elements are considerably smaller than the atoms of the matrix metal, interstitial solid solution forms, where the matrix solute atoms are located in the spaces between large solvent atoms (see Figure 2.34).

When the solubility of a solute element in interstitial solution is exceeded, a phase of intermediate

Interstitial solid solution

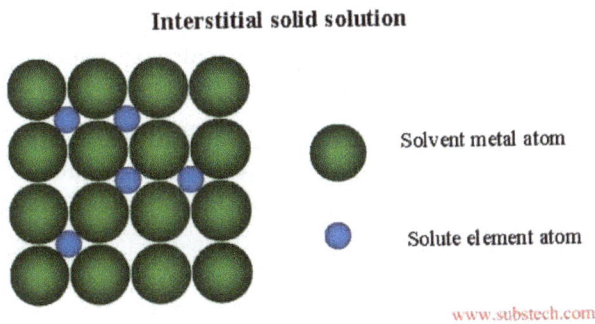

Solvent metal atom

Solute element atom

www.substech.com

Figure 2.34 Interstitial solid solution

compound forms. Often there is a strong interaction between the two types of atom, and the compound formed which will exist over a very limited composition range and will be able to be assigned a chemical formula. These compounds (TiN, WC, Fe$_3$C etc.) play an important role in strengthening steels and other alloys. Some substitution solid solutions may form ordered phase where the ratio between concentration of matrix atoms and concentration of alloying atoms is close to simple numbers like AuCu$_3$ and AuCu.

Examples of typical intermediate compounds are:

- Fe$_3$C – iron carbide
- CuAl$_2$
- Mg$_2$Si.

2.6 METALLOGRAPHY AND THE EQUILIBRIUM DIAGRAM

The need to be able to carry out a detailed metallographic examination and an interpretation of microstructures is often a barrier to gaining an understanding of steel technology. This often results from the use of confusing terminology and often unrelated names given to the different phases/aspects of the microstructure. (Austenite from Roberts-Austen, Martensite from Professor Adolf Martens, Bainite from Edgar Bain, Widmanstatten from Dr Widmanstatten.)

Maps/visual images used by the metallurgist. To simplify and assist the understanding of various process topics, both time dependent and time independent "maps" are used by the metallurgist.

Time independent:

- Equilibrium Phase Diagram
- Material selection Ashby charts (See Figure 1.8 to 1.9).

Time dependent:

- Time-Temperature-Transformations (TTT) diagrams and Continuous-Cooling Transformation (CCT) diagrams. These are important for understanding phase transformations.
- Creep mechanism maps
- Wear maps
- Fatigue S N diagram.

Equilibrium Phase Diagrams are an important tool in understanding metals technology. A phase diagram outlines the various phases present in metals and alloys. Phase diagrams can be viewed as maps similar to a map which shows geographical detail. The phase diagram shows the temperature range and compositional range particular phases are present.

For metals we use Temperature-Composition diagrams (i.e. axes are temperature and composition). The diagrams are based on the premise of achieving microstructural equilibrium. Therefore, the techniques used to establish the phase diagrams are based on slow cooling.

Therefore, it is important to remember that due to the possible range of cooling rates in the processing of real alloys and metals the metallographic appearance of commercially processed metals can be different from the actual microstructures predicted by the Equilibrium Phase Diagram.

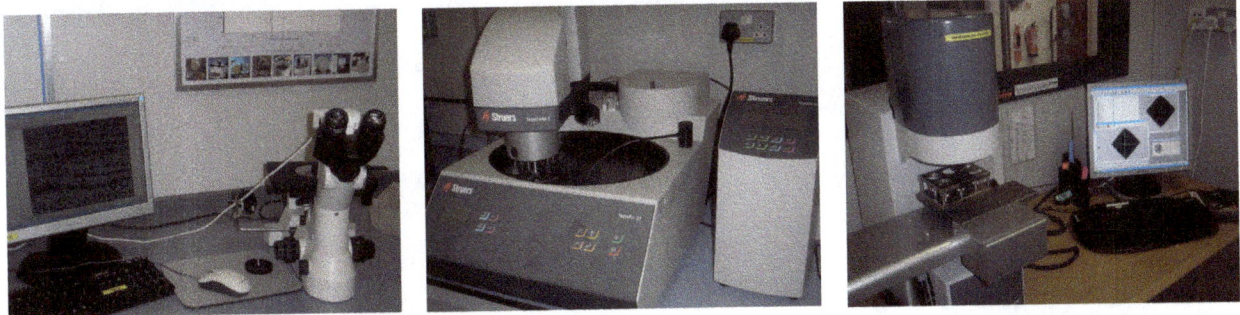

Figure 2.35 Optical metallography Automated metallurgical preparation equipment Automatic hardness

Experimental determination of an equilibrium diagram

The determination of phase diagram involves the use of a simple thermal analysis technique. The cooling curves are measured for several compositions (Figure 2.36). This allows the identification of the solidus and liquidus for various compositions. More advanced techniques would use DSC /DTA (Differential Scanning Calorimetry/ Differential thermal analysis).

An alternative method for determining the phase diagram would be the use of Thermo-Calc software. (CALPHAD – CALculation of Phase Diagrams)

The iron-carbon equilibrium diagram is the key diagram/information source that provides insight into the relationship between carbon level, temperature and the phases present in the microstructure. The equilibrium diagram allows metallurgists to provide an interpretation of any microstructure of a metal or alloy and to confirm that it has been processed correctly. There are several books covering the topic of process control by the use of metallography.

The equilibrium diagram also allows insight into the microstructural changes associated with compositional changes and the effects of temperature. Since the change in microstructure can result in a change in the mechanical properties it is essential data. An example of the iron/carbon equilibrium diagram is shown in Figure 2.38.

The following bullet points list some of the knowledge required to use and interpret a phase diagram:

- Hume Rothery rules of solid solubility. Atoms with not more than 15% size difference, the same crystal structure, similar valency and electro-negativity will form a substitution solid solution
- Gibbs phase rule $F = C - P + 1$ (see section 1.6.7.)
- Gibbs free energy $\Delta G = \Delta H - T\Delta S$ (see section 1.6.7.)
- The lever rule is a tool used to determine fraction of each phase of a binary equilibrium phase diagram. If an alloy consists of more than one phase, the amount of each phase present can be found by applying the lever rule to the phase diagram.

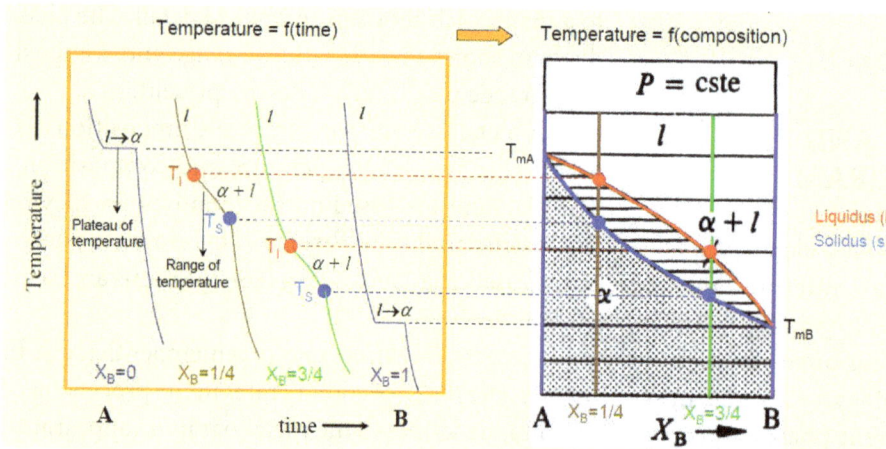

Figure 2.36 Experimental determination of an equilibrium diagram[9]

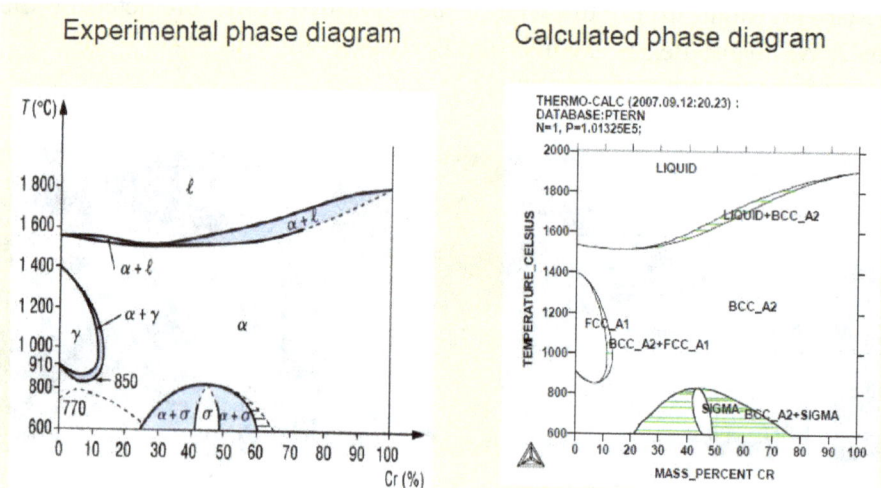

Figure 2.37 Comparison of experimental and calculated Cr-Ni equilibrium diagram[9]

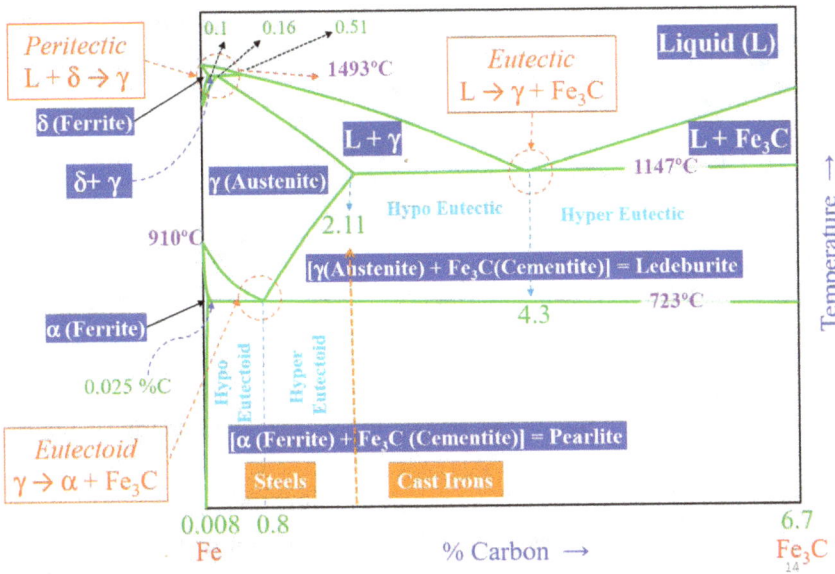

Figure 2.38 The Iron-carbon equilibrium diagram[10]

The lever rule can be explained by considering a simple balance. The composition of the alloy is represented by the fulcrum, and the compositions of the two phases by the ends of a bar.

The proportions of the phases present are determined by the weights needed to balance the system.

So,

fraction of phase 1 = (C2 – C) / (C2 – C1)

and,

fraction of phase 2 = (C – C1) / (C2 – C1).

It is important to appreciate that in Figure 2.38

- irons or cast irons are associated with the eutectic reaction on the right-hand side of the equilibrium diagram at high carbon levels and
- steels are associated with the eutectoid reaction at the left-hand side of the equilibrium diagram.

Failure to appreciate this distinction would represent a clear handicap to working with iron and steel.

The equilibrium diagram represents the summary of a significant body of knowledge and is the first diagram we use to predict the microstructure. To gain an understanding of this type of diagram there is a need to appreciate the concept of "phases", which is very important in the field of metallurgy. A phase can be defined as "*a macroscopically homogeneous body*

of matter". An example of different phases would be metal present in either the solid or liquid form. In Figure 2.38 there are 4 separate phases of iron shown. There is liquid plus three solid phases, each one given a Greek symbol (alpha BCC structure) α, (gamma FCC structure) γ and (delta BCC structure) δ.

The process by which the mechanical properties can be changed by thermal processes is known as heat treatment. This process consists of heating a metal or alloy to a specific predetermined temperature, holding at this temperature for a required time, and then cooling from this temperature.

2.7 SOLIDIFICATION OF METALS. UNDER COOLING, CONCEPT OF NUCLEATION AND GROWTH, HOMOGENEOUS & HETEROGENEOUS NUCLEATION

Solidification takes place by a nucleation and growth process. Atoms in a solid or liquid metal are never stationary. To achieve solidification the process requires the formation of a stable nucleus to form in the pool of metal. Within the pool of molten metal there is a boundary that separates a potential stable nucleus from the liquid metal. Atoms keep trying to cross the boundary and as the temperature drops (under cooling) eventually sufficient atoms cross the boundary and cluster together to form a stable nucleus.

Once stable nuclei form and become stable they can continue to grow. Initially the growth occurs at the same rate in all directions until the growth is hindered due to impingement. The formation of new nuclei

Figure 2.39 A sketch showing a schematic representation of nucleation and growth of solid nuclei during solidification from the molten liquid state. Colours denote different orientations of grains. (a) Initially there are fewer nuclei. Some of them have grown. (b) Shows that growth ceases along certain directions due to impingement. A few more nuclei have formed. All of these continue to grow. (c) Shows a state when most of the space is filled up indicating that the process is nearly complete. Grains appear to be randomly oriented. From- https://nptel.ac.in/courses/113105023/Lecture7.pdf

Homogeneous Nucleation & Energy Effects

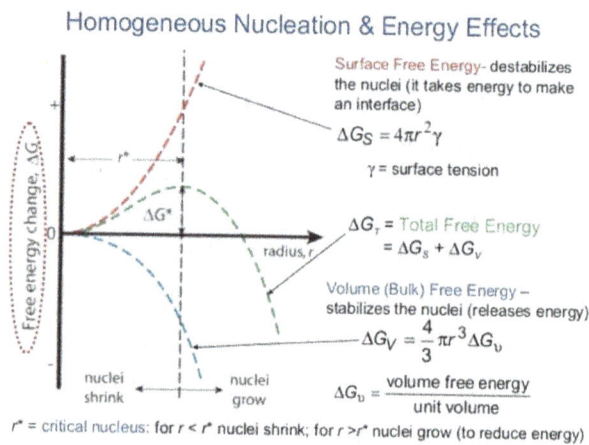

Surface Free Energy- destabilizes the nuclei (it takes energy to make an interface)

$$\Delta G_S = 4\pi r^2 \gamma$$

γ = surface tension

ΔG_T = Total Free Energy
= $\Delta G_s + \Delta G_v$

Volume (Bulk) Free Energy – stabilizes the nuclei (releases energy)

$$\Delta G_V = \frac{4}{3}\pi r^3 \Delta G_\upsilon$$

$$\Delta G_\upsilon = \frac{\text{volume free energy}}{\text{unit volume}}$$

r* = critical nucleus: for r < r* nuclei shrink; for r >r* nuclei grow (to reduce energy)

Figure 2.40 Homogeneous Nucleation Griffiths equations shown in Figure 2.28 represents energy versus crack length and it is analogous to the diagram of free energy change versus embryo size in the theory of nucleation of phase transformations shown in this Figure.

Solidification of Ingots and Castings

- Most engineering alloys are poured or cast into a ingot moulds (for steel they are made from cast iron) for cast iron and non-ferrous either sand moulds or permanent moulds.

- If the as-cast metal retains their shape afterwards and only reshaped by machining, they are called castings.

- If the cast metal id subsequently mechanically worked, e.g. by rolling, extrusion or forging, the cast metal is referred to as ingots

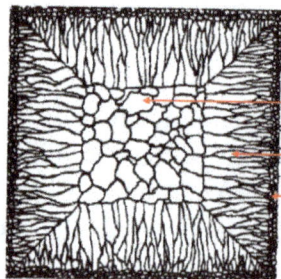

Equiaxed

Columnar

Chill zone

Figure 2.41 General macro-image of a cast metal.

Following the stable nuclei dendrite growth along specific planes <100> in cubic metals.

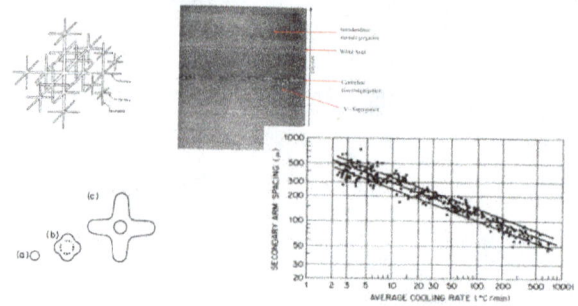

Figure 2.42 Growth for an alloy – the dendritic structure.

within the remaining liquid also continues. When the process is complete the solid is found to consist of several grains (crystals). The size of the grains may differ depending on whether it developed from a nucleus formed right in the beginning or towards the end of the process. This is shown in Figure 2.39 and also Figure 2.25 Top LHS. Each original nucleus produces a grain with its own orientation of atomic structure, separated from the neighbouring grains, which may have a different orientation, by a grain boundary. The grain boundary is a narrow transition region in which the atoms adjust themselves from the arrangement within one grain to that in the other orientation. The grain boundary will have a high level of imperfections and therefore often provides a preferential diffusion path. The grain size of a material is an important microstructural feature and all attempts are made to achieve a small grain size which will give improved mechanical properties.

This process can be represented in terms of change in free energy as shown in Figure 2.40.

The formation of a stable nucleus in a pure metal is a balance between the surface free energy and the bulk free energy. In real alloys there are impurities (**this then becomes heterogeneous nucleation**) and mould walls that assist in the formation of the stable nuclei and require less undercooling.

Following successful nucleation then growth occurs. For pure metals this would be equiaxed followed by columnar growth. For alloys this would be equiaxed followed by dendritic growth.

The classic picture of the grain structure in a cast alloy is shown in Figure 2.41.

SUMMARY

The structure of metals can be considered on a number of levels. In order to understand the engineering properties, it is necessary to consider the crystal structure and the microstructure. The crystal structure describes the way in which the atoms are arranged, whilst the microstructure describes the way in which the crystals are assembled together in the metal. Figure 2.43 shows a summary of grain development.

In addition, we have now entered the era of achieving "nano" sized grains which provide an increase in both toughness and strength. (Ultrafine grain is used here in the context of average grain sizes between 1 μm and 2 μm in diameter; submicron refers to grain sizes between 100 nm and 1000 nm; while nano-structured refers to grain sizes below about 100 nm.) http://www.dierk-raabe.com/ultrafine-grained-steels. Figure 2.44

shows details of the various numerical values used to quantify the grain size.

When metals are alloyed a number of different constituents may form. However, it is difficult to accurately predict which constituents will form under any given set of conditions. It is often necessary to determine experimentally which constituents will exist under given conditions. This information is generally presented in the form of a "phase diagrams" (or equilibrium diagrams), a topic which was introduced in Chapter 1 with the iron-carbon equilibrium diagram.

In addition, when changes in the microstructure are not in accordance with equilibrium conditions such as the heat treatment of steel, additional methods have to be used to predict the microstructure such as TTT (temperature time transformation) and CCT (continuous cooling transformation) diagrams.

Several important aspects that have been introduced in this chapter are of vital importance in

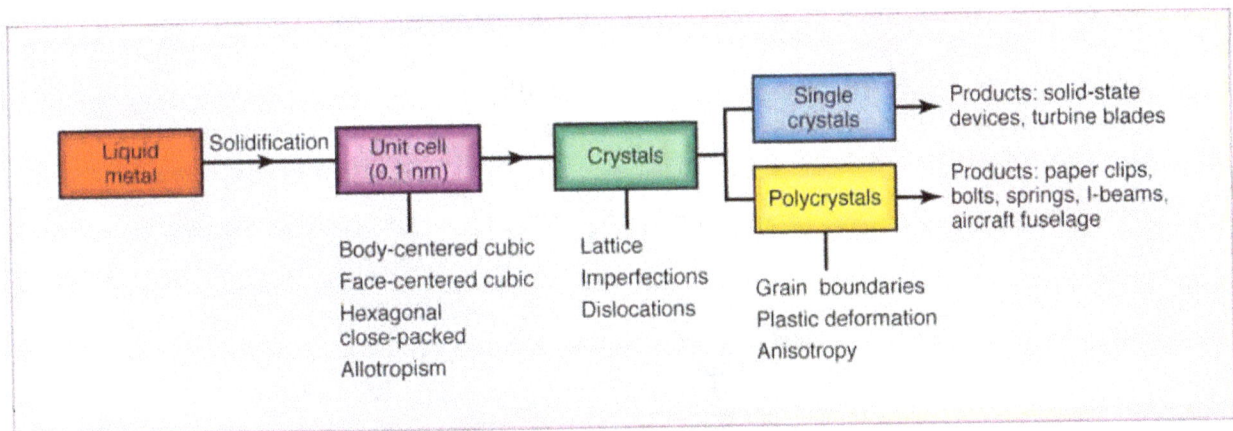

Figure 2.43 *Summary of from solidification to crystal structure showing option for single crystal.*

TABLE 1.1

Grain Sizes

ASTM No.	Grains/mm^2	Grains/mm^3
-3	1	0.7
-2	2	2
-1	4	5.6
0	8	16
1	16	45
2	32	128
3	64	360
4	128	1,020
5	256	2,900
6	512	8,200
7	1,024	23,000
8	2,048	65,000
9	4,096	185,000
10	8,200	520,000
11	16,400	1,500,000
12	32,800	4,200,000

ASTM Grain Size:

$$N = 2^{n-1}$$

where

N = Grains per square inch at 100x magnification

n = ASTM grain size number

ASTM GRAIN SIZE NUMBER	AVERAGE GRAIN DIAMETER MICRONS
0	360
1	254
2	180
3	127
4	90
5	64
6	45
7	32
8	23
9	16
10	11
11	8
12	6

Figure 2.44 *Various ASTM relationships, grains per cubic mm, ASTM grain size number relationship and the relationship between ASTM number and the average rain diameter in microns*

the understanding of topics such as solidification, homogenisation, carburising, transformations and fracture.

REFERENCES

1. The future impact of materials security on the UK manufacturing industry Future of Manufacturing Project: Evidence Paper 27 Foresight, Government Office for Science. David Parker Centre for Remanufacturing and Reuse, Oakdene Hollins Ltd October 2013. file:///C:/Users/user/Pictures/train/New%20folder/ep27-material-security-impact-uk-manufacturing.pdf

2. Materials Availability: Comparison of material criticality studies – methodologies and results Working Paper III UKERC/WP/TPA/2013/002 February 2013 Jamie Speirs et al Imperial College Centre for Energy Policy and Technology (ICEPT) file:///C:/Users/user/Downloads/Materials%20Availability%20Working%20Paper%20III.pdf

3. Metallurgy made in and for Europe The Perspective of Producers and End-Users Roadmap Edited by Achilleas Stalios Rapporteurs: Victoria Folea and Edmond Cahill https://ec.europa.eu/research/industrial_technologies/pdf/metallurgy-made-in-and-for-europe_en.pdf

4. Substitution of critical raw materials, Deliverable report D5.3 Final Roadmap Report MAY 2015. DANIELA VELTE, TECNALIA

5. Strategic and Critical Materials 2015 Report on Stockpile Requirements, Under Secretary of Defence for Acquisition, Technology and Logistics January 2015

6. Serope Kalpakjian, Steven Schmid: Manufacturing Engineering and Technology prentice Halll 2010

7. Michael Ashby, Hugh Sherdiff, David Cebon, Materials engineering science processing and design BH 2007.

8. Helmut Mehrer: Diffusion in Solids Fundamentals, Methods, Materials, Diffusion-Controlled Processes Springer Berlin Heidelberg New York 2000

9. http://iramis.cea.fr/meetings/matgen4/Presentations/MNDN-1.pdf

10. http://www.vssut.ac.in/lecture_notes/lecture1428553162.pdf

QUESTIONS

1. Based on this training document state the three levels of structure identified for understanding the properties of metals? a)? b)? c)?

2. Explain the terms 'unit cell' and 'Bravais lattice'.

   Answers
   1. The three levels of structure are:

      a) atomic structure b) crystal structure c) microstructure.

   2. The unit cell is the geometric figure which illustrates the grouping of the atoms in a crystal. The crystal consists of large numbers of atoms arranged in a regular repetitive pattern or array, i.e. the "Bravais" or space lattice.

3. Why are some dislocations called 'edge' dislocations, whilst others are called 'screw' dislocations?

   Answer
   The term "edge dislocation" is used because the distortion line lies along the 'edge' of the extra half plane of atoms. The term "screw dislocation" is used because the planes intersected by a screw dislocation are distorted into a spiral shape, similar to a 'screw' thread.

4. With reference to the previous discussion of defects in the crystal structure, give an example of a defect for each of the following classes:

   a) a point defect b) a line defect c) a plane defect.

   Answers
   a) For a point defect you could have chosen either vacancy, interstitial, interstitial impurity, or substitutional impurity.
   b) A line defect is a dislocation (edge or screw).
   c) A plane defect is a grain boundary.

SELF-ASSESSMENT QUESTIONS

1. What is the total number of Bravais lattice types?

2. For the two-dimensional lattice shown below, determine the following:

   a) the number of nearest neighbours
   b) the unit cell.

   o o o o o o
   o o o o o o
   o o o o o o
   o o o o o o

3. Sketch the following unit cells and for each case give one example of a metal which adopts that structure:

   i) f.c.c.
   ii) b.c.c.
   iii) h.c.p.

4. Describe the relevance of each of the following defects to the properties of metals:

   a) grain boundary
   b) dislocation
   c) vacancy.

5. Explain what is meant by an alloy.

6. Describe each of the following metal constituents.

7. How would you improve the strength and toughness of AISI 4130 steel that was purchased as forged?

8. Describe two differences between metallic solid solutions and intermediate compounds.

9. State a formula of a compound which can form in the iron-carbon system.

10. The supplier of a forged part has stated that the grain size is ASTM 8. What will be the average grain size in microns? Is there any further information required to understand the expected grain size in the supplied forging?

11. There are two method of nucleation of grains in a casting Homogeneous & Heterogeneous Nucleation. Which form of nucleation would give the smallest grain size? In practice how would this be carried out?

12. What is diffusion? Name a heat treatment process that uses diffusion?

13. What change in the grain size of an alloy would improve the yield strength and toughness?

14. How does work hardening occur?

15. Name three types on non-metallic inclusions that may be found in steel.

16. What could be checked on a chemical analysis test certificate to gain an indication of the steel cleanliness of the steel supplied?

17. What chemical elements would you request on a test certificate to check the inclusion content of a steel?

18. If the ferrite grain size on a sample of S235 steel plate was 10 microns and another plate was 100 microns what would be the estimated yield strength and impact transition temperature for each steel.

19. Which steel would be likely to give the best resistance to brittle fracture when used in a cold environment?

20. A casting was metallographically examined and was found to have a dendrite secondary arm spacing of 100 microns. What was the average cooling rate of the casting?

21. Name two metals that have a BCC crystal structure at room temperature.

22. What crystal structure would a sample of austenitic stainless steel have?

CHAPTER 3

Steel selection and steel quality

3.1 INTRODUCTION

The key objective involved in steel selection should be "zero risk" in terms of achieving the minimum level of specified mechanical properties and avoiding any risk of failure during the warranty or life expectancy period. In addition, there is a requirement to ensure that the quality of product achieves "fitness for service" at an economic price. This means that the steel should be both "*suitable and affordable*".

Any policy of risk reduction requires that the steel should be made by a process which has an established *robust process capability* with appropriate control procedures to ensure steel quality requirements are met. It is not expected for steel users to be experts in steel manufacture but an understanding is helpful to ensure the procurement of steel with the best combination of price and quality. Many large steel users have detailed supply specifications that include the requirement for steel maker approval by quality assurance audits and steel quality validation. However, such quality assurance approvals are expensive, and many users must rely on supply standards and test certificates.

The quality of the steel product such as a forging or machined part is dependent on the quality of the primary steel supply that is used to make the part or component. Therefore, a company that specializes in the manufacture of quality parts must have a basic understanding of how the steel making technology and heat treatment choice can affect steel product quality. There are many steel making factors that affect the quality of steel and process control requires good technical co-operation and communication between the steel user and the steel supplier to ensure that the quality that is required is delivered. This chapter aims to introduce some of the important factors which need to be controlled during the manufacturing route by the steelmaker:

1. Low Hydrogen levels – especially when low sulphur steels are required
2. Steel cleanliness – Secondary steel making and vacuum degas are needed to minimise inclusions
3. Segregation – Ingot or continuously cast need to meet the required standard
4. Control of chemical analysis – (The size effect/ hardenability) Many large cross section products require the chemistry close to the top of the maximum required and ideally with a similar consistent composition for all steel supplied.
5. As cast surface quality and freedom from cracks, porosity and seams, which could be damaging to the final products.
6. Sufficient mechanical working measured in terms of forging ratio or rolling reduction ratio.
7. Heat treatment control, Furnace control, austenitising time and temperature and control of quenchant condition temperature, agitation and composition for polymer grades.

The effective control of these factors requires a technically based audit system with the understanding "of what can go wrong". It also requires a standard of comparison that relates to "best world class standard" to allow a quantitative comparison or benchmarking to rate individual manufacturers.

3.2 MATERIAL STANDARDS

For every material there exists a standard. Based on country of origin, the standards can be categorized as.

- American standards (SAE, ASTM)
- British standards (BS)
- European standards (EN)
- German standards (DIN)
- Indian standards (IS)
- Chinese standards (GB)

ISO International Organization for Standardization https://www.iso.org ISO is an independent, non-governmental international organization with a membership of 162 national standards bodies that attempts to consolidate various national codes worldwide. The ISO (International Standards Organization) also publishes its own standards

2.3.1 American Standards

- The popular American standards are *SAE* (Society for Automotive Engineers). In 1995 the AISI turned over future maintenance of the system to SAE because the AISI never wrote any of the specifications. *ASTM* (American Society for Testing of Materials) & *ASME* (American Society of Mechanical Engineers) and API (American Petroleum Institute).

- The *SAE standards use* a designation/steel grading system in which the first two digits indicate a particular group of steel and the last two or three digits indicate the average carbon content. For example, SAE 8620 indicates a low Ni-Cr-Mo steel with average carbon content of 0.20 %. (See Table 3.1)

- ANSI American National Standards Institute www.ansi.org/ oversee all the "standards development organizations" and is USA link with ISO.

- UNS In the 1970s, a "Unified Numbering System for Metals and Alloys" (UNS) was jointly developed by ASTM and SAE to simplify material

Table 3.1 *SAE designations*

Carbon Steels		Nickel-Chromium-Molybdenum Steels (Continued)	
10xx	Plain carbon (Mn 1% max)	87xx	Ni 0.55%, Cr 0.50%, Mo 0.25%
11xx	Resulphurized	88xx	Ni 0.55%, Cr 0.50%, Mo 0.35%
12xx	Resulphurized and rephosphorized	93xx	Ni 3.25%, Cr 01.20%, Mo 0.12%
15xx	Plain carbon (MN 1.0% to 1.65%)	94xx	Ni 0.45%, Cr 0.40%, Mo 0.12%
Manganese Steels		97xx	Ni 0.55%, Cr 0.20%, Mo 0.20%
13xx	Mn 1.75%	98xx	Ni 1.00%, Cr 0.80%, Mo 0.25%
Nickel Steels		Nickel-Molybdenum Steels	
23xx	Ni 3.5%	46xx	Ni 0.85% or1.82%%, Mo 0.2% or 0.25%
25xx	Ni 5.0%	48xx	Ni 3.5%, Mo 0.25%
Nickel-Chromium Steels		Chromium Steels	
31xx	Ni 1.25%, Cr0.65%, or 0.08%	50xx	Cr 0.27% or 0.4% or 0.5% or 0.65%,
32xx	Ni 1.25%, Cr 1.07%	51xx	Cr 0.8% or 0.87% or 0.92% or 1% or 1.05%
33xx	Ni 3.5%, Cr 1.5% or 1.57%	50xxx	Cr 0.50%, C 1.0% min
34xx	Ni 3.0% Cr 0.77%	51xxx	Cr 1.02%, C 1.0% min
Molybdenum Steels		52xxx	Cr 1.45%, C 1.0% min
40xx	Mo 0.20% or 0.25%	Chromium-Vanadium Steels	
44xx	Mo 0.40% or 0.52%	61xx	Cr 0.6% or 0.8% or 0.9%, V 0.1 or 0.15 min
Chromium-Molybdenum (Chromoly) Steels		Tungsten-Chromium Steels	
41xx	Cr 0.5/0.8/ 0.95, Mo 0.12/0.2/ 0.25/0.3%	72xx	W 1.75%, Cr0.75%
Nickel-Chromium-Molybdenum Steels		Silicon-Manganese Steels	
43xx	Ni 1.82%, Cr 0.50% or 0.8%, Mo 0.25%	92xx	Si 1.4 or 2, Mn 0.65/0.82/0.85, Cr 0/0.65
43BVxx	Ni 1.82%, Cr 0.50, Mo 0.12/0.35, V 0.35%	High-Strength Low-Alloy Steels	
47xx	Ni 1.05%, Cr 0.50%, Mo 0.2% or 0.35%	9xx	Various SAE Grades
81xx	Ni 0.30%, Cr 0.50%, Mo 0.25%	xxBxx	Boron Steels
86xx	Ni 0.55%, Cr 0.50%, Mo 0.20%	xxLxx	Leaded Steels

identification. It is not widely used in industry. These codes begin with "UNS", followed by a letter and 5-digit number https://www.astm.org/BOOKSTORE/PUBS/DS56L-EB.htm

The codes for some of the more common metals of the UNS Series:

A00001 to A99999 Aluminium and aluminium alloys

C00001 to C99999 Copper and copper alloys

D00001 to D99999 Specified mechanical property steels

E00001 to E99999 Rare earth metals and alloys

F00001 to F99999 Cast irons

G00001 to G99999 AISI and SAE carbon and alloy steels (except tool steels)

H00001 to H99999 AISI and SAE H-steels

3.2.2 British standards

For Engineering steel grades, the main specification was BS 970, for weldable structural steels BS4360 but when the UK joined the European Economic Area these were superseded usually by the DIN based specifications. The document PD 970 2005 outlines the relevant European specification for each grade of steel. (See the introduction and the Annex A in PD 970 2005.)

The first edition of BS 970 specification was in 1941 to replace BS. 5005. 1924, which, until then, was the only specification for wrought steel for automobiles. The main objective was to provide a comprehensive schedule of steels for general use in the engineering industries as the nation entered the war years where there would be alloy shortages.

Revisions were subsequently issued in 1942 and 1947. The 1947 issue included the results of discussions with the Society of Motor Manufacturers and Traders, to allow the inclusion of the post-war requirements of the motor industry.

The original 'EN' or 'Emergency Number' material designation was developed during the early years of WW2 to aid the standardisation of steel reference specification to allow the production of components from the appropriate material. The EN reference has continued to be used from that time. However, in theory they are now replaced by modified nomenclature,

either the newer 970 format or the European based numbering, but the original EN designation tends to remain in use in Industry due to being specified on Engineering Drawings particularly in the rail, defence and aerospace drawings.

The EN designation was replaced by a six-digit system when the BS 970 was revised in 1970, e.g. 080A15 – although officially this has been superseded by other European standards. Several steel grades are still produced to BS970 designations. However, most carbon and alloy grades have officially been superseded by new BS EN steel standards such as BS EN 10083, BS EN 10088 and BS EN 10277.

The BS 970 numbering system was designed to clearly detail the carbon content and the specific type of steel via the numbers and letter given to the material (as shown below) following a similar trend to the SAE/AISI designation.

070 M 20
Steel Type   Letter   Carbon Content

Steel Type:

This three-digit number indicates the type of steel:

000 to 199 - Carbon manganese steel, number shows the manganese content (x100)
200 to 240 - Free cutting steel, the 2nd and 3rd digit represents the sulphur content (x100)
250 to 250 - Silicon manganese steel
300 to 499 - Stainless steels and steels resistant to heat
500 to 999 - Reserved for alloyed steels.

Letter:

The single letter will be one of four designations, A, H, M or S:

A - The steel is supplied to a chemical composition as attained from a chemical batch
H - The steel specification is 'hardenable'
M - The material is produced to certain mechanical properties
S - The steel is stainless in specification.

3.2.3 European Committee for standardization

https://www.cen.eu

- One European Standard replaces 34 national standards. They have created the *EN (European Norms)* standards which is to replace the individual national standards of the member countries – UK, Germany, France, Italy etc.
- All the materials covered under these standards have a unique *material number*. It is therefore easy to correlate the materials even if the designation is changed. E.g. 17CrNiMo6 of DIN17210 is designated as 18CrNiMo7-6 in EN10084 but the material number remains the same i.e. 1.6587.

3.2.4 German Standards

Material numbers Werkstoff Numbers

DIN 17007 In 1959 this standard was issued which developed a numerical system to identify materials suitable for computer-based systems. A material number consists of 7 digits, which are in turn divided into 5 groups as shown below:

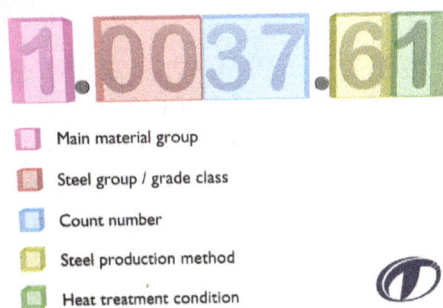

Main material group

Steel group / grade class

Count number

Steel production method

Heat treatment condition

Main material group

The first digit contains the material group and indicates the material involved. The principal group 1 covers all materials that are made of steel or cast steel. The material group 2 contains all heavy metals such as copper, zinc and lead. The third group contains the lightweight metals which include, among other things, aluminium and titanium.

Steel group / grade class

The next two numbers give the steel group of the material group or the grade number of the remaining other groups. These figures provide additional information such as the carbon content or whether it involves grades such as mild steel, pressure vessel steel, tool steel or roller bearing steel. There are also steel group numbers for steel grades with special properties such as heat and rust resistance, resistance to chemical substances or magnetic properties.

In the case of the material groups 2 and 3 the grade number is used to subdivide the materials into non-ferrous base metals such as copper, zinc, cadmium, lead, aluminium, magnesium, titanium and precious metals.

Count number

The next two numbers are count numbers (serial numbers), in order to differentiate between materials which, belong to the same steel group or, in the case of non-ferrous materials, which have the same grade number. But they do not provide any information whatsoever about the composition of the material.

3.3 STEEL QUALITY FACTORS

3.3.1 Hydrogen in steel

Hydrogen has the smallest atomic diameter (see Figure 2.5) at 0.1 nm diameter. As a consequence, in the atomic state, hydrogen can enter the metal lattice, where it has a relatively high diffusion coefficient. In ferrite steels (bcc lattice) the mobility of hydrogen is high however, in the more closely packed of austenitic steels (f.c.c. lattice) the diffusion coefficient can be four orders of magnitude smaller than in ferritic steels which results in a lower susceptibility to hydrogen cracking in the temperature range below 800C. Most of the hydrogen diffusion mechanisms suggest that hydrogen atoms combine into hydrogen molecules in the interface of second phase and matrix, where the hydrogen molecules that are formed can produce high hydrogen pressure. If the pressure of hydrogen causes local stresses that exceed the fracture stress of the metal then micro-cracks can form. (hydrogen cracking or hydrogen flakes)

The hydrogen level in a solid metal should be kept to a minimum to avoid the risk of hydrogen cracking and the possible reduction in the tensile ductility (elongation % and reduction of area %). For low sulphur steel the hydrogen levels in steel should be as low as possible. Hydrogen level is normally measured

Figure 3.1 SEM images of fracture surfaces showing evidence of hydrogen embrittlement

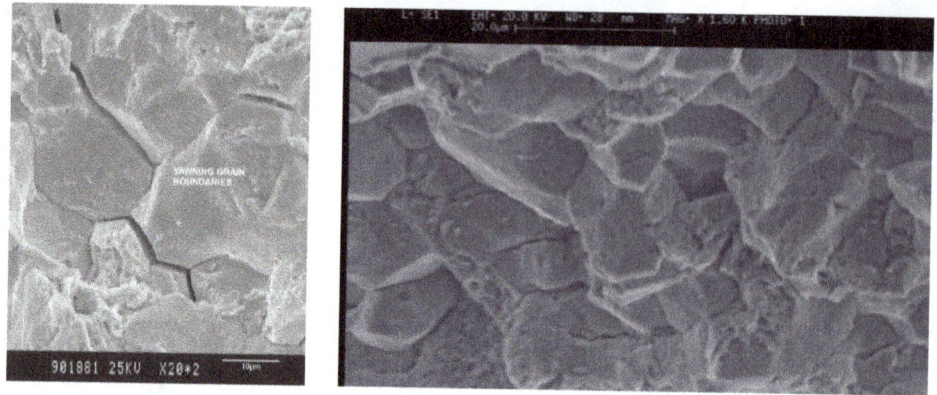

at the liquid steel stage (Hydris) and the use of vacuum degassing equipment during steel manufacture allows hydrogen removal, reducing the risk of hydrogen problems caused by hydrogen picked-up during steel manufacture

Hydrogen-induced fracture can proceed either in an intergranular or in a transgranular fracture path relative to the prior austenite grains. The transgranular hydrogen embrittlement shows separation at the prior austenite grains with additional cracks (widening grain boundaries), ductile hair lines (crow's feet) and micro pores at the grain boundaries. The ductile hair lines are formed during separation at the prior austenite grain boundaries and are caused by small areas of plastic deformation. (See Figure 3.1)

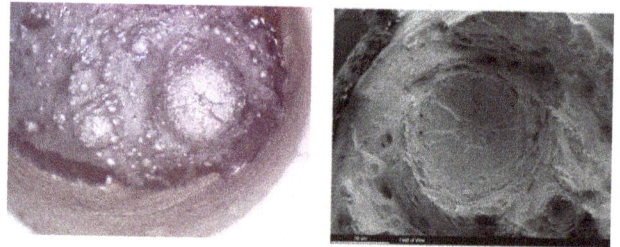

Figure 3.2 Fisheyes associated with hydrogen on an F22 tensile specimen. The elongation was 11%. Fisheyes in ferritic steels have been seen and understood for many years. However, they are now rare and their occurrence can cause serious concern to anyone unfamiliar with the effects of hydrogen. Fisheyes have been found on plasma nitrided torsion-bending fatigue specimens. LHS = Optical macro image, RHS = SEM image

3.3.1.1 Avoidance of hydrogen during steel manufacture

Any hydrogen present in steel would be in solution in the monatomic form. The hydrogen atom is small in comparison with that of iron and therefore it occupies interstitial positions within the lattice. It can diffuse both inwards and outwards at elevated temperatures. The embrittlement which occurs as a result of a high hydrogen content is generally believed to be due to the gas coming out of solid solution during cooling and accumulating as molecular hydrogen at various sites within the metal such as at voids at the tips of manganese sulphide non-metallic inclusions of or actual cracks and as cloud precipitates in the vicinity of dislocations. This diffusion to local sites can build up internal pressures which can produce small internal fractures.

As the sulphur is reduced the number of sulphide "sinks" decreases and therefore the pressure at each inclusion increases, which can result in fracture. Therefore, low sulphur steels are more vulnerable to hydrogen cracking and must be manufactured with

a lower hydrogen content. When a steel specimen with a high hydrogen content is fractured in a slow bend test, these internal fissures become apparent as small silvery areas – variously called flakes, fissures of fisheyes. Diffusion of hydrogen is known to be easier (both into the metal and out of it) when the steel is in the body-centred state and a safeguard adopted with large forgings is to anneal for long periods at a sub-critical temperature, say 600/650°C. Figure 3.3 RHS shows a graph used to predict the holding time at 650°C to give a particular hydrogen reduction. Hydrogen can enter and diffuse through a metal alloy surface at ambient or elevated temperatures. This can occur during various manufacturing and assembly operations or when the component is in service, i.e. anywhere that the metal comes into contact with atomic or molecular hydrogen.

3.3.1.2 Other sources of hydrogen pick-up

Typical industry processes that can lead to hydrogen embrittlement include: acid cleaning prior to galvanizing, phosphating, acid pickling, electroplating

Koyama et al. / Acta Materialia 70 (2014) 174–187

Damage evolution plotted against local plastic strain with and without hydrogen.

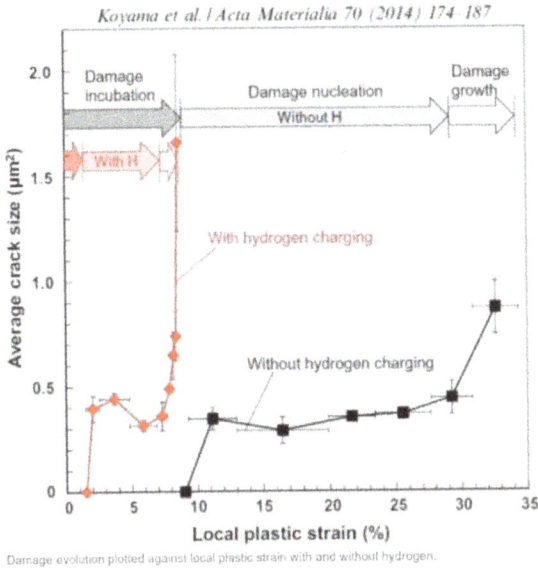

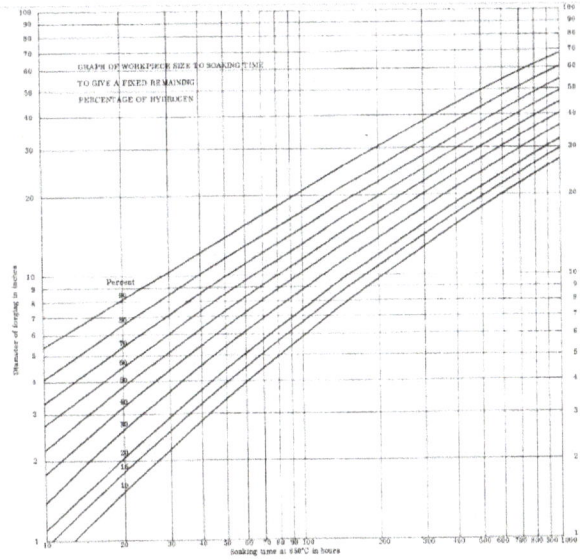

Figure 3.3 *LHS = Effect of hydrogen on crack propagation/ reduced toughness, RHS = Hydrogen removal*

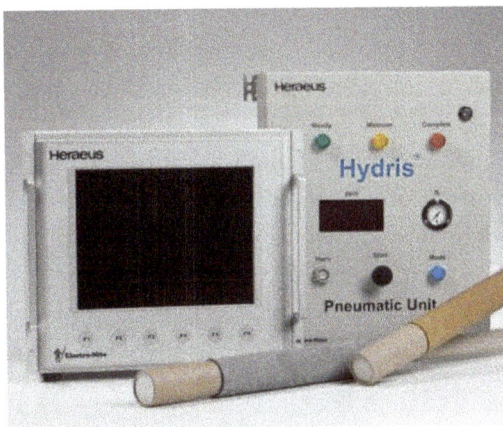

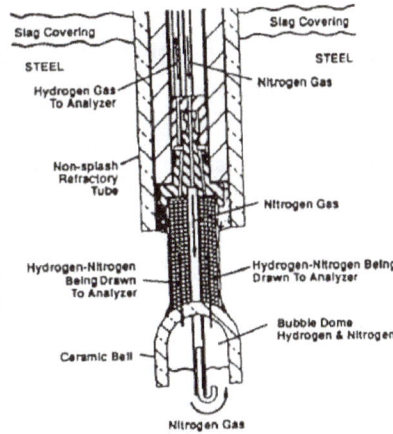

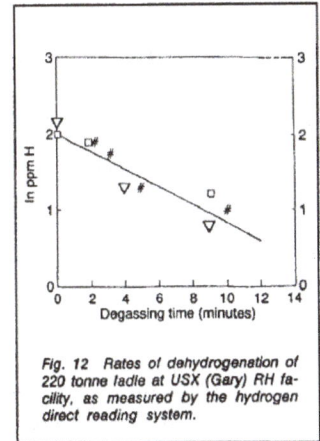

Fig. 12 Rates of dehydrogenation of 220 tonne ladle at USX (Gary) RH facility, as measured by the hydrogen direct reading system.

Figure 3.4 *Hydris "in situ" hydrogen measurement.*[1 to 3] *Probe details RH degasser and Hydrogen*

coatings such as zinc, chromium and cadmium and arc welding (HAZ cold cracking). During these processes, there is a possibility of absorption of hydrogen by the material. For example, during arc welding, hydrogen is released from moisture, for example in the coating of the welding electrodes; to minimize this, special low-hydrogen electrodes are used for welding high-strength steels. The cathodic region during aqueous corrosion can experience hydrogen pick-up.[10 to 13]

When metals are used to manufacture products that are used where corrosion can occur, hydrogen can be introduced into metal as a result of environmental corrosion issues. (For example, sour gas with H_2S related oil and gas extraction and processing can result in sulphide stress cracking.) Oil wells are classed as either sweet or sour. Sweet wells are mildly corrosive, whereas sour wells are very corrosive since they can contain hydrogen sulphide, carbon dioxide, and chlorides.

During the next decade the most severe new challenges of oil and gas exploration are in the domain of deep wells, specifically those in deep seawater. In addition to the corrosive conditions these wells experience temperatures up to 260°C and pressures up to 25000 psi (172 MPa). Alloy selection is, therefore, especially critical for such sour gas deep sea wells. As the operating conditions become more severe, material selection changes from carbon steels for sweet wells to duplex (austenitic-ferritic) stainless steel, to nickel-based alloys such as Incoloy alloys 825 and 925, Inconel alloys 725HS and 725.[4]

An important challenge relating to achieving a better understanding of the nature of hydrogen embrittlement in metallic alloys, especially in steels, is the determination of the location of the hydrogen.

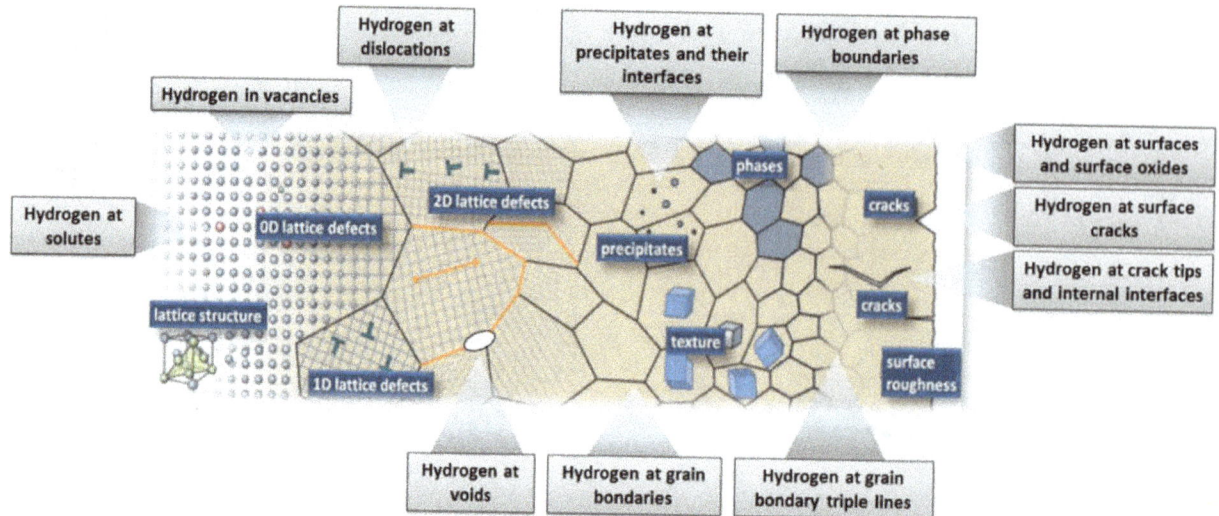

Figure 3.5 *Hydrogen can get trapped and transported at various types of lattice defects such as vacancies and certain solutes, dislocations, grain boundaries as shown the diagram. The main challenge in identifying the governing mechanisms behind hydrogen embrittlement and microstructural sensitivity*[5]

Because of the low concentrations, and high diffusivity, and the possible presence of compositional, thermal and mechanical gradients, the measurement of the hydrogen is difficult. Figure 3.5 shows an image from a recent review on measurement techniques showing locations where trapped hydrogen can be found.[5]

3.3.2 Non-metallic inclusions – a low content is essential for premium grade steels.

Steel cleanliness is one of the main quality issues in the production of steel. During the manufacturing process the steel can be contaminated by non-metallic inclusions in the product. These non-metallic inclusions are either oxides or sulphides. The sulphides in the steel are initially inherited from the sulphur in the coal/coke used in the blast furnace to produce iron. To achieve low sulphur levels requires desulphurisation of the iron prior to steel making. During the subsequent refining of the iron from the blast furnace in the "steelmaking" process, the carbon (begins around 4%) is removed with oxygen. At the end of refining with the carbon around 0.1% there is a high oxygen content in the steel. This is removed by the "deoxidation" process by either vacuum degas methods or the addition of silicon and aluminium. The oxygen in the steel forms non-metallic inclusions (Silicate, alumina, calcium oxides), some of which float out of the liquid steel (by Stokes Law) and some remain trapped in the solidified steel. The oxide inclusions that remain in the solid steel are divided into two main categories:

- Indigenous inclusions which are formed during the de-oxidation of the steel to remove the excess oxygen introduced to remove the carbon from the iron. Also, re-oxidation of the liquid steel.
- Exogenous inclusions which originate from the refractory lining of the furnace or the slag used to refine the steel.

The type and quantity of non-metallic inclusions can impact the steel's physical and mechanical characteristics such as toughness, tensile strength, fatigue limit, and can assist in failure modes that can result in service failures of engineering components.[6] There are several published papers relating to steel properties and steel cleanliness by Brian Pickering who was one of the key persons in the development of microstructure and property relationships. In 1998 Brian Pickering prepared a detailed review was on "The effects on non-metallic inclusions on the properties of steels", which has endured the passage of time and still represents the current status.[7]

Another important area where the inclusion content is important is with machineability. It is estimated that for automotive components the cost of machining represents about 40 percent of the production cost. Typically, a range of sulphur 0.02 to 0.04% is preferred for machinability. However, for high toughness and ductility very low sulphur levels are preferred. A recent study regarding the effect of inclusions on machineability was carried out by Niclas Anmark et al which shows the complexities to find steels with optimised machineability.[8]

3.3.2.1 Measurement of steel cleanliness

The type of inclusions, shape and size distribution are linked with the melting, refining and solidification practice. An EU project determined the improvements in mechanical properties for steel with low sulphur and oxides for SAE 5150, 4340, 4130 and 8620.[42] To quantify the steel cleanliness optical metallography is used. These assessments are carried out in accordance with several international standards, such as ASTM E45 and DIN 50602. These are based on chart methods where for example the worst field on a prepared microstructure at X100 magnification is compared to pictures with various sizes and area fraction of inclusions (Figures 3.6 and 3.7). An additional way to measure the level of oxides would be the total oxygen level measured by a Leco fusion method (Figure 3.17).

3.3.2.2 Inclusion cracking in Wind Turbine Gearbox bearings

A recent published study of rolling contact fatigue of wind turbine bearings based on examination of failed bearings and laboratory rolling contact work was carried out in 2018.[6] Based on the SEM examination of the samples the work presented some high-quality unique images of the effects on the non-metallic inclusion in the bearing steels (see Figures 3.8 and 3.9). The work was carried out due to wind turbine gearbox failures causing significant downtime. It was estimated that approximately two-thirds of these failures initiate in the bearings. This was found surprising since the bearings were designed to the same standards that are used in many other industrial applications.

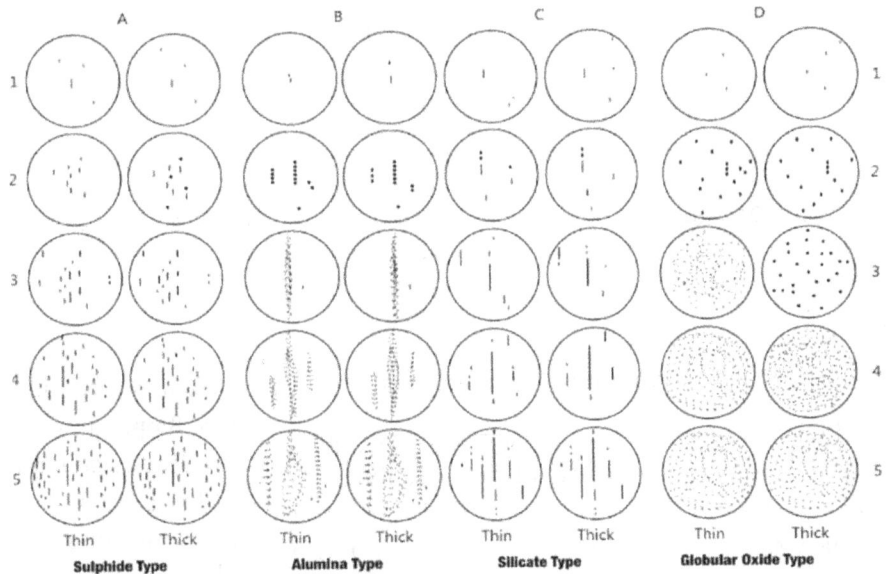

Figure 3.6 Chart used for inclusion counts to ASTM E45

Acceptance level/Quality	A-Type		B-Type		C-Type		D-Type	
	Thin	Thick	Thin	Thick	Thin	Thick	Thin	Thick
Air Melted quality	2.5	2.0	2.0	2.0			2.0	2.0
Ladle Treated/Argon Purged	2.0	1.0	2.0	1.0			1.5	1.0
Vacuum Degassed	1.0		1.0				1.0	

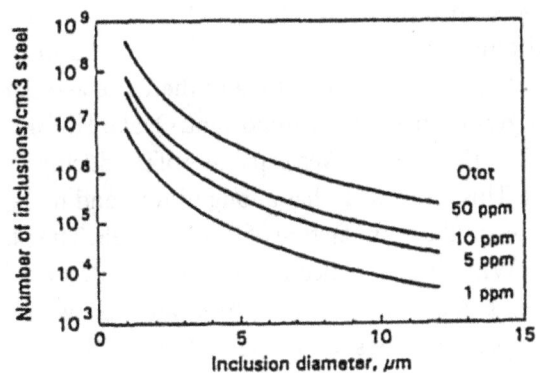

Figure 3.7 Typical values of ASTM E45 inclusion counts in accordance with method of manufacture Figure 3.7 The relationship between total oxygen, average oxide inclusion size and the number of inclusions per cubic cm

▶ *Figure 3.8* *Four different forms of subsurface-damaged MnS inclusions from a failed WTG planetary bearing: a separation of inclusion from the surrounding steel matrix; b internal cracking of inclusion; c crack, propagation into surrounding material; d WEAs attached to these cracks and/or free surfaces (6)*

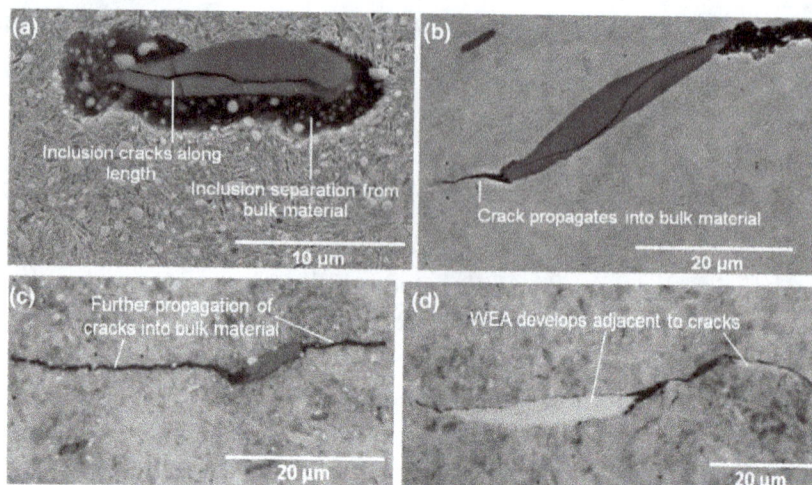

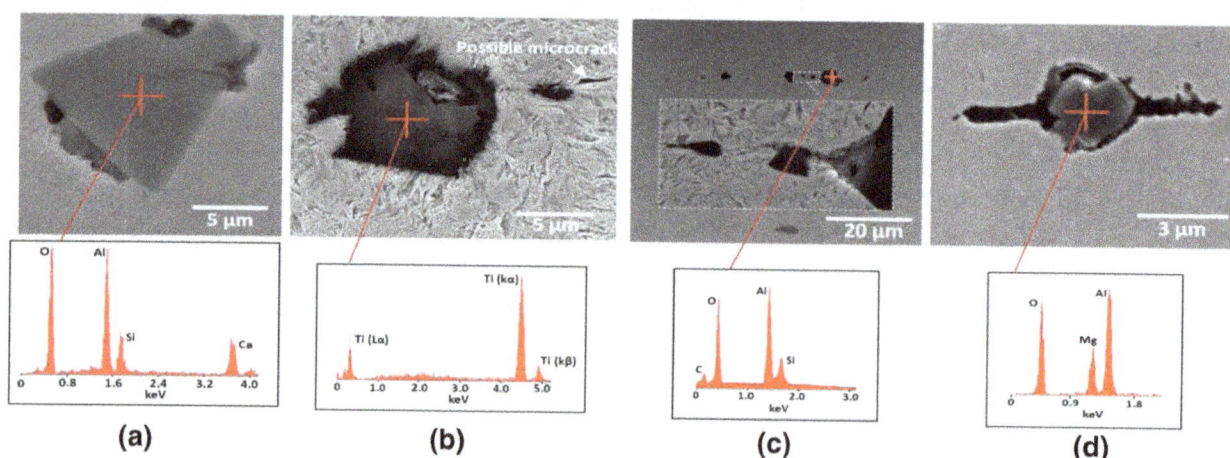

Figure 3.9 *Damage at other types of inclusions and EDX analysis: an internal cracking of alumina inclusion; b micro-crack close to titanium inclusion; c possible damage at alumina inclusion; d micro-cracks at either side of alumina inclusion*[6]

3.2.2.3 Steel cleanliness for bearing steels

Steel selection for roller bearing applications require that the steel has a high level of steel cleanliness to achieve the demanding requirement of rolling contact fatigue resistance. In addition, these steels have a high carbon level and a high alloy content (1% Carbon, 1.5% Chromium) and balls and rollers and inner and outer raceways are through hardened and tempered to around 58 to 62Rc and therefore avoidance of segregation would be important. Ovako Hofors is a steel-works with a proven capability of producing clean steel. This steel-works has a long history and has specialised in the manufacture of bearing steels. The plant uses WEU ingots which are argon shrouded during teeming. A review of some of the facilities allows a better appreciation of some of the challengers, difficulties, and danger that faces the steel maker. The Appendix 1 shows a review of the facilities and key plant and equipment prepared following a visit during 2019.

Clean steels will generally have oxygen content below 8 ppm and a sulphur content below 100 ppm. Ultraclean steels tend to have even lower oxygen content (5 ppm) and a sulphur content below 20 ppm, thus minimizing the presence of sulphide and oxide inclusions. Depending on the requirements of the application and what needs to be achieved for current or future generations, there are different levels of cleanliness that will offer different potential. Figure 3.10 illustrates schematically how defect size affects fatigue strength and what performance can generally be expected from different types of material cleanliness.

The casting process can vary from the use of various ingot sizes to relatively small concast bloom or billet. To produce clean steel a large casting size, such as ingot mould, is preferred. The advantages of ingot casting are that it is flexible, gives a good starting cross section, and because the solidification process

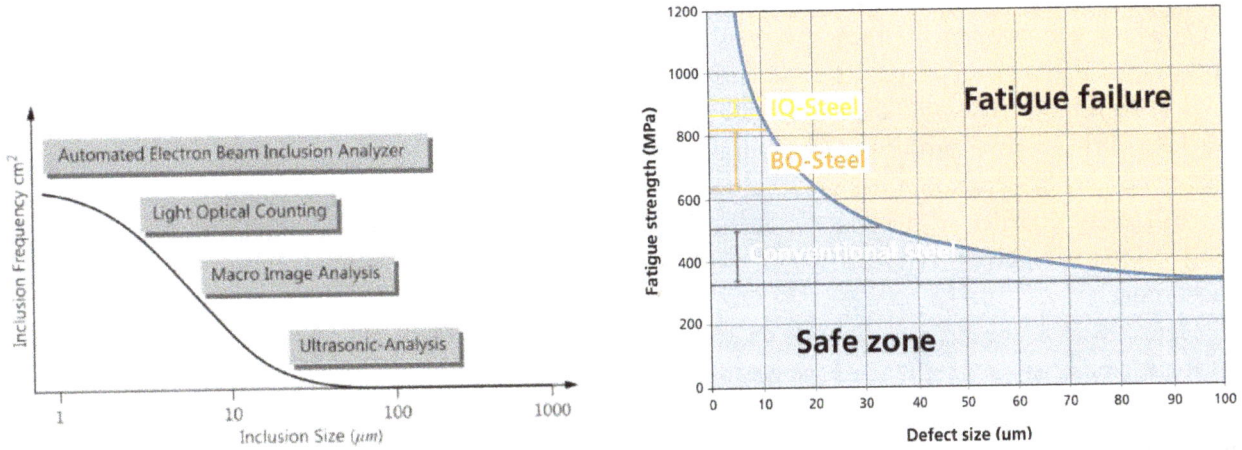

Figure 3.10 *LHS = Size of non-metallic inclusions seen by the various methods. RHS = Fatigue limit vs defect size (9) BQ=bearing quality IQ= Isotropic quality*

is inward and upward, inclusions are generally pushed towards the top. The top is then cut off and scrapped and the amount that is cropped can also be adjusted, depending on grade and quality produced. However, ingot steel has a lower liquid steel to final product yield compared to the continuously cast routes and the steel made by an ingot route has to be more expensive.

The final main production stage is rolling, including the homogenization and hot-working of the material into smaller sizes. Using an ingot results in higher reduction ratio that also gives improved material quality. An example, an ingot size of approximately 500 × 500 mm will have a reduction ratio of 65:1 for a 70mm bar.

When a continuous casting process is used, small cast sizes are produced in a cost-effective process. The typical size would be 240 × 240 mm, which gives a

small reduction ratio of approximately 15:1 for a 70mm bar. Since hot working and area reduction are needed for eliminating porosity and reducing inclusion sizes, material from small, continuous cast routes tends to have a larger inclusion population and thus a different level of material cleanliness and performance. These types of steels are widely used in the automotive industry and are suitable in applications where service loads and durability requirements are relatively low.

3.3.3 Desulphurizing and vacuum degas techniques to ensure low sulphur and oxygen

Using the desulphurizing and vacuum degas techniques, that are available, (see 3.15 and 3.16) will

Element	1950/1960's	1990's	2010's
Sulphur	100 – 300	50 – 80	Below 20
Phosphorous	150 – 300	80 – 140	50 – 100
Hydrogen	4 – 6	3 – 5	2 – 3
Nitrogen	80 – 150	< 60	< 40
Oxygen	60 – 80	< 10	Below 5

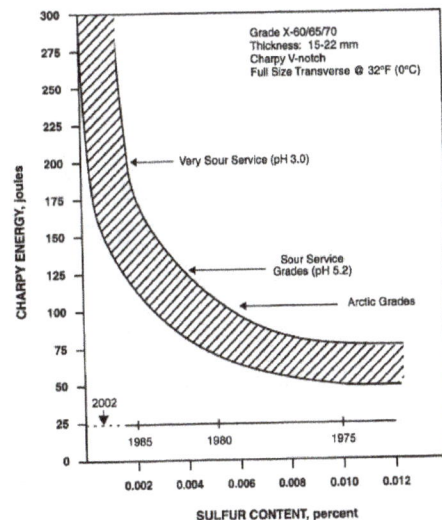

Figure 3.11 *LHS = Typical impurity level in world class tonnage steel manufacture for different period of time showing continuous improvements. RHS = Effect of sulphur on toughness*

ensure low sulphur levels and oxygen levels (see Figure 3.12) which will ensure a high Charpy shelf energy and tough ductile steel (see Figure 3.11) and a high level of endurance. This will also reduce the risk of cracking during thermal processes such as forging and heat treatment. Steel can be at risk of cracking due to thermal stresses, particularly when water quenching is used to process relatively lean hardenability medium carbon steel.

Figure 3.13 lists the suggested steel cleanliness limits for various steel grades There have been several recent reviews on the use of REM (rare earth metals). Due to China being almost the sole supplier of REM, China has taken a major role in the development of REM used in steel manufacture. In China, REM are widely used in steel metallurgy to induce a refined microstructure and modify inclusions and are being used to develop the next generation of steels. [14, 15 and 16]

The theory and practice of rare earth additions has been known for many years. Thermodynamic data predicts that rare earth elements have a very strong affinity for oxygen and sulphur and are capable of reducing the oxygen and sulphur contents of steel as well as modifying the inclusions remaining after solidification. The methods of adding rare earth metals to steel are important and methods of using plunging and encapsulation techniques should be used. Also, magnesia ladle and tundish refractories should be used.

A recent development in steel cleanliness has been called "oxide metallurgy". This is based on the premise that all inclusions cannot be removed and therefore the population of inclusions should be controlled to a small size so that they have no effect on the product. [15]

Figure 3.12 *Trend of oxygen levels in steel*

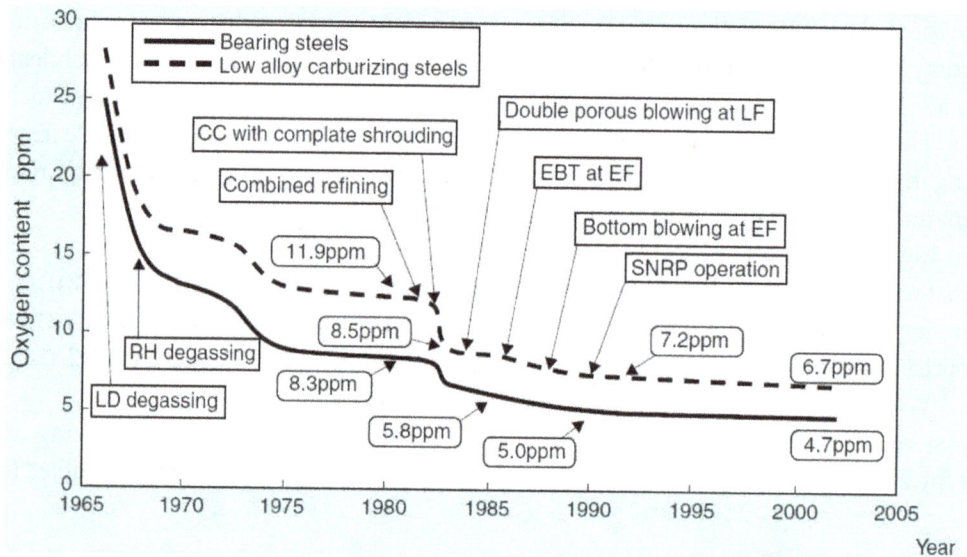

Steel product	Maximum impurity fraction	Maximum inclusion size
IF steel	[C]≤30ppm, [N]≤40ppm, T.O≤40ppm [7], [C]≤10ppm[8], [N]≤50ppm[9]	
Automotive & deep-drawing Sheet	C]≤30ppm, [N]≤30ppm [10]	100µm [10, 11]
Drawn and Ironed cans	[C]≤30ppm, [N]≤30ppm, T.O≤20ppm [10]	20µm[10]
Alloy steel for Pressure vessels	[P]≤70ppm[12]	
Alloy steel bars	[H]≤2ppm, [N]≤10-20ppm, T.O≤10ppm [13]	
HIC resistant steel (sour gas tubes)	[P]≤50ppm, [S]≤10ppm[12, 14]	
Line pipe	[S]≤30ppm [12], [N]≤35ppm, T.O≤30ppm [13], [N]≤50ppm[9]	100µm[10]
Sheet for continuous annealing	[N]≤20ppm[12]	
Plate for welding	[H]≤1.5ppm[12]	
Bearings	T.O≤10ppm[12, 15]	15µm[13, 15]
Tire cord	[H]≤2ppm, [N]≤40ppm, T.O≤15ppm[13]	10µm[13]
Non-grain-oriented Magnetic Sheet	[N]≤30ppm [9]	
Heavy plate steel	[H]≤2ppm, [N]30-40ppm, T.O≤20ppm[13]	Single inclusion 13µm[10] Cluster 200µm[10]
Wire	[N]≤60ppm, T.O≤30ppm[13]	20µm[13]

Figure 3.13 *Steel cleanliness limits for various steel grades.* [18]

3.3.3.1 Oxide Metallurgy

Non-metallic inclusions (oxides, sulphides, and nitrides) in steel damage the continuity of steel matrix and affect the material properties. Traditionally, inclusions with sizes of less than 1 μm were thought to make little difference to the surface defects and strength of steel. In addition, in the 1970s, researchers found that inclusions with sizes of around 1 μm were beneficial and induced intra-granular ferrite (IGF), and thus the steel microstructure could be refined and the strength and toughness could be significantly improved in the heat affected zone (HAZ). This phenomenon has drawn a lot of attention, because inclusions with sizes of around 1 μm, which can form during solidification and cooling, are very difficult to remove. Oxide metallurgy, which can refine grains without thermal processing, has gained research attention, especially for slab steel and structural steel, for which grain refinement cannot be realized via thermal processing. The concept of oxide metallurgy was first developed up by researchers at Nippon Steel Company in 1990.

Concept of oxide metallurgy

- Controlling the oxide distribution and properties in steel (chemical content, melting point, size, and size distribution).
- Using the oxides as nuclei for the heterogeneous nucleation to refine grains and, at the same time,

as the nuclei for heterogeneous nucleation of sulphides, nitrides, and carbides to control the segregation distribution.

- Preventing grain growth by pinning the austenitic grain boundary and using the inclusions in the austenite to affect the transformation from austenite to ferrite and induce intra-grain ferrite similar weld metal solidification.

Successful industrial applications of oxide metallurgy include JFE EWEL (Excellent Quality in Large Heat Input Welded Joins), produced by JFE Steel Corporation, and the HTUFF (Super High HAZ Toughness Technology with Fine Microstructure Imparted by Fine Particles) technique, developed by Nippon Steel. The characteristics of these techniques are shown in Figures 3.14 Lower. Both techniques improve steel properties by using existing inclusions without changing the chemical content and refining the microstructure.

3.3.3.2 Carbon and sulphur

Figure 3.15 shows a graph of steel quality requirements for Carbon and Sulphur for five different steel grades. IF (interstitial free) steel is an ultra-low carbon (ULC) steel type basically used for deep drawing applications. In most applications IF steel is surface coated (galvanized) and is used in automotive construction. Electrical steel is another ULC steel type used in transformer and generator fabrication. Due to its elevated Silicon content the electrical losses of its application can be minimized. CSP steel is basically low Carbon steel for all kind of construction applications like wheels, cylinders, welded pipes, agricultural, etc. The heavy plate shown here is for HIC resistant steel for large diameter pipeline tubes and 100Cr6 is the classic high carbon, high strength steel used for roller bearings. In order to improve machinability of this steel the sulphur content is controlled at a certain range.

The measurement of the oxygen and sulphur levels can be done using Leco analysis. Carbon/Sulphur Elemental Analysis to ASTM E 1019 / ISO 9556 / ISO 4935. Our LECO CS-244 is a simultaneous carbon/sulphur determination based on the combustion-infrared absorption method. The sample is combusted in oxygen that converts the carbon and sulphur in the sample to CO_2 and SO_2. The carbon and sulphur are then measured by infrared absorption.

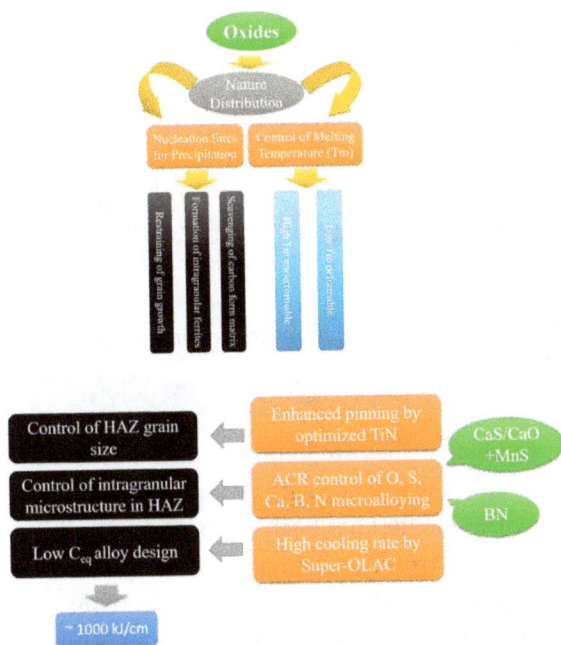

Figure 3.14 *Upper = Concept of oxide metallurgy. Lower = Characteristics of JFE EWEL OLAC: online accelerated cooling; ACR: atomic concentration ratio.*[15]

▶ **Figure 3.15** *Carbon and Sulphur Requirements for different Steel Grades*[19]

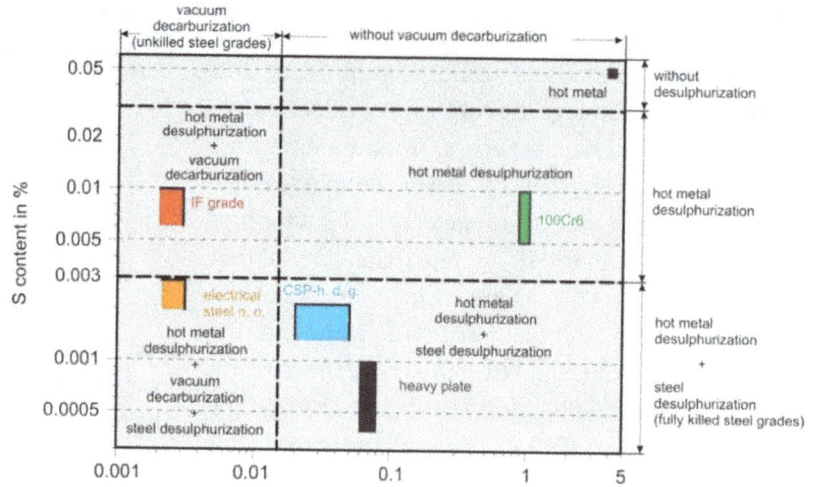

▶ **Figure 3.16** *Sulphur in Steel Removal Process for different Steel Grades*[19]

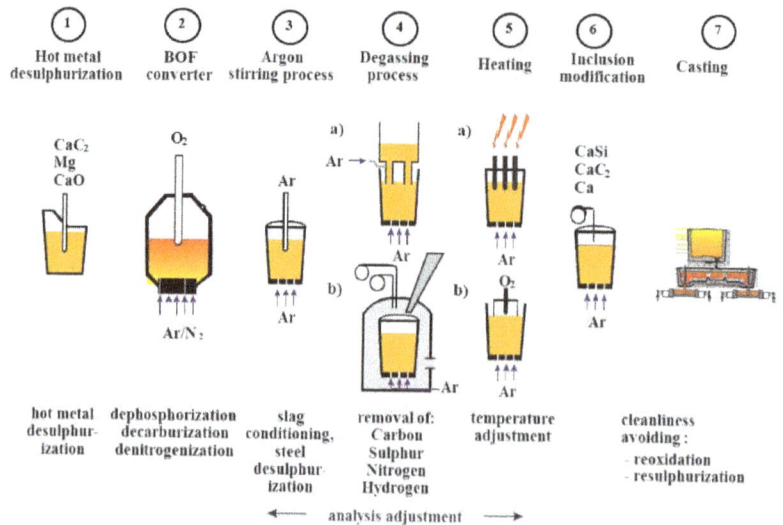

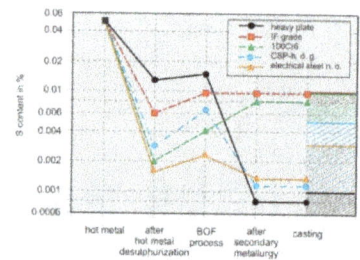

Figure 3.17 *LHS and Centre=Leco analysis equipment used to determine carbon, sulphur, oxygen and nitrogen. RHS = Sulphur removal in for the different Steel Grades*[19]

3.3.4 Avoidance of segregation to ensure uniformity of mechanical properties

There are potential chemical element segregation problems with both continuously cast steel and ingot steel, particularly for large piece items such as reactor pressure vessel (RPV) steels for nuclear power plants and steam and gas turbines. "A" segregate and "V" segregate zones are known to cause problems in ingots and centreline segregation can cause problems with steel that has been from continuously cast. The degree of segregation can depend upon the steel type, ingot size and composition, and for concast the condition of the machine and whether EMS (electromagnetic stirring) has been used.

Product quality of rolled bloom and billet can be checked using ASTM E381. An excellent outline of the background, history and use of macrostructural examination can be found in chapter 1 of a book prepared by George F Vander Voort, (Metallography Principles and Practice, ASM, 1999) The methods use standard charts to rate the macro-etch (50% HCl 50% water 15 to 30 mins at 70°C) appearance of the sub-surface condition (S), the centre segregation (C) and random condition (R).

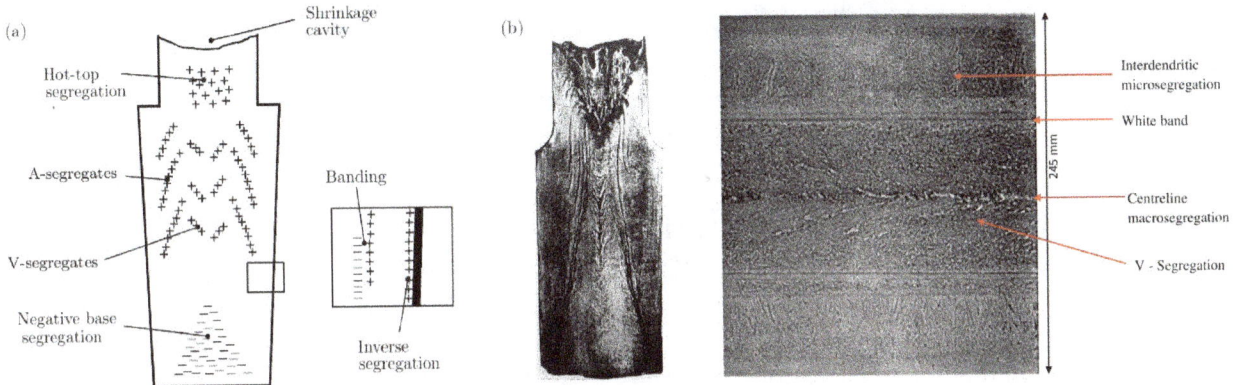

Figure 3.18 *LHS = Ingot segregation RHS = Slab centreline segregation macro-etch*

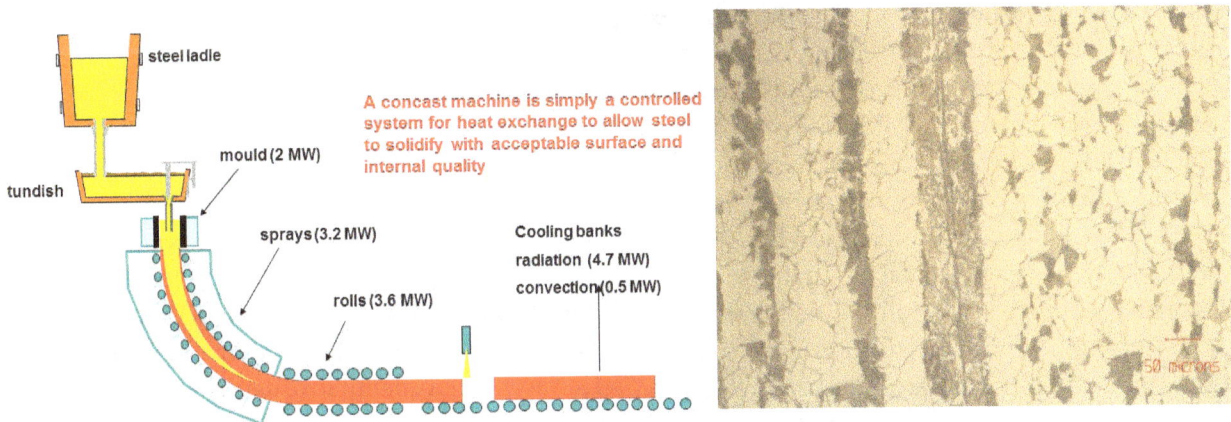

Figure 3.19 *LHS = Slab concast RHS = Hard centre-line 100-micron segregation band*

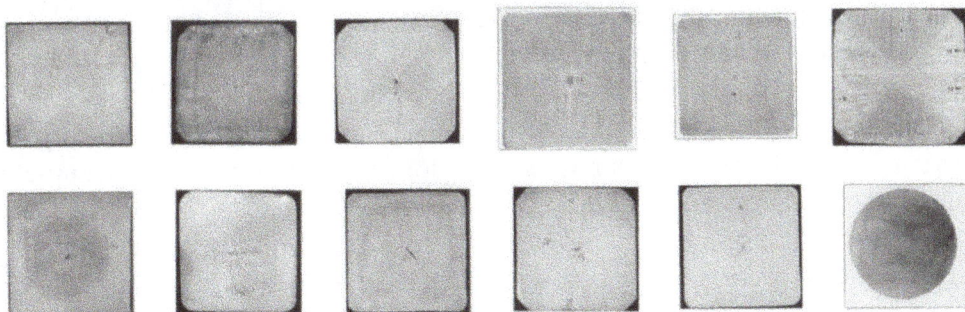

Figure 3.20 *Example of comparison charts used in ASTM E381 (plate 3)*

Figure 3.21 RHS = Macro-structure from a through thickness sample which had low forge reduction ratio from the cast to the forged product. Size of the macro-specimen was 17mm × 13mm. LHS= 180mm diameter bar centre segregation Mid radial carbon = 0.41%, Core carbon = 0.64%.

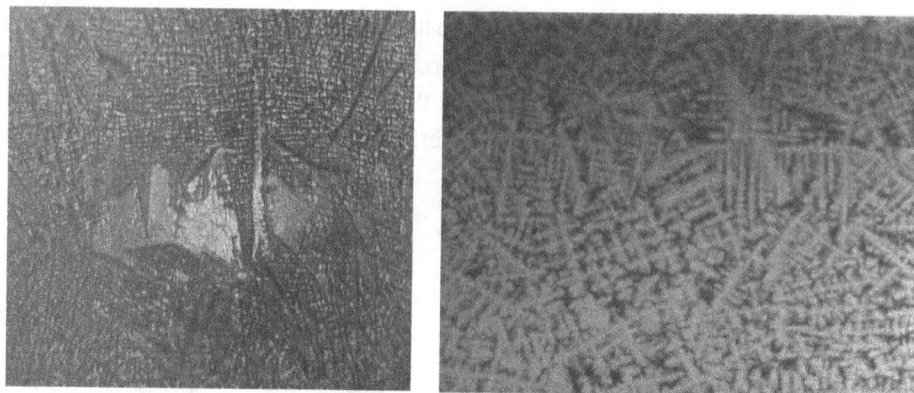

3.3.5 Appropriate forge or rolling ratio from cast shape to wrought hot worked shape

The reason for a minimum "forge ratio" is to consolidate any "as cast" porosity and imperfections and to help make any elemental segregation pattern more uniform through the thermo-mechanical treatment which allows some diffusion of elements in the steel. The mechanical deformation/working during forging also allows essential grain refinement to take place. The addition of niobium can assist this process by delaying recrystallization during the hot working (controlled rolled). In terms of control of the quality of product an important factor is the mechanical reduction from the "cast" condition to the final "wrought" condition. This reduction ratio is generally a simple calculation of the cross section prior to reduction to the cross section after deformation. There are several reviews on the required reduction ratio to provide the required quality level. If continuous forging or rolling is used a reduction ratio of 5:1 is often recommended. Work on continuous cast product concluded a 7:1 was needed and observed that centre line unsoundness was found even after 15:1 reduction.

3.3.5.1 The conclusions from key reports on the forge ratio

The report on the multi-national project carried out for the European funded project[20] commented

"Further rolling trials with continuously cast billet, to examine the change of both central porosity and the mechanical properties for area reductions up to 20:1 have shown that homogenisation of tensile ductility is achieved at an area reduction of between 4:1 and 6:1. However it cannot be assumed that central porosity has been completely eliminated because

full mechanical properties have been attained at the centreline of the stock, since remnant porosity was observed for area reductions of up to 15:1. For demanding applications of the rolled product, a more sensitive criterion than tensile ductility is probably needed."

Another report commented

"Unless the variability in the amount of centreline porosity contained within a continuously cast billet can be eliminated, it will be extremely difficult to move customers away from a safe, minimum billet area reduction of 6:1 for critical applications."

Based on several reviews the conclusion would be that the steel maker has vested interest in stating that a 4:1 should be an acceptable minimum for ingot cast steel. However, when quantitative data is examined relating to toughness and fatigue properties in low alloy steel there is a need for higher values to give suitable properties and a comfortable safety margin.[21]

3.3.5.2. Measurement and the importance of forge ratio

Forging reduction is generally considered to be the amount of cross-sectional reduction taking place during drawing out of a bar or billet. The original cross-section divided by the final cross-section is the forging ratio.

There is an equivalent reduction on upsetting for forgings being upset during forging (gear blanks, for example). In this case, the upset ratio of beginning billet length over final height is the upset ratio. This is similar in total reduction to the bar reduction. However, the uniformity of deformation from centre to edge may not be as easy to estimate as the drawing out reduction. For these applications such as closed die the ideal is to aim to get the high level of deformation in the billet supply. Grain refining is an essential part of forging and if the temperature for forging steel is above about 1200°C, then reductions of at least 15-20% are needed to ensure that the grain size from heating is refined again.

The supply of large cross sections can be compromised by insufficient grain refinement or mixed grain size that can be retained in the final product. The following example is for a large F65 forging. Figures 3.22 and 3.23 show mixed grain size and the effect of the toughness (ITT).

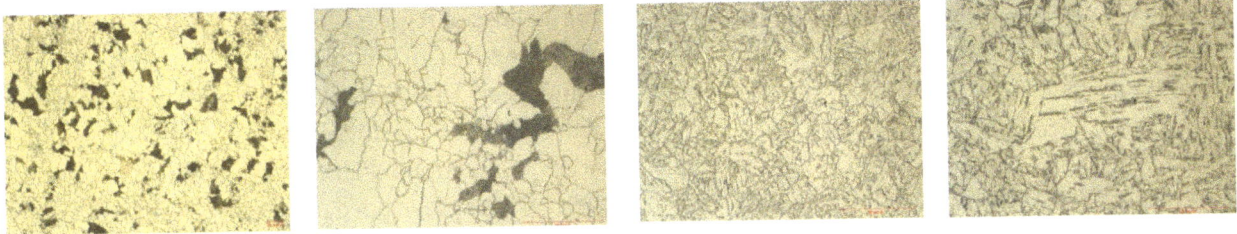

Figure 3.22 *Examples of mixed grain size in a forged part*

Sample	Fine ferrite grain size distribution	Predicted ITT	Coarse ferrite grain size distribution	Predicted ITT
Surface RT	11 microns	-50	64 microns	+5
Surface L	11 microns	-50	70 microns	+5
Surface RL	9 microns	-55	86 microns	+5
Centre RT	22 microns	-35	114 microns	+10
Centre L	31 microns	-30	109 microns	+10
Centre RL	41 microns	-25	125 microns	+25

Figure 3.23 *Grain size average measurements for fine grains and coarse grain and predicted ITT using the diagram shown in Figure 3.24 top RHS.*

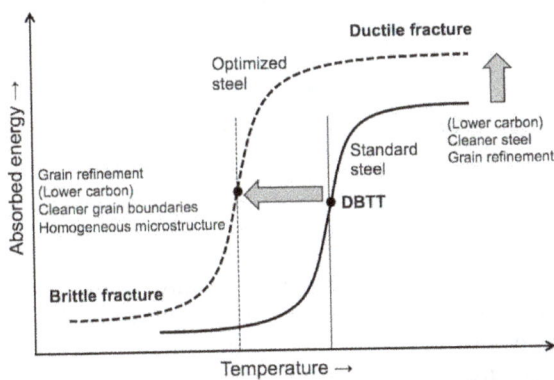

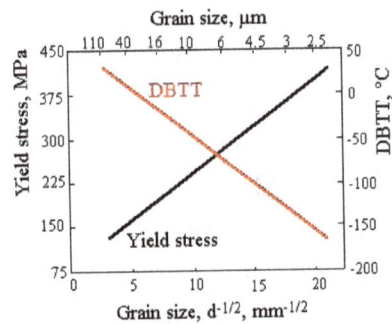

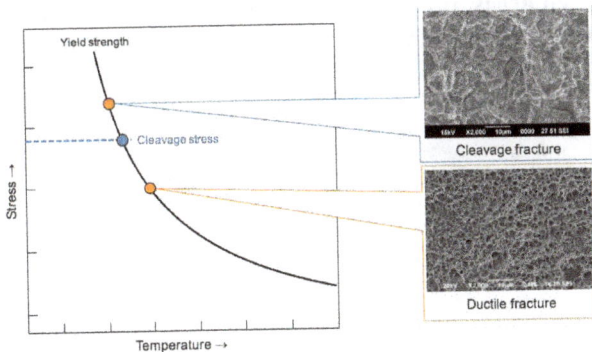

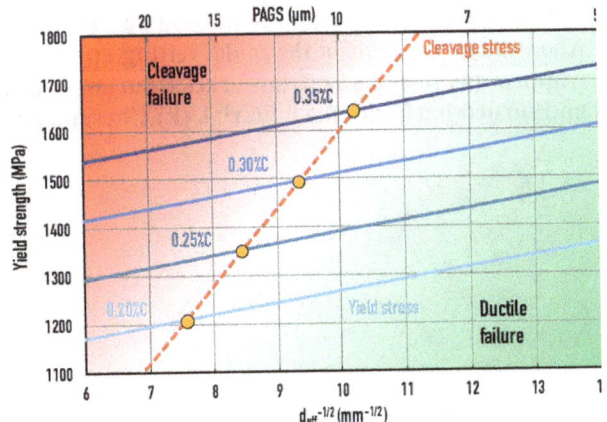

Figure 3.24 *Grain size and Impact Transition Temperature. Ductile rupture vs Cleavage[25]*

There have been several reviews of the effect of round product rolling practice (diamond -round and oval-round pass sequences) on the uniformity of grain refinement across the section.[22, 23 and 24] It is important for steel to have the correct rolling practice to ensure a fine grain size in the final steel parts.

Figure 3.24 shows the various factors that affect the impact transition temperature of steels with a ferritic microstructure (bcc).

3.3.6 Steel surface quality

During steel manufacturing the surface of the steel should be kept free from defects. In general, the types of surface imperfections include cracks, seams, embedded scale and laps. However, the specific types of surface imperfections can depend on the type of product such as slab used for making flat products of plate and strip and bloom used for billet, bar, rod and tube manufacture. The types of imperfection also depend upon whether the steel has been cast by either concast or ingot.

In addition, the types of imperfections on castings will also be specific to the casting process.

The types of cracks formed can be either due to teeming condition or due to thermal stress created during processing. Teeming cracks are usually transverse cracks and are caused by the breaking of the initial solidified skin during the solidification process. The avoidance requires a balance between the fill rate (ladle nozzle size) and the degree of superheat in the liquid steel. The thermal cracks are often associated with regions of low ductility during the cooling of the large cross sections of steel combined with the level of thermal stress developed due to none uniform cooling (corners cool faster that faces)

Seam defects that run longitudinally along the length of rolled products or helically on a tube surface due to the rotation during piercing originate from gas porosity. As the steel solidifies the gas (hydrogen, nitrogen and carbon dioxide) exits the solid steel and can form blowholes. If these blowholes are sub-surface and are exposed during subsequent heating or during rolling the blowholes can form seam defects.

Some of the known imperfections are shown in the diagram below

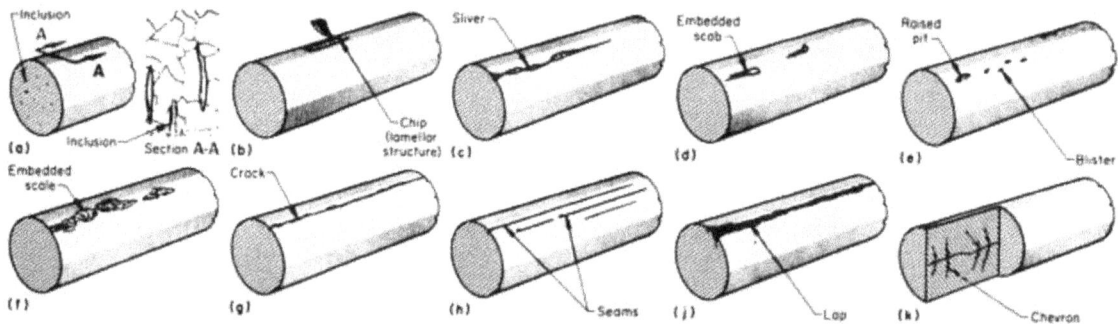

Ten different types of flaws that may be found in rolled bars. (a) Inclusions. (b) Laminations from spatter (entrapped splashes) during the pouring. (c) Slivers. (d) Scabs are caused by splashing liquid metal in the mold. (e) Pits and blisters caused by gaseous pockets in the ingot. (f) Embedded scale from excessive scaling during prior heating operations. (g) Cracks with little or no oxide present on their edges when the metal cools in the mold, setting up highly stressed areas. (h) Seams that develop from elongated trapped-gas pockets or from cracks during working. (j) Laps when excessive material is squeezed out and turned back into the material. (k) Chevron or internal bursts.

REFERENCES

1. Hydris applications in modern steelmaking, Heraeus, 1996, https://www.heraeus.com/media/media/hen/doc_hen/steel_applications/continuouscasting/Hydris_applications_in_modern_steelmaking.pdf also, R P Stone et al, Experimental determination of the accuracy of hydrogen measurement of liquid steel with Hydris system, Hydris 74th steel making conference 1991, and G Frigm et al, Accuracy of hydrogen measurement in liquid steel with in-situ microprocessor based system, Industrial Heating 1991,

2. K. VRBEK et al, CHANGES IN HYDROGEN CONTENT DURING STEELMAKING, Archives of Metallurgy and Materials, Volume 60, 2015 Issue 1 p295, http://www.imim.pl/files/archiwum/Vol1_2015/46.pdf

3. Matjaž Knap et al, Influence of process parameters on hydrogen content in steel melt, RMZ – M&G I 2013 I Vol. 60 I pp. 233–238, http://www.rmz-mg.com/letniki/rmz60/RMZ60_0233-0238.pdf

4. http://www.dierk-raabe.com/hydrogen-and-hydrogen-embrittlement/

5. Motomichi Koyama et al, Recent progress in microstructural hydrogen mapping in steels: quantification, kinetic analysis, and multi-scale characterisation, MATERIALS SCIENCE AND TECHNOLOGY, 2017 VOL. 33, NO. 13, 1481–1496. file:///C:/Users/user/Downloads/Mater%20Sc%20Techn%202017%20Recent%20progress%20in%20microstructural%20hydrogen%20mapping%20in%20steels.pdf

6. T. Bruce et al, Threshold Maps for Inclusion-Initiated Micro-Cracks and White Etching Areas in Bearing Steel: The Role of Impact Loading and Surface Sliding, Tribology Letters (2018) 66:111, http://eprints.whiterose.ac.uk/134205/1/2018-07%20Bruce%20et%20al%20ThresholdMapsForInclusion-Open Access.pdf

7. Brian Pickering, The effects of non-metallic inclusions on the properties of steel, Mechanical Working and Steel Processing Proceedings, 1989, page381

8. Niclas Anmark et al, The Effect of Different Non-Metallic Inclusions on the Machinability of Steels, Materials 2015, 8, https://www.diva-portal.org/smash/get/diva2:806314/FULLTEXT01.pdf

9. Lily Kamjou, Joakim Fagerlund, Brent Marsh and Thomas Björk, Performance and Machining of Advanced Engineering Steels in Power Transmission Applications—Continued Developments, GEAR TECHNOLOGY I May 2017 https://www.geartechnology.com/issues/0517x/steels.pdf

10. M. Möser and V. Schmidt, Fractography and the mechanism of hydrogen cracking- The Fisheye concept, file:///C:/Users/user/Downloads/FRACTOGRAPHY_AND_MECHANISM_OF_HYDROGEN_CRACKING_-_.pdf

11. J. Ćwiek, Hydrogen degradation of high-strength steels, JAMMI, Volume 37, Issue 2, December 2009. http://jamme.acmsse.h2.pl/papers_vol37_2/3722.pdf

12. M Teresa Ferraz, Steel fasteners failure by hydrogen embrittlement, Ciência e Tecnologia dos Materiais, Vol. 20, no 1/2, 2008, http://www.scielo.mec.pt/pdf/ctm/v20n1-2/20n1-2a19.pdf

13. Daniel H. Herring. Hydrogen Embrittlement, Wire Forming Technology International/Fall 2010, http://www.heat-treat-doctor.com/documents/hydrogen%20embrittlement.pdf

14. Rensheng Chu et al, Study on the Control of Rare Earth Metals and Their Behaviors in the Industrial Practical Production of Q420q Structural Bridge Steel Plate, Metals 2018, 8, 240, file:///C:/Users/user/Downloads/metals-08-00240.pdf

15. Xing LI, Zhouhua JIANG, et al, Evolution Mechanism of Inclusions in H13 Steel with Rare Earth Magnesium Alloy Addition, ISIJ International, Advance Publication by J-STAGE, https://www.jstage.jst.go.jp/article/isijinternational/advpub/0/advpub_ISIJINT-2019-094/_pdf

16. Fei Pan, Review Effects of Rare Earth Metals on Steel Microstructures, Materials 2016, 9, 417, file:///C:/Users/user/Downloads/materials-09-00417%20(1).pdf

17. Jinzhu Gao et al, Effects of Rare Earth on the Microstructure and Impact Toughness of H13 Steel, Metals 2015, 5, 383-394, file:///C:/Users/user/Downloads/metals-05-00383.pdf

18. L Zhang et al; Evaluation and control of steel cleanliness-review 85th Steelmaking Conference Proceedings, ISS-AIME, Warrendale, PA, 2002 pp. 431-452.

19. Buğra Şener İsdemir et al, Desulphurization Strategies in Oxygen Steelmaking, Abs ID: 2395 Paper No.: 30613, file:///C:/Users/user/Downloads/036_DesulphurizationStrategiesin OxygenSteelmaking_AIS Tech12_120430_2395final%20(1).pdf

20. P W. Morris et al: Extending the product size range from continuously cast sections. Final report ECSC Agreement No. 7210.EB/801

21. P. W. Morris et al: Optimization of the deformation process for continuously cast billets to provide the most appropriate material properties, Final report, Contract No 7210.EB/804 (D1-D5.5/88) 1 December 1988 to 30 November 1991

22. H W Lee et al, Numerical investigation of austenite grain size distribution in square-diamond pass hot bar rolling, Journal of Material Processing Technology, 191 (2007) 114-118.

23. Hyuck-Cheol Kwon et al, Numerical Prediction of Austenite Grain Size in Round-Oval Round Bar Rolling, ISIJ International, Vol. 43 (2003), No. 5, pp. 676–683.

24. Ho-Won Lee et al, Local Austenite Grain Size Distribution in Hot Bar Rolling of AISI 4135 Steel, ISIJ International Vol 45 (2005) No 5 pp 706-712.

25. http://www.niobelcon.com/NiobelCon/niobium/niobium_in_automotive_flat_steel/auto_martensitic_steel/tough_martensite/

SELF-ASSESSMENT QUESTIONS

1. What chemical elements would need to be controlled to produce a weldable structural grade of steel with a toughness of 200J at 0°C.?

2. Some alloy steel blooms have been rolled from steel that had a liquid steel hydrogen level of 6ppm. What problems could be expected? What should be done to avoid problems?

3. If the forging was 10 ins in diameter and the end user wants the hydrogen to be below 1 ppm how long does the forging have to be soaked at 650°C?

4. List four ways that the toughness impact transition temperature can be lowered.

5. List two ways that the Charpy ductile energy could be increased.

6. A low carbon weldable structural steel has a grain size of 40 microns what would be the expected toughness impact transition temperature and the expected yield strength?

7. Name three types of segregated zones found on a large forging ingot.

8. A request has been made to specify the steel cleanliness for a tyre cord application. What factors should be considered and what recommendations would you make?

9. A low sulphur high strength weldable structural steel slab has been ultrasonically tested 12 hours after manufacture and was found acceptable. The slab has been ultrasonically tested after 7days and has failed. What would be the probable cause for the defective slab?

10. A sample of steel has been found to have 50ppm total oxygen. The sample had a volume of about 1 cubic cm. The average inclusion size was 10 microns. Estimate how many inclusions there would be in the sample of steel.

11. What reduction ratio from concast product to rolled bar would you recommend for bolt manufacture?

12. High durability shafts 200mm in diameter are required in alloy steel. What should be specified as the minimum ingot cross section for the manufacture of these shafts?

13. Low alloy steel100mm diameter tube was found to have longitudinal surface imperfections along the length of the tube. What would be the cause of these imperfections?

14. A high-quality steel was required for high endurance roller bearing manufacture. What maximum level of total oxygen would be required in the steel?

15. Name a method used to measure the liquid steel hydrogen level.

16. A bolt has failed in service and the probable cause has been hydrogen embrittlement. The facture surface needs to be examined using SEM (Scanning Electron Microscope). What features would confirm that the fracture had been assisted by hydrogen?

17. A steel has been specified as SAE 8630. What would be the alloying elements and the typical level of alloy elements in the steel.

18. An alternative steel has been recommended as SAE 4330. Which steel will have the highest level of nickel?

19. A steel bar has been supplied as BS 970 grade 070M20. What does the letter M indicate?

20. A steel bar is supplied as BS 970 grade 817H20. What does the letter H indicate?

21. What equipment would be used to measure the oxygen content in a steel sample?

22. There two potential suppliers of steel bar. The application requires a high level of toughness at ambient temperature. Supplier A has steel with a sulphur level of 0.005%. Supplier B has similar steel but with a sulphur level of 0.025%. Which would be the best supply of steel for this application?

23. An ASTM E45 inclusion count has to be carried out on an air meted steel. What would be the maximum levels for both Thin and Thick worst fields for A-type, B-type and D-Type inclusions?

24. If the steel was a vacuum degassed grade what would be the maximum worst fields

CHAPTER 4

Heat treatment and surface engineering

4.1 INTRODUCTION

Heat treatment and surface engineering are engineering terms that have been applied to the treatment of metals for decades. Heat treatments probably date back to antiquity, whereas surface engineering as a concept was developed in the 1980s although some of the individual techniques are considerably older. Heat treatment is the application of thermal cycles such as through hardening of steel or thermo- chemical environments plus thermal cycles such as nitriding or carburizing of steel. (See Figure 4.2 and 4.3)

Surface engineering involves the concept of having different sets of properties for the surface and subsurface compared to the core of the material. Typically, the destructive mechanisms such as wear, corrosion and fracture initiation occur at the surface of a material where there is contact with other components and the environment. Therefore, providing a high durability surface layer has proven to be a cost-effective approach to creating high integrity parts.

In 1985 a definition given by Professor Bell at Birmingham University for Surface Engineering was "makes possible the design and manufacture of engineering components with combination of bulk and surface properties unobtainable in a single monolithic material"

There is not a clear demarcation between surface engineering and heat treatment and plating. In many businesses there is overlap especially where heat treatment companies have added plasma processing options such as plasma nitriding and PVD coatings.

The development of many of these technologies and innovations associated with materials technology, heat treatment and surface engineering have been customer or application led. Frequently they have been sold under a proprietary name which has subsequently been entered onto the manufacturing documents resulting in the "named" process being mandatory. This can create a monopoly of supply which only tends to break down when the culture of cost reduction takes precedence.

The main objective for improved processes has been to either achieve a commercial advantage in terms of either lower cost of production or higher integrity and product life. The end point has been a number of competing processes, many with advantages but frequently with some disadvantages. The main task for the user and designer has been the selection of the appropriate treatment. This requires a broad in-depth appreciation of heat treatment and surface engineering.

The natural progression of the development of material technology can be demonstrated by the description of the history of mankind. Initially the stone age then the bronze age followed by the iron age and then the silicon chip or space age. We could invoke several other age descriptors because of the rapid scientific developments and others that await discovery and exploitation by future generations probably associated with robotics, AI and nanotechnology.

There are several substantial books on the "heat treatment of metals". (See photos in Figure 4.1.) One approach to materials and heat treatment describes the important factors associated with material selection such as mechanical properties and their associated test methods. This is then followed with information to provide the reader with an understanding of important metallographic features that affect the mechanical properties of strength and toughness. Other texts begin by introducing microstructure and how the microstructure forms and how it is related

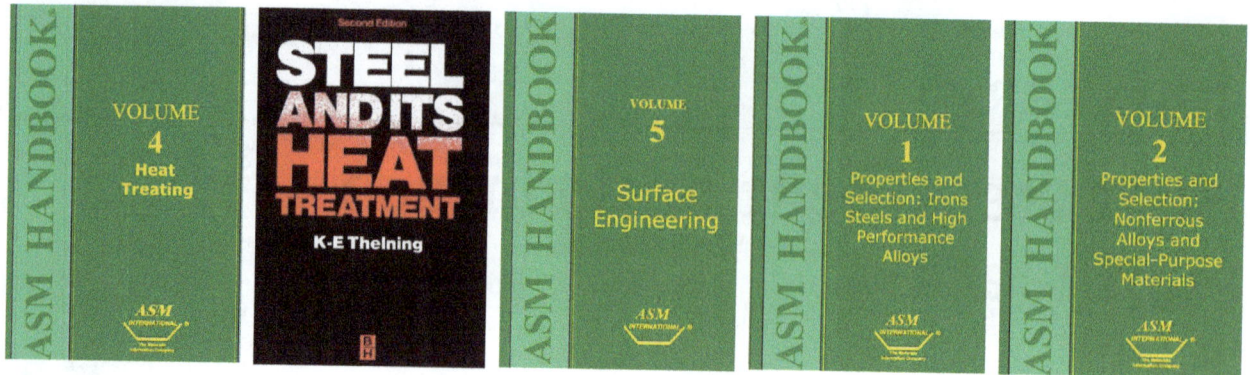

Figure 4.1 *Important heat treatment and surface engineering sources of information*

to mechanical properties. The most logical answer to this "chicken and egg" problem in my view would be to begin with the microstructure of metals. To gain an understanding of metals and alloys, it is vital to appreciate that an important factor that controls most of the engineering properties of metals and alloys is the crystal structure and microstructure. However, that relatively simple statement excludes the important effect of metal imperfections such as porosity or non-metallic inclusions introduced during casting or mechanical working or other imperfections associated with other methods of manufacture

4.2 HEAT TREATMENT – STEEL COMPOSITION, HARDENABILITY, SECTION SIZE AND QUENCH RATE

For many engineering applications the "core" mechanical properties are important. Therefore, an understanding of the key requirements of "through hardening" to obtain the best combination of strength and toughness is important. To achieve that understanding there are some important definitions that are important.

4.2.1 Definitions

- **Hardenability** is the ability of steel to partially or completely transform from austenite to some fraction of martensite at a given depth below the surface, when cooled under a given condition. Hardenability therefore describes the capacity of the steel to harden to a particular depth under a given set of conditions. The hardenability is normally determined based on 50% martensite at the centre of a bar. Two methods to determine the hardenability are outlined in ASTM 255 clause *1*. *"The two test methods include the quantitative end-quench or Jominy Test and a method for calculating the hardenability of steel from the chemical composition based on the original work by M. A. Grossman"* Other tests include "Grossman's critical diameter method" see Fig 4.47 RHS, and a fracture test if there is a clear demarcation between martensite zone and pearlite zone.

- **Equivalent Diameter** (**ED**) Is a method used to compare the hardening characteristics of heat-treating low-alloy steel of various shaped sections to a round bar equivalent.

- Recently, Terry Khaled, the Chief Scientific / Technical Advisor, Metallurgy at the Federal Aviation Administration carried out a detailed review comparing the history and background of the US term "Equivalent round", with the UK equivalent "diameter and ruling section"[1] The main conclusion was that the US equivalent round has been "carried through" into many specifications and the historical basis for the calculations have been lost. However, there are similarities between the equivalent round and equivalent diameter values. An important point was made that equivalent round could be applied regardless of the material of the cooling medium, which cannot be correct in all circumstances. In some ways this has been overtaken by FEM approach to the quench process and even the old International Harvester method of (CHAT) where a part is quenched and then sectioned and then compared to the jominy results on a similar steel and the various areas of the part are given "J equivalent position values" (J_5 mm etc) values based on hardness which can be related to a cooling rate.

- **Ideal Diameter (DI)** quantitative measure of a steel's hardenability is expressed by its DI, or ideal diameter, value. This abbreviation comes from the French phrase "diamètre idéal" and refers to the largest diameter of steel bar that can be quenched to produce 50% martensite at its centre. The quench rate of the bar is assumed to be infinitely fast on the outside; that is, it has sufficient quench severity so the heat removal rate is controlled by the thermal diffusivity of the metal and not the heat transfer rate from the steel to the quenchant. DI values are an excellent means of comparing the relative hardenability of two materials as well as determining if it is possible to harden a particular cross section (or ruling section) of a given steel.
- **Grossman Quench Severity Factor.** The effectiveness of a given cooling medium is measured by a parameter called its 'severity of quench'. It is given the symbol 'H', which is a measure of the rate of heat removal from a quenched part by the quenching medium.
- **Ruling Section:** The ruling section is the equivalent diameter of that portion of the object at the time of heat treatment that is important in relation to mechanical properties.
- **Limiting Ruling Section.** The limiting ruling section for any given steel composition is the largest diameter in which certain specified mechanical properties can be achieved after a specified heat-treatment. The chemical composition of the steel and quench determine the limiting ruling section.
- **Critical cooling rate CCR.** The cooling rate at which a fully martensitic microstructure is formed is called the critical cooling rate. A regression equation to obtain the minimal critical cooling rate (V_M) and a 100% martensitic structure based on the steel composition:

$$\log V_M = 4.5 - 2.7C - 0.95Mn - 0.18Si - 0.38Cr - 0.43Ni - 1.17Mo - 1.29\ C.Cr - 0.33\ Cr.Mo$$

where V_M is in $°Cs^{-1}$. This equation can be applied to hypoeutectoid carbon steels and low-medium alloy steel.[2]

4.2.2 Heat Treatment and Surface Engineering Processes Thermal and Thermo-chemical heat treatment processes

There are many competing heat treatment and surface engineering processes which have advantages and disadvantages, different costs and provide different levels of performance. Based on my experience I prefer to divide the processes into two groups. The two groups can then be sub-divided into a further more detailed list.

- Thermal processing. (See Tables 4.1)
- Thermo-chemical processing. (See Table 4.2)

In the thermal processing the composition of the steel needs to be appropriate and the mechanical properties and microstructure are developed through the application of heat. These include:

- Through hardening
- Thermo-mechanical processing such as controlled rolling
- Induction hardening
- Flame hardening
- Laser hardening.

For thermo-chemical processing the heat treatment takes place in a controlled atmosphere usually either carbon or nitrogen or both diffuse into the surface. The four basic heat treatment processes are shown in Table 4.2. The list below repeats these four and adds some of the established Surface Engineering processes.

- Carburizing has only carbon is added
- Carbo-nitriding has mainly carbon is added but with a little nitrogen
- Nitro-carburizing has mainly nitrogen is added with a little carbon
- Nitriding has only nitrogen is added
- Boriding
- Electro-plating
- Gas phase deposition (PVD, CVD, PACVD)
- Fusion-hot dip processes (Galvanising, Tin Plating
- Transferred plasma arc

Figure 4.2 shows a flow chart of the various **thermal processes** listed according to whether they are carried out above the AC3, (fully austenitised) or below the AC1 in the ferritic condition or whether they are limited to only surface hardening.

Figure 4.3 shows a flow chart of the various **thermal-chemical processes** listed according to whether they are a surface hardening treatment, gas phase deposition process or the application of a coating by either plating, fusion or thermo mechanical.

Figure 4.4 shows the depth and hardness of the heat treatment and surface engineering methods.

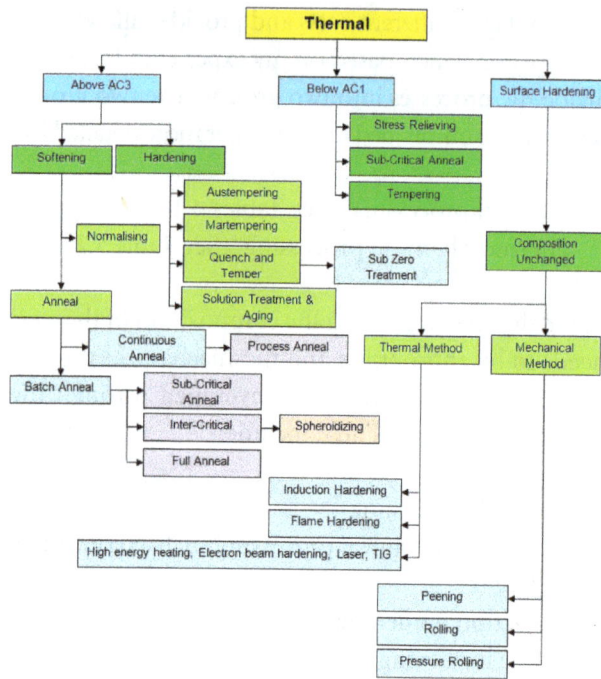

Figure 4.2 Thermal Heat Treatment processes

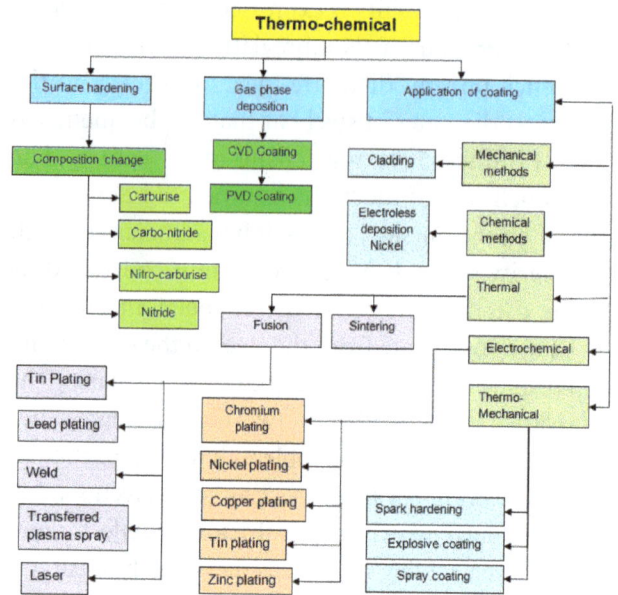

Figure 4.3 Thermo-chemical Heat Treatment and Surface Engineering processes

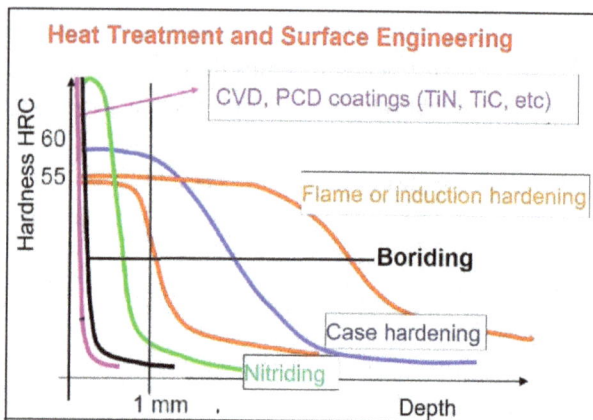

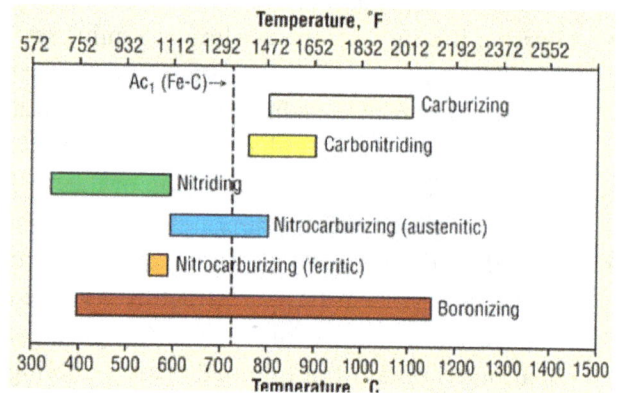

Figure 4.4 Depth and hardness of heat treatment and surface engineering methods

Table 4.1 Thermal Heat treatment processes

Process	Characteristics
Austenitizing: Transformation to austenite by heating the steel above AC3.	The optimal austenitizing temperature is 30-50°C above AC3 for hypo eutectoid steels and is 30-50°C above ACm for hypereutectoid steels. AC3 is the temperature at which the transformation of ferrite to austenite is completed. The heating rate must be limited and uniform to avoid cracking or warpage and to control thermal stresses in the range of 250-600°C. The carbon level and alloy content control the propensity for steel to crack. The holding time is dependent on geometrical factors related to the furnace (emissivities, temperature and atmosphere composition) and load (type of steel and thermo-physical properties. There are general rules such as 1 hr /inch or 1min per mm empirically established

Process	Characteristics
Annealing: Heat treatment consisting of heating and soaking at suitable temperature followed by cooling under conditions such that, after return to ambient temperature the metal will be in a structural state closer to that of equilibrium. The primary purpose of annealing is to soften the steel to enhance its workability and machinability. Also, it relieves internal stresses, restores ductility and toughness, refines grains, reduces gaseous content in the steel and improves homogenization of alloying elements.	**Full Annealing:** Heat 30-50°C (55-90°F) above AC3 for hypo eutectoid steels, then furnace cool though critical temperature range at a specific cooling rate. The aim is to break the continuous carbide network of high carbon steels. It improves machinability. **Intercritical Anneal:** Heating within the critical temperature range (AC1-AC3) followed by slow furnace cooling. It improves machinability. **Subcritical Annealing:** Heating 10-20°C below AC1 followed by cooling in still air. It can be used to temper bainitic or martensitic structures to produce softened microstructures containing spheroidal carbides in ferrite. Improves the cold working properties of low carbon steels (0.25%C) or softens high carbon and alloy steels. **Recrystallisation Annealing:** Heat the steel for 0.5 to 1h at temperature above the recrystallisation temperature ($T_r = 0.4\ T_m$) then the steel is cooled. The treatment temperature depends on prior deformation, grain size and holding time. The recrystallisation process produces strain-free grain nucleation, resulting in ductile, spheroidized microstructures. **Isothermal Annealing:** Heating the hypo-eutectoid steel within the austenitic transformation range above AC3 for a time sufficient to complete the solution process, yielding a completely austenitic microstructure. At this time the steel is cooled rapidly at a specific rate within the pearlite transformation range until the complete transformation to ferrite plus pearlite occurs and then it is cooled rapidly. **Spheroidising (Soft Anneal):** Involves the prolonged heating of steel at a temperature near the lower critical temperature (AC1) then furnace cooling. **Diffusion (Homogenizing Annealed):** Heat the steel rapidly to 1100-1200°C for 8-46h, furnace cool to 800-850°C and then cool to room temperature in still air. It is performed on steel ingots and castings to minimise chemical segregation.
Normalising: The aim is to provide a uniform microstructure of ferrite plus pearlite (small grains and finer lamellae than in annealing).	Heat the steel to 40-50°C above AC3 for hypo eutectoid steels and 40-50°C above ACM for hypereutectoid steels. The holding time depends on the size, and then the steel is cooled in still air. It produces grain refinement and improved homogenization.
Stress Relieving: It is typically used to remove residual stresses that have accumulated from prior manufacturing processes. Stress relieving may result in a significant reduction of yield strength in addition to reducing the residual stresses to some "safe" value.	Heat to a temperature below AC1 for the required time to achieve the desired reduction in residual stresses and then the steel is cooled at a rate sufficiently slow to avoid the formation of excessive thermal stresses. Below 300°C, faster cooling rates can be used. No microstructural changes occur during the stress-relief processing. The recommended heating temperature range is 550-700°C depending on the type of steel. These temperatures are above the recrystallisation temperature. Little or no stress relief occurs at temperature <260°C and approximately 90% of the stress is relieved at 540°C. The maximum temperature for stress relief is limited to 30°C below the tempering temperature used after quenching. The results of the stress relief process are dependent on the temperature and time.
Quenching: Quench severity is the ability of a quenching medium to extract heat from a hot steel workpiece.	Specific recommendations for quench media selection for use with various steel alloys are provided by standards such as SAE AMS 2759. Quench media includes water, brine, aqueous polymer, gas or air quenching and caustic quenching.
Tempering: Tempering is the thermal treatment of hardened and normalised steels to obtain desired mechanical properties which include improved toughness, ductility, lower hardness and improved dimensional stability.	The tempering process involves heating steel to any temperature below the AC1 temperature. During tempering, as quenched martensite is transformed into tempered martensite which is composed of highly dispersed spheroids of cementite (carbides) dispersed in a soft matrix of ferrite, resulting in reduced hardness and increased toughness. The objective is to allow the hardness to decrease to the desired level and then stop the carbide decomposition by cooling. The extent of the tempering effect is determined by the tempering time and process.

Table 4.2 Thermo-chemical heat treatments

Process	Temp °C	Diffusing Elements	Process Media	Steels	Case Characteristics	Process Characteristics	Applications
Carburising	900-1000	Carbon	Pack Salt Gas Fluidised Bed Vacuum Plasma	Low carbon steels, mild and low alloy.	• Medium to deep cases • Oil quenched • Typical surface hardness 57-62 HRC (650-800HV) after low temperature tempering	Case required to minimise distortion	• High surface stress conditions • Alloy steels – large sections • Mild steels – small sections (<12.5mm)
Carbonitriding	800-880	Carbon and Nitrogen (Carbon mostly)	Salt Gas Fluidised Bed	Low carbon steels, mainly mild steels.	• Shallow to medium cases • Oil quenched • Typical surface hardness 57-62 HRC (650-800HV) after low temperature tempering	Less distortion than carburising	• High surface stress conditions • Mild steels – to section sizes above 12.5mm
Nitriding	500-550	Nitrogen	Gas Plasma Fluidised Bed Vacuum Plasma	Alloy steels and some tool steels with appropriate levels of nitride forming elements (Cr+Al).	• Shallow to medium cases • No quenching required • Surface hardness in range 650-1100HV depending on steel	Very low distortion Long process time	• Severe surface stress conditions • Gives highest hardness and temperature resistance up to 200-300°C
Nitrocarburising ("Ferritic") **Austenitic Nitrocarburising**	560-570 600-780	Nitrogen and Carbon (Nitrogen mostly)	Salt Gas Fluidised Bed Vacuum Plasma	Wide range of steels, from low carbon / non alloy to tool steels.	• Thin (10-20µm) hard surface compound layer • Underlying nitrogen diffusion zone • Surface hardness depends on steel type and process route. Can range from 350-550HV1 on low carbon/non alloy steels to 1000+HV1 on some tool steels	Very low distortion No post heat treatment machining possible	• Low to medium surface stress conditions • Good wear resistance • Post oxidation/impregnation gives salt corrosion resistance (PlasOx)

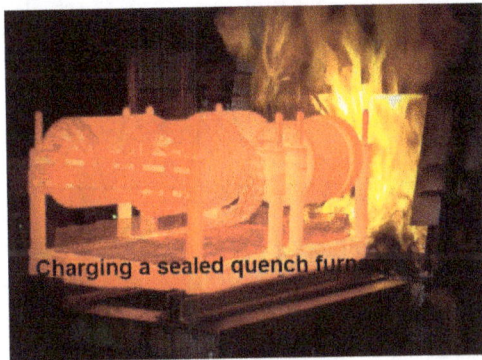

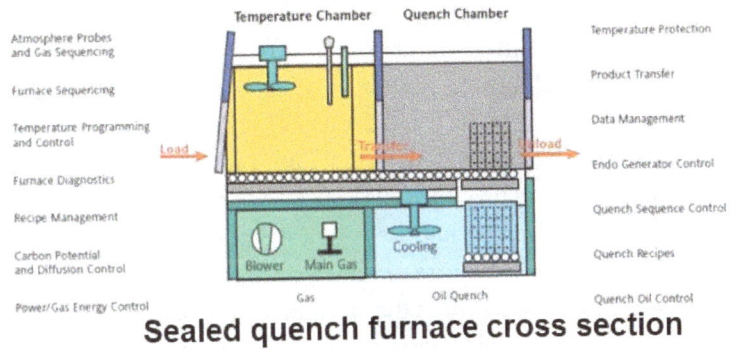

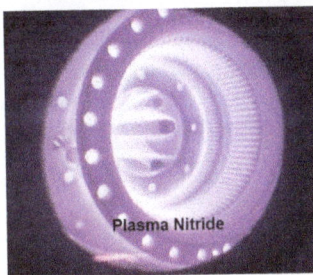

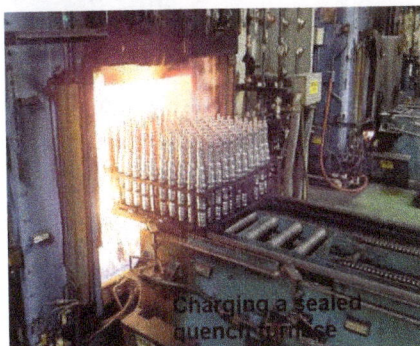

Figure 4.5 *Photographs of various heat treatment facilities*

4.2.3 The temperature time sequence diagram

There are a number of factors of importance which are to be considered when heat treating a metal or alloy.

- The temperature up to which the metal/alloy is heated
- The length of time that the metal/alloy is held at the elevated temperature
- The rate of cooling
- And the atmosphere surrounding the metal/alloy when it is heated.

These heat treatment process parameters are often summarised on a diagram known as a **temperature time sequence diagram**. Figure 4.6 shows a simple heat treatment cycle, whereas Figure 4.7 represents some complex heat treatment cycles. Figure 4.8 shows the style of diagram used by Rolls Royce and the terminology used for each part of the diagram. Figure 4.6 is the simplest possible heat treatment cycle in which the metal/alloy is heated, held at the elevated temperature for some time, and then cooled to room temperature.

Figure 4.7 shows a more complex cycle, used for the hardening of D2 tool steel where an initial stress relief and preheat stages are introduced to minimise distortion.

The heat treatment profiles shown in Figure 4.7 and 4.8 show several terms /steps. The definitions associated with these terms are as follows:

- **Preheating.** This can be used to minimise the thermal stresses and to minimise distortion. It can involve both slow heating and can include subsequent holding at one or more temperatures to allow temperature equalisation to take place (multi-stage preheating) usually initially below the AC1. The aim of preheating is to reduce stresses which may lead to distortions and to avoid the formation of heating stress cracks.
- **Time of Surface heating.** This is the time taken to reach the specified temperature in the surface/sub-surface.
- **Time for temperature equalisation,** the time taken for both the core and surface to be at the required temperature after reaching the specified temperature.
- **Time for through heating.** This is the sum of the time for the surface to achieve temperature plus the time for temperature equalisation.

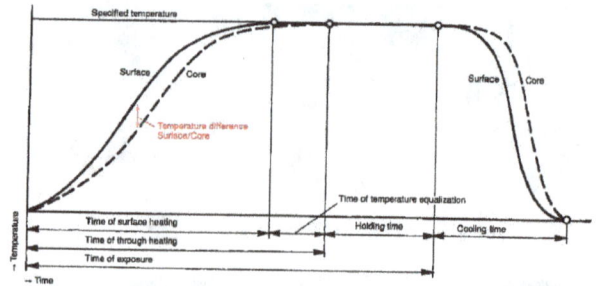

Figure 4.6 Graphical representation of a relatively simple heat treatment cycle (rate of heating

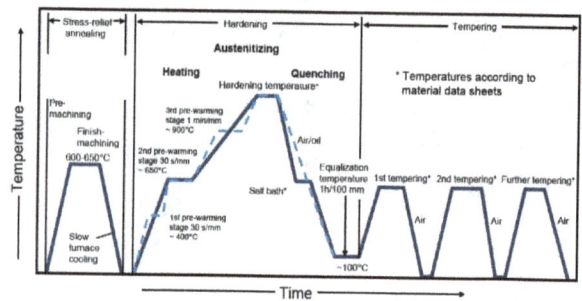

Figure 4.7 Hardening of D2 tool steel

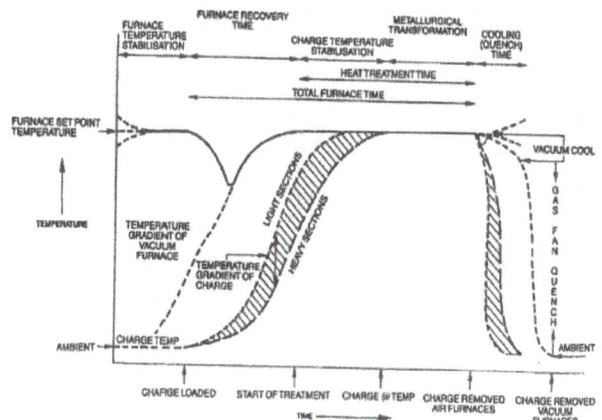

Figure 4.8 Rolls Royce diagram showing their terminology

- **Holding (soaking) time.** This is the time held at temperature to allow full temperature homogenisation, full homogenisation of composition and appropriate dissolution of alloy elements into solution in the austenite.
- **Cooling time.** This is the time taken to cool back to ambient temperature. When cooling rates faster than still air is used it is referred to as quenching. Quenching can be carried out in either water, polymer or oil.
- **The time of exposure.** (Formerly also called time of immersion, when salt-bath heat treatment was employed) I.e. the period of time elapsing between introduction of a workpiece into the furnace and

its withdrawal from the furnace, comprises the time of through heating and the time of holding.

4.2.4 Solution heat treatment and precipitation hardening – a hardening mechanism for ferrous and non-ferrous metals. (See 10.2.7 Historical aspects of aluminium alloy development.)

Figure 4.9 shows a typical heat treatment cycle suitable for a solution heat treatment and a precipitation hardening treatment. In this case, the alloy is heated and held at predetermined high temperature. This step is referred to as "solution treatment". The alloy is then cooled rapidly to room temperature by quenching. The quenched alloy is heated and held at a suitable temperature above the room temperature, followed by slow cooling. The last step, i.e. heating to and holding at a temperature is referred to as "ageing" or more specifically as "artificial ageing". This is because some precipitation hardening alloys harden even at room temperature. Such alloys are known as natural age hardening alloys referred to as "natural aging". Duralumin is a natural age hardening alloy. (see section 10.4.12 for the discovery precipitation hardening)

4.3 SURFACE ENGINEERING

Surface engineering consists of a range of technologies designed to modify the surface properties of metallic and non-metallic components for decorative and/or functional purposes. The use of surface engineering aims to improve corrosion and wear resistance to extend component life. It also targets improved aesthetics to make parts visually attractive. Surface engineering has a wide range of processes and covers the following:

4.3.1 Aqueous electrolytic

Refers to the electroplating of metal coatings such as chromium, anodising of aluminium and titanium, as well as the electro-polishing of stainless steel.

4.3.1.1 Chromium

Commercial-grade chrome plating was developed at Columbia University in 1924, based on research carried out by Dr George J. Sargent. His work was published in 1920 which had discovered that 1-part sulphuric acid to 100 parts chromic acid would make it possible to electroplate chromium, creating chrome plating. Unfortunately, the recent history has been

Figure 4.9 Section of equilibrium diagram showing decreasing solubility plus temperature time sequence diagrams.

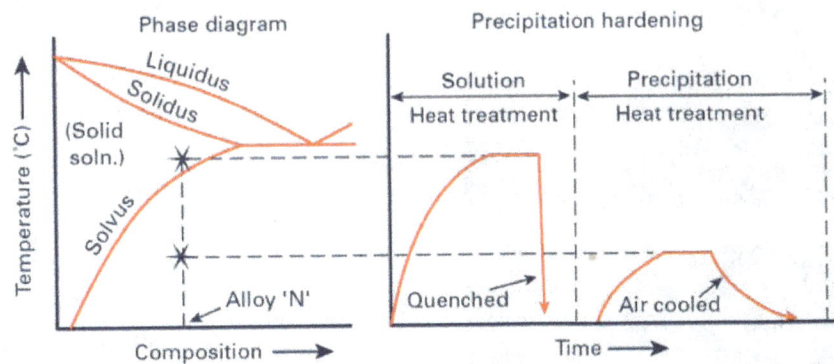

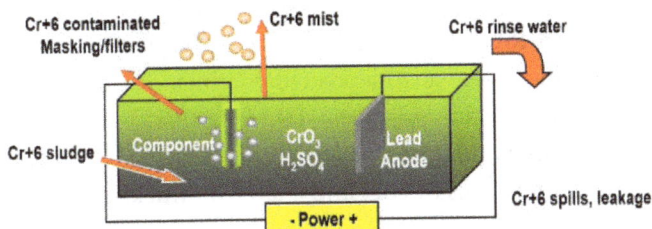

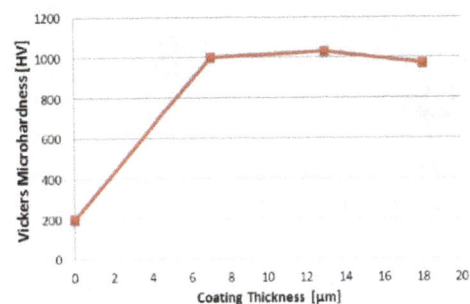

Figure 4.10 Process diagram showing the areas of Hexa-valent Chromium. RHS = Hardness (4)

concerned with environmental issues. Since 2006 the chrome plating industry has faced regulations that prevent chrome plating processes from releasing fumes, dust, or mist from its operations in the United States. This has required increased investment in extraction equipment.

The environmental problems associated with chrome plating were well publicised by the film "Erin Brockovich". The film was a dramatization of the true story of Erin Brockovich, portrayed by Julia Roberts, who fought against the energy corporation Pacific Gas and Electric Company that had contaminated the ground water with hexavalent chromium in Hinkley, California. On September 21st, 2017, the EU enacted legislation banning the use of hexavalent chrome solutions to plate decorative components. The view was that plating with trivalent chrome was a viable alternative. A recent statement by Hardide PLC gives their business a good future:

> "REACH and OSHA have tried repeatedly to restrict the use of hexavalent chromium salts, but the industry has extended the deadline a number of times as they have been unable to find a viable alternative. With Hardide (CVD Tungsten Carbide) they don't get that excuse."

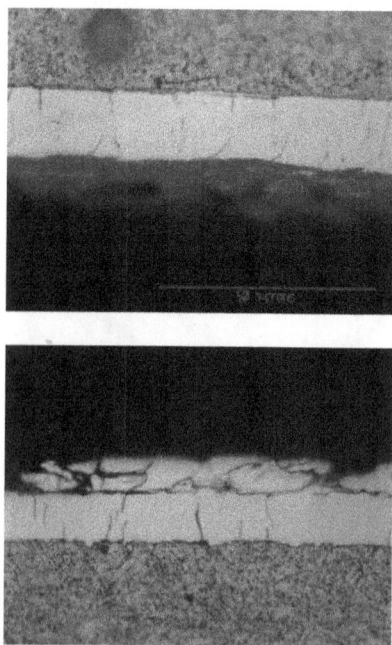

Figure 4.11 *Microstructure of chromium plating after use showing cracking and deformation*

4.3.1.2 Chromium plating specifications

BS 4641-1986 – Method for specifying Electroplated coatings of chromium for engineering purposes. This specification provided significant details regarding the theory and practice of chromium plating. It includes an appendix on "Guidance on types and uses of chromium coatings".

Electroplated coatings of chromium used for engineering purposes are much thicker than those used for decorative purposes. They are often called "hard chrome" or "heavy chromium" coatings. Chromium coatings are used for engineering purposes in industry, mainly to take advantage of one or more of the following characteristics: a) low coefficient of friction; b) wear resistance; c) corrosion resistance; d) anti-stick properties; e) anti-fret properties; f) reflectivity; g) hardness.

The thickness of chromium to be applied will depend upon the properties required and the particular application. In view of the wide variety of industrial uses of chromium coatings, it is not possible to specify thicknesses but the following is intended to indicate typical practice.

Coatings normally used in the as-plated condition. Depending upon the shape, many articles may be chromium plated to tolerances which are acceptable to the engineer. The addition of lapping, honing or polishing can often eliminate the requirement for grinding.

Examples of use are as follows.

1. Coatings up to 12 microns i) plastics moulds, where chromium provides a corrosion-resistant surface and allows easy release of the product; ii) rams for use in hydraulic machines.
2. Coatings 12 to 50 microns i) cylinder liners for the internal combustion engine; ii) slideways for machine tools to increase accurate life.
3. Coatings 50 to 75 microns i) press tools to reduce friction and improve appearance of pressings; ii) aircraft undercarriages to improve wear and corrosion resistance.

BS EN ISO 6158:2011- Metallic and other inorganic coatings — Electrodeposited coatings of chromium for engineering purposes (ISO 6158:2011)

This standard provides tables on the codes for the different types of coat and a table on thickness.

Table 1 — Symbols for different types of chromium   Typical thickness of chromium specified in engineering applications

Type of chromium	Symbol
Regular hard chromium	hr
Hard chromium from mixed acid solutions	hm
Microcracked hard chromium	hc
Microporous hard chromium	hp
Duplex chromium	hd
Special types of chromium	hs

Typical thickness μm	Application
> 2 to ≤ 10	To reduce friction and for light wear resistance
> 10 to ≤ 30	For moderate wear resistance
> 30 to ≤ 60	For adhesive wear resistance
> 60 to ≤ 120	For severe wear resistance
> 120 to ≤ 250	For severe wear, abrasion and erosion resistance
> 250	For repair

Figure 4.12 Tables from BS EN ISO 6158:2011 showing the type of chromium symbol and the coating thickness

4.3.1.3 Poor chromium appearance

Figure 4.13 shows the possible surface appearance problems. The bright deposit typically gives the harder deposit and the best wear resistance. The most common reason for poor appearance is the use of the wrong combination of current density and bath temperature. The current density is the applied current divided by the surface area. The bath temperature will be held constant by the temperature controller.

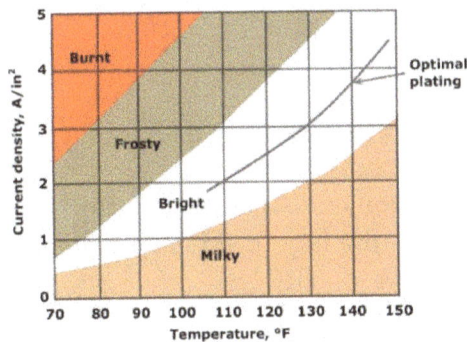

Figure 4.13 Relationship between current density, bath temperature and appearance of the chromium deposit (5 and 6)

4.3.1.4 Hardness checks

Chromium deposits 25 microns or more on a hardened steel surface are thick enough for satisfactory measurement with 50 to 100 g load. It is recommended that the plate thickness should be at least 14 times the depth of the indent. Hardness tests can be made on the cross section of the deposit if it is thick enough and well supported. If the deposit cracks around the indent, a smaller load should be used.

4.3.1.5 Corrosion

Where corrosion issues are a problem the use of an electroless nickel coating below the hard chromium plating has been used.[7 and 8]

4.3.2 PEO. Plasma Electrolytic Oxidation

Developed by Keronite International Ltd allows the surface hardening of aluminium and magnesium.[9] http://www.keronite.com

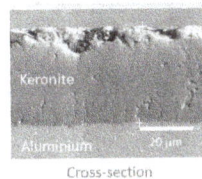

As with anodising, we have a dense, well-adhered ceramic layer, resulting from substrate oxidation. In the PEO process, however, this is modified by melting, melt-flow and re-solidification to become far harder crystalline phases such as "sapphire", and also a far more complex microstructure than the simple columnar pores of anodising.

Current applications in aerospace and defence:

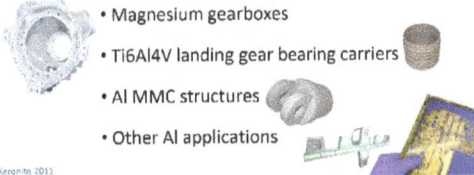

- Magnesium gearboxes
- Ti6Al4V landing gear bearing carriers
- Al MMC structures
- Other Al applications

Figure 4.14 PEO process used for hard coatings on aluminium and magnesium alloys

4.3.3 Electroless processes

For those metals, notably copper, nickel, gold and tin that can be applied by chemical reduction methods. This method avoids the use of electrical energy that is used in electroplating processes.[10 and 11] A good example of Electroless plating would be Electroless Nickel Plating (ENP). The ENP is a nickel phosphorus deposit containing 2 to 14% phosphorus and the plating depth would be around 25-75 micron. The higher the phosphorus content the greater the corrosion resistance, however the compromise on increased phosphorus content is a decrease in hardness. The deposit can be hardened by heat treatment at low temperature to around 1000HVas shown in Figure 4.15.

Figure 4.15 *Electroless nickel and Electroless nickel with PTFE filler. Lower RHS = Hardness (12)*

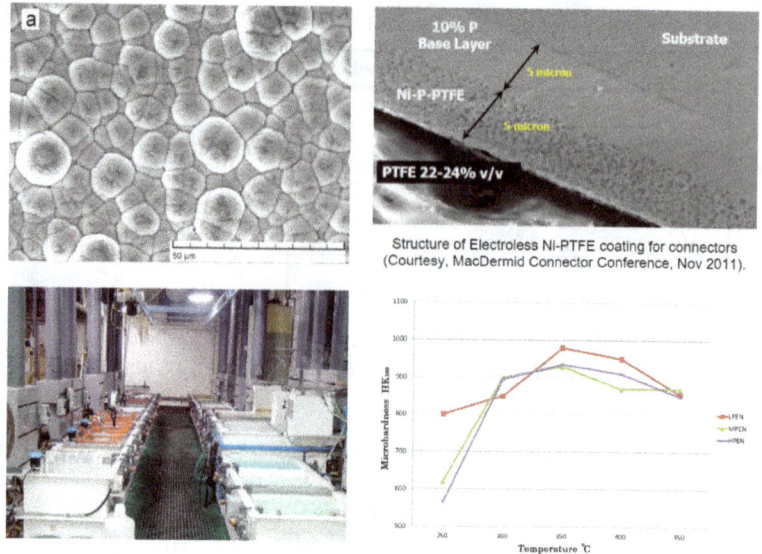

Structure of Electroless Ni-PTFE coating for connectors (Courtesy, MacDermid Connector Conference, Nov 2011).

ENP is deposited by reducing nickel ions to metallic nickel with a chemical reducing agent such as sodium hydrophosphite. ENP offers excellent corrosion resistance to common corrodents such as salt water, carbon dioxide, oxygen and hydrogen sulphide. High phosphorus deposits of ENP (10-14% phos) is also amorphous, which means that there are no grain or phase boundaries to create initiation sites for corrosion. The uniformity of ENP versus electrolytic deposits is also advantageous, creating a uniformly thick coating across the whole component.

4.3.4 Aqueous non-electrolytic

Typically cleaning, pickling, phosphating, (iron, manganese and zinc phosphate) passivation, mechanical plating or peen plating (sherardizing) and a variety of other colouring processes, e.g. "blacking" of steel. And corrosion resistance such as spray coatings of Moly Disulphide and Zylan. A treatment with manganese phosphate (pinion in Figure 4.16) has a plating thickness approximately 3 to 15 μm.

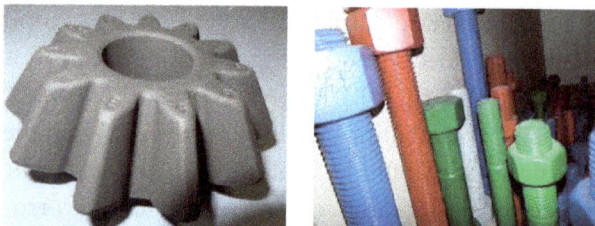

Figure 4. 16 *Left: Manganese phosphate, Right: Zylan Coating are a family of fluoropolymer coatings*

4.3.5 Organic (liquid)

Which can be solvent or water based but applies pigmented or metal containing coatings by dipping, dip-spinning, flow coating, conventional spraying, or in the case of water-based paints, by electrophoretic or auto-catalytic means.

4.3.6 Organic (powder)

The application of dry powders, usually by the process of electrostatic spraying or by fluidised bed techniques.

4.3.7 PVD and CVD (Physical Vapour Deposition and Chemical Vapour Deposition)

Physical vapor deposition (PVD) consists of a range of vacuum deposition methods which can be used to produce thin films and coatings. PVD is characterized by a process in which the material goes from a solid phase to a vapour phase and then back to a thin film solid phase.

The main processes are:

i. Evaporation techniques

    a. Vacuum thermal evaporation.
    b. Electron beam evaporation.
    c. Laser beam evaporation.
    d. Arc evaporation.
    e. Molecular beam epitaxy
    f. Ion plating evaporation.

ii. Sputtering technique

    a. Direct current sputtering (DC sputtering).
    b. Radio frequency sputtering (RF sputtering).

PVD coatings are about 5µm maximum. Since these coatings are so thin their wear life is short. PVD coatings suffer from line-of-sight problems during deposition which results in a non-uniform coating. The most common PVD coatings are TiN and CrN. PVD processes are environmentally friendly vacuum deposition techniques which involve three steps:

- Vaporization of the material from a solid source assisted by high temperature vacuum or gaseous plasma.
- Transportation of the vapour in vacuum or partial vacuum to the substrate surface.
- Condensation onto the substrate to generate thin films.

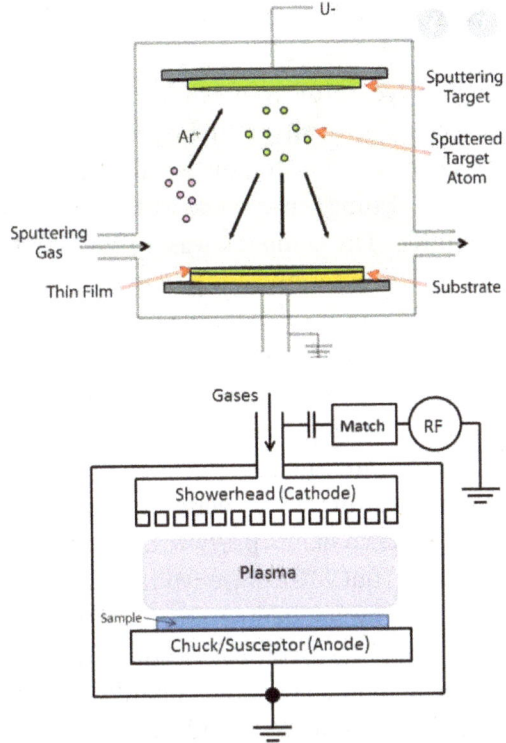

Chemical Vapor Deposition (CVD) is a process in which the substrate is exposed to one or more volatile precursors, which react and/or decompose on the substrate surface to produce the desired thin film deposit. The methods used are:

a. Low pressure (LPCVD)
b. Plasma enhanced (PECVD)
c. Atomic layer deposition (ALD)

CVD coatings are used in many industrial and consumer products. The applications range from machine tools, wear components, analytical flow path

Figure 4.17 Left: PVD Sputter Coating, Right: Plasma Enhanced Chemical Vapour Deposition PECVD

components, instrumentation. The Plasma enhanced (PECVD) technique can be used to deposit diamond like coatings.

4.3.8 (Klosterising)

Low temperature carburising of austenitic stainless steel. This can also be accomplished by low temperature plasma carburising using butane.

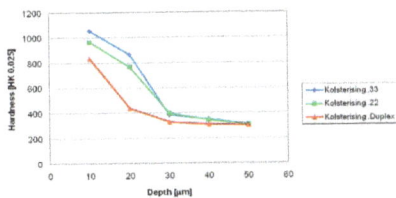

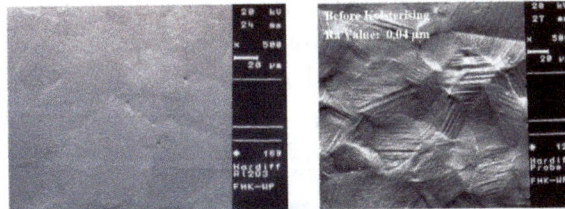

Table 1 Commercially available processes for low-temperature nitriding and carburizing of stainless steels

Entity	Process name	Interstitial hardening element	Temperature °C	Temperature °F	Method	Applications	Comments
University of Birmingham, U.K.	LTPN	N	<450	<842	Plasma	...	...
	LTPC	C	<550	<1022	Plasma	...	...
Bodycote, U.K.	Kolsterising	C	Trade secret	Trade secret	Trade secret	Hardware, watch cases	...
	Nivox2	N	<400	<752	Plasma	Control rods used in nuclear reactors	Bodycote acquired Nutrivid (France) 2010
	Nivox4 and Nivox LH	C	<460	<860	Plasma	...	...
Nihon Parkerizing, Japan	Palsonite	N + C	450–490	842–914	Cyanide salt bath	Small bore weapon components	...
Airwater Ltd., Japan	NV Super Nitriding	N	300–400	572–752	Gas	Flatware, hardware	Fluoride activation
	NV Pionite	C	<500	<932	Gas	Hardware, watch cases	Fluoride activation
Swagelok Company, U.S.A.	SAT12	C	380–550	716–1022	Gas	Tube fitting ferrules, hardware	HCl activation
Nitrex Metal Technologies, Canada	Nitreg-S	N	...	...	Gas	Piston rings, pitch gears	...
Expanite A/S, Denmark	Expanite	N + C	...	...	Gas	...	...

Figure 4.18

Low temperature plasma carburising

4.3.9 Hardide-CVD Tungsten Carbide

Following 15 years of research by scientists at Moscow State University and Frumkin Institute of Electrochemistry in Moscow a nano-structured coating made of tungsten-carbide was invented and named Hardide. The company was formed in 2000 by a team of UK-led venture capitalists (Flintstone Technologies) to commercialise the process. (http://www.hardide.com/downloads/Hardide-FT-6-6-04.pdf) The coating has also been developed as a REACH compliant replacement of Hard Chrome Plate (HCP) and can match the thickness of HCP and can exceed the hardness with better fatigue properties. www.hardite.co.uk. https://www.youtube.com/watch?v=aCq-crY0uCM&feature=youtu.be The coating is deposited by low-temperature CVD process (quoted as 500°C). The company claims that it has proven to be a good alternative to hard chromium for valve applications in several industries. The product has given better corrosion resistance compared to thermal sprayed coatings due to the low porosity and absence of a binding material.

4.3.10 Galvanizing

A process where ferrous articles are dipped into molten zinc (or an alloy of zinc) to produce a relatively thick surface layer giving protection against corrosion. Hot dip galvanising deposits a thick layer of zinc iron alloys on the surface of a steel item. The various strip grades that are used for automotive application are also galvanised and several improved grades have been developed to give better performance.

The relevant specifications are ASTM A123- Standard Specification for Zinc (Hot-Dip Galvanized) Coatings on Iron and Steel Products and ASTM A143- Standard Practice for Safeguarding Against

- Made by low Temperature low Pressure CVD;
- Process Temperature ~500°C
- Pore-free nano-structured coatings crystallised atom-by-atom from gas mixture;
- Non line of sight and complex shapes can be coated uniformly.

Figure 4.19 A Hardite process chamber

Embrittlement of Hot-Dip Galvanized Structural Steel Products and Procedure for Detecting Embrittlement. In addition, ASTM A 385- Standard Practice for Providing High-Quality Zinc Coatings (Hot-Dip)

The process consists of several steps of cleaning, rinsing fluxing and dipping into the molten zinc bath. The bath must have a minimum of 98% zinc. Other metal additions can be made to assist the process.

Surface preparation – prior to galvanizing consists of three main steps, caustic cleaning, acid pickling and fluxing.

Caustic cleaning – Often a proprietary solution of hot alkali is used to remove organic contaminants of grease and oil from the surfaces. Epoxies, Vinyl, asphalt, paint or welding slag must be removed by grit or sand-blasting or other mechanical methods.

Pickling – Scale or rust should be removed by pickling in a dilute solution of hot sulphuric acid or un-heated hydrochloric acid.

Fluxing – Fluxing is the final preparation and removes oxides and prevents any further oxidation prior to galvanizing. The method of applying depends upon whether the process uses the wet or dry galvanizing process. (In the dry galvanizing process, the item is separately dipped in a liquid flux bath, removed, allowed to dry, and then galvanized. In the wet galvanizing process, the flux floats atop the molten zinc and the item passes through the flux immediately prior to galvanizing)

Type	Hardness	Toughness	Thickness	Applications
Hardide-H "Ultra-Hard"	3000 – 3500 Hv	Satisfactory	5...10 μm	Self-sharpening blades
Hardide-M "Multi-Layer"	1200 – 2000 Hv	Good	Typically 50 μm	Erosion-resistance
Hardide-T "Tough"	1100 – 1600 Hv	Excellent	Typically 50 μm	Oil tools, pumps, valves, actuators
Hardide-A "Aerospace"	800 – 1200 Hv	Excellent	50-100 μm	Hard chrome replacement
Hardide-W	400 Hv	Excellent	Up to 300 μm	X-ray Anodes

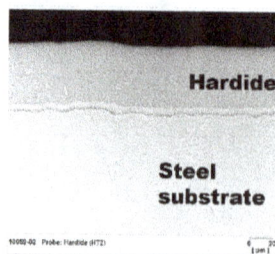

A family of nano-structured Tungsten Carbide & Tungsten Coatings;

Produced by Chemical Vapour Deposition (CVD)

Hardness varied from 400 Hv up to 3500 Hv;

Coating thickness 5...100 μm

Cross-section of 60 μm thick Hardide

Figure 4.20 Hardide Coatings

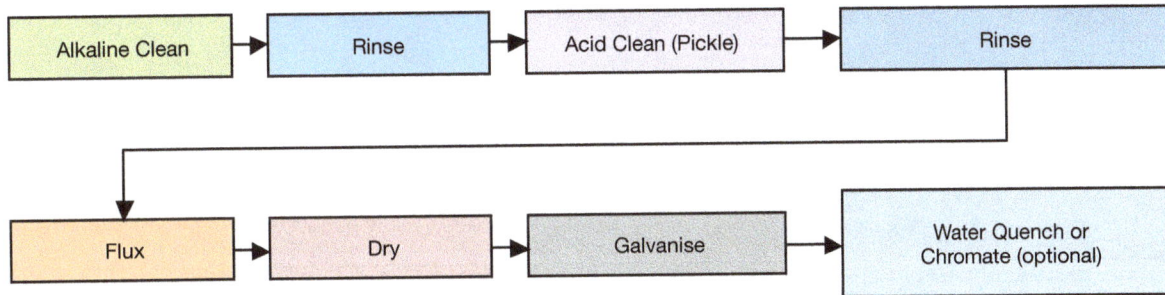

Figure 4.21 *Galvanizing Process. (13)*

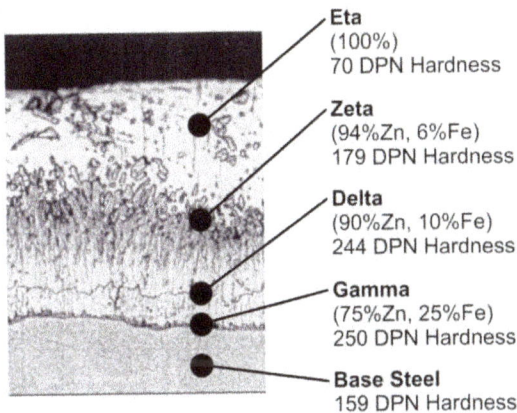

Figure 4.22 *Microstructure of a hot dip zinc coating. RHS = The effect of silicon content of the steel on zinc thickness (13)*

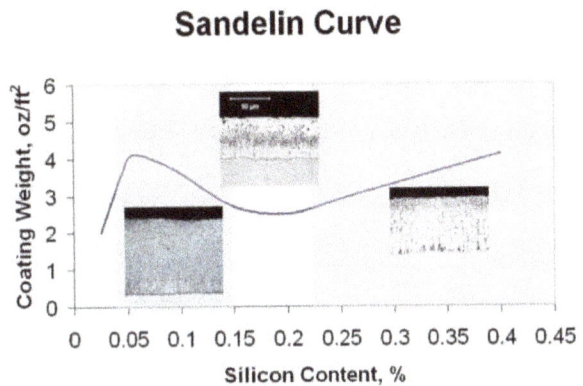

Figure 4.23 *The effect of silicon content of the steel on zinc thickness (13)*

The Metallurgical Bond

Galvanizing forms a metallurgical bond combining the zinc and the underlying steel or iron, creating a barrier that is part of the metal itself. During galvanizing, the molten zinc reacts with the iron in the steel to form a series of zinc-iron alloy layers. Figure 4.22 is a photomicrograph of a galvanized steel coating's cross-section and shows a typical coating microstructure consisting of three alloy layers and a layer of pure metallic zinc.

The Sandelin Curve

Silicon can have a major effect on the growth of galvanized coatings. (Phosphorus and manganese also increase the reactivity of the steel, and in combination with specific silicon levels, could also produce a thicker matte grey coating.) The Sandelin Curve (Figure 4.23) shows the recommended levels of silicon to produce typical hot-dip galvanized coatings. For highest-quality galvanized coatings, silicon levels should be less than 0.04% OR between 0.15% and 0.23%. Steels outside these ranges, considered

reactive steels, can be galvanized, and typically produce an acceptable coating; however, these steels often form a thicker coating, thus a darker appearance should be expected. For fasteners where thread fits are important control of the galvanised thickness is essential.

Coating Thickness

The average thickness of coating for all specimens tested shall conform to the requirements of Table 2 from ASTM A123 which is shown in Figure 4.24.

Time to First maintenance

Figure 4.25 shows a graph of the thickness of the galvanized coating against the expected time to first service of the coating under outdoor exposure conditions. This data is a compilation of many exposure tests of zinc-coated steel since the 1920s.

Today's atmosphere has substantially improved through anti-pollution measures, so the data curves represent a conservative relationship of the current performance of zinc coatings.

Coating Grade	mils	oz/ft²	µm	g/m²
35	1.4	0.8	35	245
45	1.8	1.0	45	320
50	2.0	1.2	50	355
55	2.2	1.3	55	390
60	2.4	1.4	60	425
65	2.6	1.5	65	460
75	3.0	1.7	75	530
80	3.1	1.9	80	565
85	3.3	2.0	85	600
100	3.9	2.3	100	705

A The values in micrometres (µm) are based on the Coating Grade. The other values are based on conversions using the following formulas: mils = µm × 0.03937; oz/ft² = µm × 0.02316; g/m² = µm × 7.067.

Figure 4.24 *Coating grades shown in ASTM A123*

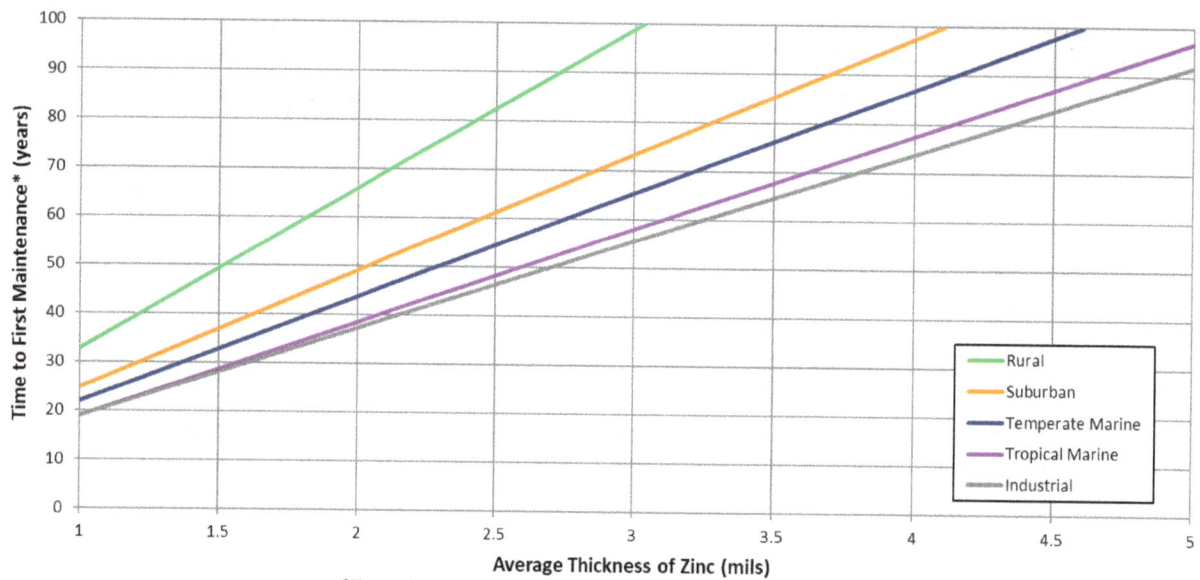

Figure 4.25 *Time to first maintenance (13)*

Cracking

The problem of steel cracking during hot dip galvanizing has been known for many years.[14] Such cracking has been associated with welding and cold working during fabrication. These processes can induce susceptibility to embrittlement in certain steels which, when combined with residual and thermal stresses, can result in cracking of the steel during galvanizing. Historically, high-strength steels have been most susceptible (>825MPa) but lower-strength steels (<620MPa) can also be prone to cracking at flame-cut sections, weldments and severely cold-worked areas such as bends and corners of cold-formed tubing. The cracking has been associated with hydrogen pick up during acid cleaning. For high strength bolt grades acid cleaning is avoided and blast cleaning is mandatory. Recently cracks have occurred on high strength automotive grades.

Aluzinc

In recent years there has been continuous improvement in hot dip coatings used in the automotive industry. These include Zn-Al (Galfan, Galvalume, Aluzinc)[17] and more recently Al-Mg which claims better life prior to developing red rust in salt spray tests. The corrosion resistant aluminium–zinc coatings were developed in the Homer laboratories of Bethlehem Steel in the 1960s and corrosion trials over 30 years optimised the composition and established that the coating achieved over 4 times the life of zinc coatings.[15] Aluzinc's composition: 55% Al, 1.6% Si, the rest zinc, was selected from a systematic study during the development of the coating in the 1960s, and provides an excellent combination of galvanic protection and low corrosion. A thin organic coating anti-finger print sealant, known as Easyfilm®, is applied to the surface. This thin transparent organic

ALUZINC - Corrosion Resistance

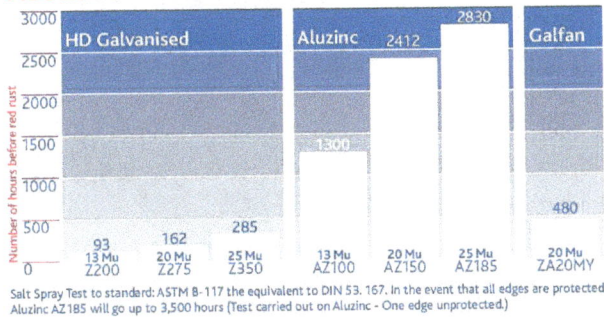

Salt Spray Test to standard: ASTM B-117 the equivalent to DIN 53. 167. In the event that all edges are protected: Aluzinc AZ185 will go up to 3,500 hours (Test carried out on Aluzinc - One edge unprotected.)

Figure 4.26 Corrosion resistance of Aluzinc/Galvalume (17)

coating offers corrosion protection, increases protection against fingerprints and improves sliding characteristics during forming operations. It can also be used as a primer coat for subsequent painting[16].

4.3.11 Metal spraying

A technique for uniquely transferring metals by the use of heat, plasma, or arc to the surfaces of prepared components. Low Pressure Plasma spray is used to deposit NiCrAlly for aeroengine parts such as nozzle guide vanes and fuel vaporisers.

4.3.12 Vitreous enamelling

The application of metallic glass containing liquids by dipping or spraying techniques onto ferrous components.

4.3.13 DLC Diamond Like Carbon Coatings

The great versatility of carbon materials arises from the strong dependence of their physical properties on the ratio of sp2 (graphite-like) to sp3 (diamond-like) bonds. There are many forms of sp2 bonded carbons with various degrees of graphitic ordering, ranging from micro-crystalline graphite to glassy carbon. The compositions of nitrogen-free carbon films are shown on the ternary phase diagram in figure 1.73. We define diamond-like carbon (DLC) as an amorphous carbon (a-C) or a-C:H with a significant fraction of sp3 bonds. It can have a high mechanical hardness, chemical inertness, optical transparency. DLC films have widespread applications as protective coatings in areas such as optical windows, magnetic storage disks, car parts, biomedical coatings and as micro-electromechanical devices.

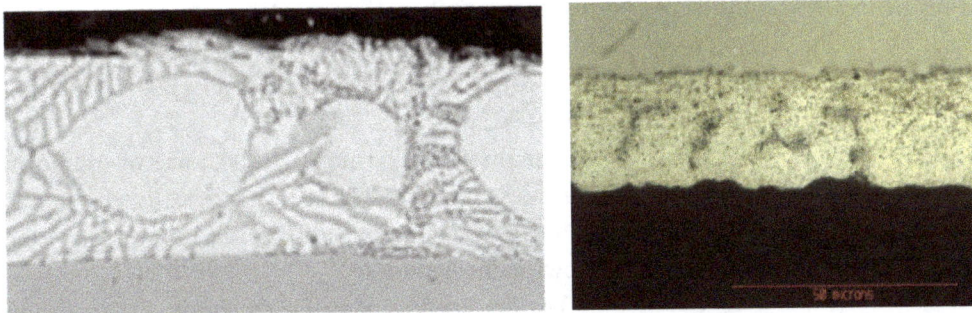

Figure 4.27 Metallographic comparison Zn Mg (LHS) and Aluzinc (RHS)

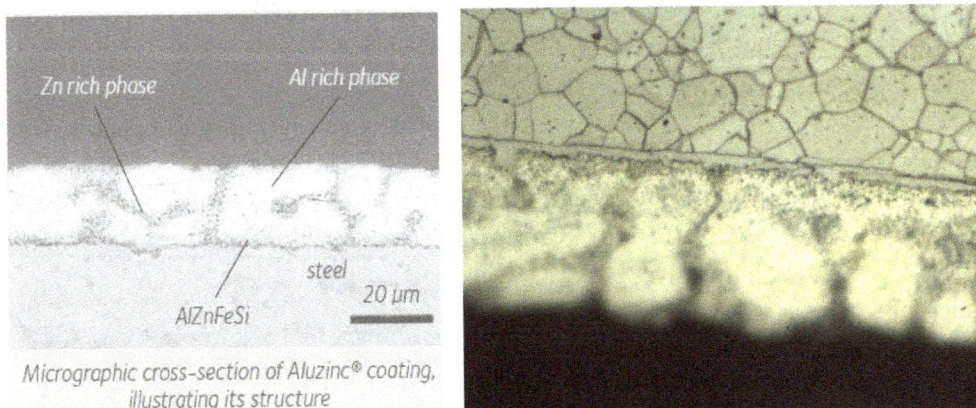

Micrographic cross-section of Aluzinc® coating, illustrating its structure

Figure 4.28 LHS = Published image of the microstructure of Aluzinc (17). RHS = Etched sample

a-C	Amorphous, non-hydrogenated carbon (**a-C**) coatings: dominated by sp^2 bonds and have typically less than 1% of hydrogen
a-C:H	Hydrogenated amorphous carbon (a-C:H) films: varying amounts of sp^3/sp^2 bonds and hydrogen content resulting in a wide range of properties
ta-C	Tetrahedral amorphous carbon (ta-C): the highest fraction of sp^3 bonds; synthesized typically from solid graphite - do not contain a much hydrogen; closest to diamond
ta-C:H	Hydrogenated tetrahedral amorphous carbon (ta-C:H): typically around 30% hydrogen content and variable fraction of sp^3/sp^2 bonds

The ratio of sp^3/sp^2 bonds and the hydrogen content in the coating determine the properties of DLC films

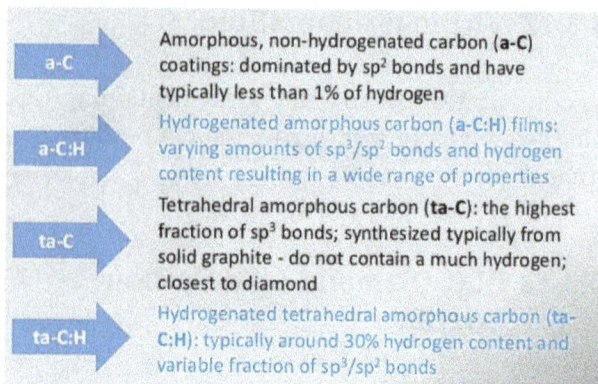

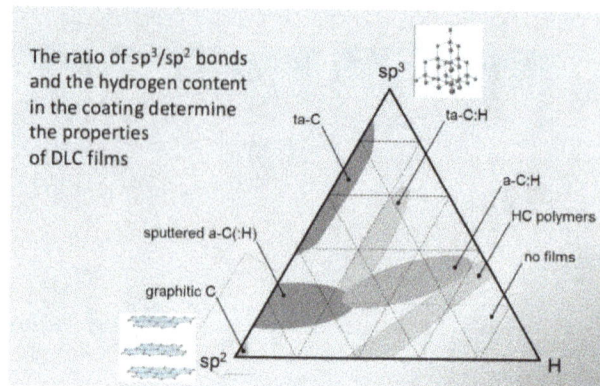

Figure 4.29 *DLC is a generic term describing a range of amorphous carbon. DLC have a mixture of sp2 and sp3 bonds. Now finding use in oil and gas applications. (Ref 18)*

4.3.14 How to select the required surface engineering technology

The selection of suitable surface treatments should be based on an understanding of the wear, service stresses and corrosion conditions that the surface of the part experiences due to the operating environment. Since the surface of the part is exposed to service conditions and possible contact stresses the surface has to survive all mechanical, thermal, chemical, and electrochemical interactions with the environment. The evaluation of the risk of these conditions causing potential failure modes should then result in a "product design specification" that outlines the main surface property requirements that are needed to function and to avoid failure. The range of surface requirements can include:

1. Corrosion resistance (plating, painting, oxide cover Cr on stainless, Al_2O_3 on aluminium, Oxide on titanium)
2. Wear resistance (high hardness, carbide content)
3. Defined tribological behaviour (frictional coefficient)
4. Optical behaviour (anti-reflection coating)
5. Decorative behaviour (PVD deposition of TiN on taps- gold colour)
6. Electronic (Impurity doping of semi-conductor devices to EMI shielding- Electromagnetic shielding)
7. Antimicrobial surfaces (copper alloys, innovative metal oxides in antimicrobial surfaces)
8. Matched interface behaviour (e.g. for joining purposes such as weld overlay).

Figure 4.30 top illustrates the main kinds of load conditions divided into volume and surface loads. Wear and corrosion are the main forms of deterioration that have to be controlled by the surface technology.

Incorrect materials selection as well as unsuitable or missing protective layers can result in damage and potential failure. Some examples are shown in Figure 4.30 centre and lower. In many cases, appropriate surface treatment can either prevent or delay damage and potential failure.

The main challenge is selecting the appropriate material and surface engineering process from the large number of competing processes. Some of the techniques have been outlined in this section but a total list is estimated at over 1000 competing processes.

A main aspect of making the choice would be an awareness of what has been historically acceptable for any particular application and a knowledge of the known failure modes.

The application of surface technology focuses on creating properties to exceed the service loads and stresses by a safety margin. To achieve this the materials properties of part surfaces are systematically modified by either:

- applying a protective coating to the part. (chromium plating, PVD, Hardite)
- modifying the surface/sub-surface of the part. (carburising, nitriding, boriding, carbo-nitriding)

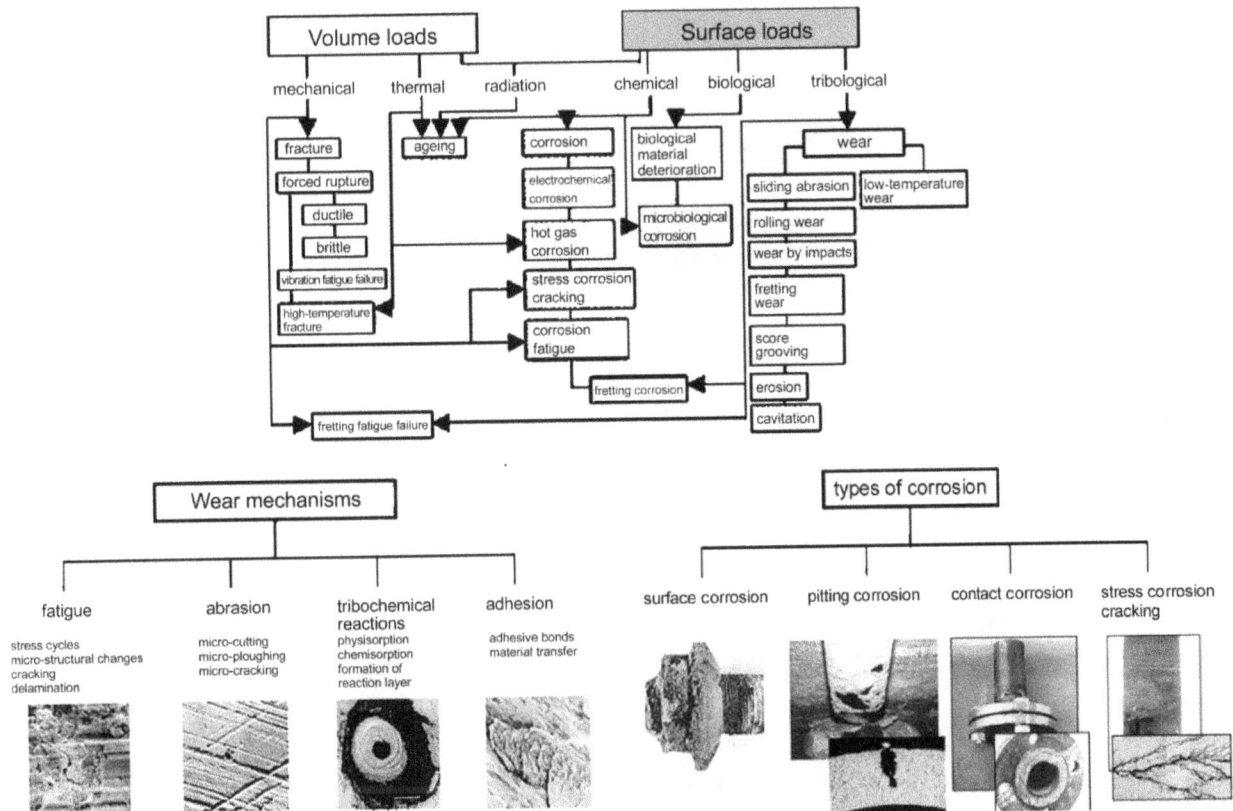

Figure 4.30 *Selection of appropriate surface engineering methods Top=service conditions, Centre=potential wear mechanisms, Lower= potential corrosion mechanisms*[19]

4.4 IMPORTANT ASPECTS OF THROUGH HARDENING STEEL. STEEL COMPOSITION, HARDENABILITY, SECTION SIZE AND QUENCH RATE

4.4.1 Hardenability and Steel selection for heat treated parts

For any particular part the steel selected must have appropriate hardenability (DI) for the section size (ED) and the quench severity (H) to allow the formation of a microstructure that will guarantee the required YS/UTS ratio and low impact transition temperature (ITT) at the key locations in accordance with the design requirements. This would normally require a minimum of 90% martensite in the as-quenched microstructure at the key location.

Methods to measure, evaluate and control hardenability include:

(A) The quantitative end-quench or Jominy Test
(B) A method for calculating the hardenability of steel from the chemical composition based on the original work by M. A. Grossman.

(C) "Grossman's critical diameter method" see Fig 4.47 RHS,
(D) A fracture test method if there is a clear demarcation between martensite zone and pearlite zone.

(A) Jominy A simple yet very effective technique that has withstood the passage of time. The Jominy end-quench test was invented by Walter E. Jominy (1893-1976) and A.L. Boegehold, in the Research Laboratories Division of General Motors Corp., in 1937. The end-quench test is the most familiar and commonly used procedure for measuring steel hardenability. This test has been standardized in ASTM A 255, SAE J406, DIN 50191, and ISO 642. A 100 mm (4 in.) long by 25 mm (1 in.) diameter round bar is austenitized, dropped into a fixture, and one end rapidly quenched with 24°C water from a 13mm orifice. The austenitizing temperature is selected according to the specific steel alloy being tested generally in the range of 870–900°C. The cooling velocity decreases with increasing distance from the quenched end. After quenching, parallel flats are ground on opposite sides of the bar and hardness measurements made at 1/16 in. (1.6 mm). Hardness as a function of distance from the quenched end is measured and plotted.

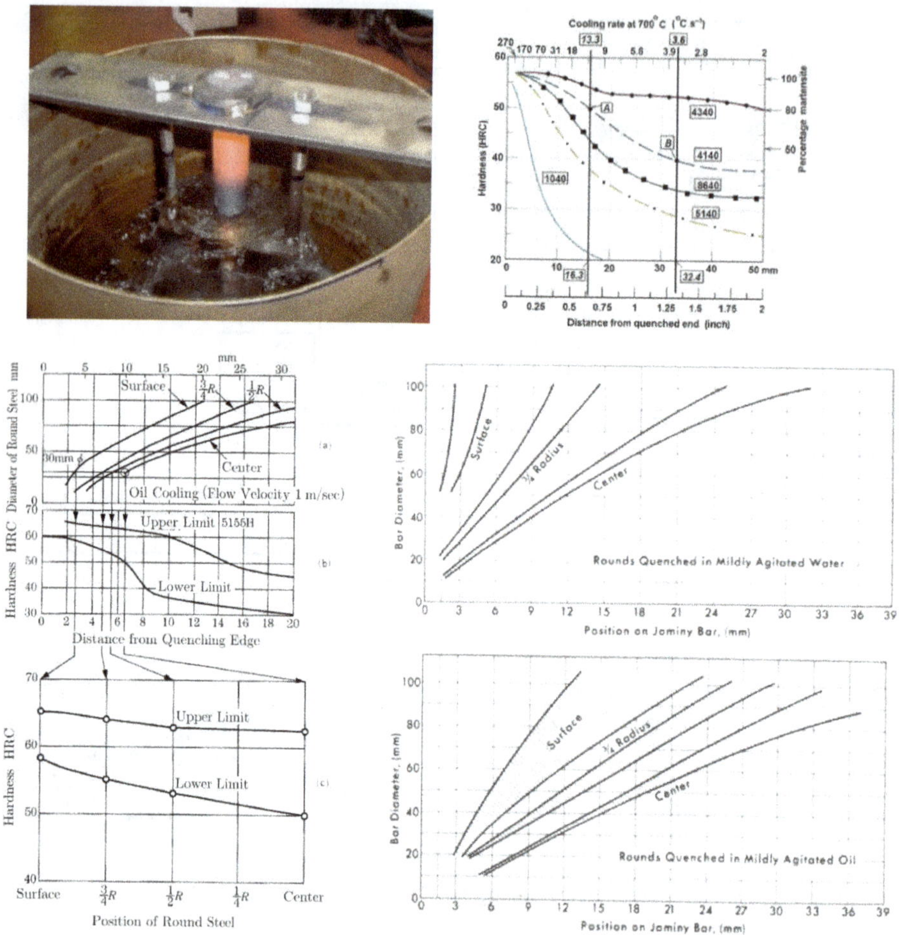

Figure 4.31 *Top = LHS Jominy sample during cooling. Top RHS = Jominy data for various steel (28). Lower LHS = Demonstrates the use of hardenability data to predict properties in a round bar Use appropriate graph on the lower RHS depending upon quench. Predict Jominy curve from chemistry. Use the graphs on the RHS to relate J values to position on a particular diameter bar*

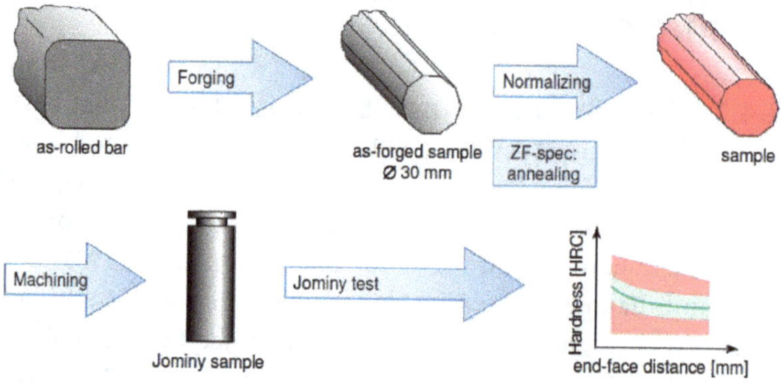

▶ **Figure 4.32** *The Jominy sample preparation*[20]

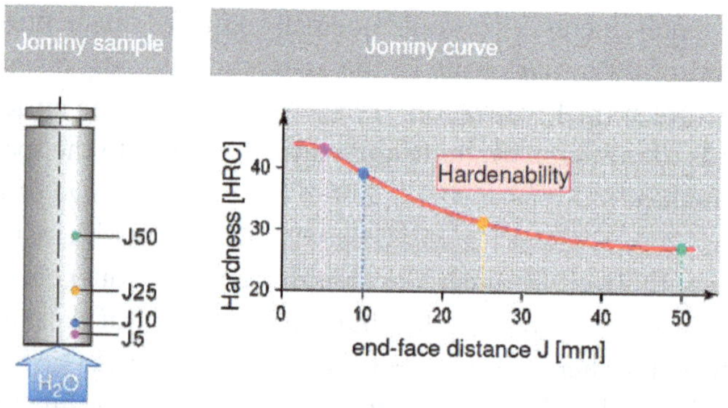

▶ **Figure 4.33** *Jominy sample hardness test positions and test data presentation*[20]

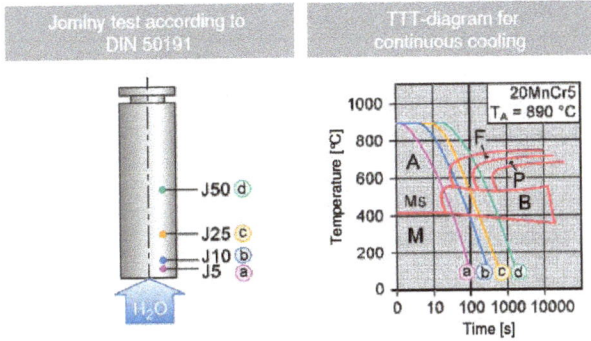

Figure 4.34 *Jominy sample hardness test locations and similar cooling curve shown on the CCT diagram. (20)*

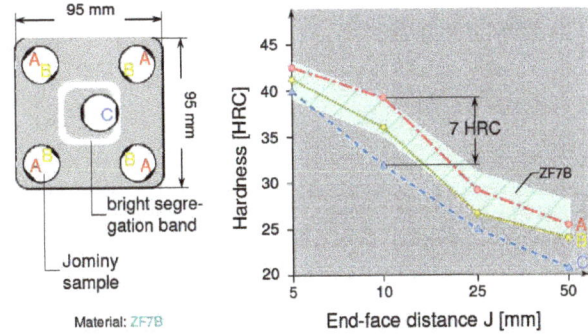

Figure 4.35 *Understanding the effect of segregation on the Jominy test and the spread of results (20)*

Characteristic parameter		Effect on J25
▪ Austenitising temperature	870 ± 20 °C	± 0,4 HRC
▪ Length of sample	100 - 5 / - 10 mm	± 0,4 HRC
▪ duration before water quenching	5 + 15 sec.	± 0,4 HRC
▪ Duration of austenitizing	60 ± 15 min.	± 1 HRC
▪ Surface roughness	R_z = 17 / 25 µm	± 1 HRC
▪ Alignment of water jet	0°/ 15°	± 1 HRC

Figure 4.36 *Effect of test conditions on results (20)*

The major symposium on hardenability in 1946 confirmed that the end quench test was reproducible and capable of producing reproducible results. It continues to be the main method to measure hardenability of steel.[22] The Jominy test has also been on non-ferrous metal to evaluate precipitation hardening.[3]

(B) Calculation of Hardenability. Grossman defined DI as the ideal diameter of a given steel that would harden to 50-percent martensite when quenched in a bath where H = ∞. This is the hypothetical infinite cooling rate that is equivalent to instantly reducing the surface temperature of the steel bar to the quenchant temperature.

The ideal diameter is a true measure of hardenability associated with a steel composition. The concept of the ideal diameter can be used to determine the critical size of steels quenched in quenchants of differing severity. The calculation of the ideal diameter steels relies on a series of multiplying factors. This base DI is determined from the grain size and carbon content and then is multiplied by the various factors from the composition:

$$DI = DI\ base \times fMn \times fSi \times fNi \times fCr \times fMo \times fV$$

These multiplying factors are tabulated in ASTM A255.

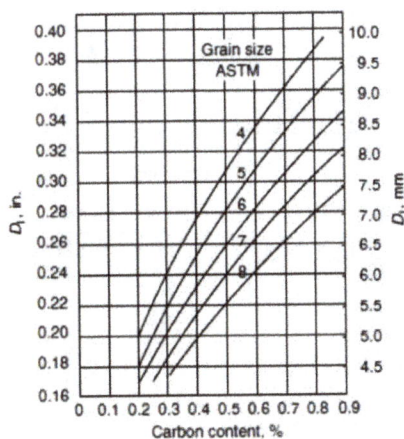

Figure 2-54 The ideal critical diameter (D_I) as a function of the carbon content and austenite grain size for plain carbon steels, according to Grossmann[251].

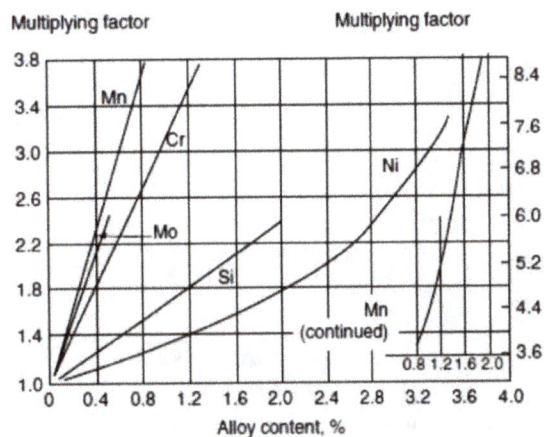

Figure 2-55 Multiplying factors for different alloying elements when calculating hardenability as D_I value, according to AISI[252].

Figure 4.37 *Grossman multiplying factors*

Figure 4.38 *Hardness of various martensite percentages*

Carbon %	Hardness (HRC)				
	99% martensite	95% martensite	90% martensite	80% martensite	50% martensite
0.10	38.5	32.9	30.7	27.8	26.2
0.12	39.5	34.5	32.3	29.3	27.3
0.14	40.6	36.1	33.9	30.8	28.4
0.16	41.8	37.6	35.3	32.3	29.5
0.18	42.9	39.1	36.8	33.7	30.7
0.20	44.2	40.5	38.2	35.0	31.8
0.22	45.4	41.9	39.6	36.3	33.0
0.24	46.6	43.2	40.9	37.6	34.2
0.26	47.9	44.5	42.2	38.8	35.3
0.28	49.1	44.8	43.4	40.0	36.4
0.30	50.3	47.0	44.6	41.2	37.5
0.32	51.5	48.2	45.8	42.3	38.5
0.34	52.7	49.3	46.9	43.4	39.5
0.36	53.9	50.4	47.9	44.4	40.5
0.38	55.0	51.4	49.0	45.4	41.5
0.40	56.1	52.4	50.0	46.4	42.4
0.42	57.1	53.4	50.9	47.3	43.4
0.44	58.1	54.3	51.8	48.2	44.3
0.46	59.1	55.2	52.7	49.0	45.1
0.48	60.0	56.0	53.5	49.8	46.0
0.50	60.9	56.8	54.3	50.6	46.8
0.52	61.7	57.5	55.0	51.3	47.7
0.54	62.5	58.2	55.7	52.0	48.5
0.56	63.2	58.9	56.3	52.6	49.3
0.58	63.8	59.5	57.0	53.2	50.0
0.60	64.3	60.0	57.5	53.8	50.7

4.4.2 Quenchant selection and control

To minimise the hardenability requirements (alloy content) the fastest cooling rate should be aimed for commensurate with avoidance of quench cracking and distortion. This usually requires a low temperature water with a high level of agitation. When steels have a high carbon level and alloy content, particularly with complex geometries, the conventional approach has been to use quenchants that give relatively slow cooling. This is normally a polymer quench or an oil quench. However, in the era of Intensive Quench

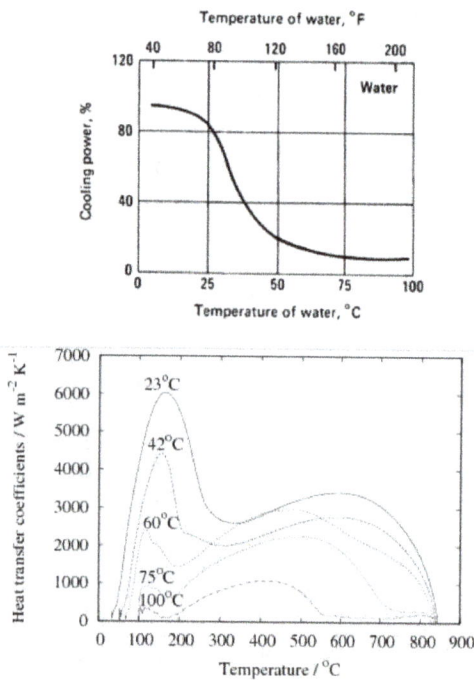

Figure 5-47: Heat transfer coefficient as a function of surface temperature during quenching at different water temperatures.

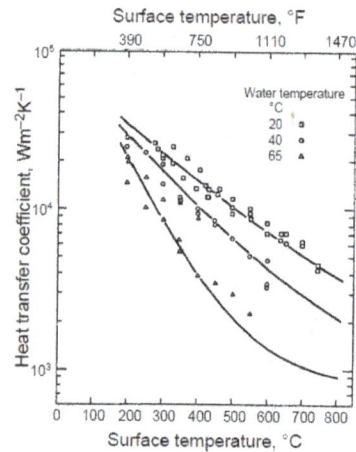

Fig. 8 Heat-transfer coefficient as function of surface temperature during immersion in water at 20, 40, and 65 °C (70, 105, and 150 °F); lines are calculated values. Source: Ref 25

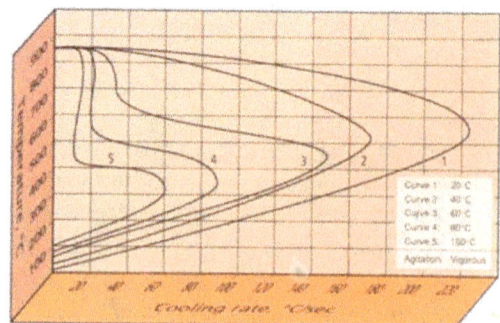

Figure 4.39 *Top LHS = The effect of water temperature on water cooling power (agitated water) Top RHS = Effect of water temperature on heat transfer Lower LHS = Effect of water temperature on heat transfer coefficient RHS = Houghton data*

Technology there are opinions that for certain sizes and shapes a rapid quench should be used even for some tool steels. For open furnace work the trend has been to polymer quenchants (PVP) due to the fire risk. When using polymer quenchants, appropriate control is needed on percentage polymer in the quench, the operating temperature and the quench tank agitation.

For water quenching the maximum cooling rate is required and the quench tank should operate at as low a temperature as possible and additional cooling (refrigerated) can often be needed during hot weather or high plant output. Figure 4.39 shows the effect of water temperature on the quench rate.

To ensure process capability the severity of quench (H) of a quench tank should be measured. This can be done in several ways:

1. Using a steel sample (approx. 125 mm diameter, 500 mm long, with limited known hardenability for that cross section so that a hardness gradient is developed from the surface to the centre. By measuring the hardness across the diameter and using Lamont and Craft's graphs the H value can be established.
2. The use of 19mm diameter C45 steel. (See Figure 4.40 centre and lower diagram.)

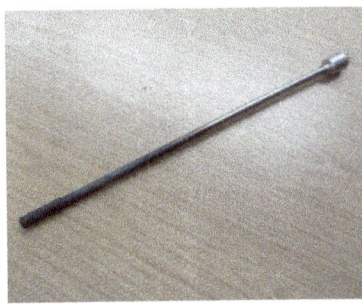

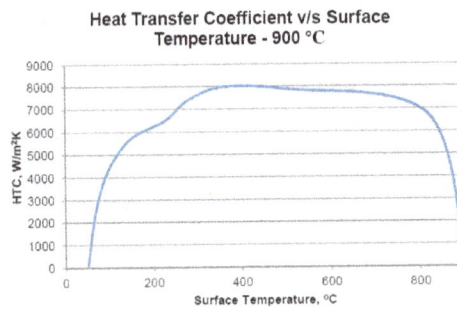

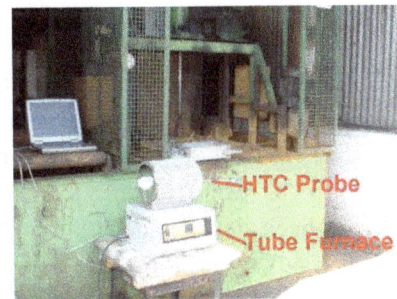

Agitation	Air	Oil	Water	Brine
No circulation	0.02	0.25-0.30	0.90-1.0	2
Mild circulation	-	0.30-0.35	1.0-1.1	2-2.2
Moderate circulation	-	0.35-0.40	1.2-1.3	-
Good agitation	-	0.40-0.50	1.4-1.5	-
Strong agitation	0.05	0.50-0.80	1.6-2.0	-
Violent agitation	-	0.8-1.10	4	5

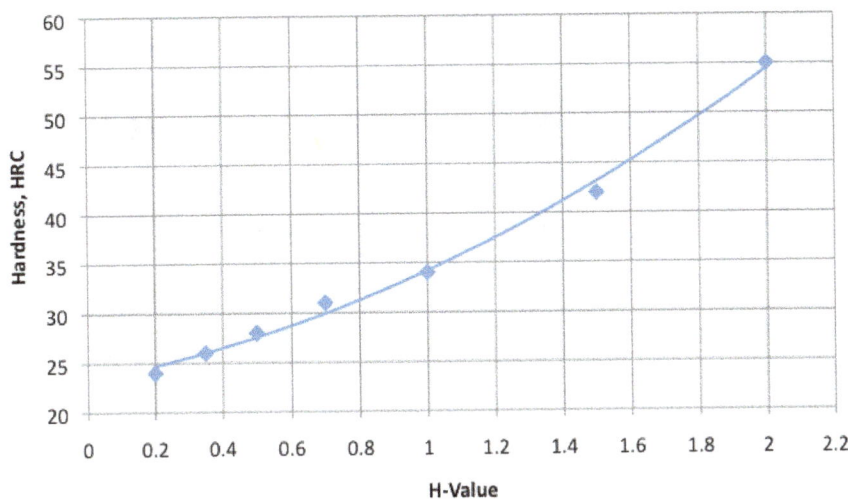

Figure 4.40 *Top=MSL Quench Characterisation Probe Centre= Grossman Quench severity values (H). RHS = Quench severity probes (19mm C45). Lower = Quench probe centre hardness for various quench severities (H values), From[23]*

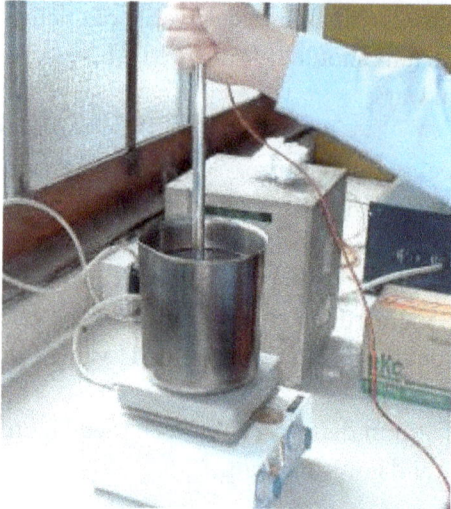

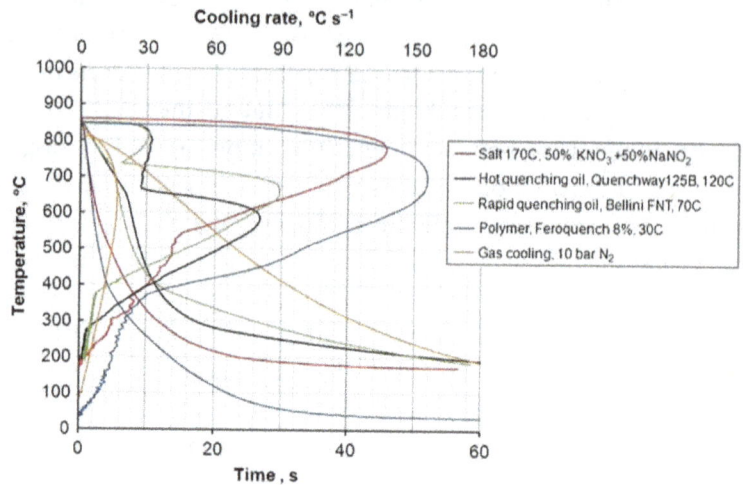

Figure 4.41 *LHS = Checking quenchant characteristics in accordance with ISO 9950 at MSL, RHS = Cooling curves and cooling rates for different quenchants. From | (23)*

3. The use of a thermal probe with twin thermocouples to quantify the heat transfer coefficient. (See Figure 4.40 top photos.) This probe, referred to as QC Probe (Quench Characterization Probe), has been developed by MSL. The main function of the QC Probe is to predict heat transfer coefficients of different quenchants based on recorded cooling curve data. The main requirement for the information obtained using the probe would be:

i. To monitor the quality of quenchants for service life and contamination, which needs to be controlled to avoid poor mechanical properties, excessive distortion or cracking
ii. The prediction of heat transfer coefficients of quench systems for input into PC based software models that predict the heat treatment such as QT Steel.

A similar probe with one thermocouple at the centre can be used to measure the quench characteristics and is carried out in accordance with ISO 9905 and ASTM 6549 (as shown in Figure 4.41).

4.4.4 Improved system for quench rate monitoring

Quench facilities are an essential part of the quench and temper process and requires regular maintenance and control. Currently the items logged are usually transit time between the furnace and quench, the quench tank temperature prior to the quench and

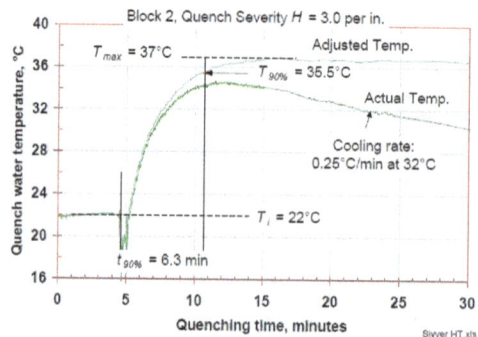

the quench tank temperature after the quench. In 2005 Canmet in Canada[24 and 25] carried out experimental logging of a quench tank and monitored the rate of release of the heat from the charge to the water. This information was then used to establish the

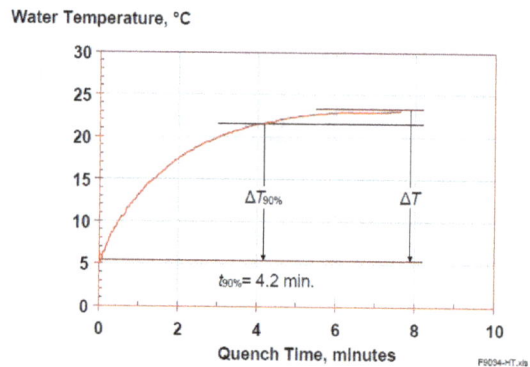

Fig. 7. Construction to determine quench time from quench water temperatures in a laboratory quench tank [3].

Fig. 8. Quench time determination in an industrial quench tank with cooling water recirculating from a cooling tower [6].

Figure 4.42 *Quench tank monitoring to measure quench rate*

effectiveness of the quench. The purpose of the work was to prevent intergranular embrittlement of high manganese austenitic steel casting. This requires a fast water quench. The basic simplicity of this concept is worthy of detailed consideration since it has not been used in commercial installations.

The concept is that the analysis of data from a suitably positioned quench tank thermo-couple (as shown in Figure 4.42) could be used to confirm that the "quenched" furnace load has had an effective quench. It would represent an additional quality assurance measure in a critical part of the heat treatment process

4.4.4.1 Furnace efficiency improvements by control of residence times

Heat treatment temperature, holding time, and rate of heating and cooling are some of the parameters that affect the heat treatment processes, and are commonly referred to as heat treatment process variables. The selected values for these process variables depend on the chemical composition, size and shape of the object and the final properties required in the metal/alloy.

For example, a component made from tool steel being heat treated in a vacuum furnace would be placed into the furnace at room temperature. This furnace would then be heated up to a required temperature. The average rate of heating can be calculated from the temperature rise divided by the total time taken.

The heating rate should be adjusted depending upon the size and shape of the parts and the thermal conductivity of the alloy. Complicated shapes, and parts with variable sections should be heated slowly. Alloys, such as high carbon steels and austenitic stainless steels, are also subjected to slow heating rates. Smaller sized and simple shaped objects can be heated with higher heating rates, and homogeneity of the structure can be ensured in such cases by increasing the holding time. Often the parts are held just below the AC1 and again above the AC1 to allow transformation to begin uniformly through the whole part and avoid the risk of thermal stresses causing distortion of the part.

The heat treatment temperature is governed mainly by chemical composition of the alloy, prior heat treatment, if any, and the final properties required. For example, for normalise or quench and temper heat treatments, steel is heated above the AC3

(upper critical temperature). This temperature is also known as austenitising temperature, and can be determined, for carbon steels, by the iron-cementite phase diagram.

For carbon steels, this temperature decreases with increasing carbon content up to eutectoid composition and again rises with increasing carbon content. Theoretically, at this temperature, steel should be fully austenitic with the smallest grain size. With rapid heating, actual austenitising temperature is raised compared to the theoretical value obtained from iron-cementite phase diagram (only relevant for rapid heating such as induction heating). The prior microstructure can also affect the austenitising temperature for induction heating.

During heat treatment the austenitising temperatures are slightly higher than the critical temperatures determined from the equilibrium phase diagram. This is usually plus 30 to 50°C above the AC3. In theory the austenite formed on heating near or slightly above the equilibrium temperature is inhomogeneous in nature. In the case of alloy steels, some alloying elements or their compounds (carbides) are slow to dissolve or elements are slow to diffuse to create homogeneity. Such steels require higher temperatures or longer times for homogenisation of austenite. Iron carbide, in steel, typically dissolves readily in austenite as compared to carbides of strong carbide-forming elements (chromium, molybdenum and vanadium).

Once the heat treatment temperature has been decided, holding time is generally provided at a rate of 1 to 2 minutes per millimetre of section thickness. For parts with variable cross section thickness the holding time should be determined on the basis of the thickest section. Parts which are heated up with high heating rates require longer holding time.

Heat treatment temperature and holding time are somewhat related in the sense that an increased heat treatment temperature could result in reduction of holding time. Similarly, lowering of heat treatment temperature demands an increase in holding time. Alloys, with high alloy content, are often kept for more time at "holding" temperature than plain carbon steels and low alloy steels to allow the alloy carbides to dissolve into solution in the austenite to ensure appropriate hardenability. Only the alloy content in solution in the austenite will contribute to the "hardenability". To allow full solution of alloying elements holding time may be increased by 25 to 40% compared to plain carbon and low alloy steels.

The choice of treatment temperature and treatment time can be a difficult and complex decision. The basic starting point should be the specified values from sources such as AMS 2759 or MIL-H 6875H or specified customer values. The aspect of selecting appropriate austenitising conditions has always been an empirical method due in part to the inability to be able to monitor thermal events at the centre of a part inside a furnace.

This has always been a topic of serious concern since fuel savings can be achieved and CO_2 reductions made by operating a practice of minimum furnace residence times. However, attempts to reduce "holding" times can increase the risk of failure to achieve the heat treatment intent, which is usually revealed by failure to achieve the mechanical property requirements such as proof strength or toughness and a full re-heat treatment has to be carried out.

One of the first projects that I carried out in the steel industry, as a young graduate, was aimed at reduced normalising times which would give fuel savings and higher furnace output. The plant where I worked had a standard cost of £20 per tonne for the normalising treatment. Other plants in the steel corporation could achieve £3 per tonne. An additional problem was that the heat treatment furnace used to carry out the normalising treatment, had been refurbished, and a substantial capital investment had been made to improve the furnace control system. Unfortunately, the new control system was of poor construction and the burners could not "turn down" when the furnace reached temperature. Therefore, the burners could only be operated at a "fixed" setting by the operators to avoid overshoot in temperature! Attempts to solve this issue was one of my first trips into "management acrimony" since I received several warnings to "leave it alone or it could affect my career" made by the senior people responsible for the acceptance of the furnace control system!

However, there are approaches to refining the temperature monitoring of the parts inside heat treatment furnaces which quote attractive savings.

In a similar way there are methods to measure and monitor the performance of the heat transfer coefficient and to apply science to the quench practice.

4.4.4.2 Zero hold time – Minimum austentising time

There are several Chinese publications that have examined the option of "Zero-time-holding" prior to quenching. This means that the samples are quenched immediately when their surface temperature reaches the austenitising temperature. The advantages compared with a conventional quenching process, are claimed as reducing the time in the heat treatment furnace, a decrease of 20-30% in the energy consumption and an increase in the labour productivity. In addition, it can also reduce or eliminate the oxidation, decarb and grain coarsening. Thus, zero-time-holding quenching can raise the quality of the product.

This topic of "time in the furnace" is still a potential R&D topic and often requires further consideration since fuel savings can be made by targeting reduced heat treatment times. The Chinese innovations that aim for "zero soak" are probably too ruthless for metallurgists with conventional views of austenitizing conditions. However, there are approaches to refining the temperature monitoring of the parts inside heat treatment furnaces which often quote attractive savings. See *Improved heat treatment process control for fuel efficiency and cycle time reduction, TCP Kamezos 2009 Penn State University*[26] In an era of environmental concern all heat treaters should review the austenitizing conditions/ soaking time required based on the furnace capability and furnace load to establish minimum furnace residence times and cost savings. Guidelines on the austenizing condition are shown in Figure 4.43 and AMS 2579.

In summary there are two areas that require greater consideration in heat treatment:

- This topic of "time in the furnace" where there are potential fuel savings and increased through-put. The initial "rules" usually originate from AMS 2759. (See Tables in section 4.4.4.2)
- The control of the process capability and monitoring of quench practice. This requires an understanding of an application of "hardenability" to steel selection to ensure that the design requirements are met at all required locations on the part.

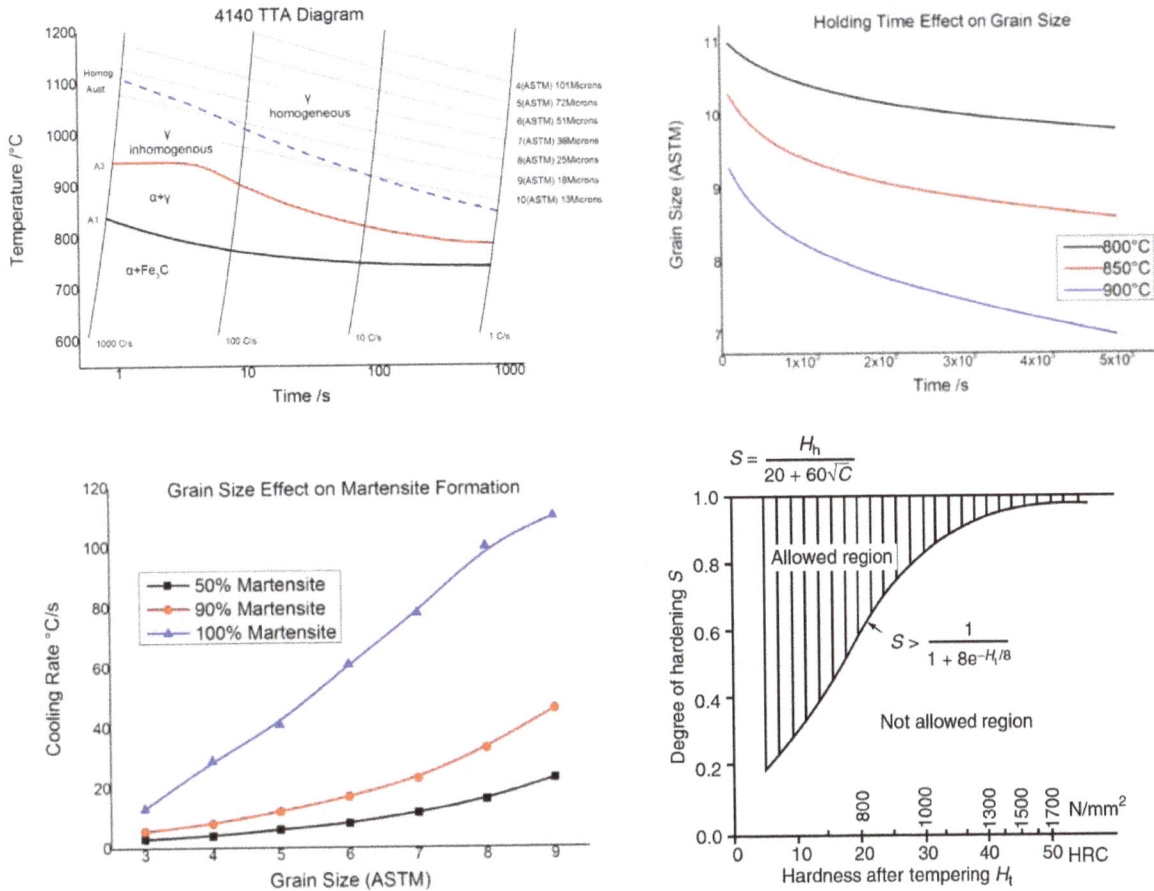

Figure 4.43 *Austenitising conditions Centre= Holding time on grain size. RHS = Effect of grain size on martensite formation (27). Lower RHS = Degree of hardening vs strength/hardness*

4.4.5 Heat treatment specifications

The following sections outlines some of the key specifications that should be used to ensure that heat treatment is carried out to world class standards

4.4.5.1 AMS 2750 Pyrometry

The use of this specification ensures that the instrumentation type and furnace class are known and tests such as TUS (temperature uniformity survey) and SAT (system accuracy tests) are carried out in accordance with defined procedures. The furnace user has the responsibility to decide and specify the required furnace class for each facility. This would depend upon the type of heat treatment carried out. The furnace control instrumentation could range from a simple controller to full over-temperature cut-out combined with temperature logging based on load thermocouples. The instrument class reflects the combination of control instrumentation, thermocouples and data recording and logging used as shown in the table below. The furnace control instrumentation type is used to determine how frequently the calibration and the furnace hot zone searches are carried out. Based on the measurement of the temperature uniformity of the furnace hot zone requirements the furnace class ranges from a "1" to a "6".

Instrumentation	A	B	C	D	E
Each control zone has a thermo-couple connected to the controller	X	X	X	X	X
Recording of the temperature measured by the control thermocouple	X	X	X	X	
Sensors for recording the coldest and the hottest spots	X		X		
Each control zone has a charge thermocouple with recording system	X	X			
Each control zone has an over-temperature protection system	X	X	X	X	

Furnace Class	Temperature Uniformity °C
1	+/- 3
2	+/- 6

Figure 4.44 *continued*

3	+/- 8
4	+/- 10
5	+/- 14
6	+/- 24

Heat treatment of steel parts-General requirements. This specification provides guidance on heat treatment procedures and the treatment temperatures and times in accordance with steel type and section size.

4.4.5.2 AMS 2759

Figure 4.44 *AMS 2750 classification of temperature control system and furnace class (Uniformity of temperature of the hot zone)*

▶ **Table 4.4**
Recommended times
AMS 2759

Thickness (1) Inches	Thickness (1) Millimeters	Minimum Soak Time (2), (3), (4), (5) Air or Atmosphere		Minimum Soak Time (2), (3), (4), (5) Salt	
Up to 0.250	Up to 6.35		25 minutes		18 minutes
Over 0.250 to 0.500	Over 6.35 to 12.70		45 minutes		35 minutes
Over 0.500 to 1.000	Over 12.70 to 25.40	1 hour			40 minutes
Over 1.000 to 1.500	Over 25.40 to 38.10	1 hour	15 minutes		45 minutes
Over 1.500 to 2.000	Over 38.10 to 50.80	1 hour	30 minutes		50 minutes
Over 2.000 to 2.500	Over 50.80 to 63.50	1 hour	45 minutes		55 minutes
Over 2.500 to 3.000	Over 63.50 to 76.20	2 hours		1 hour	
Over 3.000 to 3.500	Over 76.20 to 88.90	2 hours	15 minutes	1 hour	5 minutes
Over 3.500 to 4.000	Over 88.90 to 101.60	2 hours	30 minutes	1 hour	10 minutes
Over 4.000 to 4.500	Over 101.60 to 114.30	2 hours	45 minutes	1 hour	15 minutes
Over 4.500 to 5.000	Over 114.30 to 127.00	3 hours		1 hour	20 minutes
Over 5.000 to 8.000	Over 127.00 to 203.20	3 hours	30 minutes	1 hour	40 minutes
Over 8.000	Over 203.20	(6)		(7)	

Notes:
(1) Thickness is the minimum dimension of the heaviest section of the part.
(2) Soak time commences as specified in 3.4.2 as modified by 3.4.2.1.
(3) In all cases, the parts shall be held for sufficient time to ensure that the center of the most massive area has reached temperature and the necessary transformation and diffusion have taken place.
(4) Maximum soak time shall be twice the minimum specified, except for subcritical annealing.
(5) Longer times may be necessary for parts with complex shapes or parts that do not heat uniformly.
(6) 4 hours plus 30 minutes for every 3 inches (76 mm) or increment of 3 inches (76 mm) greater than 8 inches (203 mm).
(7) 2 hours plus 20 minutes for every 3 inches (76 mm) or increment of 3 inches (76 mm) greater than 8 inches (203 mm).

▶ **Table 4.5**
Recommended
temperatures AMS
2759Table

Material Designation	Annealing (1) Temperature, °C	Normalizing Temperature, °C	Austenitizing Temperature, °C	Hardening Quenchant
1025	885	899	871	water, polymer
1035	871	899	843	oil, water, polymer
1045	857	899	829	oil, water, polymer
1095 (2)	816	843	802	oil, polymer
1137	788	899	843	oil, water, polymer
3140	816	899	816	oil, polymer
4037	843	899	843	oil, water, polymer
4130	843	899	857	oil, water, polymer
4135	843	899	857	oil, polymer
4140	843	899	843	oil, polymer
4150	829	871	829	oil, polymer
4330V	857	899	871	oil, polymer
4335V	843	899	871	oil, polymer
4340	843	899	816	oil, polymer
4640	843	899	829	oil, polymer
6150	843	899	871	oil, polymer
8630	843	899	857	oil, water, polymer
8735	843	899	843	oil, polymer
8740	843	899	843	oil, polymer
H-11 (3)	871	-	1010	air, oil, polymer
98BV40	843	871	843	oil, polymer
D6AC (4)	843	941	885 (5)	oil, polymer
9Ni-4Co-0.20C	(6)	899	829	oil, water, polymer (7)
9Ni-4Co-0.30C	(6)	927	843	oil, polymer (7)

Notes:
(1) Cool at a rate not to exceed 111 °C per hour to below 538 °C, except 4330V, 4335V, and 4340 to below 427 °C, and 4640 to below 399 °C.
(2) 1095 parts should be spheroidize annealed before hardening.
(3) H-11 parts shall be in the annealed condition prior to the initial austenitizing treatment, except hot handled parts shall be annealed at 885 °C and furnace cooled at 28 °C per hour maximum to at least 538 °C.
(4) D6AC parts shall be in the normalized or the normalized and tempered condition prior to the initial austenitizing treatment, except that parts only normalized without tempering shall be preheated prior to austenitizing.

4.4.5.3 API 6HT Heat Treatment and Testing of Carbon and Low Alloy Steel Large Cross Section and Critical Section Components

This specification provided details of the extra control needed for the heat treatment of critical or large cross sections. For example, section 6.3.1.3 of API 6HT states:

Furnace Loading Practices

"Allow sufficient space between parts when loading a furnace to ensure that all surfaces of the parts are evenly heated and to insure a good, even quench. Do not stack or bundle parts. Fixtures may be required. Parts should not be placed directly on the furnace hearth, use a metal tray or fixture that allows the furnace atmosphere to circulate around and under the part. The refractory on the furnace floor is a large heat sink that may result in uneven heating of the part. Support long parts as needed to prevent sagging during heat treating: sagging may occur especially during the austenitizing cycle. Consider using heat treating fixtures for parts with complex geometries to prevent distortion during heat treating".

4.4.5.4 ASTM A255 Determining the hardenability of steel

The specification provides detail on how to carry out the Jominy test. It also provides the algorithms for the calculation of DI and Jominy curves. The algorithms provided in ASTM A255 has been coded in MSL Jominy software to allow predictions to compare with actual test results.

4.5 KEY GUIDANCE FOR ACHIEVING MECHANICAL PROPERTY REQUIREMENTS AND UNDERSTANDING RULING SECTION/ HARDENABILITY LIMITATION

Significant progress had been made towards an understanding of the heat treatment of steel towards the end of the1930s. In the1950s work began on the quantitative relationships between microstructure and mechanical properties which led to several well-established relationships. It was identified that the

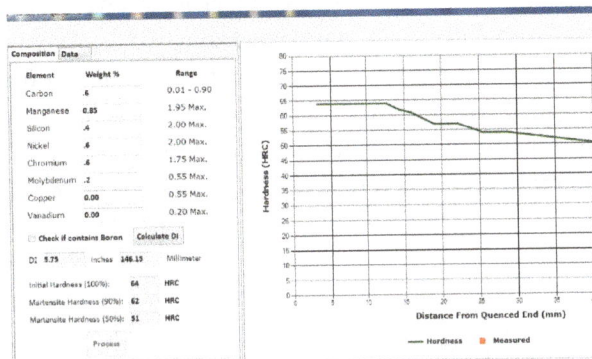

microstructure that typically gives the best combination of strength and toughness was a microstructure with the maximum amount of martensite probably consisting of 90% martensite and 10% lower bainite. Two important graphs shown in Figure 4.45 and Figure 4.46 can assist with understanding the importance of achieving the correct microstructure to produce products with the required mechanical properties in terms of minimum specified yield strength and sub-zero impact requirements. After a lifetime in steel technology I regard these as two key graphs.

- Figure 4.45 shows the effect of microstructure on the YS/UTS ratio.
- Figure 4.46 shows the effect of microstructure, grain size and strength level on the impact properties.

Figure 4.45 shows the relationship between the YS and UTS for various steels and relationship between the YS/UTS ratio and the microstructure. These relationships confirm that the YS/UTS is an important indicator of the microstructure of heat-treated steel. A 90% ratio between the yield strength and the UTS requires a microstructure of greater than 90% martensite. This microstructural requirement, which ensures a high YS, has been recognised in several national specifications:

- bolt steel standards BS EN ISO 898 requires approximately 90% martensite at the centre of the fastener in the as quenched condition

- casing steels to API 5CT where the standard specifies the requirement for 90% martensite in the as quenched product
- Spring steel specification ISO 683 requires at least 80% martensite at the centre of a spring after oil quench.

Figure 4.46 shows the effect of yield strength and microstructure on the impact transition temperature. This shows that to achieve low impact transition temperature in high yield strength steels requires a martensitic microstructure. The scatter band for the martensite curved lines also shows the effect of

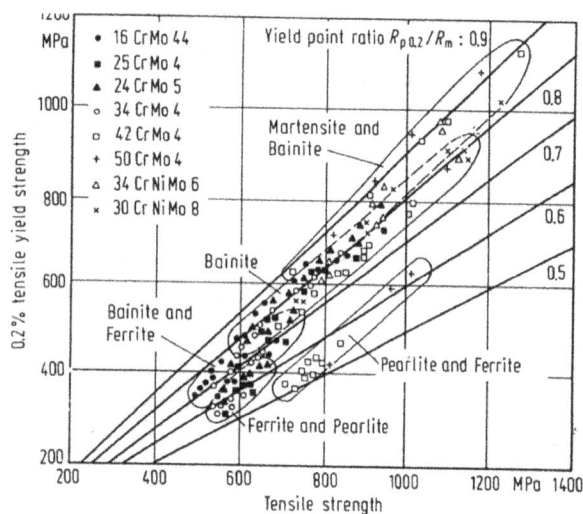

ASTM grain size. The top band has a coarse grain size of ASTM 3 and the lower line has a fine grain size of ASTM 7.

4.6 USING THE AS-QUENCHED HARDNESS AND CARBON CONTENT TO FIND THE PERCENTAGE MARTENSITE IN THE AS-QUENCHED MICROSTRUCTURE

During early studies in the USA regarding the effect of microstructure of steel on the mechanical properties the 50% martensite content was chosen to allow the determination of the "ideal diameter". This was due to a clear etching demarcation visible based on the "nital" etching response as shown in Figure 4.47.

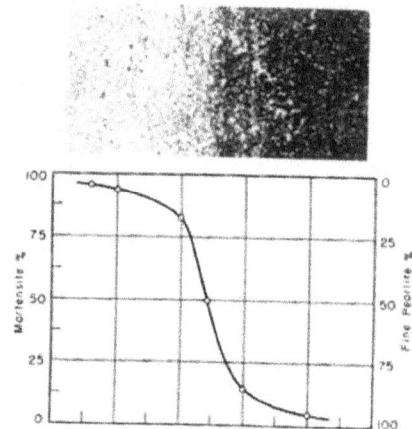

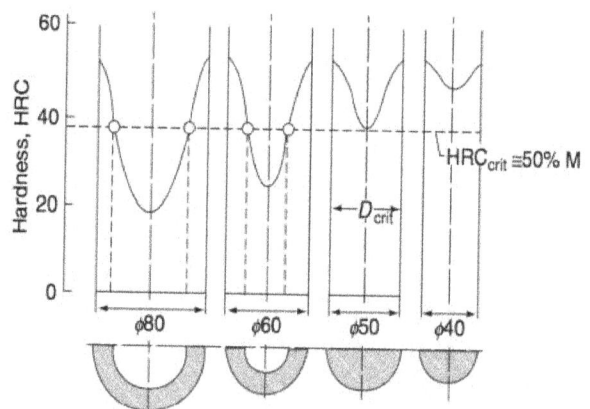

Figure 4.45 *The relationship between the YS and UTS for various steels and relationship between the YS/UTS ratio and the microstructure (From Rose, A. et al., Stahl u. Eisen, 91, 1001, 1971; Hengerer, F. et al., Stahl u. Eisen, 90, 1263, 1979)[28]*

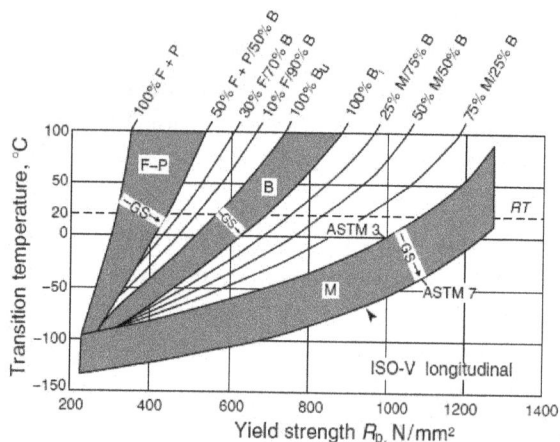

Figure 4.46 *The effect of yield strength and microstructure on the impact transition temperature. (From G. Spur and T. Stoͤferle (Eds.), Handbuch der Fertigungstechnik, Vol. 4/2, Warmebehandeln, Carl Hanser, Munich, 1987)[29]*

Figure 4.47 *LHS = A graph from the work of Grossman showing the etching response and a hardness traverse of a hardened component in the through thickness direction. The small photograph shows that when a microstructure has greater than 50% martensite it tends to etch light. When the microstructure has less than 50% martensite it etches dark. RHS = Grossman method (see Appendix 7).*

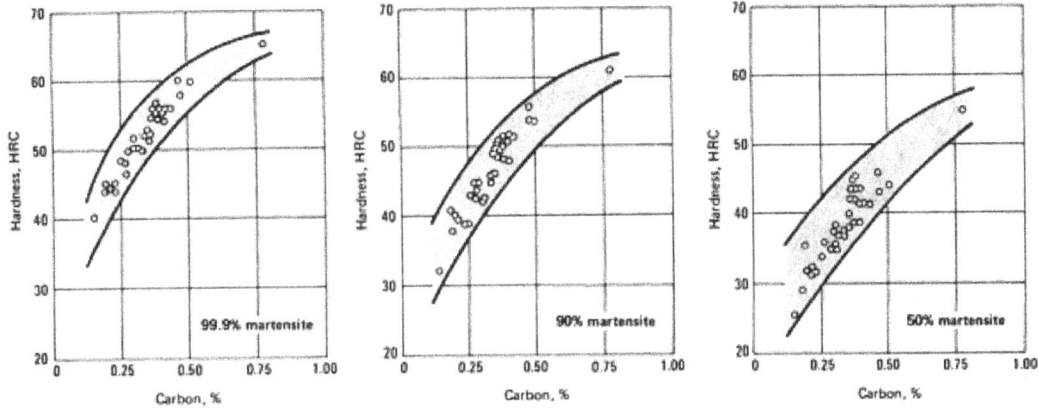

Figure 4.48 *The effects of carbon and martensite content on hardness*

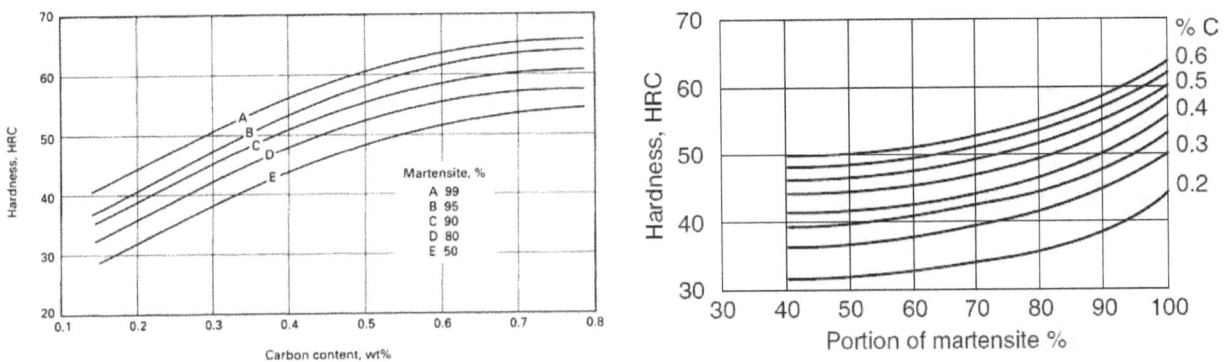

Figure 4.49 *This relationship was established by Hodge and Orehoski in 1946 and has been used in several publications in various formats. RHS = Relationship between as quenched hardness, carbon content and the percentage martensite in the microstructure*

The hardenability (DI) was defined as the maximum diameter which can be hardened with 50% martensite at the centre with an infinite quench. Obviously, this will not give the "best" microstructure for achieving the best combination of required strength and toughness and endurance properties. In general, the best properties are obtained with around 90% martensite and 10% lower bainite. However, with 50% at the centre a part may have the required level of martensite in the surface layers.

It was established in the 1940s that the maximum attainable hardness of any quenched steel depends mainly on the carbon content. Also, the maximum hardness values that can be obtained with small test specimens under the fastest cooling rates of water quenching were nearly always higher than those developed under production heat-treating conditions. This was because hardenability limitations in quenching larger sizes resulted in less than 100% martensite formation.

The effects of carbon and martensite content on hardness are shown in Figure 4.48 with the original scatter on the measurements.

SUMMARY

- To achieve the high ratio between the YS and UTS requires approximately 90% martensite in the microstructure. (See Figure 4.45)
- To achieve sub-zero impact toughness requires a high percentage of martensite in the microstructure. (See Figure 4.46)
- Hardness value can be used to indicate the "degree of hardening" and the percentage martensite in a quenched microstructure. (See Figure 4.49)

4.7 A BRIEF INTRODUCTION TO THE THEORY AND PRACTICE OF THE HEAT TREATMENT

To provide a simple understanding of the hardening of steel the following factors are important:

- Steel exists as BCC (body centered cubic) atomic structure at room temperature. The BCC can only accommodate a very small amount of carbon

- At about 910°C the atomic structure of pure iron changes to FCC (face centered cubic) which can accommodate up to 2% carbon
- The change from BCC to FCC is known as the allotropic change
- This change in atomic structure forms the basis of "transformation hardening"
- For example, the steel is heated to form the FCC atomic structure known as austenite. The carbon is taken into solution in austenite
- If the structure is cooled rapidly the carbon is trapped and the transformation occurs by a shear mechanism (no diffusion) and forms martensite which is hard
- The hardness of the martensite depends mainly on the carbon content. The graph used the most by worldwide organizations for predicting the martensite content based on the carbon content and hardness is shown in Figure 4.49
- If cooling occurs at a slower rate some carbon diffusion occurs to form a separate compound of iron carbide (Fe_3C) and the transformation creates bainite and at even slower cooling rate produces ferrite/pearlite.

4.8 THE DEVELOPMENT OF AN ED-DI-H-DC APPROACH FOR SELECTION AND THE DEVELOPMENT OF A STEEL PLANNING SYSTEM

ED – Equivalent Diameter. The conversion of the "part" size and shape to an equivalent round (ER is slightly different from ED) is recommended by API 6A. ED can be obtained by using the diagram attributed to DIN. (See Figure 4.50) (Despite years of searching I cannot find the original source. It is shown on the Edelstahl Witten-Krefeld brochure.)

DI – Ideal Diameter. This provides a quantitative value of the hardenability of the steel. This can be calculated from the steel chemistry using ASTM A255 Tables.

"H" – Grossman severity of quench. This approach to the severity of quench was developed in the 1940s. The quench severity of still water had a value of H = 1. The parameter it has linked with the heat transfer coefficient and the thermal diffusivity by the equation

$$H = \frac{h}{2\lambda}$$

Where

h = heat transfer coefficient

λ = thermal conductivity

The heat removal depends upon two factors:

h The heat transfer from the surface of the steel to the quenchant.

λ The rate of flow of heat through the steel part.

To obtain a standardized / normalized value for the hardenability of steel, Grossman measured the hardenability with reference to an infinite (or ideal) quench severity

$$h = \infty$$

This concept was based on the situation where the rate of heat removal depends **only** upon the thermal conductivity. It is assumed that the "h", the heat transfer coefficient, is a high value and therefore does not impede the heat removal. The use of the Grossman "H" value was then used to describe various quench media as shown in the table below (Table 4.6).

Table 4.6 Grossman Quench Intensity factors Figure 4.51 Average relationship between the hardenability (expressed as DI) in terms of 50, 95 and 99.9% martensitic microstructures.

Grossmann Quenching Intensity Factor H

Method of Quenching	H Value (in.$^{-1}$)		
	Oil	Water	Brine
No agitation	0.25–0.30	1.0	2.0
Mild agitation	0.30–0.35	1.0–1.1	2.0–2.2
Moderate agitation	0.35–0.40	1.2–1.3	
Good agitation	0.40–0.50	1.4–1.5	
Strong agitation	0.50–0.80	1.6–2.0	
Violent agitation	0.80–1.10	4.0	5.0

Source: Metals Handbook, 8th ed., Vol. 2. American Society for Metals, Cleveland, OH, 1964. p. 18.

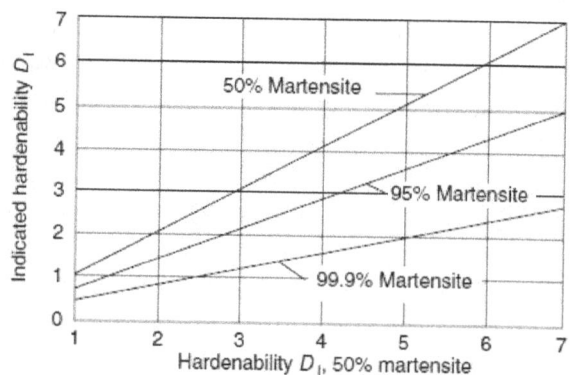

Name	Sketch of Product section	Equation for determining the appropriate heat treatment diameter
Round section	(bar) 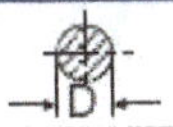	$d = D$
Square section	(bar) 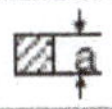	$d = 1.1 \cdot a$
Oblong section	(bar) 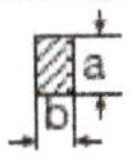	$d = 1.05 \cdot \sqrt{a \cdot b}$ $d = 1.5 \cdot b$
Disc		$d = \sqrt{h \cdot D}$ $d = 1.5 \cdot h$
Disc with hole	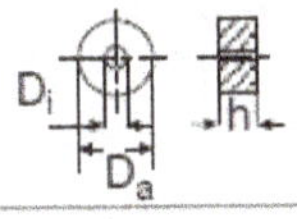	$d = \sqrt{h \cdot \frac{D_a - D_i}{2}}$ $d = 1.5 \cdot h$
Ring	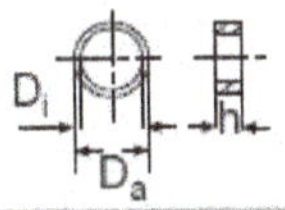	$d = 1.05 \cdot \sqrt{h \cdot \frac{D_a - D_i}{2}}$ $d = 1.5 \cdot \frac{D_a - D_i}{2}$
Tube	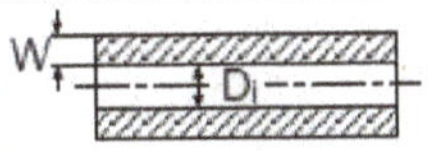	$D_i \leq 80$ mm $\quad d = 2 \cdot W$ $80 < D_i \leq 200$ mm $\quad d = 1.75 \cdot W$ $200 < D_i \quad d = 1.5 \cdot W$
one-end or double-end closed hollow body	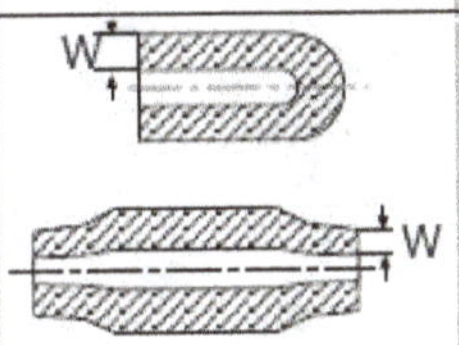	$d = 2.5 \cdot W$
End flange		$d = \sqrt{F_b^2 \cdot D^2}$
Shaft end		$d = F_d$
Centre flange		$d = \sqrt{\left(\frac{F_d - D}{4} + D\right)^2 + F_b^2}$
Schaft, roll		$d = F_d$
Triangle	(bar) 	$d = 1.03 \cdot S_w$
Dreieck	(bar) 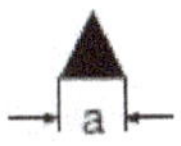	$d = 0.7 \cdot a$

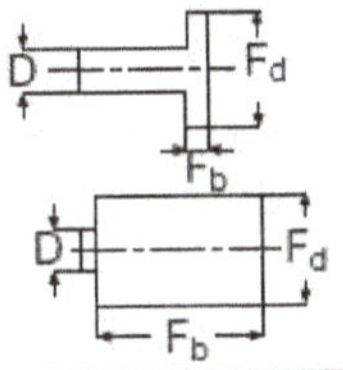

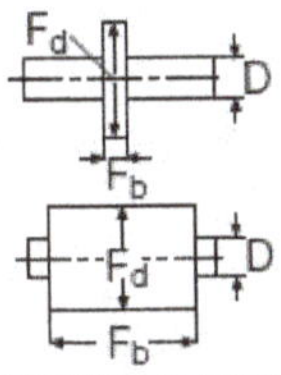

D = Diameter
D_i = Inner diameter
D_a = Outer diameter
a,b = edge length
W = wall thickness
S_w = Hexagon width
h = Height
F_d = Flange and shaft or roll diameter
F_b = Flange and shaft or roll width

Figure 4.50 *Calculation of equivalent diameter From EDELSTAHL WITTEN-KREFELD GMB, Thyrofort, Heat-treatable steels Brochure[30]*

The hardenability DI was defined as the maximum diameter which can be hardened with 50% martensite at the Centre with an infinite quench. In general, there was a view in the 1970s that the best microstructure in the as quenched condition had greater than 80% martensite with the remainder as lower bainite. Additional graphs were developed to convert the DI based on 50% martensite to different criteria (different levels of martensite (see Figure 3.46). Grossman determined the effect of the chemical composition on the hardenability. ASTM A255 provides the quantitative data needed to determine the DI. (Section 3.4.2.1)

In real quench conditions DI is converted to the actual expected diameter that will be quenched to 50% martensite at its center based upon the quench severity (Dc). This is done by using a graph with Dc on the "Y" axis and DI on the "X" axis with several lines for various H values (Grossman Severity of Quench Table 4.6)

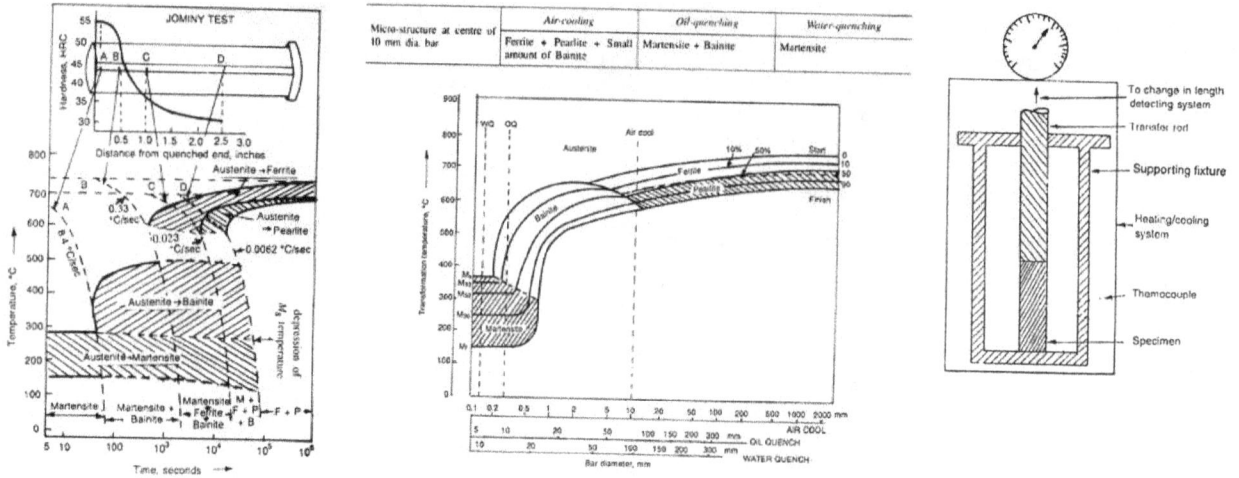

Figure 4.52 *Method used to determine CCT diagram LHS = Using Jominy data, Centre=Using 10mm bar, RHS = Using a Dilatometer*

Table 4.7 *Guidelines provided by a steelmaker for limiting ruling section. (From Corus Engineering Steels Untapped Energy Steels for oil and gas industries)*

Grade	Related Standards & Specifications	Minimum Tensile Properties				Hardness (Max) HB	Limiting Ruling Section
		UTS ksi (N/mm²)	0.2% Proof ksi (N/mm²)	Elongation %	Reduction of Area %		
Carbon							
LF2	ASTM A350	70	36	22	30	197	
X65 Flange Steel	API 5LX ASTM A694	77	65				6"
Alloy – Heat Treated							
4130	API 6A	100	80	16	35	235	2"
	NACE MR0175, ISO 15156	100	75	18	35	235	4"
	AISI 4130						
4140	API 6A	130	110	13	35	341	6"
	NACE MR0175, ISO 15156	120	100	14	30	302	7"
	AISI 4140	100	80	20	40	235	7"
4140 Mod	NACE MR0175, ISO 15156	130	110	13	35	341	8"
		100	80	20	40	235	8"
		100	70	20	40	235	11"
4145H Mod	API 7 Sections 4/5/6	140	120	13	40	341	6.7/8"
		135	110	13	40	341	12.1/4"
F22	ASTM A182 UNSK 21590 NACE MR0175, ISO15156	100	75	18	35	235	7"
9Cr1Mo	ASTM A199 ASTM A213 NACE MR0175, ISO15156	100	80	20	40	235	11"
8630 Mod	NACE MR0175, ISO15156	100	85	15	35	235	10"
		130	110	11	35	341	10"
4340	ASTM A434	135 (930)	105 (720)	14	35		7"
		130 (900)	100 (690)	14	35		9.1/2"
4330V	AMS 6427, ASTM A646	130	115	16	50		10"
	AMS 4330M	150	135	14	45		
	UNS K23080	165	150	12	40		
EN30B	BS 970 1955 835M30 – BS 970 1983	160	135	13	50		10"

Table 4.8 DI range for various low alloy steels

Steel	D_i range	Steel	D_i range	Steel	D_i range
1045	0.9–1.3	4135H	2.5–3.3	8625H	1.6–2.4
1090	1.2–1.6	4140H	3.1–4.7	8627H	1.7–2.7
1320H	1.4–2.5	4317H	1.7–2.4	8630H	2.1–2.8
1330H	1.9–2.7	4320H	1.8–2.6	9632H	2.2–2.9
1335H	2.0–2.8	4340H	4.6–6.0	8635H	2.4–3.4
1340H	2.3–3.2	X4620H	1.4–2.2	8637H	2.6–3.6
2330H	2.3–3.2	4620H	1.5–2.2	8640H	2.7–3.7
2345	2.5–3.2	4621H	1.9–2.6	8641H	2.7–3.7
2512H	1.5–2.5	4640H	2.6–3.4	8642H	2.8–3.9
2515H	1.8–2.9	4812H	1.7–2.7	8645H	3.1–4.1
2517H	2.0–3.0	4815H	1.8–2.8	8647H	3.0–4.1
3120H	1.5–2.3	4817H	2.2–2.9	8650H	3.3–4.5
3130H	2.0–2.8	4820H	2.2–3.2	8720H	1.8–2.4
3135H	2.2–3.1	5120H	1.2–1.9	8735H	2.7–3.6
3140H	2.6–3.4	5130H	2.1–2.9	8740H	2.7–3.7
3340	8.0–10.0	5132H	2.2–2.9	8742H	3.0–4.0
4032H	1.6–2.2	5135H	2.2–2.9	8745H	3.2–4.3
4037H	1.7–2.4	5140H	2.2–3.1	8747H	3.5–4.6
4042H	1.7–2.4	5145H	2.3–3.5	8750H	3.8–4.9
4042H	1.8–2.7	5150H	2.5–3.7	9260H	2.0–3.3
4047H	1.7–2.4	5152H	3.3–4.7	9261H	2.6–3.7
4053H	1.7–2.4	5160H	2.8–4.0	9262H	2.8–4.2
4063H	1.8–2.7	6150H	2.8–3.9	9437H	2.4–3.7
4068H	1.7–2.4	8617H	1.3–2.3	9440H	2.4–3.8
4130H	1.8–2.6	8620H	1.6–2.3	9442H	2.8–4.2
4132H	1.8–2.5	8622H	1.6–2.3	9445H	2.8–4.4

Reproduced from Grossmann, M. A.; Bain, E. C. *Principles of Heat Treatment*, 5th ed.; American Society for Metals: Metals Park, OH, USA, 1964. Krauss, G. Hardness and Hardenability. In *Steels: Heat Treatment and Processing Principles*; ASM International: Materials Park, OH, 1990; pp 145–178. ISBN: 0-87170-370-X (chapter 6).

as shown in Appendix 7. An additional important relationship is the link between % martensite, carbon level and hardness (see Figure 4.49). This allows a hardened steel to be hardness tested and the content of martensite in the microstructure to be estimated (assuming the carbon level is known). The relationship between DI, H and Dc is shown in Appendix 7

Ruling section. The application of risk reduction to the process of steel selection and processing requires substantial technical knowledge which is usually contained within appropriate national or international specifications. When a steel part has been heat treated for a given application it is often important to ensure that the specified mechanical properties are obtained throughout the section. This is expressed by the phrase "limiting ruling section" that describes the maximum diameter or cross-section of a bar or component in which the specified properties can be achieved by a given heat treatment. The Table 4.7 shows some guidelines provided by a steel supplier. (Corus/Tata)

4.8.2 Tempering

The martensite that forms in steel can be brittle especially for medium and high carbon steels. In addition,

there can be unwanted residual stresses in the part due to the quenching process and the thermal and transformation stresses. To achieve a more ductile microstructure the steel is reheated to a lower temperature keeping below the AC1 and avoiding the possible reformation of austenite.

Tempering is an important process to achieve the required balance of strength and toughness for the particular service conditions. The martensite is a non-equilibrium microstructure since the carbon has been trapped in the BCC crystal structure by a diffusion-less shear-type transformation. The conventional theory is that the trapped carbon atoms deform the BCC lattice by expanding one of the axes forming a body-centered tetragonal (BCT) crystal structure. These effects restrict dislocation movement and result in a high strength microstructure.

There are several microstructural changes that have been identified to occur during tempering. The diffusion of carbon to lattice defects, the precipitation of carbides, the transformation of retained austenite, and the recovery and recrystallization of the martensitic microstructure. The way in which the microstructure responds to tempering is controlled by the alloy content, the carbon level and the tempering temperature and time. The general trend is that at tempering temperatures above 150°C the microstructural changes

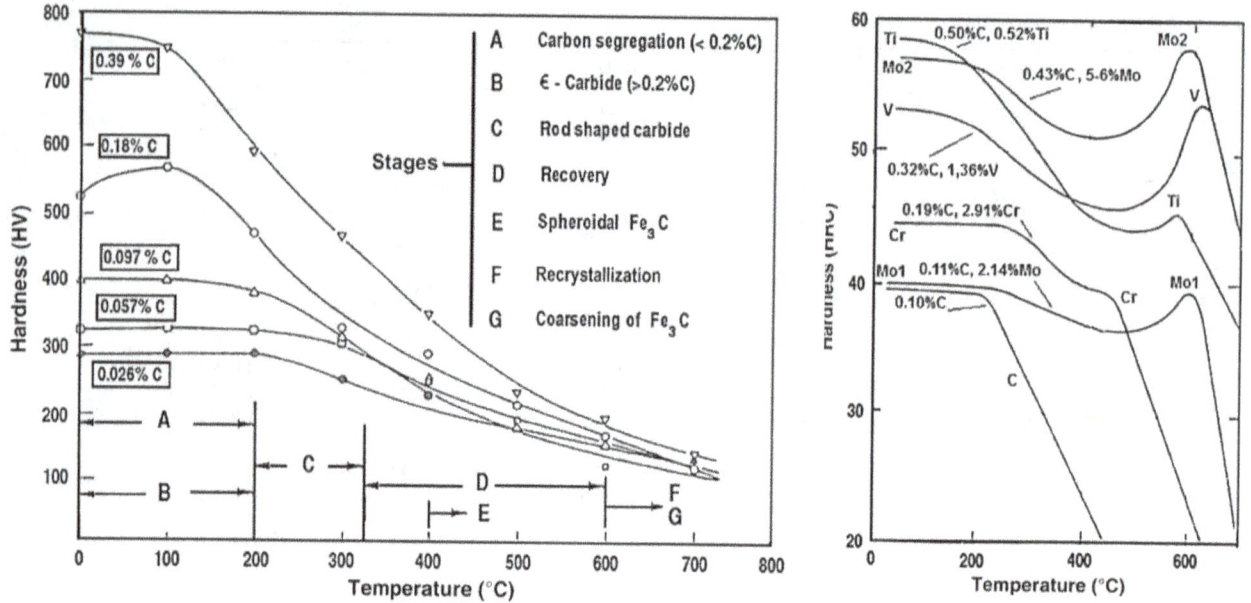

Figure 4.54 *Hardness as function of tempering temperature for plain carbon steels RHS = Tempering alloy steels. From Speich, G. R.; Leslie, W. C. Tempering of Steel. Metall. Trans. May 1972, 3 (5), 1043–1054*[31]

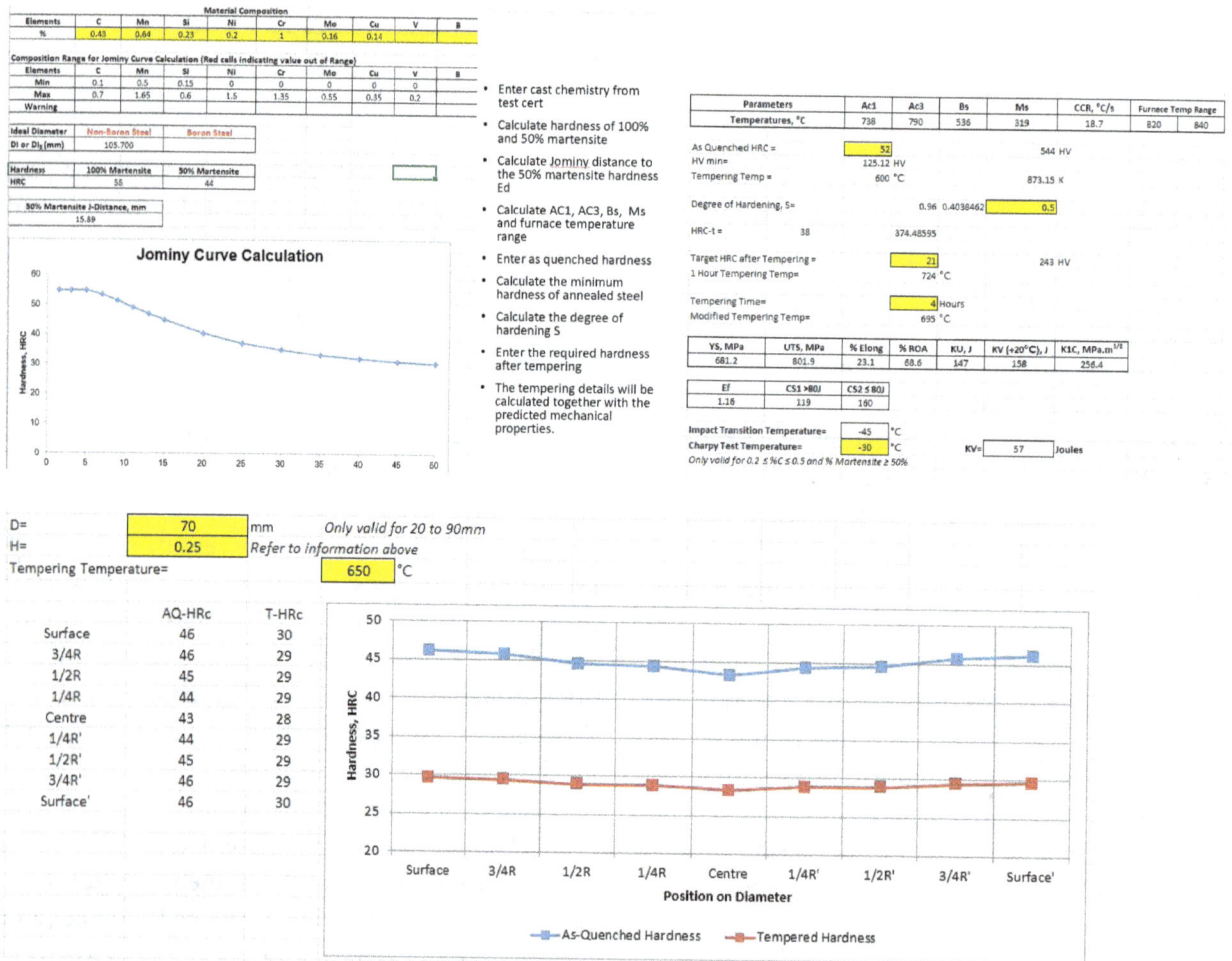

Figure 4.55 *Method to determine heat treatment parameters and to predict tempering temperature*

cause a lowering of the hardness (strength) and an increase in the ductility (toughness).

At temperatures above 250°C the carbon begins to precipitate carbides as ε – iron carbide (epsilon) $Fe_{2.4}C$ in the martensite matrix and the tetragonality of the BCT structure decreases. At a similar temperature if there was retained austenite in the microstructure it would change to bainitic ferrite and cementite (Fe_3C). Between 200°C and 350°C cementite forms on the ε – iron carbide and eventually the ε – carbide converts fully to cementite. Cementite also forms on grain and lath boundaries. At higher temperatures up to 700°C the cementite particles grow in size and form a microstructure of fine carbides in a ferritic matrix. At these high temperatures the martensite has to undergo recovery (reduction in the dislocation density) and potential grain growth. The grain growth can be prevented if the fine carbides formed effectively pin the grain boundaries. The tempering response of any bainite formed is lower than that of martensite. This is due to the much lower carbon still held in solution in the bainite. Figure 4.54 shows a summary of the changes during tempering from ASM Metals Handbook.

Sources of tempering data

1. WOLFSON HEAT TREATMENT CENTRE datasheets for the tempering of steel 1 to 26
2. Harry Chandler, Heat Treater's Guide: Practices and Procedures for Irons and Steels, ASM International, Second Edition 1995.
3. Lauralice C.F. Canale et al, A historical overview of steel tempering parameters, Int. J. Microstructure and Materials Properties, Vol. 3, Nos. 4/5, 2008 p 474, file:///C:/Users/user/Downloads/2008-96_Canale.pdf
4. B. Smoljan et al, Mathematical modelling of hardness of quenched and tempered steel, Archives of Materials Science and Engineering, Volume 74, Issue 2, August 2015, Pages 85-93
5. ASM metals Handbook Volume 4, Heat Treating

Figure 4.55 shows a spreadsheet method to determine heat treatment parameters and to predict tempering temperature and the mechanical properties.

4.8.3 Temper embrittlement – the embrittlement phenomena that occurs during tempering

Some steels suffer embrittlement during tempering in certain temperature ranges. This can be a problem when toughness requirements have to be met. Therefore, there has been considerable research and development aimed at minimizing the effect of "temper embrittlement" so that maximum Charpy impact toughness can be achieved. In addition, temper embrittlement can also occur when steels are used in service at the temperature where embrittlement occurs. The selection of steels that can resist this known type of embrittlement is a major consideration in the choice of steels for pressure vessels that operate at temperature in the range of 400 to 500°C.

There are two temperature ranges where temper embrittlement occurs

- temperature range around 300°C known as "Tempered martensite embrittlement" (TME)
- the range between 350 and 550°C (termed "temper embrittlement").

To avoid the TME (Tempered martensite embrittlement) tempering in the range 250 to 350°C should be avoided as far as possible. Temper embrittlement mainly affects steels with Mn, Cr, MnCr, CrV

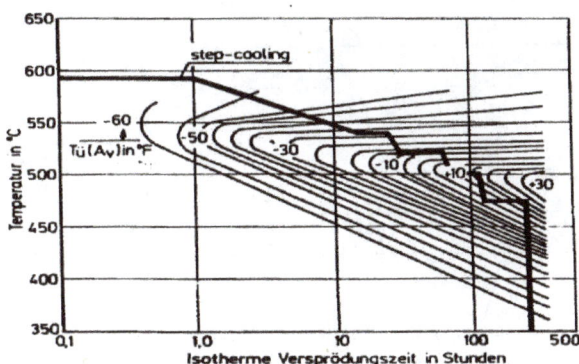

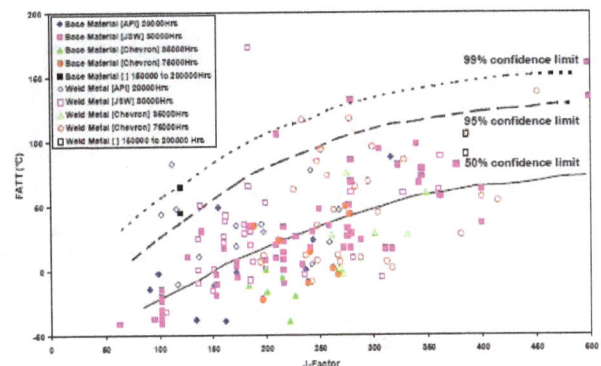

Figure 4.56 LHS = Step cooling curve, RHS = Effect of J factor on ITT (impact transition temperature)

and CrNi alloy additions when these steel types are cooled slowly after tempering above 600°C or when tempering is carried out between 350 and 550°C. The embrittlement can be reduced by low phosphorus content, using steels alloyed with molybdenum up to about 0.6 % by weight or by cooling rapidly after tempering above 600°C.

When temper embrittlement occurs, there is a decrease in toughness which is indicated by a shift of the impact transition-curve to higher temperatures. The degree of temper embrittlement can be measured by the measurement of the increase in the impact transition temperature. A standard step procedure is used (See Figure 4.56 LHS) and the difference in the ITT between step cooled specimens and naturally cooled specimens is measured.

Investigational work has been carried out to establish the mechanism of temper embrittlement. In principle, the factors that characterize this type of embrittlement for 2¼Cr1Mo steel are:

- TE impairs the toughness characteristics of steels which are held or slowly cooled down through a temperature range between 350°C and 600°C
- TE occurs in low alloy steels when tramp elements such as P, Sn, Sb or As are present.
- The intensity of TE caused by specific impurities is also influenced by other alloying elements such as Mn and Si
- The loss of toughness can be restored by reheating to temperatures above 600°C, followed by accelerated cooling
- The mode of fracture changes from trans-crystalline to inter-crystalline with increasing embrittlement with the fracture path following the former austenitic grain boundaries.

Based upon these observations various methods of assessing and avoiding potential embrittlement have been developed.

WATANABE developed the so-called J-Factor for 2¼Cr1Mo base material

$$J = (Mn+Si) * (P+Sn) * 10^4 \text{ (in \%)}$$

It is important to note that the J-Factor was originally developed for 2¼Cr1Mo base material but has been used for other steels such as CrMo with lower alloy content. In theory it should not be used for weld metal but the relationship has been used to modify the steel composition to avoid embrittlement

The X-Factor developed by BRUSCATO, for use with 2¼Cr1Mo weld metal is also often used for the base material

$$X = (10XP + 5XSb + 4XSn + As)/100 \text{ (elements in ppm) (See Figure 4.57)}$$

With increasing J-factor and X-factor the susceptibility to temper embrittlement shall also be increased. (Figure 4.56 RHS and 4.57) This is the reason why many engineering companies specify J-factors <100 and X-factors <15 ppm for 2¼Cr1Mo(V) steels.

When the tramp elements segregate to the grain boundaries in pressure vessel steels at service temperature then temper embrittlement will occur and cause reduced toughness after long-term service. It is not possible to simulate this behavior by a single mechanical test. The Step-cooling heat treatment cycle (SC) has been developed to accelerate the segregation of the tramp elements to the grain boundaries. Figure 4.56 LHS

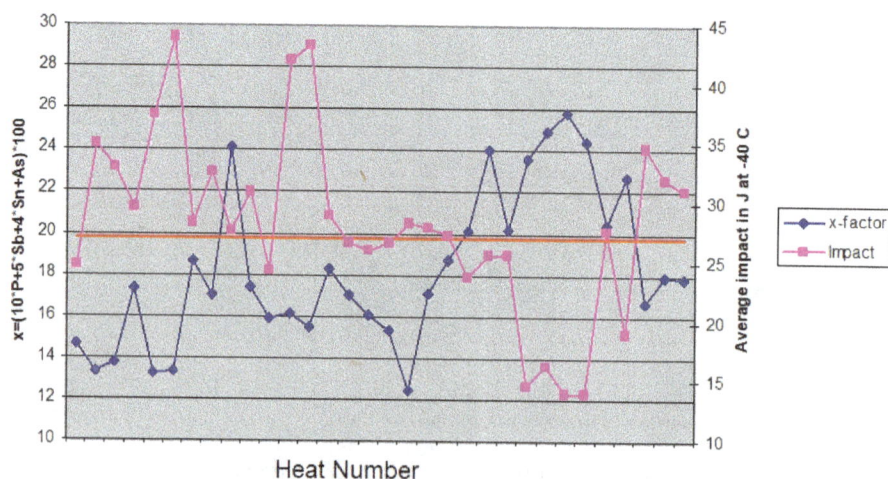

Figure 4.57 Bolt steel (SAE 4140) toughness at -40 Deg C vs X Factor. A distinct trend can be observed.

P together with other residual elements such as As, Sb and Sn all segregate to the grain boundaries by solid state diffusion During tempering these elements will also diffuse to the grain boundaries and cause embrittlement.

4.9 ERRORS AND PROBLEMS THAT CAN OCCUR DURING HEAT TREATMENT[36]

The thermal treatment of metals can be viewed as an abusive process which introduces the risk of potential cracking and distortion. Therefore, some heat treatment jobs are difficult. The main potential problems that can occur during heat treatment include:

- Lower than expected hardness after hardening (*possible cause low hardenability or slack quench*)
- The presence of variable hardness and soft spots (*low hardenability or microstructural segregation or surface decarb*)
- When heat treating in an open furnace there are problems with oxidation, decarburization and de-alloy which may or may not affect the quality of the product depending upon subsequent machining and metal removal and subsequent service conditions (*There is a need to select the correct heat treatment method with the use of controlled atmosphere furnaces for finish machined parts*)
- During the treatment of tool steels where high temperatures are required there is possibility of overheating and burning. With normal heat treatments the risk of burning is low and the greater risk would be during forging. However, overheat can result in grain coarsening and a contributing effect to quench cracking (*grain growth of the prior austenite grains*)
- Quench cracking particularly for complex geometries, variable cross sections and high carbon alloy steels (*Conventional recommendation do not water quench alloy steels with carbon above 0.3%; however, this rule is often not followed and water quench is used on several medium carbon alloy steels to maximize properties*)
- Distortion and warping occur with all thermal treatments. It can be minimized by confining the treatments to the "ferritic" based hardening mechanisms (nitriding) and avoiding the quench stresses associated with transformation and solution treatment hardening methods.

4.9.1 Reasons for low hardness

Possible reasons (excluding the presence of excess decarb) for low hardness and strength after quenching are:

- Incorrect steel composition for the size of part (*Hardenability, ruling section*)
- The use of a temperature too low for achieving the correct degree of austenitising (*Usually 30 to 50°C above the AC3*)
- Insufficient soaking time (*Wide range of guidance from 1hr/ins to 1min/mm*)
- Delayed quenching (*large transit time between furnace and quench*)
- Slower than required cooling rates due to quench too hot or lacking the necessary heat transfer coefficient (*Incorrect choice of quenchant, or inadequate cooling of quenchant or inadequate agitation*)
- Presence of large amount of retained austenite (*in high alloy steels with Mf near to or below ambient temperature*).

When low hardness occurs on steel components after surface hardening treatment such as carburizing, carbo-nitriding or nitriding, the following should be considered:

- the result of improper carburizing or nitriding atmosphere
- low heat treatment temperature or insufficient time at temperature
- the lack of case hardenability (for carburizing and carbo-nitriding)
- the presence of excessive retained austenite due to high carbon potential for the alloy content
- large volume of metal in furnace that has interfered with gas circulation or low power density with plasma treatments
- Incorrect prior heat treatment.

4.9.2 Soft spots

Sometimes, after hardening, the hardness on the surface of the component is not uniform. Hardened steels show varying hardness at different points on their surface. There are several reasons for occurrence of soft spots:

- Formation of a vapour blanket between the quenchant and component during quenching due to inadequate agitation which affects the speed of quench in certain locations

- Localized decarburization of steel during heat treatment or the presence of decarb from previous thermal treatments (forgings, castings)
- Inhomogeneity of microstructure (elemental segregation) causing different transformation rates at different locations
- Incorrect austenitising treatment to allow full alloy content to be taken into solution
- Slack quench due to high temperature of the quenchant or lack of quenchant agitation.

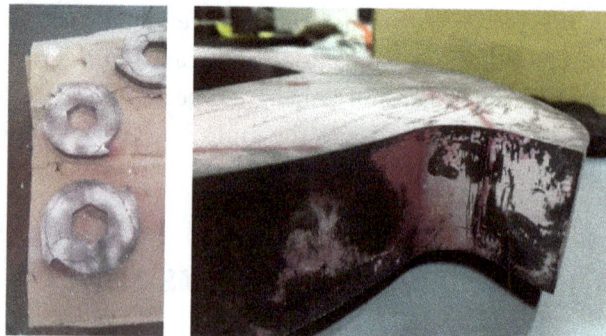

Figure 4.58 Quench cracks in SAE 4140. Cracks shown by dye-penetrant.

4.9.3 Oxidation and decarburization

When steel is heated to a high temperature in a furnace open to atmosphere, furnace gases such as oxygen, water vapour and carbon dioxide may react with the surface of the steel. This may give rise to both oxidation and decarburization.

Open furnace heat treatment is usually the lowest cost but results in oxidation and decarb. To prevent this type of imperfection there is a need to select the correct heat treatment for a particular part.

4.9.4 Overheating and burning

When steel is heated significantly above the upper critical temperature (AC3), coarsening of austenitic grains can occur. Hence, mechanical properties of steel could be adversely affected. If grain coarsening occurs, mechanical properties can be improved by a subsequent normalizing or annealing treatment. If, however, steel is heated to higher temperatures (i.e. near solidus), or it is held for a very long period, overheating and burning are likely to take place. This affects the properties of steel adversely. Consequently, there is loss of ductility and toughness. During overheating, sulphide inclusions can segregate along austenite grain boundaries and cause damage. However, fortunately the aspects of overheat are more often associated with forging rather than heat treatment.

4.9.5 Quench cracks

Quench cracks are formed as a result of stresses produced from the non-uniform transformation of austenite to martensite. The cracks may be small or large and are often visible after a steel part has been quenched. The transformation of austenite to martensite is accompanied by increase in volume. As a result, stresses are introduced and when these stresses are tensile and exceed the fracture stress of the steel, quench cracks are produced. Quench cracks are a serious problem from which there is no recovery and cracked parts are scrapped.

The task of determining the root cause of quench cracks formed during the heat treatment of engineering components involves a consideration of the metallographic evidence and the huge body of published information relating to the theory and practice of the formation of quench cracks. Heat treatment can be viewed as an abusive treatment of steel because of the thermal and transformation stresses that can be caused. If steel is not above a certain quality level it will crack during the quenching process. The quality parameters of interest would be:

1. The non-metallic inclusion content and their size, shape and orientation
2. The levels of micro-segregation and banding (*area of high chemistry= high hardness*)
3. The prior austenite grain size (*weakness at grain boundaries during the quench*)
4. The surface quality
5. The steel chemistry and the temperature of the martensite start temperature (Ms)
6. The presence of decarb.

Superimposed on the steel quality/toughness aspects are the levels of thermal and transformation stresses introduced during the quenching process. The magnitude of the stresses will be related to the heat transfer coefficient of the quenchant and the austentising temperature.

Dimensional changes

Steel has a coefficient of linear thermal expansion of 12×10^{-6} per °C. When a steel specimen is cooled from 850°C to 25°C, there is a contraction of about 0.99 percent. During the transformation of austenite to martensite, there is volume expansion. This expansion, expressed as percentage, is equal to $4.64 - (0.53 \times \%C)$. Approximately, the volume expansion is three times the linear expansion. Thus, in a 0.6% carbon steel, the linear expansion is about 1.4% as a result of transformation.

Quench sequence of events

When a steel component of an alloy steel is quenched from austenitising temperature, the surface of the component first comes into contact with the quenchant so that the temperature of the outer layer drops below the Ms temperature. Therefore, martensite forms initially preferentially at the outer surface or areas with the highest cooling rates.

The formation of martensite is accompanied by volume increase which can cause stresses to develop in the centre of the component. The yield strength of austenite is low at high temperature so that the stresses can cause plastic flow of the austenite and therefore a shape change.

As cooling progresses, the steel at the centre of the component cools below the Ms temperature. Martensite then forms at the centre. The expansion accompanying the newly formed martensite is restricted by the outer layers of martensite previously formed. The hardness and strength of this martensite is too high for the stresses to be relaxed by plastic deformation. Therefore, high internal tensile stresses can be developed which can exceed the tensile strength of the as-quenched martensite present at the outer surface of the component and can potentially initiate a quench crack.

Residual stresses arise due to temperature gradients and lack of simultaneous transformation throughout the cross-section of a component. If the cooling is slow, there will be only a small temperature gradient and transformation will occur more or less at the same time throughout the section leading to lower residual stresses. This is the established way to control the risk of quench cracking.

An alloy steel which has the required hardenability to give the required properties after an oil or polymer quench. In many cases it is important to remove the residual stresses formed during quenching. This can

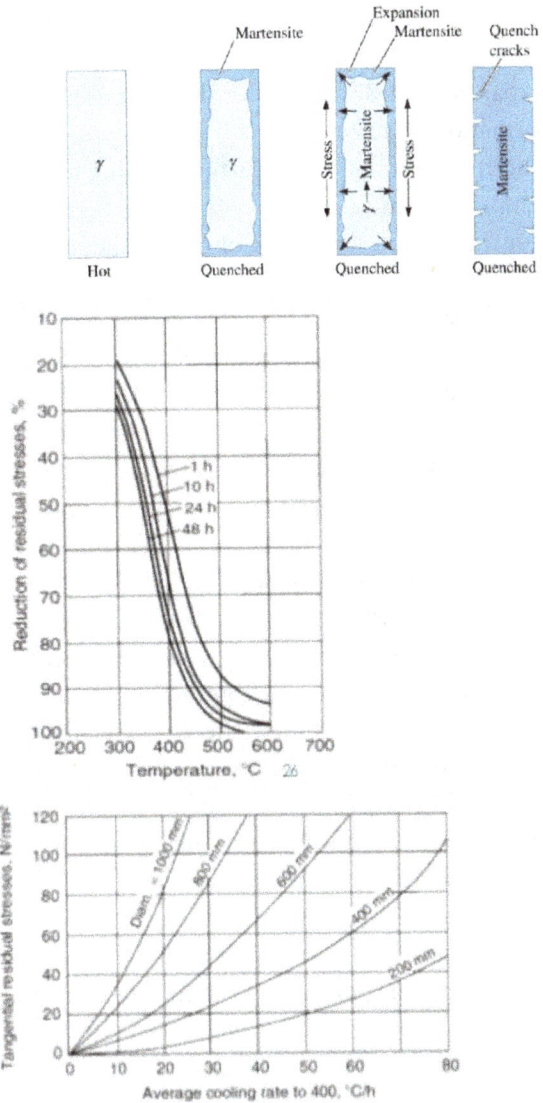

Figure 4.59 *Formation of quench cracks and temperatures needed to reduce residual stress*

be done by "tempering". At 250°C, residual stresses are reduced by about 20%. Higher temperature tempering can relieve all stresses.[33 and 34]

4.9.6 Distortion

Refers to changes in the size and shape of heat-treated parts due to thermal and transformation stresses.[35a to 35d]

Dimensional changes occurring in heat treatment include:

1. Uniform thermal expansion below the AC1 temperature
2. Contraction on transformation to austenite
3. Thermal expansion of austenite on further heating

4. Thermal contraction on cooling to the transformation temperature (Ms in hardening operations)
5. Expansion on diffusional or diffusion less transformation of austenite
6. Thermal contraction on further cooling to room temperature
7. Contraction on tempering of martensite.

These dimensional changes can give rise to distortion in steels. When steel is heated slowly to elevated temperature or cooled slowly, only small temperature gradients are developed. The temperature can be regarded as uniform throughout the part. Thermal and transformational changes then occur uniformly and simultaneously throughout the section. Hence, there is no risk of internal stresses being developed in the part. However, when steel parts are heated or cooled at a fast rate, internal stresses develop in the steel part because of differential expansion and contraction. Distortion occurs due to the combined effect of thermal and transformation stresses. The total distortion is often the result of (i) the presence of residual stresses in the part before heat treatment, or (ii) the introduction of internal stresses during heat treatment.

Distortion is one of the most difficult problems associated with heat treatment. It occurs in steel during hardening and tempering. There are two main types of distortion:

• size distortion which refers to changes in volume, and
• shape distortion which relates to changes in geometrical form of the steel part.

Size distortion occurs because of expansion or contraction in steel component, while shape distortion is manifested by changes in curvature such as twisting (warpage) or bending in steel components. The fishbone diagram in Figure 4.76 shows a summary of the many factors that can affect distortion.

4.9.7 Control of Distortion

The risk of distortion during and after heat treatment can be minimized by taking care of the following aspects: design, composition, initial condition, and machining procedures.

• Design aspects – Avoid abrupt changes in the cross section and sharp comers and thin walls

• Composition – Size distortion can be minimized by the suitable steel selection (Hardenability control)
• Initial condition. The presence of residual stresses due to previous machining or forging operation enhances the tendency for distortion. Therefore, these fabrication stresses should be relieved by subcritical annealing or normalizing operation. This is more important for intricate parts with close dimensional tolerances. The steel should have a uniform microstructure possibly similar to the final microstructure in the finished part
• Machining procedure. Follow established procedures for tool steels and for nitriding the sequence should be rough machine, stress relief followed by final machining.

Methods to Reduce Distortion during Heat Treatment.

Size and shape distortion cannot be eliminated during heat treatment. But size distortion can be minimized by proper selection of composition of steel and by adjusting the machining allowance and controlling the microstructure.

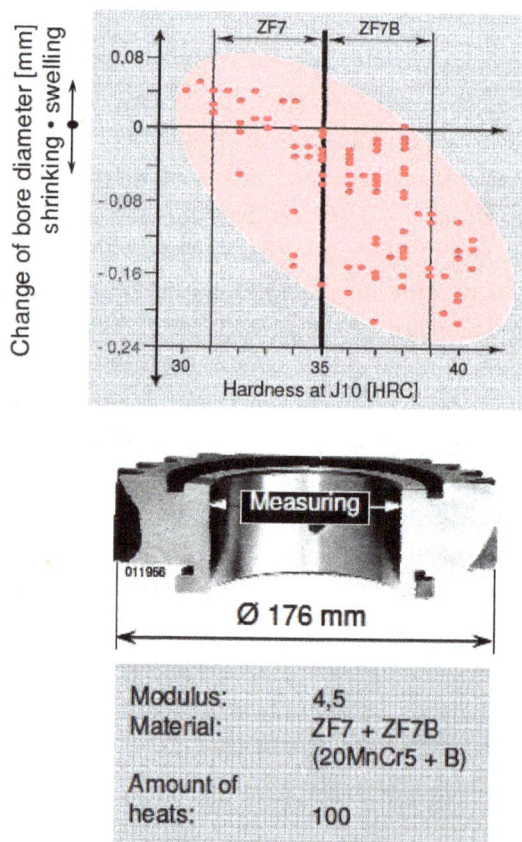

Figure 4.60 *Bore closure and the J10 hardness*[20]

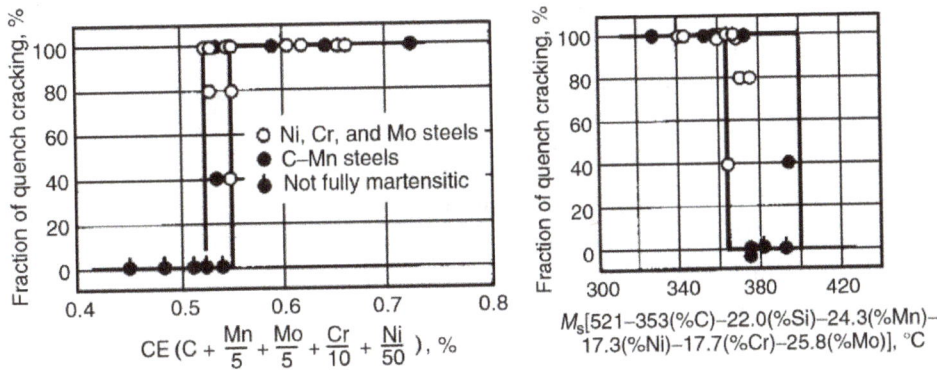

Figure 4.61 *Effect of MS temperature and carbon equivalent on the quench cracking of selected steel[35a]*

$$CE\ \left(C + \frac{Mn}{5} + \frac{Mo}{5} + \frac{Cr}{10} + \frac{Ni}{50}\ \right),\ \%$$

$$M_s[521 - 353(\%C) - 22.0(\%Si) - 24.3(\%Mn) - 17.3(\%Ni) - 17.7(\%Cr) - 25.8(\%Mo)],\ °C$$

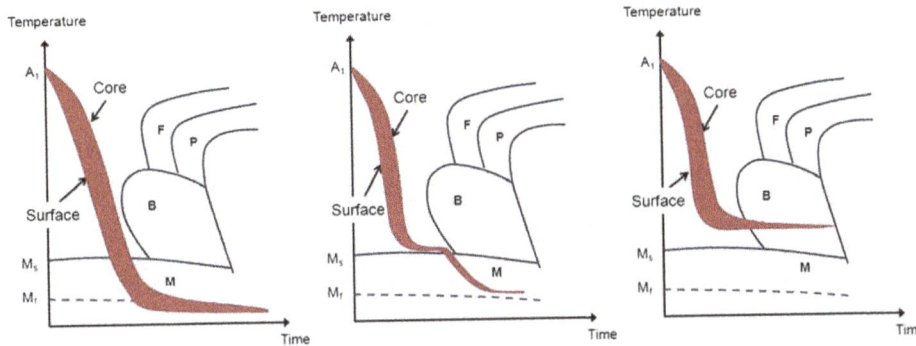

Figure 4.62 *Direct quench, martemper and austemper*

Shape distortion is also inevitable. To minimize shape distortion as well as aspects of part design, a number of factors need to be controlled. These are (i) stress relieving, (ii) heating rate, (iii) possible preheating, (iv) quenching media, (v) press quenching, and (vi) trays, fixtures and supports.

- The Heating Rate of a component should be controlled and slow with possible holds below and above the sub-critical temperatures so that the transformation occurs uniformly across the part.
- Preheating reduces shape distortion by reducing the thermal stresses produced by temperature gradients and thermal stress. Preheating can be performed between 400°C and 800°C. For a large cross-section and complicated shape parts and for high alloy steels having poor thermal conductivity, it is preferable to carry out two-stage preheating.
- Quenching Media. The basic aim in hardening of steels is to produce a martensitic structure and avoid formation of pearlite and bainite. Therefore, the rate of cooling of heated component should be fast enough for the hardenability of the steel. The ideal situation for distortion control would be to have sufficient hardenability to allow the use of marquench or vacuum quench technology.
- Press-Quenching. Press quenching or die quenching are used for precision gears. During press quenching, distortion is minimized by physical restraint of a part during quenching.

- Trays, Fixtures and Support for Complex shaped components and unsymmetrical sizes need holding trays and fixtures which maintain the shape of components during heating and rapid cooling from austenitising temperature. Trays, fixtures and supports are also used to reduce the sagging problems on heating the components to the austenitising temperature reducing the yield strength.
- Treatments for Stabilizing Dimensions Dimensional stability may be achieved by single tempering, multiple tempering, or sub-zero treatment.

4.9.8 Controlling the cooling process during the quenching of steel parts

The selection and control of quenchants is an important factor in the successful application of the quench and tempered process to steel components. Unfortunately, modern quench technology has become a complex specialist topic that requires a knowledge of the metallurgy of heat treatment and heat flow.[36 and 37]

The mechanical properties achieved by the quench and tempering process are dependent on achieving the correct heating times and temperatures (correct austenitising conditions) and the correct cooling rates throughout the parts. The first condition requires a suitable furnace with a qualified work zone (by

TUS – temperature uniformity survey) and the second statement needs the quenchant to perform in a predictable and consistent manner (correct heat transfer performance for the size of part and steel hardenability).

The use of an incorrect quenching practice can result in the potential build-up of residual stress that can cause cracking and distortion. This is often related to the non-uniformity of heat transfer due to poor circulation patterns of the quenchant around the part which can generate large temperature gradients within the part. Therefore, the control of circulation of quenchant, by the use of sufficient agitation, to assist with the uniformity of heat extraction has to be a major requirement.[38 and 39] To reduce the formation of undesirable thermal gradients and poor microstructures formed during quenching, a high level of agitation should be used to give high flow conditions to remove any stable vapour blankets. The tank agitation can be provided by various methods including a recirculation pump, a submerged spray, an impeller stirrer, an ultrasonic generator[38], and actual movement of the part itself, and submersible agitators. Among these agitation methods, the most common and cost-effective one is an impeller mixer.[40 and 41]

There are several publications that review the use of CFD software to optimise quench tank agitation and the development of methods to measure the tank flow. Some videos of submersible quench tank agitator can be viewed at the following site. http://www.dhanaprakash.com/quenchant-tank-agitation.html

In addition, it is essential to monitor and control the factors that cause the potential deterioration of the performance of a quenchant such as temperature change, oxidation damage to oil and water in oil[42, 43 and 44] and variation in concentration for polymers due to drag out and uncontrolled water make-up, and the build-up of oil, scale and damaging contaminants.[45]

Modern quenching technology makes use of several different quenchants. Quenchants range from synthetic to vegetable oils, different polymer-solutions and other water-based solutions such as brine and caustic, molten salt-baths, fluidized beds, gas backfill of nitrogen, argon or helium in vacuum furnaces and high velocity turbine driven gases. Any of these quenchants may be used in a range of different process conditions such as temperatures and agitation rates, thus creating a vast number of possible combinations that need to be optimised for a particular process route. Figure 4.64 shows the range of quench intensity factors achieved by the various quenchants.

There are also different quenching techniques: direct immersion quenching; intensive quenching; interrupted quenching; delayed quenching; press quenching; martempering; austempering; spray quenching. There are a range of furnaces from sealed quench furnace that use oil quench and similar style furnaces used for austempering that use molten salt quench. Open gas fired furnaces for bulk heat treatment that use chargers to transfer from the furnace to the quench tank, many of which now operate with polymer quenchants to reduce the fire risk and environmental fume and cleaning problems. Continuous furnace such as shaker hearth, mesh belt and cast link furnace.[46]

Some of the important aspects relating to the quenching of steel are:

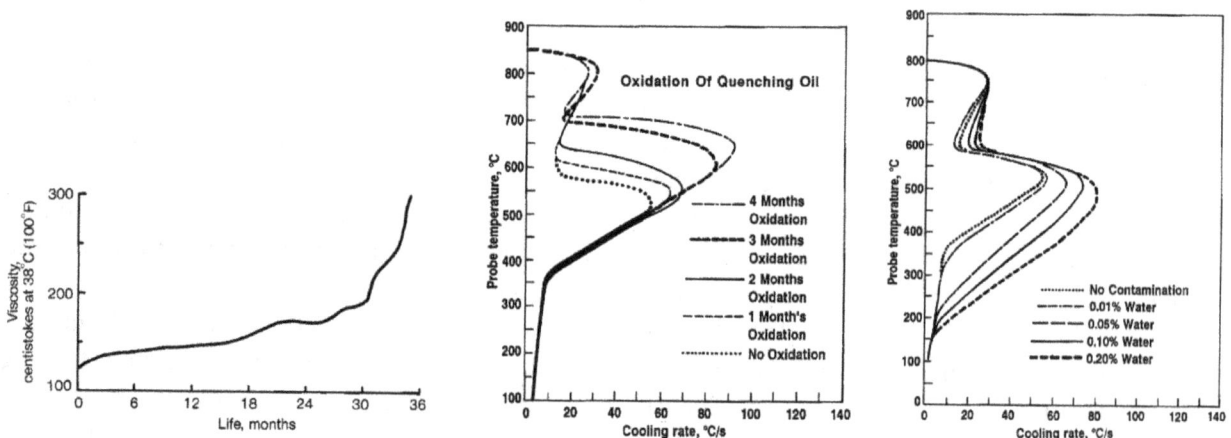

Figure 4.63 LHS = Change in viscosity of a martemper oil Centre= Oil oxidation and effect on cooling rate, RHS = Effect of water content on cooling rate[42]

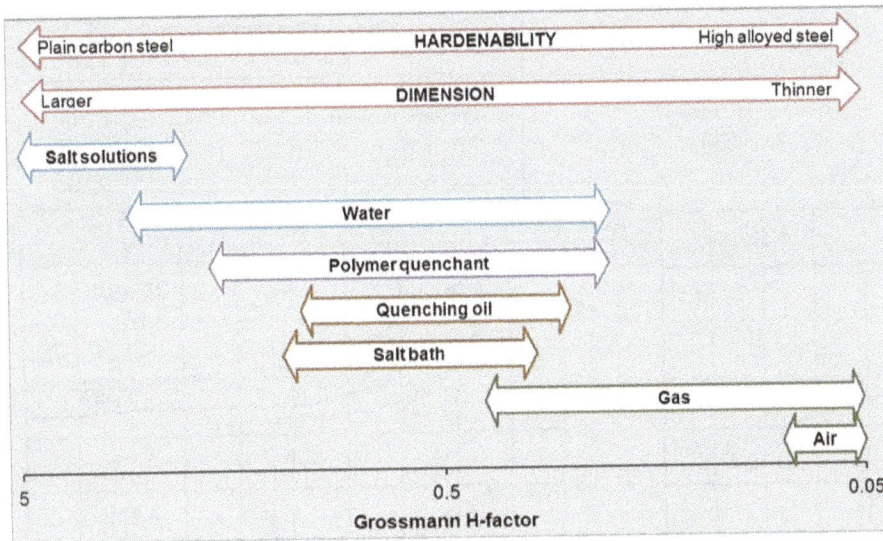

Figure 4.64 *The range of Grossman H values of different quenchants*[37]

- achieving the correct austenitising conditions
- the use of a suitable quenchant which can give the appropriate quench rate
- the use of a quenchant that can be controlled and maintained to give consistent quench rates throughout the life of the quenchant (temperature, water contamination for oil, for polymers the percentage of polymer, oil contamination and bacterial growth. Control requires sampling and testing to specifications such as ASTM A6666, BS 9950 and an understanding of cooling curve analysis)
- to have the appropriate environmental performance (control of fume, fire risk, cleaning requirements after quenching and effluent disposal).

Austenitising

There is a need to understand how to austenitise steel correctly. The starting point to determine the heating temperature and time would be AMS2759[47] and the ASM Heat Treater's Guide.[48] However, it is also important to understand that the prior condition of the steel can affect the austenitising of alloy steels. This is shown Figure 4.65 as Jominy data for various austenitising times and prior steel condition, S= spherodized A= Annealed, HR= Hot rolled, N= Normalised, Q= oil quenched.[49]

In addition, TTA diagrams can give guidance for austenitising conditions. Figure 4.66 is a TTA AISI4140 TTA diagram (time-temperature-austenitizing) for AISI 4140 steel simulated by JmatPro.[50] Steel treated with the conditions of time and temperature shown for the homogeneous part of the diagram

will have higher hardenability compared to inhomogeneous steel.

The cooling process

Steel parts are heated in a furnace to above the AC3 (generally 30 to 50°C above the AC3) held at temperature for an appropriate duration and then cooled in a quenchant suitable for the hardenability of the steel and the size of the parts. If the cooling rate is sufficiently high, the diffusionless phase transformation will occur and the steel will have a high percentage of martensite in the microstructure and the formation of the softer ferrite, pearlite or bainite phases will be avoided. The main objective is to achieve the maximum percentage of martensite after quenching, which after suitable tempering can deliver the best combination of strength, toughness and endurance.

The manufacturers of the quenchants have always provided excellent technical information on quenchants and one of the leading companies was Houghton who hold many patents.[51 to 57]

The operation of quench systems requires a knowledge and understanding of the analysis of cooling curves to allow the measurement of the heat transfer between the hot steel and the quenchant, and how to apply the knowledge of the heat transfer to the hardening process. This allows the characterization of a quenchant to allow appropriate selection and comparison between quenchants and to allow and process control measures to ensure consistent quench performance.[58]

To see how the technology has changed Figure 4.67 is of interest from 1951 where observations were made of the difference between the cooling of the surface and the cooling of the centre for different quenchants.

Jominy Curves for Various Prior Microstructures

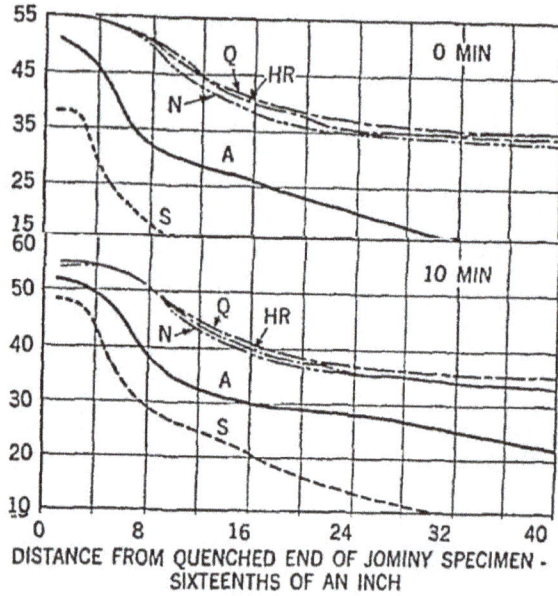

Curve	Prior Structure		Brinell Hardness
	Condition		
S	spheroidized..		152
A	annealed (lamellar + spheroidized carbides)..............		170
HR	hot rolled..		331
N	normalized from 1650 F............................		277
Q	oil quenched from 1550 F..........................		477*

* converted from Rockwell C

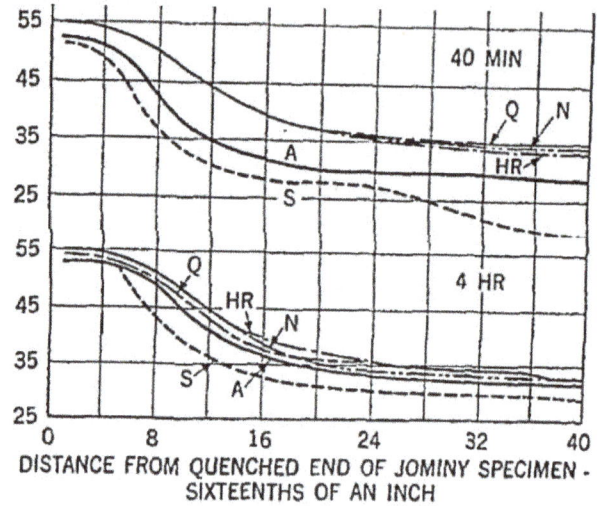

Figure 4.65 Jominy data for various austenitising times and prior steel condition, S= spheroidized A= Annealed, HR= Hot rolled, N= Normalised, Q= oil quenched[49]

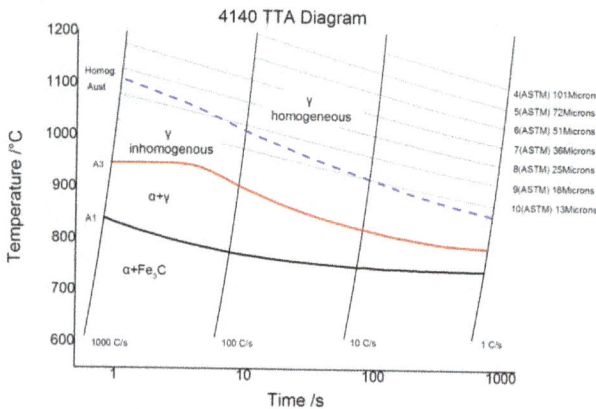

Figure 4.66 TTA diagrams for AISI 4140 steel simulated by JmatPro.[50]

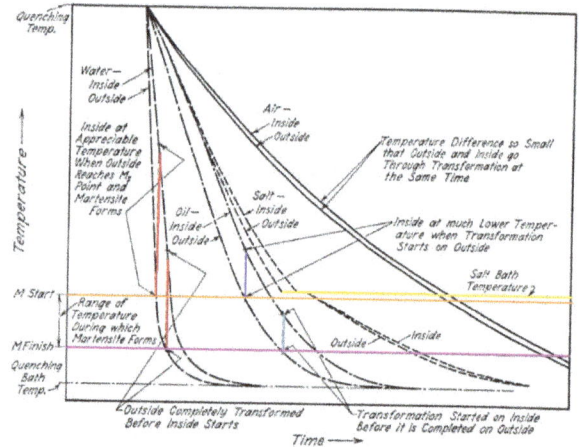

Figure 4.67 Surface and centre cooling curves of steel cylinders quenched in water, oil, hot salt bath and air[59]

During immersion quenching, several stages of heat transfer from the specimen occur. As shown in Figure 4.68 and the YouTube videos, when a hot component comes in contact with the liquid quenchant, there are normally 3 stages of quenching which are also labelled on the cooling rate diagram.

• Vapour Blanket Stage. The vapour stage happens when the hot surface of the heated part comes in contact with the liquid quenchant. The part is surrounded with a stable blanket of vapour. Heat

transfer is slow, and occurs primarily by radiation through the vapour blanket. Some conduction also occurs through the vapour phase. This removal of the blanket can be achieved by strong agitation or additives. The vapour stage can be responsible for surface soft spots. If this stage is allowed to persist other transformation products other than martensite will form.

• Nucleate Boiling Stage. The second stage encountered in quenching is the boiling stage. This is where the vapour stage starts to collapse and all

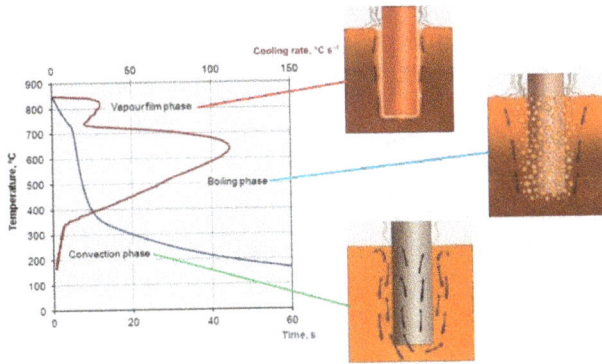

Figure 4.68 Cooling curve converted into a cooling rate diagram. The cooling rate diagram shows the 3 phases of cooling (see https://www.youtube.com/watch?v=Hfi9kpjHBlE for polymer quench) (see https://www.youtube.com/watch?v=bMN1QUVUBkw for water quench)

liquid in contact with the component surface erupts into boiling bubbles. This is the fastest stage of quenching. The high heat extraction rates are due to carrying away heat from the hot surface and transferring it further into the liquid quenchant, which allows cooled liquid to replace it at the surface. In many quenchants, additives have been added to enhance the maximum cooling rates obtained by a given fluid. The boiling stage stops when the temperature of the component's surface reaches a temperature below the boiling point of the liquid. For many distortion prone components, high boiling temperature oils or liquid salts are used if the media is fast enough to harden the steel, but both of these quenchants see relatively little use in induction hardening.

- Convection Stage. The final stage of quenching is the convection stage. This occurs when the component has reached a point below that of the quenchant's boiling temperature. Heat is removed by convection and is controlled by the quenchant's specific heat and thermal conductivity, and the temperature differential between the component's temperature and that of the quenchant. The convection stage is usually the slowest of the 3 stages. Typically, it is this stage where most distortion occurs.

The quenchant is often overlooked and neglected and given a low priority until the occurrence of failed mechanical property requirements, cracking and the distortion of parts. Under these circumstances the financial loss can be high and the acquisition of knowledge about quenchants then becomes part of problem-solving procedure. Fortunately, the tolerance for this type of situation should now be an event of the past since automotive and aerospace quality assurance requirements cover control of the quench process.[60]

The action of a polymer quench in forming a polymer-rich film during quenching depends on the structure of the polymer. For polymers such as PVP exhibiting normal solubility in water, localized evaporation occurs adjacent to the hot metal surface, producing a polymer concentration gradient which increases towards the liquid-gas interface. Then, the blanket vapour suddenly ruptures resulting in the nucleate boiling process. The polymer film formation and subsequently rupture process may occur repetitively depending on the type of polymer and the bath conditions. When the surface temperature falls, the cooling becomes slow and the heat transfer of convection is done by simple conduction through the film and the solution. The following standards for determining cooling characteristics of quenching media:

- **ISO 9950:1995**, "Industrial quenching oils – Determination of cooling characteristics – Nickel-alloy probe test method"
- **ASTM D 6200-01**, "Standard Test Method for Determination of Cooling Characteristics of Quenching Oils by Cooling Curve Analysis"
- **ASTM D 6482-06**, "Standard Test Method for Determination of Cooling Characteristics of Aqueous Polymer Quenchants by Cooling Curve Analysis with Agitation (Tensi Method)"

Hardening Power (HP) Equations

Segerberg has developed a test using ISO 9950 with an Inconel 600 cylindrical probe to quantify the quench severity of polymer and oil quenchant media. The quantitative empirical formulas of HP take into

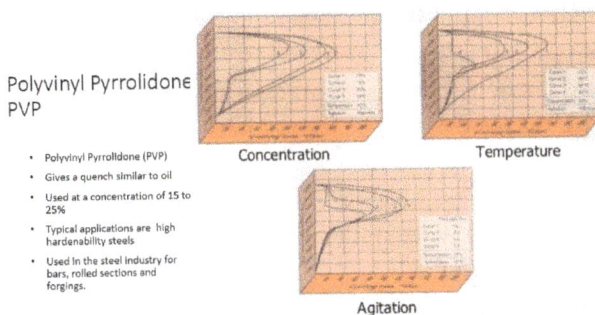

Polyvinyl Pyrrolidone PVP

- Polyvinyl Pyrrolidone (PVP)
- Gives a quench similar to oil
- Used at a concentration of 15 to 25%
- Typical applications are high hardenability steels
- Used in the steel industry for bars, rolled sections and forgings.

Concentration   Temperature

Agitation

Figure 4.69 Polyvinyl Pyrrolidone (PVP)

▶ *Figure 4.70* Cooling
parameters from cooling
and cooling rate curves of
polymer quenchant

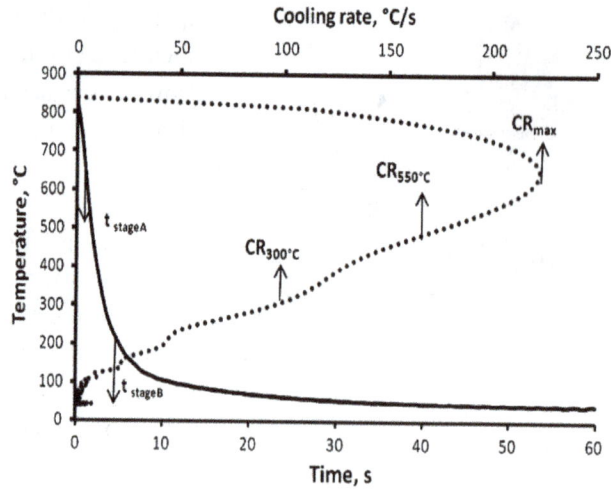

▶ *Figure 4.71* Comparison
of slow, medium and fast
quench oils

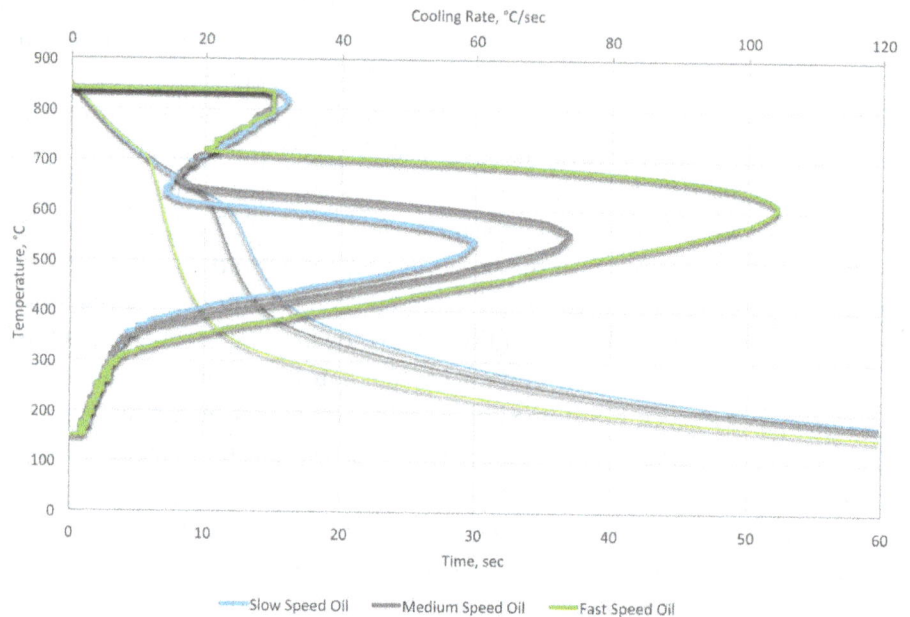

account the characteristic parameters, CRF and CRM, determined from the cooling curve, as illustrated in Figure 4.70.

The HP empirical formula, HP polymer, used for agitated polymer quenchant is:

$$HP\ polymer = 3{:}54\ CRF + 12{:}3CRM - 168$$

Where CRF is the cooling rate at 550°C (550°C is the temperature for the ferrite/pearlite nose in the CCT diagram); CRM is cooling rate at 330°C (330°C is the temperature at the beginning of the phase transformation austenite to martensite).

Hardenability aspects

Hardenability is the composition-dependent property of a steel that describes its ability to harden by martensite formation and is related to parameters such as austenitising temperature, cooling rates after austenitising, and part size and geometry. Alloying elements in general terms shift the CCT diagrams towards right, but the effect of an element on pearlitic and bainitic region is not normally the same. A cooling rate which may be sufficient to suppress the pearlitic reaction may not be large enough to prevent bainite formation. In such a case, the critical cooling rate for the 100% martensite formation is governed by the bainite reaction, and the non-martensitic product shall be bainite.

This results in two types of hardenability:

- Pearlitic hardenability
- Bainitic hardenability.

Pearlitic hardenability is based upon where the remaining 50% non-martensitic product is ferrite-pearlite, whereas, the bainitic hardenability has bainite as the non-martensitic product. A clear distinction is very important as this 50% non-martensitic product can have a significant influence on the mechanical properties such as toughness shown in Figure 4.72 The presence of bainite or ferrite pearlite is not easily distinguishable in the Jominy test, but the relative positions of the bainite start line and ferrite pearlite start line can be seen on the CCT diagram.

The hardenability can be increased by manganese, chromium, molybdenum and very small additions of boron (0.001 – 0.002%). Nickel and silicon have a smaller effect (see Figure 4.73 LHS). The hardenability is specially increased, when several alloying elements are added to steel (instead of one in a large amount).

Alloying elements affect the hardenability in a complex manner. Alloying elements affect the kinetics of austenite decomposition in a different and complex manner. Elements like Cr and Mo make bainite as the main non-martensitic product.

Also, the effect of all the alloying elements is not separable in that, for the same amount of addition of an alloying element, the effect on the hardenability would, in general, depend upon the presence, or the absence of certain other alloying elements in the steel. In spite of these problems, some correlation between effects of alloying elements on hardenability has been found based on average effects derived from a large amount of empirical data.

Thus, these elements should be used effectively in design of parts for heat treatment. The amount of carbon should be sufficient to give the required martensitic hardness. Carbon has an effect on the hardenability as shown in Figure 4.73 (RHS), but higher levels decreases the toughness, and can increase the risk of distortion and quench cracking and also weld cracking. Manganese increases the hardenability (see

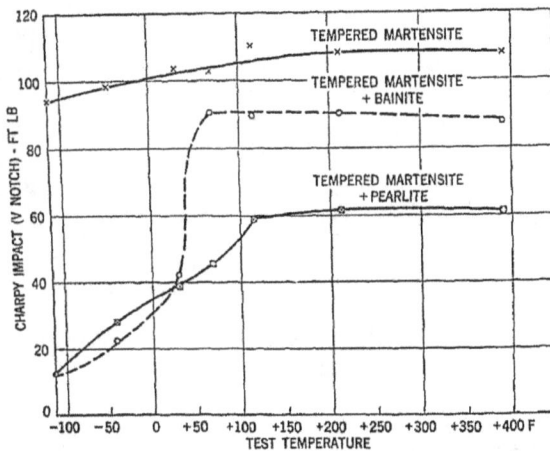

FIG 85 – Effect of Microstructure on the Low Temperature Impact Properties of a Chromium-Molybdenum-Nickel Steel (8735) at a Tensile Strength of Approximately 125,000 psi (data from Figures 3, 8 and 10 of (24))
Analysis: 0.34% C, 0.95% Mn, 0.26% Si, 0.014% P, 0.019% S, 0.56% Cr, 0.24% Mo, 0.66% Ni
Heat Treatment: (% in. rd)
Tempered Martensite: water quenched from 1650 F, tempered at 1210 F
Tempered Martensite + Bainite: water quenched from 1650 F and held for five seconds; transferred to salt pot at 750 F and held for five minutes; water quenched to room temperature; cooled in liquid nitrogen and tempered at 1095 F
Tempered Martensite + Pearlite: austenitized at 1650 F; transferred to salt pot at 1110 F and held for 115 min; water quenched to room temperature; tempered at 1160 F

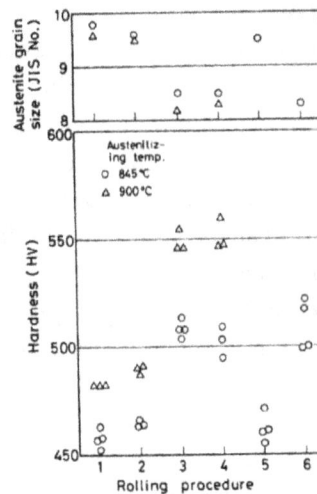

Fig. 1. Effect of rolling procedure on the hardness of specimens cooled at a rate equivalent to 15 mm distance of Jominy specimen and on the austenite grain size of specimens austenitized at 845°C or 900°C for quenching (K1 steel).

Figure 4.72 LHS = Effect of pearlite compared to bainite in the microstructure on the impact properties (14), RHS = Effect of hot working on the hardenability (the hardness of the J15 position) (69)

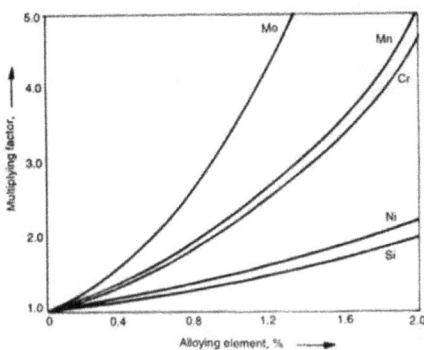

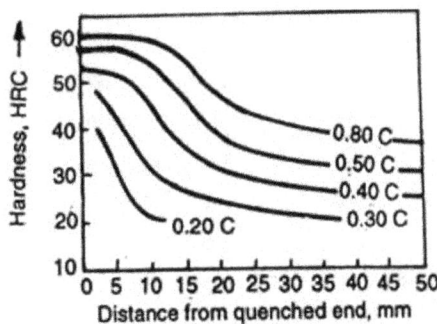

◀ **Figure 4.73**
LHS = Grossman multiplying factor, RHS = Effect of increasing carbon on hardenability 0.5% Cr, 0.5% Mn, 0.25% Mo.

Figure 4.73 LHS), particularly of plain carbon steels. Chromium and molybdenum can be cheaper effective alloying elements per unit of increased hardenability.

Intensive Quenching Technology 61 to 68

An intensive quenching (IQ) process was developed as an alternative method for hardening steel parts. The process uses highly agitated plain water instead of the environmentally unfriendly quench oils and polymers. Dr Nikolai Kobasko began work with intensive quenching in 1964 in the former Soviet Union. This early work established that the correlation between the probability of part cracking during quenching can be represented by a "bell-shaped" curve as shown in Figure 4.74. The area at the end of the right side of the "bell-shaped" curve with high cooling rates was the "surprise" discovery. It was established that at extremely high cooling rates, the probability of part cracking was zero. In addition, the parts had superior mechanical properties and performance characteristics. The explanation was the development of very high surface compressive stresses due to the fast quenching that developed uniformly around the part and prevented cracking. An additional part of the process was an interruption of intensive quench at a time when these surface compressive stresses were at their maximum value resulting in the formation of high residual surface compressive stresses. These stresses remained even after the completion of the phase transformation of the part core by air cooling.

An important quote in the documents that describe intensive quench technology was a comment by Dr George Totten, a person with substantial experience of quench technology as shown in Figure 4.75.

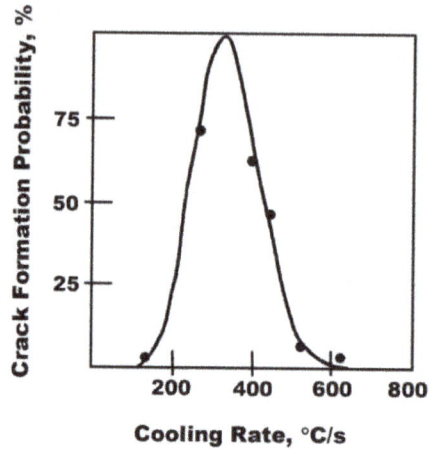

Figure 4.74 *Original data obtained by Dr Kobasko in 1964 for Ø6mm bars made of plain carbon and low alloy steels*

"My personal thoughts are that there is really no excuse to use oils, polymers, etc. Water (or water-salt) works fine, is cheaper, and better; if the system is properly engineered. This means IQ should be rightfully featured."

~ Dr. George Totten, 2013, Fellow of ASM International, SAE, IFHTFE, and ASTM; past president of International Federation for Heat Treating and Furnace Engineering (IFHTFE) and worldwide quenching authority

Figure 4.75 *Dr George Totten in 2013 suggesting that it is time to avoid oil and polymer*

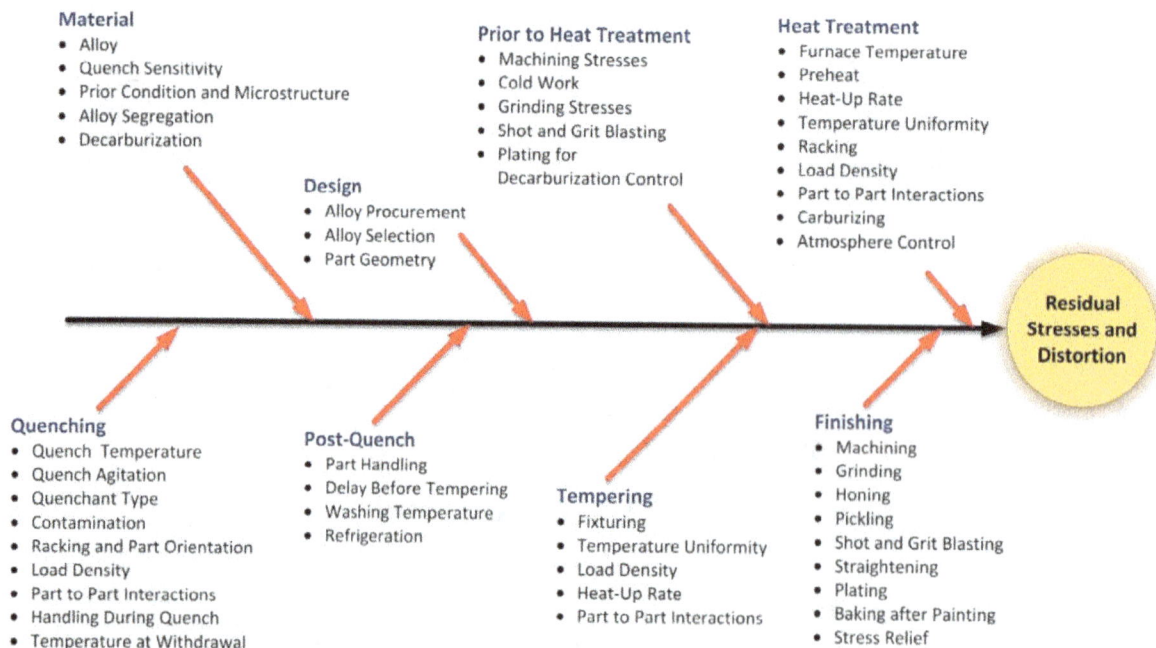

Material
- Alloy
- Quench Sensitivity
- Prior Condition and Microstructure
- Alloy Segregation
- Decarburization

Design
- Alloy Procurement
- Alloy Selection
- Part Geometry

Prior to Heat Treatment
- Machining Stresses
- Cold Work
- Grinding Stresses
- Shot and Grit Blasting
- Plating for Decarburization Control

Heat Treatment
- Furnace Temperature
- Preheat
- Heat-Up Rate
- Temperature Uniformity
- Racking
- Load Density
- Part to Part Interactions
- Carburizing
- Atmosphere Control

Residual Stresses and Distortion

Quenching
- Quench Temperature
- Quench Agitation
- Quenchant Type
- Contamination
- Racking and Part Orientation
- Load Density
- Part to Part Interactions
- Handling During Quench
- Temperature at Withdrawal

Post-Quench
- Part Handling
- Delay Before Tempering
- Washing Temperature
- Refrigeration

Tempering
- Fixturing
- Temperature Uniformity
- Load Density
- Heat-Up Rate
- Part to Part Interactions

Finishing
- Machining
- Grinding
- Honing
- Pickling
- Shot and Grit Blasting
- Straightening
- Plating
- Baking after Painting
- Stress Relief

Figure 4.76 *Fishbone diagram for residual stress and distortion during manufacture*[52]

REFERENCES

1. Terry Khaled, EQUIVALENT ROUND, EQUIVALENT DIAMETER & RULING SECTION, 5 September 2017 Report #: ANM-112N-17-03 https://www.faa.gov/aircraft/air_cert/design_approvals/csta/publications/media/equivalent_round.pdf

2. J.W. Newkirk and D.S. MacKenzie, The Jominy End Quench for Light-Weight Alloy Development, Journal of Materials Engineering and Performance Volume 9(4) August 2000 page 407

3. Thomas C.P. Karnezos, IMPROVED HEAT TREATMENT PROCESS CONTROL FOR FUEL EFFICIENCY AND CYCLE TIME REDUCTION, The Pennsylvania State University, The Graduate School, Master of Science May 2009

4. El-Amoush, Tribological properties of hard chromium. Solid State Science, 2011, Vol 13 pp 529 to 537.

5. Hard Chrome Plating Training Course, National Metal Finishing Resource Center (NMFRC) Chapter 5 available on line on the troubleshooting of Chromium plating. https://platingbooks.com/HCrSection5secured.pdf

6. http://www.sterc.org. The Surface Technology Environmental Resource Center (STERC), previously known as the National Metal Finishing Resource Center (NMFRC), provides a wealth of useful environmental compliance information to the surface finishing and surface treatment industry. STERC is the result of a new partnership with NCMS, the AESF Foundation and NASF initiated in 2016 and through continued grant funding from EPA

7. Don Baudrand, Electroless Nickel and Hard Chrome And Hard Chrome vs. Electroless Nickel http://www.plateworld.com/editorial14.htm

8. TECHMETALS,INC., ELECTROLESS NICKEL WITH A HARD CHROME OVERLAY, https://techmetals.com/pdfs/TM_111.pdf

9. http://www.materialsfinishing.org/attach/9.%20Keronite%20James%20Curran.pd f

10. https://www.youtube.com/watch?v=CVzMNzI6mV8

11. https://nickelinstitute.org/~/Media/Files/TechnicalLiterature/PropertiesAndApplicationsOfElectrolessNickel_10081_.pdf

12. Durkin B, Electroless nickel as a replacement for hard chrome. Waste Reduction Resource Centre, New Hudson.

13. Hot-Dip Galvanizing For Corrosion Protection of Steel Products, ©2013 AZZ Galvanizing Services, ©2004 American Galvanizers Association

14. Ignatius C. Okafor et al, Effect of Zinc Galvanization on the Microstructure and Fracture Behavior of Low and Medium Carbon Structural Steels, Engineering, 2013, 5, 656-666, https://www.scirp.org/pdf/ENG_2013080616165210.pdf

15. SABS, Properties of a 55 % Aluminium-Zinc Coating on Steel Sheeting, http://www.safalgroup.com/SABS-AZ-fact-sheet.pdf

16. http://industry.arcelormittal.com/catalogue/E80/EN

17. Arcelor Mittal. Metallic coated steel User manual, also https://www.metalbulletin.com/events/download.ashx/document/speaker/8353/a0ID000000X0kNrMAJ/Presentation https://www.nipponsteel.com/product/catalog_download/pdf/U008en.pdf

https://www.jfe-steel.co.jp/en/products/sheets/catalog/b1e-004.pdf

18. http://www.slideshare.net/TomaszLiskiewicz/dlc-coatings-in-oil-and-gas-production

19. W. Tillmann et al, Selecting Surface-treatment Technologies, Modern Surface Technology, 2006, Publisher Wiley-VCH Chapter

20. Stefan Hock, Ingo Kellermann, Jörg Kleff, Helmut Mallener, and Dieter Wiedmann: Relevance of Hardenability for the Machining and Application of Case Hardening Steels, Drive system technique, Volume 19, Edition 1/2005 China.

21. Callister, W. D.; Rethwisch, D. G. Applications and Processing of Metal Alloys. In Fundamentals of Materials Science and Engineering, An Integrated Approach, 3rd ed.; John Wiley & Sons Inc.: New York, NY, 2008; pp 358–413 (chapter 11

22. Nature June 21st 1947, Vol 159, page 854, Hardenability of Steel,

23. Holm, T.; Olsson, P.; Troell, E. Steel and Its Heat Treatment. ISBN: 978-91-86401-11-5, 2012, www.swereaivf.se

24. S Kuyucak et al: Quench time measurement as a process control tool. Heat treating process Jan/Feb 2005 p60

25. S Kuyucak et al: Quench time measurement as a process control tool. Heat treating process March/April2005 p42

26. Thomas C.P. Karnezos, IMPROVED HEAT TREATMENT PROCESS CONTROL FOR FUEL EFFICIENCYAND CYCLE TIME REDUCTION, The Pennsylvania State University, The Graduate School, Master of Science May 2009

27. Yuan Lu, Heat Transfer, Hardenability and Steel Phase Transformations during Gas Quenching Doctorate Thesis, WORCESTER POLYTECHNIC INSTITUTE, https://web.wpi.edu/Pubs/ETD/Available/etd-111417-121355/unrestricted/Lu.pd f

28. Rose, A. et al., Stahl u. Eisen, 91, 1001, 1971; Hengerer, F. et al., Stahl u. Eisen, 90, 1263, 1979

29. G. Spur and T. Sto¨ferle (Eds.), Handbuch der Fertigungstechnik, Vol. 4/2, Warmebehandeln, Carl Hanser, Munich, 1987.

30. EDELSTAHL WITTEN-KREFELD GMB, Thyrofort, Heat-treatable steels Brochure https://www.schmolz-bickenbach.co.za/fileadmin/files/schmolz-bickenbach.co.za/documents/heat_treatable_steel.pdf

31. Speich, G. R.; Leslie, W. C. Tempering of Steel. Metall. Trans. May 1972, 3 (5), 1043–1054 XXXXXBozidar Liscic et al, Quenching Theory and Technology Second Edition, CRC Press Taylor & Francis Group 2010, file:///C:/Users/user/Downloads/BLieseciac_InternationalFederationforHeatTreatmentandSurfaceEngineering._etal-Quenchingtheoryandtechnology-CRC2010.pdf

32. DIN 17022-1 to 5

33. R.R. Blackwood, L.M. Jarvis, D.G. Hoffman and G.E. Totten: "Conditions Leading to Quench Cracking Other than Severity of Quench" in 18th Heat Treating Society Conference Proceedings, Eds., H. Walton and R. Wallis, ASM International, Materials Park, OH, 1998, pp.575-585.

34. Kyozo Arimoto, Fumiaki Ikuta, Takashi Horino, Shigeyuki Tamura, Michiharu Narazaki, and Yo- shio Mikita: "Preliminary Study to Identify Crite- rion for Quench Crack Prevention by Computer Simulation"; Transactions of Materials and Heat Treatment, 2004, vol.25, no.5, pp.486-493.

35a. C.E. Bates, G.E. Totten, and R.L. Brennan, in ASM Handbook, Vol. 4, ASM International, Materials Park, OH, 1991, pp. 67–120

35b. http://gearsolutions.com/departments/hot-seat-the-effects-of-microstructure-and-hardenability-on-distortion/

35c. LCF Canale and L Albano, Hardenability of Steel, COMPREHENSIVE MATERIALS PROCESSING, VOLUME 12 THERMAL ENGINEERING OF STEEL ALLOY SYSTEMS, 2014 Elsevier Ltd. ISBN: 978-0-08-096532-1

35d. W.T. COOK et al, Properties and service performance EVALUATION OF CLEAN STEEL PRACTICES AND RESULTANT PROPERTY IMPROVEMENTS IN ALLOY ENGINEERING STEEL, Contract No 7210-MA/805 (1.7.1981 – 31.12.1984) FINAL REPORT 1986, EUR 10340,

36. Theory and technology of quenching, a handbook, Ed B Liscic, M Tensi and W Luty Pub Springer Verlag 1992 and Second Edition 2011 Pub CRC Press.

37. COMPREHENSIVE MATERIALS PROCESSING Volume 12 THERMAL ENGINEERING OF STEEL ALLOY SYSTEMS, Ed G Krauss, Pub Elsevier 2014

38. Lauralice de C.F. Canale and George E. Totten, Eliminate quench cracking with uniform agitation, HEAT TREATING PROGRESS • JULY/AUGUST 2004 p27

39. ALLEN, F. S, The effect of quenchant characteristics on the generation of thermal stress and strain in steel plates, Sheffield Hallam University, Doctoral dissertation 1987

40. DHIRAJ BHIKA CHAUDHARI et al, EFFECT OF QUENCHING PARAMETERS ON MATERIAL CHARACTERISTICS OF SPECIMEN- AN OVERVIEW, International Journal of Mechanical and Production Engineering Research and Development (IJMPERD) Vol. 5, Issue 4, Aug 2015, https://pdfs.semanticscholar.org/295d/9d6f65364347de2ef6b86bf5d05f999ed1ac.pdf

41. http://www.dhanaprakash.com/pdf/AGITATION-Jetmixers.pdf

42. D. Scott MacKenzie et al, Care and Maintenance of Quench Oils, https://www.houghtonintl.com/sites/default/files/resources/article_-_care_and_maintenance_of_quench_oils.pdf

43. D. Scott MacKenzie, Effect of Water on Quench Oil and Quenching, https://www.houghtonintl.com/sites/default/files/resources/article_-_effect_of_water_on_quench_oil_and_quenching.pdf

44. D.A. Wachter et al, Quenchant fundamentals: quench oil bath maintenance, http://www.getottenassociates.com/pdf_files/Quenchant%20Fundamentals%20Quench%20Oil.pdf

45. ASTM D6666, Standard Guide for Evaluation of Aqueous Polymer Quenchants, 2004

46. Daniel H. Herring, Types of HeatTreating Equipment for Fasteners, Fastener Technology International/December 2010, http://www.heat-treat-doctor.com/documents/HTEquip.pdf

47. AMS 2759 Heat Treatment of Steel Parts–General Requirements

48. Heat Treater's Guide Practices and Procedures for Irons and Steels, Editor Harry Chandler, ASM International, ISBN-13: 978-0-87170-520-4, 1995

49. R S Archer et al Molybdenum Steels and Irons, Climax Molybdenum Company 1948, https://archive.org/stream/in.ernet.dli.2015.113418/2015.113418.Molybdenum-Steels-Irons-Alloys_djvu.txt

50. Yuan Lu, Heat Transfer, Hardenability and Steel Phase Transformations during Gas Quenching, WORCESTER POLYTECHNIC INSTITUTE In partial fulfillment of the requirements for the Degree of Doctor of Philosophy, Dec 2016, https://web.wpi.edu/Pubs/ETD/Available/etd-111417-121355/unrestricted/Lu.pdf

51. Houghton on quenching, https://www.houghtonintl.com/sites/default/files/resources/article_-_houghton_on_quenching.pdf

52. D. Scott MacKenzie, Selection of oil quenchants for heat treating processes, International Heat Treatment and Surface Engineering 2014 VOL 8 NO 1 page 8

53. D. SCOTT MACKENZIE, APPLICATIONS OF HARDENABILITY, https://www.houghtonintl.com/sites/default/files/resources/applications_of_hardenability.pd f

54. D. Scott MacKenzie, History of Quenching, file:///C:/Users/user/Downloads/ASMHistoryofQuenching.pdf

55. Quenching Oils and Polymer Quenchants, Fuchs https://www.fuchs.com/fileadmin/schmierstoffe/Prospekte/Brochures_EN/Product_brochures_industry/THERMISOL_brochure_09-2018.pdf

56. Quenchants – Monitoring and Maintenance, Fuchs, 2011, https://www.fuchs.com/fileadmin/schmierstoffe/Prospekte/FTI/Quenchants-monitoring-and-maintenance.PDF

57. J Vanpaemel, HISTORY OF THE HARDENING OF STEEL : SCIENCE AND TECHNOLOGY, JOURNAL DE PHYSIQUE Colloque C4, supplement au no 12, Tome 43, decembre 1982 page C4-847, https://hal.archives-ouvertes.fr/jpa-00222126/document

58. Nicholas Anthony Hilder, The behaviour of polymer quenchants, University of Aston, Doctoral dissertation 1988, https://publications.aston.ac.uk/id/eprint/11905/1/Hilder_NA_1988.pdf

59. EARL J. ECKEL et al, AN EVALUATION OF THE HARDENING POWER OF QUENCHING MEDIA FOR STEEL, UNIVERSITY OF ILLINOIS BULLETIN Vol. 48 June, 1951 No.73, https://core.ac.uk/download/pdf/4814276.pdf

60. CQI-9-3rd-Edition-AMP-041712-Ed.pdf http://www.ampht.com/pdfs/CQI-9-3rd-Edition-AMP-041712-Ed.pdf

61. xxxxMałgorzata Przyłecka and Wojciech Gestwa, The Possibility of Correlation of Hardening Power for Oils and Polymers of Quenching Mediums, Advances in Materials Science and Engineering Volume 2009, https://www.hindawi.com/journals/amse/2009/843281/

69. By A.V. Reddy et al, Simple method evaluates quenches, Heat treating process June 2001,

70. Lauralice de Campos Franceschini Canale and George E. Totten, Quenching Technology: A Selected Overview of the Current State-of-the-art, Materials Research, Vol. 8, No. 4, 461-467, 2005, http://www.scielo.br/pdf/mr/v8n4/27623.pdf

71. Hala Salam Hasan, Evaluation of heat transfer coefficient during quenching of steels, Cambridge University, Doctoral dissertation, September 2009, https://www.phase-trans.msm.cam.ac.uk/2010/Hala_Thesis.pd f

61. Joseph Powell, How IntensiQuench® Works, http://www.akronsteeltreating.com/docs/default-source/default-document-library/how-intensiquench-works.pdf?sfvrsn=2

62 Phase I Report, Cooperative Agreement Award:, W15QKN-06-2-0105, Intensive Quenching Technology for Advanced Weapon Systems, http://www.airflowsciences.com/sites/default/files/docs/AWS_Report_Phase_1_Final_redacted.pd f

63. Michael Aronovet al, Intensive Quenching Processes Basic Principles, Applications and Commercialization, IFHTSE Conference Paper May 2014, file:///C:/Users/user/Downloads/IFHTSE2014Paper02.21.14.pdf

64. Dr. Michael A. Aronov, Final Technical Report Intensive Quenching Technology for Heat Treating and Forging Industries DOE Award Number: DE-FC36-03ID 14463 Project Period: April 2003 – September 2005

65. D. Scott MacKenzie, Discussing the various methods — as well as the pros and cons — of selecting a quench oil for various applications, Thermal Processing July 2019, http://thermalprocessing.com/wp-content/uploads/2019/07/0719-IF-1.pdf

66. D. Scott MacKenzie, Selection of oil quenchants for heat treating processes, International Heat Treatment and Surface Engineering 2014 VOL 8 No 1, file:///C:/Users/user/Downloads/SelectionofOilQuenchants%20(1).pdf

67. S. R. Elmi Hosseini et al, Cooling Curve Analysis of Heat Treating Oils and Correlation With Hardness and Microstructure of a Low Carbon Steel, Materials Performance and Characterization, Vol. 3 / No. 4 / 2014 page 427.

68. N. I. Kobasko et al, Cooling Capacity of Petroleum Oil Quenchants as a Function of Bath Temperature, Materials Performance and Characterization, Vol. 2, No. 1 2014 Page 468

69 Hirooki NAKAJIMA, et al, Effect of Hot Working on the Hardenability of Steels, Tetsu-to-Hagane, Volume 68 Issue 2 Pages 284-291

81. 1998, https://www.jstage.jst.go.jp/article/tetsutohagane1955/68/2/68_2_284/_pdf/-char/en

82. DOUGLAS V. DOANE, Application of Hardenability Concepts in Heat Treatment of Steel, J. HEAT TREATING, VOLUME 1, NUMBER 1 page 6, http://www1.diccism.unipi.it/Valentini_Renzo/slides%20lezione%20met.%20meccanica/hardenability.pdf

BIBLIOGRAPHY

1. H C Child, Surface Hardening of Steel, Engineering Design Guide 37, Published by the Design Council, Oxfrod University press

2. D K Bullens, Steel and Its Heat Treatment, Volume 1 Principles, Fifth Edition, John Wiley & Sons 1943.

3. George Krauss, Steels, Processing, Structure, and Performance, ASM International, 2015, ISBN-13: 978-1-62708-083-5

4. KARL-ERIK THELNING, Steel and its Heat Treatment Bofors Handbook, Pub Butterworths, ISBN 0 408 70934 0 1975

5. G. PARRISH et al, PRODUCTION GAS CARBURISING, PERGAMON PRESS, ISBN 0-08-027312-2 Hardcover 1985.

6. Jon L. Dossett and Howard E. Boyer, Practical Heat Treating, Second Edition, ASM International, ISBN: 0-87170-829-9, 2006

7. Harry Chandler, Editor, Heat Treater's Guide Practices and Procedures for Nonferrous Alloys, ASM International, ISBN: 0-87170-565-6, 1996

8. Harry Chandler, Editor, Heat Treater's Guide: Practices and Procedures for Irons and Steels, 2nd Edition, ASM International, ISBN-13: 978-0-87170-520-4, 1996.

9. Geoffrey Parrish, Carburizing Microstructures and Properties, ASM International 1999

10. Edited by Bozidar Lisci´ c Hans M. Tensi Lauralice C. F. Canale George E. Totten, Quenching Theory and Technology Second Edition, CRC Press, International Standard Book Number: 978-0-8493-9279-5 (Hardback), 2010.

11. George F. Vander Voort, Atlas of Time-Temperature Diagrams for Irons and Steels, ASM International, ISBN: 0-87170-415-3, 1991,

12. John Holloman and Leonard Jaffe, Ferrous Metallurgical Design, Design Principles For Fully Hardened Steels, Pub John Wiley and Sons, 1947

13. COMPREHENSIVE MATERIALS PROCESSING, Volume 12, THERMAL ENGINEERING OF STEEL ALLOY SYSTEMS, Pub Elsevier, ISBN: 978-0-08-096532-1, 2014.

14. George E Totten, Steel Heat Treatment Handbook, Second Edition Pub CRC Taylor & Francis, 2007

SELF-ASSESSMENT QUESTIONS

1. Explain the term heat treatment. Give an example of how heat treatment can change the mechanical properties of an alloy?

2. Explain the term surface engineering.

3. What are the three solid phases of iron called?

4. Explain the term transformation hardening. What carbon level would be required to give an as quenched hardness of 45Rc when a fastener was hardened to achieve a microstructure with 90% martensite?

5. Give three examples of why a heat treatment would be carried out.

6. Name three heat treatment process variables. Discuss the significance of holding time at heat treatment temperature.

7. Explain why some steels are heated in stages to heat treatment temperature; by using hold periods of time at certain temperatures.

8. Outline the precautions that are needed when heat treating a component of complicated

shape having sharp corners and variable section sizes.

9. The heat treatment of alloy steels could involve a slow heating rate. Why would this be carried out?

10. For identical shape and size, compare the heat treatment schedule for high carbon, medium carbon and low carbon steels.

11. Based on the knowledge gained from this chapter, what methods would you use to determine a CCT diagram?

12. Based on the knowledge gained from this chapter, what methods would you use to simulate a specimen with the similar microstructure to the coarse-grained heat effected zone of a weld?

13. How could the grain size of an austenitic stainless steel be refined?

14. How could the grain size of a ferritic stainless steel be refined?

15. How could the grain size of a carbon steel open die forging be refined?

16. If the carbon steel forging (with 0.2% carbon) was then water quenched and 80% martensite was achieved, what would be the expected hardness?

17. A forged shaft has been made from AISI 4140 (42CrMo4) and the mechanical properties after heat treatment are UTS=800MPa, YS= 540MPa, Elongation = 22% RA= 45% What will be the main phase present in the microstructure? Use (Figure 4.45) What will be the approximate ITT for this steel? (use Figure 4.46). Will this steel be expected to pass the impact requirements at minus 46°C?

18. What element in steel has the greatest effect on the hardness of a small as-quenched part?

19. What crystal structure would a carbon steel have at 950°C? What is it known as?

20. A forging has been water quenched and the as-quenched hardness has been measured as 40Rc. The steel used was AISI 4130. What is the percentage of martensite in the surface of the part? What would be the expected YS/UTS ratio?

Would this steel part achieve impact property requirements at minus 60°C?

21. Using the Grossman factors in ASTM A255 the AISI4130 steel supplied by the customer has a DI of 3ins. The ED (effective diameter) for this component is 2ins. Will this steel provide impact requirements at minus 20°C at a yield strength of 600MPa?

22. Approximately what percentage of carbon can enter into interstitial solid solution when at medium carbon steel is held at temperature in the austenite phase?

23. A heat treatment company has measured the "quench factor" of their quench facility and has found the quench severity "H value" of 1.0. Would you personally recommend this facility as a key supplier for the processing of AISI 4140 forgings? Would this facility be appropriate to heat treat AISI4130 at 150mm diameter to grade X85 strength and toughness requirements?

24. What crystal structure would a sample of pure iron have at 50°C?

25. What is the phase change in steel that occurs in pure iron at around 910°C called? Why is this phase change important in the processing of steel? What phase would be present above 910°C?

26. Using the Jominy curves shown in Figure 4.31 top RHS determine the as quenched hardness for a 60mm diameter shaft at the centre, ¾ Radius and the surface (using graphs 4.31 lower RHS to relate the J distances to the cooling rates for the 60mm shaft use the worst condition since a range is shown). Do this exercise for 3 steel types SAE 4340, SAE 8640 and SAE1040 for both water quench and oil quench. Use Figure 4.49 to estimate the percentage martensite at the surface and centre of each shaft.

27. What would be the appropriate austenitising temperature and furnace residence time for a 60mm diameter shaft made from 4140 steel? Use the tables shown in section 4.4.5.2. and compare with Figure 4.43 top LHS to confirm that homogenous austenite has been achieved.

CHAPTER 5

Mechanical and Chemical Characterization of Metals

5.1 BACKGROUND

The initial chapters have introduced the discovery of important investigational techniques that allowed an understanding of the behaviour of metals (chapter 1), the important features associated with the microstructure of metals (chapter 2), and the important aspects of steel quality (Chapter 3) the heat treatment and surface engineering (chapter 4). The subject of how the mechanical properties of alloys and metal can be related to their microstructure has been outlined and emphasis was placed on the important aspects of the microstructure such as grain size. The microstructure to mechanical property relationships are often the key to being able to control and predict the behaviour of metals during manufacture and in service. In this chapter the aspects of how the mechanical properties are measured will be considered. A summary of empirical and theoretical equations developed over the years can be found on the web, "Steel Forming and Heat-Treating Handbook" 2018 by Antonio Gorni who has spent many years up-dating this document. (http://www.gorni.eng.br/e/Gorni_SFHTHandbook.pdf) Ref1

In this chapter the important mechanical properties are examined in detail together with the basic test methods. Mechanical tests are carried out for routine quality checks, for material validation and for final release testing to certify that the manufactured products achieved the specified mechanical property requirements.

A closely associated requirement in metals validation involves ensuring that the chemical analysis of the metal or alloy is in accordance with the specified

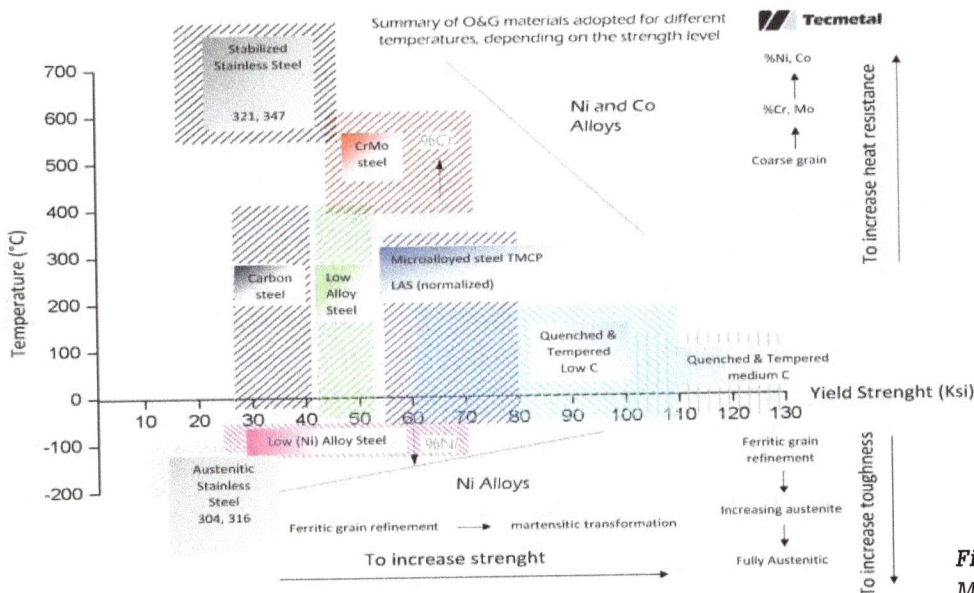

Figure 5.1 Steel Properties Map from Antonio Gorni[1]

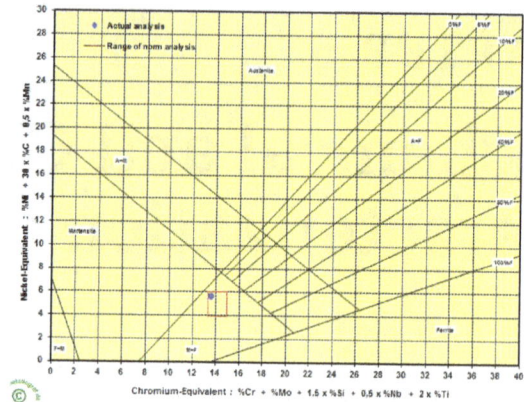

Figure 5.2 Excel spreadsheet for display of the Schaeffler diagram

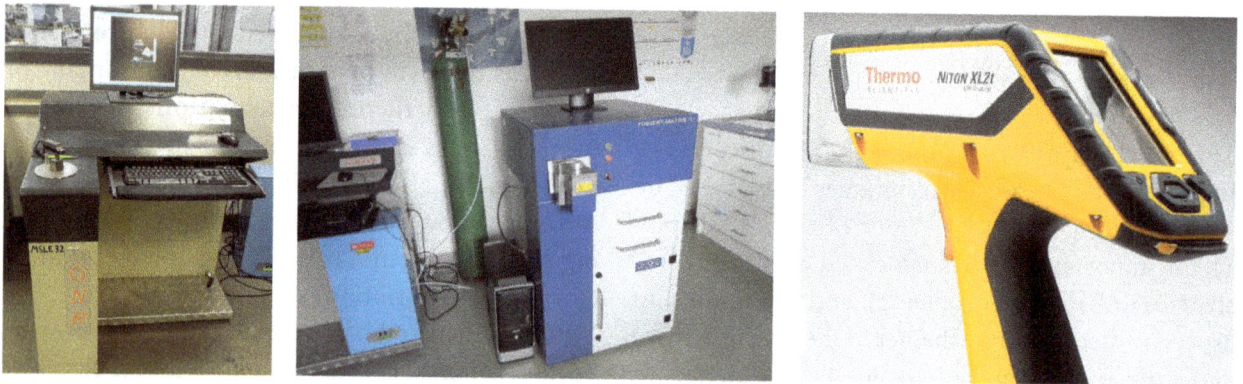

Figure 5.3 LHS = Optical Emission Spectroscopy supplied by GNR Italy (15), Centre= Oxford Instruments Foundry-Master Pro (16), RHS = XRF (X-Ray Fluorescence) supplied by Niton USA, used for PMI (Positive Material Identification) (17)

requirements. Knowledge of the exact chemical analysis would be essential to the ability to carry out the following:

- Calculate weldability CEV, CE = C + Mn/6 + (Cr + Mo + V)/5 + (Ni + Cu)/15, and PCM Ref2
- Corrosion resistance of stainless PREN = %Cr + 3.3 × (%Mo + 0.5%W) + 16 × %N[3&4]
- Predict stainless microstructure Schaeffler diagram (see Figure 5.2)[5]
- Calculate hardenability DI, Jominy Curves and Heat Treatment parameters[6]
- Check for the risk of temper embrittlement using the Wantanabe Factor J and Bruscato Factor BF or X Factor[7 & 8]
- Check residual elements in secondary aluminium grades "Tramp or Trace"[9, 10 & 11]
- Check for deleterious elements in cast irons.[12, 13 and 14]

The two basic methods OES (optical emission spectroscopy) and XRF (X-ray fluorescence) are outlined since they often provide important and vital information to support the material characterization process. Figure 5.3 shows the equipment used for chemical analysis of metals.

5.1.1 Service Loads and the need for mechanical properties to resist these loads

Galileo's bending theory as shown in Figure 5.4 RHS was believed to be the historical starting point for the consideration of load and of "elasticity". This was known as "Galileo's problem".

The knowledge and understanding of the service loads is an important requirement to allow an evaluation of the future integrity of a component. This can be an expensive and difficult task. This involves examination of the "load path" through the component to try to identify the weakest link. The loads experienced by a component are the result from the forces acting on a component. Components may be subjected to many different loads which may involve combinations of tension, compression, bending, shear and torsion.

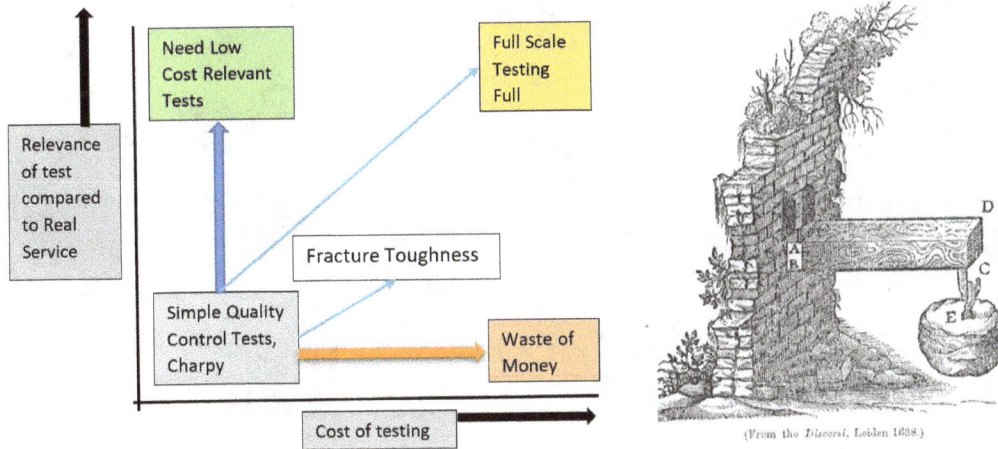

Figure 5.4 LHS = *Ideal test – It should mimic service conditions at low cost. RHS = Galileo's bending theory, known as "Galileo's problem" began detailed consideration of elasticity*[44]

The performance of the any material used to manufacture a part, component or structure must be able to resist the applied loads.

There is often a need to carry out test work to validate the choice of material, which can be complex and expensive task. Figure 5.4 shows a diagram with axis of "test relevance" vs "cost of testing". The simple quality tests are low cost but often lack relevance to the final service performance of a part or component. Quality tests are carried out to check that materials that are used conform to the specified requirements and quality test work completely fulfil that role but they would not give full guarantee that a complex shape made from the material would be acceptable under service condition.

The task of the test engineer is to rise vertically on this diagram and establish low cost very relevant tests. For the automotive industry the proven answer is the yellow box and full-scale testing of prototype vehicles on test tracks

5.1.2 Experimental Stress Analysis (ESA)

The analysis of the mechanical stress state in materials, which is usually performed at the design stage by FEA and then using experimental methods when prototype products are available. In the uniaxial tensile test, the method used to determine the yield point when there is no clearly identifiable drop-in stress, immediately following yielding, then the offset method is used. An offset at 0.2 percent (0.2 percent corresponds to 2000 micro strain). Clearly at 2000 micro-strain a part would have yielded and would be beyond the elastic region. For parts to survive there is a need to keep stress/strain as low as possible, ideally below 1000 micro-strain. Figure 5.5 shows the various techniques used to evaluate the levels of strain in component or structure. Ideally the challenge is to ensure elastic behaviour and to keep strains below 1000 micro-strain. Figures 5.6 to 5.10 show examples of stress measurement.

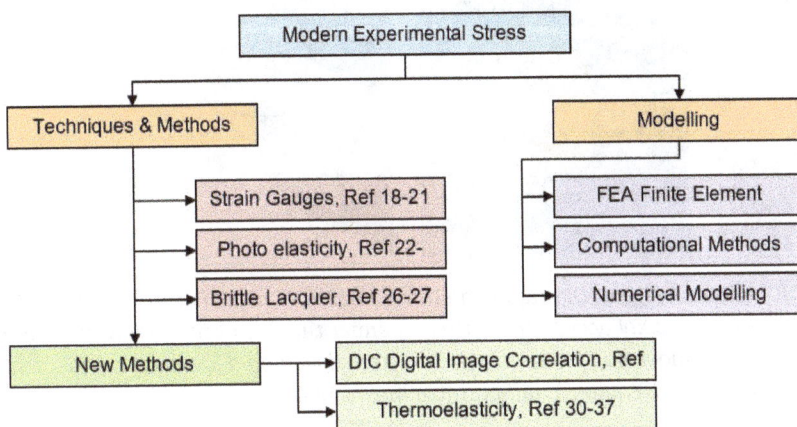

Figure 5.5 Modern Experimental Stress analysis – measurement of strain

Figure 5.6 *Specimen Alignment Test This application used a micro measurements 8 channel strain gauge conditioning with 8 gauges on a tensile specimen to measure the machine alignment. The electrical resistance strain gauge was invented by US engineers Edward E. Simmons and Arthur C. Ruge in 1938. The first application of the strain gauges was a measurement of the water tank wall deformation to define applied stresses. In this first experience, ordinary cigarette paper was glued on the tank, and a small wire with end connections glued on the paper was used with wire glued onto a cigarette paper. Now made from foil. Based on the concept discovered by Lord Kelvin, in 1856, that the resistance of a wire changes as it is stressed. The change in resistance is small and a Wheatstone bridge circuit, invented by Samuel Hunter Christie in 1833 and improved and popularized by Sir Charles Wheatstone in 1843, is used to measure the mv. Strain gage conditioning units are required to supply the accurate 5V to the bridge and to amplify the output mvs.*

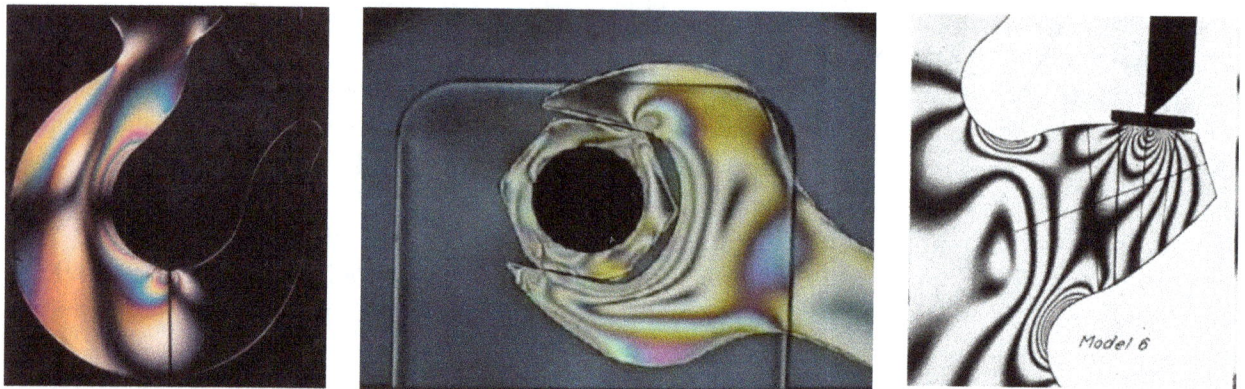

Figure 5.7 *Photoelasticity LHS = Crane Hook, frequently used photo to demonstrate photoelasticity. It was taken in the 1980s by John A Driver, a technician in the Department of Mechanical Engineering at the University of Sheffield, Centre= Spanner tightening a nut (Indian Institute of Madras). RHS = Image created in 1944 at the university of ILLINOIS showing gear tooth contact (tooth bending stress and contact stress). The initial technique involved maching parts from plastic sheet followed by paint systems.*

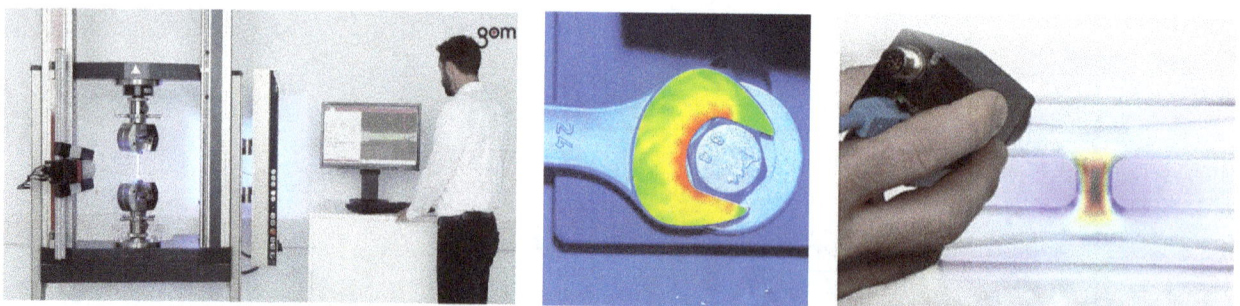

Figure 5.8 *Two Modern methods LHS = DIC Aramis system (Digital Image Correlation) DIC uses pattern and/or point markers to track the surface of the materials with subpixel accuracy, Centre= spanner turing nut from GOM web site, RHS = MiTE, Developed in Australia, MiTE is a thermoelastic stress analysis system based on a low-cost microbolometer detector technology to create an affordable, rugged, compact and portable means of imaging stress in dynamically loaded structure.*

5.1.3 Determining Mechanical Stress

Stress is defined as the physical response (deformation) in materials caused by force. It usually occurs as a result of an applied force (mechanical stress), and above the elastic limit can cause the material to deform. Stress can also often be due to the effects of force within a material (residual stresses) or stresses from other parts of a mechanically connected system. Measurement of loads or stresses involve the measurement of "strain" and then using the modulus converting to stress.

Figure 5.9
Vehicles that were instrumented using strain gauges to obtain data for fatigue life assessment during the 1990s

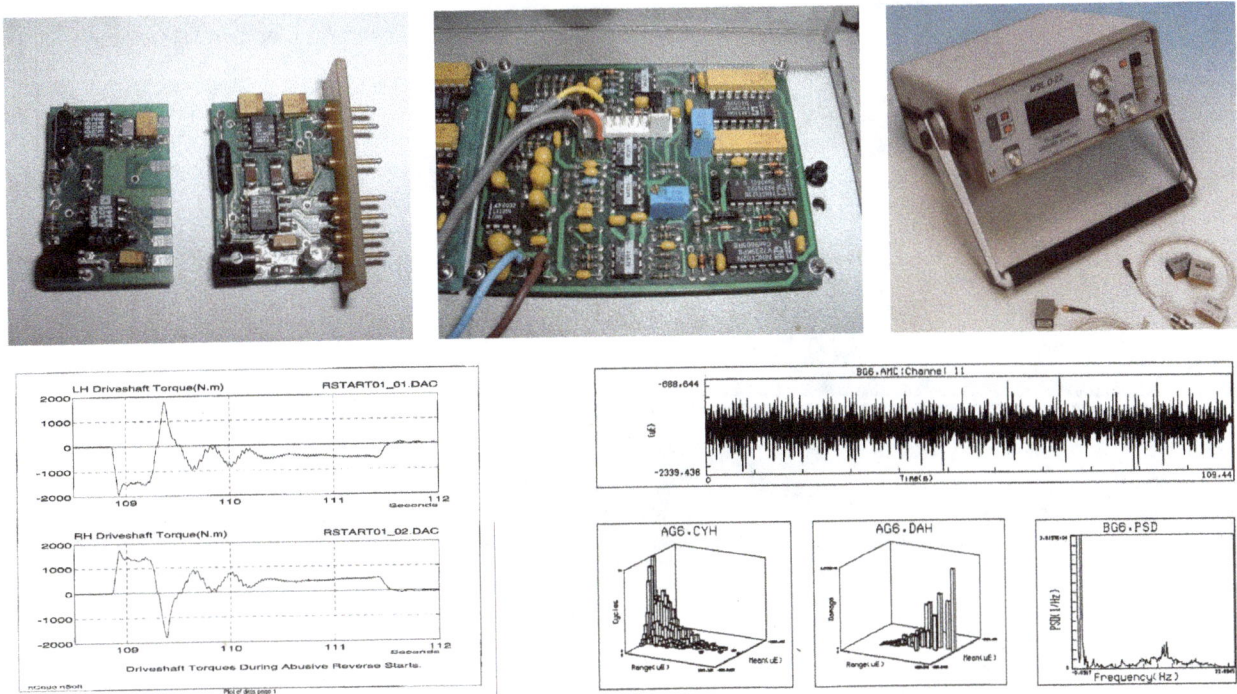

Figure 5.10 *Top LHS = Telemetry on shaft oscillator and small strain gauge amplifier, Centre Electronics in the telemetry ground station, RHS = telemetry system, Bottom LHS = Drive shaft data obtained with MSL telemetry, RHS = Load service data from a bus structure for fatigue life assessment*

Stresses are subdivided as follows:

1. **Type:** normal stresses and shear stresses
2. **Origin:** tension, compression, bending, torsion, residual and thermal stresses
3. **State:** uniaxial, biaxial, triaxial or spatial.

Mechanical tests provide quantitative data on the deformation and fracture properties of metals. There are two particular aspects of deformation that are of interest to engineers, one relating to "fitness for service" and the other the "fitness for manufacture". This can be summarized as:

Fitness for service – To ensure that parts/components have sufficient integrity and durability to resist the plastic deformation that occurs when service loads operate near to or exceed the yield strength or proof strength especially in areas of stress concentration

Fitness for manufacture – To ensure the ability of metals to deform plastically (i.e. permanently) during manufacturing in order to allow parts to be manufactured by the commonly used shaping processes, e.g. hot forging, cold forging, rolling, bending and sheet metal forming.

5.1.4 An important test laboratory microscope-the SEM

Another important metallurgical tool is the SEM (scanning electron microscope) which allows fracture surfaces to be examined to establish the fracture mode and using EDX (Energy-dispersive X-ray spectroscopy) to determine the chemical analysis of small areas of the microstructure.

When components and structures have been manufactured it is essential to guarantee the ability of the metal or alloy to retain its durability and integrity throughout its service life. This will include factors such as strength, resistance to fracture and fatigue, resistance to environmental corrosion mechanisms.

5.1.5 Training Aims

1. To outline the basic test methods that have been developed to characterize the mechanical properties of metals and alloys (stress, strain, Young's modulus, toughness, hardness)
2. To demonstrate the wealth of information available in ASTM, BS and ISO test standards and how these standards supply all the information needed to evaluate the mechanical properties of metals and alloys.

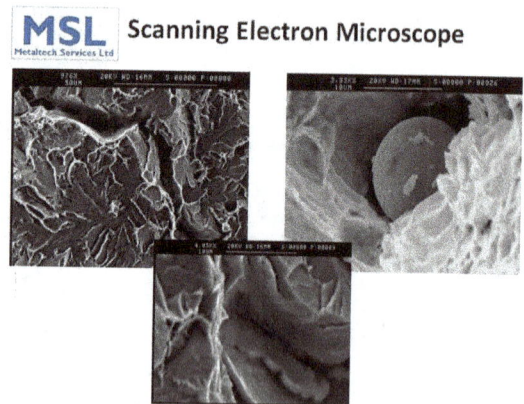

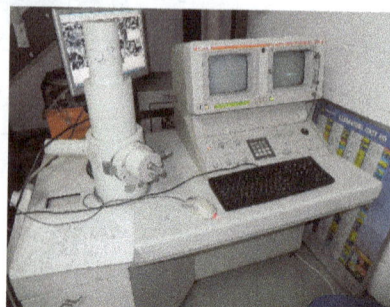

Fig 5.11 Scanning electron microscopes at MSL and images of fracture surface

3. To introduce the basic methods of chemical analysis of metals

4. To introduce the fracture modes that occur in metals

5.2 MECHANICAL PROPERTIES OF METALS – TENSILE TEST

The evaluation of the mechanical properties of a metal usually involves simple test methods such as Tensile test, Hardness, Charpy-toughness, Bend test: all of which have developed and evolved over the past hundred years. For strength measurement it is usual to carry out a tensile test. An excellent introduction to the development and history of the tensile test can be found in a report prepared by Malcolm S Loveday on the "Tensile testing of metallic materials: A review".[38] The review was carried out during 2001 to 2004 as part of the EU Tenstand project. The project formed part of the review of EN 10002-1 to try to lower the cost of material testing by allowing faster testing. The other parts of the project are outlined in (Ref 39 to 42).

Figure 5.12 shows one of the tensile testers used at MSL is a Zwick 1200kN Screw type tensile test machine, a Dartec 500kN Servo-hydraulic equipped

to carry out fracture toughness and screen shot of the control software used on the three systems. Figure 5.13 shows a 10 kN Shimadzu with a Messphysik non-contact extensometer. Figure 5.14 shows photos of a DMG 200kN single screw tensile and images of the PC control software.

Tensile Test

When a load or force is applied to a material it will deform/extend due to a reaction to the load. This is the simple principle for the tensile test method used to quantify the tensile properties of a material. Using a prepared tensile specimen in accordance with the required National or International Standards (BS EN ISO 6892-1:2016, ASTM E8/E8M-16a, ASTM A370-18) the yield or proof strength, the UTS (Ultimate tensile strength) and the tensile ductility, the percentage elongation (%El) and the reduction of area (%RA) can be measured. If the strains are accurately measured the Modulus of Elasticity can be determined (ASTM E111) although the preferred option would be the use of strain gauges or the velocity of sound.

Consider the diagrams on page 145 that are representative of a tensile test. The images show the information that can be obtained.

Figure 5.12 Zwick 1200kN Screw type tensile test machine with software control. RHS = Instron 500kN Servo-Hydraulic with Dirlik Controls Tensile and Fracture Toughness software

Figure 5.13 A 10 kN Shimadzu with a Messphysik non-contact extensometer

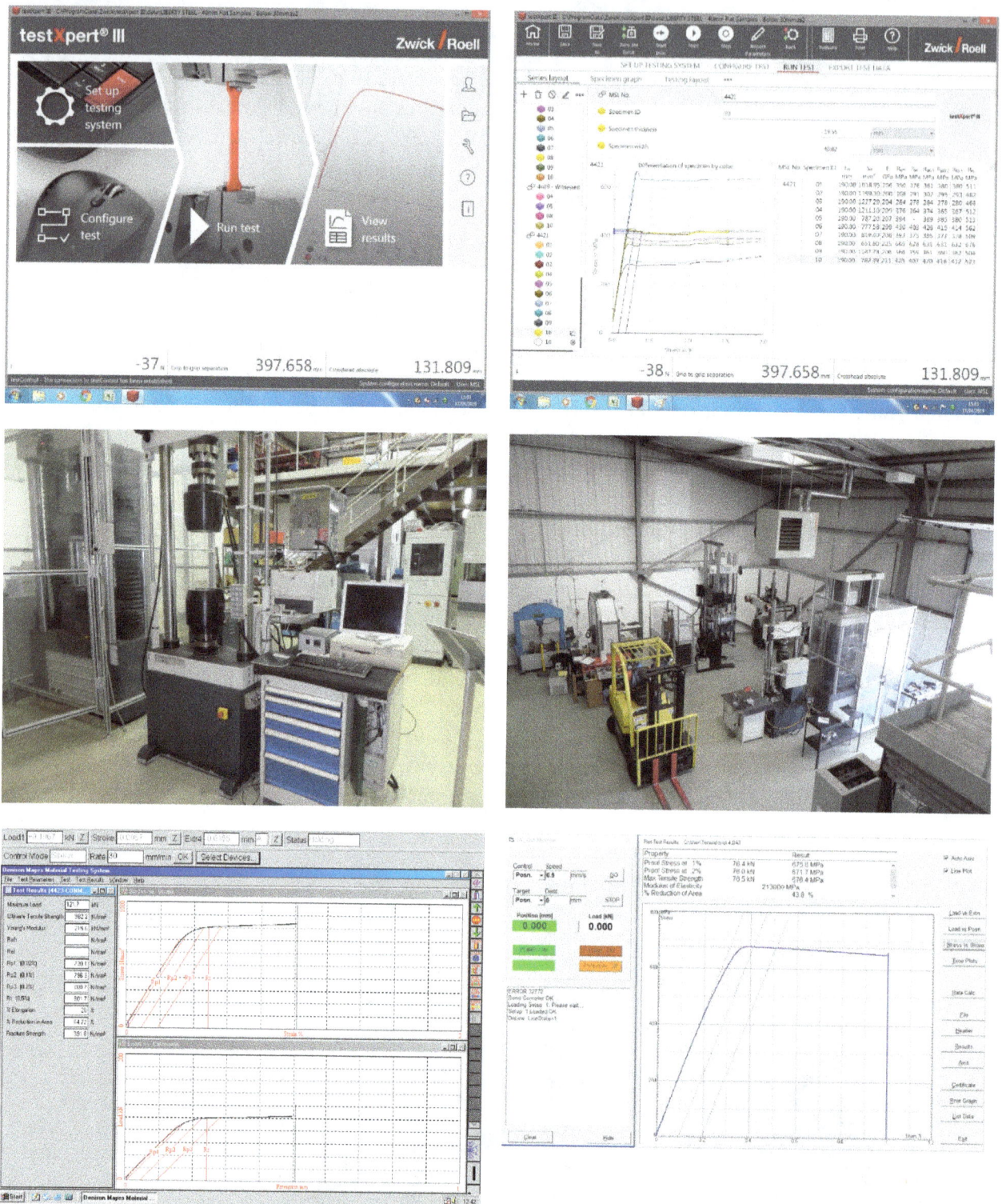

Figure 5.14 Top = Zwick Roell TestXpert III control software Centre LHS = DMG 200kN Single screw tensile with hot tensile oven and extensometer, RHS = MSL test area, Bottom LHS = DMG Rubicon Software, RHS = Dirlik Controls Software

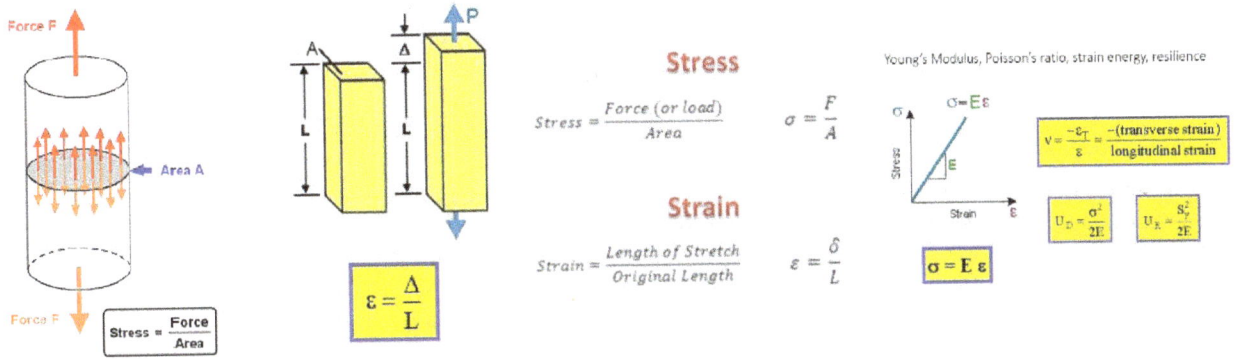

Figure 5.15 *Uni-axial loading*

The first mathematician to consider the nature of the resistance of solids to rupture was Galilei (1638) – see Figure 5.4 RHS. The two relevant discoveries that followed were the discovery of Hooke's Law in 1660 (see Section 5.2.2) and Young's Modulus (1807).[43-45]

It was concluded by J E Gordon in his book "New Science of Strong Materials: Or Why You Don't Fall Through the Floor" that there was a delay in the understanding of modulus of over one hundred years due to using deformation and force from Hooke's original concept. It took Thomas Young to point out that stresses and strains should be used instead of deformation and forces.

Thomas Young was born in Somerset in 1773 and died in London in 1829. He had an extraordinary intelligence, which was noted at a very early age: he could read at the age of 2 and had read the Bible twice through before he was 4 years old. He began learning Latin when he was five. By the time he was 13 he had also learnt Greek, Hebrew, Italian, and French and later he would study near-eastern languages such as Arabic, Persian, Chaldee, Syriac, and Samaria and he had also developed an interest in optics and telescope making. In the years that followed, he read French, Italian and English classics, mathematics, astronomy, natural philosophy, botany, chemistry, and medicine. He deciphered the ancient Greek texts of the Rosetta Stone, an engraved stone from Egypt, dated to 196 BC, and other inscriptions. He developed the wave theory of light which upset Newton's claim of light being composed of particles,

Thomas Young 1773–1829

first to understand that energy of a moving body increases as the square of velocity, first to determine the diameter of a molecule, first to use surface tension to calculate molecular tension, first to create a modern view of heat as molecular motion instead of a flow of a caloric fluid into a body, and first to imagine a spectrum of radiation from ultraviolet to infrared. See – The last man who knew everything by Andrew Robinson 2006.

The effect of specimen size can be examined using the following example

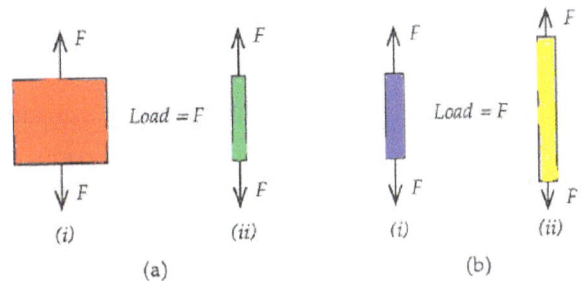

Which of the following statements do you think are true?

1. For the tests shown in (a), (i) should extend more than (ii).
2. For the tests shown in (b), (i) should extend more than (ii).

Neither of the statements is true, as we shall see in the following section.

In order to avoid the problems associated with specimen dimensions we can consider the mechanical properties of the alloy in terms of:

$$Stress, MPa\ (\sigma) = \frac{Load\ (N)}{Cross-Sectional\ area\ (mm^2)}$$

And

$$Strain\ (\varepsilon) = \frac{Extension\ (mm)}{Original\ Length\ (mm)}$$

These two relationships are fundamental and must be understood and memorized – stress equals load over area – strain equals ΔL over L.

Normally test specimens conform to a standard design, although since the results are now reported in terms of stress and strain, we should be examining material properties and not simply features of the specimen shape.

We should now be able to examine our previous examples in more detail.

If the load applied is 6000N and the geometry of the specimens is as shown below:

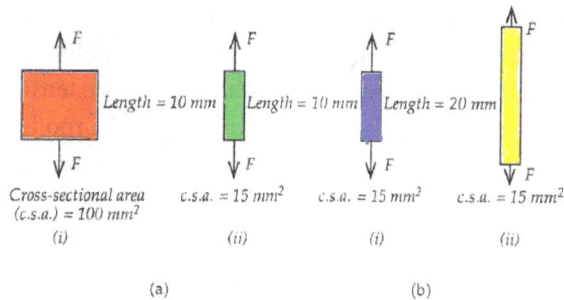

(a)   (b)

For the two specimens in (a) the stress applied

(i) $\frac{6000}{100} = 60$ (ii) $\frac{6000}{15} = 400$

The much greater stress on (ii) will produce a greater extension.

For the two specimens in (b) the applied stress is the same. However, specimen (ii) could be thought of as being made up of two specimens having the same design as specimen (i). Each of these specimens would elongate the same amount as specimen (i), hence the extension would be twice as much in specimen (ii).

See if you can calculate the stresses and strains for the following.

(a) The stress acting on specimen (a) (i) for an applied load of 10,000 N.

(b) The stress acting on specimen (a) (ii) for an applied load of 10,000 N.

(c) The extension produced in specimen (b) (i), under a strain of 0.10 (or 10%).

(d) The extension produced in specimen(b) (ii), under a strain of 0.10 (or 10%).

Answers

(a) $Stress = \frac{10000\ MPa}{100} = 1000\ MPa$

(b) $Stress = \frac{10000\ MPa}{15} = 666\ MPa$

(c) $0.10 = \frac{extension\ (mm)}{10}$, *Hence extension* = 1.0mm

(d) $0.10 = \frac{extension\ (mm)}{20}$, *Hence extension* = 2.0mm

5.2.1 Yield or Proof stress

If a metal is subjected to a uniaxial tensile load, such as is illustrated in the previous examples, and the extension is measured as the load is increased, then a stress-strain curve, similar to that shown in Figure 5.16 LHS, can be determined.

The initial part of the curve is linear, whereas the final section is non-linear. Furthermore, it is found that if the metal is stressed to a level in the linear region and then the stress is removed, the specimen will return to its initial dimensions, i.e. it has undergone elastic, or recoverable, deformation. In metals, this elastic behaviour is associated with the stretching of the atomic bonds (as we saw in chapter 2), such that when the stress is removed the bonds will return to their initial state, hence 'recoverable' deformation.

However, if the specimen is deformed into the non-linear region and then unloaded, a permanent change in shape will have occurred, i.e. it has undergone plastic or non-recoverable deformation. As we

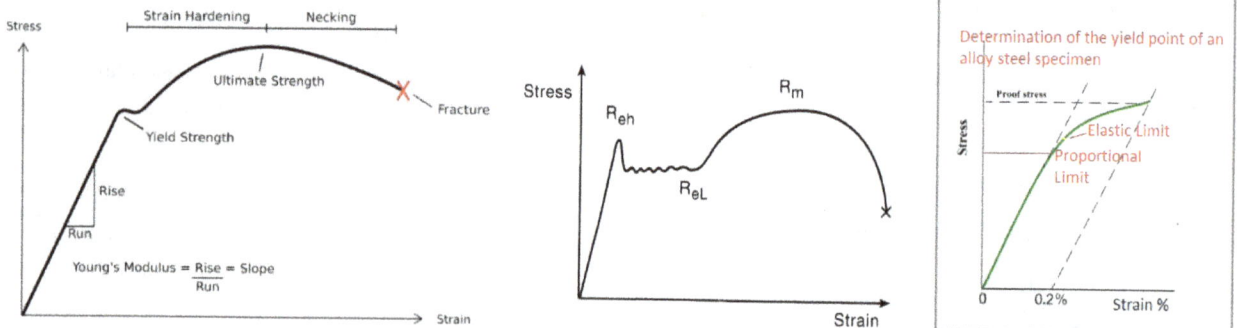

Figure 5.16 *LHS = Uniaxial tensile test Centre= Upper and Lower YS, RHS = 0.2% Proof strength*

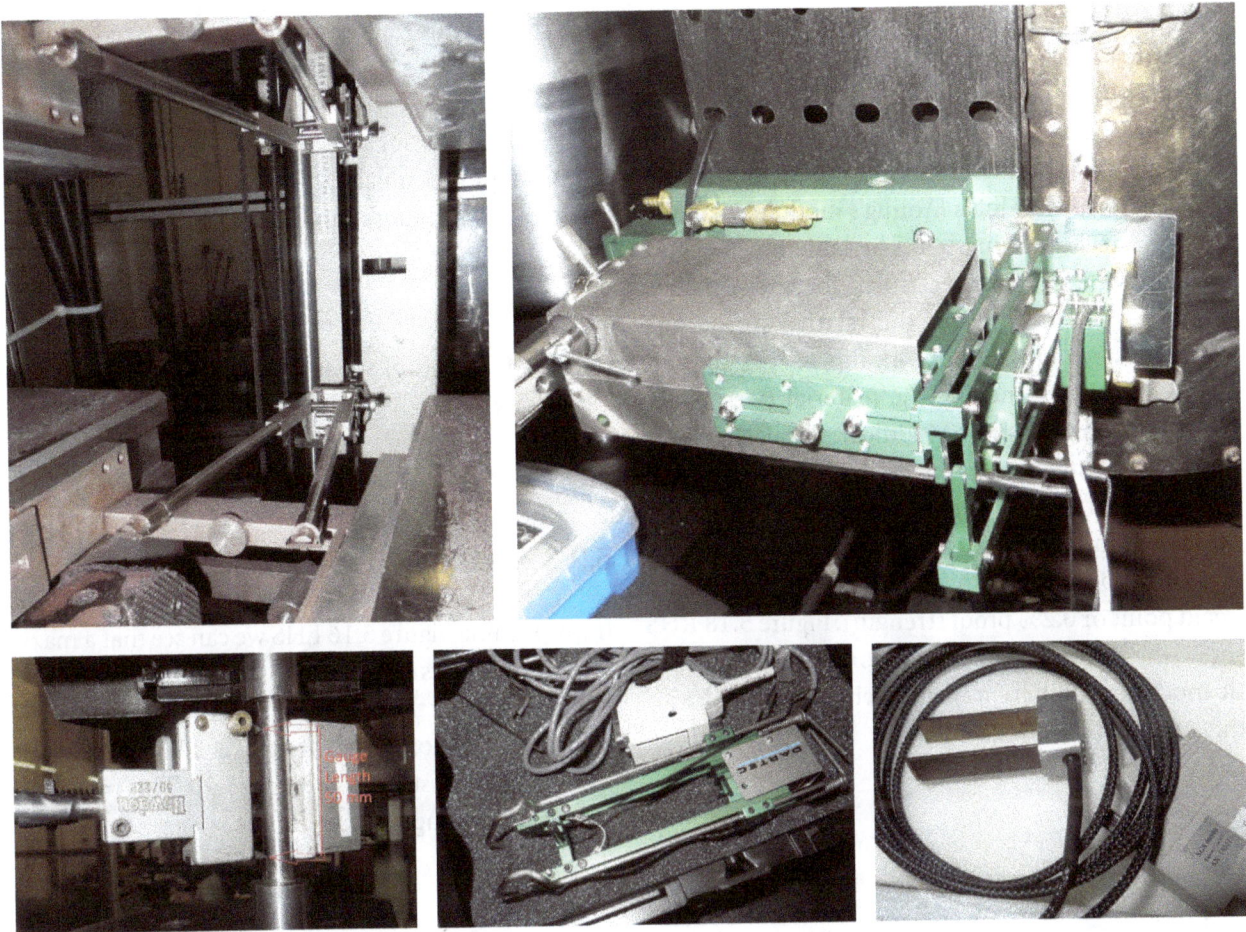

Figure 5.17 *Extensometers. Top LHS = Zwick Roell Macro extensometer. RHS = Epsilon High temperature ceramic rod. Bottom LHS = Clip on extensometer, Centre, Epsilon 25mm gauge length extensometer. RHS = Fracture toughness clip gauge*

saw in Chapter 2, this type of deformation is associated with **the movement of dislocations**. The point at which the linear elastic behaviour ceases is known as the elastic limit. As the load increases beyond the elastic limit the yield point or the 0.2% proof stress would be reached.

In some materials, such as mild steel (e.g. Fe – 0.1% C), the yield point is clearly identifiable and shows a drop-in stress immediately following yielding. This leads to the appearance of an **upper yield stress** (UYS) and a **lower yield stress** (LYS), as shown in Figure 5.16 Centre. With some materials, particularly high strength steels, the transition from elastic to plastic deformation is gradual and it is consequently difficult to identify the yield point. In these cases, a different parameter, known as the **proof stress,** is used. The proof stress is the stress needed to cause a given amount of plastic deformation (e.g. 0.1% or 0.2%) and is calculated as illustrated in Figure 5.16 RHS. The 0.2% proof strength is measured by the use of an extensometer attached to the specimen during testing.

5.2.2 Elastic modulus

Robert Hooke discovered the law of elasticity in 1676, but did not publish immediately. He protected his discovery by announcing it in the form of an anagram: "ceiiinosssttuv." Two years later the solution to the anagram was given in the Latin phrase "ut tensio, sic vis", which translated means "as the extension, so the force".

Hooke's statement expressed mathematically is, F = ku

where

F is the applied force (and not the power, as Hooke mistakenly suggested),

u is the deformation of the elastic body subjected to the force F, and

k is the spring constant (i.e. the ratio of previous two parameters).

If we examine the elastic behaviour, we can see that a measure of the rigidity or flexibility of the metal can be obtained from the behaviour in the elastic region. Consider the range of material shown in Figure 5.18. Each line shows a different slope representing a different Elastic modulus (Young's modulus) for aluminium, steel and tungsten. The table on in Figure 5.19 shows the Modulus of elasticity for several metals.

Figure 5.18 LHS shows the relative stiffness of tungsten, steel and aluminium; the steeper the line the greater the stiffness of the material (i.e. a steep line means a high Young's Modulus). The dashed lines show that the stress in the materials would continue to rise linearly if the strain were increased. The increase would be limited and terminated due to the presence of dislocations and the onset of plastic deformation (yield point or 0.2% proof strength). Figure 5.18 RHS shows the atomistic explanation for the different elastic modulus of different metals being associated with the strength of atomic bonds.

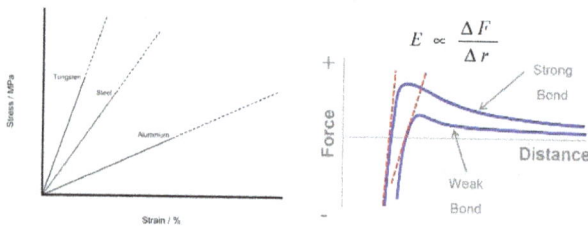

Figure 5.18 *Materials with different modulus of elasticity*

Metal/Alloy	Modulus of Elasticity GPa	Shear Modulus GPa	Poisson's Ratio
Aluminium	69	25	0.33
Brass	97	37	0.34
Copper	110	46	0.34
Magnesium	45	17	0.29
Nickel	207	76	0.31
Steel	207	83	0.30
Titanium	107	45	0.34
Tungsten	407	160	0.28

Figure 5.19 *Materials with different modulus of elasticity*

We can now see the key tensile parameters on the following diagram.

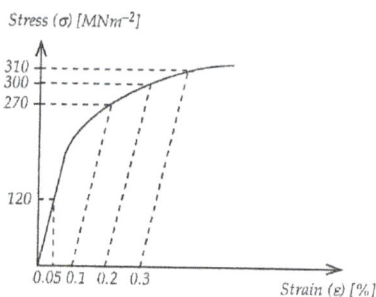

What are the values of the following parameters shown in the above diagram?

(i) The 0.2% proof stress?
(ii) Young's modulus? (N.B. You must use fractional strain for Young's modulus, not percentage.)

Answers

(i) The 0.2% proof stress is 300 MN m^{-2}.
(ii) 0.05% is 0.05 / 100 as a fraction.

$$Young's\ Modulus = \frac{120 \times 10^6}{\frac{0.05}{100}} = \frac{120 \times 10^6 \times 100}{0.05} = 240 \times 10^9\ Nm^{-2} = 240\ GNm^{-2}$$

5.2.3 Tensile strength (UTS)

If we examine Figure 5.16 LHS we can see that a maximum stress exists on the stress-strain curve. It is found that at this point the specimen begins to 'thin' down in a particular region rather than over the whole of the specimen, as illustrated in Figure 5.20. This process is known as necking.

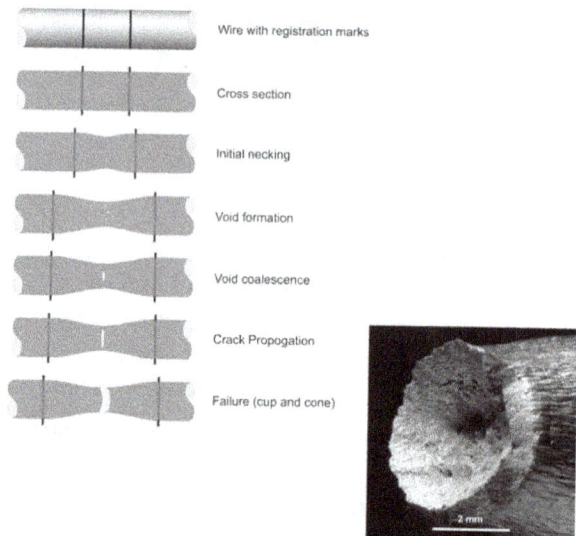

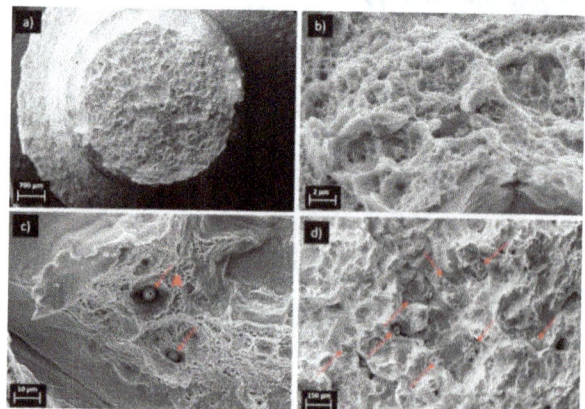

Figure 5.20 *Deformation and fracture of a tensile test specimen, https://www.youtube.com/watch?v=uW0PRpw-RCU*

The stress corresponding to the onset of necking (i.e. the maximum stress in the curve) is known as the tensile strength or the ultimate tensile strength (UTS). This parameter is often used as a measure of the strength of a metal. However, many think that this has limitations since it only refers to uniaxial stresses which are unusual to find in service environments and that once parts begin to yield then failure has already occurred. It took many years for design specifications to change from being based on a percentage of the UTS to being based on a percentage of the 0.2% proof strength.

5.2.4 Tensile ductility

The amount of deformation prior to fracture can be measured by either the percentage elongation to fracture, or the percentage reduction in area at fracture and this quantity is referred to as the tensile ductility. Figure 5.21

$$\% \, Elongation \, at \, Fracture = \frac{Final \, Gauge \, length \, at \, Fracture - Initial \, Gauge \, Length}{Initial \, Gauge \, Length} \times 100$$

$$\% \, Reduction \, of \, Area = \frac{Initial \, Area - Final \, Area \, at \, Fracture}{Initial \, Area} \times 100$$

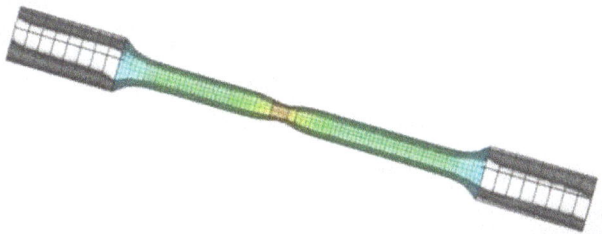

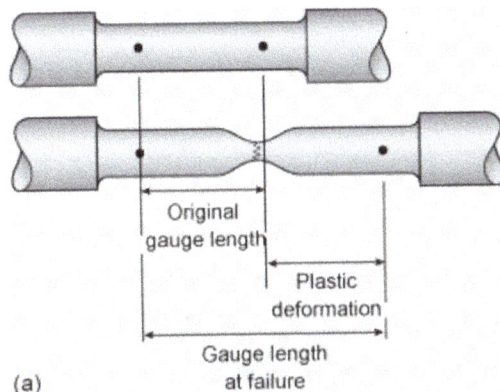

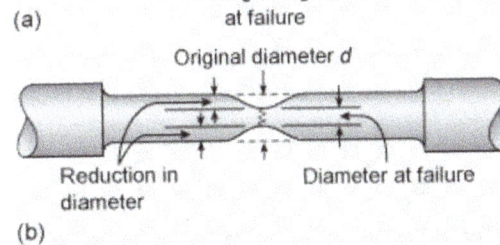

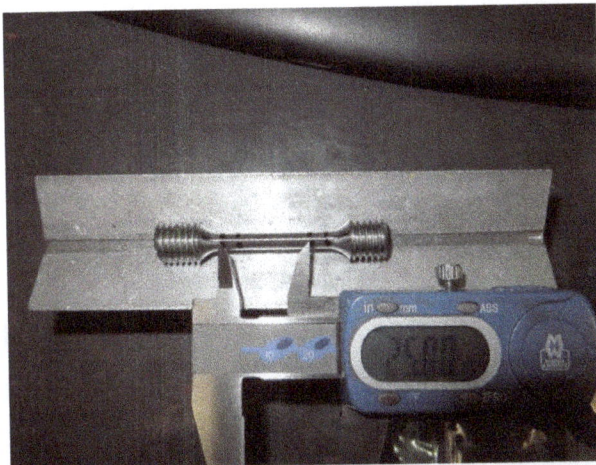

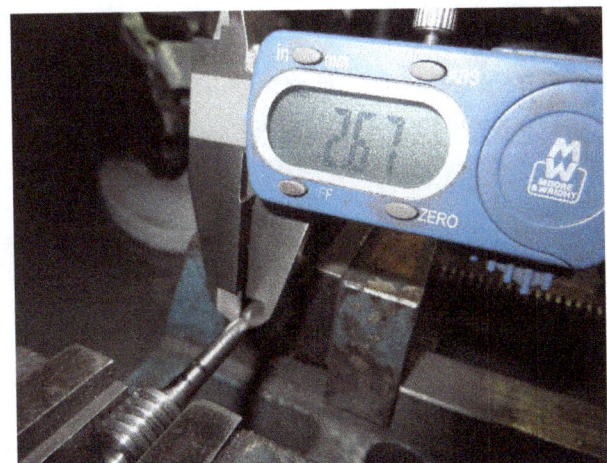

Figure 5.21 *Top= Marking the gauge length Lower LHS = Measurement of elongation, RHS = Measurement of reduction of area*

A summary of the value of elongation and reduction of area

%Elongation – %Elongation is the sum of the "uniform" elongation and the "local" elongation formed after necking. The main part is the uniform elongation which is dependent on the strain-hardening capacity of the material.

Reduction of Area – Reduction of area is more a measure of the deformation required to produce failure and its chief contribution results from the necking process. Because of the complex state of stress in the neck, values of reduction of area are dependent on specimen geometry, and deformation behaviour and they should not be taken as true material properties. RA is the most structure sensitive ductility parameter and useful in detecting quality changes in the materials.

Now see if you can calculate the values required from the following tensile test data.

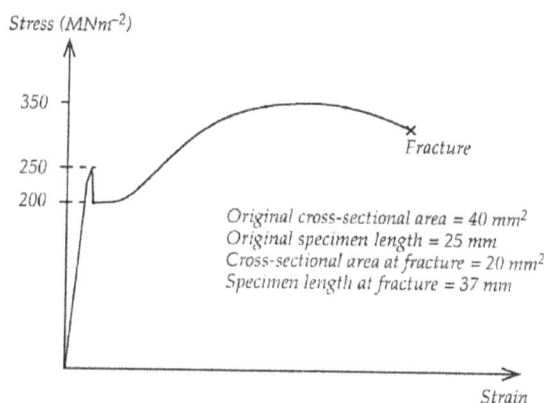

Figure 5.22 *Tensile test curve*

(i) Upper Yield Stress
(ii) Tensile strength
(iii) Lower Yield stress
(iv) Percentage elongation at fracture
(v) Reduction in area at fracture

Answers

(i) 250 N/mm^2
(ii) 350 N/mm^2
(iii) 200 N/mm^2
(iv) 48%
(v) 50%

Two supplementary considerations with tensile testing are:

5.2.5 The Bauschinger Effect[46-49]

The Bauschinger effect was observed by Bauschinger in 1881. He noted that after plastic deformation the elastic limit of a metal was lower for subsequent loading in the reverse direction compared with the loading in the same direction.

One of the key areas where this effect has been observed are formed pipe, springs and thin films.

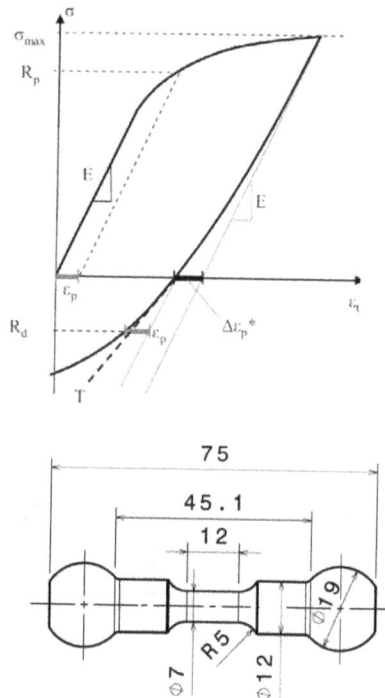

Figure 5.23 *LHS = Schematic drawing of Bauschinger parameters, RHS = Geometry of tension-compression test specimens*

A metallic specimen as shown in Figure 5.23 is deformed plastically in tension up to a tensile stress of + σ max and is then subjected to a compressive strain. The specimen will first contract elastically and then, instead of yielding plastically in compression at a stress of – σ Rp as might have been expected, it is found that plastic compression starts at a lower stress (– σ Rd) this phenomenon known as the Bauschinger effect (BE)

The BE occurs because, during the initial tensile plastic straining, internal stresses accumulate in the test-piece and oppose the applied strain. When the direction of straining is reversed these internal stresses now assist the applied strain, so that plastic yielding commences at a lower stress than that seen in tension.

The BE may be encountered in the measurement of the yield strength of some line-pipe steels.

The measured yield strength in such specimens can be significantly lower than that obtained on the (undeformed) plate from which the pipe-line is manufactured. The plate material thus has to be supplied with extra strength to compensate for this apparent loss in yield strength

5.2.6 The Considère construction.

In the engineering stress strain plot the engineering stress after the UTS is achieved, falls due to the presence of the necking. If the true stress, based on the actual cross-section (A) of the gauge length, is used, the stress–strain curve increases continuously to fracture, as indicated in adjacent plot of true stress-strain curve. (lower diagram)

The presence of a neck in the gauge length at large strains introduces local triaxial stresses that make it difficult to determine the true longitudinal tensile stress in this part of the curve and correction factors have to be applied. True stress-strain curve illustrates the continued work hardening but unlike the engineering stress–strain curve, the strain at the point of plastic instability and the UTS are not apparent.

However, these values may be readily obtained from

$$d\sigma/de = \sigma u/(1 + eu)$$

Where σu is the true stress at the UTS and eu is the strain at the point of plastic instability. This allows identification of these values by means of the **Considère construction**, whereby a tangent to the true stress– strain curve is drawn from a point corresponding to –1 on the strain axis.

Additionally, the intercept of this tangent on the stress axis will give the value of the UTS. In addition, the Considère tangent to the true stress–strain curve identifies both the point of plastic instability, the true stress at the UTS and the value of the UTS itself

5.3 HARDNESS

Hardness is a measure of the ability to resist indentation. In metals it is determined by pressing an indenter in the form of a carbide ball or a pyramid or cone type diamond into a flat surface of the metal. The hardness is a stress (force/area) but for historical reasons the

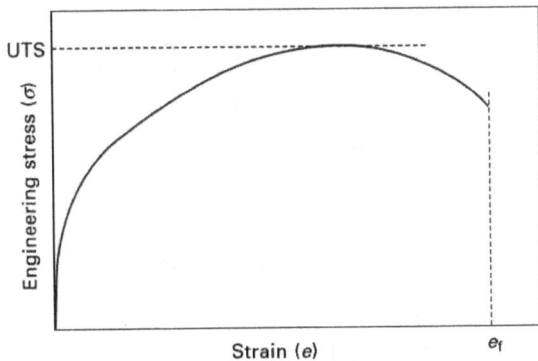

Engineering stress–strain curve.

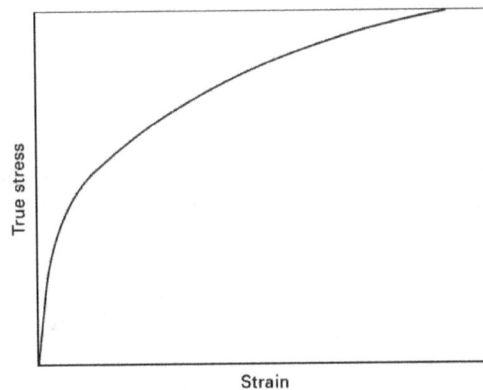

True stress–strain curve.

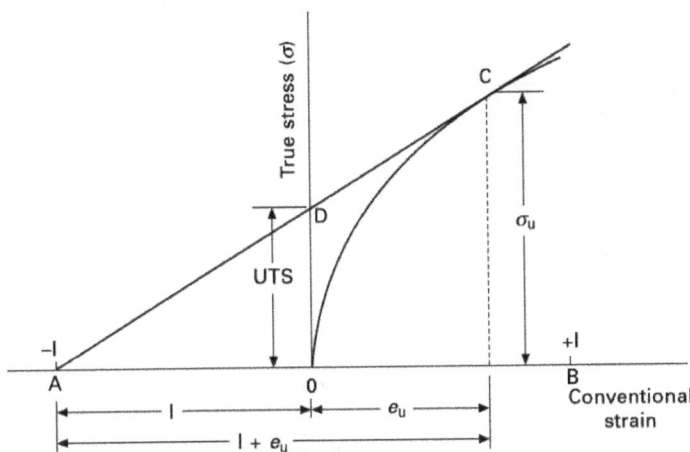

Figure 5.23a

area used for the calculation is not the projected area but on the surface area of the indentation.

The state of stress under the indenter represents a complex 3D stress system with a strong hydrostatic component which constrains the yielding. Due to this constraint the hardness of a metal is roughly three times the metal's yield strength. HV= 3 × YS (MPa) HB=UTS/2 (ksi). In 1951 the progress in the understanding of hardness was reviewed by D Tabor.[50] Figure 5.24 summarises the current understanding.

A more recent technical update can be obtained from the book by Konrad Herrman, Hardness Testing Principles and Applications, ASM International.[52 and 53]

Hardness test methods may be classified into two types: static and dynamic. The static methods involve the measurement of dimensions of the indentation (Brinell, Vickers, Knoop) or the measurement or the depth of indentation (Rockwell, Nano-indentation). The static hardness tests may also be classified as the traditional tests, which measure one contact area or penetration depth at a prescribed load (Brinell, Vickers, Knoop, Rockwell), and the recent Instrumented Indentation Tests (IIT or BIT "Ball Indentation Test"), which allows for a continuous measurement of load and displacement.

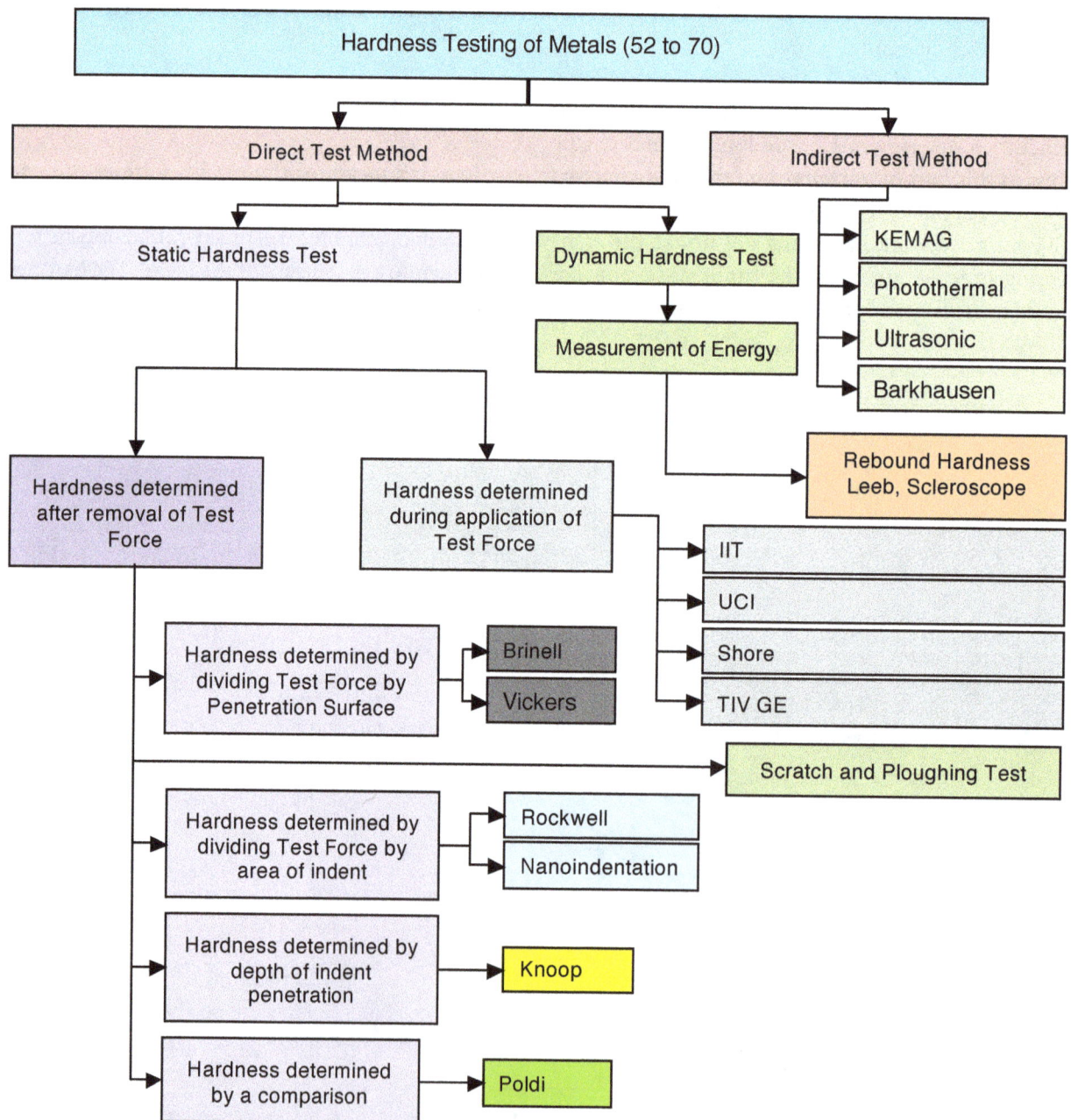

Figure 5.24 *Summary of the current status of hardness testing, the main innovation is macro IIT.*

In the dynamic tests, the indenter, usually spherical or conical, is allowed to bounce off the surface of the material to be tested, and the rebound height of the indenter is used as a measure of hardness. The Shore Scleroscope (1907) is the most well-known test of this type. In recent years the Leeb (Equotip) and the UCI (Microdur) can all be classified as dynamic test methods. The Poldi test, which was named after the Poldihütte (steel plant in Kladno near Prague),

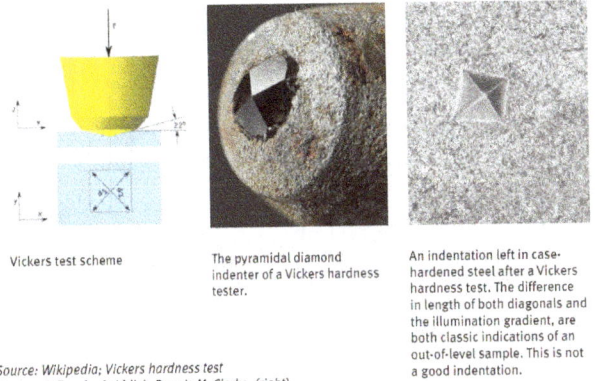

Vickers test scheme

The pyramidal diamond indenter of a Vickers hardness tester.

An indentation left in case-hardened steel after a Vickers hardness test. The difference in length of both diagonals and the illumination gradient, are both classic indications of an out-of-level sample. This is not a good indentation.

Source: Wikipedia; Vickers hardness test
Photos: R. Tanaka (middle); Dennis M. Clarke, (right)

Figure 5.25 *Vickers hardness test*

EN ISO 6506-1	Metallic materials – Brinell hardness test – Part 1: Test method
EN ISO 6506-2	Metallic materials – Brinell hardness test – Part 2: Verification and calibration of Brinell hardness testing machines
EN ISO 6506-3	Metallic materials – Brinell hardness test – Part 3: Calibration of reference blocks
EN ISO 6506-4	Metallic materials – Brinell hardness test – Part 4: Table of hardness values
ASTM E 10	Standard test method for Brinell hardness of metallic materials

The results are expressed in the following format:

600 HBW 1/30/20

which contains the following information:

600 .... Brinell hardness value

HB .... Hardness symbol

W .... indication of indenter type; tungsten carbide[1]

1 .... ball diameter in mm

30 .... approximate kgf equivalent value of applied test force where 30 kgf = 294.2 N

20 .... duration of test force (20 s) if not within the specified range (10 s to 15 s)

EN ISO 6508-1	Metallic materials – Rockwell hardness test – Part 1: Test method (scales A, B, C, D, E, F, G, H, K, N, T)
EN ISO 6508-2	Metallic materials – Rockwell hardness test – Part 2: Verification and calibration of testing machines (scales A, B, C, D, E, F, G, H, K, N, T)
EN ISO 6508-3	Metallic materials – Rockwell hardness test – Part 3: Calibration of reference blocks (scales A, B, C, D, E, F, G, H, K, N, T)
ASTM E 18	Standard test methods for Rockwell hardness of metallic materials
ISO/TR 10108	Steel – Conversion of hardness values to tensile strength values
ISO 18265	Metallic materials – Conversion of hardness values
ISO 14577-1	Metallic materials – Instrumented indentation test for hardness and materials parameters – Part 1: Test method
ISO 14577-2	Metallic materials – Instrumented indentation test for hardness and materials parameters – Part 2: Verification and calibration of testing machines
ISO 14577-3	Metallic materials – Instrumented indentation test for hardness and materials parameters – Part 3: Calibration of reference blocks
ASTM E 140	Standard hardness conversion tables for metals relationship among Brinell hardness, Vickers hardness, Rockwell hardness, superficial hardness, Knoop hardness, Scleroscope hardness and Leeb hardness

EN ISO 6507-1	Metallic materials – Vickers hardness test – Part 1: Test method
EN ISO 6507-2	Metallic materials – Vickers hardness test – Part 2: Verification of testing machines
EN ISO 6507-3	Metallic materials – Vickers hardness test – Part 3: Calibration of reference blocks
EN ISO 6507-4	Metallic materials – Vickers hardness test – Part 4: Tables and hardness values
ASTM E 384	Standard Test Method for Knoop and Vickers Hardness of Materials

Ranges of test force, F N	Hardness symbol	Designation
$F \geq 49.03$	$\geq HV5$	Vickers hardness test
$1.961 \leq F < 49.03$	$HV0.2$ to $< HV5$	Low-force Vickers hardness test
$0.09807 \leq F < 1.961$	$HV0.01$ to $< HV0.2$	Vickers micro-hardness test

Macro range	Micro range	Nano range
$2\,N \leq F \leq 30\,kN$	$2\,N > F$; $h > 0.2\,\mu m$	$h \leq 0.2\,\mu m$

Elastic resp. elastic-plastic properties	Plastic properties
Martens hardness HM	indentation hardness H_{IT}
Martens hardness, derived from the slope of the force-depth-curve HM_s	indentation creep C_{IT}
elastic indentation modulus E_{IT}	plastic deformation work W_{plast}
indentation relaxation R_{IT}	yield strength σ_y
elastic deformation work W_{elast}	
mechanical deformation work W_{total}	

Tables 5.1 *Top LHS = Specifications for Brinell, Top RHS = Specifications for Vickers Next LHS = How to express a Brinell result, Next RHS = Definitions for the Vickers hardnes load ranges, Next LHS = Specifications for Rockwell, Next RHS = ISO 14577 specified IIT (instrumented indentation test) ranges. Bottom LHS = General hardness specifications, Bottom RHS = Properties that can be determined from an IIT hardness test.*

where it was invented around the 1920s, has remained popular throughout history for the testing of large pieces of metal despite the recent development of a range of electronic hardness testers.

The main conventional hardness test methods are:

- Vickers Hardness Test (1925) (BS EN ISO 6507-1:2018, ASTM E92-17)
- Brinell Hardness Test (1900) (BS EN ISO 6506-1:2014, ASTM E10-18
- Rockwell Hardness Test (1920) (BS EN ISO 6508-1:2016, ASTM E18-19)

In the Vickers and Brinell test methods the indenter (a square based diamond pyramid and a hard tungsten carbide ball respectively) is applied to the metal being tested using a fixed load and for a fixed time. The diameter of the indentation is then measured and converted to a scale number using standard conversion tables, i.e. Vickers Hardness Number (VHN) or Brinell Hardness Number (BHN).

The Rockwell test is slightly different from the Vickers and Brinell tests in that the hardness is assessed from the depth of penetration of either a diamond pyramid or ball indenter. A summary of the

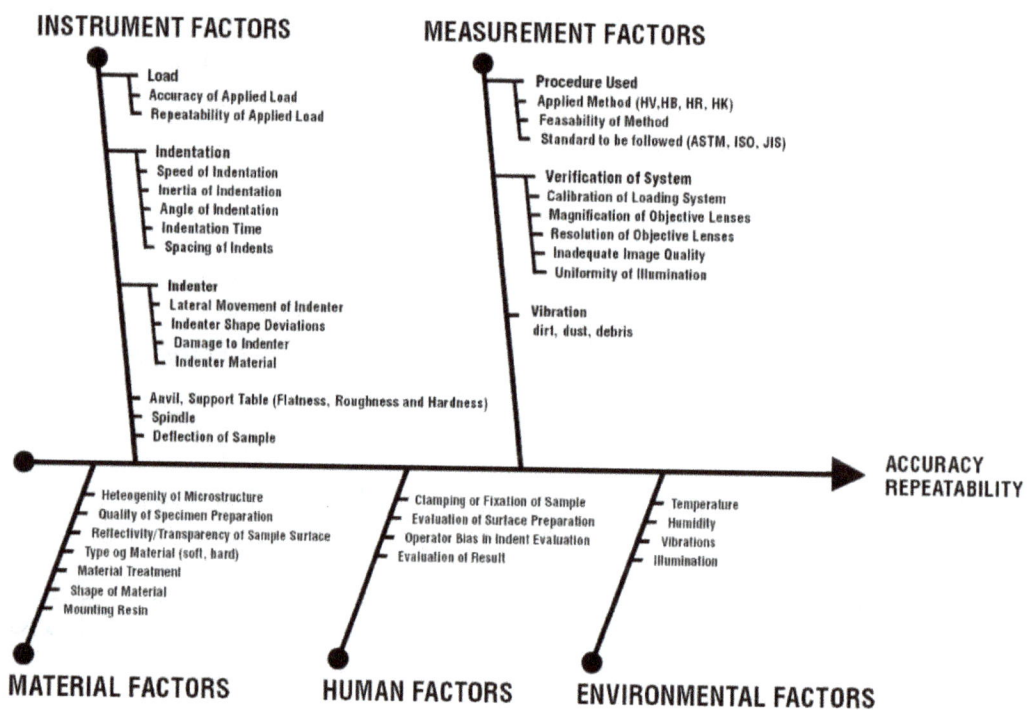

Figure 5.26 *Fish bone diagram showing factors affecting hardness accuracy. From https://www.struers.com/en-GB/Knowledge/Hardness-testing#*

Rockwell & Brinell

According to ASTM and ISO Standards: The distance between the centers of two adjacent indentations shall be at least three times the diameter (d) of the indentation.

The distance from the center of any indentation to an edge of the test piece shall be at least two and a half times the diameter of the indentation.

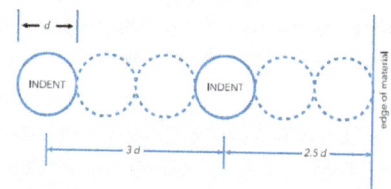

Vickers

According to ASTM Standards: The distance between two indents or an indent and the edge of the test piece shall be at least two and a half times the diagonal (dV) of the indentation.

According to ISO Standards: The distance between the centers of two indents shall be at least three times the diagonal (dV) of the indent for steel, copper and copper alloys, and at least six times for light metals, lead and tin and their alloys.
The distance between the center of an indent and the edge of the test piece shall be at least two and a half times the diagonal (dV) for steel, copper and copper alloys, and at least three times for light metals, lead and tin and their alloys.

Figure 5.27 *Recommended spacing for hardness indents. From https://www.struers.com/en-GB/Knowledge/Hardness-testing#*

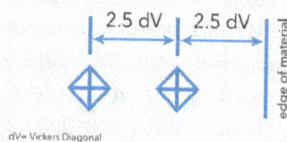

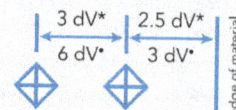

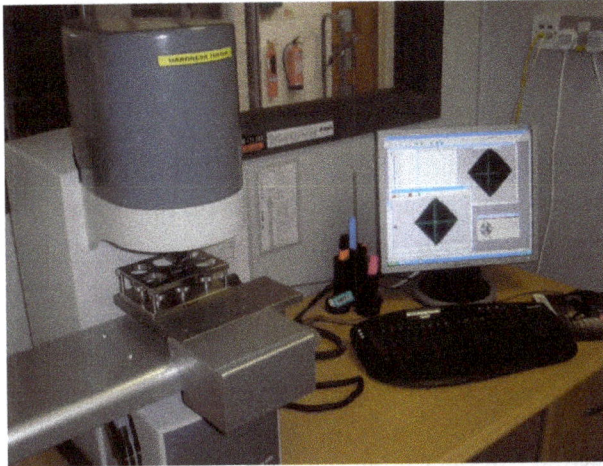

Figure 5.28 Fully automated Vickers hardness tester

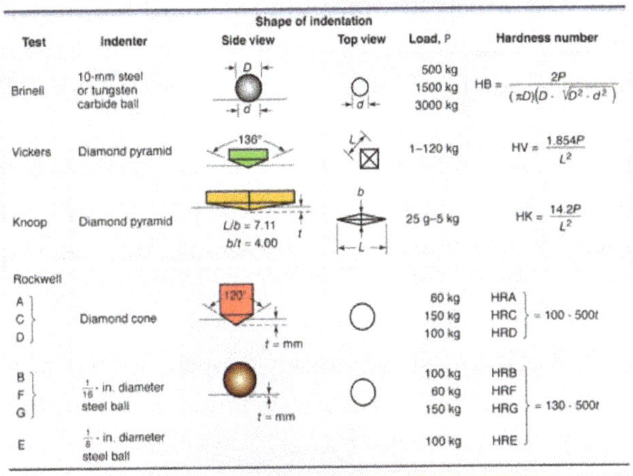

Figure 5.29 A summary of the various test methods

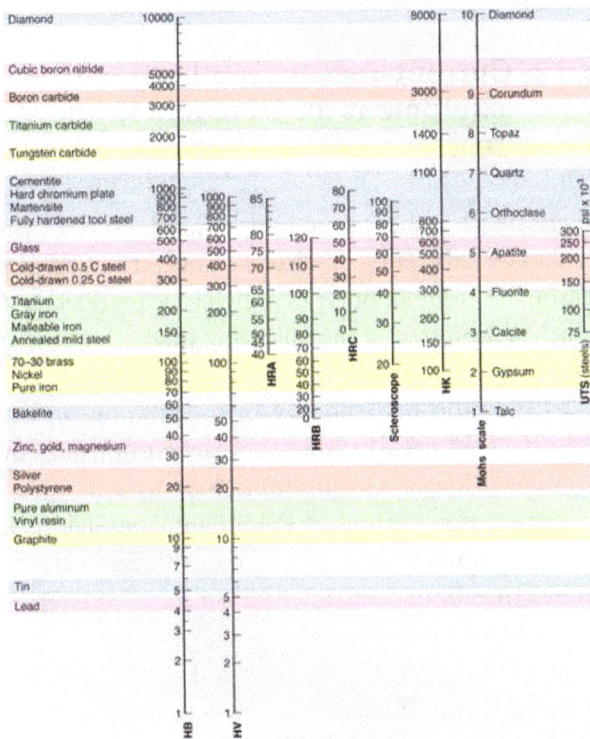

Figure 5.30 Comparison of hardness ranges and the expected hardness levels of various materials

various test methods is shown in Figure 5.29. Figure 5.30 shows an approximate range of applicable hardness and the approximate relationship between the various hardness scales. The scale on the RHS shows the tensile strength. These hardness conversions are available in ASTM 140 and ASTM 370 specifications. Often hardness tests are used for the quality control confirmation that products have met the required tensile strength.

5.3.1 A Brief History of Hardness Testing

Table 5.2 Timeline of important events relating to hardness testing

Date	Event
1889	A Martens reported on a procedure of scratch hardness
1898	A Martens proposed a device for the instrumented indentation test with mechanical, hydraulic depth measurement.
1900	J A Brinell developed the ball indentation test
1907	P Ludwik reported on a cone probe used both statically and dynamically A F Shore created the rebound hardness measurement method E Meyer set up the power law for the ball indentation test
1920	S R Rockwell developed the test force method named after him
1925	R Smith and G Sandland developed the Vickers method which allowed micro-hardness testing
1926	In the USA the first hardness test standards for the Brinell test method were introduced
1937	In Germany the first hardness reference blocks were produced
1939	F Knoop developed the knoop test method
1940	First DIN standard on the Vickers test method
1943	K Myer built the first hardness standard measuring machine
1970	D Leeb developed the Equotip, the rebound hardness testing method

The first widely accepted and standardized indentation-hardness test was proposed by J. A. Brinell in 1900. August Brinell (21 November 1849 – 17 November 1925), a Swedish Mechanical Engineer, began his career as an Engineer at the Lesjöfors Ironworks and in 1882 became chief engineer at the Fagersta Ironworks. In 1903 he became Chief Engineer at Jernkontoret, the Swedish Iron Industry's Trade Association. He remained at that post until 1914.

The Brinell test was quickly adopted as an industrial test method soon after its introduction, but several limitations also became apparent.

- The long test duration,
- the large size of the impressions from the indent,
- and the fact that high-hardness steels could not be tested with the Brinell method of the early 1900s.

The limitations of the indentation test developed by Brinell prompted the development of other macro-indentation hardness tests, such as the Vickers test introduced by R. Smith and G. Sandland in 1925, and the Rockwell test invented by Stanley P. Rockwell in 1919.

The Vickers hardness test was designed to be a more precision test for "engineering components" compared to as cast or as forged components where the Brinell test was well established. The Vickers test uses the same principle as the Brinell, that of a regulated impression on the material, but instead utilized a pyramid shaped diamond rather than the Brinell ball indenter. This resulted in a more consistent and versatile hardness test.

Later, in 1939, an alternative to the Vickers test was introduced by Fredrick Knoop at the US National Bureau of Standards. The Knoop test utilized a shallower, elongated format of the diamond pyramid and was designed for use under lower test forces than the Vickers hardness test, allowing for more accurate testing of brittle or thin materials. Both the Vickers and Knoop tests continue as popular hardness analysis methods today.

Although conceived as an idea in 1908 by a Viennese professor, Paul Ludwik, the Rockwell indentation test did not become of commercial importance until around 1914 when brothers Stanley and Hugh Rockwell, working from a manufacturing company in Bristol, Connecticut, expanded upon the idea of utilizing a conical diamond indention test based on displacement and applied for a patent for a Rockwell tester design. The principal criterion for this tester was to provide a quick method for determining the effects of heat treatment on steel bearing races. One of the main strengths of the Rockwell was the small area of indentation needed. It is also much easier to use as readings are direct, without the need for calculations or secondary measurements. The patent application was approved on February 11, 1919.

Rockwell hardness testing is the most widely used method for determining hardness, primarily because the Rockwell test is fast, simple to perform, and does not require highly skilled operators. By use of different loads (force) and indenters, Rockwell hardness testing can determine the hardness of most metals and alloys, ranging from the softest bearing materials to the hardest steels.

5.3.2 Current Testing Methods and Equipment for Dynamic Hardness Test

There are a number of methods of hardness test with dynamic loads. These methods can be arranged in a reasonable number of types of dynamic hardness test which is expressed in the following ways:

- as the ratio between size of indentation on the surface of a test sample compared to the size of indentation of a calibration block with the same dynamic force of impact: Poldi, Baumann-Steinrück
- as a product between a constant value and elastic height rebound of indenter on tested, material, Shore hardness
- as a ratio between a constant and a ratio of elastic recoil energy of indenter on the tested material and impact energy, Leeb methods (US patent expired, there are many Chinese instruments).

In theory dynamic hardness values do not accurately correspond with that statically determined hardness values. However, there is usually extensive correlation work carried out so that for most industrial applications the methods are regarded as having adequate accuracy.

5.3.3 Poldi

The name "Poldi "comes from the steel plant of the same name in the Czech Kladno, where this testing method was invented. The Poldi Hardness Tester is suitable for measuring the Brinell Hardness of Steel, Cast iron, Brass Aluminium, Copper. This Poldi Hardness Tester is useful to test large parts too big for normal test methods. Any small components should be placed on top of a large mass of metal prior to testing. Tests have shown that the results are within 5% of the actual hardness. The accuracy being dependent on the accuracy of reading the two indents. The test would not be suitable for testing very hard material.

The area to be tested should be ground or filed flat to produce a smooth flat surface. A high-quality polished surface is not required; a medium surface condition would be adequate so long as the hardness indent can be accurately measured. The hardness tester must be held vertically over the sample. Load is applied by a hammer blow on the sample being tested and a standard test bar inserted and positioned on the other side of the 10 mm dia. steel ball. The impact load on both the sample and the standard test bar is the same.

The diameter of indentation on the specimen and the test bar are measured. The harder the sample being tested, the bigger the difference in the diameter of the indent of the standard test bar and the sample.

The hardness of the specimen can be determined, by referring to charts supplied with the hardness tester or calculated from the following formula:

$$Ha = Hs \left(\frac{Ds}{Da}\right)^2$$

Ha = Hardness of sample
Hs= Hardness of calibrated bar
Da= Diameter of indent on sample
Ds= Diameter of indent on calibrated test bar

This test method is still popular despite the availability of other accurate electronic portable hardness testers due to the "Brinell type" indentation.

5.3.4 UCI (Ultrasonic Contact Impedance) MIC 10 The UCI hardness measuring method complies to ASTM A1038

A UCI probe essentially consists of a Vickers diamond attached to the end of a metal rod. This rod is excited into longitudinal oscillation by piezoelectric transducers. The diamond is then pressed into the surface of the component.

The frequency shift is proportional to the size of the test indentation produced by the Vickers diamond. Therefore, the diagonals of the test indentation are not optically determined for the hardness value, as is usually done, but the indentation area is electronically detected by measuring the frequency shift taking just a few seconds.

This is the method of UCI hardness testing, that the frequency shift is proportional to the size of the Vickers test indentation.

5.4 IMPACT TOUGHNESS TESTING – CHARPY AND IZOD

The impact resistance of a metal is important in assessing the ability of a metal to resist shock loading. It is a measure of the "energy of fracture". This is the energy required to break the atomic bonds in the

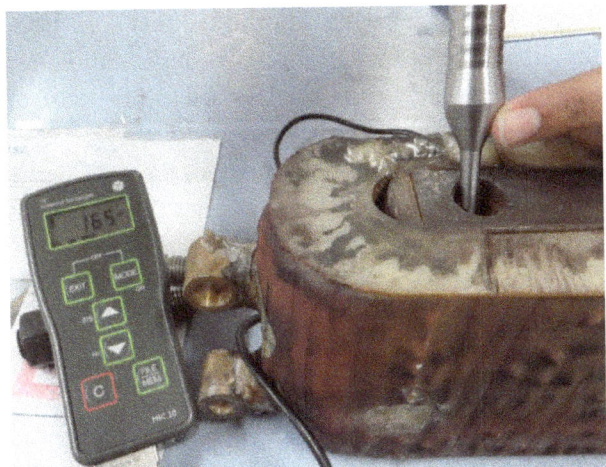

Figure 5.31 LHS = Poldi hardness tester, RHS = MIC 10.

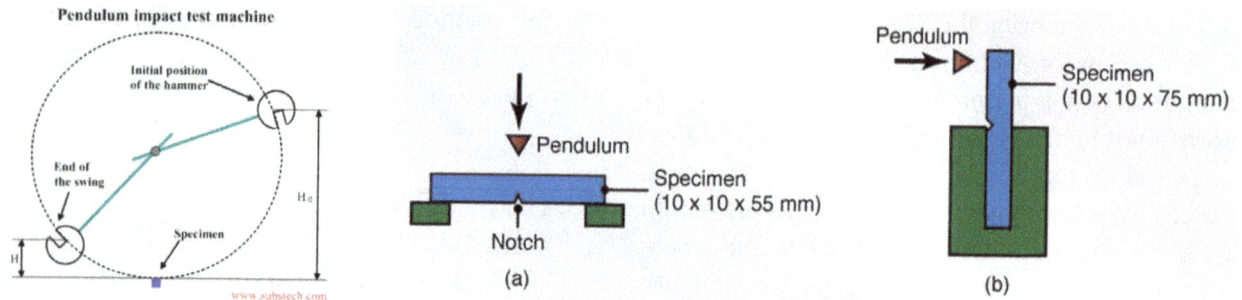

Figure 5.32 *LHS = Specimen position during the impact test for Charpy, RHS = Izod*

Figure 5.33 *MSL's Zwick Roell 450J Pendulum Impact tester.*

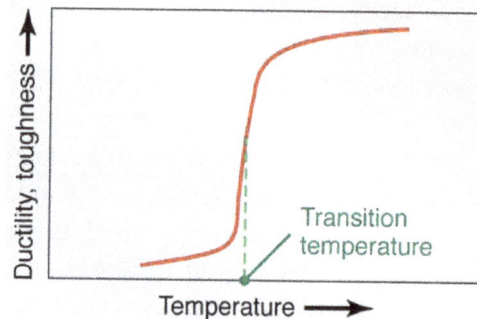

Figure 5.34 *The impact transition temperature (ITT)*

crystalline microstructure. If we wish to build a structure which could experience shock loading in service, e.g. a high-pressure gas pipeline we do not want this structure to shatter like glass when it is impact loaded, either due to a gas pressure surge or an inadvertent collision. Hence, we need a certain amount of impact resistance. The early general rules for a minimum requirement would be based on 27J at 20°C below the operating temperature. However, where the consequences of failures are high, the minimum specified energy would generally increase to 42J. This property has traditionally been measured by using a swinging pendulum to hit and break a specimen of the metal under study. The energy absorbed is calculated from the difference in pendulum height before and after impact. This energy absorbed is used as a measure of the impact resistance.[71]

The two well established methods are the Charpy and the Izod, with the Charpy being the dominant test method in many industrial areas. The Charpy being named after Georges Augustin Albert Charpy, a Frenchman, who was born in 1865 and graduated from the École Polytechnique in 1887 with an engineering degree majoring in marine artillery.

The Izod test is named after the English engineer Edwin Gilbert Izod (1876–1946) who developed the machine when working on gun barrel bursts. Izod

found that when specimens from a burst gun barrel were notched and put into a vice and hit with a hammer it broke in a brittle manner. A similar sample from an acceptable gun barrel bent plastically. Izod presented the concept of his machine in 1903.

The test is similar to the Charpy impact test but uses a different specimen size and arrangement of the specimen under test. The Izod impact test differs from the Charpy impact test in that the sample is held in a cantilevered beam configuration as opposed to a three-point bending configuration for the Charpy as shown in Figure 5.32.

5.4.1 Ductile to Brittle Transition—the Impact Transition Temperature

This impact test shows the relationship of ductile to brittle transition in absorbed energy at a series of temperatures. All metals with a BCC (body centred cubic) crystal structure undergo a transition from ductile fracture mode (which take a high energy of fracture) to brittle mode of fracture (which occurs at a relatively low energy of fracture).

The basic explanation of the transition from ductile to brittle failure as the temperature is decreased is that, as the temperature is decreased, it becomes too difficult to move dislocations (as quantified by the critical resolved shear stress) relative to the stress required to propagate a crack (as quantified by the tensile breaking stress).

Since ferritic based steel is body-centred cubic and undergoes a transition from ductile behaviour at higher temperatures to brittle behaviour at lower temperatures, this test is a basic requirement for the quality control of a number of important steel products, particular at sub-zero temperatures, for steel hull plate for ships, industrial process plant, nuclear plant pressure vessels, forgings for electric power plant generator rotors.

The task of processing steel to give a low ITT (impact transition temperature) presents a major challenge to engineers and will be considered in detail in later chapters.

One feature worthy of note was the move to cleaner steels and the effect on the Charpy shelf energy during the 1970s (see Figures 5.35 RHS and 5.38). Once the shelf energy went to high values it appeared to delay the onset of the change to brittle behaviour.

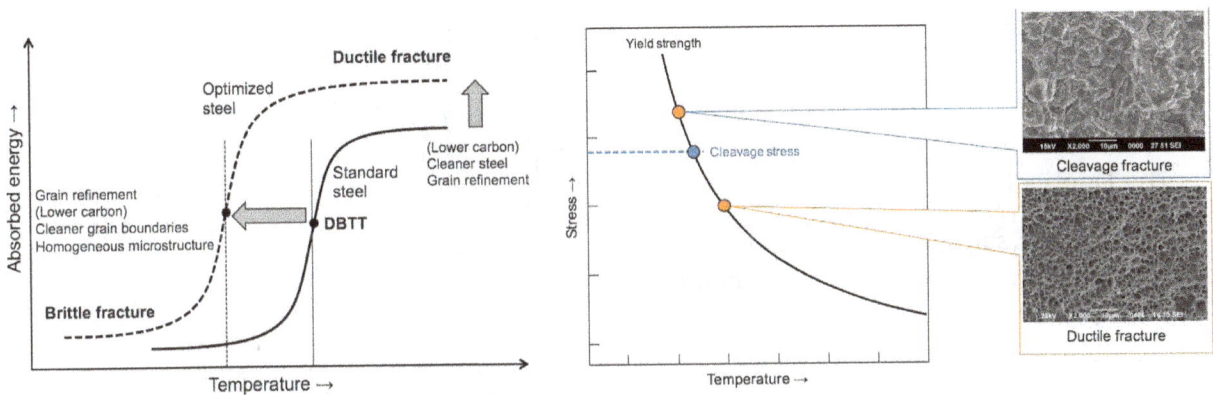

Figure 5.35 *LHS = How to improve the impact transition temperature, RHS = Reason for ductile to brittle*

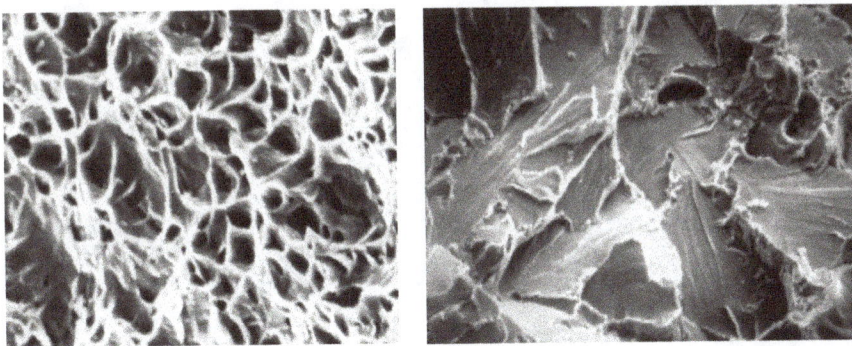

Figure 5.36 *SEM images of Charpy fracture surfaces LHS = ductile mode RHS = brittle mod*

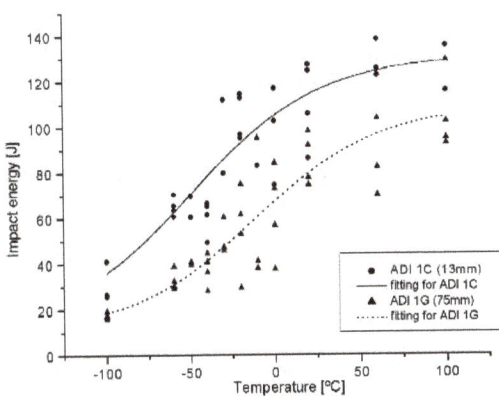

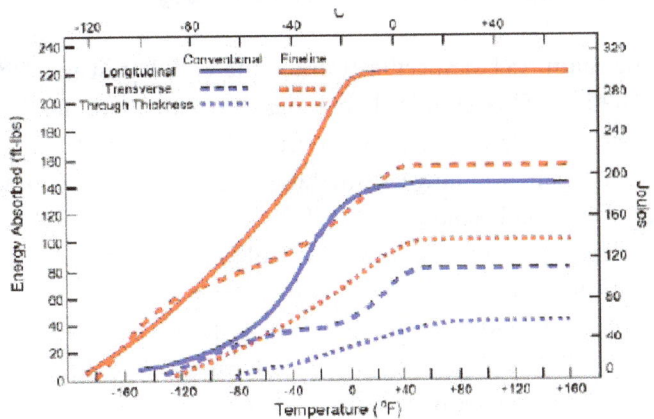

Figure 5.37 *LHS = An example of real data on an austempered ductile iron, RHS = Charpy data*

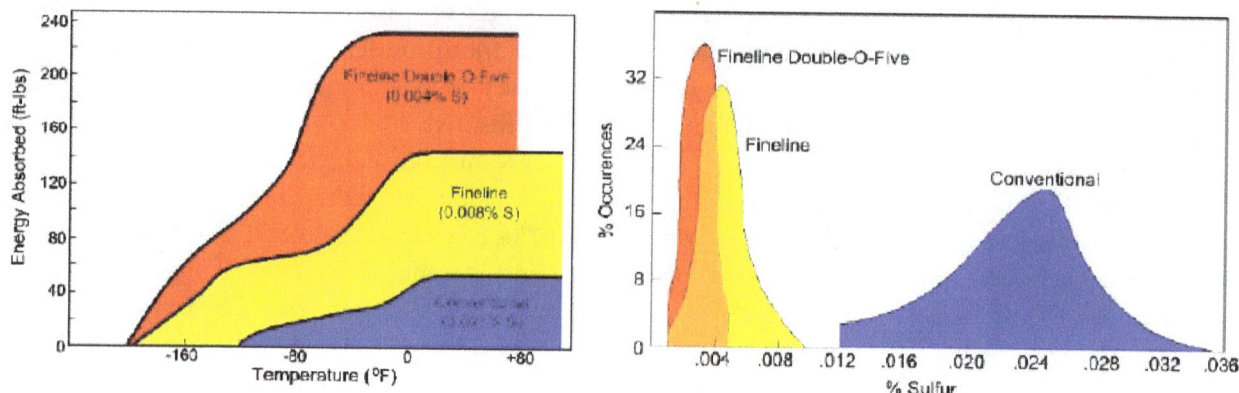

Figure 5.38 *LHS = Effect of sulphur on Charpy energy, RHS = Range of sulphur levels for conventional and lower sulphur grade*

To understand the factors which affect the toughness measurements using the Charpy method, the ITT curve can be split into the ductile shelf energy and the impact transition temperature.

Charpy Ductile Shelf Energy

- Steel cleanness, sulphur level and oxygen level.
- The carbon levels. Steels with low carbon levels have superior toughness.

The impact transition temperature (ITT)

- Free nitrogen and therefore the aluminium content in the steel is important.
- Grain size and Strength level.
- Low sulphur levels improve the toughness.
- Precipitation hardness causes a rise in the ITT.
- Microstructure. Figure 5.39 LHS shows an old graph (1948) which shows that the importance of the microstructure has been known for many years.

5.4.2 Fracture Toughness testing

The main ASTM specification for the CTOD test is E1820.[72] This standard includes procedures that allow the determination of the fracture toughness of metals using the range of methods- K, J, and Crack-Tip Opening Displacement (CTOD).

The fracture toughness is calculated by using a specimen with a notch that has been pre-cracked by fatigue loading. The specimen is then stressed in 3-point bending while accurately measuring the amount the crack opens. To allow the displacement measurement a clip gauge is used kept in position by

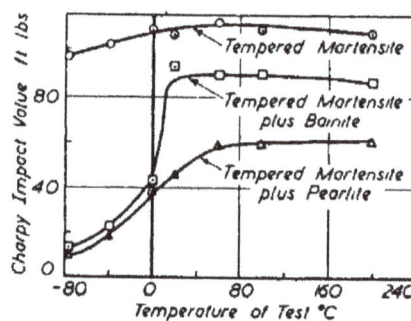

Figure 18. Impact properties of NE8735 steel at tensile strength of 55 tons/sq in in relation to partial pearlitic and intermediate transformation

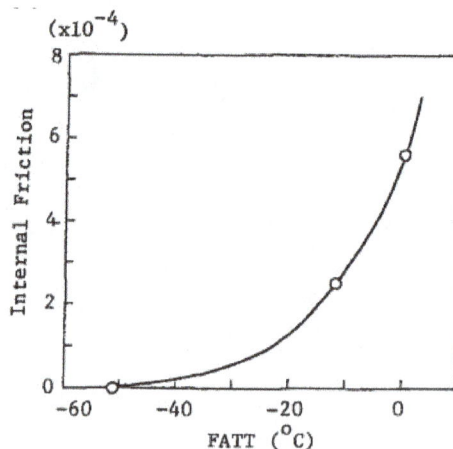

Figure 5.39 *Top = Effect of microstructure on the ITT form 1948, Bottom = Effect of free nitrogen on the ITT*

knife edges fastened either side of the mouth of the machined notch.

There are three possible ways for the specimen to fail:

- fracture instability which results in a single point-value of fracture toughness,
- stable tearing which results in an R-curve, and

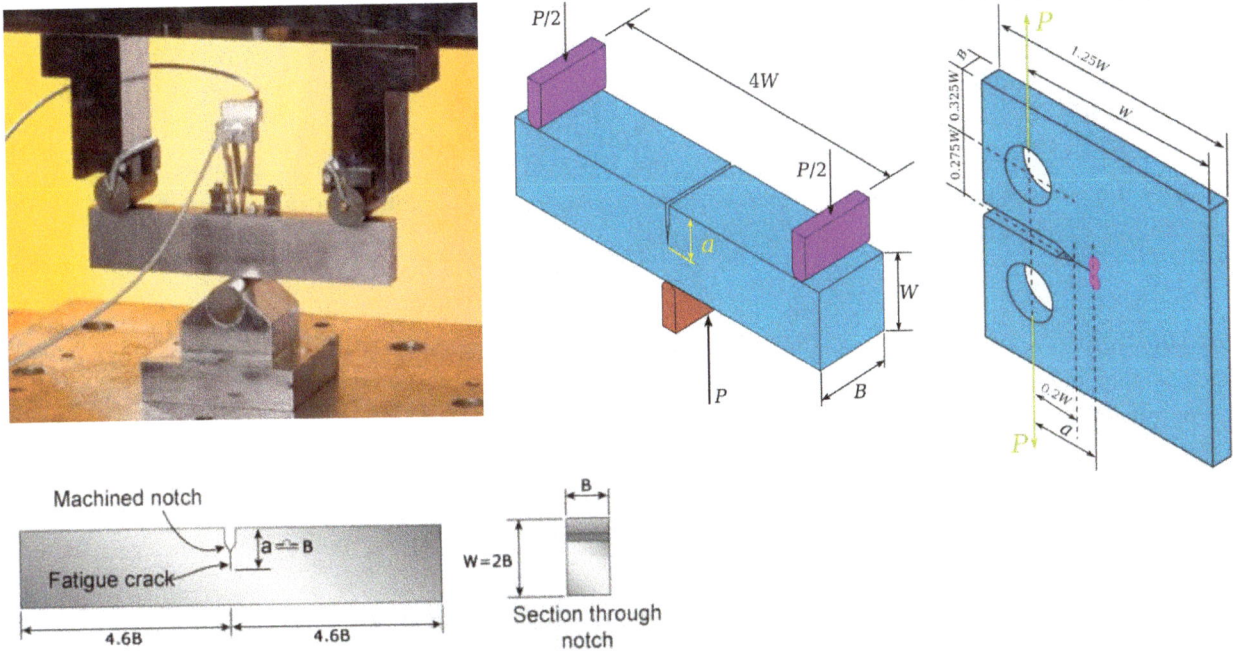

Figure 5.40 Fracture toughness test and specimens

- stable tearing interrupted by fracture instability which results in an R-curve that ends at the point of instability.

In the CTOD test the specimens are proportional. If the thickness is represented by 'A', then the depth will either be 'A', for a square cross section, or, '2A' for a rectangular cross section with the standard length being '4.6A'.

Examination of the fatigue crack surface is necessary to determine the success or failure of the test. The length of the crack itself is accurately measured. If the length of the crack is not within the specified limits the test is invalid. If the crack is not in a single plane, or at an angle to the machined notch, or, if the crack is not in the proper region, the test is invalid.

5.2.3 Hot tensile testing

Hot tensile testing is carried out in accordance with BS EN 6892 (Metallic Materials- Tensile Testing- Part 2 Method of testing at elevated temperatures) or ASTM E21 (Standard Test Methods for Elevated Temperature Tension Tests of Metallic Materials)

The Hot Tensile Test (HTT) allows the measurement of the strength and tensile ductility at high temperature. The elevated temperature properties of

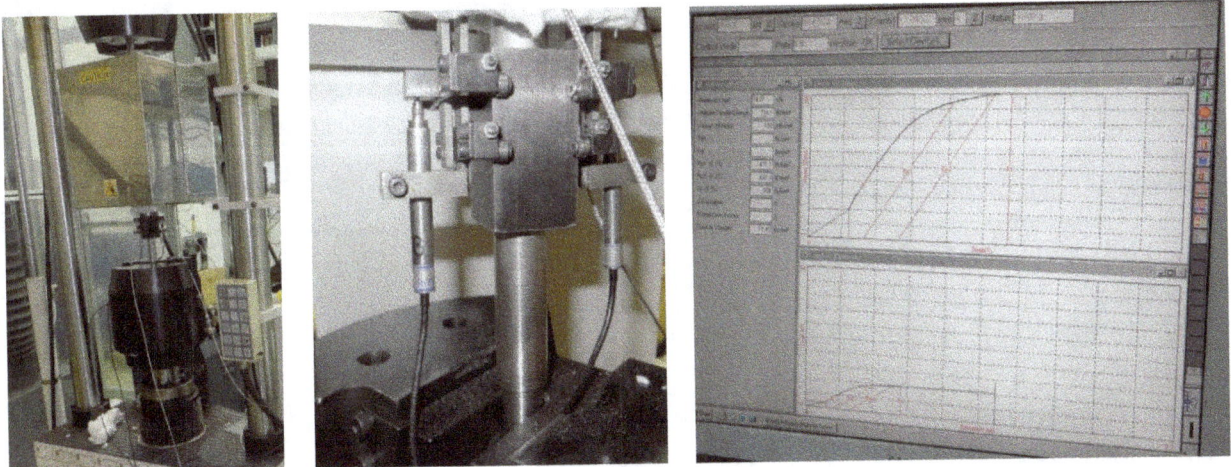

Figure 5.41 LHS = Tensile load frame with an oven, Centre A dual averaging extensometer, RHS = Software display of test result

materials are important in many applications such as oil and gas, power generation plant, aerospace, and automotive. The oil and gas industry have future plans to exploit oil and gas which involves deep wells with higher temperatures and pressures. It will be essential to carry out hot tensile tests on steels used for these applications. The tests on these materials will involve temperature that range from 50 to 250°C.

The hot tensile test is carried out by using a furnace/oven fitted onto the tensile testing load frame that allows the test specimen to be placed into the hot zone and to be tested. To measure the proof strength an extensometer is fitted to the specimen. Testing to ASTM E21 requires a dual averaging extensometer as shown Figure 5.41. A typical test would consist of heating to temperature and holding for 10 mins after the extensometer has stabilised prior to testing. There should be a minimum of two thermocouples attached to the specimen and the temperature should be within +/- 2°C

5.2.4 Fatigue Testing

Materials used to manufacture parts that experience cyclic tensile service stresses are required to have an appropriate level of endurance and durability. Fatigue tests are carried out to establish the endurance limit based on either test specimens machined from the material or actual product endurance test work. The required test standards are shown in the Table 5.3

The fatigue properties such as the endurance limit can be determined using either rotating bending methods based on similar principles used for the early work carried out by Wöhler on rail axles. August Wöhler (1819–1914) was a German engineer. His work was regarded as the first to systematically characterize the fatigue behaviour of materials using the S-N Curves. He developed a machine for repeated loading of railway axles, and showed that fatigue failure occurs by crack growth from surface defects until

the load can no longer be supported. A small machine for this type of test work is shown in Figure 5.42

To carry out endurance work in the tension-tension requires a tensile load frame with either a servo-hydraulic actuator or resonant Amsler type machines. Typically, hydraulic machine can cycle up to a maximum frequency of 10Hz whereas the resonant machines can operate up to 150Hz and can therefore complete a large number of cycles in a shorter time. Figure 5.43 shows a Zwick Roel Ambler 20kN fatigue tester

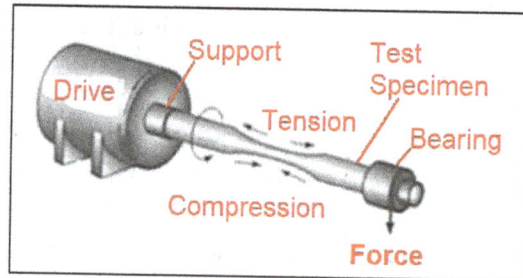

Figure 5.42 LHS = A rotating bending fatigue machine, RHS = Schematic of a rotary bending fatigue test machine

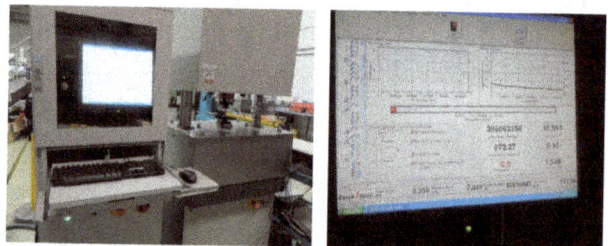

Figure 5.43 A Zwick Roell Amsler 20kN fatigue tester

Table 5.3 Test standards for Fatigue Testing

Test Method	BS EN ISO	ASTM	ISO	DIN
Axial load fatigue test Force control	1099:2017	E466	1099:2017	
Axial load fatigue test-Strain control	12111:2011	E606	12106:2017	
Axial load fatigue test da/dn	12108-2018	E647	12108-2018	
Rotating bending fatigue test	1143:2010		1143:2010	DIN 50113 E
Torsional stress test	1352:2011		1352:2011	
Dynamic force calibration of axial load fatigue testing machines	23788:2012	E467	23788:2012	

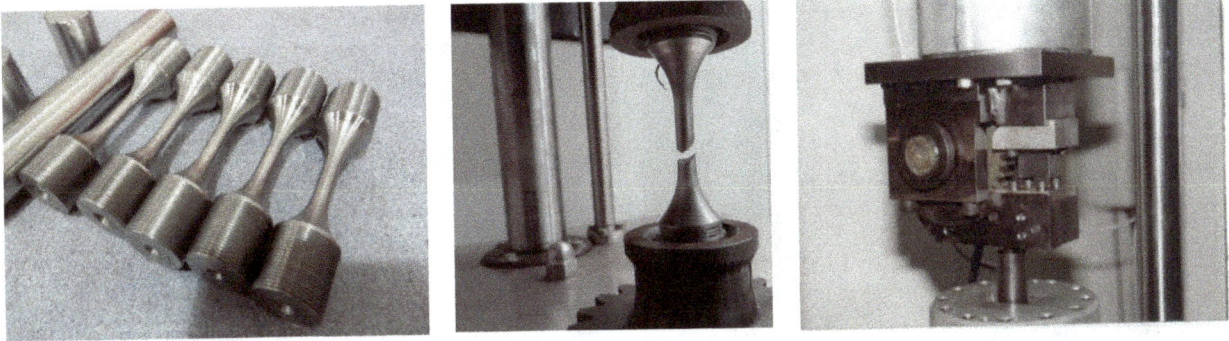

Figure 5.44 LHS = Fatigue specimens, Centre = failed specimen, RHS = Gear tooth bending fatigue test

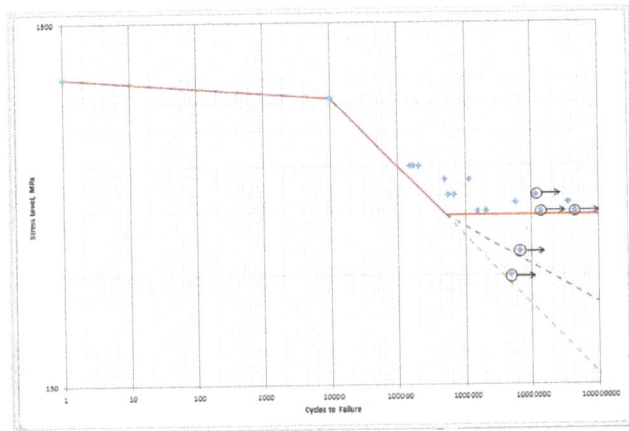

Figure 5.45 A 250kN Servo-hydraulic fatigue for testing parts, RHS = Endurance test data for a cast steel

5.5 CHEMICAL ANALYSIS

5.5.1 Spark Emission Optical Emission Spectroscopy

Why Use Spark OES Analysis?

- Quick, reliable and reproducible analysis technique
- Can analyse wide range of elements
- Usually used for metals\alloy analysis, such as:

  - Steel, cast iron and high alloyed steels
  - Non-ferrous metals and their alloys
  - AI: wrought alloys, casting alloys, etc.
  - Cu: bronze, brass, cupronickel, etc.
  - Mg, Zn alloys, solders
  - Nitrogen in steel
  - Ultra-low carbon analysis.

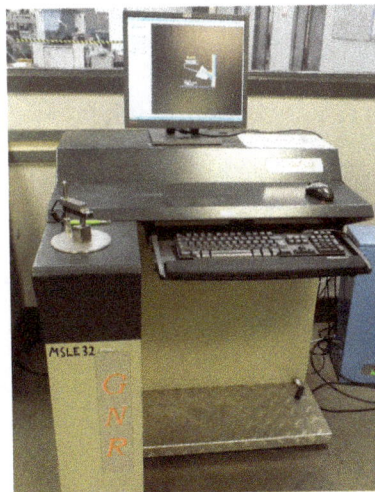

Figure 5.46 Metaltech's OES Supplied by GNR Italy. RHS = Supplied by Oxford Instruments Foundry-Master Pro

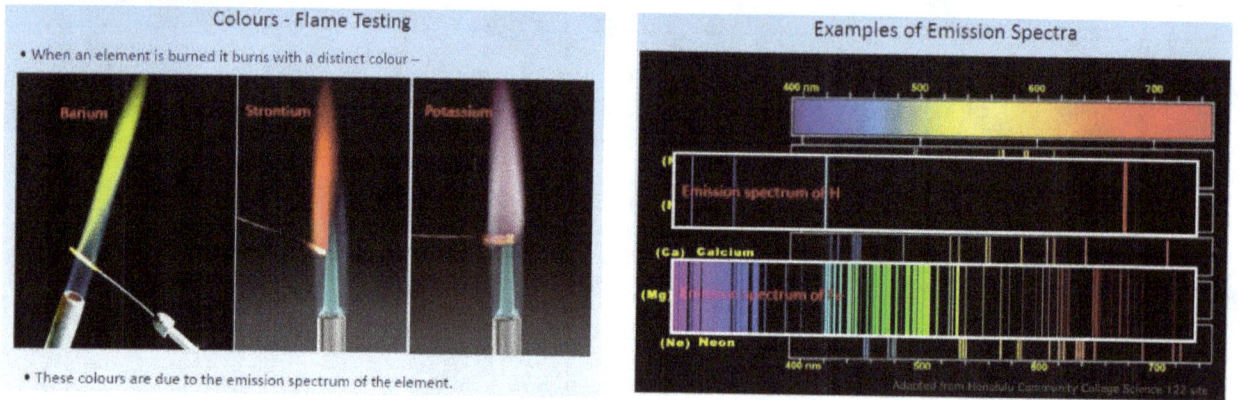

Figure 5.47 *LHS = Effect of elements of flame colours, RHS = Spectra*

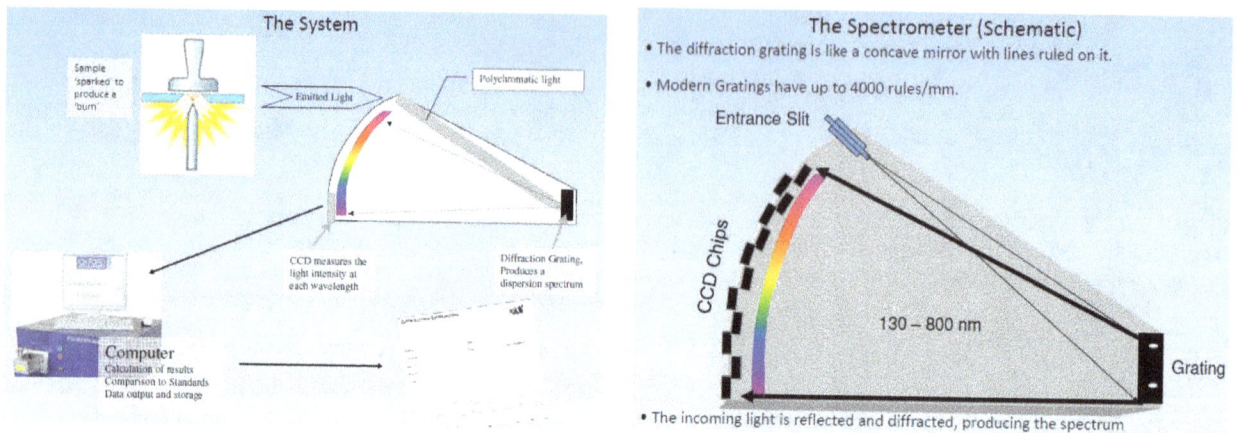

Figure 5.48 *LHS = Spark emission, RHS = The spectrometer*

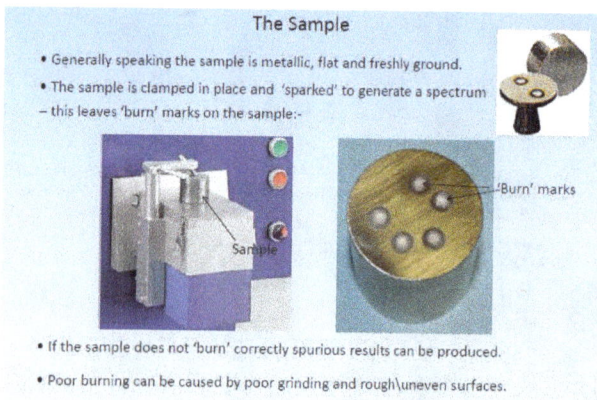

Figure 5.49 *The sample after sparking in an OES*

5.5.2 XRF – The technique used for on-site PMI (positive material identification)

When exposed to external X-rays of a sufficient energy, each of the individual elements present in a sample will produce a unique set of characteristic fluorescent X-rays that are essentially a "fingerprint" for that specific element, Figure 5.49.

At the atomic level, an element produces a fluorescent X-ray when an external X-ray generated by either a miniature X-ray tube source or sealed radioisotope within the analyser strikes an atom, thereby dislodging an electron from one of the atom's orbital shells, causing it to enter a state of excitation. As the atom regains stability, it fills the electron vacancy with an electron from one of its higher energy orbital shells. This electron drops the lower energy state by releasing a fluorescent X-ray, the energy of which is equivalent to the difference between the two quantum states of the electron.

Figure 5.52 illustrates this principle using the classic Bohr model of an atom. These characteristic X-rays can then be categorized and counted when they contact an X-ray detector within the XRF instrument. The X-ray detector, a small semiconductor, receives the emitted X-ray, and passes it to the instrument processor as an electronic signal. By counting the individual emission events from each element over a period of time, this information can be modelled to produce a chemical composition of the sample being analysed.

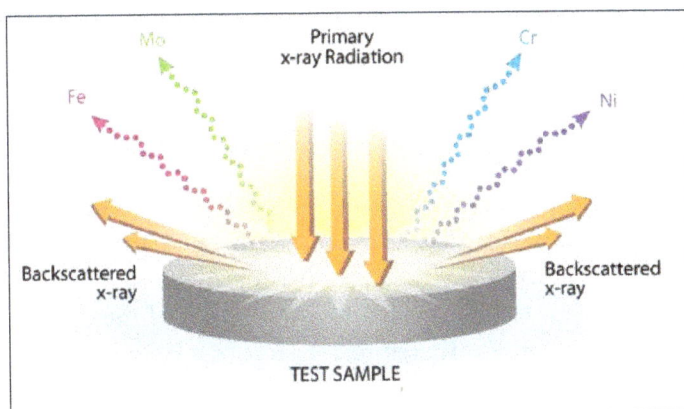

Figure 5.50 *A Niton XRF The basic principle behind XRF technology*

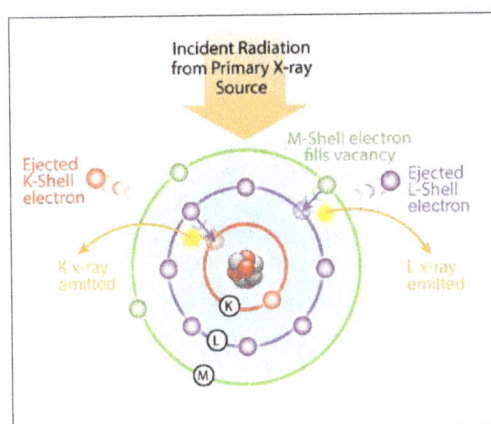

Figure 5.51 *The Bohr model of an atom illustrates the principle behind XRF technology.*

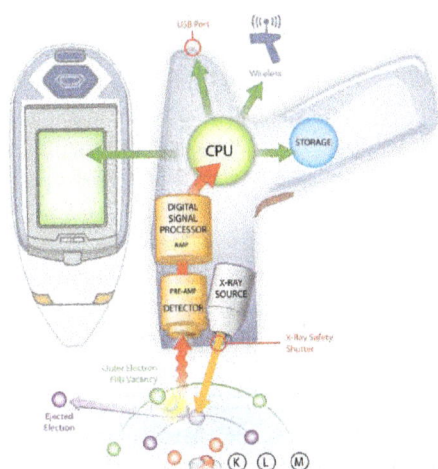

Figure 5.52 *An illustration of what occurs between the analyser and the sample during a measurement using an XRF instrument.*

Figure 5.53 *A comparison of the active area of the detector between GOLDD and standard technology*

1. Primary X-rays are released and directed at the sample
2. Primary X-rays cause elemental atoms in the sample to enter a state of excitation
3. The elements in the sample release characteristic X-rays that are captured by the X-ray detector
4. The detector converts the X-rays to electrical pulses, and sends them to a preamp
5. The signals from the preamp are sent to a digital signal processor for classification and counting
6. The digitized elemental intensity data are processed through mathematical algorithms to convert to chemical composition data, which are then used to determine the alloy grade based on values from an internal library of metals and alloys.
7. The alloy composition values and identification of the metal/alloy are displayed on the instrument screen.
8. The measurement data are stored for later recall or download to a PC.

Advances in application of microprocessor technology, X-ray detectors, and the miniaturization of X-ray tubes have all contributed to the development of the hand-held, lab quality XRF instruments. The operational aspects of the instrument are shown in Figure 5.52.

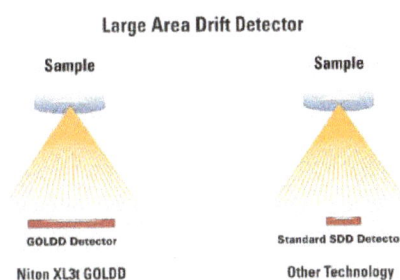

A new type of solid-state detector has been introduced that has improved the accuracy of the hand-held XRF analysis. Named the geometrically optimized large area drift detector (GOLDD™) technology increases analyser performance up to 10 times over the previous

Si PiN detector technology. To take advantage of this high-capacity detector, the analyser has been designed to generate a sufficient number of fluorescent X-ray events from a sample, and position the detector to collect as many as possible.

The silicon drift detector (SDD) has been designed to process more than 200,000 X-ray events per second and there is a 50-kV X-ray tube, which provides the excitation power necessary to take advantage of the detector capabilities. Therefore, the larger active area of the detector, which allows fewer X-rays to slip past undetected as shown in Figure 5.53.

REFERENCES

1. Antonio Gorni, Steel Forming and Heat-Treating Handbook, 2018, http://www.gorni.eng.br/e/Gorni_SFHTHandbook.pdf

2. Atlas CE Calculator – Atlas Steels 2013, www.atlassteels.com.au/documents/CE%20calculator%20rev%20July%202013p.xlsx

3. Atlas PREN Calculator – Atlas Steels Feb 2011, www.atlassteels.com.au/documents/PREN_calculator_rev_Feb_2011.xls

4. Claus Qvist Jessen, STAINLESS STEEL AND CORROSION, Damstah Stainless Solutions 2011, ISBN 978-87-92765-00-0, http://www.damstahl.com/Files/Billeder/2011/PDF/BOOK/book.pdf

5. Possible microstructures according to the Schaeffler diagram, www.hegesztesportal.hu/letoltes/schaeffler-eng-1.xls

6. ASTM A255-10 2018

7. R. Viswanathan, Damage Mechanisms and Life Assessment of High-Temperature Components, Page 50, ASM INTERNATIONAL, 1993. ISBN: 0-87170-358-0 https://www.moys.gov.iq/upload/common/Damage_Mechanisms_and_Life_Assessment_of_High_Temperature_Components.pdf

8. Dr. Ingo Detemple and Rainer Kraut, Avoid Temper Embrittlement, Step by Step, Dillinger Hüttenwerke, Germany, https://www.dillinger.de/imperia/md/content/dh/.../avoid_te_step_by_step.doc

9. The aluminium automotive manual, Version 2002 © European Aluminium Association, https://www.european-aluminium.eu/media/1531/aam-materials-1-resources.pdf

10. S. Bell, B. Davis, A. Javaid and E. Essadiqi, Final Report on Effect of Impurities in Aluminum, Report No. 2003-20(CF) file:///C:/Users/user/Downloads/2003-20CFClimateChange.pdf

11. Dr. Subodh K. Das, Designing Aluminum Alloys for a Recycling Friendly World, Materials Science Forum Vols. 519-521 1241 2006, http://web.mit.edu/sis07/www/das.pdf

12. Kandula Ankamma, Effect of Trace Elements (Boron and Lead) on the Properties of Gray Cast Iron, J. Inst. Eng. India Ser. D (January–June 2014) 95(1):19–26, https://link.springer.com/content/pdf/10.1007%2Fs40033-013-0031-3.pdf

13. Keivan A. Kasvayee et al, Effect of Boron and Cross-Section Thickness on Microstructure and Mechanical Properties of Ductile Iron, Material Science Forum, ISSN: 1662-9752, Vol. 925, pp 249-256 2018, file:///C:/Users/user/Downloads/Effect_of_Boron_and_Cross-Section_Thickness_on_Mic.pdf

14. EFFECT OF ELEMENTS IN GRAY – Roy Lobenhofer www.lobenhofer.com/Tensile%20Properties/effect_of_elements_in_gray.htm

15. OES GNR Metlab Pro, http://www.gnr.it/products/optical-emission-spectrometry/s7-metal-lab-plus/

16. OES Foundry Master Pro, https://kontrollmetod.se/content/files/products/pdf/foundry-master-pro_brochure.pdf

17. Niton XL2t XRF, https://assets.thermofisher.com/TFS-Assets/CAD/brochures/Niton-Product-Line-Brochure.pdf

18. SW. GEE and L. CONDER, A STRAIN GAU6E MANUAL, Structures Technical Memorandum 378, DEPARTMENT OF DEFENCE, DEFENCE SCIENCE AND TECHNOLOGY ORGANISATION, AERONAUTICAL RESEARCH LABORATORIES, Melbourne, Victoria, Australia 1984 https://apps.dtic.mil/dtic/tr/fulltext/u2/a151802.pdf

19. John Vaughan, Application of B & K Equipment to STRAIN MEASUREMENT, October 1975, https://www.bksv.com/media/doc/bn1148.pdf

20. Introduction to Strain Gages, KYOWA, https://www.kyowa-ei.co.jp/english/products/gages/pdf/whats.pdf

21. The Technical Staff of Measurements Group, Inc. Strain Gage Based Transducers Their Design and Construction, Published by Measurements Group, Inc. Raleigh, North Carolina, 27611, USA 1988, http://amet-me.mnsu.edu/UserFilesShared/Equipment_Manuals/TE_110/StrainGageBasedTransVishay.pdf

22. James W. Phillips, TAM 326—Experimental Stress Analysis, Copyright© 1998 Board of Trustees University of Illinois at Urbana-Champaign http://www.ifsc.usp.br/~lavfis/images/BDApostilas/ApEfFotoelastico/photoelasticity.pdf

23. James F. Doyle et al, Manual on Experimental Stress Analysis, Society for Experimental Mechanics, Chapter 6, http://courses.washington.edu/me354/lab/photoelas.pdf

24. Introduction to Stress Analysis by the PhotoStress® Method Tech Note TN-702-2, Micro-Measurements, Document Number: 11212, Revision 29-Jun-2011 http://www.vishaypg.com/docs/11212/11212_tn.pdf

25. THOMAS J. DOLAN AND EDWARD L. BROGHAMER, A PHOTOELASTIC STUDY OF STRESSES IN GEAR TOOTH FILLETS, UNIVERSITY OF ILLINOIS BULLETIN Vol. XXXIX March 24, 1942 No. 31, file:///C:/Users/user/Downloads/engineeringexperv00000i00335.pdf

26. DARCOM-R 310-6, BRITTLE LACQUER TECHNIQUE OF STRESS ANALYSIS, 29 July i981 https://apps.dtic.mil/dtic/tr/fulltext/u2/a102509.pdf

27. Negussie Tebedge et al, MEASUREMENT OF RESIDUAL STRESSES A STUDY OF METHODS, Fritz Engineering Laboratory Report No. 337.8, 1971, http://digital.lib.lehigh.edu/fritz/pdf/337_8.pdf

28. E.J. HEARN, An Introduction to the Mechanics of Elastic and Plastic Deformation of Solids and Structural Materials CHAPTER 16, EXPERIMENTAL STRESS ANALYSIS,

http://nguyen.hong.hai.free.fr/EBOOKS/SCIENCE%20AND%20ENGINEERING/MECANIQUE/MATERIAUX/Mechanics%20of%20Materials.rar_FILES/Mechanics%20of%20Materials/Volume%201/32658_16.pdf

29. DIC MEASUREMENTS IN ENGINEERING APPLICATIONS – AUGUST 2015, DIC Algorithms and Strategy, Benchmarking, and Projects, http://image-correlation.com/assets/files/MESOCOS-DIC.pdf

30. MiTE, https://www.dst.defence.gov.au/opportunity/mite

31. Nik Rajic & David Rowlands, Thermoelastic stress analysis with a compact low cost microbolometer system, Quantitative InfraRed Thermography, 2013 Vol. 10, No. 2, 135–158, https://www.tandfonline.com/doi/pdf/10.1080/17686733.2013.800688?needAccess=true

32. K.S.Anish et al, STRESS PATTERN ANALYSIS USING THERMAL CAMERA, INTERNATIONAL JOURNAL OF ADVANCES IN PRODUCTION AND MECHANICAL ENGINEERING (IJAPME), VOLUME-2, ISSUE-5, 2016, http://troindia.in/journal/ijapme/vol2iss5/5-10.pdf

33. Pierre Brémond, Cedip Infrared Systems, New developments in Thermo Elastic Stress Analysis by Infrared Thermography. IV Conferencia Panamericana de END Buenos Aires – Octubre 2007, https://www.ndt.net/article/panndt2007/papers/138.pdf

34. J M Dulieu-Barton, Development and applications of thermoelastic stress analysis, JOURNAL OF STRAIN ANALYSIS VOL 33 NO 2, 1998.

35. STEVEN B. CHASE, Evaluation of Fatigue-Prone Details Using a Low-Cost Thermoelastic Stress Analysis System, Final Report VTRC 17-R8, http://www.virginiadot.org/vtrc/main/online_reports/pdf/17-r8.pdf

36. V. Le Sauxa et al, Performance Comparison between ImageIR® 8300 hp and ImageIR® 10300 on a Thermoelastic Stress Analysis Experiment, © InfraTec 10/2017

37. Lovre KRSTULOVIĆ-OPARA, APPLICATION OF THERMOGRAPHY IN ANALYSIS OF FATIGUE STRENGTH OF MATERIALS AND STRUCTURES, 2013, https://hrcak.srce.hr/file/219012

38. Malcolm S Loveday et al Tensile testing of metallic materials: A review, Tenstand- Works Package 1, Final Report, April 2001, http://resource.npl.co.uk/docs/science_technology/materials/measurement_techniques/tenstand/test_method_review.pdf

39. J Lord, M Loveday, M Rides, I McEnteggart, Digital Tensile Software Evaluation, NPL REPORT DEPC MPE 015, "TENSTAND" WP2 Final Report, http://libsvr.npl.co.uk/npl_web/pdf/depc_mpe15.pdf

40. J Lord, M Rides, M Loveday, Modulus Measurement Methods, NPL REPORT, DEPC MPE 016 "TENSTAND", WP3 Final Report, JANUARY 2005

41. H. Klingelhöffer, S. Ledworuski, S. Brookes, Th. May, VALIDATION OF THE MACHINE CONTROL CHARACTERISTICS, TENSTAND – WORK PACKAGE 4 – FINAL REPORT,2005 http://resource.npl.co.uk/docs/science_technology/materials/measurement_techniques/tenstand/machine_control_tests.pdf

42. Stephanie Bell, Measurement Good Practice Guide No. 11, A beginner's Guide to Uncertainty of Measurement, National Physical Laboratory, 1999

43. Thomas Young, A course of lectures on natural philosophy and the mechanical art, 2007 http://www.mm.bme.hu/~kossa/books/Young%20-%201807%20-%20Course%20of%20lectures%20on%20natural%20philosophy%20and%20the%20mechanical%20arts%20Vol%201.pdf

44. A HISTORY OFTHE THEORY OF ELASTICITY AND OF THE STEENGTH OF MATERIALS FROM GALILEI TO THE PRESENT TIME. BY THE LATE ISAAC TODHUNTER, D.Sc., RR.S. EDITED AND COMPLETED FOE THE SYNDICS OF THE UNIVERSITY PEESS BY KARL PEAESON, M.A. http://www.mm.bme.hu/~kossa/books/TodHunter%20-%201893%20-%20A%20history%20of%20the%20theory%20of%20elasticity%20Vol%202%20Part%201.pdf

45. A TREATISE ON THE MATHEMATICAL THEORY OF ELASTICITY BY A. E. H. LOVE, M.A. FELLOW AND LECTURER OF BT JOHN'S COLLEGE, CAMBRIDGE VOLUME I. CAMBRIDGE: AT THE UNIVERSITY PRESS. 1892, https://hal.archives-ouvertes.fr/hal-01307751/document

46. A. Ellermann, The Bauschinger Effect in Different Heat Treatment Conditions of 42CrMo4, INTERNATIONAL JOURNAL OF STRUCTURAL CHANGES IN SOLIDS – Mechanics and Applications Volume 3, Number 1, February 2011, pp.1-13

47. Andrii Gennadiovych Kostryzhev, BAUSCHINGER EFFECT IN Nb AND V MICROALLOYED LINE PIPE STEELS, A thesis submitted to The University of Birmingham for the degree of DOCTOR OF PHILOSOPHY, April 2009

48. F.S. Silva, THE BAUSCHINGER EFFECT AND FATIGUE CRACK GROWTH, 10th PORTUGUESE CONFERENCE ON FRACTURE – 2006,

49. THIAGO SOARES PEREIRA, BAUSCHINGER EFFECT IN MACRO AND MICRO SIZED HIGH STRENGTH LOW ALLOY PIPELINE STEELS, A thesis submitted to the University of Birmingham for the degree of DOCTOR OF PHILOSOPHY School of Metallurgy and Materials, University of Birmingham, August 2015

50. D Tabor, The Hardness of Metals 1951 https://ia601606.us.archive.org/30/items/TaborHardnessOfMetals/Tabor%20-%20Hardness%20of%20Metals.pdf

51. Guidelines on the Estimation of Uncertainty in Hardness Measurements, EURAMET cg-16, Version 2.0 (03/2011), file:///C:/Users/user/Downloads/EURAMET_cg-16__v_2.0_Hardness_Measurements.pdf

52. Konrad Herrman, Hardness Testing Principles and Applications, ASM International 2011

53. Konrad Herrmann, NEW DEVELOPMENTS AND APPLICATIONS IN HARDNESS METROLOGY, HARDMEKO 2007 Recent Advancement of Theory and Practice in Hardness Measurement 19-21 November, 2007, Tsukuba, Japan, https://www.imeko.org/publications/tc5-2007/IMEKO-TC5-2007-KL-002.pdf

54. Adrien TRILLON et al, Magnetic Barkhausen Noise for hardness checking on steel, 18th World Conference on Nondestructive Testing, 16-20 April 2012, Durban, South Africa, https://www.ndt.net/article/wcndt2012/papers/569_wcndtfinal00568.pdf

55. Freddy A. Franco et al, Relation Between Magnetic Barkhausen Noise and Hardness for Jominy Quench Tests

in SAE 4140 and 6150 Steels, J Nondestruct Eval (2013) 32:93–103, https://core.ac.uk/download/pdf/37522967.pdf

56. Xiaoyu Luo et al, Non-destructive hardness measurement of hot-stamped high strength steel sheets based on magnetic barkhausen noise, 11th International Conference on Technology of Plasticity, ICTP 2014, 19-24 October 2014, Nagoya Congress Center, Nagoya, Japan.

57. Hardness Testing with Zwick Roell, Brochure file:///C:/Users/user/Downloads/15_303_Haerte_FP_E.pdf

58. Dr. Stefan Frank, Mobile Hardness Testing Application Guide for Hardness Testers, GE Inspection Technologies Brochure, https://www.gemeasurement.com/sites/gemc.dev/files/hardness_testing_application_guide_english_0.pdf

59. D. TABOR, The physical meaning of indentation and scratch hardness, BRITISH JOURNAL OF APPLIED PHYSICS, VOL.-7, MAY 1956, Page 159

60. Per-Lennart Larsson, On Plowing Frictional Behavior during Scratch Testing: A Comparison between Experimental and Theoretical/Numerical Results, Crystals 2019, 9, 33, file:///C:/Users/user/Downloads/crystals-09-00033%20(1).pdf

61. Ing. TOMÁŠ KADLÍČEK, SCRATCH TEST, http://ksm.fsv.cvut.cz/~nemecek/teaching/dmpo/clanky/2015/Kadlicek_zav%20prace_Scratch%20test.pdf

62. Dr.-Ing. Erhard Reimann, Martens Hardness – More Than Just Hardness Testing, https://pdfs.semanticscholar.org/4dac/9970041a5205cee968f6e170ce78f134fe52.pdf

63. Professor Giulio Barbato, "ZERO-POINT" IN THE EVALUATION OF MARTENS HARDNESS UNCERTAINTY, http://staff.polito.it/fiorenzo.franceschini/Pubblicazioni/ZeroPoint%20in%20the%20Evaluation%20of%20Martens%20Hardness%20Uncertainty.pdf

64. Michael Griepentrog, INSTRUMENTED INDENTATION TEST FOR HARDNESS AND MATERIALS PARAMETER FROM MILLINEWTONS TO KILONEWTONS, Federal Institute for Materials Research and Testing BAM Project

group VIII.2901 Reference Materials and Coatings for Surface Technologies Unter den Eichen 44-47, D-12203 Berlibn

65. Instrumented indentation testing (IIT), https://wiki.anton-paar.com/en/instrumented-indentation-testing-iit/

66. TRIALS USING THE INSTRUMENTED INDENTATION TECHNIQUE (IIT) (MARCH 2005) TWI, https://www.twi-global.com/technical-knowledge/published-papers/trials-using-the-instrumented-indentation-technique-iit-march-2005

67. Kamal Sharma, Application of Automated Ball Indentation for Property Measurement of Degraded Zr2.5Nb, Journal of Minerals & Materials Characterization & Engineering, Vol. 10, No.7, pp.661-669, 2011, https://file.scirp.org/pdf/JMMCE20110700008_66130518.pdf

68. Fahmy M. Haggag, ATC's Standard Haggag Test Method (HTM) also known as ABI® Test Method for Measuring Tensile and Fracture Toughness of Ferritic Steels, https://static1.squarespace.com/static/514c9dcde4b0b45af33e225b/t/54ad5e3be4b01e05d9c607f0/1420647995205/Haggag+Test+Methods+(HTM)-2015.pdf

69. Fahmy M Haggag, Estimating Fracture Toughness Using Tension or Ball Indentation Tests and a Modified Critical Strain Model, The American Society of Mechanical Engineers PVP — Vol. 170, Innovative Approaches to Irradiation Damage, and Fracture Analysis Eds.: D.L. Marriott, T.R. Mager, and W. H. Bamford Book No. H00485 1989, https://www.researchgate.net/publication/239869316_Estimating_fracture_toughness_using_tension_or_ball_indentation_tests_and_a_modified_critical_strain_model

70. Frontics, Technical Offer AIS2100, 2010, file:///C:/Users/user/Downloads/frontics_data_ais2100_eng.pdf

71. http://www.boulder.nist.gov/div853/Publication%20files/NIST_CharpyHistory.pdf

72. E1820-11 Standard Test Method for Measurement of Fracture Toughness

SELF ASSESSMENT QUESTIONS

1. Explain the difference between strength, hardness and toughness.
2. What is "Superficial Rockwell Hardness Testing" and when should it be used?
3. Determine the Tensile Strength of a steel sample with a Rockwell C reading of 39.
4. What is the primary difference between Charpy and Izod Impact Testing?
5. Qualitatively relate hardness to impact energy.
6. A steel wire 3mm in diameter will take a stress of 450MPa. What was the associated force on the wire?
7. Sketch a load extension diagram for low carbon steel and show the following points: (a) Elastic limit (b) Yield point (c) Ultimate tensile strength.
8. A carbon steel has hardness of 42HRC. When hardened it is 62 HRC. Which specimen would have the greater (a) wear resistance (b) toughness (c) strength?
9. Describe the Charpy test.
10. Name three types of hardness test.
11. Describe the Rockwell hardness testing machine, and how a test is carried out.

12. A steel component has been tested at a temperature which caused ductile fracture but the energy was only 60Joules. The design requirement was an energy level of 150 Joules. Can you recommend changes to allow the design requirement to be met?

13. A sample of AISI 4130 has been impact tested at minus 60°C by the Charpy test method. The energy was 10 Joules. What type of fracture would be observed on the fractured specimens?

14. A structural component has to operate at temperatures as low as minus 20°C and will undergo significant impact loading. What recommendations would you make for the Charpy test requirements?

15 A casting has been Brinell hardness tested and the hardness was found to be 180 BHN. The specification range was 197 to 237 BHN. After grinding 1mm from the surface a re-test was carried out and the hardness was found to be 224 BHN. Can you provide an explanation for this effect?

16. A supplier has reported a Brinell hardness of 324 BHN on the test certificate. A re-test at goods inwards has reported a value of 280 BHN. Describe a measurement and reporting procedure that could prevent the problems associated with this type of event.

17. The inspection department have chromium plated samples from several different suppliers, and they want to do a hardness test on the plating to see which supplier has a better hardness on the plating. What hardness test method would you recommend?

18. What is XRF? What is it used for? Explain the operating principles.

19. What is OES? What is it used for? Explain the operating principles.

20. What is an extensometer used for?

21. Explain the term Impact Transition Temperature.

CHAPTER 6

Fracture

6.1 INTRODUCTION

This chapter provides a general introduction to the field of "Fracture Mechanics". Fracture mechanics describes the concepts, equations and procedures required to determine how cracks grow and how cracks affect the final "strength" of a part and forms the basis of damage tolerance analysis.

My initial introduction to the subject was at university in the late 1960s and it was a topic of high significance with the development of space exploration and nuclear power where failure due to fatigue and fracture were to be avoided at all cost. This was followed by time spent in the steel industry gaining an appreciation of manufacture and testing of products to meet the mechanical property requirements associated with the prevention of fracture. (For Pipelines, Bridges, Pressure vessels, LNG Storage, Fighting Vehicles, Ship-building, Offshore Oil Rigs, Rail Infrastructure.)

Based on this experience my view is that the best introduction to the subject is again via the timeline of the evolution and development of ideas and the eventual clarity that was brought to the understanding of the key aspects of the science. This understanding led to the introduction of methods for control of the fracture of engineering materials and to establish acceptance standards for material with imperfections.

6.2 ELASTICITY CONSIDERATIONS AND LIMITATIONS FOR FRACTURE CONTROL

Ever since the Industrial Revolution, and for a long time afterwards, engineering design was based only on avoidance of failure by plastic collapse. The material property specified in the codes of design was initially the tensile strength which subsequently changed to the flow stress usually the yield strength or the 0.2% proof strength. The allowable design stress was the calculated stress to cause "plastic collapse" divided by a safety factor. Table 6.1 shows the early values of factors of safety applied for various parts.

Class of service	a	b	c	d	Safety factor F= a.b.c.d
Boilers	2	1	1	2.25-3	4.5-6
Piston and con rods for double acting engines	1.5-2	3	2	1.5	13.5-18
Shaft with bandwheel, flywheel or armature	1.5-2	3	1	1.5	6.75-9
Mill shafting	2	3	2	2	24
Steel work in building	2	1	1	2	4
Steel work in bridges	2	1	1	2.5	5
Cast iron wheel rims	2	1	1	10	20
Steel wheel rims	2	1	1	4	8
Material	a	b	c	d	Safety Factor
Cast iron and other castings	2	1	1	2	4
Wrought iron or mild steel	2	1	1	1.5	3
Oil-tempered or nickel steel	1.5	1	1	1.5	2.25
Hardened steel	1.5	1	1	2	3
Bronze and brass, rolled or forged	2	1	1	1.5	3

a = the UTS/YS (or 0.2%PS) ratio

b = 1 for dead load, 2 for load cycling 0 to +x

c = 1 for gradually applied load, 2 for sudden applied load

d = a factor of ignorance to provide against unknown conditions; varies between 1.5 and 3 but may be as high as 10

Table 6.1 Factor of safety for various application (1)

This approach to design did not consider other types of failure such as "brittle fracture". There was a general belief that the generous safety factor as well as covering for many unknowns would also cover for the occurrence of low stress brittle fracture. Unfortunately, a number of failures cast doubt on the reliability of design based on plastic collapse and a safety factor. Additional considerations were needed which initially became the requirement for a toughness such as Charpy energy at a specific temperature based on empirical service performance data. The development of fracture mechanics allowed the control of imperfections by NDT techniques and an understanding of the need to remove tensile residual stresses inherent in welds and castings that could cause "brittle fracture".

6.2.1 Stress based approach Maximum nominal stress, brittle materials – 1850 Rankine, Maximum shear stress 1868 Tresca, Maximum distortion energy 1913 von Mises, Material Models

Engineering components often experience complex loads due to a combination of various service stresses **such as tension, compression, bending, torsion, shear or pressure**. Stress analysis methods (using Mohr's circle) can enable the summation of the stresses to allow calculation of maximum principle stress and maximum shear stress. If the sum of these stresses exceeds a certain value it can result in material yielding and plastic deformation or actual fracture. To establish safe limits for use of a material under combined stresses requires the specification and application of a failure criterion. The two specific failure criteria would be:

- failure by yielding (yield criteria)
- or failure by fracture (fracture criteria).

This type of approach to providing a guarantee to the performance of a part would be described as a stress-based approach. A stress-based approach would involve the calculation of the combined stresses and then comparison of the value to material properties obtained for uniaxial tension.[2 and 3]

Many mechanics of materials books have sections on strength and failure criteria. Predicting the yielding of ductile materials or the fracture of brittle materials depends on a "Theory of Failure".

A good failure theory required:

- Well proven empirical data of a large number of test results in different load conditions
- A plausible theory explaining the microscopic mechanisms consistent with the observations

If these conditions were met then there was a high level of confidence that it would apply to other conditions that have not been studied or tested in detail. Various theoretical criteria have been proposed to obtain adequate correlation between the estimated component life and the actual life achieved in service.

The development of these "failure theories" ocurred over the last 200 years and the main theories were:

1. Rankine or Maximum Principal stress theory
2. Saint venant or Maximum Principal strain theory
3. Tresca or Maximum shear stress theory
4. Von Mises or Shear strain energy theory
5. Haigh or Total strain energy per unit volume theory
6. Mohr-Coulomb failure theory – cohesive-frictional solids

Usually the von Mises and Tresca criteria are the main theories applied to ductile materials. These are based on standard formula to calculate the equivalent shear stress and these methods are often used in structural finite element analysis. Tresca is referred to as the maximum shear stress criterion and is often used to predict the yielding of ductile materials. Yield in ductile materials is usually caused by the slippage of crystal plane along the maximum shear stress surface.

Figure 6.1 LHS shows shows the permanent deformation of a single crystal under a tensile load. Note that the slip planes align in the direction of the tensile force. This behaviour can be simulated with a pack of card wrapped with a rubber band (b) Shows twining in a single crystal in tension. Figure 6.1 in the centre, shows a schematic image of slip lines and slip bands in a grain due to the shear stress. A slip band consists of a number of slip planes. The grain in the upper image is show surrounded by several other grains. Figure 6.1 RHS shows the stress required to initiate slip in a pure and perfect single crystal, the critical resolved shear stress (CRSS) is a constant for a material at a given temperature. This rule, known as Schmid's Law.

A part in service is considered safe as long as the maximum shear stress at a critical point is under the yield shear stress obtained from a uniaxial tensile test.

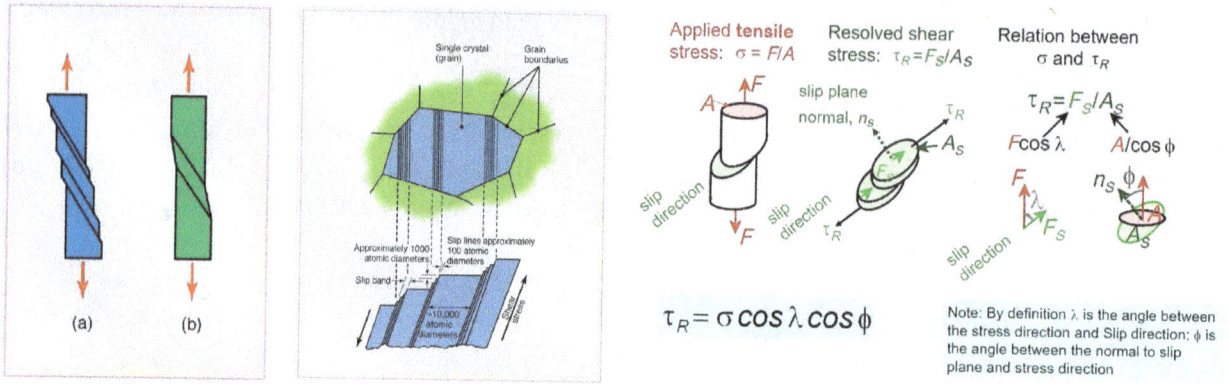

Figure 6.1 *LHS = This shows the permanent deformation of a single crystal under a tensile load. Centre=This shows a schematic image of slip lines and slip bands in a grain due to the shear stress. RHS = stress and dislocation motion.*

Rankine 1850

• Rankine had observed, that brittle failure, hence failure was not proceeded by plastic deformation

• When the first principal stress reaches the tensile strength of the material failure occurred

• The fracture then formed perpendicular to the largest principal stress

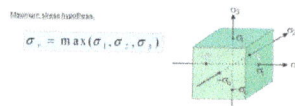

Tresca 1868

• The maximum stress hypothesis was a big success, but even Rankine knew, that it was only valid for cleavage fractures.

• Tresca proposed that yielding occurs when the largest shear stress reaches the uniaxial yield value.

• For failure, the largest principal stress difference, must be two times the maximum shear stress

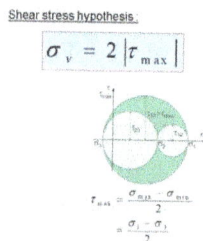

Von Mises 1913

• An alternative proposal to Tresca was given by Von Mises based on the idea that plastic deformation was not the result of one stress component.

• Von Mises preferred that an energy per unit volume was responsible to give plastic deformation

• This was then called «maximum distortion energy theorem» or von Mises criterion

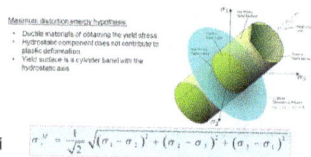

Figure 6.2 *Key features of Rankine, Tresca and von Mises*

The von Mises Criterion (1913), also known as the maximum distortion energy criterion, octahedral shear stress theory, or Maxwell-Huber-Hencky-von Mises theory, is often used to estimate the yield of ductile materials. In addition to bounding the principal stresses to prevent ductile failure, the von Mises criterion also gives a reasonable estimation of fatigue failure, especially in cases of repeated tensile and tensile-shear loading.

Elasticity theory provided guidance of failure stress by allowing the maximum shear stresses and tensile stresses to be determined and allowed a comparison to the uniaxial yield strength but does not extend to the process of fracture. **The plasticity that follows beyond the elastic region during failure was often regarded as an addition "safety" margin.**

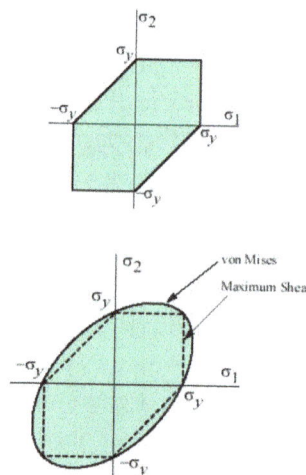

With respect to 2D stress, the maximum shear stress is related to the difference in the two principal stresses (often using Mohr's Circle). Therefore, the criterion requires the principal stress difference, along with the principal stresses themselves, to be less than the yield shear stress.

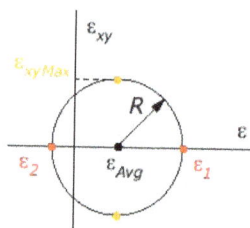

Figure 6.3 *Tresca, von Mises, Mohr's circle*

Techniques for Failure Prevention

Computational software has magically transformed the field of stress analysis with the widespread use of FEA. It is common practice to verify design stresses with finite element (FEA) packages. A criticism now directed at users of FEA is that they simply look just for red spots in the simulation output, without really understanding the stress systems that are operating. By using the old basic rules of thumb danger points can be anticipated and avoided without total reliance on computer simulation.

Loading Points	Maximum stresses are often located at loading points, supports, joints, or maximum deflection points.
Stress Concentrations	Stress concentrations are usually located near corners, holes, crack tips, boundaries, between layers, and where cross-section areas change suddenly. Sound design avoids rapid changes in material or geometrical properties. For example, when a large hole is removed from a structure, a reinforcement composed of generally no less than the material removed should be added around the opening.
Safety Factors	The addition of safety factors to designs allow engineers to reduce sensitivity to manufacturing defects and to compensate for stress prediction limitations.

6.2.2 The Era of Engineering with cracks – Fracture Mechanics and Defect Tolerance

Once it was accepted that all metal components are likely to contain imperfections/cracks, especially cast and welded structures, it was also accepted that in the "real world" the possibility of fracture of parts at stress levels below the yield strength could occur. There was clear evidence for this event since it has caused bridges to fail, ships to sink, pipelines to leak and aircraft failures. There were also some major surprises, for example in January 1943 SS Schenectady, one of the "Liberty" ships, broke in two whilst still in the fitting out dock and at nominal stress levels of around 60MPa!

Fracture control of materials with defects and imperfections during the first half of the 1900s was still in the "dark ages" and we can find comments that confirm the difficulties and the limitation of knowledge at that moment in time. Some of which was due to the lack of a need. The developments in Space, Nuclear Energy, large welded structure and Military needs would eventually create the financial motivation.

In 1907 Karl Weighardt carried out stress analysis on a failed roller bearing and in a publication with the title "On the splitting and cracking of elastic bodies" stated:

"Knowledge of the theoretical stress distribution does not allow the evaluation of crack initiation upon exceedance of the loading with certainty and it is not at all possible to determine the path of further cracking".

It was pointed out by Rossmanith who translated the Karl Weighardt publication that he came near to the idea of "stress intensity factor" by suggesting the important parameter was the gradient of stress near the crack tip and not the actual stress at the crack tip.

Love's treaties on the mathematical theory of elasticity first published in 1892 with a 4th edition in 1927 concluded that

"The properties of rupture are but vaguely understood".

The theory of elasticity allowed the development of closed-form of analytical procedures in order to develop constitutive equations for predicting failure of crack-free solids. But, when solids contain flaws or cracks, the theory of elasticity cannot consider the **stress singularity phenomenon** near a crack tip. This gap in knowledge could only be filled by the development of "Fracture mechanics": the study of mechanical behaviour of cracked materials subjected to an applied load.

Irwin is regarded as the founder of fracture mechanics which was based on the work of Inglis[4], Griffith[5], and Westergaard.

Essentially, fracture mechanics deals with the process of rupture due to nucleation and growth of cracks. It was also appreciated that the formation of cracks was often a complex fracture process, which was dependent on:

- The microstructure that plays a very important role in a fracture process due to dislocation motion, and imperfections such as precipitates, inclusions, grain size, and type of phases making up the microstructure

- The applied loading which may be cyclic or static and which can create plane stress or plain strain conditions at the crack tip
- The environment was also important since some materials are sensitive to certain corrosion environments and there can be environmentally assisted cracking.

6.3 TOUGHNESS AND SIMILITUDE – "THE SIZE EFFECT"[6]

Another important factor that plagued the attempts to find ways to "design" against fracture was the lack of similitude [or scaling] of toughness measurements. It was found that the toughness behaviour was different from strength.

The strength results from different size specimens from similar microstructures and quality level were similar. However, this was not true for toughness and for a similar microstructure and quality level toughness measurements changed with the size of the component. This can be seen in the measurement of toughness with LEFM (linear elastic fracture mechanics) and the importance of the change from plane stress to plane strain conditions as the thickness increased, on the measured fracture toughness. [Figure 6.4]

The key parameter developed in the theory of fracture mechanics; a quantity called the **stress intensity factor** (K). The **stress intensity factor** can be viewed as a "scaling factor" that characterizes the severity of the crack situation as affected by crack size, stress, and geometry. In defining K, the material is assumed to behave in a linear-elastic manner, according to Hooke's Law, so that the approach to fracture control is referred to as linear-elastic fracture mechanics (LEFM).

A given material can resist a crack without brittle fracture occurring as long as this K (SIF) is below a critical value Kc, which is a property of the material called the fracture toughness. Values of Kc vary widely for different materials and are affected by temperature and loading rate, and also by the thickness of the part under consideration. (K = **stress intensity factor** < Kc = material fracture toughness)

The universally accepted fracture mechanics approach is based on the simple assumption that cracks can have **the same cracking driving force if they have similar stress intensity factors at the crack tip**. This enables small laboratory specimens to duplicate the response of large structures with similar stress intensity factors.

The effect of size also affects the use of subsidiary sized Charpy specimens. When testing material where the standard 10mm × 10mm specimens cannot be taken, smaller sized specimens 7.5mm × 10mm or 5mm × 10mm are used. The energy values obtained from the various samples can be related to the fractured area. However, it is also found that the ITT [Impact Transition Temperature] measured using different size Charpy specimens can also change. These are shown in Figure 6.5.

6.4 THE EFFECT OF MAJOR FAILURES ON DEVELOPMENT OF FRACTURE CONTROL METHODS

The Liberty ship failures in the 1940s and Comet in the 1950s emphasized the inadequacy of the "safety factor" based design approach [outlined in chapter 1]. The metallurgical based examination to find the

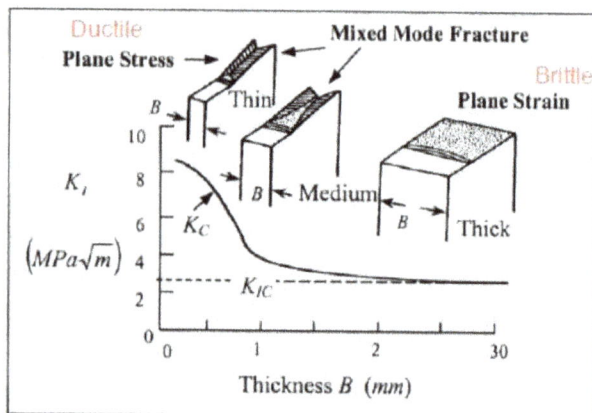

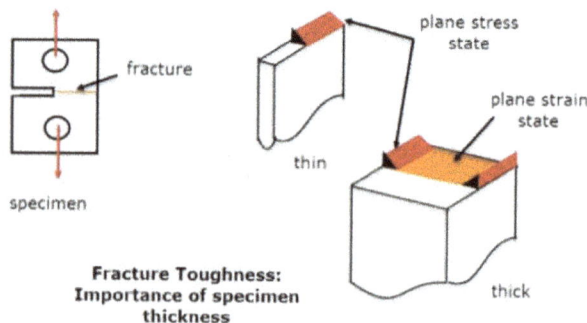

Figure 6.4 *The variation of fracture toughness with sample thickness*

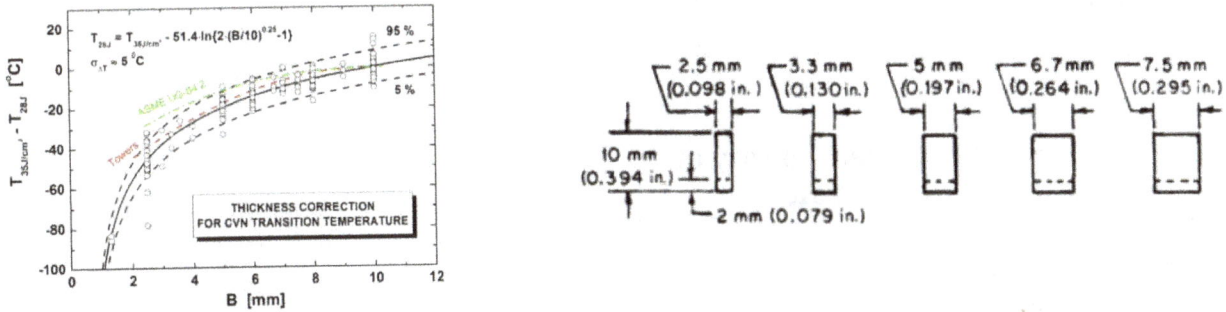

$$T_{28J} = T_{36J/cm^2} - 51.4 \cdot \ln\{2 \cdot (B/10)^{0.25} - 1\}$$

$$\sigma_{\Delta T} = 5\,°C$$

THICKNESS CORRECTION FOR CVN TRANSITION TEMPERATURE

TABLE 9 Charpy V-Notch Test Acceptance Criteria for Various Sub-Size Specimens

Full Size, 10 by 10 mm		¾ Size, 10 by 7.5 mm		⅔ Size, 10 by 6.7 mm		½ Size, 10 by 5 mm		⅓ Size, 10 by 3.3 mm		¼ Size, 10 by 2.5 mm	
ft-lbf	[J]	ft-lbf	[J]	ft-lbf	[J]	ft-lbf	[J]	ft-lbf	[J]	ft-lbf	[J]
40	[54]	30	[41]	27	[37]	20	[27]	13	[18]	10	[14]
35	[48]	26	[35]	23	[31]	18	[24]	12	[16]	9	[12]
30	[41]	22	[30]	20	[27]	15	[20]	10	[14]	8	[11]
25	[34]	19	[26]	17	[23]	12	[16]	8	[11]	6	[8]
20	[27]	15	[20]	13	[18]	10	[14]	7	[10]	5	[7]
16	[22]	12	[16]	11	[15]	8	[11]	5	[7]	4	[5]
15	[20]	11	[15]	10	[14]	8	[11]	5	[7]	4	[5]
13	[18]	10	[14]	9	[12]	6	[8]	4	[5]	3	[4]
12	[16]	9	[12]	8	[11]	6	[8]	4	[5]	3	[4]
10	[14]	8	[11]	7	[10]	5	[7]	3	[4]	2	[3]
7	[10]	5	[7]	5	[7]	4	[5]	2	[3]	2	[3]

Figure 6.5 *Effect of Charpy size on the ITT and ASTM values for Sub-Size specimens*

cause of the failure also highlighted the metallurgical problems of brittle fracture with the Liberty ships and fatigue with the Comet failure. In addition, recent information on the Comet failure also stressed the importance of control of the manufacture process by quality assurance procedures. There was a reported controversial death-bed confession of a "cover-up" which appeared as a sensational "news paper headline", which was subsequently denied. The approved and validated design was a press fit for the port holes; however, this method was too slow for the manufacturing engineers and the method of installation was replaced by the use of rivets. The rivet holes added addition stress concentration.

A major outcome of these failures was that the search began for improved methods to control the risk of catastrophic failures caused by "fracture". This became essential as the world embarked on nuclear power, space travel and the extraction and movement of oil and gas from inhospitable climates.

6.4.1 LEFM (Linear Elastic Fracture Mechanics)

The development of LEFM took place in the USA at NRL [Naval Research Laboratory] based on defence finance. This was based on the initial Griffiths equation for brittle materials modified in two ways, one based on energy with an added term for the energy involved in plastic movement of atoms on either side of the fracture and the second changing to a stress-based approach.

- **Surface Energy** The term for the surface energy required for fracture was modified to include a term for the plastic deformation during the fracture (γp)
- **Stress Intensity Factor.** Based on Irwin's review of the work by Westergaarde on the elastic solution of the stress distribution at a sharp crack, Irwin created the concept K the stress intensity factor (the stress field ahead of the crack) which constitutes the "driving force" and Kc the material's fracture toughness or the "resistance force" or the critical value at failure. This is a material property analogous to yield strength used in strength applications. The value of fracture toughness being determined with specimens and loading that ensures plain strain conditions which allows the lowest value to be determined.[7]

6.4.2 EPFM (Elastic Plastic Fracture Mechanics)

The early development of EPFM took place in the UK at the British Weld Research Association (now TWI), which was a government owned organization. In 1963 the COD [Crack Opening Displacement] was introduced, which eventually matured into CTOD [Crack Tip Opening Displacement] test method.[8]

6.4.3 Comparison between Charpy and Fracture Toughness values

The Fracture Toughness test can cost over ten times the cost of a Charpy test and therefore there were many attempts to establish quantitative relationships between Fracture Toughness values and Charpy values. In addition, since we are all familiar with the expected Charpy energy values with various metals, some guidance on the relationship between Charpy and Fracture Toughness would be helpful. Figure 6.6 shows a comparison of several published relationships by Professor Kim Wallin during the early part of his career in 1992. In addition, there are several published equations linking Charpy values with fracture toughness values as shown in Table 6.1.

Table 6.1 *Fracture toughness vs CVN correlations*

K_{IC}-CVN correlations for lower shelf regions and different zones found in published literature.

	Transition temperature region	
Barsom and Rolfe [33]		
$\frac{K_{IC}^2}{E} = 2(CVN)^{3/2}$	$ksl\sqrt{in}$, ksi, ft-lb	40-250 ksi, 4-82 J
Marandet and Sanz [34]		
$K_{IC} = 19(CVN)^{1/2}$	$MPa\sqrt{m}$, MPa, J	303-820MPa 43-118ksi
Sailor and Corten [7]		
$\frac{K_{IC}^2}{E} = 8(CVN)$	$psi\sqrt{in}$, CVN=ft-lbf E= psi	268-923MPa 39-134 ksi

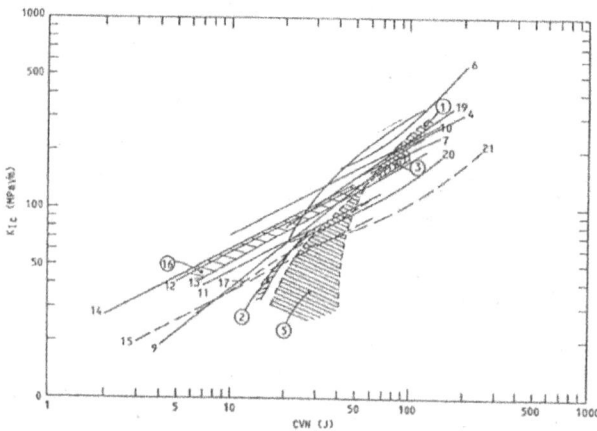

Figure 6.6 *Comparison of different published empirical Charpy-V vs KIc correlations. From (9)*

6.4.4 Introduction to the use of fracture mechanics and defect tolerance

To apply the "fracture mechanics/defect tolerance" requires a different approach compared with the simple stress approach. With the stress-based approach applied stresses experienced by parts are matched by some percentage of the uniaxial yield strength, together with the selection of a material that shows some uniaxial tensile ductility and a "safety factor" to prevent brittle fracture.

The fracture mechanics approach has three variables that must be considered which include the applied stresses, defect sizes and fracture toughness. This is shown diagrammatically in Figure 6.7.

A useful additional diagram is shown in Figure 6.8 which allows engineers to gain a "feel" for the quantitative values and also shows the ability to predict the critical defect size based on the material fracture toughness and service stress.

The next important stages in the development of fracture mechanics was practical application and applying the technology to industry. This resulted in the:

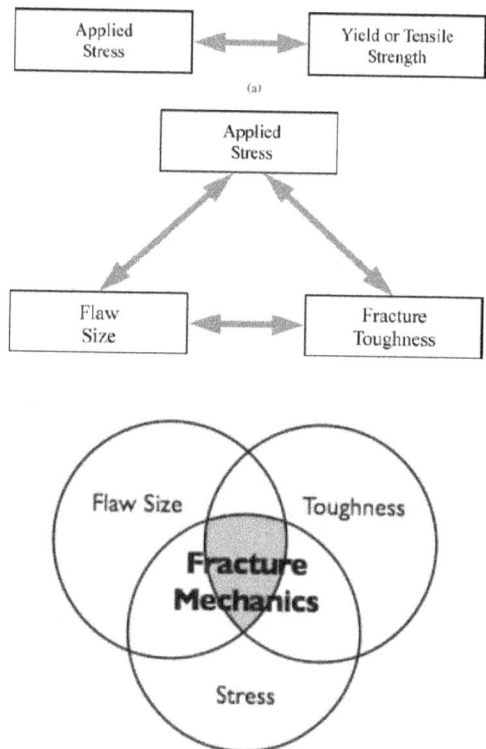

Figure 6.7 *Comparison of the fracture mechanics approach to design with the traditional strength of materials approach (a) the strength of material approach and (b) the fracture mechanics approach. RHS = Venn diagram*

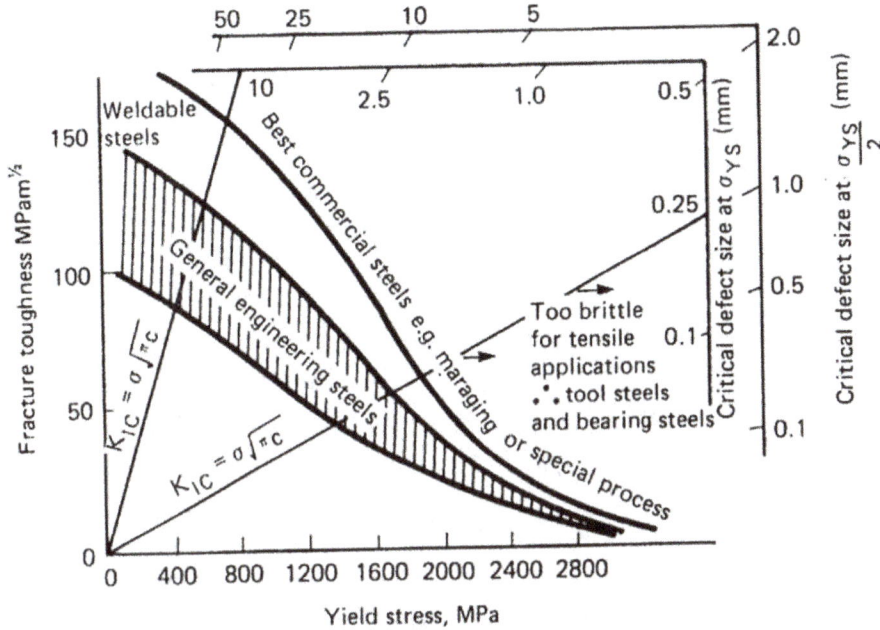

Figure 6.8a Ratio analysis diagram. From (10)

The bullet points to the right of the diagram read:

- The X axis represents the steel's yield strength

- The Y axis shows the steel's Fracture toughness

- A vertical line is drawn from the YS and a horizontal line drawn from the KIC

- A line is drawn from the "0" through the intersection of the lines and extrapolated to the outer scales which shows the critical defect size

- The inner scale represents service loads at YS and the outer service loads at half YS

- CTOD design curve "approach" in 1971, In 1971, Burdekin and Dawes applied several ideas proposed by Wells several years earlier and developed the CTOD design curve, a semi-empirical fracture mechanics methodology for welded steel structures. The cost of Wide -plate testing was high and therefore correlating the results for Wide-plate tests was an economic design strategy. The CTOD design curve was adopted in PD6493. The CTOD design curve was a simple and conservative way to specify weld flaw acceptance levels and it worked well for many years.

- R6 code by the CEGB (Central electricity generating board) in 1987

- PD 6493 in 1980 by British Standards followed by the first edition of BS 7910 in 1999 (now 3rd edition 2013)

- "Guidelines for Fracture-safe and Fatigue-reliable Design of Steel Structures" Nov 1983 by W.S. Pellini[11]

- API 579 used for the inspection and repair of pressure vessels. The areas of damage considered is shown in below. See RP 571 for detailed description and photos of expected failures

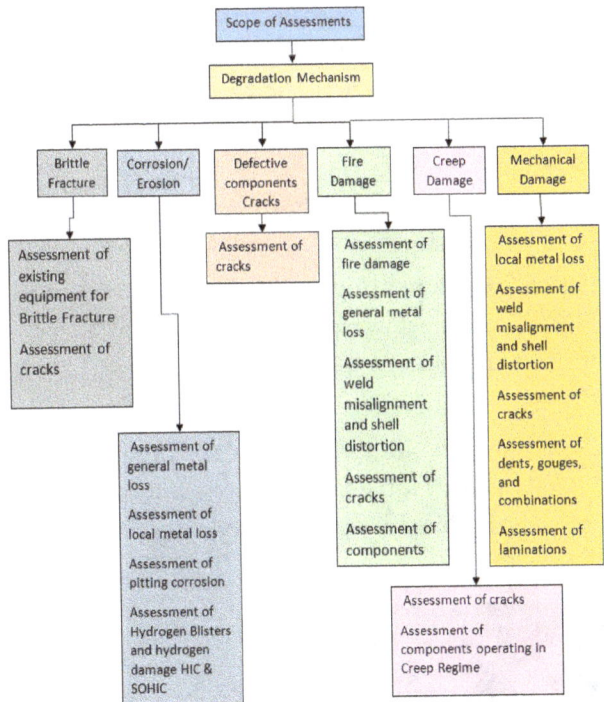

Figure 6.8b API 579 used for the inspection and repair of pressure vessels. The areas of damage considered is shown in below

6.5 TIMELINE SUMMARY

Approx. 1500	Leonardo da Vinci – failure stress of iron wires depends on length i.e. on the probability of flaw
1638	Galileo Galilei
1856	Mr Roebling in The Engineer referenced by Dr Biggs in 1960
1901	Charpy Swinging hammer tests
1903	Izod cantilever impact test
1907	Weighardt Rupture not related to stress at a point but stress over a small portion of the part
1909	Ludwik explained impact transition temperature on the effect of temperature on YS
1913	Inglis–elastic stress field around elliptical hole
1920	Griffiths equation for brittle materials
1930	Obreimoff's experiment
1938	Westergaarde elastic solution of the stress distribution at a sharp crack
1945	Constance Tipper and the liberty ships. Recognition of the ductile to brittle transition
1945	Orowan using X-rays identified the plastic zone below cleavage facets
1956	Irwin – development of the concept of energy release rate
1957	Irwin NRL [Naval Research Laboratory, USA]: K parameter. (Stress Intensity Factor) Kc Fracture toughness
1958	Wells wide plate test. Professor Alan Wells 1924-2005
1960	Paris law relating the crack growth rate to the stress intensity factor
1960-61	Irwin: Small scale yielding (plasticity correction factor)
1963	A. A. Wells (1963): CTOD (crack tip opening displacement)
1965	Cambridge Instruments introduces a commercial scanning electron microscope
1968	Proposal of the J-integral by Rice
1971	Burdekin & Dawes at Welding Institute: CTOD design curve. The Welding Institute's CTOD design curve was developed by the Welding Institute
1976	Shih and Hutchinson establish the theoretical basis of the J-Integral and link it to the CTOD
1976	Harrison et al CEGB in the UK: R-6 method
1980	PD 6493 introduced by British Standards
1999	BS 7910 introduced by British Standards. "Fitness for service"
1999	SINTAP Procedure
2000	API 579 introduced by API
2005	2nd edition BS7910. "Fitness for service"
2013	3rd edition BS7910 "Fitness for service"

The major review of fracture "Fracture and Life" prepared by Brian Cotterell[12] includes a timeline Figure 6.9.

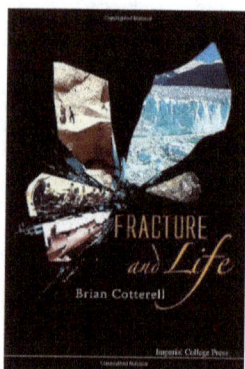

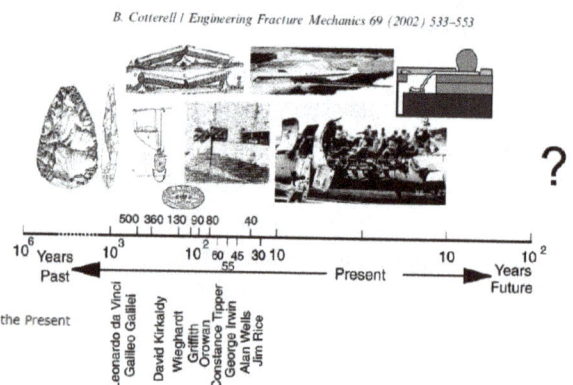

Figure 6.9 Shows the views of Dr Brian Cotterell regarding the history of fracture. Brian Cotterell obtained his PhD from Cambridge and began his career at the British Weld Research Association in 1959 during a similar time as Alan Wells. This was followed by 30 years at Sydney University. In the 1990s he became a professor at the University of Nanyang in Singapore. He retired in 2007 and took on a role as visiting professor at the University of Sydney until 2010. He completed a book in 2010 on "Fracture and Life".

Design against fracture has evolved over many years and has been a very active area of research. The following section provides a brief introduction to some important contributions.

6.5.1 Leonardo da Vinci

Around 1500, Leonardo da Vinci studied the relationship between failure stress and length under test using iron wires. The basic concept was that the probability of flaw in the wire would depend on length tested. An important aspect of fracture mechanics is the effect of scaling. Strength based on tensile tests can give similar results irrespective of the specimen size; however, it is now known that toughness tests are dependent upon specimen dimensions. Leonardo da Vinci (1452–1519) was the first to record an understanding of the scaling of fracture and the illustration in Figure 6.10 comes from one of his note-books and shows how he carried out strength tests on iron wires.

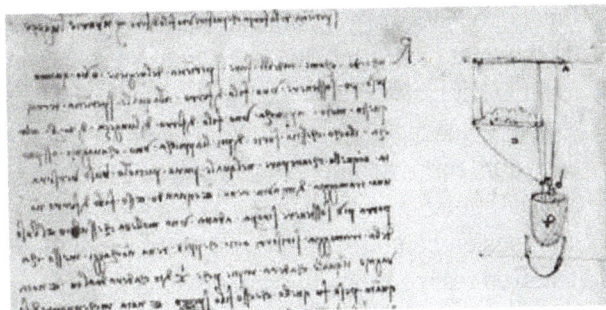

Figure 6.10 *Leonardo da Vinci's fracture of metal wires.*

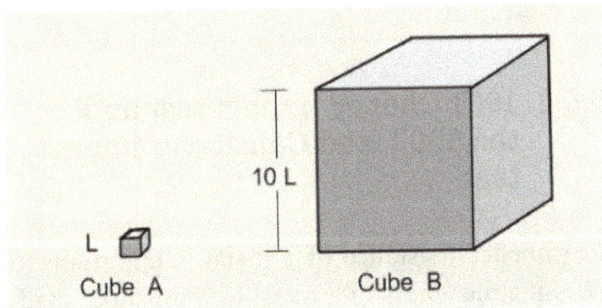

Figure 6.11 *Area up X100 but volume up by X1000*

6.5.2 Galileo Galilei (1638) and the "square-cube" law

In 1638 Galileo prepared a document "Dialogues Concerning Two New Sciences", in which he was the first to give the correct scaling laws for bars under tension and bending and he explained the reasons why objects cannot be any arbitrary size.

He gave the first answer to the question "does size matter" or more specifically, can a rock, a person, a car, an aeroplane, or a dinosaur exist at any arbitrary size? The size effect is important in fracture. Galileo saw that this effect placed a limit on the size of structures, both man-made and natural, which made it impossible to build ships, or temples or other man-made structures of infinite size. For similar reasons nature limits the size of people and animals (bones break under their own body weight) or trees of extraordinary size because their branches would break and fall under their own weight.

Galileo explained, when an object is scaled up its *area increases by the square of the multiplier while the volume increases by the cube of the multiplier.* (See Figure 6.11.) Since the ratio between the area and the volume is changing with size then the properties of the object are changing with size. He demonstrated the reasons that a large structure proportioned in exactly the same way as a smaller one must necessarily be weaker in accordance *with the square-cube law.*

"From what has already been demonstrated, you can plainly see the impossibility of increasing the size of structures to vast dimensions either in art or in nature; likewise the impossibility of building ships, palaces, or temples of enormous size in such a way that their oars, yards, beams, iron-bolts, and, in short, all their other parts will hold together; nor can nature produce trees of extraordinary size because the branches would break down under their own weight; so also it would be impossible to build up the bony structures of men, horses, or other animals so as to hold together and perform their normal functions if these animals were to be increased enormously in height; for this increase in height can be accomplished only by employing a material which is harder and stronger than usual, or by enlarging the size of the bones, thus changing their shape until the form and appearance of the animals suggest a monstrosity."[13]

6.5.3 1856 Mr Roebling in The Engineer referenced by Dr Briggs in 1960

Dangerous Brittleness in Steel

The Engineer magazine observed:

Between Preface and Introduction of a book on The Brittle Fracture of Steel by Dr. W. D. Briggs, to be reviewed soon in our columns, we were intrigued to see printed the following quotation from THE ENGINEER, of June 18, 1861: *"Effects of percussion and frost upon iron . . . we need hardly say that this is one of the most important subjects that the engineers of the present day are called upon to investigate. The lives of many persons, and the property of many more, will be saved if the truth of the matter discovered – lost if it be not."*

THE ENGINEER

Friday, January 18, 1861

THE EFFECTS OF PERCUSSION AND FROST UPON IRON

Up to the present moment it has been very generally supposed, even by metallurgical authorities, that when wrought or malleable iron, even of good fibrous quality, is acted upon by long-continued percussion, or by tremulous motion to which it is subjected in railway work, it necessarily loses its fibrous nature, and acquires a crystalline character. In order to examine the fitness of various qualities of iron for the manufacture of wire rope, Mr. Roebling undertook, during the hard winter of 1856, at his establishment, at Trenton, a series of experiments when the thermometer was five to ten degrees below zero.

Part of the way through the text we find the description of his test procedure

The samples for testing, about one foot long, were reduced in the centre to exactly three quarters of an inch square, and their ends left larger, were welded to heavy eyes, making in all a bar of three feet long. Thus prepared, they were thrown outside of the mill, covered with snow and ice and left exposed for several days and nights. Early in the morning, before air grew warmer, a sample enclosed in ice, would be put into the testing machine and at once subjected to a strain of 26,000lbs, the bar being suspended in a vertical position, left free all around. A stout mill-hand armed with a billet of one and a half inches in diameter, and two feet long, then struck the sample horizontally a number of blows, hitting the reduced section as hard as he could. The blows were counted and continued until rupture took place. Care was given to maintain a tension of 26,000lbs during this test by screwing up the lever, while the sample kept stretching. Other means for producing vibration were attempted, but none were proved so effective as the hitting with an iron bolt. He remarks that most of these irons would support 70,000-80,000lb per square inch; and that good samples of three-quarters of an inch square would support a strain of 26,000lb for a whole week without visible stretching, provided all vibration and jarring were avoided. But the least jar would produce a permanent elongation. Without going into details of these interesting and instructive experiments, we will only state the number of blows which the different samples resisted, when encased in ice ranged from three to one hundred and twenty. Inferior qualities of a crystalline texture would break at the third or fourth blow. Good samples of refined puddle bar resisted very well, and went up to sixty blows, while the better qualities of hammered charcoal iron, supported up to one hundred and twenty blows, stretching and drawing all the time. Indeed, the tension being reduced to 20,000lb, some good samples resisted the almost incredible number of three hundred blows before breaking. Such qualities of iron may, says Mr. Roebling, be depended upon for the construction of wire cables and car axles.

6.5.4 1901 Charpy impact test and the 1903 Izod Cantilever impact test

The impact resistance of a metal is important in assessing the ability of a metal to resist shock loading. It is a measure of the "energy of fracture" which is the energy required to break the atomic bonds in the crystalline microstructure when cleavage fracture occurs. For ductile rupture the microstructural features such as grain size and second phase particles are responsible for the toughness. If we wish to build a structure which needs to withstand shock loading in service, e.g. a high-pressure gas pipeline, we would not want this structure to shatter like glass when it is

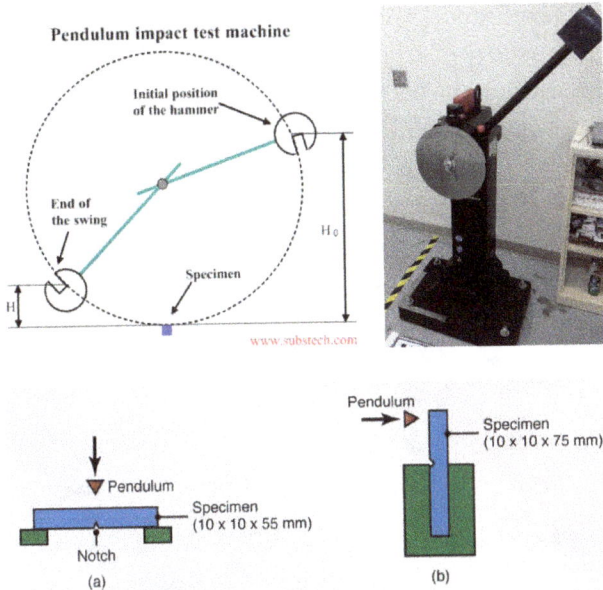

Figure 6.12 Charpy and Izod test equipment and specimens (See section 5.7)

impact loaded, either due to a gas pressure surge or an inadvertent collision. Hence, we need to specify a certain amount of impact resistance. The early general rule from an analysis of "Liberty Ship" failures was 27J at 20°C below the operating temperature. However, as the consequences of failure increased and welded construction was involved, the minimum specified energy would generally increase to above 40J.

The toughness has been traditionally measured since the beginning of the 1900s by using a swinging pendulum to hit and break a specimen of the metal under study. The energy absorbed is calculated from the change in potential energy based on the difference in pendulum height before and after impact. This energy absorbed is used as a measure of the impact resistance.[14]

The two well established methods are the Charpy and the Izod with the Charpy being the most popular test method that has been chosen by Industry. The Charpy was named after Georges Augustin Albert Charpy, a Frenchman, who was born in 1865 and graduated from the École Polytechnique in 1887 with an engineering degree majoring in marine artillery. The Izod test is named after the English engineer Edwin Gilbert Izod (1876–1946) who developed the machine when working on gun barrel bursts.

The evolution of impact testing may be divided into the following four periods:

1. Early developments: up to the time of standardisation of testing procedures

2. The stage of brittle fracture: period up to the beginning of the 1950s including the brittle-fracture story (Tipper) and the transition-temperature concepts (Liberty ships)

3. The development of fracture mechanics: up to the early 1980s including the correlation between the absorbed energy measured with CVN and other fracture-mechanics parameters LEFM (linear elastic fracture mechanics) EPFM (elastic-plastic fracture mechanics). Fracture mechanics allowed a quantitative approach to fracture especially in the approach to design and the link between NDT inspection to quantify the size of defect in the structure to evaluation of the allowable defect which can be accepted without the occurrence of risk of brittle fracture in service

4. A recent stage: including instrumented impact testing, testing on sub-size specimens, and a better understanding of the toughness in the Ductile-to-Brittle Transition Region.

6.5.5 Weighardt 1907

Wieghardt in 1907 rejected the idea of stress or strain related criteria for the onset of rupture. He was concerned with the paradox that the stresses at the tip of a sharp crack in an elastic body are infinite no matter how small is the applied stress. This fact led him to argue that rupture does not occur when the stress at a point exceeds some critical value, but only when the stress over a small portion of the body exceeds a critical value.

The concept of a critical stress intensity factor slipped through Wieghardt's fingers.

6.5.6 Ludwik 1909

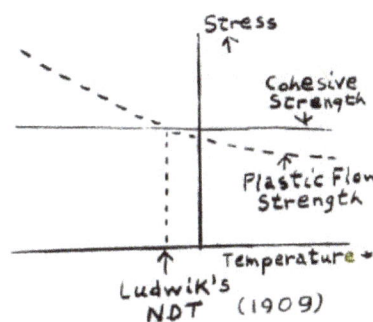

Figure 6.13 Sketch by G R Irwin

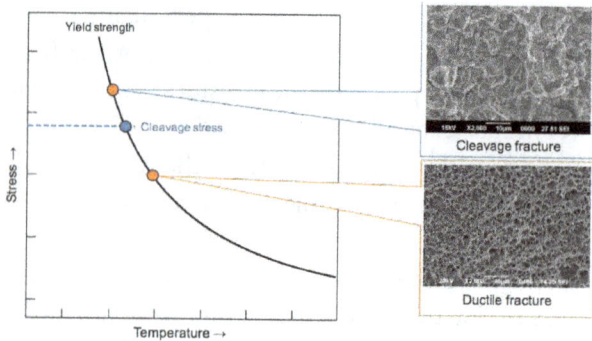

Figure 6.14 Current understanding

As early as 1909, Ludwik explained a possible reason for the transition from ductile to cleavage behaviour in ferritic steels. He suggested that the cohesive strength was little affected by temperature, but there was a marked increase in the yield strength of low carbon steel as the temperature decreased so that a particular temperature cleavage fracture became easier than yielding.

The effect of a notch on the transition temperature was seen to be primarily due to a constraint on yielding and Orowan, using Ludwik's concept, showed how a notch would increase the transition temperature.

6.5.7 Stresses at Elliptical Holes – Inglis's Solution (1913)[15] Inglis's original publication[4]

In 1913 Sir Charles Edward Inglis demonstrated how stresses were concentrated around sharp localized changes in shape. He modelled holes and notches in uniform stress fields to produce a two-dimensional stress analysis. The results are shown in Figure 6.15:

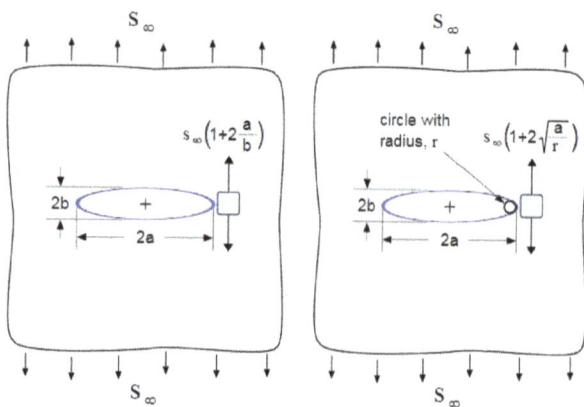

Figure 6.15 Inglis solution for the stress concentration due to an elliptical hole

The simplicity of Inglis's equation, together with the quantity of corners and notches manufactured in industry, in the early 1900s, guaranteed its popularity. Experimental evidence of the concept was also confirmed by photo elastic tests.

The final sentence in the Wikipeda page on Inglis quotes, "Inglis has been described as the greatest teacher of engineering of his time". [16]

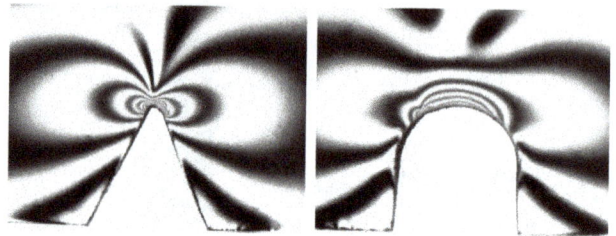

Figure 6.16 Photo elastic models demonstrate that stresses are more concentrated at sharp notches

6.5.8 1920 A A Griffiths equation for brittle materials

Many people that review fracture consider the work of Griffiths as a breakthrough event although at the time it was not considered as such a major event. A A Griffiths abandoned elasticity and stress approach and proposed an energy-balance of fracture. The basic approach was that the strain energy had to provide the energy to create the new surface created by the fracture. Figure 6.17 shows an extract from a 1920s'

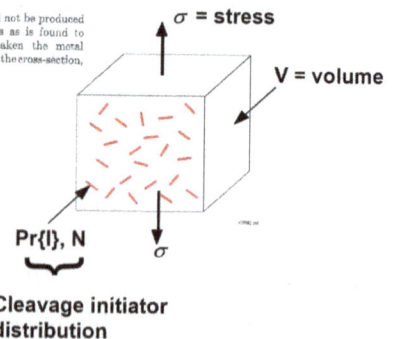

Figure 6.17 LHS = Extract from a 1920s' book showing a cartoon of Griffiths theory. RHS = The modern-day equivalent – Basis of the general statistical model – material in front of the crack contains a distribution of possible cleavage fracture initiation sites i.e. cleavage initiators. From [17]

book showing a cartoon of Griffiths theory. Figure 6.18 shows the proposed energy approach the strain energy creating the surface energy.

Griffith's theory

- Theory of unstable crack growth
- Theory establishes a relationship for the occurrence of unstable crack growth
- Basic underlying principle was an Energy balance (strain energy converting into new surface energy)
- The basic approach considers the introduction of a crack in a stressed / loaded component and determination of the release of strain energy as crack grows
- The crack growth is seen by a reduction in stiffness
- What happens to change in strain energy? or released strain energy?
- An infinite body Linear elastic – subjected to stress
- Initially there was no crack
- Strain energy stored

$$U_o = \frac{\sigma^2}{2E} \cdot Volume$$

- Crack size = 2a
- Material above and below the crack will be stress-free to some extent
- The stress-free portion would be assuming triangular distribution with height of triangle = β
- Griffith carried out extensive calculations and experiments and established that β = ϖ
- Strain energy after introducing crack = Strain energy before introducing crack – Strain energy loss

$$U = U_o - \frac{\sigma^2}{2E} \cdot Volume$$

$$\Delta U = U_o - U = \frac{\sigma^2}{2E}(2 \cdot \frac{2a\beta at}{2})$$

$$\Delta U = \frac{\sigma^2 \beta a^2 t}{E} = E_R \quad \begin{array}{l}\textit{Strain energy decreased}\\ \textit{due to crack extension}\end{array}$$

- When a body is cracked, material bonds are broken and the energy comes from a reduction in the strain energy
- Two key aspects in crack growth. How much strain energy is released when crack advances? What is the minimum energy required for the crack advance in forming two new surfaces?
- Some external work can also be involved which would increase in strain energy.

Surface energy gives the crack growth
   Assume a crack of '2a' size exists

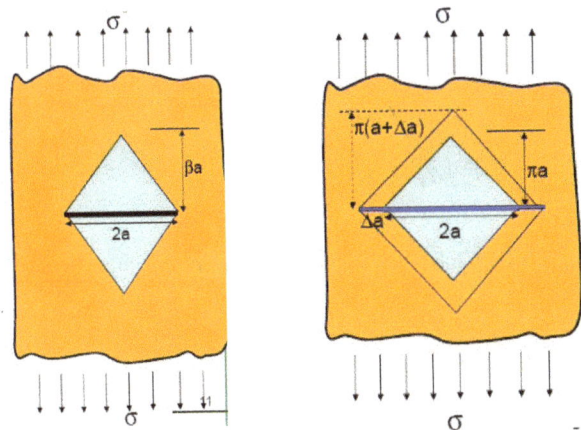

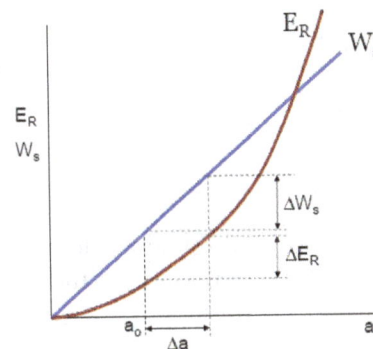

Figure 6.18 *LHS = Diagram 1, Centre= Diagram 2, RHS = Plot E_R and W_s wrt crack length 'a'*

Initial stress-free area (Figure 6.18 LHS Diagram 1)

$$A1 = 2 \cdot \frac{2a \cdot \pi a}{2} = 2\pi a^2$$

Crack grows by Δa on both sides
   Stress-free area after crack growth (Figure 6.18 LHS Diagram 2)

$$A2 = 2 \cdot \frac{2(a + \Delta a) \cdot \pi(a + \Delta a)}{2} = 2\pi(a + \Delta a)^2$$

Change in stress-free area

$$A1 - A2 = 2\pi a^2 - 2\pi(a + \Delta a)^2$$

$$A1 - A2 = 2\pi\left[(a + \Delta a)^2 - a^2\right]$$

$$\Delta a = 2\pi(2a \cdot \Delta a) = 4\pi \cdot a \cdot \Delta a$$

$$\Delta V = 4\pi \cdot a \cdot \Delta a \cdot t$$

Change in Strain Energy

$$\Delta U = \frac{\sigma^2}{2E} \cdot Volume$$

$$\Delta U = \frac{\sigma^2}{2E} \cdot 4\pi \cdot a \cdot \Delta a \cdot t = \Delta E_R$$

$$\frac{\Delta E_R}{\Delta a} = \frac{\sigma^2 . 2\pi . a . t}{E}$$

Surface energy required to create new area

$$\Delta W_s = \gamma_s . (4. \Delta a . t)$$

$$\frac{\Delta W_s}{\Delta a} = 4.\gamma_s . t$$

Unstable crack growth takes place, if

$$\frac{d E_R}{d a} \geq \frac{d W_s}{d a}$$

$$\frac{2.\pi.\sigma^2.a.t}{E} = 4.\gamma_s.t$$

$$2.E.\gamma_s = \sigma^2.\pi.a$$

'a' existing crack length – Stress required to grow crack

$$\sigma^2 = \frac{2.E.\gamma_s}{\pi.a}$$

Given the loading on a part this gives the "Stress". The above formula can be used to calculate the maximum allowable crack size before the crack becomes unstable, a_c. This gives the approach to "Damage tolerant design". If NDT can give the maximum size of imperfection a calculation can be made on whether the part is "fit for service".

$$a_c = \frac{2.E.\gamma_s}{\pi.\sigma^2}$$

γ_s = *Specific Surface Energy – Surface Energy per unit area of crack surface*

Area of crack = 2.(2.a . t) = 4.a . t

Surface Energy stored ≥ 4.a . t . γ_s

Energy balance

$$E_R \geq W_s$$

$$\frac{\pi.\sigma^2.a^2.t}{E} = 4.a.t.\gamma_s$$

Plot both E_R and W_s wrt crack length 'a' Crack growth takes place, if $\Delta E_R \geq \Delta W_s$
Therefore:

• Not satisfying above inequality – crack remains dormant
• Surface energy required can be obtained from external sources as well – increase in applied stress – No change in ΔW_s
• Strain energy used to break the bonds – creates surface energy
• Crack growth – energy conversion process
• When a crack propagates – strain energy gets reduced and surface energy increases for a constant displacement case.

Figure 6.18 RHS shows the net summation of the strain energy release and the new surface area created.

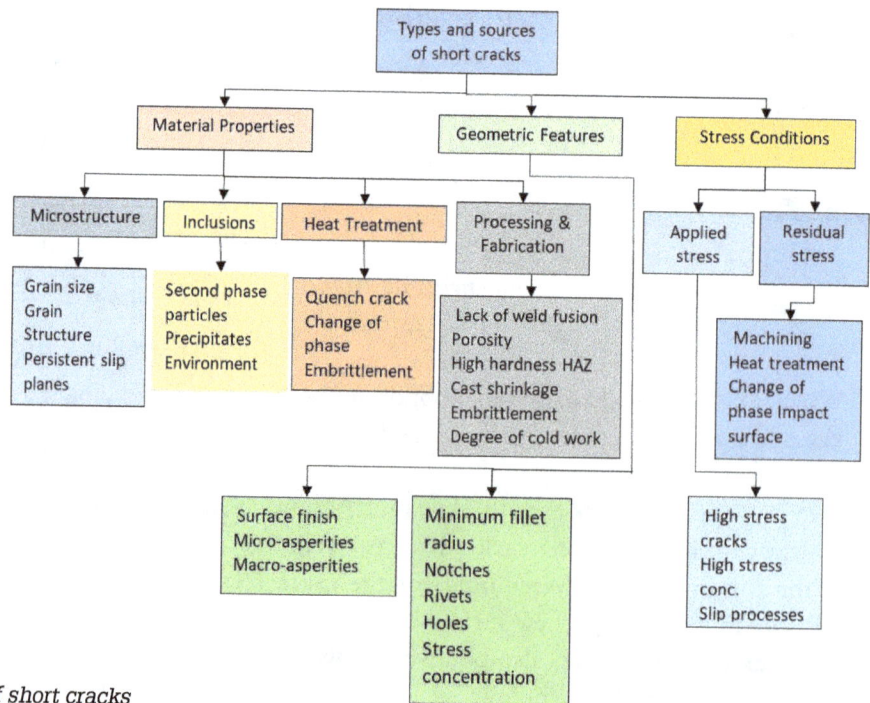

Figure 6.19 *The possible origin of short cracks*

Once the strain energy release is greatest then a critical crack length has been achieved and brittle fracture will occur.

6.5.9 1930 Obreimoff's wedging experiment

This experiment gave support to Griffiths by pushing a wedge to cause delamination

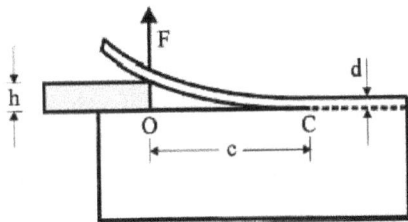

Figure 6.20 Wedge pushed beneath a layer of thin material

$$U - U_E = \frac{Ed^3h^2}{8a^3}$$

The surface energy needed to grow the crack is

$$U_S = 2a\gamma \quad \text{where } \gamma \text{ is the surface energy.}$$

Equating the elastic energy to the surface energy gives an equilibrium crack length a_o of:

$$a_o = \sqrt[4]{3Ed^3h^2/16\gamma}$$

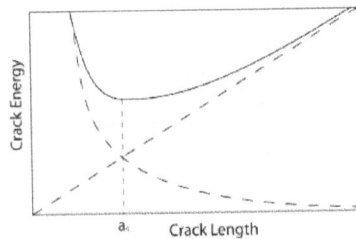

Figure 6.21 Energy in bent beam and crack length increase

6.5.10 Westergaard elastic solution of the stress distribution at a sharp crack

H. M. Westergaard, working on bearing contact pressures in 1938, developed elastic solution of the stress distribution at a sharp crack.

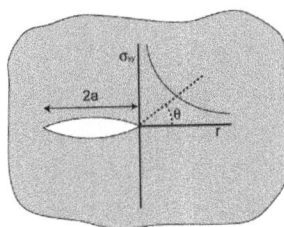

$$\sigma_{xx} = \frac{\sigma_o\sqrt{\pi a}}{\sqrt{2\pi r}}\cos\left\{\frac{\theta}{2}\right\}\left[1-\sin\left\{\frac{\theta}{2}\right\}\sin\left\{\frac{3\theta}{2}\right\}\right]$$

$$\sigma_{yy} = \frac{\sigma_o\sqrt{\pi a}}{\sqrt{2\pi r}}\cos\left\{\frac{\theta}{2}\right\}\left[1+\sin\left\{\frac{\theta}{2}\right\}\sin\left\{\frac{3\theta}{2}\right\}\right]$$

$$\tau_{xy} = \frac{\sigma_o\sqrt{\pi a}}{\sqrt{2\pi r}}\sin\left\{\frac{\theta}{2}\right\}\cos\left\{\frac{\theta}{2}\right\}\cos\left\{\frac{3\theta}{2}\right\}$$

Irwin reviewed this approach and identified that perpendicular to the cracking plane ahead of tip and close to the crack tip that the local tensile stress has a distribution given by

σL= K/ /(2πr) where r= the distance ahead of the crack tip. K is a constant determining the level of the stress distribution. The constant K became known as the stress intensity factor.

6.5.11 Constance Tipper – Work carried out on the Liberty Ships identified the importance of the ductile to brittle impact transition

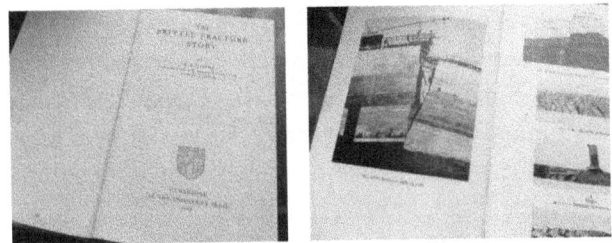

Figure 6.22 Tipper's book "The brittle fracture story" (18)

Constance Fligg Tipper (born 1894-1995) carried out work on the failed Liberty ships. She concluded that the cause of the fractures in the ships was not totally attributed to the welding imperfections and hard HAZs, as many engineers had thought, but because of the steel itself. Structure plate steel made in the early 1940s had Charpy transition temperatures, where the fracture behaviour changes from ductile to brittle, around plus 20°C. It was well known at the time that steel produced in the late 1800s by the early Bessemer process had brittle behaviour of steel due to high nitrogen levels caused by blowing air through the molten metal.[19, 20] These problems were recognized and documented by test engineers such as David Kirkaldy. However, in riveted structures brittle fractures rarely caused catastrophes because a fracture was usually arrested at the edge of the plate in which it initiated.

It was in 1943 that Tipper began her important work on the cause of brittle fracture in Liberty ships, after the appointment of J.F. Baker (later Lord Baker) as Professor of Engineering at Cambridge. The Liberty ships were the merchant ships, constructed with great haste at the beginning of the war, which brought vital supplies across the Atlantic to Britain. A number of these ships, complete with their cargoes, had been lost

Figure 6.23 Details of the Tipper test.

Particulars of specimen	Method of load	Criteria for assessing transition temperature	Remarks

Tipper notched tensile 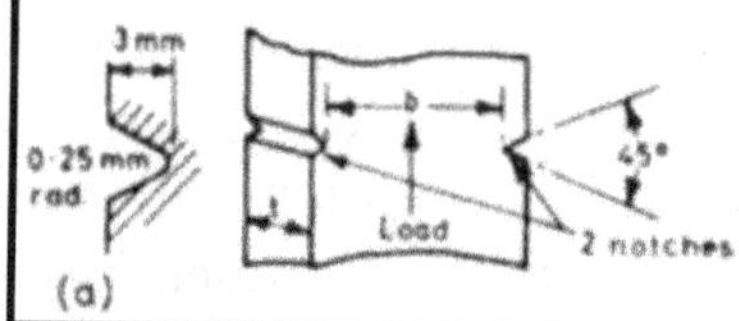	Static tension	1. Fracture appearance
		2. Ductility –
		  a) Extension over notch
		  b) Reduction in thickness
		3. Work done when record of load/ extension available
			Uniform rectangular strips of full plate thickness (unless limited by capacity of testing machine). Width 'b' greater than thickness 't': preferably b = 2·5 t. Length to suit grips.

in heavy seas, breaking up with a fast crack running instantaneously right round the ship. Tipper discovered that in the cold of the North Atlantic, the type of steel used for the ships became brittle, were susceptible to brittle fracture.

The work she carried out allowed modifications to be made to the ships allowing the vital supply route to remain open throughout the war. She was also the first person to use a scanning electron microscope to examine metals, and developed the Tipper Test which is still used to determine the brittleness of steel used in construction and manufacture.

Welded structures needed steel with adequate toughness to avoid brittle fracture. Welding provided high residual stresses equal to the yield strength, a heat affected zone adjacent to a weld with a much higher transition temperature than the parent plate, and crack-like defects. Since a welded structure is continuous an unstable brittle fracture could propagate through the structure and cause a catastrophic failure.

6.5.12 Orowan 1945 Identified the presence of plastic strain below the cleavage facets

Orowan in 1945 studied the depth of plastic strain beneath cleavage facets in low carbon steel using X-ray scattering. Irwin in 1948 noted that the energy expended in this plastic straining could be estimated from Orowan's results. This fracture energy, Cp (plastic) for low carbon steel around 0°C, turns out to be roughly two thousand times the surface energy, Cs (surface) used in the Griffith model. Irwin concluded that Grifftth's theory could be used if the plastic work plus surface energy was substituted for the surface energy. Orowan presented the same idea in 1949. However, it was Irwin who grasped the engineering significance of the extension of Griffith's work and went on to develop LEFM.

6.5.13 1956 Irwin NRL [Naval Research Laboratory, USA]

Based on the observation by Orowan in 1945 of the plastic straining below the cleavage facets Irwin realized the magnitude of the energy dissipated due to the plastic straining. Irwin developed the concept of energy release rate Gc similar to surface energy term in the Griffiths equation; replacing the term 2γ in the Griffith equation with Gc which represented the total energy required including the energy dissipated in plastic deformation.

To move from   To   Took 36 years!

$$\sigma = \sqrt{\frac{2\gamma_s E}{\pi a}} \qquad \sigma = \sqrt{\frac{(2\gamma_s + \gamma_p)E}{\pi a}}$$

6.5.14 1957 Irwin NRL [Naval Research Laboratory, USA]: K parameter. (Stress Intensity Factor)

Irwin developed the concept of stress intensity factor based on Westergaard's work. Irwin applied the findings of Westergaard to show that the stresses and displacements near the crack tip could be described by a single parameter which later became known as the "stress intensity factor". Mathematically it could be proven that the stress intensity approach was related to the energy release rate method.

6.5.15 1958 Wells wide plate test. Professor Alan Wells 1924-2005

This test consisted of a full thickness 3ft square plate made up of two halves as shown in Figure 6.16.

The plates had an appropriate weld profile with a saw-cut into the edge of midpoint to create an artificial defect. The two plates are then welded together with a butt weld. After the test piece has cooled it was loaded in tension parallel to the butt weld. This

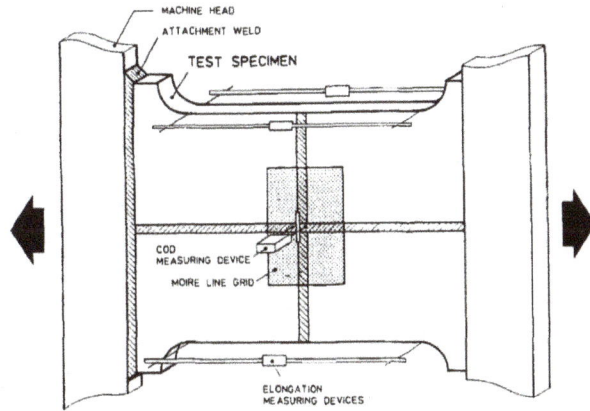

Figure 6.24 LHS = Wells wide plate test, RHS = Instrumentation used for Wide-plate test

test demonstrated the transition effect from ductile to brittle behaviour as the temperature was lowered. Below a certain temperature which was dependent on the weld procedure and defects, present fractures occurred at relatively low levels of stress. Above the transition temperature the loading had to be above general yielding before fracture occurs.

This test has provided useful information on the significance of residual stress and of defects and has allowed the benefits of stress relief to be demonstrated.

A comment made by Alan Wells after his retirement reported a hazardous incident associated with the test:

"Testing the plates required compact 600 tonne capacity equipment . . . none of us will forget the first successful fracture test," he recalled. "About half a tonne of equipment flew out of the end of the hut."

6.5.16 1960 Paris law relating the crack growth rate to the stress intensity factor[6]

After the initial universal acceptance of the "Irwin" approach there were a number of applications to real engineering problems. For example, Wells used the method to show the fatigue cracks on the Comet propagated under the service stresses and then after reaching a critical crack size caused complete fracture of the fuselage.

Paris applied fracture mechanics approach to fatigue crack growth. Paris postulated that the range of stress intensity factor could be used to quantify the sub-critical crack growth under cyclic fatigue

loading in a similar way that the stress intensity factor predicted the occurrence of fast fracture. Paris examined a number of metals and found that plots of crack growth rate against range of stress intensity gave straight lines on log-log scales. This implies that:

$$\log\left(\frac{da}{dN}\right) = m\log(\Delta K) + \log C$$

Taking out the logs gives:

$$\frac{da}{dN} = C\Delta K^m$$

Figure 6.25 LHS = Paris equation, RHS = Data for austempered ductile iron in air

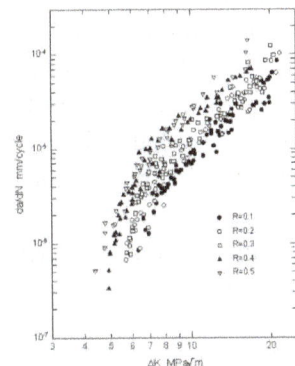

Paris made it possible to make a quantitative prediction of crack growth under cyclic service stresses. In addition, the fracture toughness could be used to predict the maximum crack before final failure.

Figure 6.25 RHS shows data for austempered ductile iron in air, as a function of stress ratio (minimum stress in cycle divided by maximum stress in cycle which is a measure of mean stress in the fatigue cycle).

6.5.17 1960-61 Irwin: Small scale yielding [plasticity correction factor]

The principles of linear elastic fracture mechanics (LEFM) can only be applied in situations where the stress field outside the plastic zone is dominated by the stress intensity factor.

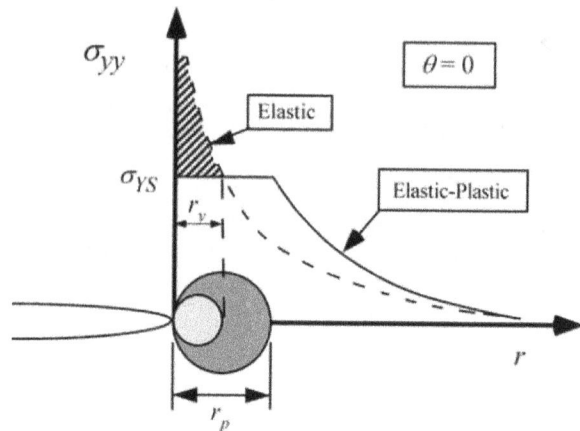

Figure 6.26 Plastic zone

LEFM is valid in practice for high strength steels, alloy steels and aluminium alloys, i.e. valid results can be obtained with the use of small test pieces. However, for the lower strength forging steels large specimens are needed to satisfy the measurement requirements.

The problem becomes critical when applied to metals such as structural steels at room temperature, i.e. those of lower yield strength. For example, A533B steel (0.23%C, 1.5%Mn, 0.5% Mo) which finds use in welded pressure vessels in nuclear power stations. For such steel, the yield strength is approximately 500 MPa and the fracture toughness at room temp is 180 MNm-3/2.

Applying the standard requirement for specimen size from

$$B, a \geq 2.5 \left(\frac{K_{IC}}{\sigma_{YS}} \right)^2, \quad 0.45 \leq a/W \leq 0.55$$

This would equate to a weight of approximately 1500 Kg as a CTS specimen and would require a fracture load of 5MN.

The thickness is not unrepresentative of the practical service application (very large, thick vessels) but the size of the test-piece that would be required to give a valid fracture toughness test is far too large for a routine test.

Since A533B is used in the nuclear industry, there was a requirement to assess the effects of neutron irradiation on toughness. It became clear that there was a need to measure toughness parameters on small test-pieces. Consideration was directed towards the effect of plasticity on the crack-tip stress field to produce a solution. This resulted in Plasticity "Correction Factors".

6.5.18 1963 A. A. Wells (1963): CTOD (crack tip opening displacement)

During the testing of many metals and alloys the assumption of small-scale plasticity is not valid, especially when the plastic zone sizes are large compared to the crack size or specimen dimensions.

The first approach developed to test low strength metals was the crack opening displacement (COD). The COD is the amount of crack opening before crack extension. Originally developed from research at British Welding Research Association (became TWI) during the 1960s. It was proposed independently by Cottrell (1961) and Wells (1961) as a parameter for characterizing the stress–strain field ahead of the crack tip. In plane stress, this is given by Smith; and Burdekin and Stone:

$$\delta = \left(\frac{8\sigma_{ys}a}{\pi E} \right) \ln \sec \left(\frac{\pi a}{2\sigma_{ys}} \right)$$

Expanding this equation and taking the first terms gives the COD under plane stress conditions as

$$\delta = \frac{K^2}{E\sigma_{ys}}$$

Rice (1974) has also obtained the following expression for the COD under plane strain conditions, using finite element analyses:

$$\delta = \frac{K^2}{2E\sigma_{ys}}$$

The COD is generally measured with a crack-mouth clip gauge. Using similar triangles (**Figure 6.27**), accurate measurements with this gauge can be related easily to the crack-tip opening displacement (CTOD). (Figure 6.27) Values of the CTOD measured at the onset of fracture instability correspond to the fracture toughness of a material. Hence, the CTOD is often used to represent the fracture toughness of materials that exhibit significant plasticity prior to the onset of fracture instability. Guidelines for CTOD testing are given in the ASTM E-813 code. And BS EN ISO 7448.

CTOD has been the most widely used fracture toughness parameter within the oil and gas industry for nearly 50 years. CTOD was an ideal parameter for characterizing the fracture toughness of medium-strength carbon-manganese steels used in the

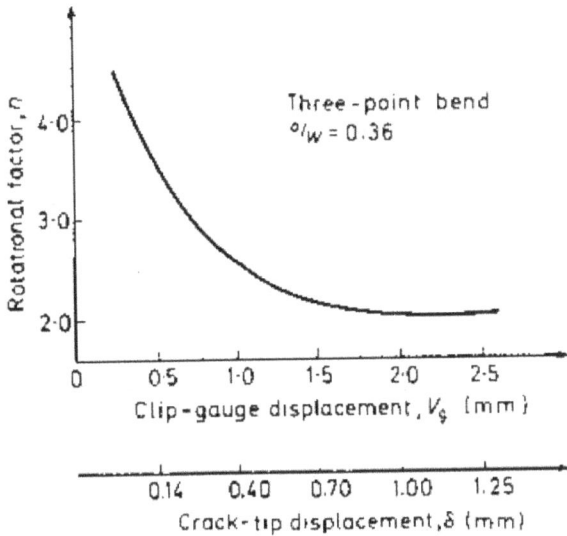

Figure 6.27 Diagram showing relationships between crack-tip displacement and knife-edge displacement

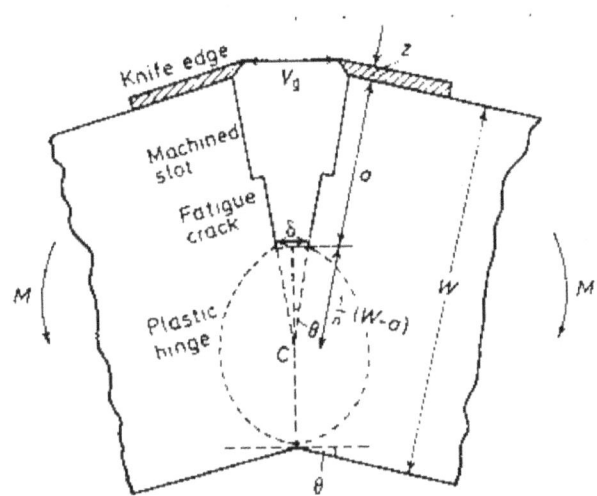

Figure 6.28 Shows a diagram of the test specimen showing the machine slot, the fatigue crack

manufacture of pressure vessels which displayed too much ductility to allow the application of linear-elastic fracture mechanics (LEFM).

The development of North Sea oil and gas from the 1970s onwards hastened the application of CTOD testing and analysis concepts for application to the construction of steel-jacket production platforms, and pipelines. The fracture-toughness testing of single-edge-notched bend specimens (or the 'CTOD test' as it is sometimes called) is the standard method to measure it.

Wells supposedly developed the crack-tip-opening displacement (CTOD) model of fracture mechanics from an observation of the movement of the crack faces apart during plastic deformation of notched test pieces. He showed that fracture would take place at a critical value of COD, and for calculations below general yield, this was proportional to the square of the critical stress intensity factor divided by the yield strength.

Furthermore, he showed that the critical value of COD determined in bend specimens and wide-plate specimens (representing structural components) of the same thickness were equivalent. Thus, he was able to demonstrate transferability of fracture toughness determined from test specimens to other structural geometries. This was to have far-reaching implications on the development of fitness-for-service concepts for welded structures for the avoidance of fracture. As a result of this, the CTOD parameter was used extensively in the UK for elastic-plastic fracture mechanics (EPFM) analysis of welded structures from the 1960s especially once the development of North

Sea oil reserves in the 1970s was driving much of the fracture research at that time.

6.5.19 1965 SEM Cambridge Instruments introduces a commercial scanning electron microscope

The SEM microscope was essential for the study of fracture surfaces and modes of fracture and therefore useful for failure investigations and the ability to understand the fracture process. Failure analysis deals with the determination of the causes of failures. This involves an analysis of factors influencing failures and finally deciding the "root cause" that allow the elimination or prevention of such failures in the future.

The practical approach needed to study the cause of failure, frequently requires the fractographic examination of the fracture surface. The scanning electron microscope (SEM) has become a major tool especially when combined with EDS chemical analysis. The fracture surface provides information on the "mode" of fracture which indicates the stresses, level of defects present and the material toughness. Failure analysis is currently becoming an important skill for engineers whose work is relevant to the design and operation of machines, components, or engineering structures.

6.5.20 1968 Proposal of the J integral by Rice

J-integral is a more comprehensive approach to fracture mechanics of lower-strength ductile materials. J-integral can be interpreted as the potential energy difference between two identically loaded specimens having slightly different crack lengths. Testing is carried out in a similar manner to fracture toughness KIC but using a series of identical specimens (the multi-specimen approach) or a single specimen.

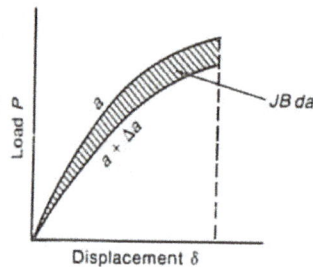

Figure 6.29 Physical interpretation of the J integral

6.6 FRACTURE OF METALS

The types of fracture can be divided into two broad categories, which are cleavage (brittle) or ductile rupture (ductile fracture).

Cleavage fracture has long been regarded as the most troublesome and destructive. Cleavage mode of fracture appears as brittle fracture which was often unpredictable and was responsible for loss of life and substantial financial loss and disruption. The early failures were associated with welded parts which often contain hard HAZ microstructures, defects and residual stress.

Ductile rupture in the absence of crack-like defects would only be experienced in the cases of accidental overload which exceed the normal design stresses. When it does occur the fracture, surface shows rough and dull appearance with evidence of plastic deformation. Since this type of failure is less damaging it has gained less metallurgical attention compared to brittle failure.

During any failure investigation that involves fracture an understanding of the "mode" of fracture can assist with the "root cause" analysis and allow appropriate actions to prevent future similar failures.

The types of fracture can also be described based on whether they occur through the "grain" structure or through the grain boundaries. For example:

1. Through the grain structure – **Transgranular fracture**, which can be classified into

   a) Brittle cleavage fracture
   b) Ductile rupture fracture
   c) Fatigue fracture

2. Through the grain boundaries – **Intergranular fracture**, which can be classified into

   a) Intergranular fracture without micro voids
   b) Intergranular fracture with micro voids.

6.6.1 Brittle cleavage fracture

Figures 6.31 (a) and (b) show fracture surfaces of tensile specimens, which have been failed in ductile and brittle manners respectively. Figure 6.31(b) shows

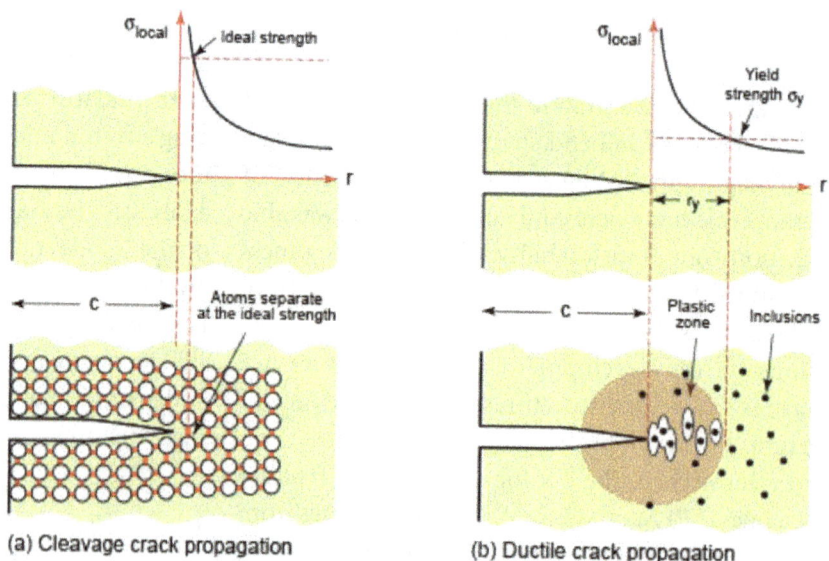

Figure 6.30 Comparison of cleavage and ductile rupture

(a) Cleavage crack propagation

(b) Ductile crack propagation

Figure 6.31 Ductile fracture and brittle cleavage fracture observed from tensile fracture surfaces.

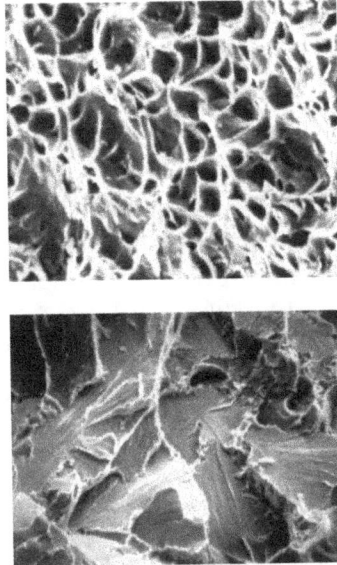

the fracture surface shows a river line pattern or stress line pattern in which its direction points toward the origin of the crack.

When the crack is initiated possibly at for example an inclusion, or large carbide, the metal grains then readily cleave along the low-fracture-energy crystallographic plane.

The river lines or the stress lines are steps between cleavage or parallel planes, which converge in the direction of local crack propagation as seen from Figure 6.32.

Transgranular cleavage fracture is usually associated with defects such as cracks, porosity, inclusions or second phase particles which, can obstruct dislocations movement. Stress is therefore concentrated in front of these defects, initiating a crack of a critical size. The propagation of this crack then finally causes the failure with very little plastic deformation as illustrated in Figure 6.32. As the defects are known as the cause of the failure, minimizing these imperfections will avoid stress concentration, and increase fracture resistance of the materials. This can be achieved by control of the quality of the products in manufacturing processes.

The crack initiation point can be identified by observing the fracture surfaces of the specimens or components by visual inspection or by means of magnification. An example of the crack initiation site which can be observed by visual examination of the fracture surfaces of metal samples is illustrated in Figure 6.33. There are apparent stress lines pointing to one end of the plate on both halves as depicted in Figure 6.33 (a) and an initiation site located in the middle of the top surface as seen in Figure 6.33 (b).

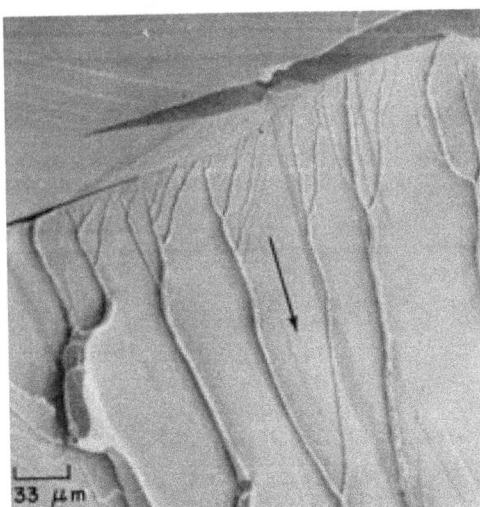

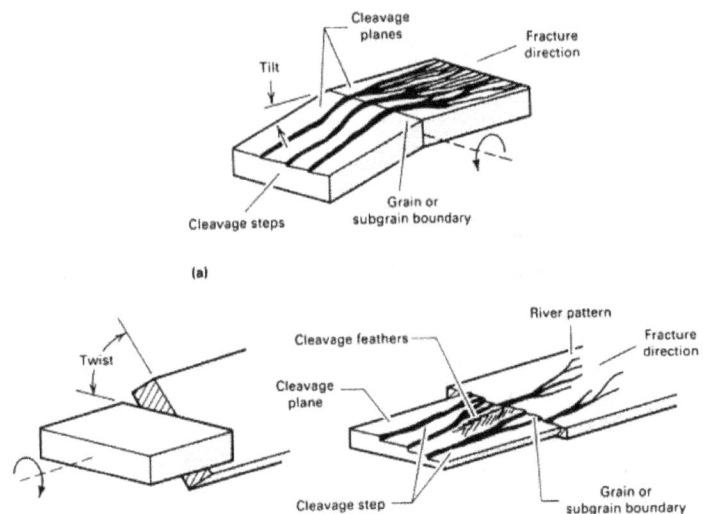

a flat surface with limited plastic deformation (zero reduction of area) which are features of tensile specimens that have been failed in a brittle failure mode.

Figure 6.31(a) shows the fracture surface of the tensile specimen that failed in a ductile mode and show the "classical" "cup and cone" type fracture, with plastic deformation (significant reduction of area).

Brittle materials exhibit flat fracture surfaces consisting of transgranular cleavage facets. The fracture energy associated with this type of fracture is low. Figure 6.31(b) shows a classic characteristic of brittle facets with their size similar to the grain size of the material. The brittle facets originate by fast propagation of the crack moving in a transgranular direction across the grains. When examined at a higher magnification using a scanning electron microscope (SEM),

Figure 6.32 River line pattern observed from brittle cleavage fracture surfaces.

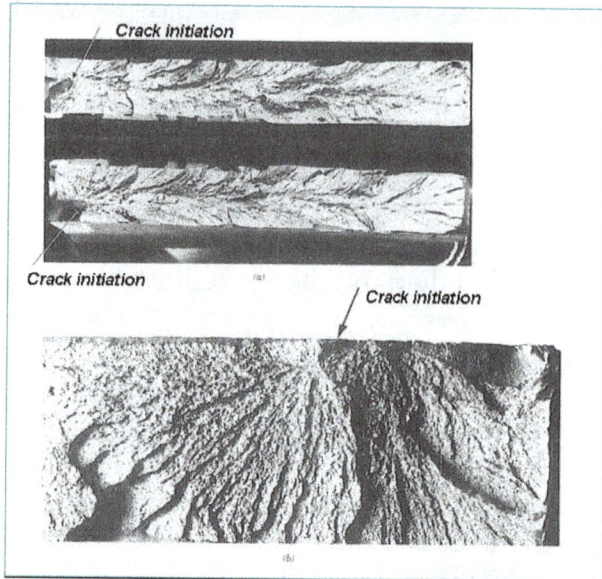

Figure 6.33 *Fracture surfaces with features pointing to the initiation point*

6.6.2 Ductile fracture

A diagrammatic illustration of a tensile sample loaded beyond the maximum tensile strength is shown in Figure 6.34. Necking has occurred along the specimen within the gauge length. Inside the specimen, within the necked area, micro-voids form around any second phase particles (inclusions, carbides). As the load increases, these micro-voids/cracks expand and join together to create a crack with its plane perpendicular to the axis of the tensile force applied. As this crack grows, the shear plane of approximately 45 degrees to the tensile direction develops around the specimen edge and merges with the existing crack, resulting in the classic cup and cone fracture

surfaces as shown in the final stage of fracture process in Figure 6.34.

Ductile materials normally exhibit plastic deformation during fracture, providing rough fracture surfaces with relatively high surface areas as has been illustrated in Figure 6.31 top. Ductile fracture therefore requires higher energy to create two new fresh fracture surfaces in comparison to energy required to cleave flat brittle surfaces. Investigation under higher magnifications shows ductile fracture surface consisting of many of ductile dimples or micro-voids. The mechanism of micro-void formation is also illustrated in the diagram in Figure 6.35, which is similar to Figure 6.34 showing the influence of inclusions or second phase particles as micro-void initiation sites. During loading, these particles undergo de-cohesion from the matrix, forming a micro-void/cracks. Generally, these micro-voids are normally observed to be centred on inclusions or other second phase particles as shown in Figure 6.35(d). Materials such as ductile cast iron, aluminium or alloys operating at high temperature exhibit this type of fracture surface.

The shape of the micro-voids observed in a microscopic scale can be used to determine the type of the force applied onto the specimen. If the micro-voids are in equi-axed shape, as shown in Figure 6.36(a), the applied force are uniaxial tensile loading. Parabolic-shaped micro-voids with its tip pointing in the opposite direction when seen on both halves of fracture surface are due to shear force, see Figure 6.36(b). If the micro-voids appear in the parabolic shape with the tip of each micro-void orientated in the same direction on both fracture surface halves, the tensile tearing is applied in this case as shown in Figure 6.36(c).

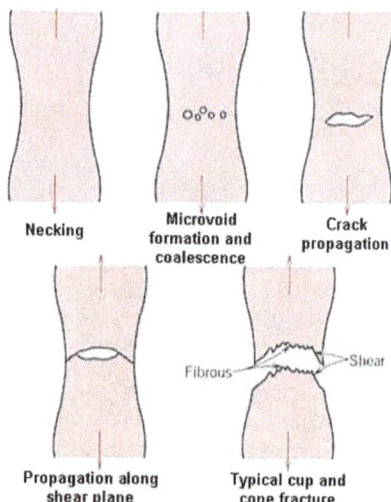

Figure 6.34 *Cup and cone fracture*

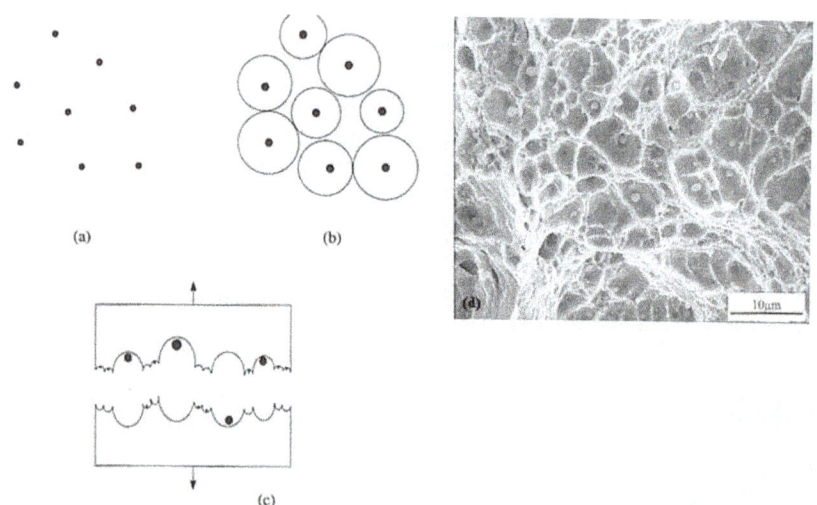

Figure 6.35 *Ductile rupture*

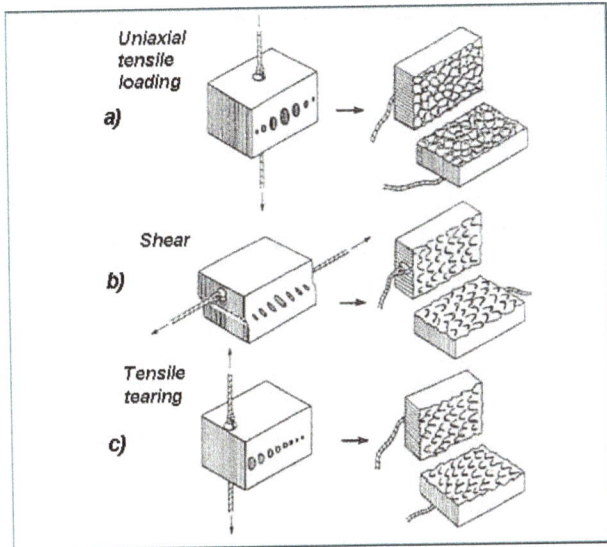

Figure 6.36 *Different characteristics of micro voids observed under (a) Uniaxial tensile loading, (b) shear and (c) tensile tearing.*

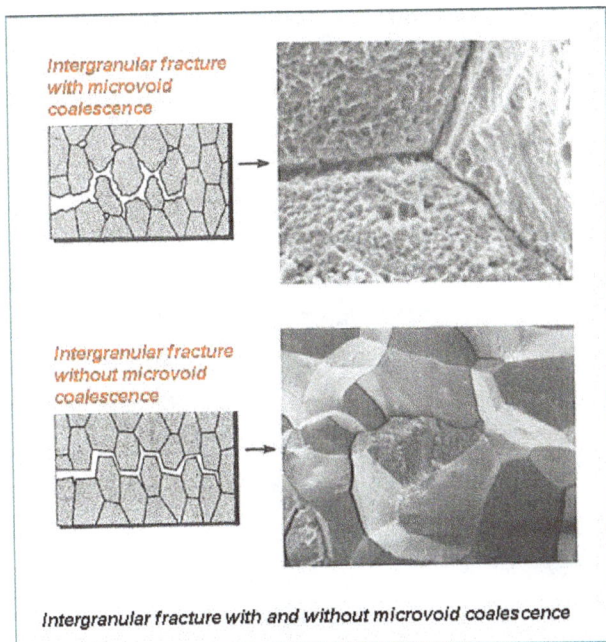

Figure 6.37 *Fracture surfaces of (a) Intergranular fracture with micro void coalescence and (b) without micro void coalescence*

6.6.3 Intergranular fracture

Intergranular fracture is normally associated with service conditions which are corrosive or have occurred at high temperature. These conditions result in precipitation of the second phase particles along the weakened grain boundaries. Figure 6.37 illustrates two types of intergranular fracture: a) intergranular fracture with micro void coalescence and b) without

micro void coalescence. Although the former exhibits areas of ductile dimples similar to those observed from typical ductile fracture surfaces, these ductile dimples are relatively shallow and require much less energy involved in the fracture process. These ductile dimples or micro voids are formed around particles precipitated in some cases as a network along the grain boundaries. This considerably reduces the bonding between the particles and the matrix, which in turn allows micro void coalescence to occur more readily. As a result, the material has significantly reduced fracture energy which promotes a rapid crack propagation path along the grain boundaries.

Temperature is one of the most important factors influencing intergranular fracture. High temperature not only facilitates ductile failure (increased amount of plastic deformation) but also promotes metallurgical changes. If the bonding between the new phases and the matrix is quite poor, responses of materials to external loads will be significantly affected, which can subsequently alter the fracture mode of the materials. In general, oxidation or combustion reactions are involved when materials are subjected to high temperatures, such as in the case of turbine blades. Oxide scales on the metal surface are normally brittle and heterogeneously formed, leading to a crack formation on the surface. If the load continues at high temperature, oxidation would progress even further into material interior through the existing surface cracks. This might result in the formation of brittle oxidation products within the grain and especially along the grain boundaries. If these grain boundary oxidation products are interconnected and poorly bonded with the matrix, intergranular fracture will result. Therefore, it can be seen that although high temperature promotes ductile failure, metallurgical changes might alter the fracture mode from transgranular ductile fracture to intergranular fracture due to poorly bonded grain boundary particles. This embrittlement effect results in noticeable reductions in tensile strength and ductility. The utilization of the materials at high temperatures or strong oxidation will considerably affect mechanical properties of the materials. Material selection therefore becomes significantly important in this case. Nickel based alloys are a good alternative for service conditions at high temperature and oxidation. Furthermore, material which is subjected to corrosive environment where corrosion products are observed along the grain boundaries will give comparable results.

6.6.4 Fatigue fracture

Fatigue fracture surfaces display unique characteristics with flat-surfaces with limited plastic deformation and in some cases show visible "beach marks" as shown on the shaft subjected to fatigue failure in Figure 1.5 or Figure 1.20 top LHS. The surface finish/condition is a key factor in controlling surface fatigue crack initiation. Crack-free or defect-free surfaces greatly help to improve fatigue life of the components. Rough surfaces can result in stress concentration, leading to easy fatigue crack initiation.

A recent Royal Society published paper (2018) by Haël Mughrabi[22] provides an interesting summary of the work of Wöhler and others combined with metallographic presentation of the key factors involved in fatigue crack initiation.

August Wöhler, an engineer working on the Prussian railway system in the 1850s, was concerned about the number of railway axles failures. Wöhler decided to carry out experimental work and was the first to carry out systematic fatigue studies under conditions of rotating bending. The most important result of his work was the identification of a stress amplitude (now called fatigue limit (see Figure 6.38)) below which failures would not occur. In this review, Haël points out that Wöhler tabulated his results and did not use the graphical display that now bears his name!

The Wöhler plot of fatigue life data consists of stress amplitude $\Delta\sigma/2$ ($\Delta\sigma$=stress range) versus number of cycles to failure, Nf. These plots are also referred to as or S–N diagrams (S, stress; N, number of cycles). Fatigue life plots have been extensively used by engineers to establish safe designs based on factual test data for components.

The work of Wöhler was followed by Bauschinger and he identified the main effect that following a tensile loading beyond the yield point the compressive yield point occurred prematurely: the effect which now bears his name the "Bauschinger Effect". Bauschinger also developed very sensitive extensometers to measure strains as low as some 10^{-5} and was able to show that, during cyclic loading, very small inelastic, i.e. plastic, micro-strains occurred.

The main break through regarding the actual mechanism was delayed until the beginning of the twentieth century, when Ewing and Humfrey observed slip bands in which microcracks had developed on the surface of a fatigued steel. The main conclusion from the work of Bauschinger and Ewing & Humfrey was:

- that fatigue damage results from the accumulation of many very small irreversible plastic cyclic micro-strains
- that the accumulation of plastic strains leads to the initiation of cracks.

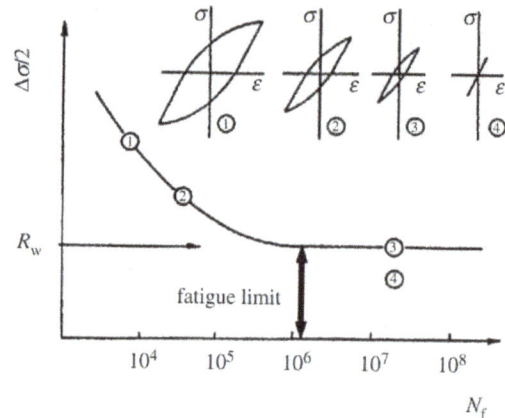

Figure 6.38 Schematic Wöhler curve with a fatigue limit, with indication of hysteresis loops (σ,stress; ε,strain) at different stress levels[22]

Cyclic slip irreversibility and damage evolution

The physical origin of the initiation of fatigue damage in ductile crystalline materials is intimately related to

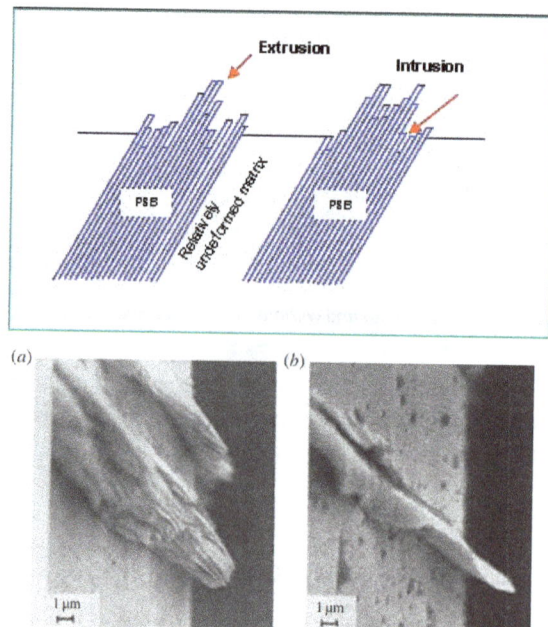

Figure 6.39 SEM micrographs of PSB surface profiles in fatigued copper single crystals (a) after fatigue at 77K and (b) after fatigue at 403K. Stress axis vertical. Primary Burgers vector lies approximately in the plane of the figure, parallel to the PSBs. Note voluminous roughened extrusions at low temperature and thinner extrusions at elevated temperature[22]

the occurrence of slip irreversibility in the core, resulting in unreversed slip steps at the surface. The link between the slip irreversibility and fatigue life relates to the mechanisms of cross slip of screw dislocations, mutual annihilation of dislocations, random to-and-fro glide of dislocations (leading to surface roughening), cutting of shear able precipitates etc. At the surface, unreversed slip steps and microcracks are left behind which have now been seen by scanning electron microscopy (SEM), replicas or, more recently, by atomic force microscopy (AFM). The study of cyclic slip irreversibility in the core are more difficult and can only be inferred from indirect observations by transmission electron microscopy (TEM).

The stages of cyclic deformation and evolution of fatigue damage

Based on metallographic observations fatigue failure is not a sudden event that occurs unexpectedly. It should be seen as the final event that has been preceded by a sequence of cyclic deformation and fatigue damage processes. In the case of ductile FCC metals, four stages of cyclic deformation and fatigue can be identified, as shown in Figure 6.40. Cyclic hardening/softening, cyclic saturation, the evolution of fatigue damage caused by cyclic slip irreversibility, initiation and subsequent growth of cracks leading to failure. In addition, intercrystalline fatigue damage can occur at grain boundaries. In heterogeneous materials containing inclusions or pores, these defects can be sites of surface or subsurface fatigue damage. The latter can become life-controlling in the UHCF regime. Figure 6.42, 6.43 and 6.44.

If we study the stage II fatigue fracture surface using SEM technique, striations orientated normal to the fatigue crack growth direction as shown in Figure 6.45 (a) will be observed. Each striation is due to plastic blunting process as demonstrated in Figure 6.45(b). When the tensile loading progresses, the crack opens and allows the slips to operate at the top and bottom ends to produce local plastic deformation as shown by the arrows. At a higher tensile loading, plastic blunting occurs, which leads to an increase in the fatigue crack length. During unloading, the fatigue crack is then closed and the slips are now operating in the opposite direction. Therefore, after one cycle, the fatigue crack now arrives at the original stage of crack closing with an increase in one fatigue striation.

The fatigue failure can be found to occur in conjunction with corrosive and high temperature environment, which are called corrosion fatigue and thermal fatigue respectively. Corrosion and high temperature accelerate the rate of the fatigue crack

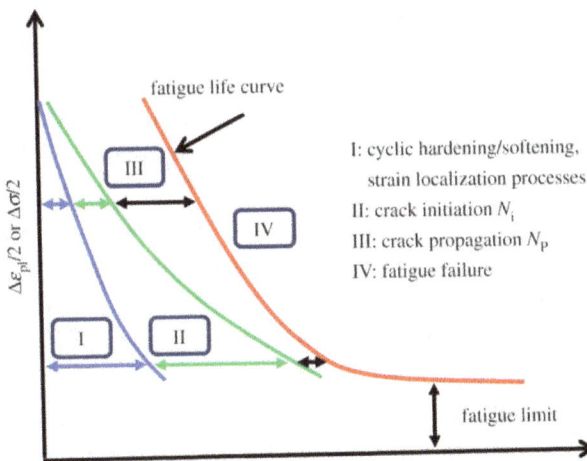

Figure 6.40 Schematic illustration of the four stages of fatigue in ductile metals until failure

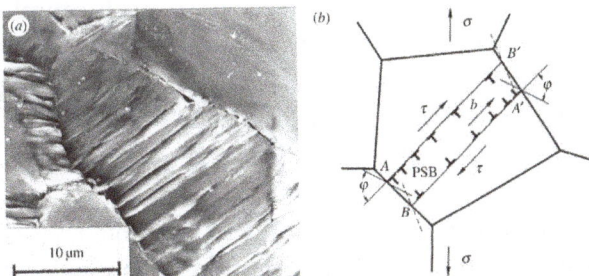

Figure 6.41 Intercrystalline PSB-GB cracks induced by PSBs impinging against grain boundaries. (a) SEM micrograph of PSB-GB cracking in fatigued copper polycrystal. (b) Dislocation model of action of PSBs against grain boundaries[22]

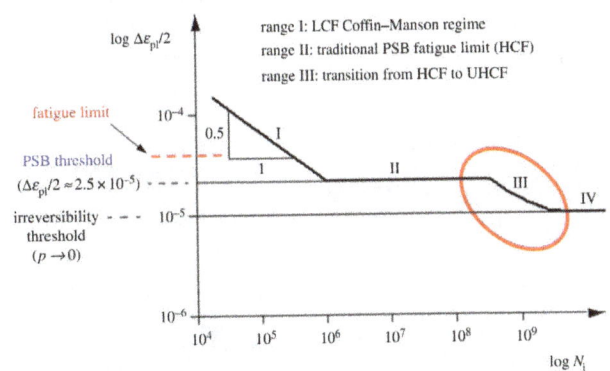

Figure 6.42 Schematic fatigue crack initiation diagram for type I materials, extending over LCF, HCF and UFG ranges, indicating reduced threshold stress for cyclic strain localization (in PSBs) in transition to UHCF range

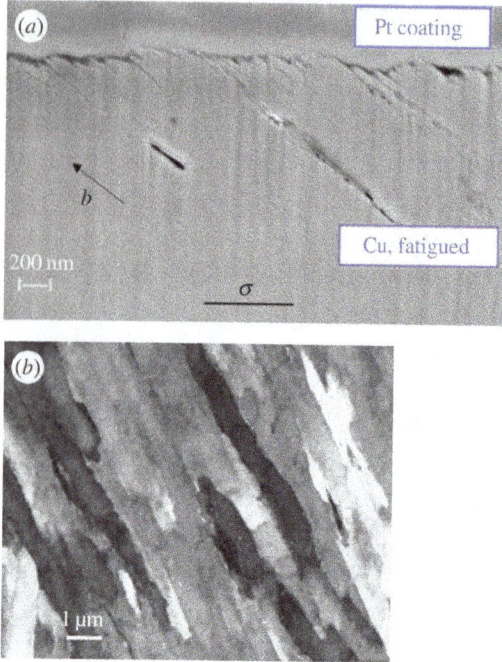

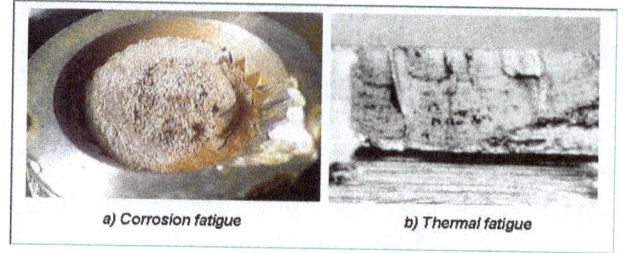

Figure 6.46 *Fracture surfaces of specimens failed in a) corrosion fatigue and b) thermal fatigue*

propagation and promote severe fatigue failures. An example of corrosion fatigue shows obvious areas of rust on the fatigue surface observed in an automobile shaft as depicted in Figure 6.46(a). This significantly reduces the fatigue life of the automobile shaft. Thermal fatigue illustrated in Figure 6.46(b) shows a beach mark which indicates the fatigue crack initiation to have started at the top surface.

Figure 6.43 *Copper polycrystal fatigued ultrasonically at room temperature (some MPa below PSB threshold) to N=1.59 × 10^10 cycles. Stress axis horizontal. (a) Surface roughness profile and stage, I crack initiation, as seen in a FIB section. (b) Lamellae of cyclic slip localization, observed by SEM back scattered electron contrast*

6.6.5 Application of fracture toughness to engineering applications

The level of fracture toughness required for a particular part can depend on several factors including the severity of the operating conditions, and the consequences of a failure. In practice because of the "similitude" effect in toughness (the toughness can vary depending upon the specimen size – unlike strength) the specified toughness requirement for components to ensure "fitness for service" has to be adjusted depending upon several factors such as:

- Operating temperature range particularly the lowest temperature especially for for ferrous materials
- The stress system constraint caused by the size of parts especially plain strain conditions
- The location, type and size of flaws
- The applied, the residual stresses such as weld residual stresses and any unknown stress
- Any heat treatments such as stress relief during manufacture or cold deformation
- Potential embrittlement in service.

The application of these considerations would represent a complex task. However, this has been simplified since the eventual understanding of fracture ended with two main approaches to fracture control:

a) Quality control-based toughness requirements Charpy, Izod, DWT, DWTT (based on experience

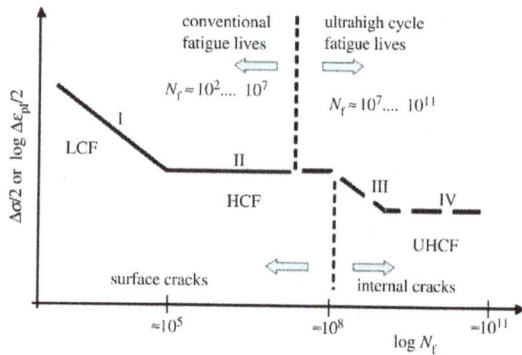

Figure 6.44 *Schematic multistage fatigue life diagram for type II materials, showing LCF, HCF and UHCF fatigue ranges and transition from surface to subsurface cracks*

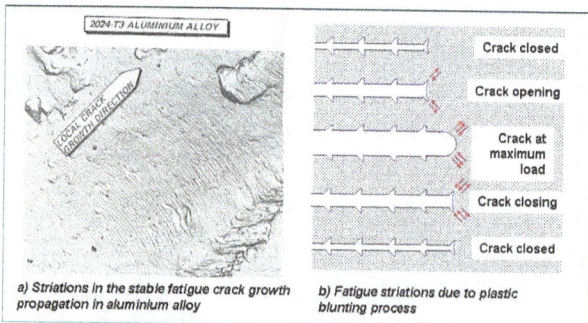

Figure 6.45 *a) Fatigue striations in aluminium and b) Fatigue striation due to plastic blunting process*

and correlation with fracture mechanics tests, such as wide plate tests)

b) Fracture assessment-based toughness requirements K_{1c}, CTOD, J and R curve.

Quality Control Based Toughness Requirements

With "quality control-based method" the toughness is measured using small scale toughness tests such as:

- **The Charpy impact test.**
  The Charpy impact test is a simple test introduced in 1900 and the most widely used fracture toughness test. This test consists of a rectangular specimen $10 \times 10 \times 55$mm or a sub-size specimen with a machined notch which is impact loaded by a pendulum hammer. V and U notches are used, although the V notch is more frequently employed. Results obtained from breaking from Charpy impact specimens are quantified in three ways;

  a) the impact energy required to fracture the specimen,

  b) the lateral expansion,

  c) and the percentage shear fracture appearance at the specified test temperature.

The measurement of the lateral expansion and the percentage shear are intended to measure the ductility.

The specified requirements initially came from studies of casualty material such as the Liberty Ships. A Charpy energy of 15ft pounds (20 Joules) at the failure temperature seemed to minimise the risk of failure. To improve margins of safety, 20ft pounds (27 Joules) became a common specification for low strength steels. In many international codes the adoption of a specified Charpy energy at a given temperature based on experience and linked to the minimum operating temperature has become accepted practice to minimise the risk of brittle failure. For pressure vessels in the early 1960s the Charpy transition temperatures, measured in the parent plate, was correlated with the temperature at which survival was obtained in welded notched wide plate specimens at an applied strain approximately 0.5% plastic strain. There is a general rule that the 27J or the 42J value should be specified at 20° below the lowest operating temperature. The use of the quality control

methods combined with the good workmanship requirements in most fabrication codes has historically ensured that structures rarely fail.

- **The Pellini drop weight test (DWT).**
  The DWT specimen consists of a rectangular beam with a brittle weld bead and saw-notched to provide a crack initiation site which is impact loaded by a falling weight. The falling weight causes a crack to initiate and to propagate through the test specimen. The DWT test was intended to establish the nil-ductility-temperature, the maximum temperature at which the specimen breaks.

- **The Drop weight tear test**
  The drop weight tear test (DWTT), which consists of a full-thickness rectangular plate with a pressed notch, impact loaded by a falling hammer, resulting in complete fracture. The DWTT is used to establish the temperature transition curve in terms of the percentage shear fracture appearance. Codes in the gas and pipeline industry (such as British Gas and API) require a specified percentage shear fracture area (typically 50 or 80%) at a specified test temperature. In this way, the DWTT is used to ensure that the fracture made is mainly ductile.

- **Other procedures**,
  These include bend tests and weld ductility tests. Bend and weld ductility tests involve straining a fabricated test weld to a given amount, with the requirement that no cracking occurs. These tests provide a qualitative measure of the weld ductility, ensuring that the weld is not too brittle at the test temperature.

Fracture Assessment Based Toughness Requirements

The quality control-based toughness requirements Charpy impact tests, Izod, DWT and DWTT cannot allow for the effects of applied stress, constraint, flaw location and size on the integrity of components. However, for components that require a high integrity a fracture mechanics-based assessment procedure has been proven to be more appropriate. The ASME III (Appendix G) and ASME XI (Appendices A and G) and API approaches are based on linear elastic fracture mechanics principles. For design purposes, a maximum planar flaw size is postulated. If flaws are detected during inspection their effect on the structural integrity of the component can be assessed. In addition, damage tolerance method which use NDT methods to measure the size, shape and orientation of imperfections and then fracture mechanics can be used to evaluate their significance regarding the structural integrity. Fracture mechanics compares the driving force and the material resistance. BS 7910 permits fracture mechanics assessment procedures to be employed. There are several different testing standards that are used to determine parameters that quantify the materials fracture resistance (ASTM E 399, ASTM E 1820, BS 7448, ESIS P2 etc.). Unfortunately, this has led to several different parameters such as static and dynamic K, CTOD, J and R-curve testing which has in many ways made the use of fracture assessments a complex task.

Fracture Toughness of Ferritic Steels in the Ductile-to-Brittle Transition Region

When the test temperature gives high values of toughness associated with the ductile upper shelf there would be little concern regarding the variation of energy values. However, when the values of toughness and fracture appearance indicate that the test temperature is in the "transition zone" then there are concerns regarding the potential variation and the need to find the "worst" values to ensure a high degree of confidence that the steel meets the specified requirements. In the last 30 years detailed work has been carried out to understand and quantify the toughness in the transition temperature range. The characterization of fracture resistance of ferritic steels in this region is problematic due to scatter in results, (bimodal behaviour) as well as size and temperature dependences. There were two explanations to the size effect. One was based on **constraint effects**, while the other made use of **statistical weakest link** concepts to explain the probability to find a cleavage initiator site at the crack front.

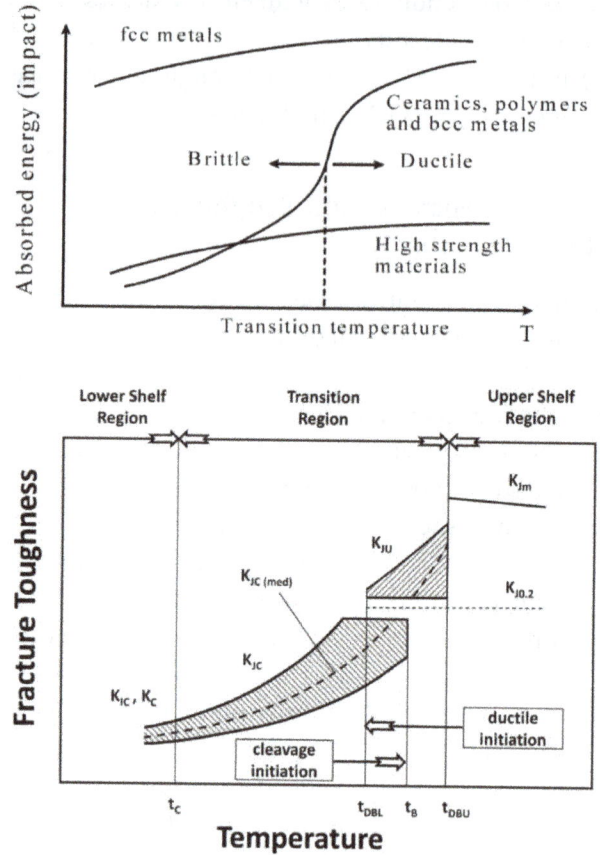

Figure 6.47 LHS = Fracture behaviour vs temperature for different materials[17], RHS = Schematic representation of fracture toughness-temperature dependence[23]

One explanation is that there are small areas of low toughness or weak links (initiators of cleavage) randomly distributed in the crack front, so that the brittle fracture could be viewed as a statistical event. The cleavage fracture is a local fracture process controlled by a critical stress, and it will occur when the critical stress is reached in one of these weak links. The load required to produce the fracture will depend upon

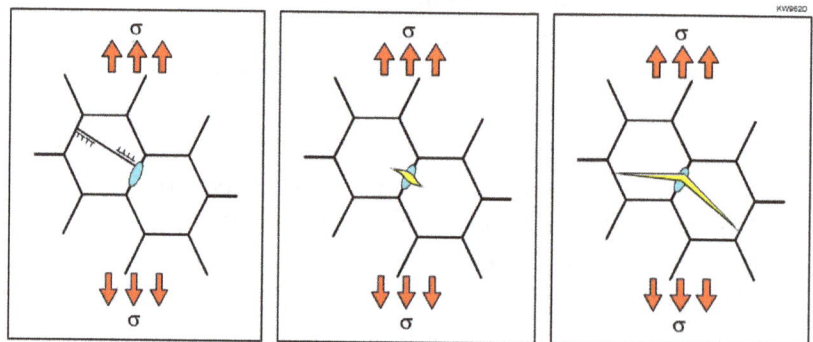

Figure 6.48 Cleavage crack initiation, "cleavage fracture initiation is a statistical event"[17]

Local stress produces a dislocation pile-up which impinges on a grain boundary carbide.

Cracking of the carbide introduces a microcrack which propagates into the matrix.

Advancing microcrack encounters the first large angle boundary.

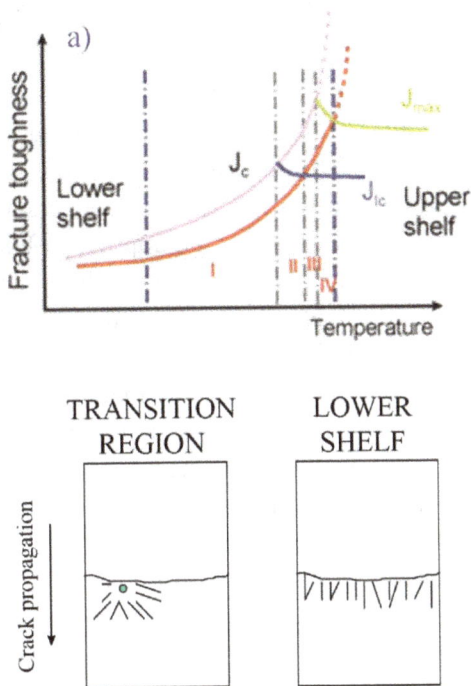

Figure 6.49 *Various zones in the transition region and RHS = lower shelf lots of crack initiators[17]*

I: All specimens fracture by cleavage without any stable crack growth.

II: Some specimens fracture by cleavage without any stable crack growth, while others fracture by cleavage after some amount of stable crack growth.

III: No cleavage without stable crack growth occurs. All the specimens fracture by cleavage after some amount of stable crack growth, or

III: some specimens fracture without stable crack growth, others with stable crack growth, and others reach the maximum load condition and do not present instability

IV: Some specimens fracture after some amount of stable crack growth, while others reach the maximum load condition and do not present instability.

For higher temperatures, no cleavage occurs and this behaviour corresponds to the upper shelf.

the location of the weak link and its critical stress. A Weibull statistical approach led to improved understanding and the development of the "master curve" approach and test method ASTM E1921.

In addition to providing an explanation of the scatter in energy values that occurs in the transition region, the weakest link model also explains the effect of specimen size, since an increase in the length of the crack front enlarges the highly stressed volume of material at the tip of the crack, also increasing the likelihood of finding a weak link.

These concepts were developed using Weibull probability methods by Professor Kim Wallin at the Finnish VTT laboratories which led to a master curve

(MC) methodology and a lower bound conservative value of the toughness at a particular temperature. Wallin identified that in the ductile-brittle transition region micro-mechanisms of cleavage fracture cause the toughness data to be highly scattered when compared to the lower shelf region. Rather than single value of toughness at a particular temperature, the material has a toughness distribution. Research over the past three decades on the fracture of ferritic steels in the ductile-brittle transition region has led to two important conclusions:

1. Scatter in fracture toughness data in the transition region follows a characteristic statistical distribution that is very similar for all ferritic steels.
2. The shape of the fracture toughness vs. temperature curve in the transition range is virtually identical for all ferritic steels. The only difference between steels is the absolute position of this curve on the temperature axis.

This was achieved by establishing the T_0 value. The T_0 value is the lowest temperature that a 1-inch specimen

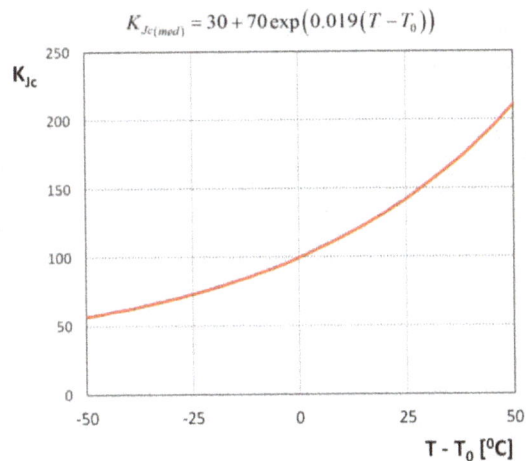

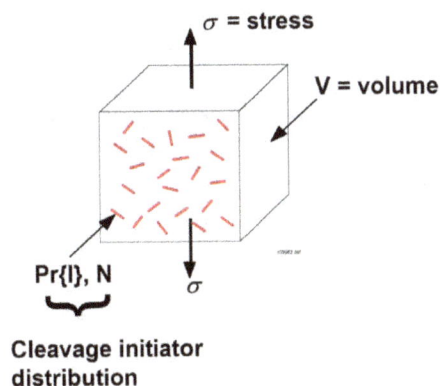

$$K_{Jc(med)} = 30 + 70\exp\left(0.019\left(T - T_0\right)\right)$$

Figure 6.50 *LHS = Master curve from Ref, RHS = Probalistic model for cleavage initiator distribution[17]*

can give a median fracture toughness value of 100MPa m$^{-1/2}$. This is determined in accordance with ASTM E1921.

The lower-bound 3% and 5% probability curves and the upper-bound 95% probability curve can also be set up. These three curves are given by the following expressions:

$$K_{Jc(3\%)} = 24.6 + 32.2\exp\left[0.019(T-T_0)\right].$$

$$K_{Jc(5\%)} = 25.4 + 37.8\exp\left[0.019(T-T_0)\right].$$

$$K_{Jc(95\%)} = 34.6 + 102.2\exp\left[0.019(T-T_0)\right].$$

Due to the high cost of test work to ASTM E1921 there have been several attempts to correlated Charpy test results with the T_0 temperature.[17 to 29]

Two of the first Charpy-V notch (CVN) correlations that were published correlated T_0 and the 28J CVN transition temperature and To and the 41J CVN transition temperature. The first correlation is used in the SINTAP structural integrity assessment procedure and the standard BS 7910. The second correlation was developed by ORNL between, which is used in nuclear surveillance work. These relationships for pressure vessel steels (A508Cl3, A302B, A 533B) are shown below.

$$T_0 = T_{28J} + 3 \ [^\circ C]$$

$$T_0 = T_{41J} - 1 \ [^\circ C]$$

An important application of this work has been for life extension of USA nuclear power plants from the design life of 40 years to 60 years.[28 and 29] (see Figure 6.51). This was of vital importance to allow time to replace the decommissioned facilities due to the delay in the building of nuclear power stations Unfortunately, this cannot be applied to UK AGRs because of the condition of the graphite that is used as a moderator. However, a graphite irradiation research program named Blackstone project was launched in 2006 in support of ageing management of Advanced Gas-Cooled Reactors (AGRs) in the UK. And on the 16 February 2016 EDF Energy announced that the scheduled closure dates for its Heysham 1 and Hartlepool plants had been extended by five years to 2024, while those of Heysham 2 and Torness had been extended by seven years to 2030. The announcement followed life extensions at EDF Energy's other AGR power plants.[31]

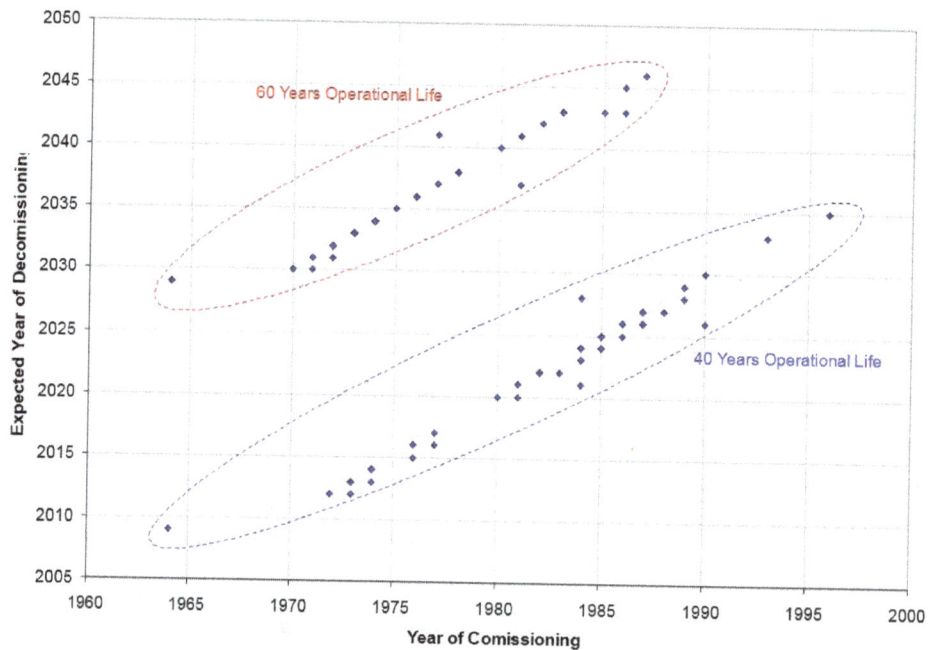

Figure 6.51 Life extension of USA nuclear power plants based on the master curve

REFERENCES

1. W. J. Jackson, FRACTURE TOUGHNESS IN RELATION TO STEEL CASTINGS DESIGN AND APPLICATION, https://www.sfsa.org/publications/misc/Fracture%20Toughness.pdf

2. Robert Juvinall And Kurt M Marshek, Fundamentals of machine design, second edition Pub John Wiley and sons, 1991 page 222

3. Norman E Dowling, Mechanical Behavior of Materials, Pub Prentice-Hall International Editions, 1993, page 233,

4. C. E.Inglis, Stress in a plate due to the presence of cracks and sharp corners, http://imechanica.org/files/1913%20Inglis%20Stress%20in%20a%20plate%20due%20to%20the%20presence%20of%20cracks%20and%20sharp%20corners_0.pdf

5. A A Griffths, The Phenomena of Rupture and Flow in Solids http://rsta.royalsocietypublishing.org/content/roypta/221/582-593/163.full.pdf

6. Adrian Demaid, Fail Safe, The Open University 2004 page 38, http://www.fracturetraining.com/downloads/fail-safe2006.pdf?LMCL=cKCqVB

7. Practical Fracture Mechanics, T357, Structural Integrity designing against failure, Block 2, Part 1, The open University.2009.

8. Elastic – Plastic fracture mechanics, T357, Structural Integrity designing against failure, Block 2, Part 5, The open University.2009.

9. Kim Wallin, GUIDELINES FOR DERIVING FRACTURE TOUGHNESS ESTIMATES FROM NORMAL AND MINIATURE SIZE CHARPY-V SPECIMEN DATA, Rakenteiden Mekaniikka, Vol 25 No 3 1992, ss . 24 -40, http://rmseura.tkk.fi/rmlehti/1992/nro3/RakMek_25_3_1992_2.pdf

10. J A Charles and F A A Crane, Selection and use of engineering materials, Second edition, Pub Butterworth, 1989

11. W.S. Pellini, Guidelines for Fracture-safe and Fatigue-reliable Design of Steel, Publisher: Welding Institute (1983)

12. Brian Cotterell, Fracture and Life, ISBN: 9781848162822, Imperial College Press (2010)

13. http://www.dinosaurtheory.com/scaling.html

14. http://www.boulder.nist.gov/div853/Publication%20files/NIST_CharpyHistory.pdf

15. http://www.fracturemechanics.org/ellipse.html

16. https://en.wikipedia.org/wiki/Charles_Inglis_(engineer)

17. Iradj Sattari-Far and Kim Wallin, SKI Report 2005:55, https://www.osti.gov/etdeweb/servlets/purl/20752439

18. Constance Tipper, The Brittle Fracture Story, University Press (1962)

19. H Tholander, The presence of nitrogen in steel Iron, 29 November 1889 p461 – from Enterprise and Technology Ulrich Wenenroth, 1994 ISBN 978-0-521-38425-4.

20. F B Pickering, Some beneficial effects of nitrogen in steel, Institute of Metals 1989

21. https://www.iom3.org/iron-steel-society/steel-heroes-constance-fligg-tipper

22. Haël Mughrabi, Microstructural mechanisms of cyclic deformation, fatigue crack initiation and early crack growth, https://royalsocietypublishing.org/doi/full/10.1098/rsta.2014.0132?url_ver=Z39.88-2003&rfr_id=ori:rid:crossref.org&rfr_dat=cr_pub%3dpubmed

23. Jan Dzugan, Pavel Konopik and Martin Rund, Fracture Toughness Determination with the Use of Miniaturized Specimens, https://www.intechopen.com/books/contact-and-fracture-mechanics/fracture-toughness-determination-with-the-use-of-miniaturized-specimens

24. Carl von Feilitzen et al, Implementation of the Master Curve method in ProSACC, Report number: 2012:07 ISSN: 2000-0456 Available at www.stralsakerhetsmyndigheten.se https://www.stralsakerhetsmyndigheten.se/contentassets/3c70c8c3a94b4235b250040f9a8e4b2f/201207-implementation-of-the-master-curve-method-in-prosacc

25. P. R. Sreenivasan et al, NOVEL CHARPY-FRACTURE TOUGHNESS CORRELATIONS FOR PREDICTING REFERENCE TEMPERATURE AND MASTER CURVE, file:///C:/Users/user/Downloads/10522-40800-1-PB%20(2).pd f

26. Kim Wallin, Master curve analysis of ductile to brittle transition region fracture toughness round robin data The "EURO" fracture toughness curve, VTT PUBLICATIONS 367

27. file:///C:/Users/user/Downloads/Master_Curve_Analysis_of_Ductile_to_Brittle_Transi.pdf

28. Mark T. Kirk, The Technical Basis for Application of the Master Curve to the Assessment of Nuclear Reactor Pressure Vessel Integrity, Mark Kirk USNRC Rockville, Maryland 18th December 2009, https://www.nrc.gov/docs/ML0935/ML093540004.pdf

29. DANIEL J. COGSWELL, Statistical Modelling of the Transition Toughness Properties of Low Alloy Pressure Vessel Steels Volume 1: Main Body, A thesis submitted to the Department of Metallurgy and Materials, School of Engineering, The University of Birmingham July 2012

30. Graphite research to support AGR life extensions, 22 February 2016

31. http://www.world-nuclear-news.org/C-Graphite-research-to-support-AGR-life-extensions-2202164.html

Books

1. Richard W. Hertzberg Richard P. Vinci Jason L. Hertzberg, Deformation and Fracture Mechanics of Engineering Materials, Fifth Edition John Wiley & Sons, Inc. 2013,

2. T L Anderson, FRACTURE MECHANICS Fundamentals and Applications, Third Edition 2005 Pub Taylor and Francis.

3. AKDas, Metallurgy, of Failure Analysis, McGraw-Hill, 1997.

4. Alan F. Liu, Mechanics and Mechanisms of Fracture: An Introduction, ASM International, 2005.

5. P. G. FORREST, Fatigue of Metals, PERGAMON PRESS, 1962.

6. David Broek, The Practical Use of Fracture Mechanics, Kluwer Academic Publishers, Dordrecht, 1988. 522 pp.

7. Norman E Dowling, Mechanical Behavior of Materials, Pub Prentice-Hall International Editions, 1993, page 233,

8. Robert Juvinall And Kurt M Marshek, Fundamentals of machine design, second edition Pub John Wiley and sons, 1991 page 222

9. Kim Wallin, Fracture Toughness of Engineering Materials, ISBN: 978 0 9552994 6 9, EMAS Publishing, 2011.

SELF-ASSESSMENT QUESTIONS

1. If a linear elastic fracture toughness test was carried out using a 5 mm thick specimen and another test uses a 30 mm thick specimen which specimen would give the highest fracture toughness value?

2. Explain the reason for a possible difference in toughness.

3. In the example given in question 1, which value of fracture toughness would be nearest to a valid value?

4. Based on ASTM requirements if the required test value for a full size 10mm × 10mm Charpy was 54J what would be requirement for a 5 mm × 10 mm specimen and a 2.5 mm × 10 mm specimen?

5. What was the important change the Irwin made to the Griffiths model of fracture?

6. A steel was found to have a Charpy energy of 50J. What would be the expected K_{IC} of the steel?

7. Cleavage fracture appears (a) Bright (b) Dull (c) Difficult to identify (d) None.

8. What are the units of fracture toughness?

9. What would be the expected fracture toughness for a premium grade steel with a yield strength of 800MPa

10. How could the largest flaw size on a part be found?

11. Would a larger imperfection reduce the maximum allowable stress on a structure?

12. Does an increase in the yield strength increase the fracture toughness?

13. What type of fracture would be expected on a vehicle crack shaft?

14. The old traditional design procedure would use the allowable maximum stress to be based on the material yield strength adjusted by a safety margin. How does the Fracture Mechanicals procedure work?

15. Name three potential causes for short cracks during manufacture and fabrication?

16. Explain the difference between a Charpy test and an Izod test?

17. Name three factors that have to be allowed for when specifying the toughness requirement to ensure that a part is "fit for service".

18. Name the two terms used to describe a fracture surface.

19. What type of fracture would we expect on a Charpy specimen that failed with an energy of 10J at minus 60°C?

20. What type of fracture would we expect on a Charpy specimen that failed with an energy of 110J at +20°C?

21. Name two test methods that are classed as elastic-plastic fracture toughness measurement methods?

22. What is a The Pellini drop weight test (DWT)?

23. What is the Paris Law and why is it important?

24. Explain the technical justification for a life extension of USA nuclear power plants from the design life of 40 years to over 60 years.

25. What statistical approach has been used to gain an understanding of the toughness in the transition zone?

26. The depth of plastic strain beneath cleavage facets in low carbon steel has an associate energy. How many times greater was the estimate for this plastic work energy compared to the new surface energy term in Griffiths model.

CHAPTER 7

Corrosion

7.1 INTRODUCTION

Corrosion is the result of a chemical reaction at the interface between the material and the operating environment. Most metals are found in nature combined with either oxygen or sulphur and undergo extraction and refining processes to separate the metal. To achieve this separation, energy in the form of heat and chemical potential are supplied.

In service metals frequently revert back to the combined state by the processes known as corrosion. Therefore, corrosion can be considered as "extraction metallurgy" in reverse. For example, iron and steel are made from iron ores, and they revert to iron oxides when corrosion occurs. (Figure 7.1 top)

Therefore, if we can effectively separate or protect the metal from the environment by the use of paints and coatings then the corrosion can be prevented. The use of paint systems to protect products in many different environments is a mature science. However, R&D work is still undertaken by major paint manufacturers to validate their paint systems for various applications. An example of the attention to detail needed to achieve "fitness for service" with a paint system can be demonstrated by the fact that Nissan dedicated six man-years to the effect of different bird droppings throughout the world on the paint systems used on their cars. Because birds feed on different fruits and deposit the digested waste on cars in different temperature and humidity regimes, it was necessary to validate the paint systems in all conditions where their vehicles may be sold.

The corrosion reactions are chemical reactions, and by the application of chemical thermodynamics and reaction kinetics predictions can be made on the rate, likelihood and extent of a corrosion reaction. Consequently, corrosion engineering has a strong scientific basis that is combination of both theoretical and an empirical approach. The are several

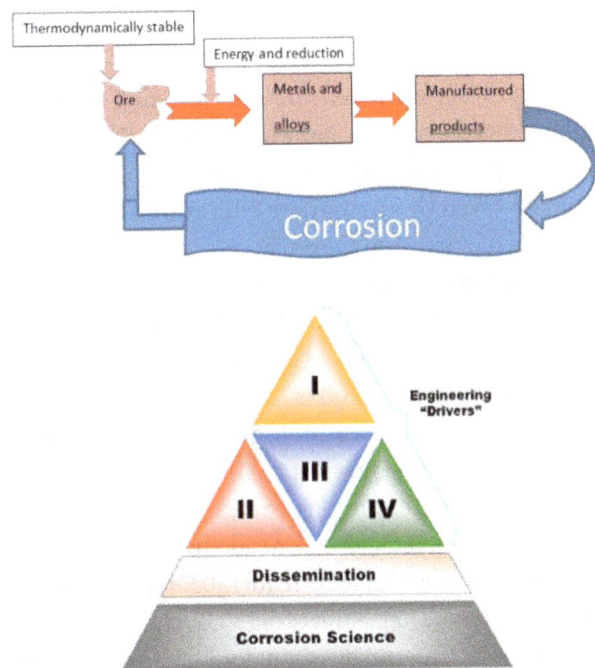

Figure 7.1 Top = Ore to metal to manufactured parts and back to oxide. Bottom = Hierarchy of the four corrosion challenges identified by the USA National Academy of Sciences committee ref1 See text for I, II, III and VI

reference documents that show the corrosion rates for a wide range of metals in an extensive range of environments. Most environments are corrosive, but the actual corrosion rate varies for a given metal and for a given location. In general terms rural inland environments cause the lowest corrosion rates whereas coastal, industrial environments would provide the worst corrosion rates.

Government funds have been applied to both quantifying the cost of corrosion and methods to mitigate its effects. One of the first estimates of the cost of corrosion was done by Robert Hatfield in 1922 when he estimated that the rusting of worldwide iron and steel would cost £700 million. During the 1970s governmental surveys were carried out in several countries to estimate the cost of corrosion issues.

Country	Year	Cost	Percentage of the GNP (%)
USA[1]	1949	USD$5.5 billion	2.1
West Germany[9]	1969	USD$6 billion	3
UK[4]	1971	£1365 billion pounds	3.5
Australia[10]	1972	AUD$900 million	3.5
Australia[11]	1974	AUD$470 million	1.5
Japan[2]	1977	USD $9.2 billion	1.8
USA[5]	1978	USD $70 billion	4.5 (GDP)
Australia[6]	1983	AUD$2 billion	1.5
Kuwait[7]	1995	USD $1 billion	5.2 (GDP)
USA[8]	1998	USD $276 billion	3.1 (GDP)
Japan[3]	1999	3.9 trillion Yen (Uhlig) 5.3 trillion Yen (Hoar) 9.7 trillion Yen (input/output)	0.77 (Uhlig) 1.02 (Hoar) 1.88 (input/output)

Figure 7.2 *Summary of reviews of the estimated cost of corrosion*[2]

These projects identified the cost as being between 2% and 4% of the gross national product. (See Figure 7.2) It was also estimated that at least 25% of this cost could be saved by appropriate education and known methods of corrosion prevention.

One of the challenging aspects of corrosion science is that it has to be an interdisciplinary subject and difficult for one person to hold all the knowledge. Corrosion science involves chemistry, metallurgy, material science, micro-biology, electro-chemistry, surface chemistry, physics of solids and environmental chemistry. An interesting review of corrosion was carried out by the National Academy of Sciences in 2011. https://www.nap.edu/read/13032/chapter/1#xii and https://www.nap.edu/read/13032/chapter/2

There was a hierarchy of the four corrosion grand challenges identified by the committee and was presented diagrammatically[1] (See Figure 7.1 Bottom) These were:

CGC I: **Development** of cost-effective, environment-friendly, corrosion-resistant materials and coatings.

- Materials with inherently high corrosion resistance that also possess other important characteristics demanded by the applications of interest.
- Design and modelling of new corrosion-resistant alloy chemistries and structures.
- Durable, environmentally compatible, and cost-effective protective coatings that eliminate or significantly reduce corrosion.
- Materials that biodegrade in a predictable and benign manner.
- Determination of properties and design parameters/rules.

CGC II: High-fidelity **Modelling for the prediction** of corrosion degradation in actual service environments.

- Ability to predict effects of corrosion and the lifetimes of materials subjected to a wide range of service environments.
- Computer modelling of material surfaces at the nanoscale, where corrosion initiates and propagates and where corrosion resistance must be imparted.
- Increasing the fundamental understanding of new corrosion science and its utilization.

CGC III: **Accelerated corrosion testing** under controlled laboratory conditions that quantitatively correlates to long-term behaviour observed in service environments.

- Smart accelerated corrosion testing that accurately predicts performance under a range of exposures, ensuring durability, and early detection of unforeseen corrosion-related failure mechanisms.
- Development of an "environmental corrosion intensity factor" that facilitates quantification of the acceleration provided by test conditions and enables prediction of performance, based on exposure time, in any combination of field environments.
- Hypothesis-driven models to increase fundamental understanding of corrosion science, improve prediction of structural lifetimes, and optimize maintenance programmes.

CGC IV: Accurate forecasting of remaining service time until major repair, replacement, or overhaul becomes necessary—i.e., **Corrosion prognosis.**

- Accurate and robust sensors that track and monitor corrosion damage and protection.
- Automated defect-sensing devices for quality inspections.

- Remaining life prediction "reasoners" based on measured corrosion deterioration and knowledge of a material's capability in the particular environment.

Sources of information relating to corrosion, www.icorr.org www.nace.org www.npl.co.uk www.iso.org https://efcweb.org

7.2 CORROSION CLASSIFICATION AND COMMON TYPES OF CORROSION

There are several ways to classify the corrosion that results from real service and environmental conditions. The advantage of trying to group together corrosion that occurs from similar causes allows the introduction of common preventative measures to be developed and applied. However, it can often represent a challenge to corrosion engineers that have to place an observed corrosion into a specific category. Typical groupings can be based on:

- Low temperature or high temperature corrosion
- Dry corrosion or wet corrosion (atmospheric, industrial gases, aqueous solutions, liquid chemicals
- Chemical corrosion and electrochemical corrosion
- Types of metal steels, aluminium alloys, ceramics
- Types of environment, sulphuric acid, alkalis, marine.

Wet corrosion often involves electrochemical corrosion – taking place in the presence of an electrolyte such as water, sea water, acids etc. In electrochemical corrosion part of the metal behaves as an anode and introduces ions into the electrolyte (corrodes) and part of the metal or a metal in electrical contact acts as a cathode and accepts and disposes of electrons by a cathodic reaction such as hydrogen evolution. The major cause of corrosion damage is achieved by this mechanism.

Sequence of Events Involved in the Corrosion Process of Steel

Consideration of the series of events that result in corrosion damage can give a better understanding of the process. These events are typically described as follows:

1. Ions move from the steel and into solution (usually water)
2. Oxygen is known to take part in the process and needs to be available
3. The steel needs to dispose of electrons to allow corrosion reaction to occur
4. A compound is formed which could form a protective barrier (oxide) or may proceed to a further reaction
5. For the corrosion to occur the above sequence of events needs to happen and needs to be thermodynamically and kinetically favourable.

The main value of this analysis is that if any of the sequence of events can be prevented the corrosion will be terminated. Appendix 8 shows some corrosion terms.

A classification of industrial metal corrosion would involve different mechanisms:[3 and 4]

1. Uniform or general corrosion
2. Galvanic corrosion
3. Crevice corrosion
4. Pitting corrosion
5. Intergranular corrosion
6. Selective leaching
7. Stress corrosion cracking
8. Corrosion fatigue
9. Hydrogen damage
10. Oxidation
11. High temperature corrosion
12. Preferential weld corrosion
13. Microbial corrosion
14. Stray current corrosion
15. High temperature corrosion.

7.2.1 Uniform or general corrosion[5-7]

This is the form of corrosion that most metals experience in the atmosphere. Corrosion rates and expected service life are well documented together with protective measures by paints and coatings. Figures 7.3 to 7.6 show examples of a uniform corrosion.

Although the products that result from uniform corrosion are unsightly, metal loss is normally predictable which allows calculation of the necessary thickness required for a particular service life. Uniform corrosion can be slowed or prevented by using the five basic methods:

Figure 7.3 Uniform corrosion

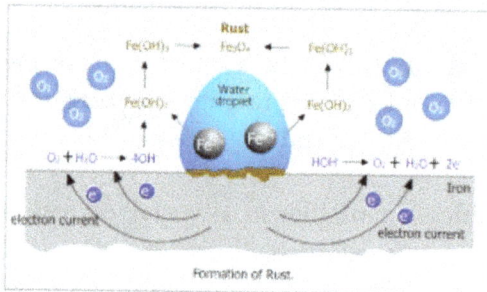

Figure 7.4 Formation of rust

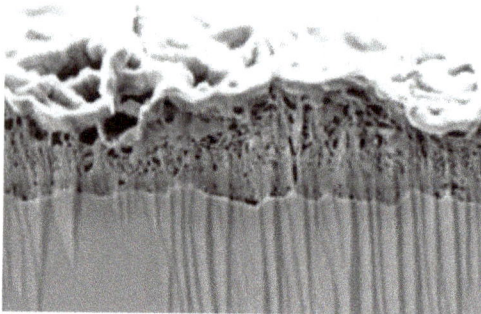

Figure 7.5 Zinc corrosion

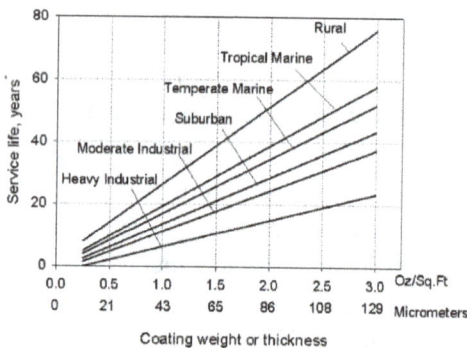

Figure 7.6 Effect of environment on zinc corrosion

1. Slow down or stop the movement of electrons (prevent the cathodic reaction)

    a) Coat the surface with a non-conducting medium such as paint coating.
    b) Reduce the conductivity of the solution in contact with the metal, an extreme case being to keep it dry. Wash away conductive pollutants (salt solutions) regularly.
    c) Apply a current to the material (cathodic protection such as zinc coating).

2. Slow down or stop oxygen from reaching the surface. This would be difficult to do completely but coatings can help.
3. Prevent the metal from giving up electrons by using a more corrosion resistant metal higher in the electrochemical series. Use a sacrificial coating which gives up its electrons more easily than the metal being protected (galvanize coating).
4. Select a metal that forms an oxide that is protective and stops the reaction. Control and consideration of environmental and thermal factors is also essential.

7.2.2 Galvanic corrosion[8 and 9]

When dissimilar metals are in electrical contact in an electrolyte the less noble metal acts as an anode (corrodes) and the noble metal has a degree of protection (see Figures 7.7 to 7.9). Three special features of this mechanism need to operate for corrosion to occur:

- The metals need to be electrically connected together
- One metal needs to be better at giving up electrons than the other
- A path for ion and electron movement must exist.
- To prevent galvanic corrosion the three factors, need to be prevented. This is accomplished by:
- Prevent the electrical contact between the two metals using plastic insulators or coatings
- Use metals that are close together in the galvanic series
- Prevent ion movement by coating the junction with an impermeable material, or ensure environment is dry and liquids cannot be trapped.

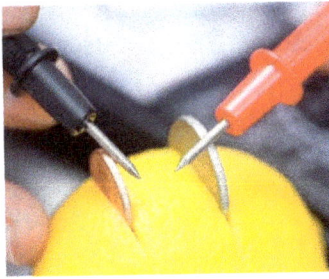

Figure 7.7 *A lemon battery*

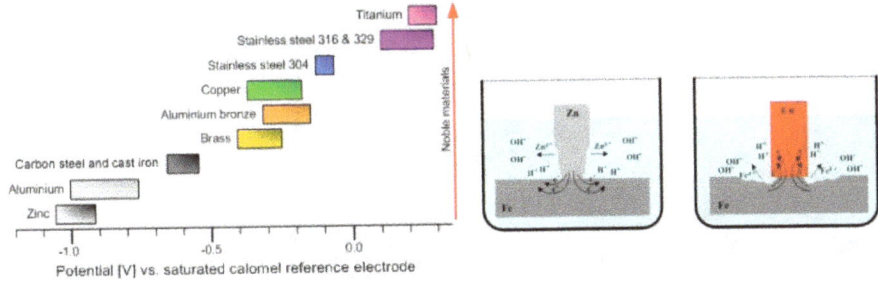

Figure 7.8 *Galvanic series in sea water and Galvanic corrosion*

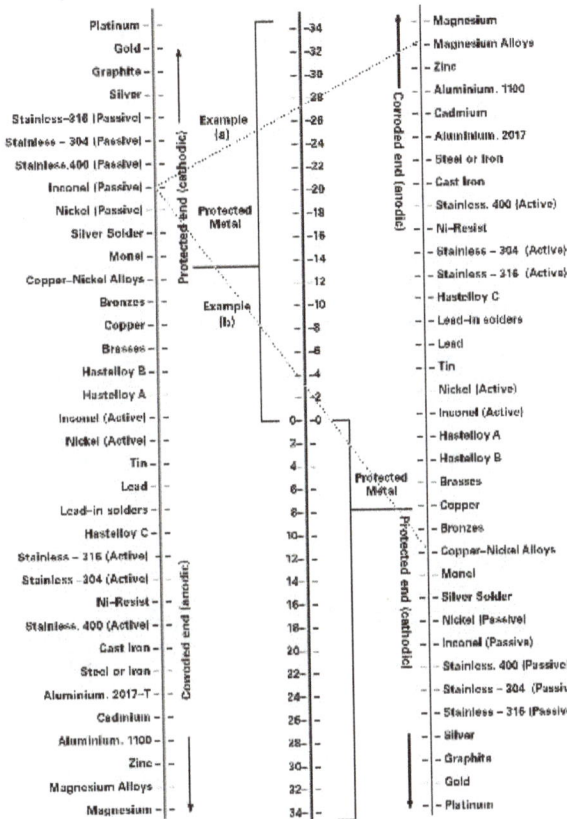

Figure 7.9 *Corrosion indicator. Join the metals and the closer the line to zero when it crosses the central scale the lower the potential corrosion. Examples line (a) = high corrosion, line (b) = low corrosion.*

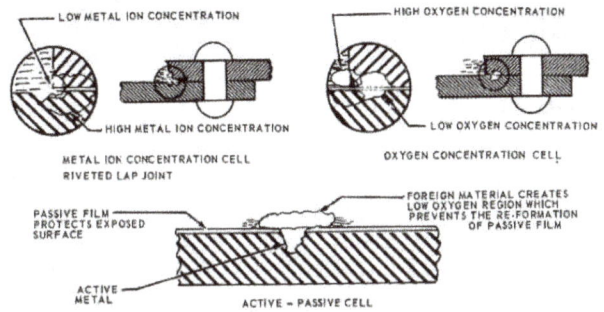

Figure 7.10 *Examples of concentration corrosion*

The potential for corrosion can be reduced by:

- Avoiding sharp corners and designing out stagnant areas
- Use of sealants
- Use welds instead of bolts or rivets
- Selection of resistant materials.

7.2.4 Pitting corrosion[11 to 14]

Pitting corrosion occurs in materials that have a protective film such as a corrosion product or when a coating breaks down. The exposed metal gives up electrons easily and the reaction initiates tiny pits with localised chemistry which can subsequently cause rapid attack.

- Control can be ensured by:
- Selecting a resistant material
- Ensuring a high enough flow velocity of fluids in contact with the material or frequent washing
- Control of the chemistry of fluids and use of inhibitors
- Use of a protective coating
- Maintaining the material's own protective film.

Note: Pits can be crack initiators in stressed components or those with residual stresses resulting from

7.2.3 Concentration cell corrosion (also referred to as differential aeration)[10]

If two areas of a component in close proximity differ in the amount of reactive constituent available the reaction in one of the areas is speeded up. An example of this is crevice corrosion which occurs when oxygen cannot penetrate a crevice and a differential aeration cell is set up. Corrosion occurs rapidly in the area with less oxygen.

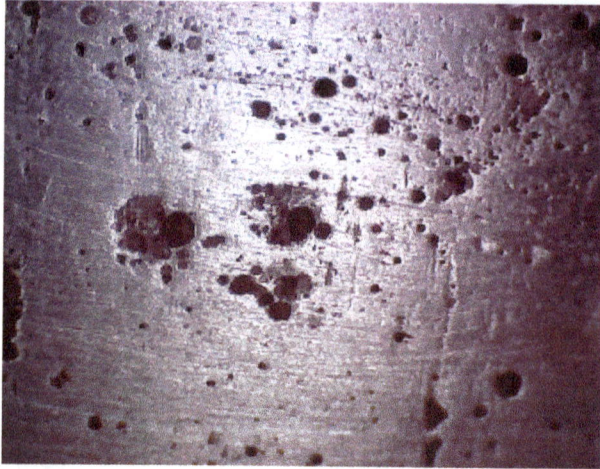

Figure 7.11 Pitting corrosion

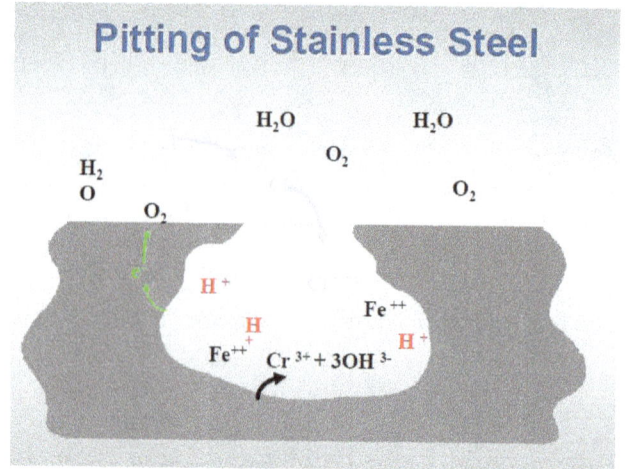

Figure 7.12 Pitting in stainless steel

	Co	Mo	N	PREN
2.4819 / Hastelloy C-276	15	16	-	67,8
1.4547 / 254 SMO	20	6,2	0,2	43,7
1.4410 ("superduplex")	25	4,5	0,3	43,0
1.4539 / 904L	20	4,5	-	34,5
1.4462 (UNS S32205)	22	3,0	0,15	34,3
1.4435	17,0	2,5	-	25,3
1.4436 / 4432	16,5	2,5	-	24,8
1.4362 (duplex 2304)	23	-	0,10	24,6
1.4162 (lean duplex 2101)	21	0,1	0,20	24,5
1.4401 / 4404 / AISI 316(L)	16,5	2,0	-	23,1
1.4571 / "AISI 316Ti"	16,5	2,0	-	23,1
1.4521 / AISI 444	17	1,8	-	22,9
1.4301 / 4307 / AISI 304(L)	17,5	-	-	17,5
1.4509 / AISI 441	17,5	-	-	17,5
1.4016 / AISI 430	16	-	-	16,0
1.4034 / AISI 440B (0,43-0,50 C)	14	0,5	-	15,7
1.4057 / AISI 431 (0,12-0,22 C)	15	-	-	15,0
2.4816 / Inconel 600	14	-	-	14,0
1.4021 (0,16-0,25 C)	12	-	-	12,0
1.4003 / AISI 410	11	-	-	11,0

PREN table containing some of the most commonly found grades of stainless steel. The higher the PREN, the better resistance towards pitting corrosion. The colours of the rows refer to the metallurgical structure:

Austenit	Ferrit	Duplex	Martensit	Nikkelleg.

Figure 7.13 PREN (Pitting Resistance Equivalent Number) of stainless grades

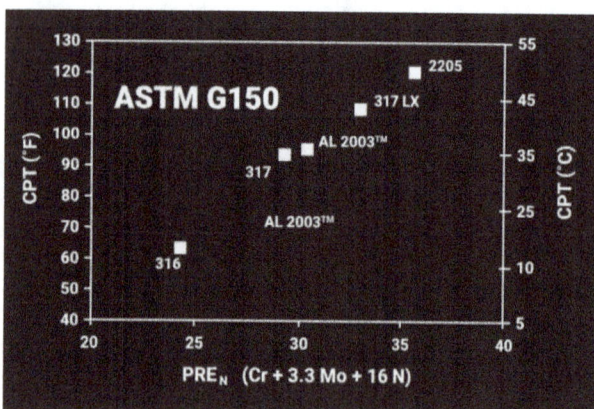

Figure 7.14 PREN vs CPT (critical pitting temperature) determined by ASTM G150 (From Rolled Alloys)

forming operations. This can lead to stress corrosion cracking.

The PREN is a way to rank the grades of stainless in accordance with their pitting resistance and provides a useful way to compare the grades. The formula is empirical and there is no theoretical basis. It reflects practical experience. It was initially developed for austenitic grades, but it is now used for ferritic and duplex grades. The formula is simply based on the chemistry and therefore does not allow for heat treatment or surface condition. Figure 7.14 shows a relationship between PREN and CPT.

7.2.5 Dezincification (selective attack)[15 and 16]

This occurs in alloys such as brass when one component or phase is more susceptible to being attacked than another and corrodes preferentially leaving a porous material that crumbles. This occurs in brasses which contain less than 85% copper. The zinc is selectively corroded to leave a porous residue of copper with little structural strength. It is best avoided by selection of a resistant material, but other means can be effective such as:

• Coating the material
• Reducing the aggressiveness of the environment
• Use of cathodic protection
• Use a dezincification resistant alloy.

ISO 6509 specifies a method for the determination of dezincification depth of copper alloys with zinc exposed to fresh, saline waters or drinking water. The method is intended for copper alloys with a mass fraction of zinc more than 15%.

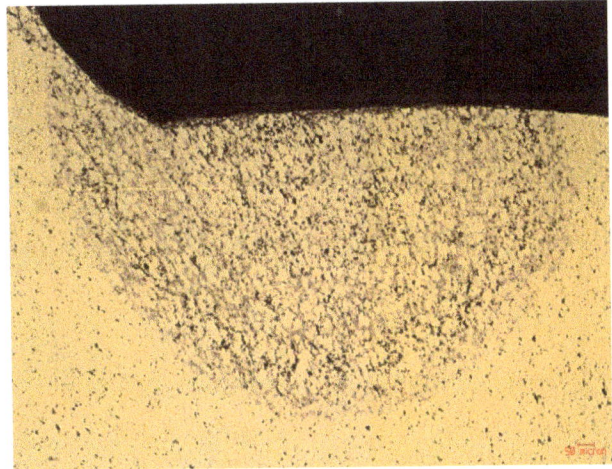

Figure 7.16 Area of plug dezincification X50 Higher magnification of dezincification X500

7.2.6 Intergranular corrosion[17-21]

This occurs when grain boundaries become more chemically active than the body of the grain and corrode preferentially. The classic example is the sensitizing of austenitic stainless steel due to carbide precipitation and thus chromium depletion in the Heat Affected Zone (HAZ) of welds.

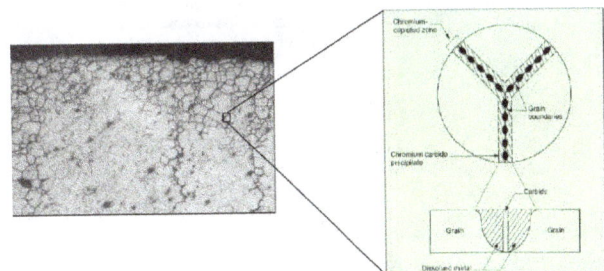

Figure 7.17 Intergranular corrosion

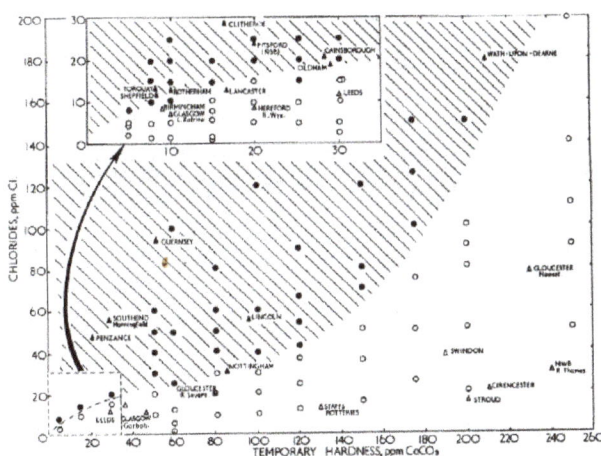

Figure 7.15 Dezincification. Bottom = Turner's diagram for predicting meringue dezincification

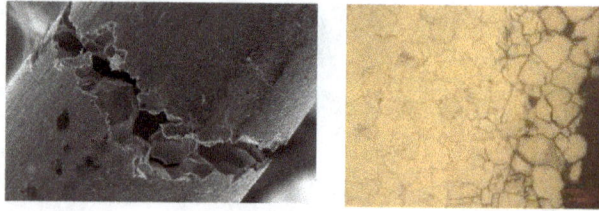

Figure 7.18 Intergranular corrosion LHS SEM image RHS microstructure

7.2.7 Stress corrosion cracking[22 to 24]

This corrosion mechanism results from the combined action of a static tensile stress and corrosion which forms cracks and eventually catastrophic failure of the component. This is associated with specific to a metal and a specific environment. Prevention can be achieved by:

- A reduction in the overall stress level and designing out stress concentrations
- Selection of a suitable material not susceptible to the environment
- Design to minimise thermal and residual stresses
- Developing compressive stresses in the surface of the material
- Use of a suitable protective coating.

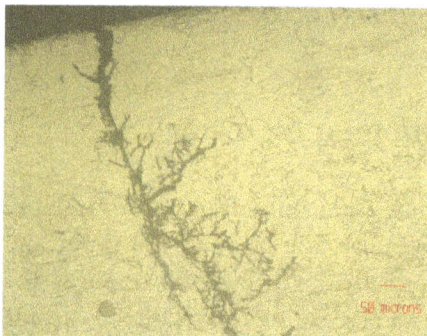

Figure 7.19 Stress corrosion cracking

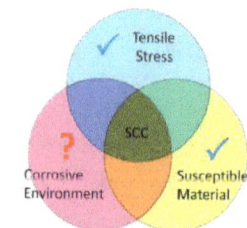

- 304 and 316 Stainless steels are susceptible to chloride stress corrosion cracking (SCC)
  - Sensitization from welding increases susceptibility
  - Crevice and pitting corrosion can be precursors to SCC
  - SCC possible with low surface chloride concentrations
- Welded stainless steel canisters have sufficient through wall tensile residual stresses for SCC
- Atmospheric SCC of welded stainless steels has been observed
  - Component failures in 11-33 years
  - Estimated crack growth rates of 0.11 to 0.91 mm/yr

2/3 of the requirements for SCC are present in welded stainless steel canisters

Figure 7.20 The 3 required conditions

7.2.8 Erosion-corrosion the combination of mechanical wear and corrosion[25]

Erosion-corrosion is the combination of corrosion and erosion and is caused by the fast flow of any unstable fluid on a metal surface. The corrosive damage can be either chemical or electrochemical action. The erosion damage is purely mechanical. The combined effect of these two processes that act together in aqueous environments is recognized as erosion-corrosion. The combined effect of the erosion and the corrosion typically produces damage greater than the sum of their separate effects. This is because of the enhancement of erosion by corrosion and enhancement of corrosion by erosion. However, it is always difficult to clearly differentiating whether erosion has enhanced corrosion or corrosion has enhanced erosion.

Figure 7.21 Erosion-corrosion

Erosion Corrosion

- It is the result of relative movement between the corrosive fluid and metal surface
- All types of equipments exposed to moving fluids are subjected to erosion corrosion.
- Surface chemistry can play a role in erosion corrosion due to mechanochemical effects.

Figure 7.22 Erosion-corrosion

Flow-Assisted Corrosion Definitions

- Erosion-corrosion

Occurs when the velocity of the fluid is sufficient to remove protective films from the metal surface.

- Impingement Corrosion

Localized erosion-corrosion caused by turbulence or impinging flow.

- Cavitation Corrosion

Mechanical damage process caused by collapsing bubbles in a flowing liquid.

Figure 7.23 Other flow assisted corrosion mechanisms

Erosion Corrosion Remedies

- Change the metallurgy (adding chrome and moly 316 SS)
- Alter the design   (provide for large sweeps, eliminate 90 elbows)
- Lower the velocity (less than 2 meters per second)
- Raise the pH   (pH above 9 helps)
- Lower the temperature
- Deaerate
- Filter
- Weld overlays or metallic coatings
- Rubber linings

Figure 7.24 *Prevention*

Grade	UNS number	Nominal composition (wt. pct.)				CCT (°C)	CPT (°C)
		Cr	Ni	Mo	N		
304L	S30400	18	8	-	-	(<-2.5)	2.5
316L	S31603	16	11	2	-	-2.5	10.0
317L	S31703	18	12	3	-	0.0	(28.8)
317LM	S31725	18	13	4.3	-	4.2	(37.7)
904L	N08904	20	25	4.5	-	12.2	42.5
254 SMO®	S31254	20	18	6	0.20	37.5	72.9
654 SMO®	S32654	24	22	7.3	0.50	75	>102.8
SAF 2304™	S32304	23	4	-	0.10	(-4.0)	20.0
2205	S31803	22	5	3	0.16	15.0	34.4
Alloy 255	S32550	25	6	3	0.20	25.0	(55.5)
SAF 2507™	S32750	25	7	4	0.27	37.5	78.9
Alloy G	N06007	22	44	6.5	-	24.5	(71.1)
Alloy 625	N06625	20	60	9	-	40.0	93.9
Alloy C-276	N10276	15	77	15	-	56.6	≥102.8

( ) = estimated

Figure 7.27 *Critical pitting temperature and critical crevice temperature*

7.8.9 Crevice corrosion[26, 27 and 28]

If two areas of a component in close proximity differ in the amount of reactive constituent available the reaction in one of the areas is speeded up. An example of this is crevice corrosion which occurs when oxygen cannot penetrate a crevice and a differential aeration cell is set up. Corrosion occurs rapidly in the area with less oxygen. The potential for crevice corrosion can be reduced by:

- Avoiding sharp corners and designing out stagnant areas
- Use of sealants
- Use welds instead of bolts or rivets
- Selection of resistant materials.

7.2.10 Fretting corrosion[29]

The relative motion between two surfaces in contact by a stick-slip action causes the breakdown of protective films or welding of the contact areas, allowing other corrosion mechanisms to operate. Prevention is possible by:

- Designing out vibrations
- Lubrication of metal surfaces
- Increasing the load between the surfaces to stop the motion
- Surface treatments to reduce wear and increase friction coefficient.

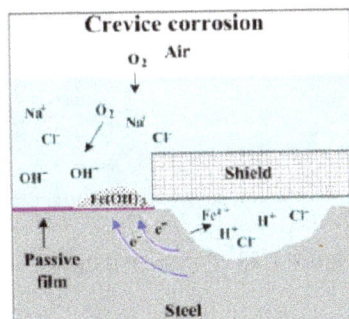

Figure 7.25 *Crevice corrosion*

Fretting Corrosion Definition

Crevice Corrosion

- Occurs more readily than pitting
- Crevices can result from design
- Crevices can result from service (under deposits)
- Crevice geometry is critical (deep and tight crevices are more detrimental)

Figure 7.26 *Causes of crevice corrosion*

- Fretting corrosion is defined as metal deterioration caused by repetitive slip at the interface between two surfaces in contact.
- It is not corrosion due to rotation or erosion.
- More exactly.. motion from vibration and corrosion effects during that time.

Figure 7.28 *Fretting corrosion damage*

7.2.11 Hydrogen damage[30 to 33]

The hydrogen atoms are very small and hydrogen ions even smaller and can penetrate most metals. Hydrogen, by various mechanisms, embrittles a metal especially in areas of high hardness causing blistering or cracking especially in the presence of tensile stresses. The expected problems are cracking, lower toughness, porosity and blisters.

This problem can be prevented by:

- Using a resistant or hydrogen-free material
- Avoiding sources of hydrogen such as cathodic protection, pickling processes and certain welding processes
- Removal of hydrogen in the metal by baking.

Pipeline steels for sour oil and gas containing hydrogen sulphide (H_2S) generally suffer from other important damaging mechanisms of hydrogen-induced cracking (HIC) or sulphide stress corrosion cracking (SSC).

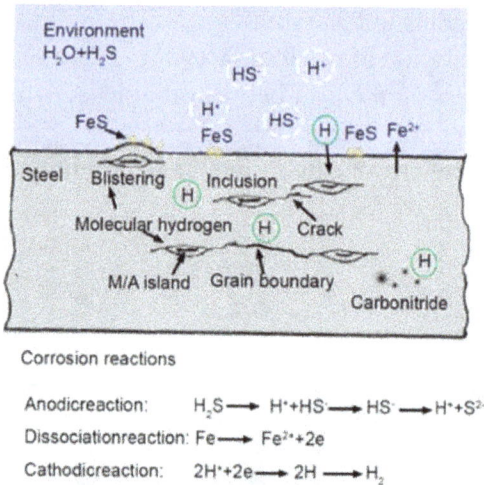

Figure 7.31 *Intergranular crack, grain boundary widening*

Hydrogen Effects

- Hydrogen is one of four or five elements (H, C, N, B, and, possibly, O) with a sufficiently small atomic diameter to dissolve interstitially in most metals.
- The introduction of hydrogen into construction steels has three major deleterious effects:
  - (1) Hydrogen embrittlement,
  - (2) Hydrogen porosity, and
  - (3) Hydrogen cracking

Figure 7.32 *Hydrogen in steel*

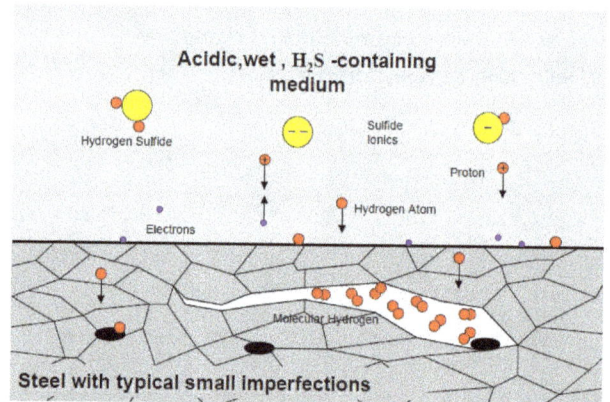

Figure 7.29 *Hydrogen in H_2S environment*

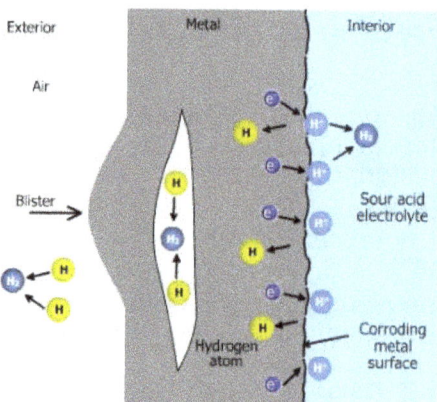

Figure 7.33 *Hydrogen damage in H2S environment*

Figure 7.30 *Hydrogen blistering*

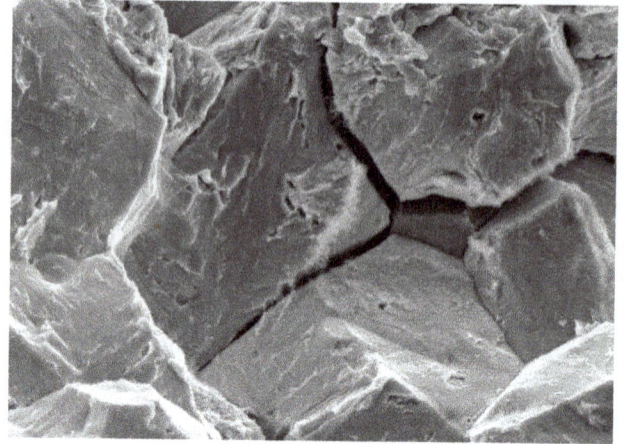

Figure 7.34 *Hindenburg*

In general, corrosion on the steel surface produces hydrogen atoms. These atoms will combine together to form hydrogen gas and escape to the environment. The hydrogen sulphide that deposits on the surface acts as poison reagent and retards the recombination reaction of hydrogen atoms to molecular hydrogen. This effect accelerates diffusion of atomic hydrogen into the steel. Internal pressure theory of HE based on hydrogen accumulation proposed by Zapffe et al. can be used to explain the HIC cracking.

7.2.12 Corrosion fatigue

The combined action of cyclic stresses and a corrosive environment reduce the life of components below that expected by the action of fatigue alone. This can be reduced or prevented by:

- Coating the material
- Good design that reduces stress concentration
- Avoiding sudden changes of section
- Removing or isolating sources of cyclic stress.

An infamous example of corrosion fatigue occurred in 1988 on an airliner flying between the Hawaiian Islands. This disaster, which cost one life, prompted the airlines to look at their aeroplanes and inspect for corrosion fatigue.

Figure 7.35 Boeing 737-200 "island hopping"

Corrosion Fatigue

- Corrosion Fatigue is a special case of stress corrosion caused by the combined effects of cyclic stress and corrosion.
- Control of corrosion fatigue can be accomplished by either lowering the cyclic stress or by corrosion control.

Figure 7.36 Fatigue plus corrosion

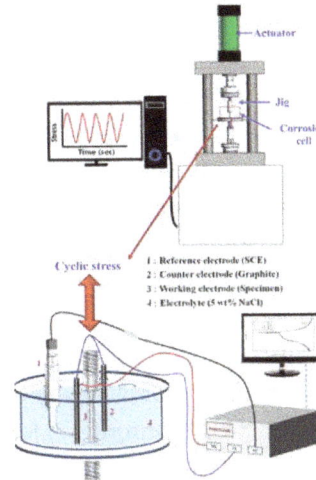

Figure 7.37 Laboratory Test rig

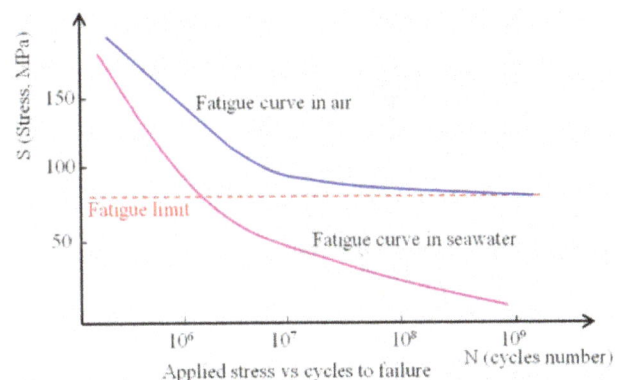

Figure 7.38 For steel endurance limit disappears

DAMAGE MECHANISM

- Corrosion fatigue
- Thermal mechanical fatigue
- Creep fatigue
- Flow induced vibration fatigue
- Flow accelerated corrosion
- Acid phosphate corrosion, caustic gouging, hydrogen damage
- internal pitting
- Thermal fatigue
- Acid dew point corrosion
- Stress corrosion cracking
- Short term heating
- Long term overheating /creep
- Low temperature creep cracking
- Fly ash erosion
- Fire side erosion
- Soot blower erosion
- Chemical cleaning damage
- Coal particle erosion

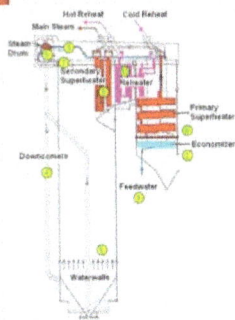

Figure 7.39 Power station failure modes

CORROSION FATIGUE

- **Where damage occurs.**
- Water touched components especially the economiser
- Steam touched tubing containing condensate during operation transients
- welded connection ,bends and attachments with high thermally induced force and bending movements.
- **Locations**
- Tube to header welds
- Scallop bar attachments
- U bend to drain line weld
- Riser and down comer tube

Figure 7.40 Power station corrosion fatigue

7.2.13 Microbial corrosion

This is the new rising star in destructive corrosion despite sulphur reducing bacteria (SRB) being identified in 1934[34] as being responsible for metal corrosion under anaerobic conditions. The mechanism proposed was of anaerobic corrosion which was referred to as the cathodic depolarization theory. Although their effect has been known for many years there has been a difficulty in diagnosis. Microbial corrosion covers the degradation of materials by bacteria, moulds and fungi or their by-products. It can occur by a range of actions such as:

- Attack of the metal or protective coating by acid by-products, sulphur, hydrogen sulphide or ammonia
- Direct interaction between the microbes and metal which sustains attack.

Prevention can be achieved by:

- The use of resistant materials
- Frequent cleaning
- Control of chemistry of surrounding media and removal of nutrients
- Use of biocides
- Cathodic protection

Figure 7.41 *MIC*

Figure 7.42 *Background*

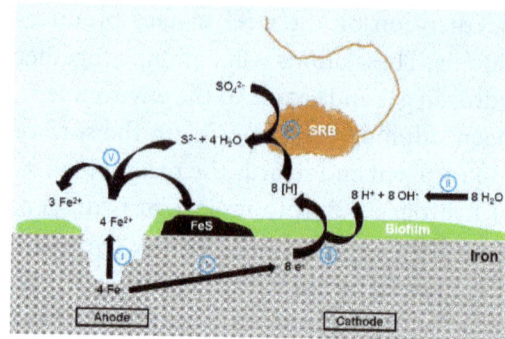

Scheme of iron corrosion by SRB based on reactions as suggested by the cathodic depolarization theory. I, iron dissolution; II, water dissociation; III, proton reduction; IV, bacterial sulfate reduction and V, sulfide precipitation. Source: Mechanisms of Microbiologically Influenced Corrosion: A Review

Figure 7.43 *SRB*

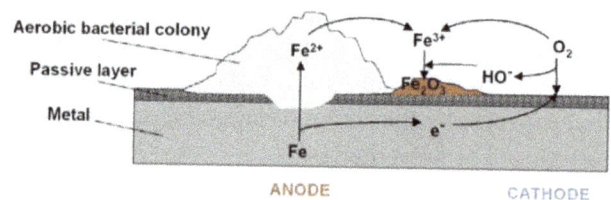

Figure 7.44 *Aerobic bacterial colony*

7.2.14 Stray current corrosion

When a direct current flow through an unintended path and the flow of electrons supports corrosion. This can occur in soils and flowing or stationary fluids. The most effective remedies involve controlling the current by:

- Insulating the structure to be protected or the source of current
- Earthing sources and/or the structure to be protected
- Applying cathodic protection
- Using sacrificial targets.

Figure 7.45 *Damage on a trawler shaft*

Stray current corrosion

- Stray current corrosion occurs where the current from an external source leaves the metal surface, by either inductive, capacitive or via a direct path and enters into the surrounding electrolyte.
- The external power source is a driving force for the corrosive reaction therefore damage can be severe.
- Stray current corrosion is different from other types of corrosion since it is caused by an externally induced current.
- The corrosion tends to be independent of other external factors such as concentration cells, resistivity, pH and galvanic corrosion.

Figure 7.46 Stray current corrosion

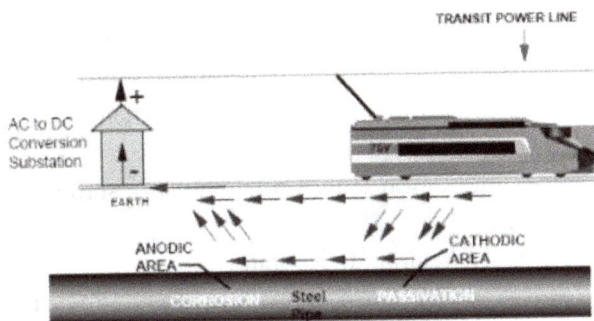

Figure 7.50 Rates of oxidation

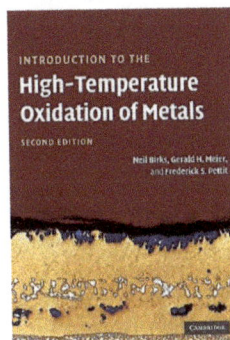

Figure 7.47 Rail to pipe potential damage

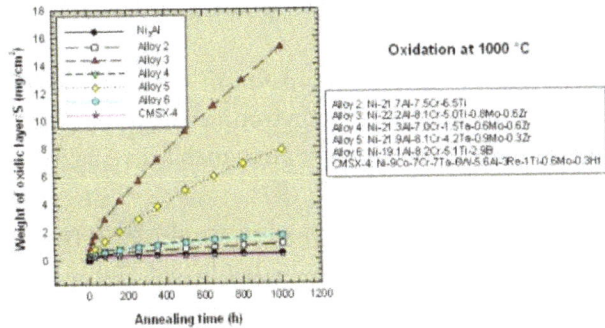

- **Stray Current**: Cathodic protection of underground structures results in the intentional introduction of direct currents in the ground.
- Following the laws of electricity, these currents choose paths of lowest resistance and sometimes interfere with intended cathodic protection or with corrosion control of other structures.
- Pipelines, especially those joined by welding, offer paths of much lower resistance than the earth.
- Pipe areas where current enters become cathodic, and areas where the current leaves become anodic, and the cell is complete.
- Unlike naturally occurring corrosion cells, voltages of such cells can be quite high and anodes correspondingly more active.

Figure 7.48 Pipeline damage

Figure 7.51 Book on oxidation and contents

7.2.15 High temperature corrosion

High temperature corrosion is a phenomenon that occurs in components that operate at very high temperatures, such as gas turbines, jet engines and industrial plants. Engineers are constantly striving to understand and prevent this type of corrosion by the use of protective oxide scales and coatings.

7.3 CORROSION SCIENCE OF THE MECHANISMS INVOLVED IN METAL CORROSION

The basic aspects of corrosion science, include:

- the electrochemical mechanism of corrosion
- thermodynamic and kinetic aspects of corrosion,
- the use of Pourbaix and Evans diagrams.

This knowledge is needed to understanding the corrosion of metals immersed in water, buried in soil, exposed to the atmosphere, used in reinforced concrete, in the human body and in petrochemical plants, or at risk of high-temperature corrosion

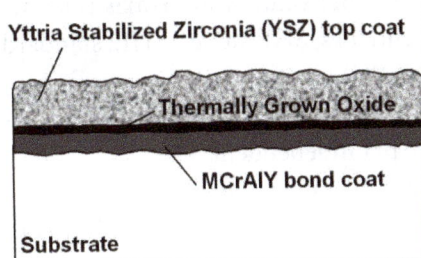

Figure 7.49 MCrAlY thermal barrier coat

7.3.1 The driving force for corrosion

To predict the risk of corrosion of a metal surface exposed to an aggressive environment an important question would be " Is there an available driving force for the corrosion process".

The answer to this question would require the consideration of the thermodynamic likelyhood of a chemical reaction based on the change in Gibbs free energy. If thermodynamics considerations excludes the risk of corrosion then the answer to the question would be "No". "No" would be answer for gold and is why gold nuggets can be found in Australia. It would be the case for silver, copper and their alloys if all contact with oxygen was avoided such as outer space.

If the thermodynamic analysis of the potential chemical reactions was positive which, happens for most metals used in industry, then there may be significant corrosion or limited corrosion, depending on the intervention of other factiors which slow down corrosion processes such as the formation of a protective films which, occurs with aluminium, titanium and stainless steels.

There would also be a need for a second question which would be "how fast would the corrosion occur".

The answer to this question would be based on the kinetics of the reaction. This would require an understanding the driving force and how fast the corrosion would proceed. This can be based on empirical data from previous experience. Or, indications of the rate at which corrosion can occur can be determined by potentiostatic studies.

Aqueous corrosion can be examined as four basic processes anodic reaction, cathodic reaction, electron transport and current transport. These reaction should take place at the same time and at the same rate

$$I_a = I_c = I_m = I_{el} = I_{corr}.$$

The corrosion rate is determined by the slowest part of the processes. If the slowest process was the anodic, because of passivation, the corrosion would be under anodic control. If the cathodic was the slowest it would be under cathodic control and if ohmic resistance in the electrolyte retarded the corrosion it would be under ohmic control

7.3.2 Pourbaix Diagram

In 1945 Marcel pourbaix proposed in "the atlas of electriochemical equilibria in acqueous solutions"

the potential-pH diagram. The diagrams are now called "Pourbaix Diagrams". Pourbaix diagrams use thermodynamic considerations to define potentials corresponding to equilibrium states of all possible reactions between a given element, its ions and its solid and gaseous compounds in aqueous solutions as a function of pH.

This is a useful diagram for examination of the risk of corrosion based on pH and potential. The Pourbaix diagram presents potential-pH information in a diagram. This represents the stability of a metal as a function of potential and pH. The diagram is constructed based on Nernst equations and metals and its species solubility in equilibrium state. In this diagram, there are different regions which are:

- Immune region (thermodynamically stable, no corrosion)
- Corrosive region (active state)
- Passive region (forming passive layers, inhibiting corrosion)

The Pourbaix diagram would:

- Provides information on which ranges of pH and potential will prevent corrosion
- The diagram allows an estimate of the corrosion product compositions at different pH and potential
- Provides information on the directions of various reactions (that are reversible) at different pH and potential

The details/characteristics of Pourbaix diagram:

1. Hydrogen and oxygen lines are drawn in dotted line.
2. The Pourbaix diagram is shown for equilibrium conditions at 25°C.
3. The concentration of metal ions is assumed to be 10^{-6} mol. (Note: lower concentration shouldn't cause corrosion)
4. X-axis is showing pH value while Y-axis is showing redox potential vs, SHE value. (SHE: standard hydrogen potential)
5. Horizontal lines represent electron transfer reaction which are pH independent.
6. Vertical lines are pH dependent.
7. Sloping lines give redox potentials of a solution in equilibrium with hydrogen and oxygen, respectively.
8. Oxidizing power increases with increasing potential.

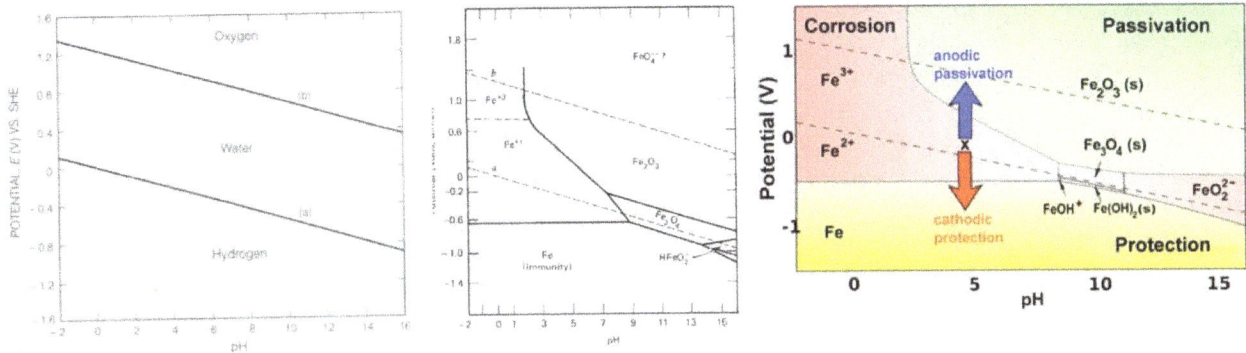

Figure 7.52 *Pourbaix Diagrams*

Pourbaix for water

The redox reactions for water can be referred from EMF series. One of the reactions is $1/2O_2 + 2H + 2e => H_2O$, which has a potential of 1.23V.

By using this potential value, knowledge from pH applicability of Nernst equation, the Nernst equation becomes: $E = 1.23 - 0.059pH$.

Another redox reaction for water is $2H + 2e => H_2$ with a potential of 0V. Using similar technique as above, the Nernst equation becomes: $E = -0.059pH$.

These two equations are then plotted on the Pourbaix diagram. The diagram is shown in Figure 7.52 LHS and shows that:

- above line (a), water is stable and any H_2 present is oxidised to water
- below line (b), water is stable and any O_2 present is reduced to water
- above line (b), water is unstable and oxidize to give O_2
- below line (a), water is unstable and reduce to H_2

The lines for the metal which, in this example will be Iron, are formed by using the same method as above in addition with Gibbs free energy calculation (which for simplicity is not shown here). Region of stability for iron is at the bottom part of the diagram, which lies about below -0.6V. Several important observations from the diagram are:

1. Iron will be unstable in water, no matter what are the pH and potential. That is, iron will corrode in water.
2. Iron can be protected by passivation [coating with Fe_2O_3 or cathodic protection (connected with more active metal that has larger negative potential).

The diagram can relate to real world conditions. For example, a bridge with part of the structure submerged in water or mud will corrode more than other bridge structure in air. This occurs because the mud or water has potential(voltage) that is close to H_2 line, where iron, can corrode to Fe_2. The where corrosion will occur are shown in a Pourbaix diagram in Figure 7.52 RHS.

The Pourbaix diagram has limitations but does often give the general trend of what to expect

1. The diagram is only applicable for 25°C. The diagram cannot predict corrosion at higher temperatures.
2. The diagram provides information that corrosion will happen but not the corrosion rate.
3. The diagram would only be applicable for pure metals. The diagram would not be applicable for alloys or impurity effects.

7.3.3 Electrochemical Basis of Corrosion

Aqueous corrosion is an electrochemical process of oxidation and reduction reactions. As corrosion occurs, electrons are released by the metal (oxidation) and metal ions enter the solution. The electrons are disposed of by reacting with ions in the corroding solution (reduction).

The flow of electrons (current) in the corrosion reaction can be measured and controlled electronically in a corrosion cell with a potentiostat. A potentiostat, in a three-electrode configuration, with a reference electrode, a working electrode and a counter electrode can be used to characterize the corrosion properties of metals in combination with various electrolyte solutions. The corrosion characteristics are unique to each metal/solution system.

To carry out a corrosion test a polarization cell is setup consisting of an electrolyte solution, a reference

Figure 7.53 *LHS = Diagram of a Potentiostat RHS = Polarization curve (green) with Evan's diagram (blue)*

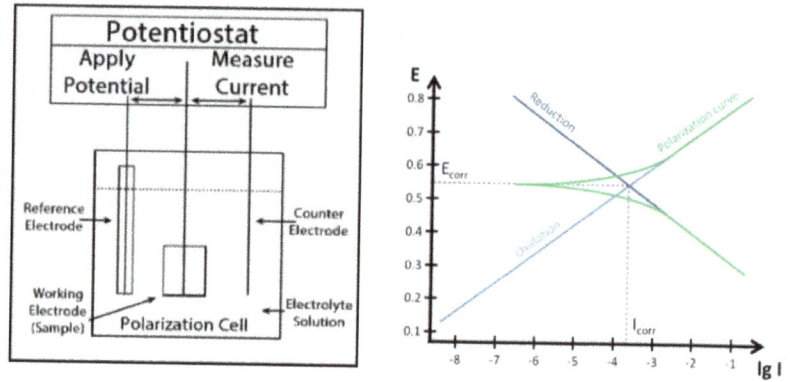

electrode, a counter electrode(s), and the metal sample of interest connected to a specimen holder. (The sample is called the working electrode.) The electrodes are connected to the potentiostat. The working, reference, and counting electrodes are placed in the electrolyte solution, generally a solution that most closely resembles the actual application environment of the material being tested. In the solution, an electrochemical potential (voltage) is generated between the various electrodes. The corrosion potential (E_{CORR}) is measured by the potentiostat as an energy difference between the working electrode and the reference electrode.

7.3.4 Tafel Plot and Evans Diagrams

The potentiostat provides details of the anodic and cathodic reactions in terms of voltage and current. These are plotted as a polarization curve and are known as Tafel lines as shown in Figure 7.53. Based on the polarization curve and the lines for the two reactions the corrosion current and the corrosion potential are determined by the point where the two Tafel lines of the reduction reaction and oxidation reaction meet. Plotting the two Tafel lines (or more) into one plot is an Evans diagram (see Figure 7.53 RHS).

7.4 IMPORTANT ORGANIZATIONS RELEVANT TO CORROSION SCIENCE AND CORROSION TECHNOLOGY

NACE https://nace.org/home.aspx Institute of Corrosion https://www.icorr.org/

European Federation of Corrosion. EFC. https://efcweb.org/ NPL http://www.npl.co.uk/

7.4.1 NACE National Association of Corrosion Engineers

Established in 1943 by 11 pipeline engineers, NACE has grown into one of the world's premier technical societies for corrosion professionals, with more than 30,000 members worldwide.

NACE International has slowly evolved as the global leader in developing corrosion prevention and control standards, certification and education. The members of NACE International still include engineers, as well as numerous other professionals working in a range of areas related to corrosion control. The society operates via several technical committees as shown in the chart below.

NACE Technical Committees https://www.nace.org/uploadedFiles/Committees/TCC-Organization-Chart.pdf

These committees produce technical specifications and practices which are prepared in accordance with NACE International Technical Committee Publications Manual Approved by the Technical and Research Activities Committee April 2018. https://www.nace.org/uploadedFiles/Committees/TechPubMan%20January%202017.pdf

This document outlines the basic objectives of NACE such as to ensure the use of good practice to prevent financial loss or environmental damage. (Clause 3.2 NACE Publications manual)

Abbreviations used on NACE documents:

- **SP – Standard Practice.** NACE SPs shall be methods of selection, design, installation, or operation of a material or system when corrosion is a factor. This class of standard may provide details of construction of a corrosion-control system; methods of treating the surface of materials to reduce requirements for using corrosion-control devices; criteria for the proper operation and maintenance of a corrosion-control system etc (Clause 3.3.1.1 NACE Publications manual)
- **TM – Standard Test Methods.** NACE TMs shall provide test methods related to corrosion prevention and control. This class of standard may give the method of conducting tests of any type to ascertain the characteristics of a material, design, or operation. TMs shall not include pass/fail criteria for the material, design or operation being evaluated by the TM. These criteria may be covered in SPs or MRs (Clause 3.3.1.2 NACE Publications manual)
- **MR – Standard Material Requirements.** NACE MRs shall be standards that define the necessary or recommended characteristics of a material when corrosion is a factor in the selection, application, and maintenance of the material. This class of standard may include chemical composition of the material, its mechanical properties, its physical properties, material selection, and other aspects of its manufacture and application (Clause 3.3.1.3 NACE Publications manual)
- RPs – (Not outlined in the publications manual) Recommended Practices – It contains recommendations prepared to encourage the use of uniform and industry proven requirements and methods
- STA – Specific Technology Group
- PSIG gage pressure referenced to Atmos
- PSIA absolute pressure Atmos = 14.7 psis
- PSIS sealed pressure
- HIC – Hydrogen induced cracking (H_2 no strain needed) NACE TM 0284-2003
- SSC – Sulphide stress cracking (H_2 and strain) NACE TM 0177
- SOHIC – Stress orientated hydrogen induced cracking (H_2 and strain) NACE TM 0103 /2003
- CRA Corrosion resistant alloy
- SZC Soft zone cracking.

Most widely used specifications to ensure reliable plant in the presence of H_2S and other threats are:

- MR0175 also known as ISO 15156 parts 1-3 upstream (oil and gas production) (Revised 2015)
- MR0103 downstream (refining and gas production) 01=version /03 = year (Revised 2016).

7.4.2 European Federation of Corrosion. EFC. https://efcweb.org/

Its website introduction outlines that the EFC is a federation of 38 organisations (Member Societies and Affiliate Members) with interests in corrosion based in 25 different countries within Europe and beyond. Taken together, its Member Societies represent the corrosion interests of more than 25,000 engineers and scientists.

Figure 7.54 *LHS = Up-stream-- MR0175, RHS = Down-stream—MR0103*

It was founded in 1955, its aim is to advance the science of the corrosion and protection of materials by promoting cooperation in Europe and collaboration internationally. The EFC accomplishes its most important activities through twenty-one active working parties devoted to various aspects of corrosion and its prevention.

EUROCORR is the EFC's annual conference; it is the flagship event of the European corrosion calendar.

1. 2009, 6 – 10 September 2009 Nice, France, "Corrosion from the Nanoscale to the plant"
2. 2010, 13 – 17 September, Moscow, Russia, "From the Earth's Depths to Space Heights"
3. 2011, 4 – 8 September, Stockholm, Sweden, "Developing Solutions for the Global Challenge"
4. 2012, 9 – 13 September, Istanbul, Turkey, "Safer world through better corrosion control"
5. 2013, 1 – 5 September, Estoril, Portugal, "Corrosion Control for a Blue Sky"
6. 2014, 8 – 12 September, Pisa, Italy, "Improving materials durability: from cultural heritage to industrial applications"
7. 2015, 6 – 10 September, Graz, Austria, "Earth, Water, Fire, Air, Corrosion happens everywhere"
8. 2016, 11 – 15 September, Montpellier, France, Advances in linking science to engineering
9. 2017, 3 – 7 September, Prague, Czech Republic, Corrosion Control for Safer Living
10. 2018, 9 – 13 September, ICE Krakow, Poland, Applied Science with constant Awareness
11. 2019, 9 – 13 September, Seville Spain, New Times, new materials, new corrosion challengers

The working parties are shown in the following list. These working parties have prepared and published several books of aspects of corrosion which are frequently consulted. https://efcweb.org/Publications/List+of+EFC+Publications.html

Working Parties and Task Forces

The main work of the Federation is conducted through Working Parties of which 21 are currently active.

- EFC Working Party 1: Corrosion and Scale Inhibition
- EFC Working Party 3: Corrosion by Hot Gases and Combustion Products
- EFC Working Party 4: Nuclear Corrosion
- EFC Working Party 5: Environment Sensitive Fracture
- EFC Working Party 6: Surface Science and Mechanisms of Corrosion and Protection
- EFC Working Party 7: Corrosion Education
- EFC Working Party 8: Physico-chemical Methods of Corrosion Testing
- EFC Working Party 9: Marine Corrosion
- EFC Working Party 10: Microbial Corrosion
- EFC Working Party 11: Corrosion of Steel in Concrete
- EFC Working Party 13: Corrosion in Oil and Gas Production
- EFC Working Party 14: Coatings
- EFC Working Party 15: Corrosion in Refinery and Petrochemistry
- EFC Working Party 16: Cathodic Protection
- EFC Working Party 17: Automotive Corrosion
- EFC Working Party 18: Tribo-Corrosion
- EFC Working Party 19: Corrosion of Polymer Materials
- EFC Working Party 20: Corrosion and Corrosion Protection of Drinking Water Systems
- EFC Working Party 21: Corrosion of Archaeological and Historical Artefacts
- EFC Working Party 22: Corrosion Control in Aerospace
- EFC Working Party 23: Corrosion Reliability of Electronics
- EFC Task Force: CO2-Corrosion in CCS-Applications
- EFC Task Force: Atmospheric Corrosion

7.4.3 The Institute of Corrosion

Their website describes the Institute of Corrosion (ICorr) as a Learned Society and Registered Charity that has been serving the corrosion science, technology and engineering community since 1959 in the fight against corrosion, which costs the UK around 4% of GNP per annum. Key to this fight are the establishment and promotion of sound corrosion management practice, the advancement of cost-effective corrosion control measures, and a sustained effort generally to raise corrosion awareness at all stages of design, fabrication and operation.

Initially known as the British Association of Corrosion Engineers (BACE), it was inaugurated at a meeting held on the 21st May 1959. In 1966 BACE was renamed the Institution of Corrosion Technology.

A number of key Members of Council and Officers lobbied MPs about the inactivity of the country generally and industry in particular to the devastating costly effects of corrosion in all areas of industry. The result of this lobbying was brought to the attention of the Minister for Technology, the Rt. Hon. Anthony Wedgwood Benn MP, who in 1969 asked the Department of Trade and Industry to set up a Committee to assess the cost of corrosion to the UK, and to look at ways in which this could be overcome.

The Committee was formed in March 1969 and Dr TP Hoar, who was head of the Department of Metallurgy at the University of Cambridge, was appointed as Chairman. There were over 20 Members appointed to this Committee, some from the Institution of Corrosion Technology and others were Members of the Corrosion and Protection Association (CAPA), which represented the Scientific and Academic interests. The DTI report was published in February 1971 and a major public conference was held by the Institute of Mechanical Engineers in April 1971, when the report was publicised and presented in detail. It was a comprehensive document, comprising 130 pages.

During the early 1970s after the launch of the report, meetings of the Councils of the Corrosion and Protection Association and Institution of Corrosion Technology merged to form the Institute of Corrosion Science & Technology (ICorr S&T) which was formed on the 1st January 1975.

The Institute of Corrosion deals with training, certification, qualification and continuing Professional Development in the UK.

7.4.4 National Physics Laboratory. Run by the Department for Business, Innovation and Skills. Has a National Corrosion Service
http://www.npl.co.uk/science-technology/advanced-materials/national-corrosion-service/

Their website states The National Corrosion Service (NCS) provides a Gateway to Expertise on many aspects of materials degradation. Because of its major impact on the economy, corrosion continues to figure prominently in the remit of this service. A wide range of NCS products and services, as well as publications, including the very popular Corrosion Guides, is available. Advice can now be obtained from the experts at NPL, and other sources, on problems resulting from a wide range of deterioration mechanisms, including the effects of wear, fatigue and high temperature exposure, and other processes specific to polymers, composites and ceramics.

http://www.npl.co.uk/upload/pdf/
Checklist_for_corrosion_control.pdf

7.4.5 Corrosion in the oil and gas industry

There has been a long history of corrosion problems in the oil and gas industry. In recent years there have been some helpful published documents which give a detailed view of the extent and complexity of corrosion in the oil and gas industry[35-39] and the potential future challenges.[37] It is easy to appreciate why codes and standards have been essential to the dual objectives of risk reduction and the extraction of oil and gas at an economic price.

As the conventional oil and gas reserves dwindle there will be a need to extract reserves that are more difficult to acquire and, therefore, the future will present many more challenges. The most difficult to extract reserves are associated with the presence of high pressures and high temperatures (HPHT). These can be in deep-waters with depths greater than approximately 800 to 1800m. Developing oil and gas with the associated HPHT problems has increased the level of risk.

Extraction of these reserves has become an even more daunting task in view of the recent failure of BP facilities in the Mexican Gulf ($65 billion loss) and the pipe leaks caused by high H_2S at Kashagan in the Caspian Sea ($4 billion loss), which sent unprecedented warnings to the whole engineering community especially the oil and gas sector.

The Kashagan project was a joint project of seven of the top oil companies. In 1998, the consortium partners were: ENI's subsidiary Agip (Italy), BG (UK), BP Amoco (UK), ExxonMobil (USA), Shell (UK), Total FinaElf (France), Phillips Petroleum Co. (USA), Statoil (Norway), and Inpex Masela Ltd. (Japan). In 2001 BP and Statoil chose to leave the consortium and to sell their shares, possibly due to the level of H_2S and the associated difficulties.

The main challenges were:

- Deep reservoir – 5,000m
- High reservoir pressure – 800bar

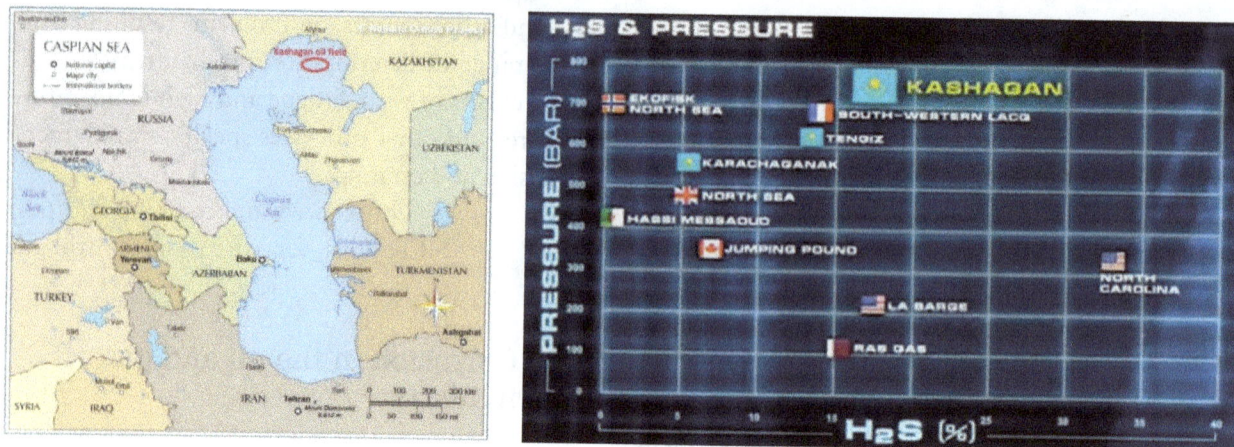

Figure 7.55 Location of the Kashagan oil field and the top of the envelope pressure and H_2S %.

- High H_2S (Hydrogen Sulphide) content (16-20%)
- Management of by-products, such as sulphur
- Use of sour gas re-injection into the reservoir.

It is therefore remarkable with such a distinguished list of operators that pipe leaks, associated with H_2S occurred within two months of start-up which were so severe that they resulted in closure of the well. This negligence also put lives at risk. H_2S has been referred to as the "knock down gas" because inhalation of high concentrations can cause immediate loss of consciousness and death at concentrations of over 500-1000 parts per million (ppm).

A report worth reading was prepared by an environmental activist group[40] titled "The Kashagan Oil Bubble" which describes the project as resembling the script from "The House of Cards": in-fighting, bribery and corruption and a veritable cash guzzler project. One amusing statement that was made due to the unending cost increases was that some of Shell's partners nicknamed Kashagan project "Cash All Gone". These events have highlighted the consequences of failure and endorsed the need for the return of the prudent and safe approach that achieved success in the past. In addition, they have stimulated detailed review of the design procedures and specification for oil and gas and focused serious management attention on the need for **corrective and preventative actions** for all failures and identified non-conformance processes and products.

UK government changes to refocus oil and gas strategy – In the UK a major review in 2013 has allowed a refocus on ways to maximise revenue from the North Sea Oil and Gas. (UKCS maximising recovery review: final report by Sir Ian Wood 24th February 2014.)

https://editor.ogauthority.co.uk/media/1014/ ukcs_maximising_recovery_review.pdf

This has allowed the creation of the "Oil and Gas Authority". The role of the OGA is to regulate, influence and promote the UK oil and gas industry in order to maximise the economic recovery of the UK's oil and gas resources. The OGA is largely funded by an industry levy introduced on 1 October 2015. OGA has their headquarters in Aberdeen with another office in London. The OGA regulates the exploration and development of the UK's offshore and onshore oil and gas resources and offshore carbon storage, offshore gas storage and offshore gas unloading activities. One of the main activities is the support of MER strategy. (THE MAXIMISING ECONOMIC RECOVERY STRATEGY FOR THE UK.)

https://editor.ogauthority.co.uk/media/3229/ mer-uk-strategy.pdf, https://editor.ogauthority. co.uk/media/5727/oga_may_2019_v1_artwork.pdf

7.4.6 Sweet and Sour and possibly with Salt

There are four main corrosive elements in oil and gas production:

- hydrogen sulphide (H_2S) (Sour)
- carbon dioxide (CO_2) (Sweet)
- dissolved oxygen, particularly in surface facilities, dissolved oxygen (O_2) but also reinjected fluids
- dissolved Chloride (Salt).

The corrosion products that are formed help to identify the cause of corrosion and can assist in failure analysis. Bruce Craig provided a guide to the interpretation and analysis of corrosion deposits in 2002.[41] The modes of corrosion damage associated with the H_2S and CO_2 have been known in oil and gas for over 70 years. The first identified cases of SSC due to high H_2S (known as sour conditions) in the oil and gas industry were found in the tubing and casing of gas wells in the United States and Canada in 1950–1951.[42]

On February 2, 1975 the deadly toxicity of H_2S was demonstrated when a ruptured pipe connection on a gas well, near Denver City, released hydrogen sulphide (H_2S). Eight people were killed while trying to evacuate their nearby homes. A worker checking that well was also killed.[43] CO_2 or "sweet" corrosion is by far the most prevalent form of corrosion encountered in oil and gas production. Early cases of CO_2 corrosion were reported in gas wells located in Texas in the 1940s. The event caused major concern amongst oil companies and, fearing governmental intervention and consequential delays to their business, realised that a solution had to be found quickly.[44]

This was similar to the Macondo accident, 60 years later, where fear of project delays resulted in significant financial investment that has allowed significant progress in a "new" approach to HTHP technology.

7.4.7 The problems with sour gas H_2S, sweet gas CO_2 and Chlorides

A number of key factors play a role in influencing the extent of CO_2 corrosion. These factors are broadly summarised into the following categories as shown in Figure 7.57.

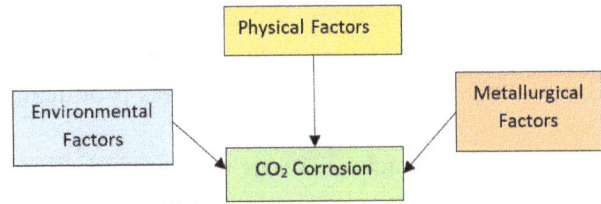

Figure 7.57 Factors that influence CO_2 corrosion

1. Temperature (major influence). Temperature has the effect of increasing corrosion rates and this is due to the reasoning behind the Arrhenius and Kinetic theories which enable faster collisions between reacting species and speeding up reaction rate
2. Pressure (determines partial pressure of CO_2). The higher the CO_2 partial pressure, in general, the greater the corrosion risk, as this directly translates to higher concentrations of carbonic acid and lower pH-values of the solution surrounding the metal surface
3. Velocity (can stimulate mass transfer and break down corrosion product layers). Higher fluid flow-rate velocities imply greater corrosion risks due to the mechanical washing away of the protective film on the metal surface
4. pH and ferrous ion concentration (determine precipitation of $FeCO_3$).

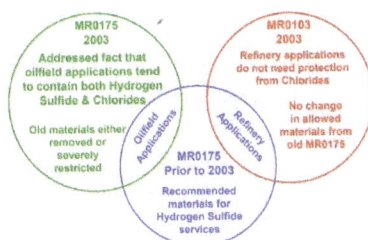

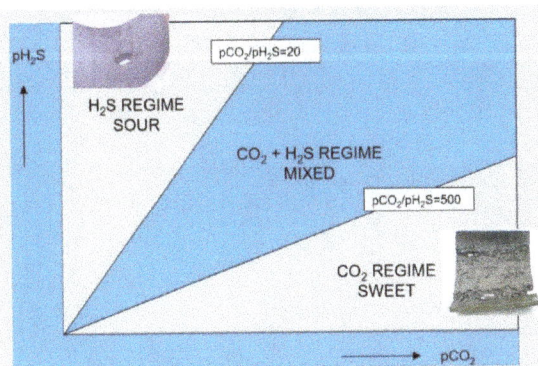

Figure 7.56 Carbon Dioxide CO_2 Major Factors Influencing CO_2 Corrosion

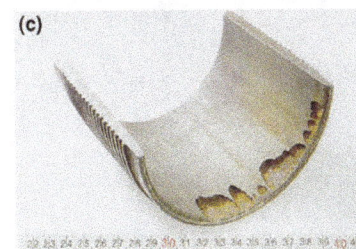

Figure 7.58 CO_2 Corrosion Damage L (a) Pitting attack (b) Mesa attack (c) Flow-induced localized attack

Pitting – This is a form of localised corrosion that usually occurs in low-velocity fluid flow environments around dew point temperatures in a gas-producing well.

Mesa-type Attack – This is a type of localised corrosion that takes place in low to medium fluid-flow environments, where the protective film forms; however, it is unstable and unable to withstand the intensity of the operating regime. Low to medium velocity fluid flow is still capable of washing away any protective films formed on the metal surface.

Flow-induced Localised Corrosion –This form of corrosion usually starts from pits/troughs that had previously been sites of localised mesa attack. This form of corrosion is solely dependent on high fluid flowrates.

Hydrogen Sulphide

Oil and gas can contain high H_2S (known as sour conditions) which can damage pipeline steels and cause cracks to form due to hydrogen-induced corrosion and cracking. SSC Cracking of a metal alloy involves corrosion and tensile stress in the presence of water and H_2S. As a result of laboratory testing and field experience, NACE MR-0175:2003 details the parameters of acceptable chemical composition, physical properties, manufacturing processes, and fabrication processes that will yield a material acceptable for use in a NACE defined sour environment.

Corrosion on the steel surface produces hydrogen atoms. These atoms can combine together to form hydrogen gas, but some will enter into the steel especially when hydrogen sulphide is present which retards the recombination reaction of hydrogen atoms to molecular hydrogen. When the hydrogen atoms diffuse into the steel, they are trapped by potential sites and cause a reduction in the toughness of the steel. Under the tensile stress cracks grow which results in fracture.

H_2S problems and MR0175

1. H_2S causes Sulphide Stress corrosion Cracking (SSC). Most metals are susceptible to SSC.
2. SSC is defined as cracking of a metal under the combined action of a tensile stress and corrosion in the presence of water and H_2S. SSC is a form of hydrogen stress cracking resulting from absorption of atomic hydrogen that is produced by the sulphide corrosion reaction on the metal surface.
3. Metallurgical Variables –

   • Higher hardness increases SSC susceptibility
   • >22 HRC is a rule for most carbon and low alloy steels
   • Higher YS and TS correlate with hardness and increases SSC
   • Finer grain size and a higher percentage martensite in the microstructure increases resistance to SSC.

4. Environmental Variables

   • Lower temperatures (<175°F) are more severe
   • Higher PH_2S increases SSC
   • Higher fraction of water in systems promotes wetting of the equipment and increases SSC
   • Lower pHs increase SSC.

5. Why was hardness testing chosen to control the conformance to specification?

   i) Hardness equates fairly well with the strength of a metal

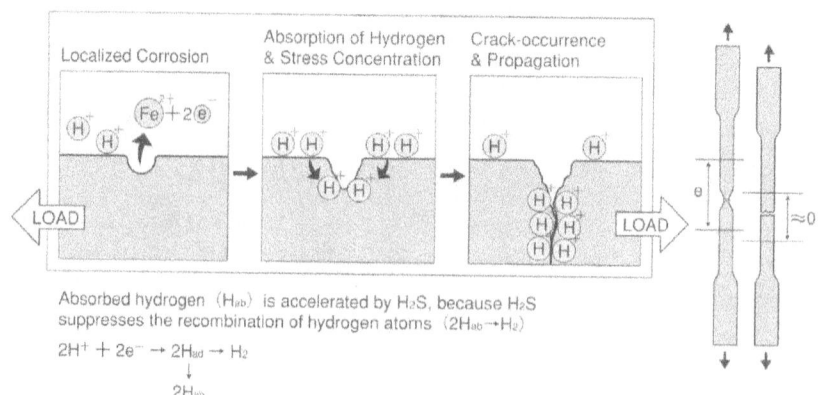

Figure 7.59 SSC

Reference - Sumitomo Metals: OCTG Materials & Corrosion

Lack of ductility is typical of SSC failures
P110 Failure
Paper 0512, CORROSION/2005

Quasi Cleavage SSC Failure SSC
fracture surface – brittle with little
ductility.
Paper 05116, CORROSION/2005

Figure 7.60 Brittle fracture caused
by SSC

ii) Easy to do and it is a non-destructive test

iii) Higher strength metals are more susceptible to SSC and Cl⁻ SCC

iv) The number and location of hardness tests are not specified except for weld testing.

6. MR0175 specified materials

i) Some materials were accepted based on prior proven performance

ii) Since MR0175, alloys have to pass testing per TM0177 MR0175 identifies acceptable alloy categories. Some alloys are only suitable for specific components.

▶ *Figure 7.61* Material selection based on Sumitomo procedure (Sumitomo Website)

7.4.8 Material selection and Software models are used for material selection

The main factors that influence the material selection for OCTG are H_2S pressure, CO_2 pressure, temperature, and Cl– content.[45 and 46]

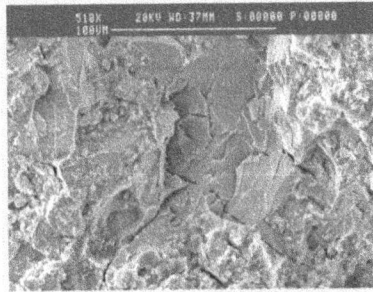

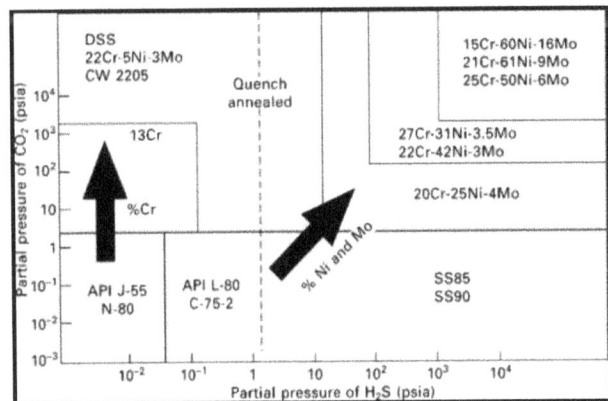

Table 1. Nominal composition of representative carbon and low alloy steels as well as CRA for oilfield applications

Alloy designation	Standard nominal composition (wt%)												SMYS
	Cr	Mo	Ni	W	N	Fe	Nb or (Nb + Ta)	Ti (Al)	Cu	C	Si	Mn	MPa (ksi)
Carbon and low alloy steels													
API 5L-X65Q (PSL 2)	–	–	–	–	–	bal.	§	§	–	0.18 (max.)	0.45 (max.)	1.70 (max.)	450 (65)
ASTM A694 F65	–	–	–	–	–	bal.	–	–	–	0.30 (max.)	0.15–0.30	1.60 (max.)	450 (65)
ASTM A508 Gr. 4	1.50 to 2.0	0.40 to 0.6	2.80–3.90	–	–	bal.	–	–	–	0.23 (max.)	0.40 (max.)	0.20 to 0.40	690 (100)
UNS K32047	1.50 to 1.90	0.50 to 0.65	3.00–3.50	–	–	bal.	–	–	–	0.14 to 0.20	0.15–0.38	0.10 to 0.14	690 (100)
10GN2MFA	0.30 (max.)	0.40 to 0.70	1.80–2.30	–	–	bal.	–	–	–	0.08 to 0.12	0.17–0.37	0.80 to 1.10	414 (60)
UNS K21590	2.00 to 2.50	0.90 to 1.10	0.25 (max)	–	–	bal.	–	–	–	0.11 to 0.15	0.10 (max.)	0.30 to 0.60	517–586 (75–85)
UNS G43200	0.40 to 0.60	0.20 to 0.30	1.65–2.00	–	–	bal.	–	–	–	0.17 to 0.22	0.15 to 0.35	0.45 to 0.65	414 (60)
Precipitation-hardened low alloy steels													
ASTM A707-L5	0.60 to 0.90	0.15 to 0.25	0.70 to 1.00	–	–	bal.	–	–	1.00 to 1.30	0.07 (max.)	0.35 (max.)	0.09 (max.)	517 (75)
Solution annealed nickel-based alloys													
UNS N06625	20.0 to 23.0	8.0 to 10.0	58.0 (min.)	–	–	5.0 (max.)	(3.15 to 4.15)	–	–	0.10 (max.)	0.50 (max.)	0.50 (max.)	290–414 (42–60)ª
Precipitation-hardened nickel-based alloys													
UNS N07718	17.0 to 21.0	2.80 to 3.30	50.0 to 55.0	–	–	bal.	(4.87 to 5.20)	0.80 to 1.15	0.23 (max.)	0.045 (max.)	0.010 (max.)	0.35 (max.)	827-965 (120–140)
UNS N07725	19.0 to 22.5	7.00 to 9.50	55.0 to 59.0	–	–	bal.	2.75 to 4.00	1.00 to 1.70	–	0.030 (max.)	0.20 (max.)	0.35 (max.)	827 (120)
UNS N07716	19.0 to 22.0	7.00 to 9.50	59.0 to 63.0	–	–	bal.	2.75 to 4.00	1.00 to 1.60	0.23 (max.)	0.030 (max.)	0.20 (max.)	0.20 (max.)	827-965 (120–140)
UNS N06059	22.0 to 24.0	15.0 to 16.5	bal.	–	–	1.50 (max.)	–	(0.1 to 0.40)	–	–	0.10 (max.)	0.50 (max.)	450 (65ksl)ᵇ
UNS N06680ᶜ	20.5	6.5	bal.	6.5	–	0.1 (max.)	3.5	1.5	–	0.010 (max.)	–	–	550–665 (80–95)
UNS N06686	19.0 to 23.0	15.0 to 17.0	bal.	3.0 to 4.0	–	5.0 (max.)	–	–	–	0.010 (max.)	0.08 (max.)	0.75 (max.)	760 (110)ᵈ
Duplex and super duplex stainless steels													
UNS S32205	21.0 to 23.0	2.50 to 3.50	4.50 to 6.50	–	0.08 to 0.20	bal.	–	–	–	0.03 (max.)	0.2 to 0.70	2.0 (max.)	450 (65)
UNS S32750	24.0 to 26.0	3.0 to 5.0	6.0 to 8.0	–	0.24 to 0.32	bal.	–	–	–	0.03 (max.)	0.8 (max.)	1.2 (max.)	550 (80)
UNS S32760	24.0 to 26.0	3.0 to 4.0	6.0 to 8.0	0.50 to 1.0	0.20 to 0.30	bal.	–	–	0.5 to 1.0	0.03 (max.)	1.0 (max.)	1.0 (max.)	550 (80)
UNS S39274	24.0 to 26.0	2.50 to 3.50	6.0 to 8.0	1.5 to 2.5	0.24 to 0.32	bal.	–	–	0.20 to 0.80	0.03 (max.)	0.8 (max.)	1.0 (max.)	550 (80)
Austenitic and highly alloyed austenitic stainless steels													
UNS S31603	16.0 to 18.0	2.0 to 3.0	10.0 to 14.0	–	–	bal.	–	–	–	0.03 (max.)	1.0 (max.)	2.0 (max.)	182 (27)ª
UNS S31254	19.5 to 20.5	6.0 to 6.5	17.5 to 18.5	–	0.18 to 0.22	bal.	–	–	0.50 to 1.0	0.020 (max.)	0.80 (max.)	1.0 (max.)	310 (45)
Martensitic and precipitation hardened-martensitic stainless steels													
UNS S41000	11.5 to 13.5	–	–	–	–	bal.	–	–	–	0.15 (max.)	1.0 (max.)	1.0 (max.)	550 (80)ª
UNS S17400	15.0 to 17.5	–	3.0 to 5.0	–	–	bal.	0.15 to 0.45	–	3.0 to 5.0	0.07 (max.)	1.0 (max.)	1.0 (max.)	724 (105)ᶠ

§ Ni + V + Ti < 0.15 wt%
ª Solution annealed
ᵇ ERNICrMo-13
ᶜ ERNICrMoWNbTi-1
ᵈ INCO-WELD-686CPT (Tensile Strength)
ᵉ Double tempered; Hardness 22HRC (max.)
ᶠ H1150-D

Figure 7.62
Range of
materials for
Oil and Gas
equipment[47]

	Category	MR0175 / ISO 15156	MR0103
Carbon Steel	Max Hardness (HRC):	22	22
	Conditions:	Annealed Normalised Normalised & Tempered Quench & Tempered Stress Relieved	Annealed Normalised Normalised & Tempered Quench & Tempered
	Welding:	Vickers Hardness Survey regardless of whether PWHT or not Suggestion regarding "other controls" detected	Vickers Hardness Survey regardless of whether PWHT or not Suggestion regarding "other controls" detected
Alloy Steel	Max Hardness (HRC):	22	For P numbered steel
	Conditions:	Annealed Normalised Normalised & Tempered Quench & Tempered Stress Relieved	Annealed Normalised & Tempered Quench & Tempered
	Welding:	Vickers Hardness Survey regardless of whether PWHT or not	Vickers Hardness Survey regardless of whether PWHT or not
Austenitic Stainless	Max Hardness (HRC):	22	22
	Conditions:	Solution Heat Treated	Solution Heat Treated
	Welding:	Vickers Hardness Survey	No specific control
Martensitic Stainless	Max Hardness (HRC):	22	22
	Conditions:	Annealed Normalised Normalised & Tempered Quench & Tempered Stress Relieved	Quench & Double Tempered
	Welding:	Vickers Hardness Survey	No specific control Suggestion regarding "other controls" detected

Figure 7.63 Comparison of MR1075 and MR0103 Material requirements

Nace MR0175 scope and layout

Scope ANSI NACE MR0175, ISO 15156 gives requirements and recommendations for the selection and qualification of metallic materials for service in equipment in the oil and gas production and in natural gas sweetening plants in H2S containing environments, where the failure of such equipment pose a risk to health and safety of the public and personnel or to the environment. It can be applied to help avoid costly corrosion damage to the equipment itself.

It supplements, but does not replace, the materials requirements given in the appropriate design codes, standards or regulations.

ANSI NACE MR0175, ISO 15156 addresses all the mechanisms of cracking that can be caused by H2S, including sulphide stress cracking, Hydrogen induced cracking, Hydrogen stress cracking, Stress corrosion cracking, Stress orientated hydrogen induced cracking, Step wise cracking, Sulphide stress cracking, Soft zone cracking and galvanically induced stress cracking.

Steel Type	Grades Included	Comments
Ferritic	405, 430, 409, 434, 436, 442, 444, 445, 446, 447, 448.	Hardness up to 22 HRC
Martensitic	410, 420	Hardness up to 22 HRC
	F6NM	Hardness up to 23 HRC
	S41425	Hardness up to 28 HRC
Austenitic	201, 202, 302, 304, 304L, 305, 309, 310, 316, 316L, 317, 321. 347, S31254(254SMO), N08904(904L) N08926(1925hMo)	Solution annealed, no cold work to enhance properties, hardness up to 22HRC
	S20910	Hardness up to 35 HRC
Duplex	S31803 (1.4462) S32520 (UR 52N+), S32750 (2507), S32760) (Zeron 100), S32550 (Ferralium 255)	PREN> 30 solution annealed condition, ferrite content 35% to 65% or 30% to 70% in welds. Note that the general restriction of 28HRC in previous editions is not found in this latest edition of the standard. There is a specific restriction on HIP'd S31803 to 25HRC. For some applications cold worked material is allowed up to 36HRC
Precipitation Hardening	17-4PH	33 HRC Age Hardening at 620°C
	S45000	31 HRC Age Hardening at 620°C
	S66286	35 HRC

Figure 7.64 *Summary of MR 0175 Requirements, A wide range of materials is covered by the standard including most groups of stainless steel. The table shows some of these grades. This summary gives a general view of this complex standard and decisions should only be made based on the actual standard.*

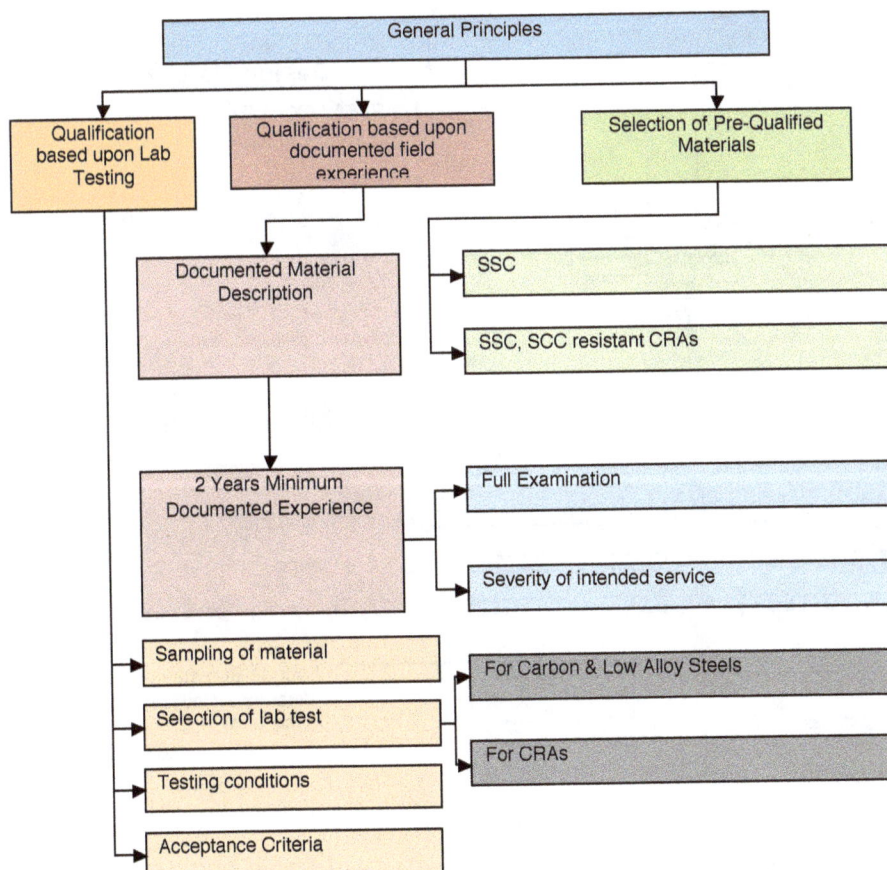

Abbreviated Terms
CRA= Corrosion resistant alloy, HIC= Hydrogen induced cracking, HSC= Hydrogen stress cracking, SCC= Stress corrosion cracking, SCHIC= Stress orientated hydrogen induced cracking, SWC=Step wise cracking, SSC= Sulphide stress cracking, SZC= Soft zone cracking

Figure 7.65 *Using MR0175*

7.4.9 API 17TR8 for the "High-Pressure High-temperature (HTHP) design guidelines."

The risks associated with oil and gas extraction equipment used for sub-sea applications represent a challenge to all companies associated with oil and gas equipment supply. This has been uniquely summarised in a slide in a Statoil presentation as shown in Figure 7.66.

Why do anything at all?
"Systems Thinking just Complicates Things."

Figure 7.66 Statoil all problems must be shared

This graphic illustrates that everyone is in the same boat and if one company fails to get it correct the repercussions will be felt by all and therefore there has been a clear team spirt and openness with the development of HTHP capability. The SC17 committee of API has attempted to lay the technological foundations needed to achieve the necessary quality level. One recent report has been API 17TR8 for the "High-Pressure High-temperature (HTHP) design guidelines". The Bureau of Safety and Environmental Enforcement (BSEE) in a recent presentation on the approval process for HPHT projects in the Gulf of Mexico (2015) stated that "BSEE believes that API 17TR8 is one of the best available guidance documents for the construction of HTHP oil field equipment at the present time". However, BSEE was also critical pointing out that a "Technical Report" is not an "Engineering Standard" and then proceeds to outline a plan for the requirements to achieve technology qualification for the Gulf of Mexico. The key feature of API 17TR8 recommended design process is that the main requirements are outlined in an easy to follow "Flow Chart" and covers 5 aspects of the design process. These processes require some detailed support mechanical and corrosion testing. The process is

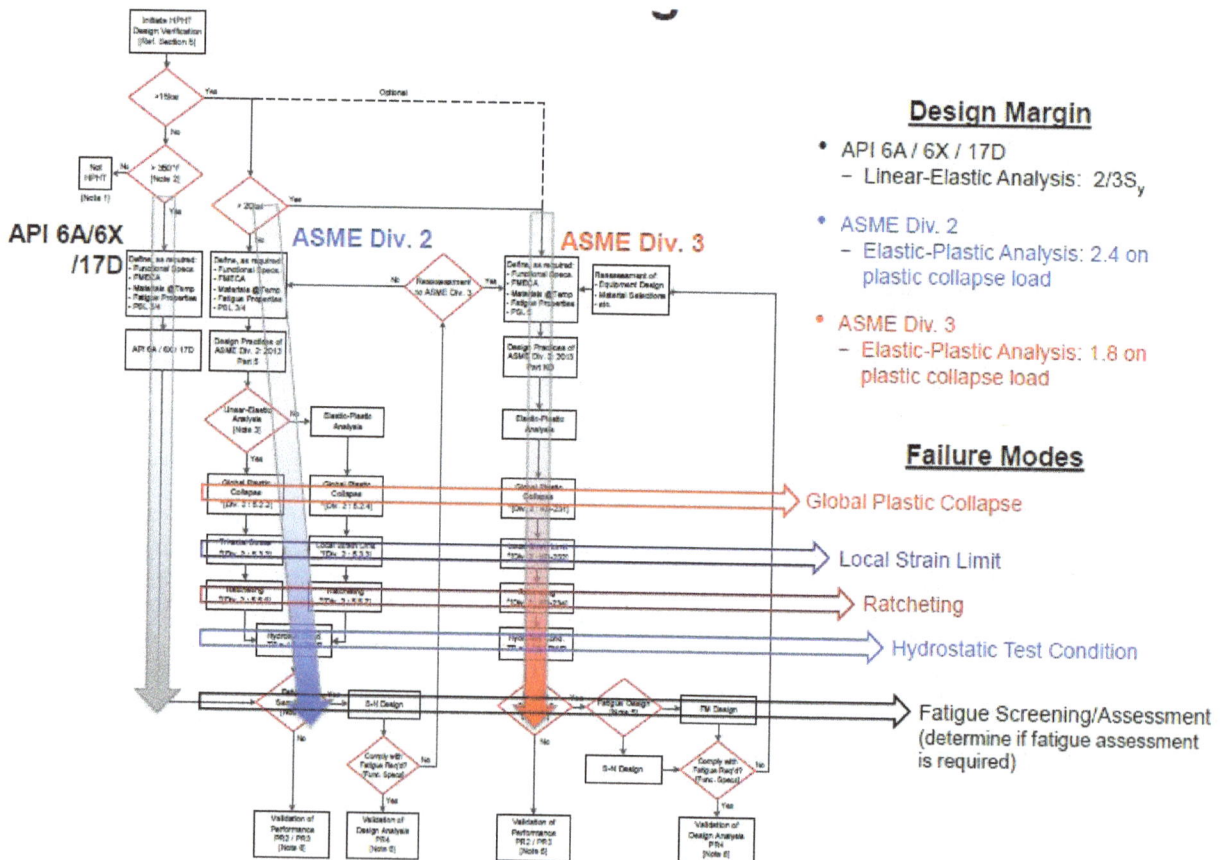

Figure 7.67 Pathway through the flow diagram for each of the codes

based on actual material results rather than test book assumptions. This specification mandates EAC testing to quantify the susceptibility of the materials to the environment and to obtain engineering design parameters such as allowable stress, fracture toughness and crack growth rates.

This has been summarised in a presentation by Man Phan of BP (2015) (Figure 7.56) which shows the pathway through the flow diagram for each of the codes.

7.5 CORROSION TESTING

7.5.1 Definitions and mechanisms

Sulphide Stress Cracking (SSC): SCC is a type of hydrogen embrittlement cracking caused by the absorption of hydrogen produced by acid corrosion of the metal surface. High hardness materials and hard HAZ weld zones are susceptible to SSC. In SSC H_2S is the main source of hydrogen which differentiates the process from other hydrogen embrittlement mechanisms. The presence of H_2S and CO_2 can lower the pH to values below the depassivation pH of the alloy increasing the rate of proton discharge. The depassivation also occurs in pits and crevices, therefore localised corrosion can initiate the failure. The presence of sulphide also prevents the formation of hydrogen molecules and for more hydrogen to be absorbed into the metal. SSC is a cathodic process and can be accelerated by an applied cathodic polarisation. Therefore, SSC can be made worse if coupled to a less noble metal.

Hydrogen-Induced Cracking (HIC): Stepwise internal cracks that connect adjacent hydrogen blisters on different planes in the metal, or to metal surface (also known as stepwise cracking). In steels, the development of internal cracks (sometimes referred to as blister cracks) tends to link with other cracks because of internal pressure resulting from the accumulation of hydrogen. The link-up of these cracks on different planes in steels is often referred to as "stepwise cracking" to characterize the nature of the crack appearance. HIC is commonly found in steels with (a) high impurity levels that have a high density of large planar inclusions and/or (b) regions of anomalous microstructure produced by segregation of impurity and alloying elements in the steel. No externally applied stress is needed for the formation of HIC.

Stress-Oriented Hydrogen-Induced Cracking (SOHIC): Arrays of cracks in steels, aligned nearly perpendicular to the applied stress, that are formed by the link-up of small HIC cracks in the steel. Tensile stress (residual and/or applied) is required to produce SOHIC. SOHIC is commonly observed in the base metal adjacent to the heat-affected zone (HAZ) of a weld and is oriented in the through-thickness direction. SOHIC may also be produced in susceptible steels at other high stress points such as from the tip of mechanical cracks and defects and from the interaction between HIC on different planes in the steel.

Hydrogen Blistering: The formation of subsurface planar cavities, called hydrogen blisters, in a metal resulting from excessive internal hydrogen pressure. Growth of near surface blisters in low-strength metals usually results in surface bulges.

As in SSC, hydrogen blistering in steel involves the absorption and diffusion of atomic hydrogen produced on the metal surface by the sulphide corrosion process. The development of hydrogen blisters in steels is caused by the accumulation of hydrogen that recombines to form molecular hydrogen at internal

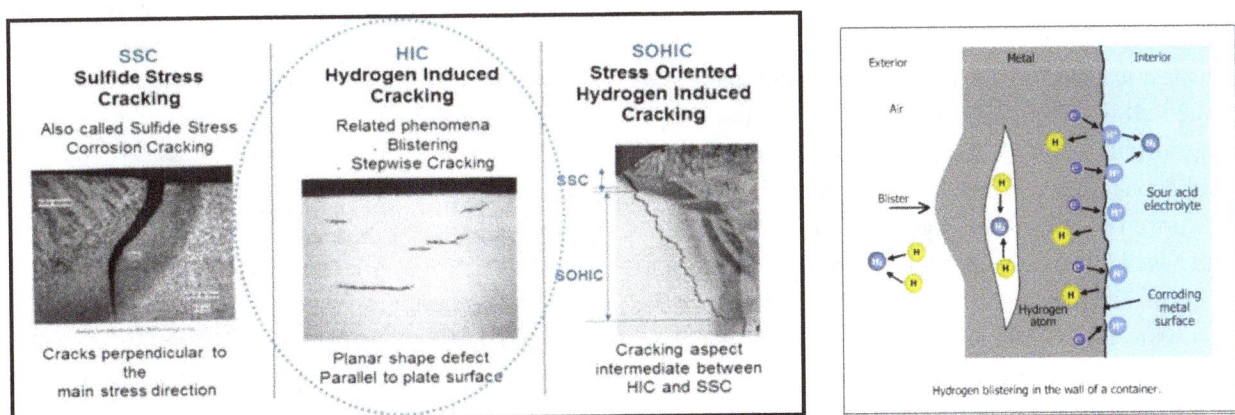

Figure 7.68 Corrosion related problems

sites in the metal. Typical sites for the formation of hydrogen blisters are large non-metallic inclusions, laminations, or other discontinuities in the steel. This differs from the voids, blisters, and cracking associated with high temperature hydrogen attack.

Stress Corrosion Cracking (SCC): Cracking of a material produced by the combined action of corrosion and tensile stress (residual or applied). Stress corrosion cracking is a form of localised corrosion involving the dissolution due to mechanical or residual micro-creep depassivation, SCC is an anodic process that requires the presence of a passive film. SCC is controlled mainly by the stability of the passive film and it is therefore sensitive to the pH temperature and the halide anion content of the environment.

7.5.2 Laboratory corrosion test techniques for assessment of pitting corrosion

Laboratory tests are an effective and established method for measuring the rate of corrosion. These methods are used for quality control purposes and also for the study of a corrosion mechanism.

There are several test methods that range from a simple immersion test such as ASTM G48 to an electrochemical test such as ASTM G61. Both these test methods are based on an accelerated test method.

7.5.2.1 ASTM G48 Ferric Chloride Test

ASTM G48 Test Method, "Standard Test Methods for Pitting and Crevice Corrosion Resistance of Stainless Steels and Related Alloys by Use of Ferric Chloride Solution".

Method A is Ferric Chloride pitting test.

This test is carried out in accordance with ASTM G48. This test is used for the testing of stainless steels and related alloys (including Ni-base alloys containing a large amount of Cr). Test is for determining pitting (and crevice) corrosion resistance property. Material is exposed to a 6% by weight $FeCl_3 \cdot 6H_2O$ solution, which is highly oxidizing, concentrated metal chloride solution. Testing time is 24 to 72 hours. Temperature for this test can be room temperature (22±2° C) or higher temperature (50°±2° C).

Mechanism of the test is based on the fact that the ferric chloride forms Fe_3+/Fe_2+ redox couple that acts

Figure 7.69 ASTM G48

as a chemical potentiostat. Potential of this couple is +0.45 V (SCE). Solution contains a high concentration of ferric ion. A reduction reaction of ferric to ferrous ion occurs on exposed metal surface. This is cathodic reaction. Other parameters that accelerated aggressiveness of solution are high chloride concentration, high temperature, very low pH of a solution (approximately around 1.3). When the potential of the solution exceeds the pitting potential of tested material pits form.

Ferric chloride test can be modified with changing temperature or exposure time.

100g of reagent grade ferric chloride $FeCl_3 \cdot 6H_2O$ was dissolved in 900 ml of distilled water (6% $FeCl_3$ by mass). The volume required to give at least 5 ml/cm^2 of surface area for specimen tested. The pH was maintained during the test in accordance to G48. Each test used a new solution.

Evaluation of pitting corrosion after Ferric Chloride Test

ASTM G46 provides guidance for the examination and evaluation of pitting corrosion. The examination consists of visual examination or low power microscope to inspect the material surface. The size, density and shape of the pits are determined and recorded. Metallographic examination can also be used to determine the cause of the pitting. The degree of pitting can be evaluated by measuring the change in weight or by the measurement of pit depths.

7.5.2.2 ASTM G61 Electrochemical testing

ASTM G61 "Standard Test Method for Conducting Cyclic Potentiodynamic Polarization Measurements for Localized Corrosion Susceptibility of Iron-, Nickel-, or Cobalt-Based Alloys".

There are several advantages of using electrochemical testing. It is an efficient method; corrosion can be studied in solution of interest rather than in less relevant environment and useful information can be collected for critical potential for initiation of pitting corrosion (or other localized corrosion).

Cyclic Potentiodynamic Polarization Test

Procedure for this test is described in ASTM G 61. This test is performed using a potentiostat which has three electrodes in the cell. The procedure according to ASTM G61 was 34g of Sodium Chloride was dissolved in 920ml of distilled water (3.56% by weight).

- Working electrode – this is metal under test
- Axillary electrode – this supplies the current to the working electrode
- Reference electrode.

In cyclic potentiodynamic polarization, the potential applied on working electrode increases with time and the current is measured. When analysing the cyclic polarization curves attention is given on two features: the pitting (breakdown) potential Epit and protection (re-passivation) potential Ep. Potential at which anodic current increases significantly is called pitting potential.

7.5.2.3 ASTM A262 Methods for detecting the susceptibility of an Austenitic Stainless Steels to Intergranular Attack

These practices cover the following five tests and the initial screening is done by the first test "practice A".

ASTM A262 is the specification which outlines five practices used to determine if an austenitic

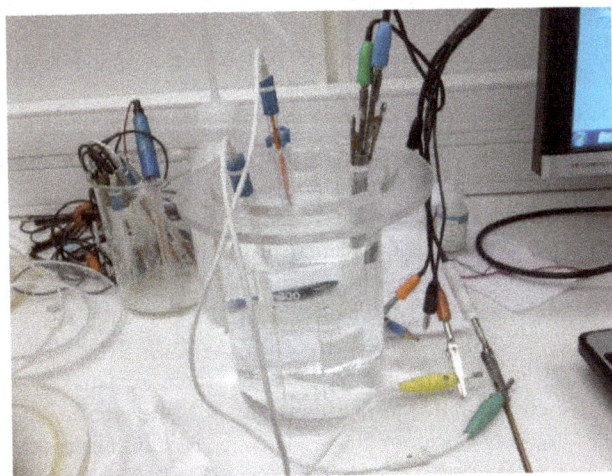

Figure 7.70 *ASTM G61*

micro-structure is susceptible to intergranular corrosion. This is not for determining resistance to Stress Corrosion Cracking (SCC) which is a transgranular attack. There are several causes for SCC, Chloride is the leading cause of transgranular cracking or attack. This specification is for detecting susceptibility to intergranular attack (IGA). Intergranular corrosion is the proper use; though the specification calls it attack, corrosion is what is taking place. Intergranular simply means that the corrosion is taking place between the grains or crystals, which is where sigma phase or chromium carbides are going to form, which makes the material susceptible to IGA. All austenitic stainless steels should meet this requirement, if a proper manufacturing route has been followed.

Practice A – Oxalic Acid Etch Test for Classification of Etch Structures of Austenitic Stainless Steels. Practice A is a rapid screening examination of the microstructure to quickly determine if the structure is certain to be free of susceptibility to rapid intergranular attack. The samples are etched after metallographic preparation for cross-sectional examination which is thoroughly viewed with a traverse from inside to outside diameters of rods and tubes, from face to face on plates, and across all zones such as weld metal, weld-affected zones, and base plates on specimens containing welds. Typical examination magnification is 200X to 500X. Classification of structure then provides either acceptance or further testing required, which typically moves you to one of the next practices. If the structure is acceptable no additional testing is required. Based on the microstructural appearance and standard photographs in the standard the specimen can be graded.

Extra-low-carbon grades, and stabilized grades, such as 304L, 316L, 317L, 321, and 347, are tested after sensitizing heat treatments at 650 to 675°C (1200 to 1250°F), which is the range of maximum carbide precipitation. These sensitizing treatments must be applied before the specimens are submitted to the oxalic acid etch test. The most commonly used sensitizing treatment is 1 h at 675°C (1250°F).

Practice B – Ferric Sulphate–Sulphuric Acid Test for Detecting Susceptibility to Intergranular Attack in Austenitic Stainless Steels.

Practice C – Nitric Acid Test for Detecting Susceptibility to Intergranular Attack in Austenitic Stainless Steels.

Practice E – Copper–Copper Sulphate–Sulphuric Acid Test for Detecting Susceptibility to Intergranular Attack in Austenitic Stainless Steels.

Practice F – Copper–Copper Sulphate–50 % Sulphuric Acid Test for Detecting Susceptibility to Intergranular Attack in Molybdenum-Bearing Cast Austenitic Stainless Steel.

Final Perspective – An interesting final perspective presented by Alan Weisman in his 2007 book "The World Without Us", (Alan Weisman, The World Without Us, Thomas Dunne Books, New York, 2007.) regarding planet Earth without us, suggested that copper and its alloys are the structural materials most likely to survive for thousands of years in a world suddenly depopulated of human beings. Some short-term indicators are the Chernobyl and the DMZ in Korea, although these have only been uninhabited for a mere 30 and 70 years. This conclusion, based on a strictly thermodynamic criterion (copper being the most noble structural metal and can be found un-combined in nature). This may not be correct since some invented materials, like stainless steel, will also endure for thousands of years because their surface is protected by passivating oxide films.

REFERENCES

1. RESEARCH OPPORTUNITIES IN CORROSION SCIENCE AND ENGINEERING, United States National Academy of Sciences 2011, http://www2.me.rochester.edu/projects/QGroup/assets/Documents/Research%20Opp%20in%20Corrosion%20by%20NAP.pdf

2. Baorong Hou et al, The cost of corrosion in China, npj Materials Degradation (2017), https://pdfs.semanticscholar.org/cc2b/e08884cd54d45ad0c77694f9e956b85e16fd.pdf

3. http://www.corrosion-doctors.org/InternetResources/NPL.htm#Beginners

4. Corrosion Factors and Corrosion Cells, https://pdfs.semanticscholar.org/03a6/569ccf7d024ca453586cb130fe61e8e7846c.pdf

5. Hilti Corrosion Handbook, https://www.hilti.pt/medias/sys_master/h34/h30/9157274632222/Hilti_Corrosion-Handbook_EN.pdf

6. Frederick Pessu et al, Localized and general corrosion characteristics of carbon steel in H2S environments, NACE International Corrosion Conference and Expo 2019, 24-28 Mar 2019, Nashville, Tennessee, USA. NACE, International. ISBN 9781510884670, http://eprints.whiterose.ac.uk/144402/1/Localized%20and%20general%20corrosion%20characteristics%20of%20carbon%20steel%20in%20H2S%20environments.pdf

7. Robert E. Melchers, A Review of Trends for Corrosion Loss and Pit Depth in Longer-Term Exposures, Corros. Mater. Degrad. 2018, 1, 42–58, file:///C:/Users/user/Downloads/cmd-01-00004%20(1).pdf

8. A. de Rooij, Bimetallic Compatible Couples, ESA Journal 1989, Vol. 13 page 199, http://esmat.esa.int/publications/published_papers/bimetallic.pdf

9. Hosking, Niamh C., Next generation corrosion protection for the automotive industry. PhD thesis, University of Nottingham. 2008 http://eprints.nottingham.ac.uk/14514/1/585527.pdf

10. J. De Gruyter et al, Short Communication Corrosion due to differential aeration reconsidered, Journal of Electroanalytical Chemistry 506 (2001) 61–63, file:///C:/Users/user/Downloads/Mertens_JEC_2001_differential_aeration.pdf

11. G. S. Frankel, Pitting Corrosion of Metals A Review of the Critical Factors, Journal of the Electrochemical Society, Vol. 145, No. 6, 1998, pp. 2186-2198, https://pdfs.semanticscholar.org/85ee/a0856f78d89a54ce34539c0f857dae54d43a.pdf

12. Fong-Yuan Ma, Corrosive Effects of Chlorides on Metals, Chapter 6 Open access peer-reviewed Edited Volume, Pitting Corrosion, Edited by Nasr Bensalah, https://www.intechopen.com/books/pitting-corrosion

13. A. Prateepasen, Pitting Corrosion Monitoring Using Acoustic Emission, Chapter 3 Open access peer-reviewed Edited Volume, Pitting Corrosion, Edited by Nasr Bensalah, https://www.intechopen.com/books/pitting-corrosion

14. Shashanka Rajendrachari, Investigation of Electrochemical Pitting Corrosion by Linear Sweep Voltammetry: A Fast and Robust Approach, file:///C:/Users/user/Downloads/63585.pdf

15. Jamal Choucri et al, Corrosion Behavior of Different Brass Alloys for Drinking Water Distribution Systems, Metals 2019, 9, 649, file:///C:/Users/user/Downloads/metals-09-00649.pdf

16. Corrosion for Engineers, Dr. Derek H. Liste, Selective Leaching, https://canteach.candu.org/Content%20Library/20053208.pdf

17. Mehmet Emin Arıkan, Determination of Susceptibility to Intergranular Corrosion of UNS 31803 Type Duplex Stainless Steel by Electrochemical Reactivation Method International Journal of Corrosion, Volume 2012, http://downloads.hindawi.com/journals/ijc/2012/651829.pdf

18. Bo-Hee Lee, Intergranular Corrosion Characteristics of Super Duplex Stainless Steel at various Interpass Temperatures, Int. J. Electrochem. Sci., 10 (2015) 7535 – 7547, http://www.electrochemsci.org/papers/vol10/100907535.pdf

19. W. D. France, PREDICTING THE INTERGRANULAR CORROSION OF AUSTENITIC STAINLESS STEELS, 21st NACE Conference,

20. St. Louis, Missouri, March 15-19, 1965, https://apps.dtic.mil/dtic/tr/fulltext/u2/624017.pdf

21. Corrosion for Engineers, Dr. Derek H. Lister, Chapter 7: Intergranular Corrosion, https://canteach.candu.org/Content%20Library/20053207.pdf

22. Chloride stress corrosion cracking in austenitic stainless steel Assessing susceptibility and structural integrity HSE, http://www.hse.gov.uk/research/rrpdf/rr902.pdf

23. Mohammed Al-Rabie, OBSERVATIONS OF STRESS CORROSION CRACKING BEHAVIOUR IN SUPER DUPLEX STAINLESS STEEL 2011 School of Materials, A thesis submitted to The University of Manchester for the degree of Doctor of Philosophy in the Faculty of engineering and physical science, https://www.research.manchester.ac.uk/portal/files/54509978/FULL_TEXT.PDF

24. Stress Corrosion, Guides to Good Practice in Corrosion Control NPL 1982, https://www.iims.org.uk/wp-content/uploads/2014/03/stress_corrosion_cracking.pdf

25. Roshan Kuruvila, A brief review on the erosion-corrosion behaviour of engineering materials, Corros Rev 2018, file:///C:/Users/user/Downloads/CorrosionReviews%20(1).pdf

26. Maija Raunio, Basic approaches and goals for crevice corrosion modelling, RESEARCH REPORT VTT-R-02078-15, 2015, https://www.vtt.fi/inf/julkaisut/muut/2015/VTT-R-02078-15.pdf

27. Ryo MATSUHASHI, Estimation of Crevice Corrosion Life Time for Stainless Steels in Seawater Environments, NIPPON STEEL TECHNICAL REPORT No. 99 SEPTEMBER 2010 https://www.nipponsteel.com/en/tech/report/nsc/pdf/n9911.pdf

28. N. Ebrahimi et al, The Role of Alloying Elements on the Crevice Corrosion Behavior of Ni-Cr-Mo Alloys, CORROSION—Vol. 71, No. 12 page 1441, https://www.surfacescienceswestern.com/wp-content/uploads/co15_shoesmithds.pdf

29. Douglas Godfrey, Fretting Corrosion or False Brinelling? DECEMBER 2003 TRIBOLOGY & LUBRICATION TECHNOLOGY, https://www.stle.org/images/pdf/STLE_ORG/BOK/LS/Bearings/Fretting%20Corrosion%20or%20False%20Brinelling_tlt%20article_Dec03.pdf

30. P. Sofronis, HYDROGEN EMBRITTLEMENT OF PIPELINE STEELS: CAUSES AND REMEDIATION, https://www.energy.gov/sites/prod/files/2014/03/f12/09_sofronis_pipe_steels.pdf

31. Jonathan A. Lee, Hydrogen Embrittlement, NASA/TM-2016–218602, https://ntrs.nasa.gov/archive/nasa/casi.ntrs.nasa.gov/20160005654.pdf

32. Marina Cabrini, Hydrogen Embrittlement Evaluation of Micro Alloyed Steels by Means of J-Integral Curve, Materials 2019, 12, 1843, file:///C:/Users/user/Downloads/materials-12-01843.pdf

33. G. Gobbi et al, A cohesive zone model to simulate the hydrogen embrittlement effect on a high-strength steel Frattura ed Integrità Strutturale, 35 (2016) 260-270, https://re.public.polimi.it/retrieve/handle/11311/983490/106685/1652-6196-1-SM.pdf

34. Rawia Mansour et al, Role of Microorganisms in Corrosion Induction and Prevention, BBJ, 14(3): 1-11, 2016; Article no.BBJ.27049, https://www.researchgate.net/publication/305344331_Role_of_Microorganisms_in_Corrosion_Induction_and_Prevention

35. Corrosion Control in the Oil and Gas Industry, Sankara Papavinasam, 2014, ISBN: 978-0-12-397022-0,

36. Overview of Corrosion in the Oil and Gas Industry, James Skogsberg (modified by R C John) April 26, 2017

37. Materials and corrosion trends in offshore and subsea oil and gas production, Mariano Iannuzzi et al, npj Materials Degradation (2017) 2,

38. Corrosion for everybody, Alec Groyman, Springer 2010.

39. Corrosion and Degradation of Metallic Materials, Francois Ropital, IFP Publications, 2010.

40. THE KASHAGAN OIL BUBBLE THE CASE OF AN OFFSHORE FIELD DEVELOPMENT IN KAZAKHSTAN. The 2017 Report prepared by Crudeaccountability. http://crude-accountability.org/wp-content/uploads/ENG_Kashagan_report_Final1-1.pdf

41. Corrosion Product Analysis – A Road Map to Corrosion in Oil and Gas Production, Materials Performance, August 2002, p2.

42. Becky L. Ogden, SULFIDE STRESS CRACKING – PRACTICAL APPLICATION TO THE OIL AND GAS INDUSTRY, Southwest Petroleum Short Course, Texas Tech University, 2005 https://www.halliburton.com/content/dam/ps/public/multichem/contents/Papers_and_Articles/web/Sulfide-Stress-Cracking-SWPSC-2005.pdf

43. https://www.lubbockonline.com/local-news/2010-09-15/denver-city-remembers-h2s-tragedy

44. http://bayanbox.ir/view/9011601610622885916/NACE-MR0103-MR0175-A-Brief-History-and-Latest-Requirements.pdf Page 6

45. Modelling CO2 Corrosion of Pipeline Steels Muhammad Hashim Abbas 2016, https://theses.ncl.ac.uk/dspace/bitstream/10443/3530/1/Abbas%2C%20M.H%202016.pdf

46. Material selection research into casings in natural gas wells in a high-corrosion environment, http://journal.it.cas.cz/62(2017)-LQ/199%20Paper%20Yifei%20Yan.pdf

47. Mariano Iannuzzi et al, Materials and corrosion trends in offshore and subsea oil and gas, Norwegian University of science and technology

BOOKS

1. Philip A. Schweitzer P.E, Fundamentals of Metallic Corrosion, Atmospheric and Media Corrosion of Metals 1st Edition 2006

2. Edited by D. Féron, European Federation of Corrosion Publications NUMBER 50, Corrosion behaviour and protection of copper and aluminium alloys in seawater 2007.

3. Harold M. Cobb, The History of Stainless Steel, ASM International 2010

4. Sankara Papavinasam, Corrosion Control in the Oil and Gas Industry, Elsevier, 2014

5. BP'S PIPELINE SPILLS AT PRUDHOE BAY: WHAT WENT WRONG? Hearing Before subcommittee on oversight and investigations of the committee on energy and commerce of the house of representatives. September 7th 2006

6. European Federation of Corrosion Publications, NUMBER 26, Advances in Corrosion Control and Materials in Oil and Gas Production, Papers from EUROCORR '97 and EUROCORR '98, Edited by P. S. JACKMAN AND L. M. SMITH, Published The Institute of Materials.

7. H.M. Shalaby A. Al-Hashem M. Lowther J. Al-Besharah (Editors) INDUSTRIAL CORROSION AND CORROSION CONTROL TECHNOLOGY Published by Kuwait Institute for Scientific Research, 1996.

8. Pierre R Roberge, Corrosion Engineering Principles and Practice McGraw Hill 2008

9. H Uhlig, Corrosion and corrosion control, Wiley 2008.

10. Stress Corrosion Cracking, Guides to Good Practice in Corrosion Control, NPL

11. Bimetallic Corrosion, Guides to Good Practice in Corrosion Control, NPL

12. Volkan Cicek, Cathodic Protection Industrial Solutions for Protecting Against Corrosion, Pub. WILEY

QUESTIONS

1. What is corrosion?

2. What is the cause of the corrosion current flow that occurs due to a corrosion reaction?

3. What is uniform corrosion?

4. What is pitting corrosion?

5. What is crevice corrosion?

6. What metals can be found on the planet as a metal?

7. What is stress cracking?

8. On average what is the annual loss due to corrosion as a percentage of a countries GDP?

9. One form of corrosion that occurs with austenitic stainless steel involves chromium carbide precipitates forming at grain boundaries. What is this known as? Explain this problem and how it is avoided?

10. What is the most common metal that can be used as sacrificial anode?

11. Name 3 metals or alloys that form an oxide scale that stops further corrosion?

12. What operational and manufacturing factors can result in to Sulphide Stress Cracking (SSC)?

13. In accordance with the requirements of MR 0175 what would be the maximum hardness for F6MN steel forging sullped to operate in accordance with NACE requirements?

14. A riser tube has to operate with a high level of are H_2S pressure what material would be the preferred choice?

15. What is dezincification?

16. How can dezincification be avoided?

17. The Kashagan project was a joint project of seven of the top oil companies. What went wrong? What was the estimate financial loss?

18. How can Microbial corrosion be prevented?

19. What factors influence CO_2 corrosion of a pipeline?

20. A pipeline has been operating for several years at a high level of H_2S. What problems could occur?

CHAPTER 8

Welding

8.1 INTRODUCTION

Welding processes have what appears to be a simple task of joining two or more materials together by the use of heat or pressure or both. However, the actual difficulty can be seen by the many competing process technologies that have evolved in an attempt to solve known problems and to find the ideal welding process. The processes used have not always delivered "fitness for service" and the solutions for many of the problems experienced have involved a considerable detailed metallurgical knowledge.[1 to 7]

Figure 8.1 shows a classification of the welding/joining processes. Figure 8.2 and Figure 8.3 the methods used to apply welding technology ranging from simple manual to fully adaptive control systems.

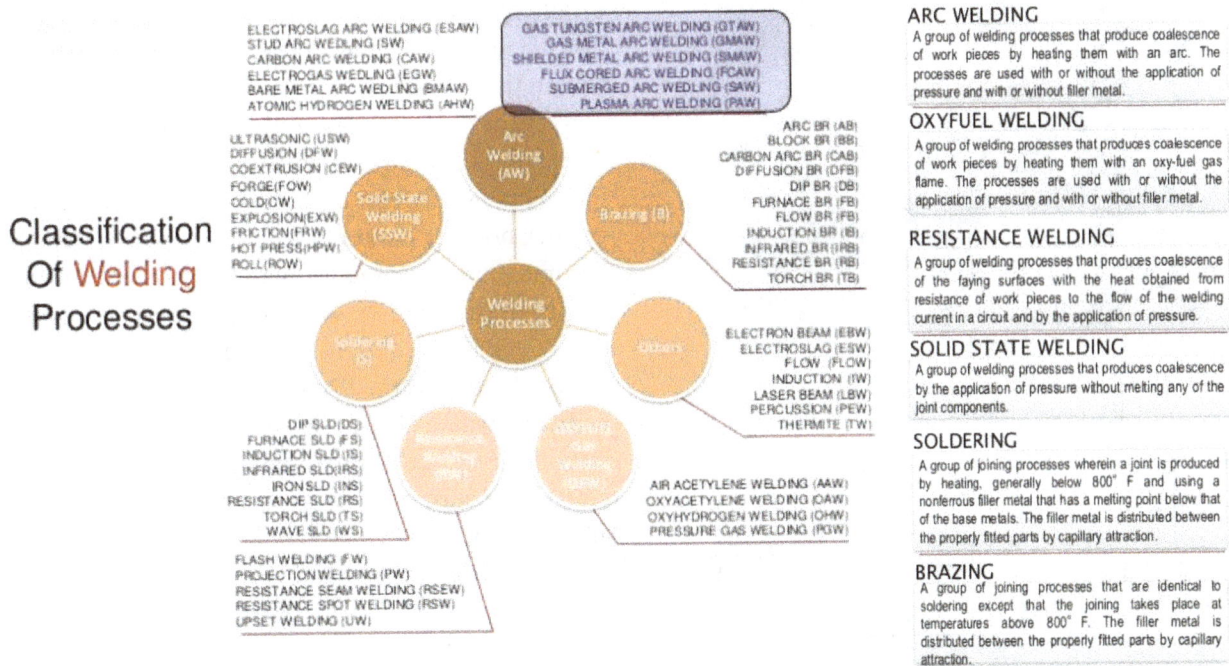

Figure 8.1 Classification of welding processes

MANUAL WELDING (MA)
Welding with the torch, gun, or electrode holder held and manipulated.

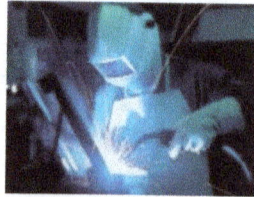

SEMI-AUTOMATIC WELDING (SA)
Manual welding with equipment that automatically controls one or more of the welding conditions.

MACHINE WELDING (ME)
Requires manual adjustment of the equipment controls in response to visual observation, with torch, gun or electrode holder by a mechanical device.

AUTOMATIC WELDING (AU)
Requires only occasional or no observation of the welding and no manual adjustment of the equipment controls.

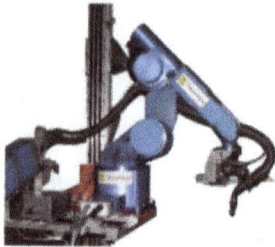

ROBOTIC WELDING (RO)
Welding that is performed and controlled by robotic equipment.

ADAPTIVE CONTROL WELDING (AD)
Welding with a process control system that determines changes in welding conditions automatically and directs the equipment to take appropriate action.

Figure 8.2 *Methods of welding application*

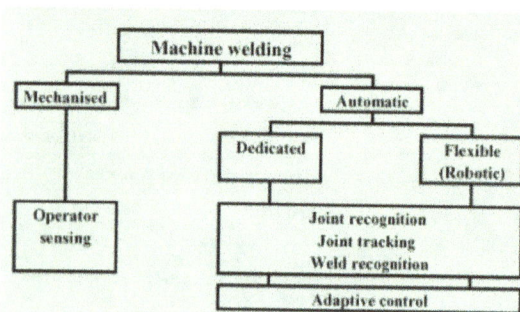

Figure 8.3 *Meta Vision Systems Digital Laser Scanner*

8.2 THE CONCEPT OF "WELDABILITY"

The American Welding Society (AWS) defines "weldability" as "the capacity of a metal to be welded under the fabrication conditions imposed into a specific, suitably designed structure and to perform satisfactorily in the intended services". Although this definition may not be universally acceptable it is worthy of note, because it refers to factors other than merely "the capacity of a metal to be welded". The terms "fabrication conditions imposed", "suitably designed structure", and "to perform satisfactorily in the intended service" are all relevant.

Obviously, for instance, it is not economical to apply preheat to every joint in the welding of a ship or bridge. It is therefore important for the steelmaker to be aware that simply "improving the capacity of a metal to be welded" will not solve all fabrication problems and that due regard should be given to other factors which may well be of greater importance (strength, toughness corrosion resistance). From a practical point of view a weldable material is one where the costs and difficulties involved in welding it is acceptable from an economic and delivery standpoint. Choosing materials of construction, therefore, becomes a matter of judgement between basic material costs, alternative welding processes available, and the quality requirements and strategic nature of the final fabrication. The industry prefers a holistic approach where all aspects are considered. For steels Graville developed a simple graphical approach which give a clear summary of the weldability of steel; this graph is shown in Figure 8.5.

Zone I steels have high hardenability, but carbon content is so low that even the hardest microstructure is not susceptible to cracking.

Zone II are shallow hardening but can develop sensitive microstructures because of their increased carbon content. Preheating the higher carbon/high CEV level is an effective means of reducing the cracking tendency.

Steels in Zone III combine high hardenability with high carbon content, compounding the problem considerably. In this instance extremely careful procedures, which include control of hydrogen, controlled cooling and post-weld heat treatment, may be required. This work further endorsed the use of the

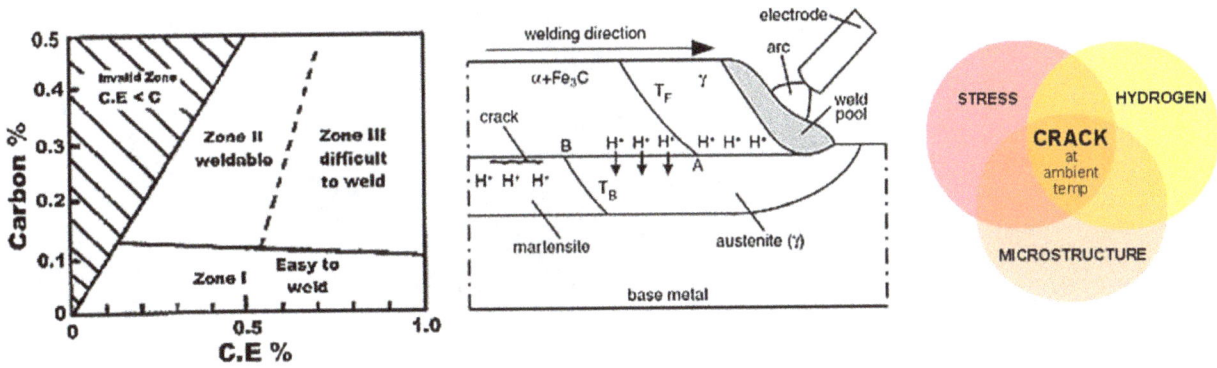

Figure 8.5 *LHS = Graville diagram showing weldability of steels with various carbon and alloy content. Centre = Schematic illustration of hydrogen diffusion in steels. RHS = Venn diagram*[8 and 9]

Table 1 *Carbon equivalent formula*[10 to 13]

Carbon equivalent formula	Application range according to Talas et al. [11]	Application range according to Yurioka et al. [3]	Reference
Group A			
$CE_{IIW} = C + \frac{Mn}{6} + \frac{Ni+Cu}{15} + \frac{Mo+V}{5}$	C-Mn steels with high CE content	$C \leq 0.08\%$	13
$CE_{WES} = C + \frac{Si}{24} + \frac{Mn}{6} + \frac{Ni}{40} + \frac{Cr}{5} + \frac{Mo}{4} + \frac{V}{14}$			14
Group B			
$CE_{DNV} = C + \frac{Si}{24} + \frac{Mn}{10} + \frac{Ni+Cu}{40} + \frac{Cr}{5} + \frac{Mo}{4} + \frac{V}{10}$	Steels with lower CE contents	$0.08\% \leq C \leq 0.12\%$	15
$CE_T = C + \frac{Mn}{10} + \frac{Cu}{20} + \frac{Ni}{40} + \frac{Cr}{20} + \frac{Mo}{10}$			16
Group C			
$P_{cm} = C + \frac{Si}{30} + \frac{Mn}{20} + \frac{Cu}{20} + \frac{Ni}{60} + \frac{Cr}{20} + \frac{Mo}{15} + \frac{V}{10} + 5B$	Pipeline steels	$C \leq 0.12\%$	17
$CE_{PLS} = C + \frac{Si}{25} + \frac{Mn}{16} + \frac{Cu}{16} + \frac{Ni}{60} + \frac{Cr}{20} + \frac{Mo}{40} + \frac{V}{15}$			18
$CE_{HSLA} = C + \frac{Mn}{16} - \frac{Ni}{50} + \frac{Cr}{23} + \frac{Mo}{7} + \frac{Nb}{5} + \frac{V}{9}$			19
Group D			
$CE_N = C + f(C) * \left(\frac{Si}{20} + \frac{Mn}{6} + \frac{Cu}{15} + \frac{Ni}{20} + \frac{Cr+Mo+Nb+V}{5} \right)$ $f(C) = 0.75 + 0.25\tanh[20(C-0.12)]$	All steels	$C \leq 0.3\%$	20

Pcm* (index of crack susceptibility), initially developed by Ito-Bessyo, for steels in Zone I and Zone III.

The use of this understanding led to the development of "crack-free" steels which could be successfully welded under the most adverse conditions.

8.3 WELDING PROCESSES

Welding processes now in use have developed over a period of 136 years since the first arc welding with a carbon electrode on the 1880s. There have been many innovations along the development curve with the development of Plasma welding to fabricate the aluminium structure of the Space Shuttle without excessive distortion and the development of Friction Stir Welding at the TWI in the 1990s is now the main process for aluminium fabrication.[14 and 15] Many recent developments have concentrated on quality and performance in production. There is now more emphasis on the "intelligent welding

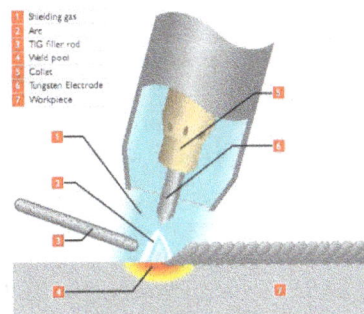

Figure 8.6 *TIG Tungsten Inert Gas process. GTAW was invented by Russell Meredith working at Northrop Aircraft Company in 1939-1941*

Figure 8.7 *MIG/MAG 1953 carbon dioxide was found to act as a protective atmosphere (providing the wire had deoxidising elements). Plus, fluxed core without gas 1954*

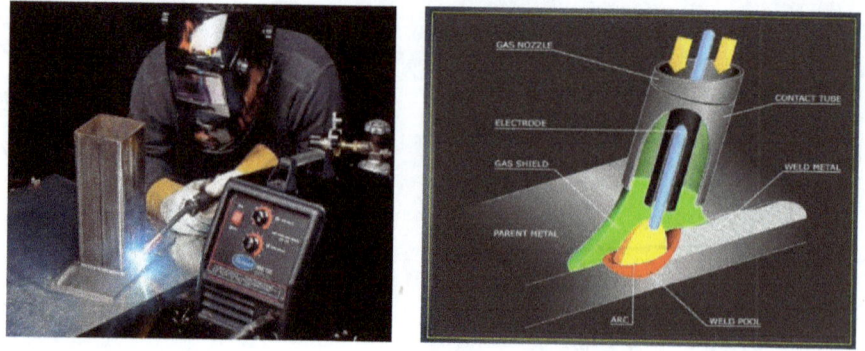

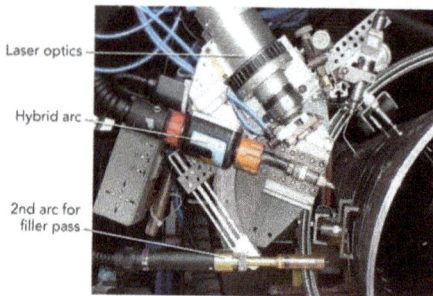

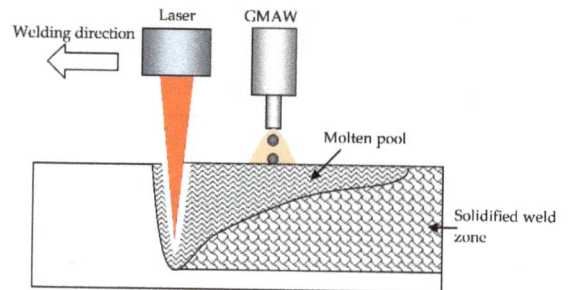

Figure 8.8 Hybrid laser-arc Bill Steen published a paper titled "Arc Augmented Laser Processing of Materials" in the Journal of Applied Physics as early as 1981. Early development work was slow because of the power limitation of the laser

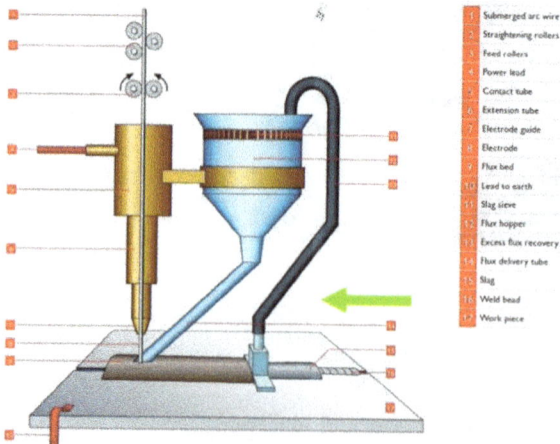

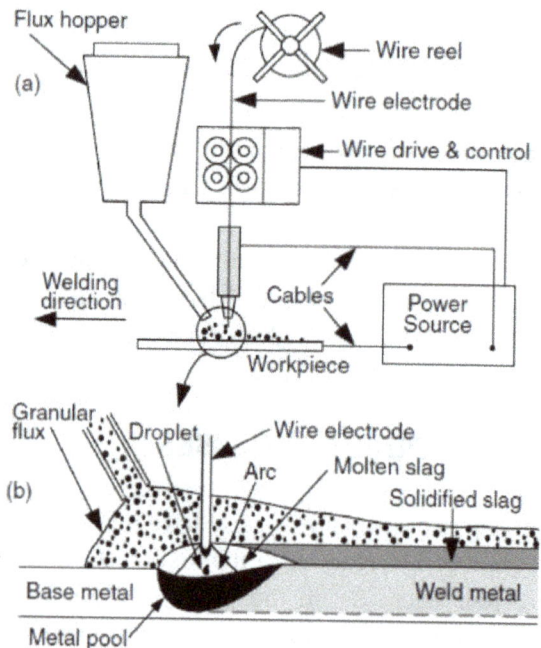

Figure 8.9 Submerged arc welding. Introduced in the USA in 1936. The process is normally limited to the flat or horizontal-fillet welding positions (although horizontal groove position welds have been done with a special arrangement to support the flux). Deposition rates approaching 45 kg/h have been reported — this compares to ~5 kg/h (max) for shielded metal arc welding

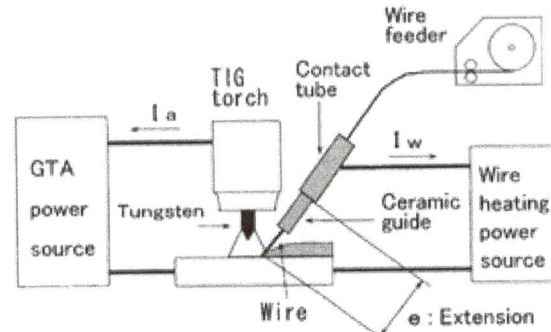

Figure 2. Principles of hot wire TIG

From Katsuyoshi, et al, 2003

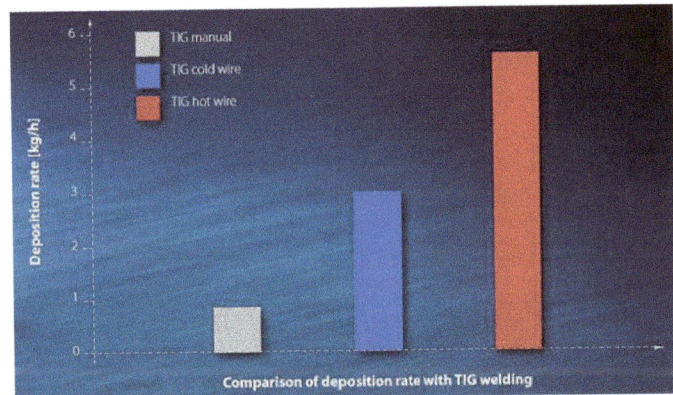

Figure 8.10 *Hot wire TIG Hot wire GTAW technology was invented in 1964 by A.F. Manz and developed by Linde. Hot wire technology is readily mechanized and automated and increased deposition rates are realized. The independent control of the arc and wire feed variables allow for flexibility in setting parameters*

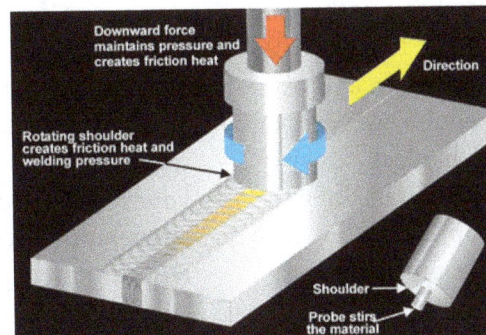

Figure 8.11 *Friction Stir Welding (FSW), invented by Wayne Thomas at TWI Ltd in 1991, overcomes many of the problems associated with traditional joining techniques. FSW is a solid-state process which produces welds of high quality in difficult-to-weld materials such as aluminium*

system". Productivity targets with greater deposition rates has favoured the development of the continuous electrode arc processes and many newer specialised processes which are suitable for automated operation.

8.4 THE METALLURGICAL EFFECTS OF FUSION WELDING

The fusion welding process has proven to be a very convenient method to join metal structures and is now an accepted method of fabrication for many metals in many industries. Codes and Standards have

played an important role which have resulted in a systematic approach to both weld procedure and welder approvals. This has allowed a closed loop control of both the technical aspects of welding and welder competency by introducing weld procedure specifications and procedure qualification records. The codes and standards therefore represent the main framework through which weld quality and integrity has improved, and development and progress has occurred.

The fusion welding process is relatively simple. The arc or laser heat source locally melts the parts to be joined together, followed by solidification to

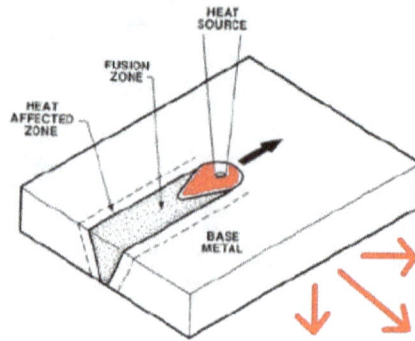

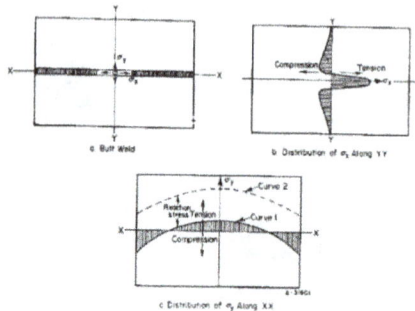

Figure 12.29 The development of the heat-affected zone in a weld: (a) the structure at the maximum temperature, (b) the structure after cooling in a steel of low hardenability, and (c) the structure after cooling in a steel of high hardenability.

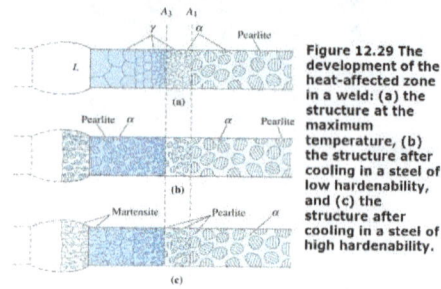

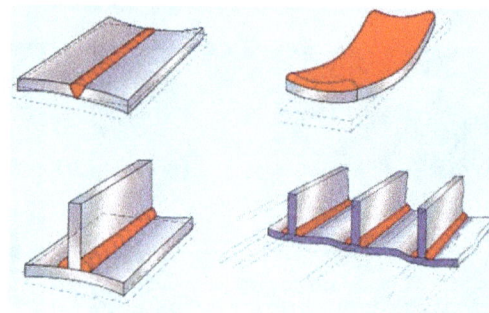

Figure 8.12 *The weld thermal input (Top LHS) causes microstructural changes and thus mechanical property changes (Top RHS), residual stresses (Lower LHS) that can assist brittle fracture, fatigue and environmentally assisted corrosion tendencies, and distortion (Lower RHS) which can result in rejection or addition stresses during subsequent assembly of for example sub-assemblies.*

produce metallurgical continuity between the two metal parts. However, the practical execution of a weld can involve several metallurgical mechanisms that can lead to a defective or imperfect weld. For example, if the shielding is not adequate, hydrogen and oxygen can be picked-up by the weld pool. The hydrogen and oxygen can reduce the mechanical integrity of the weld. The rapid re-solidification of the weld zone or weld metal can result in an "as cast" microstructures with all the associated cast metal integrity problems.

Fusion welding process contains four phases which include solid, liquid, gas and plasma. The plasma (charged particles) has a maximum temperature of 10,000K. The temperature of metal to be welded would be typically ambient temperature. The temperature of weld pool could reach a temperature of 3,000K. Therefore, there are large temperature differences available to cause distortion and residual stresses. In addition, the base metal on either side of the weld undergoes a local heat treatment which in the case of high hardenability base metal can cause high hardness values. In addition, the zone nearest the weld zone grain growth can occur and a subsequent reduction in toughness. The main effects to consider are:

- Thermal effects – Differential thermal expansion and contraction resulting is distortion and residual stresses – Computation methods and modelling can be used.
- Weld pool and shielding requirements gases, fluxes
- Weld metal solidification, composition and strength
- Four really five classical heat affected zones, High temperature, Coarse grained HAZ, Fine grained HAZ, Inter-critical HAZ, Sub-critical HAZ, Low temperature HAZ
- Weld imperfections-shape, fit-up, degree of penetration, slag, porosity.

Figure 8.13 Top LHS shows the various zones on in the HAZ that are easy to identify I a metallographically prepared sample through the weld zone. In addition, the examination of the weld metal in similar detail shows variation from the fusion zone into the parent metal. This is shown in Figure 8.13 Top RHS

Therefore, the weld metal region of the weld that is completely melted and solidified the microstructure will be dependent on weld metal composition including dilution effects and the solidification conditions. There can be local variations in composition and the weld may show three regions

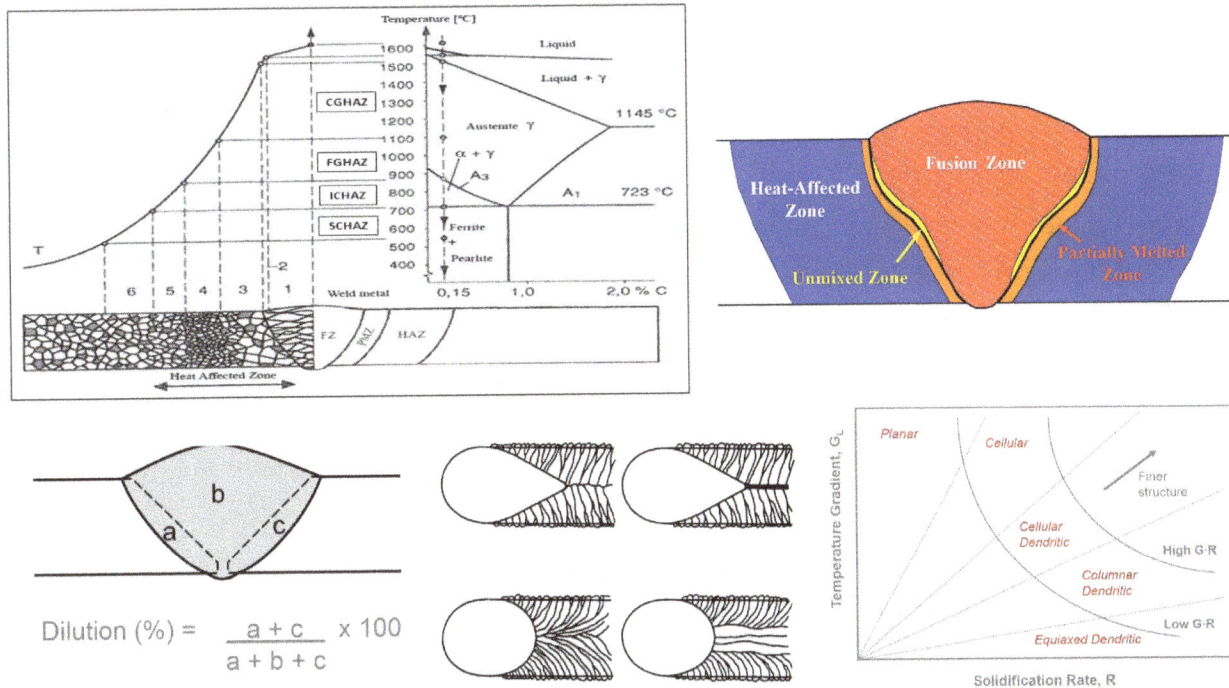

Figure 8.13 Top LHS = How the microstructural changes in the HAZ can be linked with the phase changes in the equilibrium diagram. RHS = The weld metal zone consists of Fusion zone, Composite zone and Unmixed zone (UMZ) The Heat-Affected Zone (HAZ) consists of a Partially-melted zone (PMZ) and the True heat-affected zone (T-HAZ). Lower LHS = Dilution, Centre = Weld Pool Shape, RHS = Weld metal solidification

- Composite zone
- Transition zone
- Unmixed zone

The other five classical heat affected zones are, High temperature, Coarse grained HAZ, Fine grained HAZ, Inter-critical HAZ, Sub-critical HAZ, Low temperature HAZ all have different problems.

- High temperature, Coarse grained HAZ (GCHAZ). Coarse grain size results in lower toughness. The grain coarsening that takes place will depend upon the grain coarsening characteristic of the material and the time at temperature.
- Fine grained HAZ (FGHAZ). This zone achieves a temperature of up to 50 to 100 degrees above the critical temperature AC3. This would have a refined that grain size, but it is also an area that has been austenitised and therefore the final hardness in the zone and the GCHAZ would depend upon the cooling rate. Techniques to slow the cooling would be pre-heat and weld heat input.
- Inter-critical HAZ (ICHAZ). This area has been heated between the AC1 and the AC3 and would therefore be partly transformed and likely to result in a dual phase structure or partly normalised microstructure structure.

- Sub-critical HAZ (SCHAZ). This part of the weld will undergo a tempering treatment and possibly a loss of hardness.
- Low temperature HAZ (LTHAZ). This part of the weld could have a problem of strain age embrittlement if the "free" nitrogen was at a high level. The mechanical strain and the low temperature would allow the nitrogen to "lock" dislocation and embrittle the metal.

Weld metal dilution

Figure 8.13 Lower LHS

Weld metal dilution occurs due to the amount of base metal that melts and mixes with filler metal This is typically expressed as percent base metal dilution of the filler metal. An autogenous weld would be 100% and in typical weld using filler metal would be 10 to 40%. This can have an effect on the weld metal microstructure and should be controlled by the weld joint design and welding process parameters

Weld pool shape

Figure 8.13 Lower centre

The weld pool shape depends upon the melting point and thermal conductivity, surface tension

(Marangoni effect) and therefore process parameters such as heat input and travel speed.

Weld metal solidification Homogeneous or Heterogeneous

Figure 8.13 Lower RHS

Homogeneous-A critical radius size which requires undercooling. Coarse microstructure would be expected (see Chapter 2 section 2.8.4.)

Heterogeneous-Nucleation occurs from existing substrate or particle and therefore little or no undercooling required. A fine microstructure would be expected.

Based on the above information it is easy to conclude that the thermal heat input from the arc can result in damage to the material being welded. Therefore, there is a need to control the variables that cause the damage. The control of these variables within a weld procedure by manufacturing a weld and then mechanical testing the weld is the main approach used by codes and standards.

There are a number of ways that the damage induced by the weld process can be mitigated:

*   Begin with such a high level of mechanical properties so that the drop in mechanical properties does not have significant effect on the service performance
*   Repair the damage by some method of post weld heat treatment
*   Control the heat input, pre-heat temperature and hardenability (CEV) to minimise the damage (welding nomograms).

8.5 WELD IMPERFECTIONS

Some of the types of weld imperfections that can occur are shown in Figure 8.14 and in the bullet points below:

*   Gas porosity (in weld deposit)
*   Solidification cracking (in weld deposit)
*   Liquation cracking (in weld deposit)
*   Lamellar tearing (HAZ)
*   Cold cracking or hydrogen inducted cold cracking susceptibility (HAZ)
*   Reheat cracking (Weld deposit and HAZ).

Gas porosity

Weld porosity results when gas is absorbed in molten metal and released as it solidifies. This gas can come from a number of sources such as being produced when contaminants on the weld surface are heated or shielding gas that gets trapped in the weld pool.

Nitrogen and oxygen absorption in the weld pool usually caused by poor gas shielding. As little as 1% air entrainment in the shielding gas will cause distributed porosity and greater than 1.5% results in gross surface breaking pores. Leaks in the gas line, too high a gas flow rate, draughts and excessive turbulence in the weld pool are frequent causes of porosity. Hydrogen can originate from a number of sources including moisture from inadequately dried electrodes, fluxes or the workpiece surface. Grease and oil on the surface of the workpiece or filler wire are also common sources of hydrogen.

The most common causes of porosity are atmosphere contamination, excessively oxidized work piece surfaces, inadequate deoxidizing alloys in the wire,

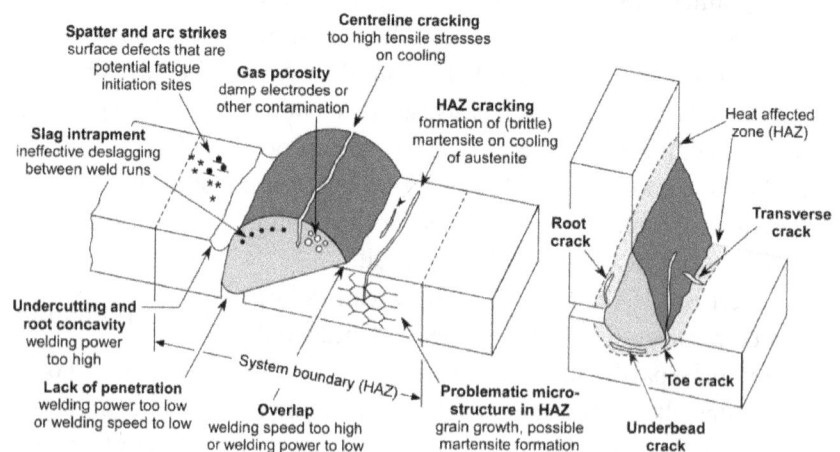

Figure 8.14 Potential weld imperfections

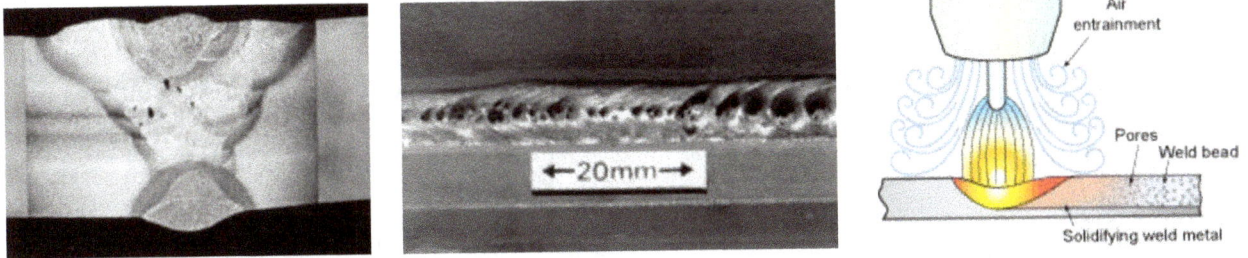

Figure 8.14a Uniformly distributed porosity, Surface breaking pores Formation during weld solidification From Weld porosity TWI job knowledge 042 https://www.twi-global.com/technical-knowledge/job-knowledge/defects-imperfections-in-welds-porosity-042

and the presence of foreign matter. Atmospheric contamination can be caused by (1) Inadequate shielding gas flow. (2) Excessive shielding gas flow. This can pull air into the gas stream. (3) Severely clogged gas nozzle or damaged gas supply system (leaking hoses, fittings, etc.) (4) An excessive draught in the welding area. This can disrupt the gas shield.

Solidification cracking

The use of higher heat input welding such as submerged arc can result in weld metal solidification cracking. This can have a serious impact of the cost of weld structures, especially where there is high restraint and high parent metal dilution, particularly in the welding of the second side of a double welded joint. The main controlling factors are the degree of weld restraint and the chemical analysis. The important factor is the carbon content of the weld pool which in the root pass can be significantly affected by the parent metal dilution. The influence of chemical composition on solidification cracking susceptibility can be assessed by using the cracking susceptibility formula (UCS):

$$UCS = 230C\% + 190S\% + 75P\% + 45Nb\% - 12.3Si\% - 5.4Mn\% - 1$$

C% represents the weld metal carbon level, unless lower than 0.08%C. Below 0.08 the value of 0.08 should be used. Experience has shown that UCS values greater than 19 to 25 for "T" fillet and butt welds respectively normally indicate potential solidification cracking problems with large structures.

Liquation cracking

The causes of liquation cracking are well understood and are associated with grain boundary segregation and the melting of grain boundaries near the fusion line. The presence of high residual stresses during weld cooling can rupture the impurity weakened grain boundaries.

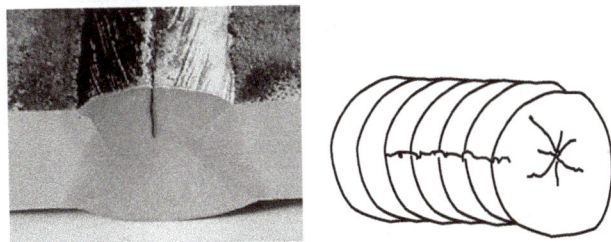

Figure 8.15 Solidification cracking – Contraction strains cause rupture of the weld at the point where the last material solidifies. Controlled by Solidification range Weld pool size and shape

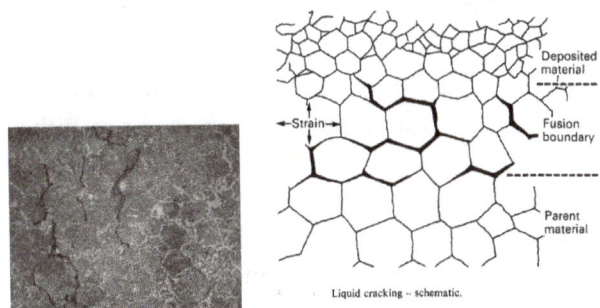

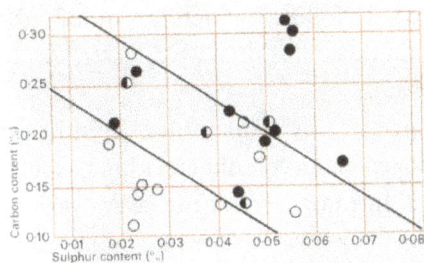

Figure 8.16 In high temp HAZ local melting/weak phase can be formed at the grain boundaries and crack due weld stress formed during weld cooling

Lamellar tearing

Caused by the weld stresses and the presence of planar non-metallic inclusions parallel to the plate surface. Lamellar tearing occurs as a result of low shot transverse tensile ductility.

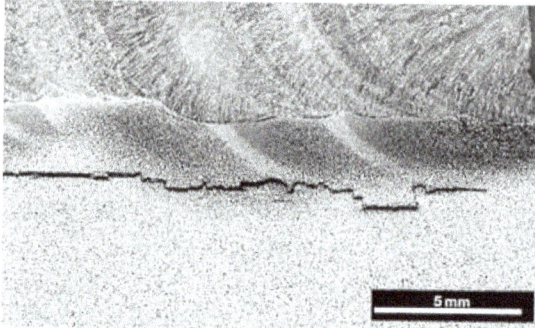

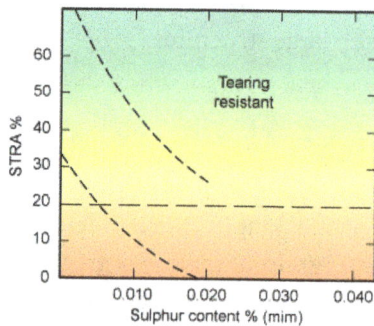

Figure 8.17 LHS = Lamellar tearing, RHS = Low sulphur needed to avoid lamellar tearing

Cold cracking

In ferritic steels the phase changes which accompany welding may lead to hardened structures in either the weld metal, HAZ or both. These hardened structures have low ductility, which is further reduced by the possible pick-up of hydrogen, unavoidably introduced into the weld during most fusion welding processes. Unless precautions are taken to modify the microstructure, reduce hydrogen level or both, cracking may occur following welding, under the influence of strain due to residual welding stresses or any applied stress.

The importance of chemical composition in the control of hydrogen-induced cold cracking has long been recognised. In practice over the years, various formulae have been evolved for the maximum compositional limits of steel which can be welded without risk of cracking and without preheat with normal welding conditions. One such formula was established by Dearden and O'Neill while working for British Rail who stated that if C+ Mn/6 was equal to or less than

0.43 then no cracks were formed during welding. In the UK the original formula has undergone modifications in the light of experience and the formula contained and referred to in British Standards is as follows:

CEV (Carbon Equivalent Value)

$$\mathbf{CEV} = C + \frac{Mn}{6} + \frac{Cr + Mo + V}{5} + \frac{Cu + Ni}{15}$$

This formula considers elements which may be present at a residual level and which are regarded as detrimental to weldability in their cumulative effect.

The choice of welding conditions for such steels can be made by reference to standards and a level of preheat may be required to avoid cracking depending on plate strength, joint combined thickness, CEV, heat input and expected hydrogen level in the weld metal. Welding nomograms can be used to predict preheat requirements.

With reference to cold cracking in higher tensile steels for ship structures:

When the CEV is > 0 .41	No special precautions are needed,
When the CEV < 0 .41 to 0.45	Use of low hydrogen electrodes is required,
If the CEV is >0.45	Use of low hydrogen electrodes and preheat are required.

Because of the restrictions to productivity inherent in the use of low hydrogen electrodes and the expense of using preheat, steel for shipbuilding is effectively restricted to a CEV of 0.41 maximum.

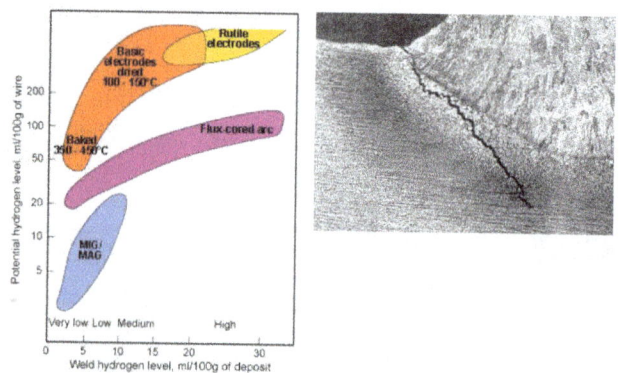

Figure 8.18 Hydrogen levels in welding processes and cold cracking

8.6 STRUCTURAL STEEL

During the 1960s significant progress was made in the UK on understanding the metallurgical effects of the chemical composition and microstructure on the mechanical properties and weldability of high strength structural steels. One of the main constraints to the use of the high strength steels was whether designers and fabricators would be able to make effective use of the higher strength which typically had reduced weldability. The answer to this point was sought at a number of UK conferences during the 1970s and 80s. The outcome has been that whilst such steels are important in specific applications including earth-moving equipment and pressure vessels, the main advance in terms of commercially produced tonnage steels halted at the 560MPa tensile strength, 320 MPa yield strength carbon/manganese steels. This can be compared with mild steel, found suitable to previous generations, at 410MPa tensile strength and 240 MPa yield strength.

This is in some part was the result of the considerable difficulty in utilising much higher strength, but more importantly because of the paramount importance of weldability where the cost of fabrication delays can be immense and the potential cost of structural failure unacceptable. Codes and standards form an essential agreement between all parties involved in fabrication and therefore represents the main framework through which development and progress can occur. Generally National specifications evolve gradually and cautiously in accordance with developments in manufacturing capability.

In the earliest specifications BS 548: 1934 the maximum allowable carbon was so high that the steel could hardly be considered easily weldable by today's standards. This was corrected in 1942 with the introduction of BS 968 which imposed restrictions on carbon and other elements. Subsequent to that date a number of catastrophic brittle fracture failures led to considerable work which enumerated the important factors controlling such failures and led to the introduction of impact property requirements in 1962.

More recent versions gave further refinements to a wider choice of impact properties, improving weldability by reduced carbon equivalent values. The pace of change in proprietary specification, however, can be rapid and is sensitive to the changes faced by designers and fabricators in the construction industry, i.e. nuclear plant, chemical process plant and off-shore oil and gas.

In the late sixties the decision to extract oil from the North Sea necessitated the construction of working platforms. These were 400-600 feet high and capable of withstanding wave heights of over 100 feet, winds of up to 100 mph and air temperatures down to -25°C.

The North Sea is now entering decline and several platforms have been built and in operation for many years. The initial specifications were based on BS 4360 Grade 50D, but gradually further restrictions were imposed such as impact properties down to -40°C, a carbon equivalent of 0.40 maximum and sulphur levels at 0.005 maximum, together with though thickness ductility guarantee of 35% reduction of area on the node steel grades.

8.7 ALLOY AND LOW ALLOY STEELS

The addition of alloying elements such as Ni, Cr and Mo, combined with suitable heat treatment low-alloy can provide better mechanical properties than carbon steels. Low alloy steels typically contain up to 5 percent total alloy content. The addition of molybdenum improves the hardenability and strength; nickel provides hardenability and improves the toughness; and chromium also adds to the hardenability and corrosion resistance.

The Graville diagram on Figure 8.5 remains one of the best indicators of the weldability of low alloy steels. With carbon levels below 0.1% the steels are classed as weldable. At higher carbon levels the CEV value needs to be taken into consideration and aspects of weld preheat and post weld heat treatment are essential. Given the chemical analysis and the selection of an appropriate weld technique and welding consumable welding procedures should be prepared and qualified.

Common Uses of Low-Alloy Steel

Low-alloy steels are used by many industries. Applications for low-alloy steels range from military vehicles, earthmoving and construction equipment, and ships to the cross-country pipelines, pressure vessels and piping, oil drilling platforms, and structural steel.

Several common groupings of low-alloy steels, beginning with HY 80, HY 90, and HY 100 steels, are used for building ship hulls, submarines, bridges, and

Figure 8.18 *The "Angel of the North" made with Corten grade of steel that develops a protective oxide https:// www.gateshead.gov.uk/article/5347/Technical*

off-highway vehicles. These low-alloy steels contain nickel, molybdenum, and chromium, which add to the material's weldability, notch toughness, and yield strength.

Another type of low-alloy steel – high-strength, low-alloy (HSLA) – is different from other low-alloy grades in that each type has been created to meet specific mechanical requirements rather than a given chemical composition. HSLA applications include warships, structural steel, and others known for their strength. Designed for strength, toughness at low temperatures, and ductility, ASTM A514, A517, steels are quenched and tempered and used in applications such as heavy equipment manufacturing and boiler and pressure vessel fabrication.

Weathering steels such as ASTM A242, A588, and A709 Grade 50W rely on certain alloys to produce a protective, corrosion-resistant layer. This layer also gives a weathered look to the finished steel and was first introduced as COR-TEN®. Weathered steels are popular in artwork, bridges, and as a facing material on buildings to achieve specific aesthetics.

One of the local metal structures where the weathering steel "Corten" was used was the "Angel of the North". Work began on the project in 1994 and cost £800,000. Most of the project funding was provided by the National Lottery. The Angel was finished on 16 February 1998. The sculpture was built at Hartlepool Steel Fabrications Ltd. The Company went out of business in 2002.

For oil and gas application the SAE 4130, 4140, 8630 modified and F22 are frequently the alloys of choice. In many thick cross sections, the standard compositions do not provide adequate hardenability and the base compositions are "modified". Currently ASTM is preparing a specification to cover these modified compositions.

Weld metal consumable selection

The weld metal consumables used to weld low-alloy steels (regardless of the specific type) are selected to match the chemical composition and mechanical properties of the base metal. Low-alloy weld metal consumables are classified by their tensile strength in ksi. The ideal weld metal consumable for low-alloy steels should match or exceed the base metal's tensile and yield strengths, as well as its elongation and toughness (Charpy V-notch) properties. A perfect match may not always be possible, however, so it is necessary to find the closest one possible. A joint that requires post weld heat treating (PWHT) benefits from a weld metal consumable alloyed with molybdenum to ensure that the deposited weld metal keeps its strength. Such applications include the PWHT of pressure vessels, which helps improve impact or toughness properties and reduce any residual stresses in the weld that could cause it to fail prematurely.

In application where high fatigue resistance is required such as earthmoving equipment, a weld metal with higher toughness would be beneficial. A weld metal alloyed with nickel provides greater resistance to cyclic loading and fatigue in such a situation, while also offering higher strength and better toughness than mild steel at low temperatures.

8.8 CURRENT KNOWN PROBLEMS

In service cracking of 625 overlay on SAE 8630 modified.

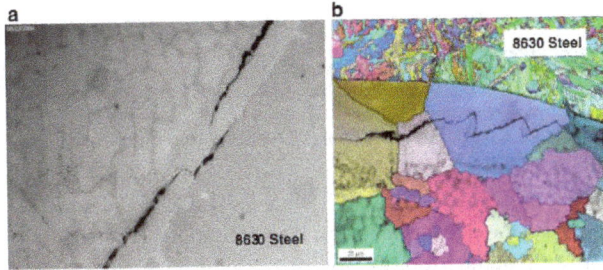

Figure 8.19 *Overlay cracking*

Stubborn tempering response some grades to obtain the NACE 22Rc F22, CA6MN

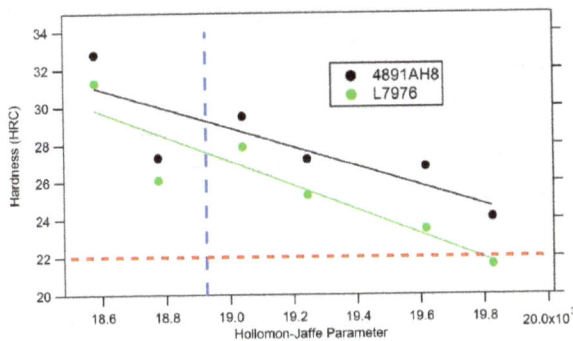

Figure 8.20 *Comparison of softening of two F22 grades*

Table 8.1 *Chemical analysis of the two F22 grades*

Element	Heat No.	
	4891AH8	L7976
C	0.167	0.161
Mn	0.42	0.75
P	0.011	0.008
S	0.011	0.012
Si	0.24	0.20
Ni	0.19	0.29
Mo	1.01	0.94
Cr	2.22	2.11
Cu	0.079	0.18
Al	0.005	0.023
V	0.015	0.006
Ti	0.011	0.002
Nb	0.001	0.001
B	0.0001	0.0003
Ca	0.0001	0.0048
O	0.0021	0.0051
N	0.017	0.017

8.9 EVALUATION OF STEEL WELDS

The established methods used to control weld quality involve the drafting of a WPS (Weld procedure specification). A weld would then be made using that procedure followed by weld testing. A typical flow diagram outlining the test work for ASME IX is shown below

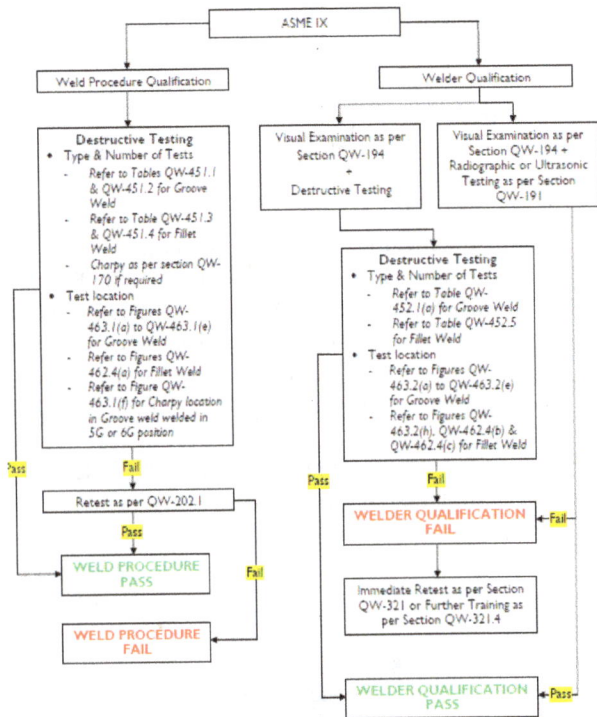

ASME IX

REFERENCES

1. P H M Hart Weld Metal Hydrogen Cracking in Pipeline Girth Welds, Proc. 1st International Conference, Wollongong, Australia, 1-2 March 1999. Published by Welding Technology Institute of Australia (WTIA), Silverwater, NSW, Australia, 1999, https://www.twi-global.com/technical-knowledge/published-papers/hydrogen-cracking-its-causes-costs-and-future-occurrence-march-1999

2. Peter Bernasovský, CASE STUDIES OF WELDED STEEL STRUCTURE FAILURES, Acta Metallurgica Slovaca – Conference, Vol. 3, 2013, p. 159-170 https://pdfs.semantic-scholar.org/ef01/de225b9fe0d9db83efc2236e0e001f10fec0.pdf

3. D. K. BHATTACHARYA, Failures of welded joints, PROCEEDINGS : COFA-1997 ©NML JAMSHEDPUR; pp. 212-220. http://eprints.nmlindia.org/1585/1/212-220.PDF

4. Dr Geoff Booth, Lessons from Catastrophic Weld Failures, Eastern Counties Branch of the Welding Institute lecture, Wednesday 16 March 2016. file:///C:/Users/user/

Downloads/G%20Booth%2016%203%202016%20-%20 Catastrophic%20Weld%20Failures.pdf

5. Göran Alpsten, Causes of Structural Failures with Steel Structures, http://www.stbk.se/1662c-paper34-iabse-2017-01-24.pdf

6. Christopher T. Mgonja, The Consequences of cracks formed on the Oil and Gas Pipelines Weld Joints, International Journal of Engineering Trends and Technology (IJETT) – Volume 54 Number 4 December 2017, file:///C:/Users/user/ Downloads/The_Consequences_of_cracks_formed_on_ the_Oil_and_G.pdf

7. The 50 Major Engineering Failures (1977-2007) Part-1 to 5. https://abduh137.wordpress.com/2008/05/05/ the-50-major-engineering-failures-1977-2007-last-part/

8. R. Kurji & N. Coniglio, Towards the establishment of weldability test standards for hydrogen-assisted cold cracking, – International Journal of Advanced Manufacturing Technology – Vol. 77, p.1581-1597 – 2015, https://hal. archives-ouvertes.fr/hal-01134401/document

9. Yong-Yi Wang et al, WELDABILITY OF HIGH STRENGTH AND ENHANCED HARDENABILITY STEELS, Proceedings of IPC2004 International Pipeline Conference October 4-8, 2004, Calgary, Alberta, Canada, IPC2004 526, http://citese-erx.ist.psu.edu/viewdoc/download?doi=10.1.1.823.4570&re p=rep1&type=pdf

10. Wesley Wang, The Great Minds of Carbon Equivalent Part lll: The Evolution of Carbon Equivalent Equations. EWI https://ewi.org/wp-content/uploads/2016/06/Great-Minds-of-Carbon-Equivalent_Part-3-Wang.pdf Also parts I and II

11. www.atlassteels.com.au/documents/CE%20calculator%20 rev%20July%202013p.xlsx

12. Doordarshi Singh et al, Inter-Relationship Between Weldability And Hardenability, Proceedings: International Conference on Advances in Mechanical Engineering-2006 (AME 2006), December 1-3, 2006,

13. P R Kirkwood, MODERN NIOBIUM MICROALLOYED STEELS AND THEIR WELDABILITY CBMM Technical Briefing, Copyright © 2016 CBMM, http://www.beta-metallurgy.co.uk/public/images/publications/htp/CBMM_ Technical_Briefing_Weldability_December_2016_External_ Version_FINAL.pdf

14. Mistry Hiten J et al, A REVIEW PAPER ON: FRICTION STIR WELDING (FSW), www.ijaresm.net, ISSN : 2394-1766, file:///C:/Users/user/Downloads/1.pdf

15. Terry Khaled, AN OUTSIDER LOOKS AT FRICTION STIR WELDING, REPORT #: ANM-112N-05-06 JULY 2005, Federal Aviation Administration, https://www.faa.gov/air-craft/air_cert/design_approvals/csta/publications/media/ friction_stir_welding.pdf

BOOKS

1. H Granyon, Fundamentals of welding metallurgy, Abington Publishers, 1991

2. Jean Cornu, Advanced Welding Systems, Fundamentals of fusion welding technology. IFS publications 1988

3. API Standard 1104, Twentieth Edition October 2005, Welding of pipelines and related Facilities

4. ASME Boiler and Pressure Vessel Code IX Welding and Brazing Qualifications, 2007 Edition

5. AWS D1.1/D1.1M:2010 An American National Standard, Structural Welding Code— Steel, 22nd Edition, Approved by the American National Standards Institute March 11, 2010

6. Sindo Kou, WELDING METALLURGY, SECOND EDITION, Pub John Wiley & Sons 2003

7. Klas Weman, Welding processes handbook, WOODHEAD PUBLISHING LIMITED, 2003

8. John Hicks, Welded design – theory and practice, Abington Publishing Woodhead Publishing Limited, 2000

9. A guide to weld inspection for structural steel, BCSA Publication No 54/12 Tata Steel

10. Kenneth Easterling, Introduction to the Physical Metallurgy of welding, Butterworth Heinemann, Second Edition 1092

11. John A. Goldak et al, COMPUTATIONAL WELDING MECHANICS, Pub Springer, 2005

12. Steven E Hughes, A Quick Guide to Welding and Weld Inspection, ASME Press

13. Nasir Ahmed Editor, New developments in advanced welding, WOODHEAD PUBLISHING LIMITED, 2005

14. ASM International, Welding Brazing and Soldering, Volume 6, 1996

15. API RECOMMENDED PRACTICE 577, SECOND EDITION, WELDING PROCESSES, INSPECTION, AND METALLURGY, 2013.

SELF-ASSESSMENT QUESTIONS

1. Alloy steels have added alloy elements. Answer the following questions:

    a) Name four alloy elements added to steel

    b) What is the difference between an alloy steel and a carbon steel?

    c) Why is manganese added to steel?

    d) How do you access the weldability of an alloy steel?

2. Define: Welding Metallurgy, Weldability, HAZ, Welding Reliability, Weld-Dilution, Residual Stresses, Preheating, Peak Temperature, Weld decay

3. Discuss briefly:

   i) Residual Stress and distortion
   ii) Weld imperfections
   iii) Differences between hot and cold cracks
   iv) Differences between crystalline changes and phase changes in steels.

4. Why is there excessive grain growth in the heat affected zone of a welded structure even if it is made of micro alloyed steel?

5. What are the precautions that need to be taken to weld creep resistant steel tubes having 0.15C, .5Mn, 0.54Si, 2.25Cr, 1.0Mo (this corresponds to ASTM T 22 grade steel)?

6. There is no chance of forming martensite in heat affected zone while welding austenitic steel. Should therefore all grades of austenitic steels be easily weldable? Give reasons for your answer.

CHAPTER 9

Metal Casting

9.1 INTRODUCTION

Near Net Shape Manufacture

A metal casting can be defined as a metal object formed when molten metal is poured into a mould which contains a cavity of the desired shape and is allowed to solidify. The casting process is used to create complex shapes that would otherwise be difficult or impossible to make using conventional manufacturing practices. The process can be adapted to use "cores" to allow internal detail to be cast followed by removal of the core. The process can be classified as near net shape manufacture and therefore should be one of the most competitive methods for part manufacture

A current common theme of value analysis (cost reduction) is the use of castings to replace fabrications both large and small and to replace steel with ductile and austempered ductile iron castings. Aluminium castings are gaining importance with the automotive industry due to the weight saving driven by potential fuel savings due to reduced vehicle weight.

A Brief History of Castings

An intriguing aspect of cast metal technology is that there is evidence that early Egyptian technology used investment casting technology. One important use of "investment casting" technology is that it is the main method of casting aerospace heat/creep resisting parts that cannot be made by other methods. Therefore, some speculate that this "know how" had been introduced by some alien travellers doing space craft repair!

One of the first castings on record is a copper frog from Mesopotamia, dated to around 3200 BC. Iron, a more common material cast today, was discovered in China around 2000 BC and first used iron in a casting process around 600 BC. Cast steel is a more recent development, although around 500 AD cast crucible steel was first produced in India; the process was lost until its rediscovery in England in 1750. Crucible steel was the first steel that was fully molten with a uniform composition in the melt. With the metal in a molten state this also allowed for alloy steel to be produced. The metal casting process has evolved since its first discovery. The industrial revolution brought about many new innovations. And since then melt practices, quality control, mould making practices and alloying have improved dramatically.

The Cast Metals Industrial base

The lean modern generation of Foundries are well automated and maintain high environmental standards. The plants typically include moulding, core making, melting, cleaning, maintenance, pattern making, quality and engineering departments. An additional important aspect of cast metals is that a significant part of the foundry industry has emigrated from the UK. Figure 9.1b shows the output from India and China. This can result in long chains of procurement which can result in potential production delays in the event of the delivery of non-conforming product. Therefore, initial appraisal of a foundry and validation of a process route are mandatory aspects of business security.

The Global Market in castings

Today metal castings are produced on a global scale. The following tables gives a view of the global metal casting market today, with China and India playing a major role in the production of metal castings.

Table 1 *World output of individual cast metals*

World Totals (metric tons)

Gray Iron	Ductile Iron	Malleable Iron	Steel	Copper Base	Aluminum	Magnesium	Zinc	Other Nonferrous	Total
45,995,817	25,167,222	1,275,473	11,299,044	1,743,817	14,051,924	226,673	587,947	486,764	100,834,681

Table 2 *Output from various European countries*

Europe (metric tons)

Country	Gray Iron	Ductile Iron	Malleable Iron	Steel	Copper Base	Aluminum	Magnesium	Zinc	Other Nonferrous	Total
Austria	39,700	93,000	-	17,258	-	123,865	5,687	12,871	-	292,381
Belgium	36,500	6,400	-	31,474	-	790	-	-	-	75,164
Bosnia and Herzegovina	10,942	2,058	-	4,973	-	6,905	-	-	-	24,878
Croatia**	22,107	17,375	-	1,313	459	11,652	-	230	661	53,797
Czech Republic	179,608	52,911	9,240	94,929	5,367	73,165	-	8,268	870	424,358
Denmark*	31,800	47,400	-	-	1,273	3,172A	-	-	290	83,935
Finland	24,553	38,431	-	15,637	3,008	3,619	-	259	-	85,507
France	657,700B	675,700	-	102,200	17,688	324,509	-	20,064	2,295	1,800,156
Germany	2,392,654	1,641,528	31,679	217,197	77,330	802,501	16,444	34,772	9	5,214,114
Hungary	49,000	31,000	11	3,535	1,745	96,128	189	4,367	124	186,099
Italy	626,435	416,805B	-	72,184	12,727	717,213	6,790	56,846	50,680	1,959,680
Norway	13,400	36,400	-	3,000	-	5,575	-	-	-	58,375
Poland	486,000	141,000	10,000	51,500	5,500	330,500	3,300	8,000	1,000	1,036,800
Portugal	35,043	73,884	-	7,982	9,206	18,940	-	1,027	-	146,082
Romania	31,669	2,910	637	24,853	4,878	45795	5,050	20	6	115,818
Serbia	37,251	15,162	10,328	9,050	2,220	4,958	-	-	7,528	86,497
Slovakia*	2,700	18,200	-	4,100	-	46,000C	-	-	-	71,000
Slovenia	100,200	24,900	-	33,900	1,052	30,065	-	2,250	-	192,367
Spain	328,600	580,700	5,900	76,100	11,760	112,384	-	8,639	601	1,124,684
Sweden	153,900	51,100	-	23,400	10,300	37,800	2,600	4,300	-	283,400
Switzerland	16,200	29,700	-	2,000	2,347	17,970	-	1,235	-	69,452
Ukraine	420,000	140,000	40,000	530,000	45,000	260,000	20,000	35,000	42,000	1,532,000
United Kingdom	128,000	191,000	3,300	74,000	10,000	104,500A	-	8,500	1,000	520,300

* 2011 data   ** 2010 data   A) Includes magnesium   B) Includes malleable iron   C) All nonferrous

Cast metal capability plays an important role in the making of components for a vast range of manufactured products. It therefore represents an essential part of all countries' manufacturing capability. Table 1 shows some recent total output values for individual metals and for the total worldwide output. The total market capacity is around 100 million tonnes. The major volume of the metal castings being iron based castings (about 70%). Within the United Kingdom the percentage of iron castings is similar at 62%.

Table 2 shows the output from European countries which shows that the total output for the United Kingdom is around 0.5 million tonnes or 0.5% of world output.

Figure 9.1 Upper images shows the cast metal split between nonferrous and ferrous and comparison of grey iron castings that are made flake graphite and the quantity made spheroidal graphite (Ductile iron) Figure 9.1 Lower images shows the output from India and China.

The impact of the growth of cast metal in "developing" low wage countries has been significant. It is interesting to contrast a 1998 US report "Metalcasting Industry Technology Roadmap"[1], which was a detailed plan with over 100 R&D projects recommended with the US International Trade report in 2005[2] which showed the cost disadvantages of the US foundry industry and the approach from big industry buyers that wanted the lowest cost casting.

"Purchasing Patterns and Practices – Price is a significant factor in purchasing decisions, with nearly one-half of responding purchasers indicating that they usually buy castings at the lowest possible price."[2]

▶ *Figure 9.1 Upper – World cast output showing the split between flake and ductile iron Lower – Comparison of India and China*

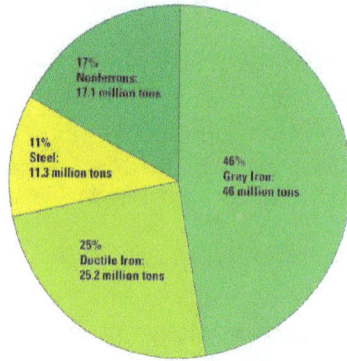

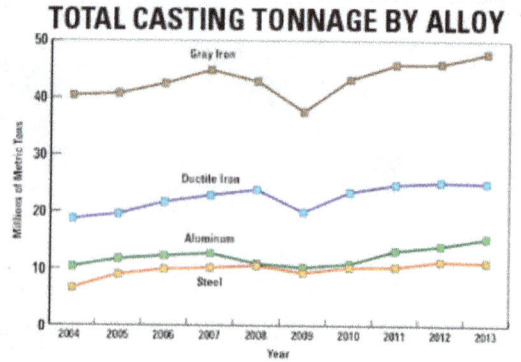

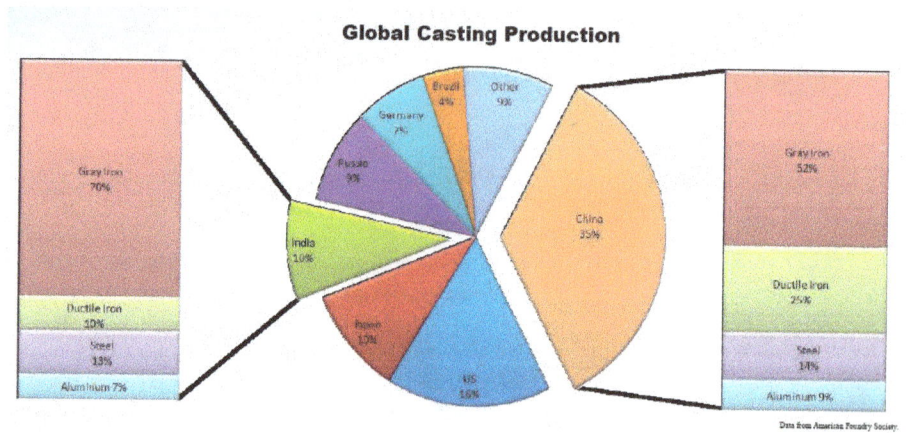

▶ *Figure 9.1b Indexes of hourly compensation costs for production workers in manufacturing, 2003[2]*

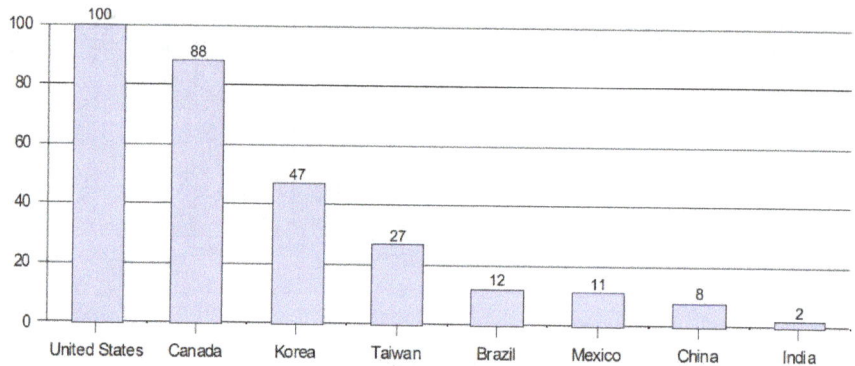

9.2 WORLD CRUDE STEEL OUTPUT AND ASSOCIATED CASTING PROCESSES

The cast metal sector which involves significant tonnage is the world crude steel output. The production of crude liquid steel is followed by a casting process followed by a hot deformation process of forging or rolling. This steel would be cast as either continuously cast product or ingot cast product.

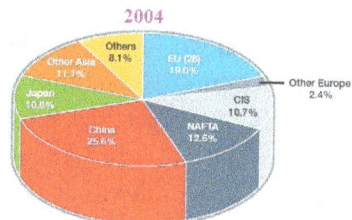

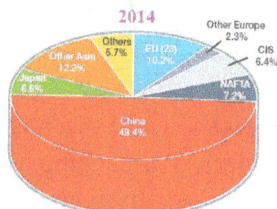

Figure 9.2 Worldwide output of crude steel by output areas 2004 compared with 2014

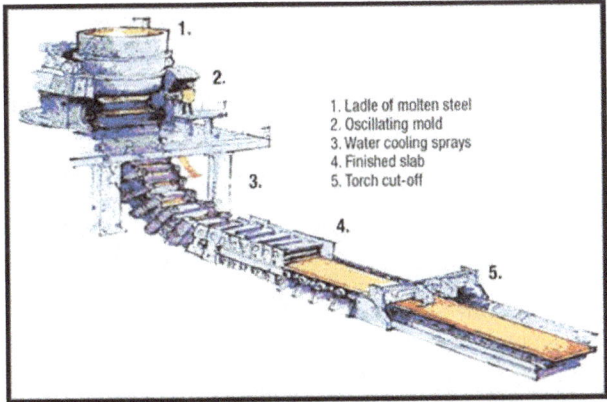

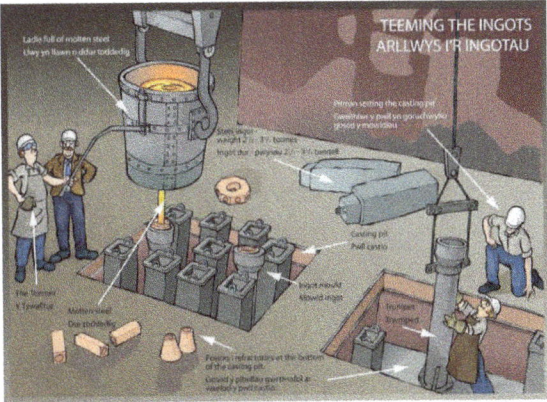

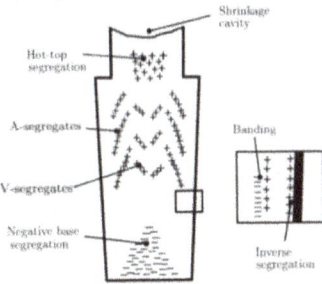

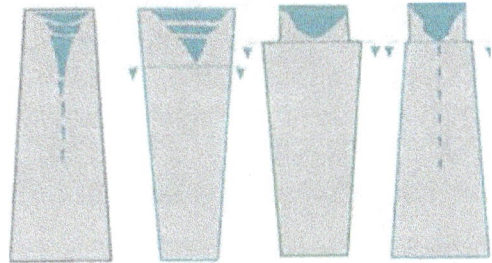

Figure 9.3 Top LHS = Slab concast, RHS = Ingot casting, Centre LHS = Vertical concast, RHS = Horizontal cast (irons), Lower LHS = Forged ingot, Centre = Ingot segregation, RHS = Ingot types

9.2.1 Continuously cast steel is cast as slab, bloom or billet

The process has been commercialised for over sixty years and the process parameters and potential problems are well understood and yet there can still be problems. Section 9.2.1.1 outlines the problem of inter-columnar cracking

9.2.1.1 Continuously cast slab inter-columnar cracking

Intercolumnar cracking of slabs has been a major problem associated with slab produced by the continuous casting process. There are published accounts of several detailed studies over many years.[3 to 8] Process control and sampling/testing are required to prevent the production or use of non-conforming product.

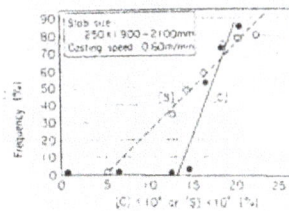

Figure 9.4 Published data on the problem of inter-granular cracking with continuously cast steel

Figure 10 : Influence de la composition en carbone et en soufre de l'acier sur le taux de criques internes (d'après [4])

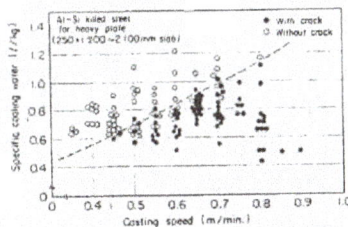

Figure 9 : Influence de la vitesse de coulée et du taux d'arrosage secondaire sur des criques internes (d'après [4])

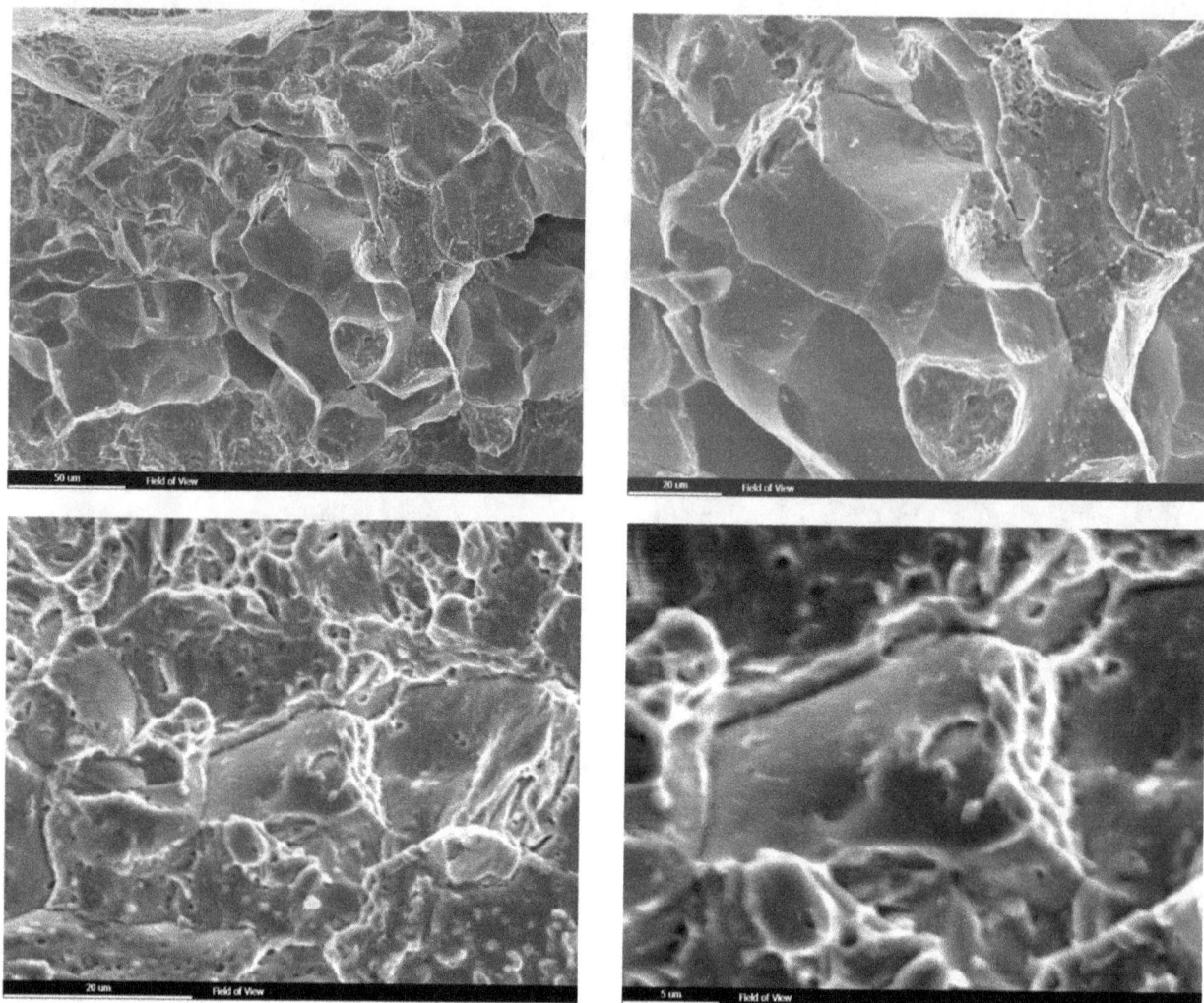

Figure 9.5 SEM images of the fracture surface of a through thickness bend test showing evidence of slab intercolumnar cracks that have not closed during plate rolling

The main reasons identified for the occurrence of intercolumnar cracking are the carbon content, cooling intensity and casting speed. (Figure 9.4) The cracking has also been related to the "maintenance" condition of the casting machine particularly relating to the occurrence of bulging between the rollers due to the lack of adequate roller support.

9.2.1.2 Closure of cracks during subsequent rolling or forging

The production of a commercial quality steel plate relies upon the ability to completely weld any inter-columnar cracks inside the slabs.[9] This depends upon the crack severity and whether the cracks have been exposed to atmosphere and oxidised and also the rolling deformation. Failure to "close" the cracks during rolling would result in poor-quality plate and a tendency for cracking in service. If the centre imperfections go undetected there would be high risk of

failure in service. A slow bend test of affected steel can show the signs of inadequate repair as shown in Figure 9.5.

9.2.2 Ingot cast two options WEU (wide end up) or NEU (narrow end up)

The use of WEU moulds gives a process that controls the segregation providing the mould taper and H/D ratio are optimised (6% taper. Maximum H/D). The optimisation was part of an EU project.[10, 11 and 12] With the correct taper bridging and secondary segregation can be avoided. With NEU multiple bridging can occur but can be repaired during rolling to a certain quality level. The other option for ingot is either direct team or uprun team (Figure 9.2 Top RHS). The source of steel for heavy plate for oil and gas and nuclear is big WEU ingot and there are now recent studies of all the conventional areas of ingot segregation on steel hardenability and toughness capability. ("A" segregate, bottom cone segregation, primary segregation.)[13 and 14]

http://www.msm.cam.ac.uk/
phase-trans/2014/evolution.pdf

9.3 CAST METAL TECHNOLOGY

The relatively simple task of filling a shaped container with a fluid forms the basic principle of cast metals technology as shown in Figure 9.6. Unfortunately, this simple task is more complex in practice due to the fact that the fluid is molten metal and the mould would be made from refractory sand. It is an easy task to prepare a list of what can go wrong!

The cast process ranks as a "near net shape" manufacturing route and therefore there can be a reduced requirement for machining and other finishing processes and therefore cost advantages. In addition, it is applicable to material that cannot be made by wrought methods of manufacture or fabrication process. These include the range of cast iron such as ductile iron and the austempered ductile iron having a capability of matching the mechanical properties of alloy steels in some applications. High chromium white irons and austenitic high manganese steel (Hadfields Steel) that provide the wear resistance can only be made by a cast metal route and provides wear resistant casting for power station and mineral handling.

There has always been competition between cast product and wrought product. Cast metal suppliers claim a product with isotropic mechanical properties whereas wrought products develop directionality in mechanical properties due to the mechanical working. However, the problems of directionality can be minimised by the manufacture of cleaner steels by vacuum degas and low sulphur levels.

Products made by metal deformation have the known benefits of removal of porosity and unsoundness that may develop during casting. In addition, the metal deformation if carried out correctly can effectively refine the grain size and redistribute any segregation.

Cast metals have associated problems of coarse grain size and cast defects. However, there have been recent developments that can assist with the manufacture of quality castings.

- The introduction of software to simulate the solidification and assist in establishing the methoding aspects particularly the feeding to ensure directional solidification and allow unwanted segregation to be removed in the riser.
- Filters placed in runners to remove non-metallic inclusions.
- Mould coatings to improve the surface quality.
- Micro-processor control of thermal analysis.
- Automation which gives repeatability.

The main foundry challenge is the application of foundry and engineering skills to design a system which will allow the manufacture of a sound (pore free) casting, free from defects (sand inclusions, slag, cracks, etc.), with the correct dimensions and with mechanical properties in accordance with the design requirements. The procedure to achieve this end point was once all skill based but now there are many aspects of computer aided input ranging from almost expert systems to the ability to simulate solidification. A recommended source of information would be Foundry Technology by P R Beeley 1972, Butterworths.

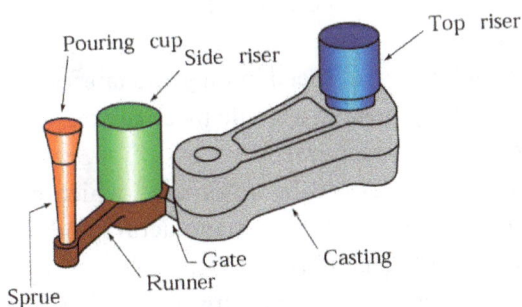

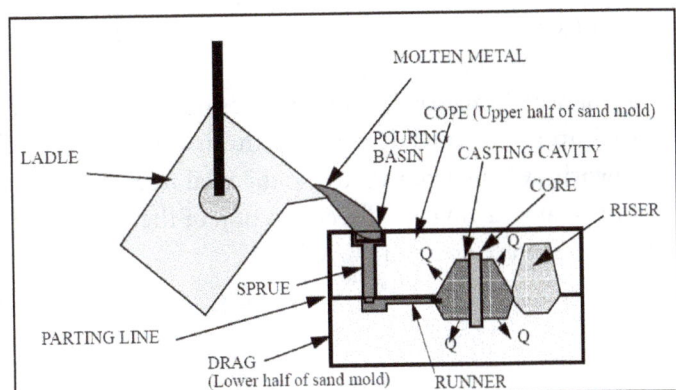

Figure 9.6 LHS = A sand mould, RHS = A side riser and a top riser

A structured approach would involve the following steps:

1. The design of the metal entry into the mould cavity avoiding gas, slag or sand pick-up. With reference to Figure 9.6 this involves the pouring basin/cup, sprue and runner and final entry into the mould cavity.

2. The design of an appropriate metal feed (called "riser" in the foundry industry but the actual purpose is "feeding") to counteract the shrinkage during casting. Liquid metals contract as they cool and solidify. The riser may have to provide up to 5 – 7% by volume for the casting as it solidifies.

Table 9.3 Volume change during casting

Metal	Solidification Shrinkage %	Solid thermal contraction %
Aluminium	7.0	5.6
Aluminium alloy (typical)	7.0	5.0
Grey cast iron	1.8	3.0
Grey cast iron, high carbon	0	3.0
Low carbon cast steel	3.0	7.2
Copper	4.5	7.5
Bronze (Cu-Sn)	5.5	6.0

3. The control of the direction of heat flow (Q in Figure 9.6 LHS) so that the last region to solidify can still be fed by the riser (called riser because metal can be seen entering and rising during filling the mould but its purpose is a feeder). This may require adjustment of the size or position of the riser or the use of more than one riser Figure 9.6 shows a side runner and a top riser. (Directional Solidification Required.)

4. In addition, the heat flow should be controlled to produce a sound casting by allowing the feeding of any inter-dendritic shrinkage regions and the avoidance of excessive build-up of residual stress. An important consideration is the use of the Chworinov Rule from 1938 that the total freezing time of any casting is a direct function of the volume to surface area $t = k(V/A)^2$

9.4 NEAR NET SHAPE MANUFACTURE

In line with the principles of world class manufacture the removal of need to machine parts can provide a lower cost of production. Casting along with powder metallurgy methods of manufacture are identified as a near net shape manufacturing method. The metal casting process is the most direct route to a near net shape product, and is often the lowest cost route. The manufacture of a cast metal product involves the pouring of molten metal into a mould. For any particular casting there are a number of competing casting techniques. The choice of a process will be dependent upon the size of casting, the number of parts required, the surface finish, the tolerances and the metal or alloy involved.

In general terms all metal products begin life at the molten stage. This needs to happen whether the method of manufacture was forging, powder metallurgy, metal injection moulding or 3D manufacturing methods.

- Forgings and wrought parts require ingot moulds or continuous casting.
- Powder metallurgy requires the production of powder by liquid metal atomization, either as an alloy (pre-mixed) or as a pure metal for later mixing.

The main advantages of castings are:

- Intricate detail both internal (by the use of cores) and external can be cast
- Some metals which are impossible to "hot work" can only be produced by casting. The best example being cast iron which provides low cost useful casting. Cast iron is the material of choice for hydraulic valve bodies which require metal with high fluidity to create the complex internal passages
- Can avoid the use of metal joining and fabrication techniques by making a single piece casting (one of Boothroyd principles)
- Casting processes can be automated which can improve the economics of manufacture and the overall process capability
- Very large items can be cast. Large pump housings
- Some mechanical properties may be more favourably distributed in the as cast condition:

1. Machineability and damping capacity of cast irons
2. Better directionality of mechanical properties
3. Good bearing properties are obtained in cast metal bearing alloys
4. A wide range of compositions and properties can be produced.

9.5 MAJOR CASTING PROCESSES

There are a large number of competing industrial casting processes. These are usually classified based on:

i. The type of mould material (sand, permanent, etc.)
ii. The pressure on the molten metal during filling (gravity, centrifugal force, vacuum, low pressure, high pressure)
iii. The method of producing the mould.

Table 9.4 shows the process capability associated with some of the key casting process and Table 9.5 shows the mould material, the way in which metal enters the mould and the method of making the mould.

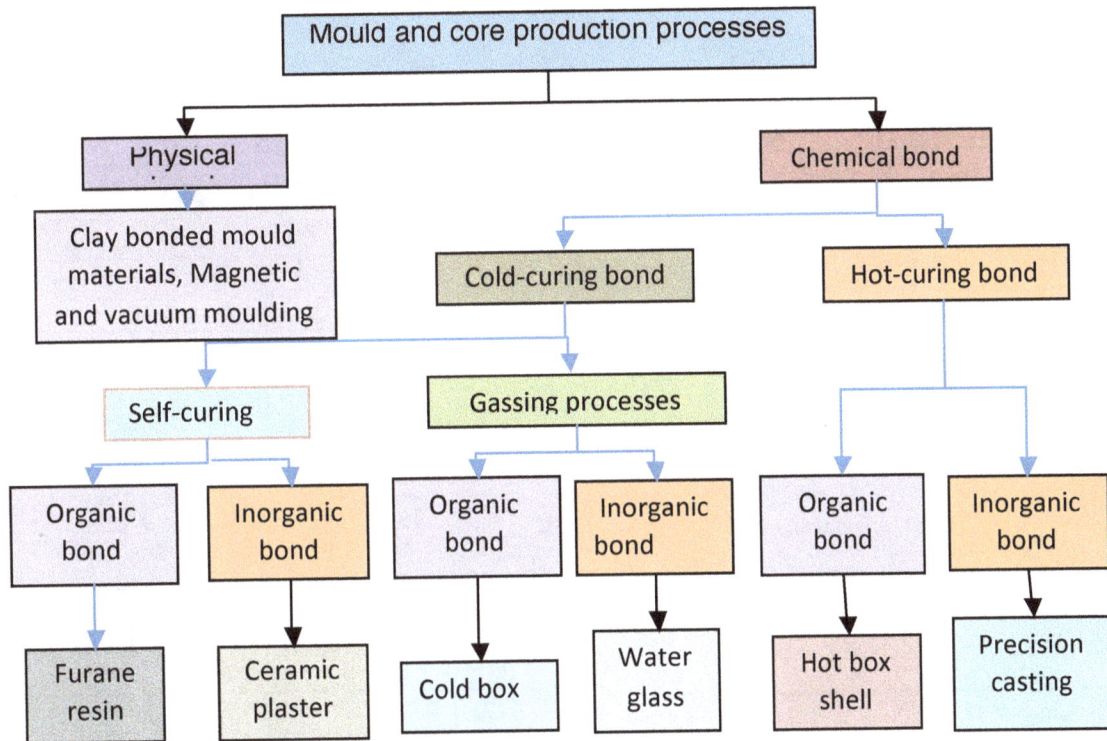

Figure 9.7 Types of moulding material/methods

Table 9.4 Capabilities of common casting processes

Attribute \ Process	Sand	Investment	Gravity Die	Pressure Die
Maximum size	several tons	up to 20 kg	up to 50 kg	up to 8 kg
Dimensional tolerance	> 0.6 mm	> 0.1 mm	> 0.4 mm	> 0.05 mm
Surface finish	> 200 RMS	> 60 RMS	> 150 RMS	> 30 RMS
Minimum thickness	> 6 mm	> 1.5 mm	> 4.5 mm	> 0.8 mm
Economic quantity	any number	> 100	> 500	> 2500
Sample lead time	2-10 weeks	8-10 weeks	8-20 weeks	12-24 weeks

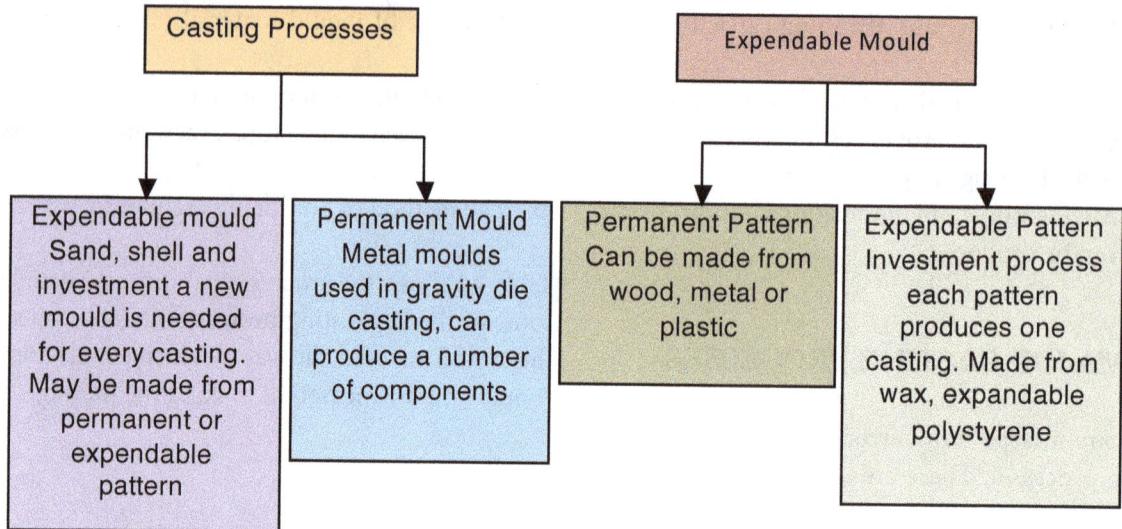

Figure 9.8 *Difference between expendable mould and permanent mould. Figure 9.8 Difference between permanent pattern and expendable pattern*

Table 9.5 *Casting Processes, and Mould Details*

Casting Process	Mould Material
Sand Casting	Sand (Bonded with clay and water or Chemicals)
Permanent Mould	Metal
Die Casting	Metal
Investment	Ceramic
Lost Foam EPC	Sand (Unbonded)
Thixocasting Rheocasting	Metal
Cosworth	Sand
V Process	Sand (Unbonded with vacuum and enclosed plastic film)
Centrifugal	Metal or Graphite
Ingot Casting Not net shape	Metal
Concast	Copper cooled mould
Metal Entry	Method of Mould Making
Gravity	Expendable
Gravity	Permanent
Pressure	Permanent
Gravity	Expendable
Gravity	Expendable
Pressure	Permanent
Vacuum	Expendable
Gravity	Expendable
Centrifugal Forces	Permanent
Gravity	Permanent
Gravity	Permanent

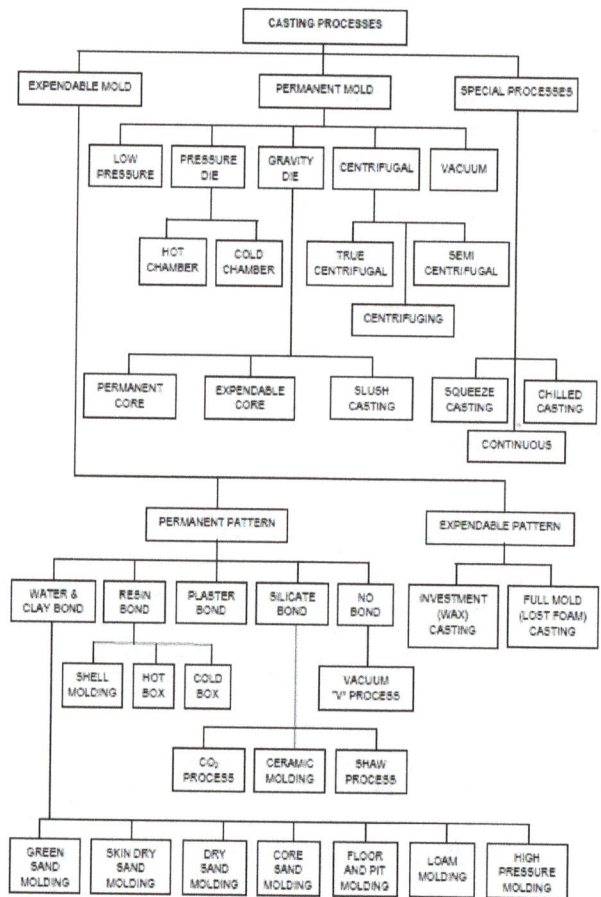

Figure 9.9 *Casting processes*

9.5.1 Sand Casting

All sand-casting processes start out with the manufacture of "tooling". The tooling consists of a pattern which is a positive replica and has the shape of the external surfaces of the casting. The tooling typically consists of two horizontally separate patterns. The top half of the pattern is called the cope and the bottom half is called the drag. Sand cores can be manufactured to create an internal surface of the casting and are made from tooling called "core boxes". The cost for the casting tooling is often quoted as a separate cost and depreciated over the life of the tooling.

The cost of the tooling will depend upon the material used to manufacture the tools which in turn depends upon whether it is for had moulding or automated moulding. Hand tooling (typically low volume production) can be made from wood, plastic or aluminium and tools for automated lines (high production runs) would be made from steel or tool steel to ensure cost effective life.

Mould Making Processes

Once the tooling has been developed and manufactured moulds can be made using the patterns. There are several different processes used to create moulds. The two most common methods are:

- Green Sand Moulding
- No Bake Moulding.

Green Sand Moulding

Green sand moulding is the most common process for making moulds for small to medium size metal castings. Typically, silica sand is used with a mixture of bentonite clay and water. This coating of clay and water allows the sand grains to stick together when compacted against a pattern surface. The sand mixture is compacted in a removable flask that is attached to the pattern. After the sand is compacted the pattern is removed revealing a cavity in the sand mould. This sand mould is used only once and sand recycled after use.

No bake moulding uses a polymer resin instead of clay and water to hold the sand in place. The term no-bake comes from the early forms of resins where heat was required to bond the sand. Current resins use a room temperature chemical reaction resulting when components in the right proportion are brought together. In no bake moulding the sand, resin

and catalyst are mixed in a continuous mixer and the mixture is delivered right into the flask. Vibration is sometimes used to compact the sand. The mixer is controlled by the operator to deliver sand when needed.

The compaction in the green sand process can either be by hand or by using machinery. Jolt squeeze moulding is a common manual process used to make moulds. This process uses a moulding machine that both jolts and squeezes the sand onto the pattern plates.

a. Jolting

Jolting is a process where the pattern plate is assembled to a flask and is filled with green sand. This assembly is then lifted and then dropped, providing the jolting action that helps to compact the sand around the pattern plate.

b. Squeezing

Squeezing is a process where the pattern plate is again assembled to a flask and filled with green sand. The assembly is then squeezed either by hydraulic or pneumatic pressure.

c. Jolting/Squeezing

Both jolting and squeezing are combined into a jolt squeeze machine. This helps to give good uniform sand compaction properties to the mould.

The cope and drag portions of the mould are created using these processes. The necessary cores are then placed into the cope or drag and the two halves of the mould are then assembled together.

Once the mould is finished, metal is poured into the mould and allowed to cool. And once the metal has cooled sufficiently the mould then goes through shakeout to remove and possibly reclaim the sand. The gating system is trimmed off, the casting cleaned up and the final product is ready for shipment to the customer.

The process shown here involves manual labour and produces horizontally parted moulds. This process is suited for low to medium production volumes. Higher volume parts are produced on automatic moulding machines. Moulds produced by these machines may be either horizontal or vertical. The use of automatic moulding machines lowers the labour required and increases the production rate.

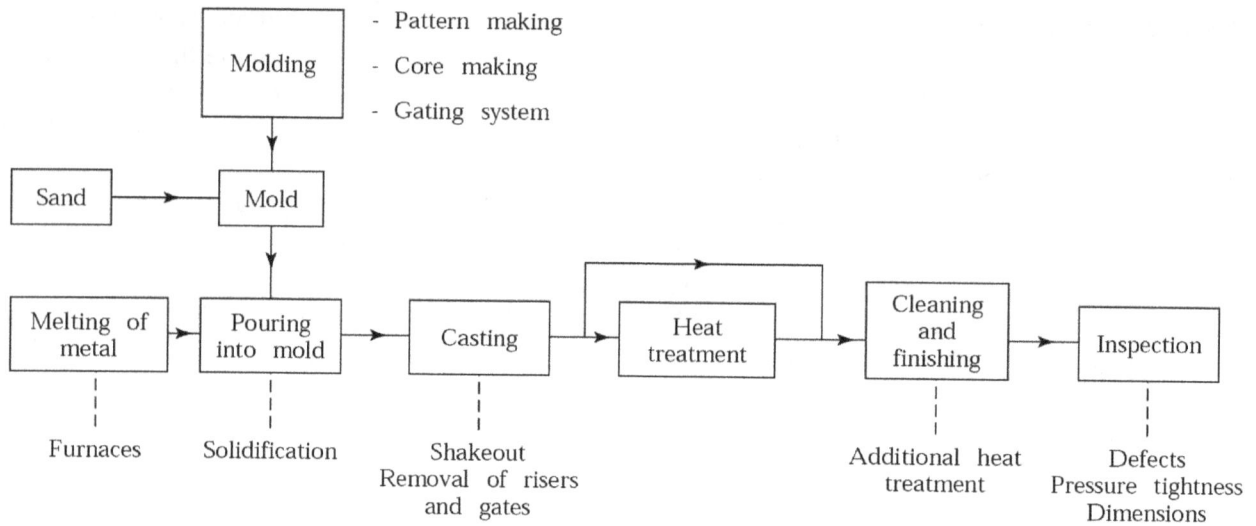

Figure 9.10 *Flow diagram for sand casting*

Steps in sand casting

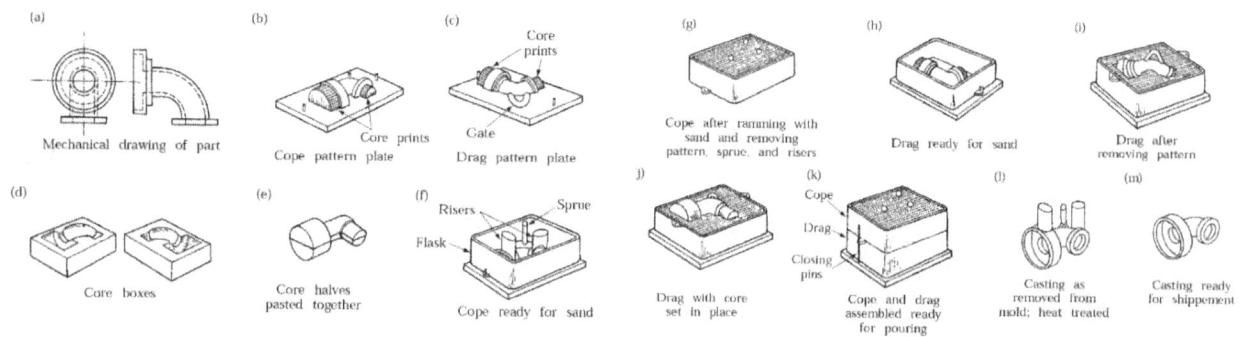

Figure 9.11 *Sequence of sand casting*

9.5.2 Investment Casting

The shell mould investment casting process starts out with the creation of a wax mould, from which the wax pattern is made. This wax mould must be sized to account for two shrink factors, one for the wax itself and the other for the steel. After enough wax patterns are produced, a pattern assembly or tree is created. Here the wax patterns are glued onto a main gating system.

Once the pattern assembly is complete it is then coated in a ceramic slurry. After each dip the wet pattern assembly is then placed in a fluidized bed of sand for coating. Another method is to rain refractory material over the wet pattern assembly.

After each dipping cycle of ceramic slurry and sand coating, the pattern assembly is allowed to dry thoroughly before the next coating. The number of coatings required is dependent on the size of the metal casting and the temperature of the alloy being poured.

Once the ceramic shell is complete, the shell is placed upside down in an auto-clave and the wax is removed and re-claimed from the shell.

The shell is then fired, to 982°C, which is necessary to burn out any remaining wax and any moisture that may be present in the shell. The shells can then be stored or immediately poured. If the shells are stored, they will need to be preheated before pouring. The shells can be gravity poured in air or under vacuum. After the castings have cooled completely, the shell is removed and the castings will be cut from the tree.

There are variations of this process across the globe. The type of refractory material used can vary from a true ceramic to a silicate or sand material. The ceramic investment cast process will yield a casting with an excellent surface finish, while the silicate investment cast will yield a casting with a rougher, sand cast like, surface finish, but with a lower part cost.

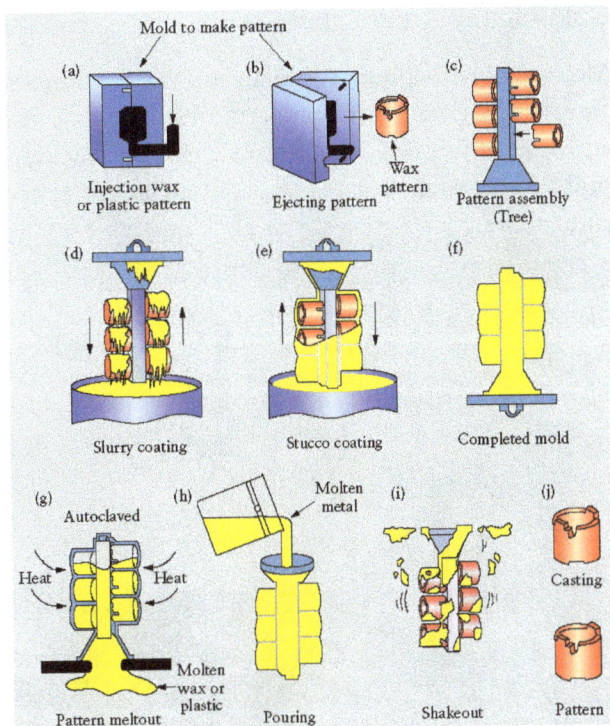

Figure 9.12 *Investment or lost wax*

9.5.3 Expendable Mould Casting – Expanded Polystyrene Mould

The lost foam casting method uses an expanded polystyrene (EPS) foam pattern that is expendable. For each casting that is poured an EPS foam pattern is burned up.

The process starts out by creating a foam tool from which foam pieces are made. The casting design can have undercuts, minimal draft, anything that can be placed into the foam tool. Also, the final foam pattern can be made from several pieces of EPS foam that are glued together allowing design flexibility not possible in other casting processes. If the part is small, a tree of multiple foams can be glued together. This final foam piece will also have a gating system attached to it for pouring.

After the foam is complete, it is coated in a refractory wash, placed into a large flask, where dry, loose sand is poured around the foam while the flask is vibrated. The vibration effectively compacts the sand around the foam and helps to fill any difficult to fill areas. The molten metal is poured into the flask and the foam is vaporised upon contact with the metal which takes the shape of the foam.

The solidified casting is removed from the flask after cooling and gating system is removed. Lost foam castings exhibit a characteristic surface finish

of the expanded beads and very little if any parting line flashing. An innovation using this technique is to place fine particles inside the foam that can assist with grain refinement or a way of adding abrasion resistant compounds to produce wear resistant parts.

Advantages

- Pattern need not be removed during mould making
- No cope /drag is needed, all features are built into the pattern
- Possible to automate the production process.

Disadvantages

- The pattern is not reusable.

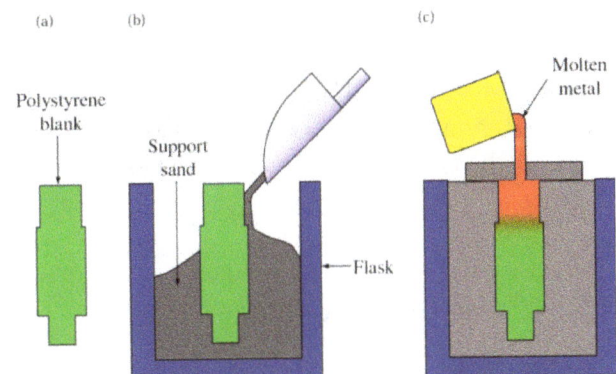

Figure 9.13 Schematic illustration of the expendable pattern casting process, also known as lost foam or evaporative casting. Pattern made from Polystyrene and vaporized when in contact with molten metal. The pattern can include the sprue and runner. No cope / drag is needed

Lost Foam Castings

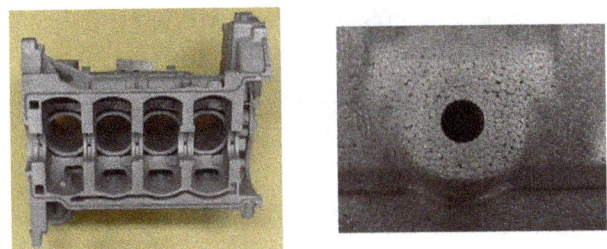

Figure 9.14 *Lost foam castings*

9.5.4 Shell moulding (Zircon sand for hydraulic spool valves)

Shell moulding is a variant of the sand moulding process and uses a heat activated resin to bind the sand together similar to cold box or no bake moulding. Although shell moulding produces a mould with much less sand and resin used. Dimensional control and surface finish are improved with shell moulding.

In shell moulding a pattern for the cope and drag are still utilized. Metal patterns are heated to a temperature around 230°C which is required to cure the resin binder. The patterns are assembled to a dump box which contains resin coated uncured sand. The dump box is tipped allowing the sand to flow onto the hot surface of the pattern and held there until a desired shell thickness is created. At which time the dump box is tipped back to its resting position and any uncured sand flows off the pattern surface and back into the dump box.

This cured mould shell is stripped from the pattern plate and the cope and drag shells, as well as any cores, are then glued, clamped or clipped together for pouring.

9.5.5 Pressure Die Casting

Molten metal is injected under pressure into a hardened steel die, often water-cooled. Metal cores are used to produce cavities and undercuts. After solidification, one half of the die is moved and the casting is pushed out by ejector pins. This process is suitable for non-ferrous castings of small to medium size, varying complexity and thin walls.

Hot-chamber machines

Metal molten in container attached to machine. Typical injection pressures are 7 to 35 MPa. The piston is subjected to the melting temperature of the metal and thus the process is often used for low melting point metals such as zinc, tin, lead or magnesium alloys.

Cold-chamber machines

Molten metal is poured into an unheated chamber from an external container. Typical injection pressures are 14 to 140 MPa. Often used for high melting point metal such as aluminium, brass, and magnesium alloys.

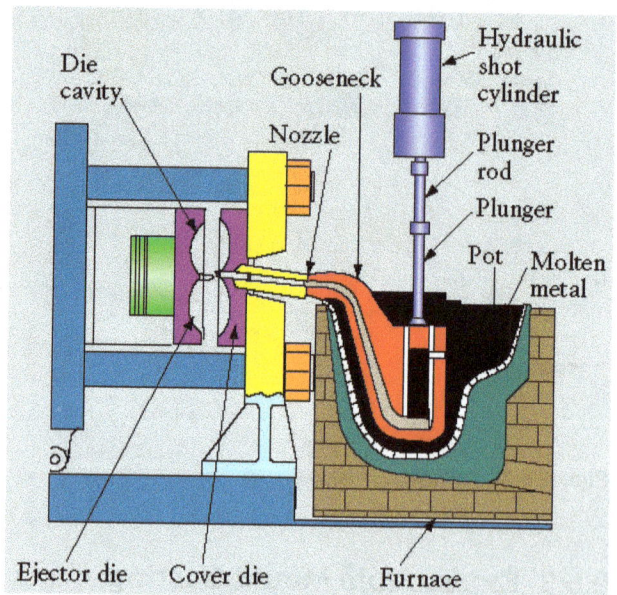

Figure 9.15 *Sequence of steps in die casting of a part in the hot-chamber process. Source: Courtesy of Foundry Management and Technology*

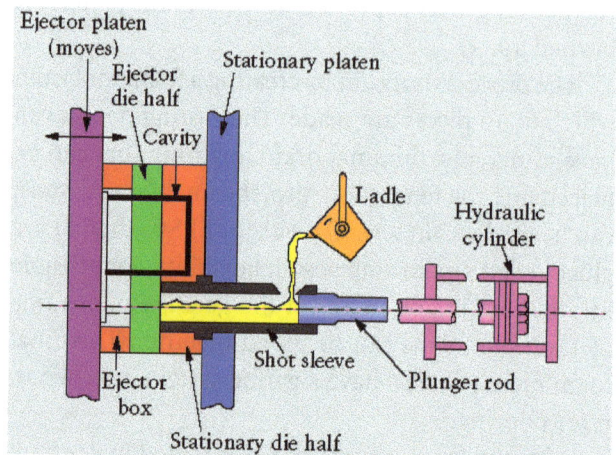

Figure 9.16 *Sequence of operations in die casting of a part in the cold-chamber process*

9.5.6 Centrifugal casting

Molten metal is poured into a horizontal rotating mould where the centrifugal force pushes the metal against the mould.

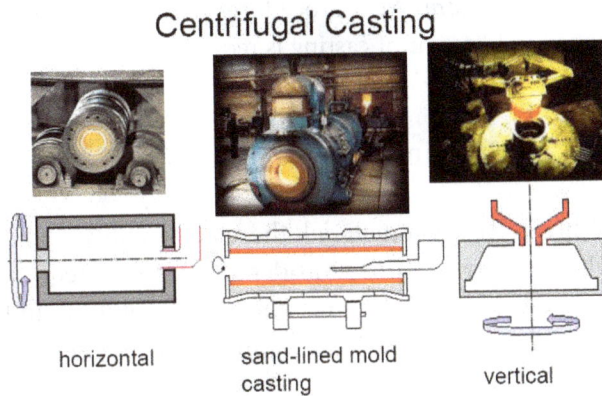

Figure 9.17 Centrifugal casting

9.5.7 Squeeze-Casting Process

Semi-solid metal is forced under pressure into a metal mould giving a fine microstructure free from a dendritic structure usually obtained in conventional casting. The mechanical properties approach those of forgings.

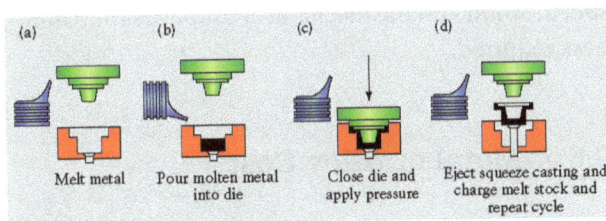

Figure 9.18 Sequence of operations in the squeeze-casting process. This process combines the advantages of casting and forging

9.6 CAST METAL SELECTION

One of the most important considerations when sourcing a casting is the metal selection. Many factors determine what material will be used for the casting which include:

- The mechanical properties, tensile strength, yield strength, elongation and hardness
- material weight – low weight has advantages for automotive parts

- material cost but casting cost is the important criteria
- machinability, wear resistance, fatigue/endurance properties, corrosion resistance and possibly weldability
- Suitability for heat treatment which would involve hardenability for alloy steels.

The choice of a metal or alloy used to make a casting is often as important to a successful service performance of a casting as the actual design of the casting. The application and use of a casting and strength requirements will determine whether a ferrous (iron base) or non-ferrous alloy is chosen. The range of non-ferrous alloys would include alloys of aluminium, zinc, magnesium, copper, lead, tin and nickel-based alloys.

Each specific alloy will have a specific combination of physical properties and the alloy choice will depend on the service requirements. For example, if a casting will be used in a saltwater environment, corrosion resistance of the alloy or a protective coating would be required. If the service stresses are of a cyclic nature, then the fatigue life would be important.

The use of cast metals has a long history and therefore there is significant material selection guidance regarding the metals that have been successful for various applications. Therefore, background research including any failures would be recommended to assist in the choice of cast metal.

A recent publication on the fracture toughness of castings outlines several important aspects of fracture toughness testing and control of the casting process to achieve high toughness.

https://www.intechopen.com/books/science-and-technology-of-casting-processes/fracture-toughness-of-metal-castings

9.6.1 Mechanical Properties

The composition of a cast metal defines and limits the achievable range of the mechanical properties. This may involve a suitable heat treatment, which would normally for steel parts be either normalise, a quench and temper heat treatment or surface hardening process. The five mechanical properties that are normally specified are the tensile strength, yield strength, elongation and hardness.

- Tensile Strength – Tensile strength is the maximum stress in uniaxial loading which a material can withstand prior to fracture.

- Yield or 0.2% Proof Strength – Yield strength is the stress at which a material extension changes from elastic behaviour to plastic behaviour. Beyond the yield point permanent metal movement takes place.
- Elongation – Elongation is the amount of permanent movement of a material as it is stressed to failure. This is expressed in a percentage of the material's original gauge length. This can be split to the uniform elongation that takes place up until the value of the UTS and the local elongation that occurs after the UTS stress has been reached during necking of the specimen.
- Hardness – Hardness is a measure of the resistance of a material to deform. This is measured by the well-established indent hardness methods such as Brinell, Rockwell and Vickers. Hardness is typically proportional to the yield and tensile strength of a material.
- Toughness – The energy of fracture determined by Charpy of Fracture toughness CTOD, K_{1C} etc.

9.6.2 Weight

In some applications the casting weight can be an important consideration in the design. This could involve a maximum weight requirement such as casting in automotive applications or aerospace applications or a weight minimum in an application such as a fork truck counterbalance weight. The lowest density of a metal and alloys are the aluminium, magnesium. Aluminium is usually the material of choice when minimum weight is a requirement. However, because of the higher yield strength of steel compared to aluminium, in some part designs it is possible for the weight of a steel casting to compete with an aluminium casting.

9.6.3 Weldability

When applications require that a part should be welded the choice would be between steel and aluminium.

The weldability of steels is well established and the weldability is evaluated based on the CEV (carbon equivalent value). The carbon content will have the major influence on the weldability of steel. The higher the carbon content the more brittle a weld will become after it cools. Typically, a material that has a carbon content at or above 0.30% may require a preheat on the parts being welded and potentially a post weld heat treatment.

Alloy steels typically have higher CEV and greater care is required. Normal procedures would be to establish and then validated by mechanical testing a weld procedure. Weld procedures are often required when weld repair of casting is required.

9.6.4 Machinability

When selecting a material, the machinability of the alloy may need to be considered. This may require a suitable anneal for steel alloys or a solution and precipitation harden for an aluminium alloy such as LM24.

9.6.5 Suitability for Heat Treatment

The selection of a material that will need a heat treatment is dependent on the hardness and heat treatment required. There are a number of competing heat treatment processes ranging from through hardening and tempering to carburising and nitriding. In addition, there are selective hardening methods such as induction and flame hardening. The choice of heat treatment would depend upon the product design specification and the mechanical properties and hardness required.

9.6.7 Metal or Alloy Cost

The cost of metal is always a consideration in the cast part design. The use of the lowest cost metal that delivers all the design specification is an essential requirement for cost effective use of castings. The metal values in order from highest to lowest cost per kilogram, would be aluminium, steel, then iron. Metal choice would be a consideration during the initial part design study.

Although aluminium is the highest cost per kilogram, its use could be justified, because of aluminium's low density, when weight savings are part of the design specification. In addition, the choice between using steel or ductile iron in a casting design, the higher strength properties of steel may allow for a thinner section in areas and thus a lower cost part.

When choosing a metal or alloy based on cost, the outcome will depend upon the design. In addition, each metal choice may require a different casting process which could also have an impact on the part cost. In addition, the relative "quality" level of the casting has a strong influence on the price. The web link shows a published review of steel cast quality carried out around 2000 by Alfred R. Buberl and Reinhold Hanus of Voest-Alpine Stahl Linz GmbH Steel Foundry, Linz, Austria (Cast steel in the competition of materials).

https://www.vgb.org/vgbmultimedia/WFC_
Budapest_98_Draft3-p-1972.pdf

The paper points out that competition takes place on price and delivery, quality and technology is specified and assumed by the customers. The paper covers the "Cost influence – quality grades of steel castings" by using a model for the relation between various quality levels and the cost of production.

The criteria which influence quality were listed and 5 quality grades (Table shown in Figure 9.19) from the lowest realistic quality up to highest quality according to the latest state of metallurgical capability. Eleven main groups of criteria (see figure 9.19) were also considered.

Each criterion the 5 quality grades were reviewed to establish the extra process costs. The expenditure was factored with regard to the extra costs to the

foundry. Finally cost associated with each grade was calculated and costs increase can be seen on the RHS of Figure 9.19.

9.7 CONSIDERATION OF CASTING QUALITY

9.7.1 Introduction

Castings are purchased to specifications which originate from several national and international organizations such as the American Society for Testing Materials (ASTM), American Society of Mechanical Engineers (ASME), Society of Automotive Engineers (SAE), American Iron and Steel Institute (AISI), and International Organization for Standardization (ISO). British standards (BS) purchasers of castings may require additional specifications for specialized applications.

Some common requirements of a casting may include the evaluation of:

* Chemical Composition
* Mechanical Properties
* Dimensional Tolerances
* Surface Roughness and Integrity

No.	Description	1	2	3	4	5
A10	Ranges	precise "point analysis" (reach the aim, no ange) ELC, Cr-, Ni- equivalent, C- equivalent	according to specification of customers	according to national and international standards	according to internal specification	only max. values of P and S are specified (e.g. S<0,05%, P<0,04%)
A20	Sulphur	< 0,003 %	< 0,010 %	< 0,020 %	< 0,030 %	< 0,050 %
A30	Phosphorus in non-alloy and low-alloy materials	< 0,010 %	< 0,015 %	< 0,020 %	< 0,025 %	< 0,040 %
A31	Phosphorus in high-alloy materials	< 0,010 %	< 0,015 %	< 0,025 %	< 0,030 %	< 0,040 %
A40	Hydrogen before pouring	< 3 ppm	3 - 5 ppm	5 - 7 ppm	> 7 ppm	not determined

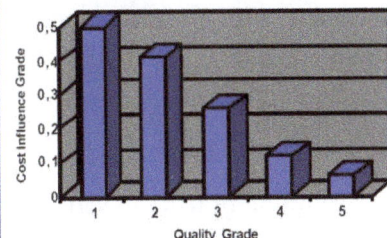

Figure 9.19 Cast steel quality grade and effect of production cost

- Internal Soundness
- Nodule and Flake shape and size (for grey and ductile irons).

9.7.2 Chemical Composition

Steel castings are commonly ordered to mechanical property specifications. Additional requirements may include heat or corrosion resistance. The chemical composition of the material is specified in order to clearly define the steel to be used to meet the service requirements. Additional requirements will also be requested which will determine how the steel should be made and treated such as the maximum and minimum level of various elements and the deoxidation practice.

9.7.3 Mechanical Properties

It is common practice to specify mandatory levels of the mechanical properties of a metal or alloy. The mechanical properties specified include tensile properties, bend properties, impact properties, fatigue properties and hardness. The material tested is taken from a representative sample, test bar of the same metal used to cast the parts. **However, it should be noted that the properties obtained from the test bar are not necessarily the same as the properties of the casting.** For example, BS 3100 "steel castings for general engineering purposes" gives the following cautionary note in section 8 on Testing:

"The mechanical properties required shall be obtained from test bars cast either separately from or attached to the casting to which they refer. The test value so exhibited represent therefore the quality of the steel from which the castings have been poured; they do not necessarily represent the properties of the castings themselves, which may be affected by solidification conditions and rate of cooling during heat treatment, which in turn are influenced by the casting thickness, size and shape."

Tensile Properties

It is common practice to verify the tensile properties of a cast material. The properties tested may include ultimate tensile strength, yield strength, elongation and reduction of area.

Bend Properties

Bend properties of a cast material are typically not measured. This is more commonly applied to weld qualification tests. Measuring the bend properties of a material is performed by taking a bar of given dimensions, bending it to a specified angle around a pin of given radius and monitoring for cracks on the surface of the material.

Impact Properties

An impact property or toughness measures the energy of fracture. The Charpy V-notch test is the common test used to do this. The higher the ductility and the lower the strength the better the toughness.

Fatigue Properties

Fatigue testing consists of cyclically stressing a specimen for a number of cycles and determining the number of cycles needed to cause failure. An S-N curve (Stress vs Number of cycles) can be developed for the material, the S representing stress and the N representing the number of cycles. Also determined from this testing is the endurance limit of the material which is the stress level where infinite life can be obtained. Fatigue testing can be used to qualify designs, processes and materials but it is unusual for this to be carried out as a standard release test for castings.

Hardness

Hardness is a measure of the resistance of the material to indentation. This is measured by methods such as Brinell, Vickers and Rockwell. Hardness tests are often used to verify the heat treatment and physical properties of the cast steel for individual furnace loads.

9.7.4 Dimensional Tolerances

A castings dimensional tolerance is affected by several process factors. The tolerance of an as cast part is dependent on the type of process used. For example, an investment cast part will give a significantly better dimensional capability compared to a sand cast part. A sand cast part that has a sand mould compacted by an automated machine will result in a better dimensional capability than a sand mould compacted by hand.

	Sand cast hand moulding	Sand cast machine mould and shell moulding	Metallic permanent mould	Pressure die casting	Investment casting
Steel	11 - 14	8 - 12	—	—	4 - 9
Grey iron	11 - 14	8 - 12	7 - 9	—	4 - 9
S. G. iron	11 - 14	8 - 12	7 - 9	—	4 - 9
Malleable iron	11 - 14	8 - 12	7 - 9	—	4 - 9
Copper alloys	10 - 13	8 - 10	7 - 9	6 - 8	4 - 9
Zinc alloys	10 - 13	8 - 10	7 - 9	3 - 6	4 - 9
Light metal alloys	9 - 12	7 - 9	6 - 8	6 - 9	4 - 9
Nickel based alloys	11 - 14	8 - 12	—	—	4 - 9
Cobalt based alloys	11 - 14	8 - 12	—	—	4 - 9

◀ *Figure 9.20*
Casting tolerances

Casting size [a]		General profile surface tolerances for tolerance grades P [b]														
above	up to, including	1	2	3	4	5	6	7	8	9	10	11	12	13	14	15
<	≤															
—	10	0,09	0,14	0,19	0,27	0,37	0,53	0,76	1,1	1,6	2,1	2,9	4,3	—	—	—
10	30	0,12	0,16	0,23	0,31	0,44	0,61	0,86	1,3	1,8	2,5	3,4	4,8	6,5	8,5	10,5
30	100	0,14	0,19	0,27	0,38	0,53	0,74	1,1	1,5	2,1	3	4,2	5,8	8,5	10,5	13
100	300	0,15	0,23	0,35	0,5	0,7	1	1,4	2	2,8	4	5,6	8	11	14	18
300	1 000	—	—	0,42	0,64	0,8	1,3	1,9	2,7	3,8	5,5	7,5	10,5	15	19	23,5
1 000	3 000	—	—	—	—	—	1,6	2,4	3,8	5,4	8	10	15	21	26	33
3000	6300	—	—	—	—	—	—	—	7	10	13	19	26	33	41	
6300	10000	—	—	—	—	—	—	—	—	11	16	23	32	40	50	

a   Diameter of smallest enveloping sphere of nominal contour in the moulded condition (space diagonal).
b   Mismatch (see ISO 10135) within profile surface tolerance.

◀ *Figure 9.21*
Casting tolerances

During the making of patterns certain assumptions regarding the size of the contraction between the pattern and the final as cast shape. If the part does not contract the predicted amount due to the part geometry or other causes then this can result in a larger than expected dimensional change on the final part. If certain dimensions are critical, a capability study of castings dimensions, using the production process, should be undertaken before regular production begins.

The size and the weight of the part will determine the tolerance required on specific dimensions. Figure 9.20 from ISO 8062 shows the casting tolerances (CT) achievable for several materials and mould making processes. Tolerance grades will differ for different casting processes. The tolerance table shows the different tolerances required for various dimensions.

In a complex mould, there may be 4 or more components in the mould that need to be assembled before pouring. This assembly of the mould will also determine the dimensional variation of the parts. Critical dimensions of the parts should be designed so that they are moulded in as few components as possible. Cleaning and heat treating will also affect a part's dimensions.

9.7.5 Surface Finish

Figure 9.22 shows the surface finish of various cast metal surfaces (from AFS). The surface finish of a casting will vary depending on the process being used. An investment cast part will yield about a 125 μinch RMS surface, a die cast part about a 200 μinch RMS surface, while a sand casting will be around 500

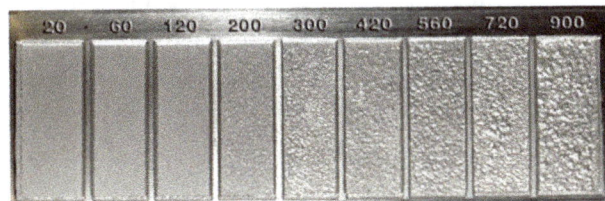

Casting Process	Surface Smoothness in RMS
Investment Casting	50-125
Shell Molding Casting	75-150
Die Casting	90-200
Lost Foam Casting	125-175
Vacuum Casting	150-200
Nobake (Resin) Sand	150-600
Green Sand Casting	250-900

Figure 9.22 Surface finish grades from AFS b) Surface finish from casting processes

μinch RMS. The surface finish will be modified in areas where a gate or riser has been removed. This can be a broken area where the gate was snapped off or a ground/cut surface where the gate was removed. The surface finish in all casting processes can be modified throughout the part by shot or bead blasting if required, but this would add to the cost of part.

9.7.6 Internal Soundness

The internal soundness of a casting can be important in some applications or at particular locations in the casting. It is now accepted that it is very difficult to cast a defect free casting. The current approach is to establish the acceptable level of defects that can be tolerated for a particular application and to base the supply specification on this acceptance level. This helps avoid over-specification of the defect level which simply leads to higher scrap rates and higher casting costs.

Section 9.8 of this chapter outlines the reasons for the many different casting defects that can occur. The three most common are porosity, inclusions and shrinkage. These will be considered in more detail.

Porosity – Definition: Porosity is a void in the casting that is characterized by smooth interior walls that are shiny or in the case of iron, are covered with a thin layer of graphite. The voids can appear in one large cavity or several small cavities dispersed throughout the casting.

Potential Causes:

1. Mechanical Gas Entrapment

   - A large amount of mould or core gas with insufficient venting from the mould cavity
   - Entrainment of air due to turbulence in the gating system

2. Metallurgical Origin

   - Excessive gas content in melt
   - In the case of steel and irons, formation of carbon monoxide by the reaction of carbon with oxygen present in the melt (exploited in old rimming steel manufacture).

Corrective practice:

- Include appropriate vents in the mould cavity to allow the escape of air

- Review gating design for turbulent areas
- Ensure that the sprue is kept full during pouring
- Reduce pouring height
- For steel, deoxidize the melt adequately. Ideally vacuum degas but this is not common in foundries
- For iron, avoid using rusty charge material which will introduce oxides into the melt
- For non-ferrous alloys, avoid excessive melt temperatures and use care in degassing.

Inclusions – Definition: Inclusions are a piece of foreign material in the cast part. An inclusion can be a metallic, intermetallic or non-metallic piece of material in the metal matrix.

Possible Causes:

1. Metallic Inclusions

   a) Charge materials which have not been completely dissolved in the melt
   b) Exposed core wires or rods

2. Intermetallic Inclusions

   a) Combinations formed between the melt material and a metallic impurity

3. Non-metallic Inclusions

   a) Loose sand in the mould
   b) Flakes of refractory coating breaking loose from the mould.

Corrective practice:

- Avoid entrainment of slag and dross from the furnace while filling the ladle
- Use siphon, teapot and bottom pour ladles
- Keep pouring basins and sprue filled during pouring
- Use strainer cores or filters[20] file:///C:/Users/user/Downloads/Wall_chart_STELEX__e_.pdf
- Create slag traps in casting gating
- Care should be taken when creating moulds to reduce the likelihood of non-metallic inclusions.

Shrinkage – Definition: Is a vacancy typically internal to the casting that is caused by a molten island of material that does not have enough feed metal to supply it. Shrinkage cavities are characterized by a **rough, dendritic, interior surface.**

Possible Causes:

- Volume contraction of the metal, either from liquid contraction of the melt or from contraction during phase change from liquid to solid
- Insufficient feed metal in defect areas
- Gating, feeding system and part design creates locally hot areas within the casting that are not fed well.

Corrective practice:

- Avoid heavy isolated casting sections that are difficult to feed
- Design the part with a progressive change in casting thickness
- Design the gating and feeding system to provide for directional solidification back to the risers
- For grey and ductile irons, increase carbon and silicon content as allowed, to decrease volumetric contraction of the metal during solidification
- Limit the pouring temperature so that the liquid contraction is minimized.

9.7.7 Inspection Methods

There are various test methods that may be employed to evaluate the soundness of a casting. These can be destructive or non-destructive tests. The more common non-destructive tests employed are magnetic particle inspection, ultrasonic, dye penetrant and radiography. The destructive methods of testing involve cutting slices in key areas to reveal any defects.

A combination of these inspection methods is employed to some degree for all castings. The inspection gives an added cost and therefore the extent of testing will be linked with the importance and integrity requirements of the casting. In addition, the test procedures are often carried out in detail during casting development and validation. Once a capable process has been established lower rates and frequency of testing are adopted. In addition, care must be taken when specifying the allowable defect severity. Over-specifying can lead to a more expensive part.

The testing requirements are typically based on some of the variables listed below:

1. The casting has a proven history with little to no issues
2. The design uses large safety factors
3. The application is not critical
4. The part can be cast with little trouble. Extra testing may be required if:

- The design is new and untested
- There are low safety factors used in the design.

9.7.8 Machining and Casting Drawings

Casting drawings will include information about the casting geometry (size, shape, draft, radii, etc.) The casting drawing will also contain the acceptable tolerance level, defect level, and surface finish. The drawing should include any inspection requirements, radiography, magnetic particle inspection, destructive testing. Material should be defined on the drawing with possibly the mechanical properties of the material. Geometric dimensions and tolerances should also be shown.

A casting can also have the part number, heat lot and foundry code cast into the part. Specification of the location and size of this lettering is required. If raised lettering causes interference, the lettering can be cast in a recessed pad. A machined casting will have dimensions for the machined features as well as dimensions relating the machined features to the cast features. The machining drawing may include a note on corner breaks required, surface finish requirements for the machined surfaces, geometric tolerances.

Foundries typically prefer two drawings, one for the casting and one for the finished machined part. The casting drawing should include the amount of finished stock required. (3mm of machine stock is typical.)

When geometric tolerances are used on a casting drawing or machined casting drawing it is good practice to establish datum points at specific locations on a cast surface for the various datum planes. When surface inconsistencies occur, this practice allows consistent inspection and layout at the foundry, machinist, outside inspection lab and customer. All tooling can be built to reference these datum points and it will help with any dimensional issues that are found with a machined casting.

9.8 CAST METAL DEFICIENCIES

There are several physical laws of nature and chemical reactions that result in adverse chemical and physical reactions which need to be controlled and avoided to allow the production of a high integrity castings free from imperfections. A picture guide on cracks is available at https://www.sfsa.org/misc/duncan/duncan.ppt[22]

The mechanical properties of cast steels are dependent upon the microstructure and the size, shape and distribution of micro-imperfections which are created during the melting, refining and the casting processes. As far back as the 1950s and through the 70s onwards, R&D work, particularly the US Steel foundry Association work, was carried out to relate the severity of micro-discontinuities on the mechanical properties and performance of castings. The general consensus of opinion is that hardness and strength are not significantly affected by the presence of micro-discontinuities but ductility, impact resistance, modulus of elasticity and fatigue properties can be reduced.

Many of the problems are common across the range of metals that are cast and the range of competing casting techniques. The common potential deficiencies can include the following:

9.8.1 Re-oxidation

The potential reaction between the molten metal or alloy and the oxygen in the surrounding atmosphere and environment (vacuum casting developed for highly reactive metals).

Figure 9.23 Sources of non-metallic inclusions

9.8.2 Metal penetration

This defect may occur with moulds made with sand (in particular green sand moulding processes), regardless of the metal, as the result of physical/chemical interactions between melt and moulding material components. Reactions between metal and moulding can cause burnt sand (sintering) and metal penetration. Metal penetrations especially occur at the edges of the sand mould or sand core where the metal remains liquid for a longer period due to the geometry of the casting and the moulds can become very hot.

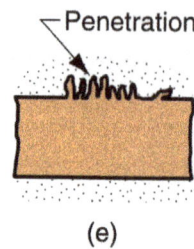

Figure 9.24 Metal penetration

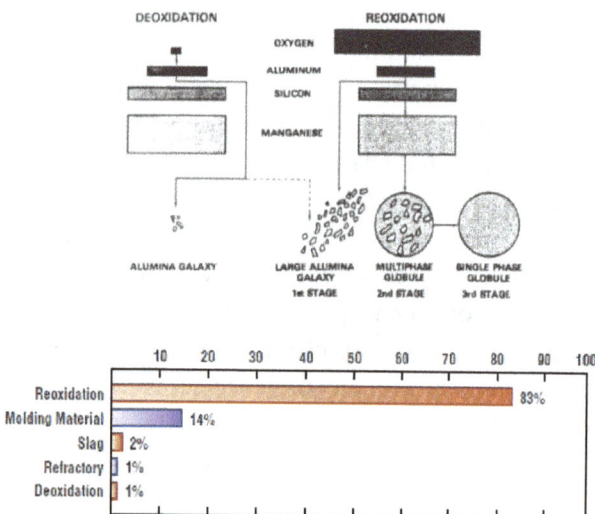

9.8.3 Shrinkage cavity

The contraction of the metal during cooling and solidification and the feeding of any last regions to solidify (interdendritic zones) (now a major dependence on software to design, the riser, and runner system and view the solidification pattern). The shrinkage defect

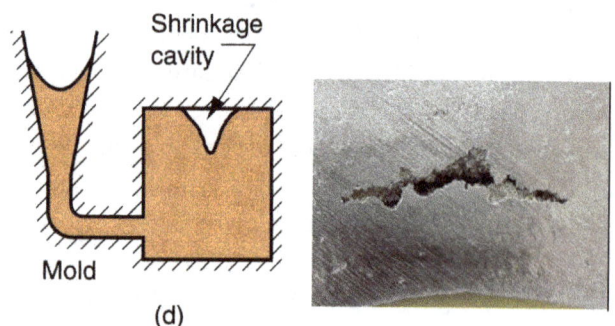

Figure 9.25 Shrinkage Cavity

is partly a function of the section thickness of the casting, but it also can be influenced by the pouring temperature, alloy purity, the riser position and size and the pouring speed.

9.8.4 Solidification crack

The potential formation of a rupture in the solid skin which forms as solidification begins due to build-up of metal pressure as the mould cavity fills as (solidification cracks) or constrain caused by mould or core hot strength.

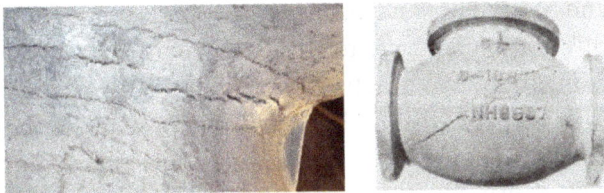

Figure 9.26 Solidification crack

9.8.5 Pinholes

The pick-up of gases (hydrogen, nitrogen and oxygen) during melting and their expulsion and potential formation of blowholes during solidification (gas porosity).

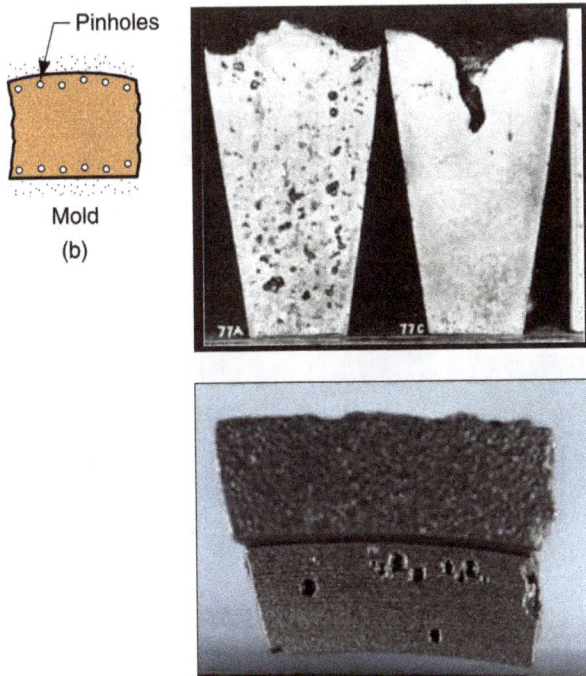

Figure 9.27 Pinhole

9.8.6 Residual stress

The generation of residual stresses due to different rates of cooling and crack formation during the occurrence of low ductility metal at various temperatures during cooling. Residual stress in a cylinder head during manufacture.

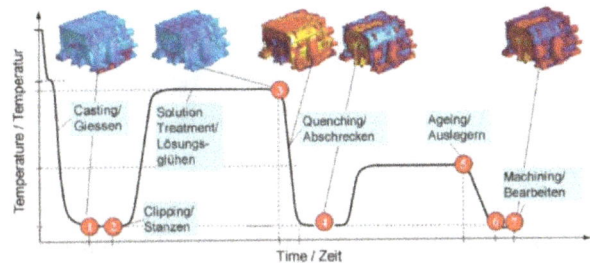

Figure 9.28 Residual Stress

9.8.7 Misrun

Incorrect mould fill due to low superheat or lack of metal fluidity.

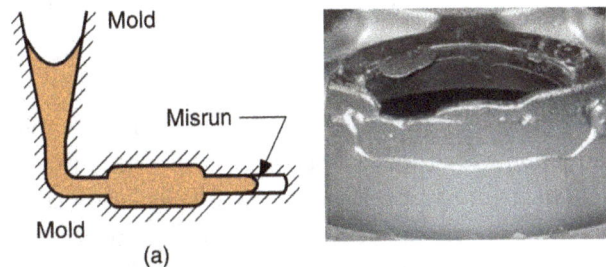

Figure 9.29 Misrun

9.8.8 Cold Shut

Two portions of metal flow together but there is a lack of fusion due to premature freezing cold shut.

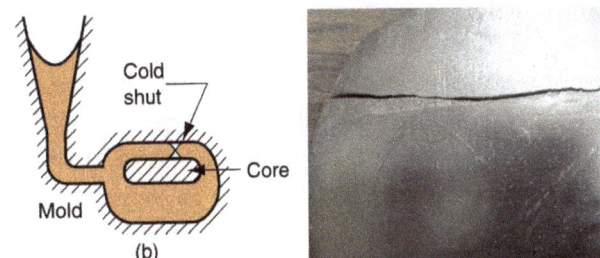

Figure 9.30 Cold Shut

9.8.9 Cold shot

Metal splatters during pouring and solid globules form and become entrapped in casting cold shot.

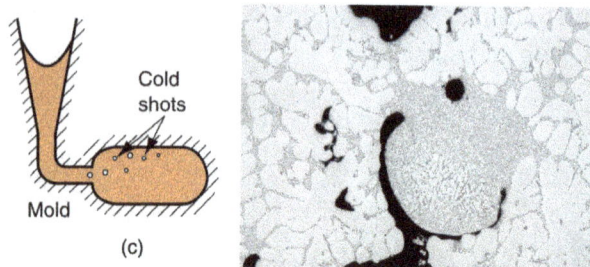

Figure 9.31 Cold shot

Cold flake is introduced into the die-cast product during the production process. During the initial stage of production, the molten aluminium is poured into the shot sleeve, and then injected into the dies by a plunger. Whilst the aluminium is being poured into the shot sleeve the molten aluminium is rapidly cooled on the inner surface of a sleeve producing a thin solidified layer. (It can be seen in the microstructure of the bowl due to its fine microstructure caused by the fast solidification rate in the sleeve.) When the melt is injected, the thin solid layer is scraped off or broken into small pieces and then pushed in the mould.

These small pieces are called as cold flakes or cold fills and have a thin oxide layer possibly also contaminated by lubricant. Poor wetting of the oxide layer with the matrix causes lack of bonding at the boundary between the cold flake and the matrix. The lack of bond introduces potential planes of weakness which reduce the strength and particularly the toughness of die-cast components.

A published R&D project sponsored by the US government identified the cause of solidified aluminium in the shot sleeve (referred to as externally solidified) which leads to the formation of the casting defect referred to as "cold flakes". The work concluded:

- The solid fraction present in the shot sleeve increases linearly with shot delay time
- The UTS, fracture elongations, and fatigue properties were dependent on the alloy pour temperatures and shot delay times
- The UTS and fracture elongation decreased as the alloy pour temperature decreased and the shot delay time increased.

The detection of cold flakes by the X-ray radiography is difficult because the cold flake and the matrix has the same transmissibility. The cold flakes have usually been detected only by destructive methods.

There are recent publications that claim cold flake in die-cast products can be detected by ultrasonic measurement. However, most current manufacturing routes are based on prevention by establishing process capability rather than inspection.

A problem associated with commercial cast aluminium alloys is that many have relatively coarse microstructure with relatively large plates of precipitates which provide easy crack paths through the microstructure. (See Optical metallography photos and SEM images.) This means that parts made from this grade may have relatively low tensile strength, elongation and fracture toughness. This type of microstructure would be acceptable for many of the commercial components that are manufactured from AISi12 alloy types. However, if the requirement is for high integrity to sustain high pressures 100bar with a

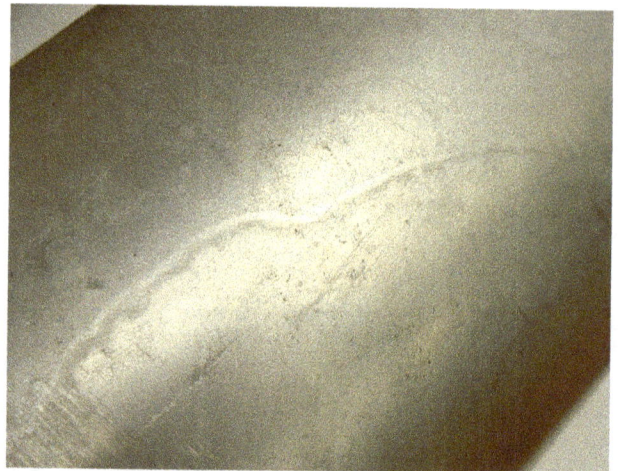

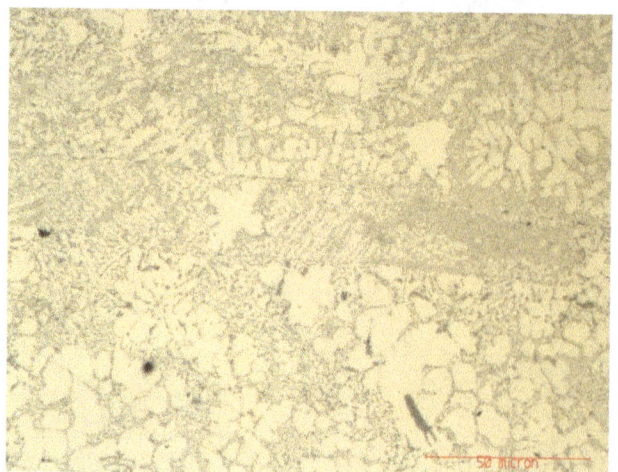

Figure 9.32 Surface indications Band of "cold flake" at inner surface x500

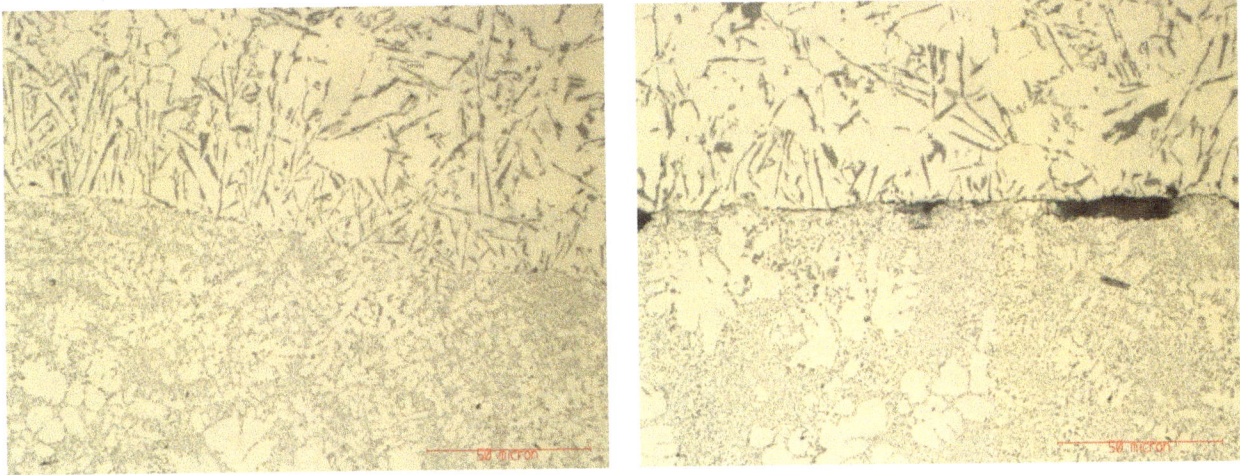

Figure 9.33 Cold flake with fine microstructure due to fast X500 Area of cold flake which has delaminated

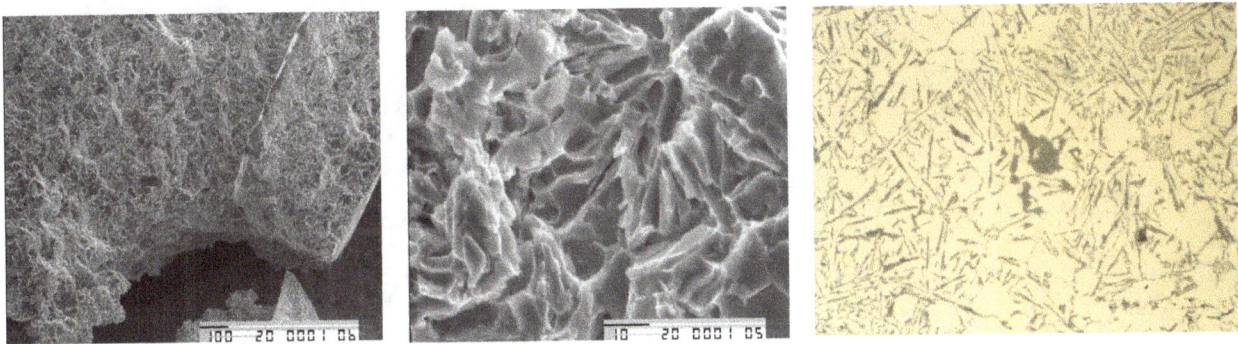

Figure 9.34 Delamination of a cold flake area Platelets that have been bent and cracked Region of coarse microstructure

5mm wall thickness, high toughness aluminium alloy would be required.

For applications that require a high integrity, high toughness, the microstructure can be refined by the use of "grain refiners" and "modifiers" in the form of Sb, Na, P and Sr are added which results in finer microstructure and improved mechanical properties.

Fracture toughness can be used to estimate the critical defect size prior to brittle fracture. Fracture toughness of cast aluminium alloys range from 7 to 8 through to 50 MPa m^1/2. High values are associated with aluminium alloys with elongation values of 10 to 15%. The low values are associated with elongations of 1 to 2%. Published data on aluminium alloys with "cold flake" gave values of toughness around 6 to 8 MPam^1/2.

9.8.10 Sand blow

Balloon-shaped gas cavity caused by release of mould gases during pouring sand blow.

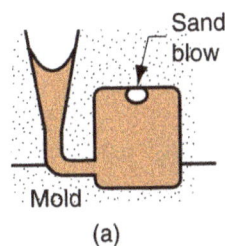

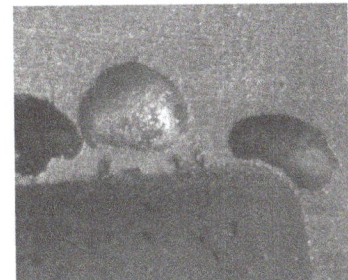

Figure 9.35 Sand Blow

9.8.11 Mould shift

A step-in cast product at parting line caused by sidewise relative displacement of cope and drag mould shift.

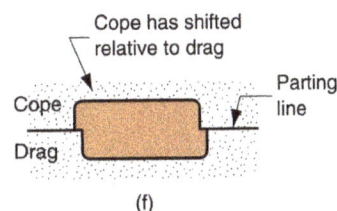

Figure 9.36 Mould Shift

9.8.12 Hot tearing

Occurs at location with high stress due to inability to shrink naturally. Resolve by mould collapsing or removing from the mould immediately after freezing.

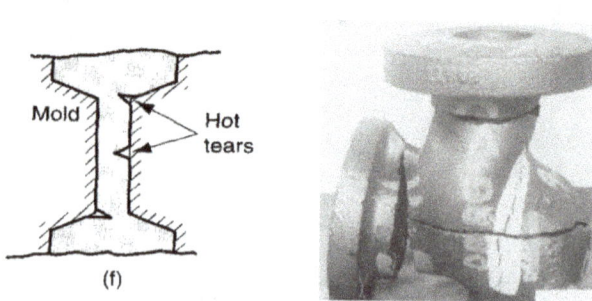

Figure 9.37 Hot tearing

9.8.13 Micro porosity

Network of small voids caused by localised solidification shrinkage. Caused by the freezing manner of the alloy.

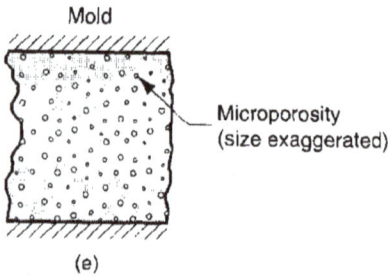

Figure 9.38 Micro porosity

9.8.14 Rock candy fracture

Caused by AlN precipitation of grain boundaries due to high aluminium and nitrogen.

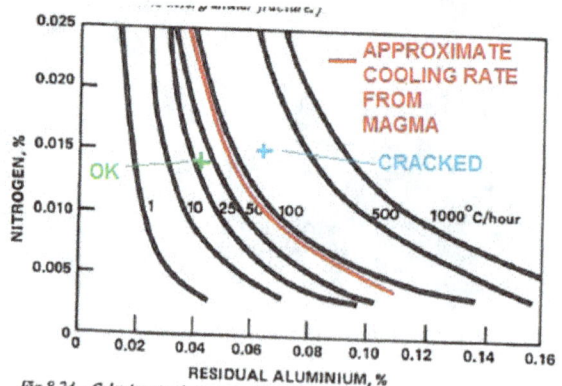

Fig. 8.24.—Calculated relationship between aluminium residual and nitrogen contents, and cooling rate, for the avoidance of intergranular fracture (after Hannerz[32]).

Figure 9.39 LHS = Rock candy fracture RHS graph shows the combinations of Al and N2 to the left of the red line are deemed acceptable for this cast part.

Rock candy fracture has been linked to coarse aluminium nitride precipitates and type II sulphide inclusions. In the worst case it can be visible on the Charpy impact specimens as shown below.

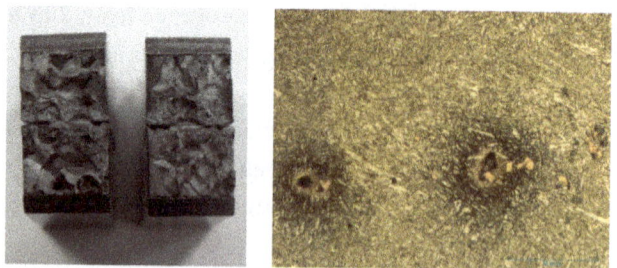

Figure 9.40 LHS. Rock candy visible on the Charpy surface. RHS. TiN inclusions in interdendritic regions. Mag approx. x500

To control the risk of rock candy embrittlement the foundry practice has been to:

- control the aluminium and nitrogen levels in accordance with the cross section of the casting
- in addition, practices have been developed to add zirconium calcium and titanium.

Both the published early R&D work carried out in the 1960s and 1970s on rock candy and recent work carried out in 2012 confirmed that titanium does not effectively reduce the risk of rock candy embrittlement.[15] In addition the titanium addition made to high nitrogen steel would result in large inclusions of titanium nitride in the inter-dendritic regions as seen in the microstructure above. The below SEM images of an inter-dendritic porosity with titanium nitride inclusions.

Many steel foundries are reluctant to use titanium for deoxidation on the grounds that it produces a "dirty" steel with an associated loss of impact strength.[16] In a recent publication on the fracture toughness of cast Cr Mo steels a similar conclusion was drawn "therefore the use of Ti in Cr Mo steels is strongly discouraged if high toughness is required"[17]. In addition, Ti will decrease the impact value by +/- 80 % and increase the transition temperature by 10 to 15 °C, compared to the effect by using aluminium.[18]

The use of a calcium addition appeared to partly modify the sulphide inclusions. There are many published endorsements of the use of ladle calcium deoxidation treatment to modify the sulphide inclusion morphology.

Previous work has evaluated the cooling rates of large castings. The cooling rates were determined using Magma software and the results compared to a SCRATA published graph.[19] The cooling rate of about 80°C per hour was determined. This was

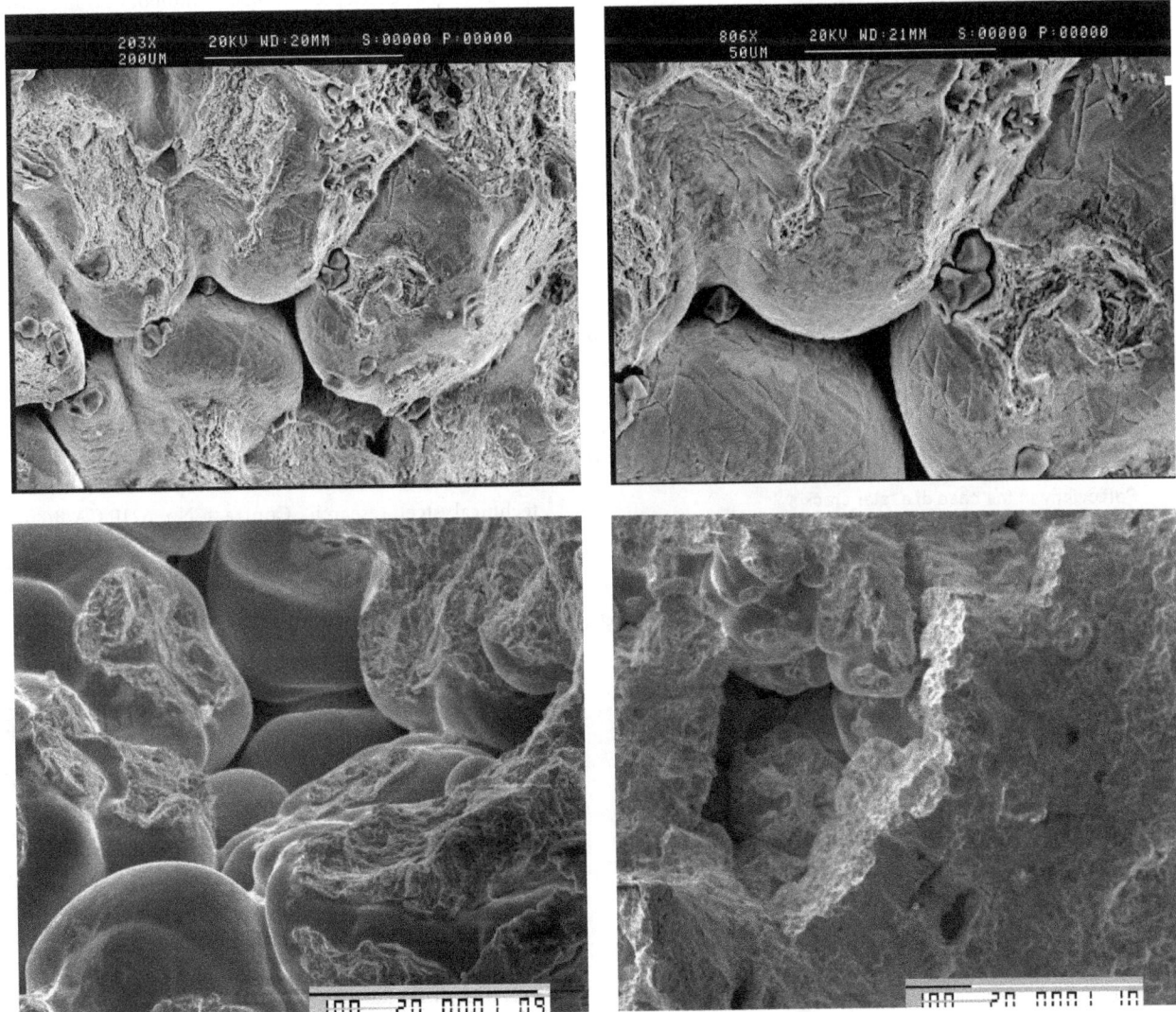

Figure 9.41 Inter-dendritic porosity on the Charpy specimens

superimposed on the SCRATA diagram (Figure 9.93) and indicated that the current maximum levels of aluminium and nitrogen were in the safe region on the SCRATA diagram. (Safe maximums would be Al=0.055% at N2= 0.015%.) It would be prudent to repeat this calculation for different sizes of the casting.

9.8.15 Riser cracks

The crack is in a riser contact surface

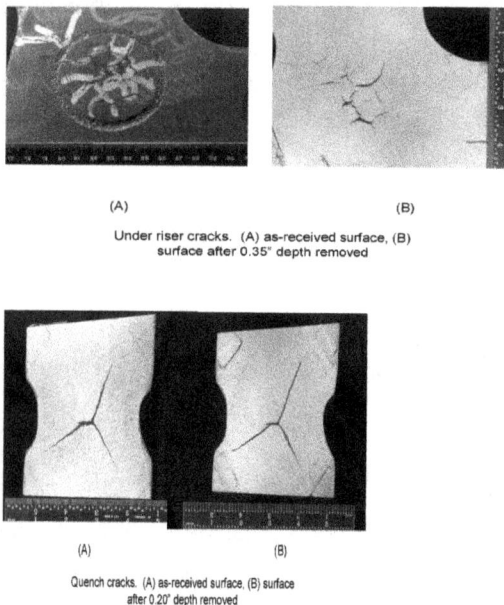

(A)   (B)

Under riser cracks. (A) as-received surface, (B) surface after 0.35" depth removed

(A)   (B)

Quench cracks. (A) as-received surface, (B) surface after 0.20" depth removed

Causes

Inadequate preheat during arc-air
Especially if alloy CE > .6
Under-riser C and Mn segregation extending into casting
Particularly in the case of a "star crack"
Secondary shrink under contact
Coarse & brittle as-cast microstructure
Grinding cracks normal to grind direction
Quench crack
Occurs only after Q&T heat-treatment

Cures

Preheat to 250F minimum before arc-air when CE >.6
Reduce segregation & secondary shrink:
Use mildly-exothermic hot-topping,
10 to 20% D riser thick
D neck > .6 D riser
Riser H/D > 1.5
Normalize @1750F prior to torch cut @ arc-air
Follow guidelines for proper quenching

Figure 9.42 Riser cracks

These potential problems frequently require additional NDT requirements compared to a wrought or forged product. In addition, the potential problems with cast metal are often used to persuade users that casting are inferior to forgings. This is frequently contested by organizations that represent foundries.

REFERENCES

1. Metalcasting Industry Technology Roadmap, Sponsored by the Cast Metal Coalition of the American Foundrymen's Society, North American Die Casting Association, Steel Founders' Society of America January 1998
2. U.S. International Trade Commission, Foundry Products: Competitive Conditions in the U.S. Market Investigation No. 332—460 Publication 3771 May 2005 https://www.usitc.gov/publications/332/pub3771.pdf
3. A Perkins: Continuous casting, casting and solidification. EUR 9086 1986.
4. J Y Lamant et al: Study of bulging of continuously cast slabs EUR 8963/111N 1985
5. B Patrick et al: Crack prevention in continuous casting EUR 18558 EN
6. S V R Raja: Control of centre-line cracking in continuously cast Corten steel slabs Steelworld 24 Oct 2008
7. K Stransky et al A transversal crack in a steel slab and its analysis P 751 GA No 106/08/0606
8. Guosen Zhu: Formation of internal cracks in continuously cast slabs Journal University Science and Technology Beijing. Vol 11 No5 October 2004 p398
9. Y U Tieu et al: Investigation of closure of cracks during rolling by FEM model considering crack surface roughness. International journal of Advanced Manufacturing Technology 75 (9-12) 1633-1640, 2014
10. P.E. WAUDBY et al, Segregation in wideendup ingots (Part I), Commission of the European Communities, 1982, EUR 7723/I
11. P.W. WATERWORTH et al, Segregation in wide-end-up ingots (Part II), Commission of the European Communities, technical steel research, Contract No 6210.CA/8/803 (1.5.1975 – 31.3.1981) FINAL REPORT 1982, EUR 7723/II
12. B.E. WATTS, Steelmaking Segregation in wide-end-up ingots (Part 3), Commission of the European Communities, Contract No 6210.CA/8/803 (1.5.1975 – 28.2.1982) FINAL REPORT 1983, EUR 7723/III
13. C-Å. Däcker et al, Innovative process technology for ingot casting by application of simulation and measuring techniques (IPTINGOT), European Commission Research Fund for Coal and Steel Grant Agreement RFSR-CT-2011-00006 1 July 2011 to 31
14. E.J. Pickering, Macrosegregation and Microstructural Evolution in a Pressure-Vessel Steel, Metallurgical and Materials Transactions A 2014, http://www.msm.cam.ac.uk/phase-trans/2014/evolution.pdf
15. J. Senberger et al, Influence of Compound Deoxidation of Steel with Al, Zr, Rare Earth Metals, and Ti on Properties of Heavy Castings. Archives of foundry engineering, Volume 12, Issue 1/2012, page 99-104

16. Nigel Howard Croft, Thesis submitted in fulfilment of Doctor of Philosophy, at the University of Sheffield. December 1981. Page 15

17. L N Bartlett et al, Dynamic fracture toughness of high strength cast steels, AFS Proceedings 2012 paper 12-054.

18. G Henderieckx, Steel casting de-oxidising. Dec 2004.

19. W J Jackson, Steelmaking for founders, Steel casting Research Association 1979, page 243 to 245

20. file:///C:/Users/user/Downloads/Wall_chart_STELEX__e_. pdf

21. https://www.sfsa.org/misc/duncan/duncan.ppt

Cast metal history

http://www.pmt.usp.br/academic/martoran/NotasFundicao/ LinhaTempoFundicao.pdf

Defects

http://maritime.org/doc/foundry/part3.htm

SELF-ASSESSMENT QUESTIONS

1. Explain the difference between an expendable mould and a permanent mould. What are the advantages and disadvantages of each?

2. Patterns are often made oversize to provide for so called "allowances". List three reasons why a pattern should be made oversize if a part is to be made as a sand casting.

3. A feeder in casting is described by which of the following (three correct answers): (a) waste metal that is usually recycled (b) gating system in which the sprue feeds directly into the cavity (c) metal that is not part of the casting (d) source of molten metal to feed the casting and compensate for shrinkage during solidification?

4. Which of the following casting processes are permanent mould operations (three correct answers): (a) centrifugal casting (b) die casting (c) expanded polystyrene process (d) sand casting (e) shell moulding (f) slush casting?

5. Which of the following metals would typically be used in die casting (three best answers): (a) aluminium (b) cast iron (c) steel (d) tin (e) tungsten (f) zinc?

6. Shell moulding is which one of the following: (a) casting operation in which the molten metal has been poured out after a thin shell has been solidified in the mould (b) Casting operation used to make artificial sea shell (c) casting process in which the mould is a thin shell of sand bound by a thermosetting resin (d) sand-casting operation in which the pattern is a shell rather than a solid form?

7. Which die casting machines usually have a higher production rate: (a) cold-chamber casting (b) hot-chamber casting (c) shell casting (d) sand casting?

CHAPTER 10

Engineering Materials

10.1 MATERIALS EVOLUTION AND SELECTION

Materials that are used to sustain mechanical loads are referred to as engineering materials. They are classified into four groups, metals and alloys, polymers, ceramics and glasses, and composites as shown in Figure 10.1.

One or more of these groups of materials are used in the manufacture of virtually all engineering products.

Although there are detailed material mechanical property data available to the design engineer, the initial design and development of engineering products still needs extensive development work and test work to guarantee a high level of reliability and long life that all users now expect. This is especially true when new materials are used. This can represent a significant part of any product development budget.

In general terms reliability can easily be achieved by the use of conservative targets of stress and the use of high-quality materials and carefully controlled methods of manufacture. A design approach adopting this practice would guarantee high reliability and low risk of failure. In the converse circumstances (**usually associated with design optimisation**) where stress levels are high and **"value analysis"** techniques dictate low material and manufacturing costs, then the risk of failure would be higher especially in designs where the details of future service loads and condition are limited. The comparison of these two design options provides the blindingly obvious solution to achieve high reliability and to prevent failures. This would be an intrinsically "safe design" with low working stresses and using techniques of manufacture and inspection that would ensure **high integrity regardless of cost**. Unfortunately, this would be a high cost strategy and normally unsustainable in a commercial environment.

In an area where we all think that cost is irrelevant, we have an interesting comment made by John Glenn regarding cost vs reliability:

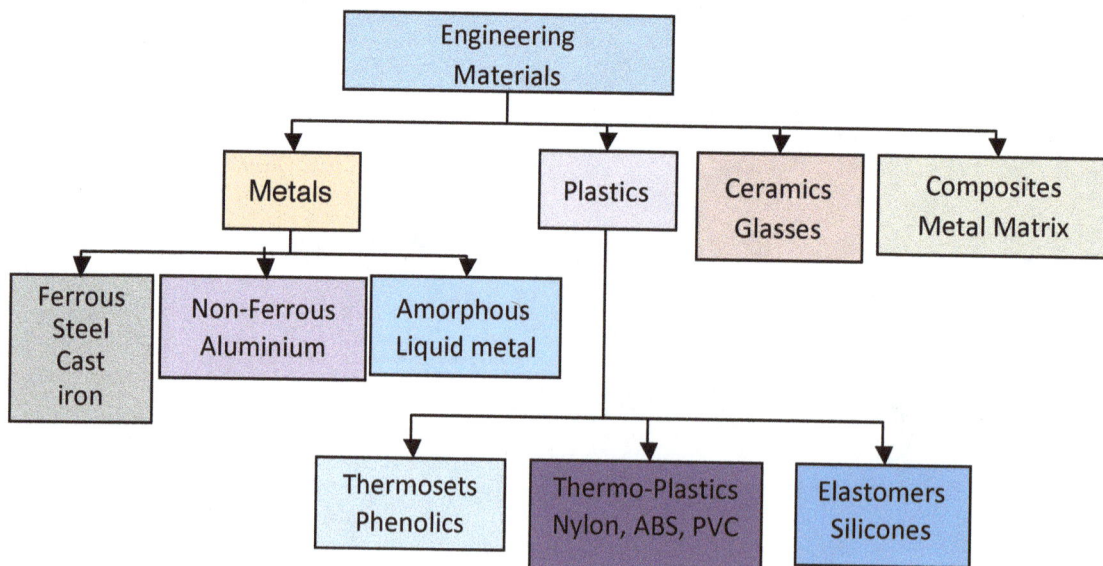

Figure 10.1 Engineering materials

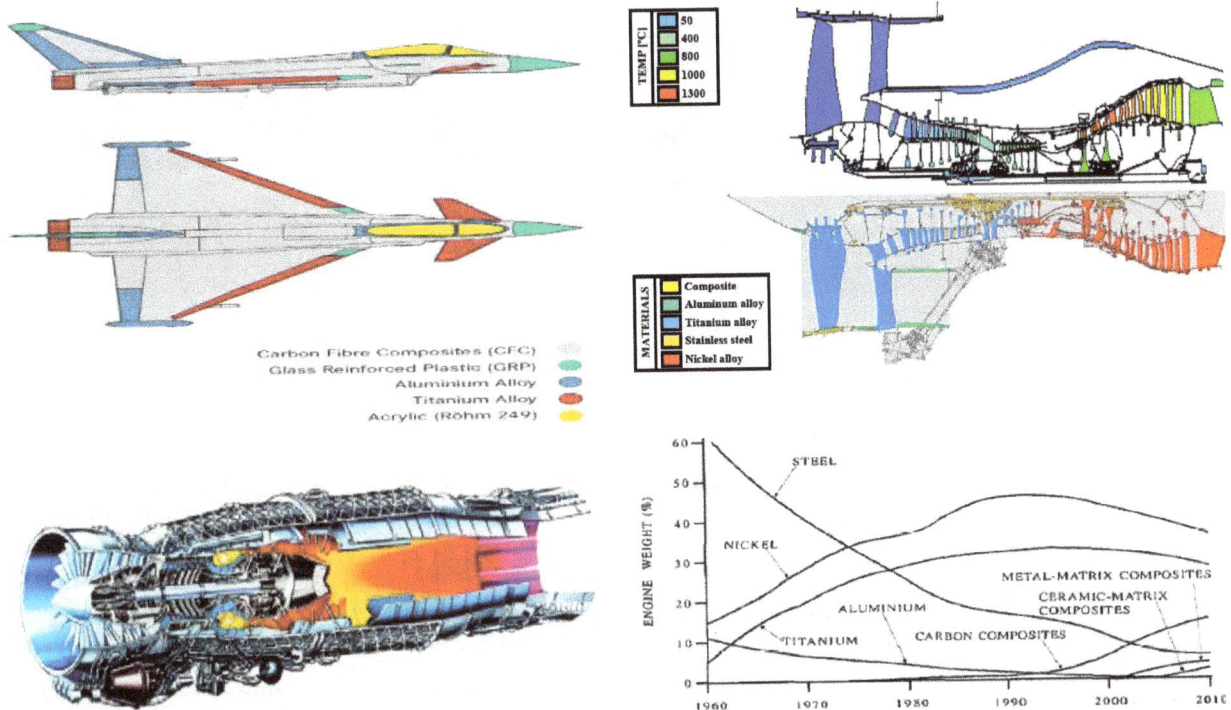

Figure 10.2 *LHS top materials used in Typhoon*[4], *RHS top Temps and materials in RR engine*[5], *LHS bottom RR EJ200 the Typhoon engine, RHS bottom Material evolution for engines*

"I guess the question I'm asked the most often is: "When you were sitting in that capsule listening to the count-down, how did you feel?" Well, the answer to that one is easy. I felt exactly how you would feel if you were getting ready to launch and knew you were sitting on top of two million parts – all built by the lowest bidder on a government contract."

– John Glenn

Each engineering item has a history of development. For example, there is a vast history of development work on Gears, Springs, Fasteners, Internal combustion engines, Pressure vessels, Storage tanks, Jet engines and each of these items has an associate "Product Design Specification" which gives a clear statement/target of the design requirements. Website http://www.tribology-abc.com/[1] together with the book for sale on this site (http://www.tribology-abc.com/book/default.htm)[2] gives an introduction to product design.

The design specification would consider the future service conditions and maintenance requirements together with the service loads and operating environments and the requirements of the methods of manufacture. Material properties cannot be considered in isolation. The need for a holistic approach has been clearly demonstrated by the work of M Ashby

and associates in the book "Materials Engineering Science, Processing and Design"[3].

An example of the materials used in the aero-space industry and how the materials have changed over time are shown in Figure 10.2. lower RHS

Figure 10.2 Top LHS shows the Euro-Fighter Typhoon aircraft where materials are selected from all four groups of engineering materials. The high temperature parts of the engine such as fuel vaporisers and high temperature turbine blades use a thermal barrier coating with a NiCrAly low pressure plasma sprayed overlay followed by a plasma sprayed Yttrium stabilised Zirconium. The body structure being made from carbon fibre composite, aluminium lithium alloy and titanium.

Most material developments have occurred through "application led" projects which have focussed on cost benefits or improved performance. It would be tempting to think that at this stage of development the laws of diminishing returns were becoming a key limitation to further improvement and that we should now begin to rationalise material specifications. This has happened to some degree in the steel sector. A key diagram used by M Ashby, H Shircliff and D Cebon in the book "Materials Engineering Science, Processing and Design"[3] is shown Figure 10.3. This clearly shows how the knowledge of materials and the number of materials has expanded. These facts

Figure 10.3 *Timeline of materials development and use[3]*

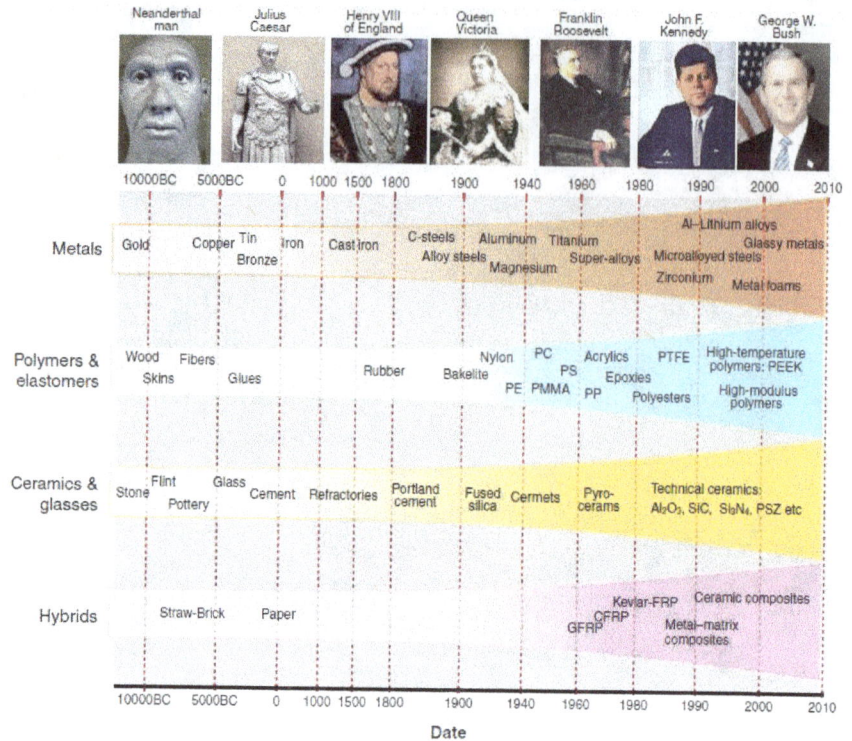

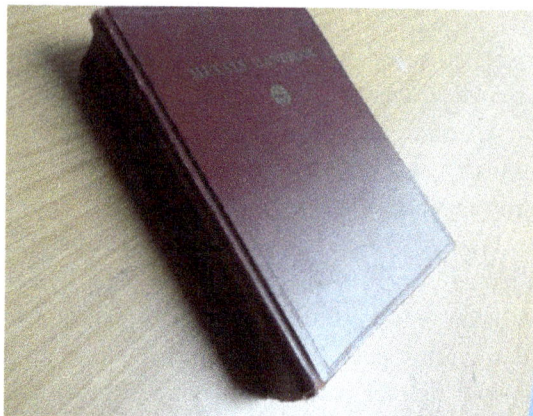

Figure 10.4 *Comparison of Metals handbook 1948 and the 9th edition electronic pdf books[6 and 7]*

are well appreciated by all elderly engineers since a significant part of this expansion has occurred in our working life.

A good example is the ASM Metals Handbook. The 1948 Edition (the year I was born) is shown in Figure 10.4 LHS and is an A4 size book about 50mm thick. The 9th edition has now 22 similar sized books and 1.5GB file size.

There are several books that have covered the topic of steel/material selection to several generations of engineers, and have provided a gradual introduction to the subject of steel selection over many years. Some examples are shown in the following pages.

The book "Steel Selection" provided several key diagrams which have been useful, for example:

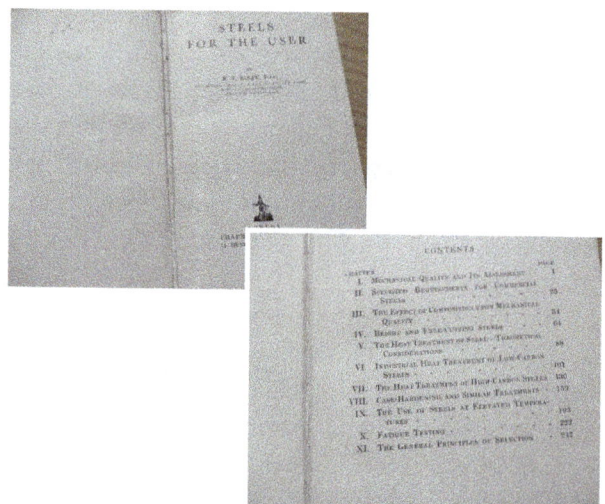

Figure 10.5 *Steels for the user. Published 1937[8]*

1 The Useful Engineering Characteristics of Steel

2 General Aspects of Proper Design for Heat Treatment: Austenitization

3 General Aspects of Proper Design for Heat Treatment: Quenching Effects

4 Commercial Aspects of the Steel Industry

5 Characteristics of Constructional Steels

6 Hardenability and Tempering Parameter

7 Compromises in Steel Selection

8 General Principles for the Selection of Direct-Hardening Steels

9 General Principles for the Selection of Steels for Case Hardening

10 Selection of Steels for Carburized Gears

11 Selection of Steels for Shafts

12 Selection of Steels for Heat-Treated Springs

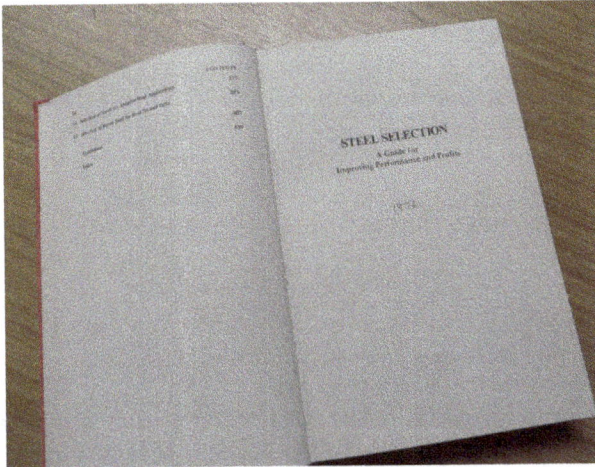

Figure 10.6 *Steel Selection 1976[9]*

Test	Unit of Measure	Typical Specified Range
Brinell (3000 kg)	Diameter of impression in millimeters (Bmm) Brinell hardness number	0.3 to 0.4 mm (or minimum only) 40 to 60 Bhn
Rockwell	C scale	5 to 6 points (or minimum only)
Rockwell	B scale	8 to 10 points
Rockwell	A scale	3 to 5 points
Rockwell Superficial	15–N	4 points (or minimum only)
Tukon	Knoop	200 to 225 points (or minimum only)
Vickers (500 g)	DPH	120 to 150 points (or minimum only)

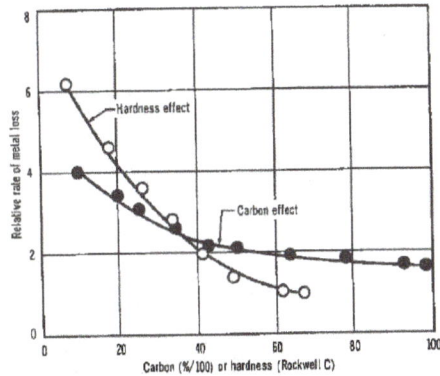

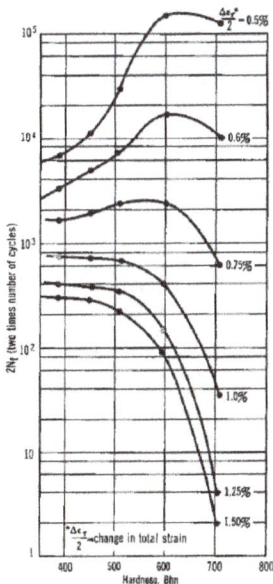

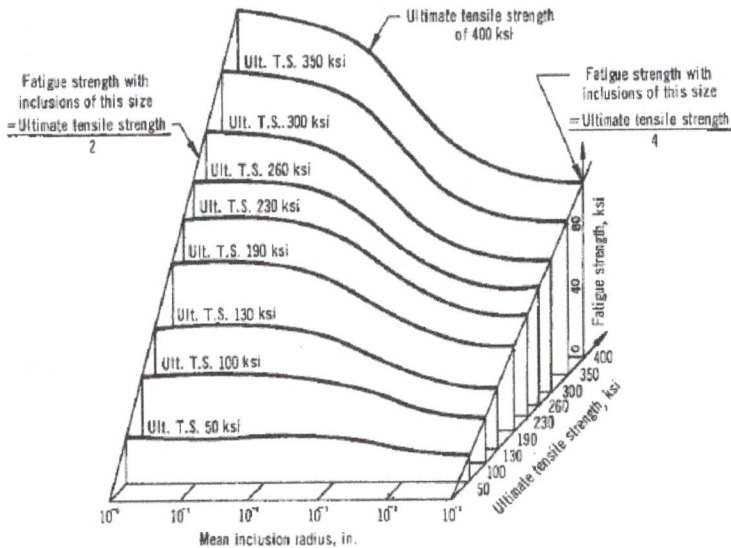

Figure 10.7 *Useful diagrams from steel selection[9]*

Steel Design rules – procedure to select a steel type.

1. Select carbon content and range to an appropriate standard (SAE or BS). Keep the carbon content as low as possible; 0.45% should be the top limit for most applications requiring direct hardening (except for spring steels for which 0.60% nominal carbon is preferred). For carburised gears it is preferable to restrict carbon to 0.25% max. For applications that are designed on yield strength Figure 10.8 may be used to establish the minimum carbon level. This nomogram relates minimum yield strength to hardness for through hardened (95% martensite) sections. In making use of this figure, it is recommended that the "as quenched" hardness be at least five points Rc higher than the final hardness.
2. For additional hardenability start with 0.80 to 1.10% Mn and avoid higher than 1.10 to 1.40% in carbon steels. For alloy steels, 0.75 to 1.00% Mn is preferable; do not exceed 1.30% Mn.
3. Set silicon content to 0.15 to 0.30% for carbon and alloy grades.
4. Sulphur content should be 0.05% max for carbon steel and 0.040% max for alloy grades.
5. Hold phosphorus to 0.040% max for carbon steels and 0.035% max for alloy grades.
6. Boron content, when specified, should be 0.0005 to 0.003%.

If an alloy steel must be designed, the following procedure will provide the best quality-cost combination based on current price books.

1. Start with molybdenum at 0.13 to 0.20%.
2. If short cycle fatigue strength is an engineering requirement, call for at least 0.40 to 0.70%Ni.
3. If additional hardenability is required call for chromium at 0.25 to 0.40% or 0.45 to 0.65%.
4. If still higher hardenability is required increase molybdenum to 0.15 to 0.25%, 0.20 to 0.30% or 0.30 to 0.40%. Note: For carburising with direct quench hardening of gears do not use more than o.30% Mo; or change heat treatment practice to reheat for hardening.
5. If further increased hardenability is required, increase chromium to 0.70 to 0.9%.
6. For maximum hardenability and /or short cycle fatigue strength increase nickel to 0.80 to 1.00%, 1.20 to 1.50% or 1.65 to 2.00%.

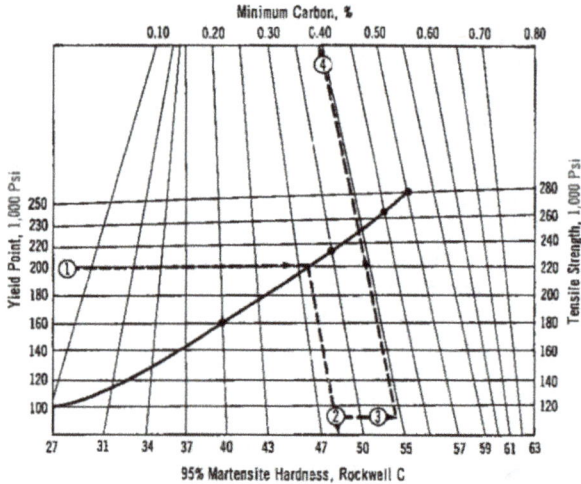

Figure 10.8 *Steel design procedure. Determination of minimum carbon content and hardness to achieve a specified yield strength or tensile strength*[9]

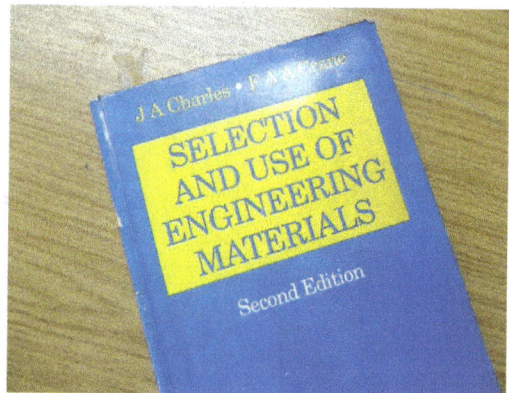

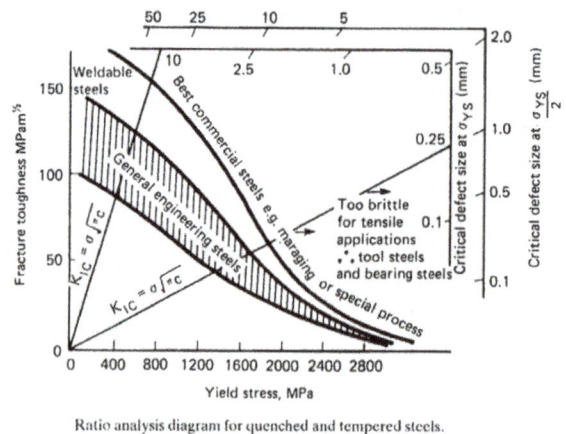

Ratio analysis diagram for quenched and tempered steels.

Figure 10.9 *Selection and use of engineering materials*[10]

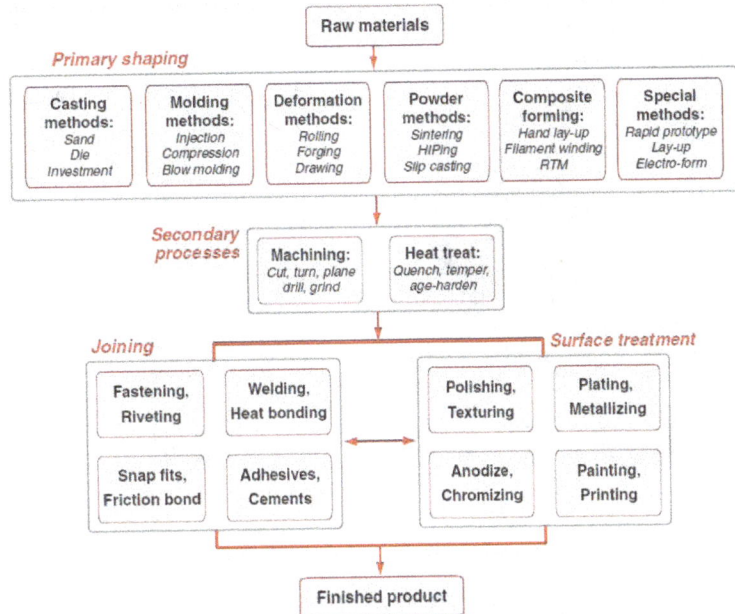

Figure 10.10 *In 2000 the guidance on materials was to focus on the holistic approach where materials, RHS = the method of manufacture and service property requirements are all considered*[3]

10.2 METALS OF INTEREST

10.2.1 Introduction to Iron and Steel—History and background

10.2.1.1 Cast iron

Cast iron is one of the oldest metals and has a long history of manufacture, being one of the first ferrous materials used. In Britain iron manufacture was carried out prior to the Roman invasion in the Weald based on local deposits of iron ore. Following the Roman invasion, the Weald supplied much of the iron for the Roman army in Northwest Europe. The Weald's iron deposits lay relatively close to the surface, and so were easily extracted by open cast mining and the local forests provided a source of charcoal.

Historical evidence suggests that iron production dropped dramatically with the breakdown of the Roman Empire in Britain. However, by late medieval times, iron production was sufficient to supply local needs and also to fulfil large military related orders. Records show that in 1242 the Archbishop of Canterbury was asked to provide 8000 horseshoes and 20,000 nails from his estates in the Weald, to be delivered to Portsmouth, almost certainly for Henry III's campaign in France that year. Throughout the history of the development of new and better materials military finance/interest has always been a major influence.

Based on carbon C14 dating an early form of iron smelting was carried out in a 'bloomery' found at Tell Hammeh, Jordan, which dates to 930 BC. The conventional date for the start of the "Iron Age" is 1000BC. This technique used a simple hole in the ground which formed a small conical furnace in which iron was smelted with burning charcoal. A goat's skin bellows was used to provide air for combustion and to achieve higher temperatures. The end point of the process was the creation of an iron rich mixture in the hearth of the bloomery. This iron 'bloom', contained a high degree of slag and non-metallics which was then either purified by hammering into a solid bar or was forged immediately into tools or weapons. (See Figure 10.11)

By the twelfth century, several continental bloomeries had increased production dramatically by using waterpower which eventually led to the permanent 'blast' furnace structures where stronger draughts were blown by the use of bigger bellows driven by a water power Figure 10.12 LHS. The improved air blast allowed higher temperatures to be achieved, which enabled the manufacture of molten pig iron.

Prior to the development of the blast furnace, the only options for steel manufacture were:

1. By modification of the conditions in the bloomery furnace such that some carbon was retained in the bloom and in this way producing NATURAL STEEL.

Figure 10.11 Bradford archaeological society replicating early iron manufacture. Bottom = The Bloom

2. By heating bloomery iron in a bed of charcoal under conditions such that carbon would diffuse into the iron. In early times such a process seems to have been confined to the treatment of finished articles so as to CASE HARDEN them.

Subsequent to the availability of cast iron, and of wrought iron derived from it, the scope became wider:

3. By enclosing small pieces of cast iron within layers of wrought iron sheet and heating them to a bright red heat (but short of fusion), out of contact with air, this being the CHINESE PROCESS.

4. By melting cast iron in a crucible and immersing bars of wrought iron in the melt; in this way the iron bars absorbed carbon from the cast iron and after a while could be taken out and forged to give BRESCIAN STEEL. Brescian steel made in Italy, enjoyed some fame in the 14th – 17th century and was used by later sixteenth-century Milanese armourers. A lump of bloomery iron ("weighing thirty to forty pounds") was supposed to be swirled about on the end of an iron rod in a bath of liquid cast iron for 4 to 6 hours, with crushed marble added, until it was carburized, and then taken out and forged into a uniform product.

5. By applying the case hardening process to layers of bar iron, interspersed with powdered charcoal in large sealed chests, BLISTER STEEL could be made by the cementation process. The blister steel could then be broken into short lengths, these made into faggots and forge welded together to give SHEAR STEEL (developed in the North East and exported to the world from Newcastle).

6. By melting the cast iron and burning out just sufficient of the carbon by what was essentially a modification of the finery process to produce STYRIAN STEEL (which was also known as GERMAN STEEL and, most confusingly, was also referred to as NATURAL STEEL).

7. By using a modification of the later puddling process and similarly burning out only sufficient carbon to produce steel, cast iron could be converted to PUDDLED STEEL.

All the routes so far described provided steel as a 'bloom' and did not involve the production of liquid steel. Full metling the steel, suitable for casting into ingots or for the production of steel castings, required the development of the Crucible process by Huntsman which became available from around 1740.[11]

The first British blast furnace referred to as the "Queenstock" furnace, was built around 1490 in the Buxted area in the Wealden District of East Sussex. Historical records show that in 1496 Henry VII sought "state of the art" weapons for his impending military campaign against Scotland. French ironworkers from the Pays de Bray in Normandy, an area renowned for its iron-smelting expertise, were invited here to cast ordnance using the Walloon system. (The pig iron was re-melted and cast, then forged using water powered hammers.) These events were nearly two hundred years before the iron manufacture began at Coalbrookdale in Shropshire and some historians were tempted to call this period the first "British industrial revolution". Many skilled immigrants arrived from France over the next fifty years or more, and much of the original French terminology has survived, e.g. 'chafery' (Blacksmith's forge) and 'tuyere' (the clay pipes for the bellows). By 1574 there were over 50 blast furnaces in the Weald.

After 1650 cheap Swedish bar iron imports seriously undercut Wealden iron prices. In addition, the demand for charcoal fuel from other industries made charcoal increasingly expensive and these factors

Figure 10.12 *LHS Blast furnace circa 1490s. RHS Bayeux tapestry showing steel armour used at the Battle of Hastings 1066. (A family feud between brothers Harold and William. 6000 dead.) Norman-French army of William, the Duke of Normandy, and an English army under the Anglo-Saxon King Harold Godwinson, beginning the Norman conquest of England*

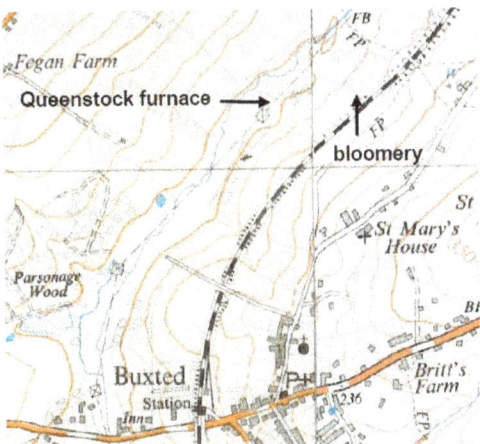

34 A shaft furnace from Greece (6th century BC) depicted on a vase showing a smith forging a bloom in front with the bellow behind (after Blumner[42])

Evidence of the efficiency of the Roman supply services was found in the enormous hoard of nails discovered at the legionary fort of Inchtuthil in Scotland.[10] This weighed over 5 t and contained nearly 900 000 nails of various sizes. As the fort was built in AD 83 and evacuated soon after AD 87, this shows that the supply services must have been operating efficiently before the period of the *Notitia*. From a metallurgical point of view there is nothing surprising about the composition or the structure of the nails. In many ways they resemble the construction of the iron cramps that key the stones of the Athenian Parthenon, i.e. they were made of forged heterogeneous bloomery iron. But they were well made to exact dimensional specifications, and the larger nails contain rather more carbon than the shorter nails (*see* Table 35). This was intentional, since the larger nails would have needed to be stronger to withstand the increased driving force and it was probably achieved by selection of suitable blooms or parts of blooms. There is no way of knowing where the iron was made: it was certainly made from low-phosphorus ores and the nickel content was also low. While such iron could have been obtained in Britain by this time, it would be shipped by sea. All one can say is that its composition is typical of the Roman period.

Table 35 Composition of nails from Inchtuthil, Scotland (after Angus et al.[13])

Group	Length, cm	Composition of heads, %				
		C	Si	S	P	Mn
A (i)	25−37	0·2−0·9	0·15	0·009	0·008	0·17
A (ii)	22−27	0·22−0·8	0·08	0·017	0·043	0·03
B	18−24	0·05−0·7	0·10	0·007	0·009	0·03
C	10−16	0·10−0·55	0·04	0·006	0·053	nil
D	7·2−10	0·05	0·08	0·01	0·035	0·03
E	3·8−7·0	0·06−0·35	0·05	0·003	0·16	nil

A (i) pyramidal heads; A (ii) flattened pyramidal heads

Figure 10.13 *Upper LHS – Buxted in the Wealden District of East Sussex. Map showing the location of the first blast furnace in Britain 1490. Upper RHS a vase showing a picture of a bloomer from the 6th century BC, Lower LHS Chemical analysis of roman nails. Above Roman nails Lower RHS description of the find of Roman nails in Scotland*

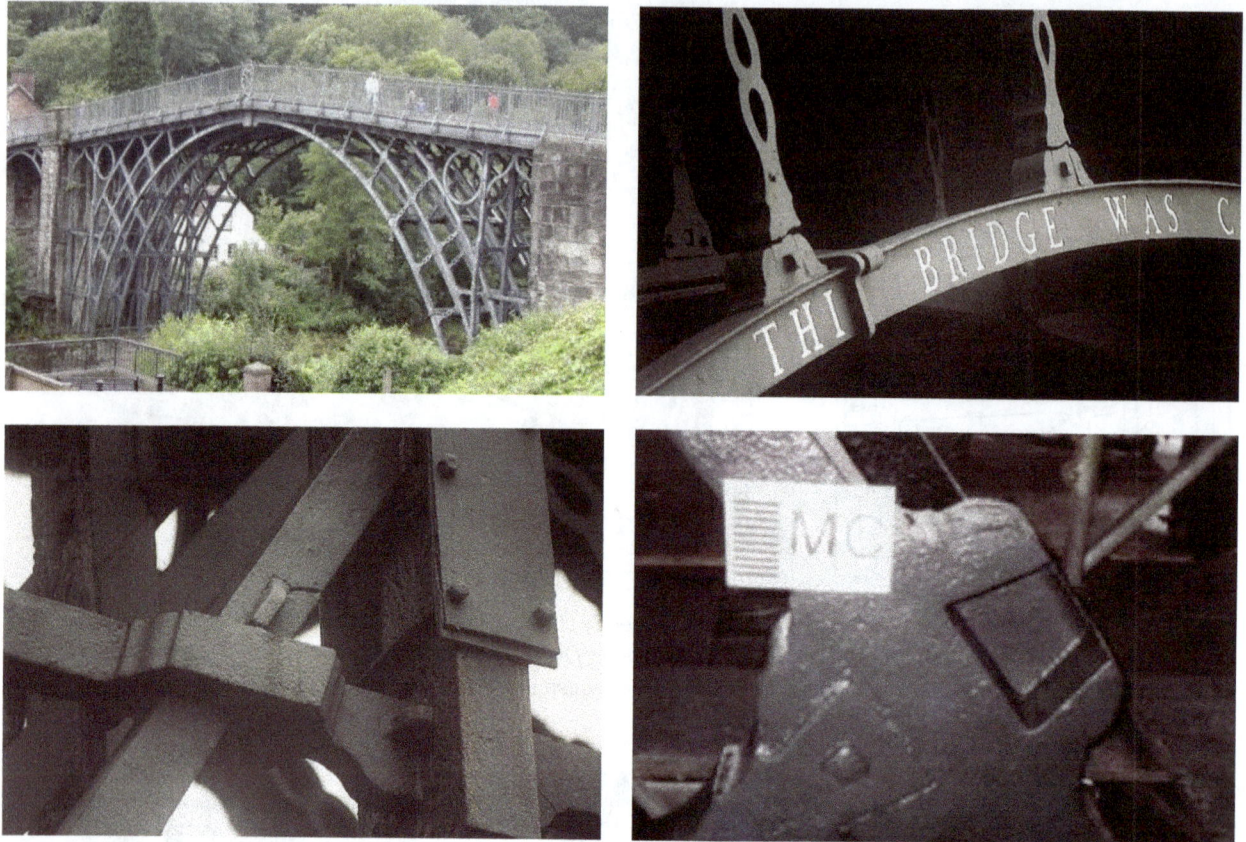

Figure 10.14 *Top= Iron Bridge. RHS Cracked parts Bottom LHS = Repairs RHS Blind dovetail joint on the Iron Bridge based on wood fabrication*

Figure 10.15 *Coke less cupola*

subsequently led to the closure of the manufacture of iron in the Weald.

Around 1709 Abraham Darby I of Shropshire finally succeeded in using local coal, converted to coke, for smelting iron, which restored the competitiveness to UK produced iron. The use of coke for iron manufacture, although initially kept secret, spread throughout Europe and began what we refer to in History as the "Industrial Revolution".

The lower cost of manufacture for cast iron made from the use of coke expanded the use of cast iron to larger structures. This allowed the construction of the

"Iron Bridge" across the River Severn in Shropshire which was the first arch bridge built with cast iron. This world's first iron bridge was erected in 1779 by Abraham Darby III. Costing over £6000, the bridge was cast in his Coalbrookdale works.

One of the final UK innovations in the last century was the "coke less cupola" which used large refractory balls to support the charge and gas or oil plus air used for the source of heat. This technology was sold to Korea in the 1990s and several European countries.

10.2.1.2 Classification of Cast Irons

The classification of cast iron has been typically based on the metallographic structure and fracture appearance. Cast iron that contains iron carbide produces a "white" fracture surface and is therefore referred to as white iron. Cast iron containing graphite produces a grey fracture appearance and is referred to as grey iron.

There are six main variables that affect the microstructure of a cast iron:

- Carbon content
- Silicon content, inoculation elements and modification elements
- The alloy and impurity content
- The cooling rate during freezing
- The cooling rate of the solidified casting
- The heat treatment after casting.

These variables control the condition of the carbon and also its physical form. The carbon may be combined as iron carbide in cementite, or it may exist as free carbon in graphite. The shape and distribution of the free carbon particles will greatly influence the physical properties of the cast iron. This results in various types of cast irons.

10.2.1.3 Steels

A comprehensive coverage of this large group of engineering alloys represents a difficult task. Steels are sometimes referenced by their microstructure and sometimes by a commercial name or application. Figure 10.18 shows the two classifications. Some of the important groups are the very low carbon automotive steel strip, low-carbon weldable structural steels, engineering steels and stainless steels, wear resistant steels, tool steels, carburizing steels nitriding steels, bearing steels, rail steel.

The technology associated with the diversity of steels can be found by reviewing "alloy steels" Special Issue Editor, Robert Tuttle, MDPI AG, First Edition 2018 file:///C:/Users/user/Downloads/Alloy%20 Steels.pdf[12]

To understand "alloy steel" it is essential to appreciate the effects of alloying elements. The purpose of adding intentional elements into steel (alloying elements), usually as ferro-alloys, is carried out to change

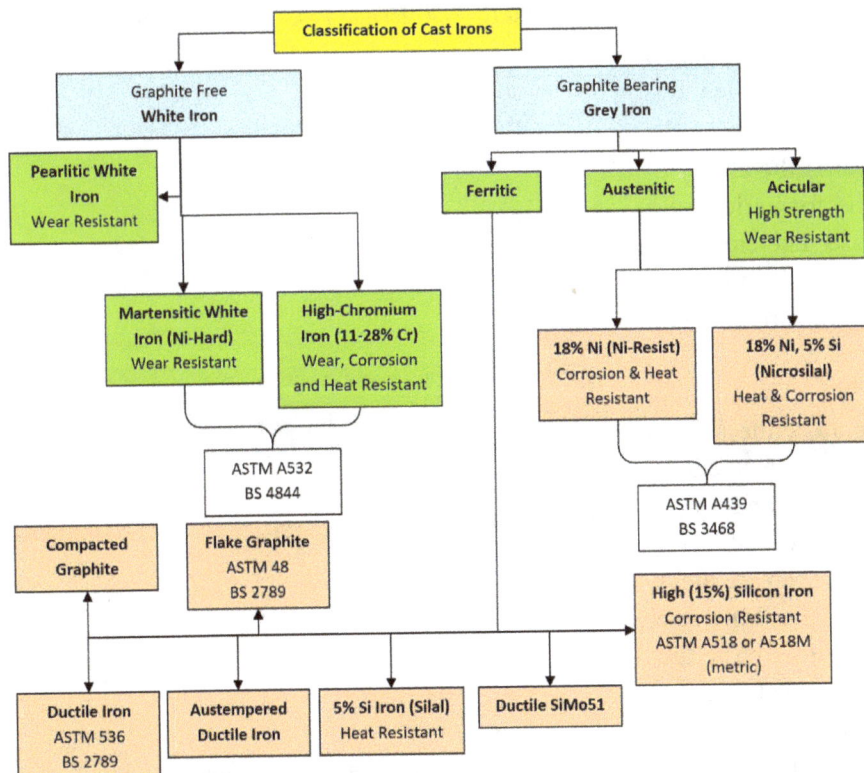

Figure 10.16 Classification of cast irons

the chemical composition with the aim of improving the steel properties as compared to plain carbon steel. Therefore, many of the elements can be classified in accordance with the intended purpose of the addition. For example:

1. deoxidation and inclusion modification (Si, Mn, Al, Ca, Ce, REM) (see Figure 10.17)
2. to combine with sulphur (Mn, Ca, REM, Zr)
3. to produce a free cutting steel (Pb, S, Te,)
4. to enable grain refining (Nb, TI, V, Zr, Al)
5. to provide hardenability (B, Cr, Mo, W, Ni, Cu, C, N)
6. to increase wear resistance (carbides-Cr, V, Mo, W)
7. to provide corrosion resistance (Cr, Al, Si Cu)
8. to improve creep resistance (Ti, Nb, Cr, Mo).

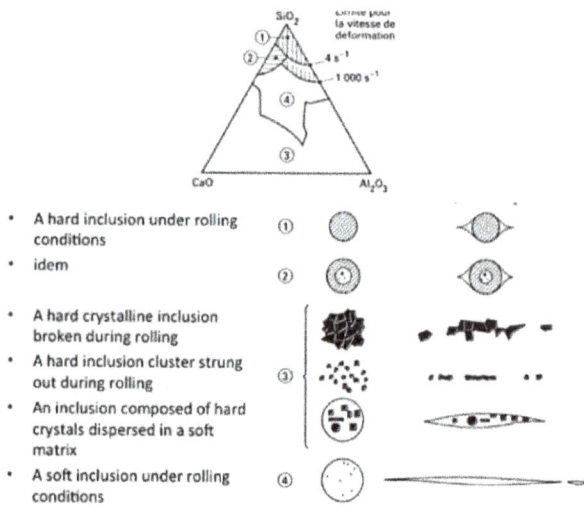

- A hard inclusion under rolling conditions
- idem
- A hard crystalline inclusion broken during rolling
- A hard inclusion cluster strung out during rolling
- An inclusion composed of hard crystals dispersed in a soft matrix
- A soft inclusion under rolling conditions

Figure 10.17 Inclusion composition and how they deform during mechanical working[13]

An additional important aspect is the ability of alloying elements to promote the formation of a certain phase or to stabilize a phase. Therefore, elements can also be grouped as austenite-forming, ferrite-forming, carbide-forming and nitride-forming elements.

Austenite-forming elements These elements have the same crystal structure as the austenite (FCC). The elements that are classed as austenite formers are nickel (Ni), manganese (Mn), cobalt (Co) and copper (Cu). These elements increase the range of stability of austenite by raising the A_4 point and decrease the A_3 temperature. Above a certain level of Ni or Mn a steel can be austenitic at room temperature. (See phase diagrams.) An example of this is the so-called Hadfield steel which contains 13% Mn, 1-2% Cr and 1% C. In this steel both the Mn and C take part in stabilizing

the austenite. Another example is austenitic stainless steel containing 18% Cr and 8% Ni.

There are several good reviews available on stainless steel[14 and 15].

Ferrite-forming elements These elements have the same crystal structure as that of ferrite (BCC). The main elements that are classed as ferrite formers are chromium (Cr), tungsten (W), Molybdenum (Mo), vanadium (V), aluminum (Al) and silicon (Si). These elements lower the A4 point and increase the A3

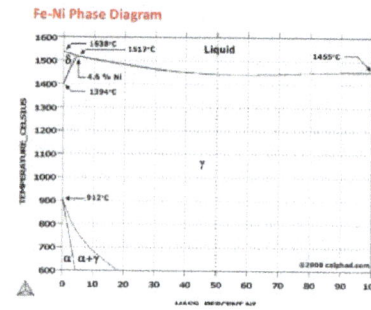

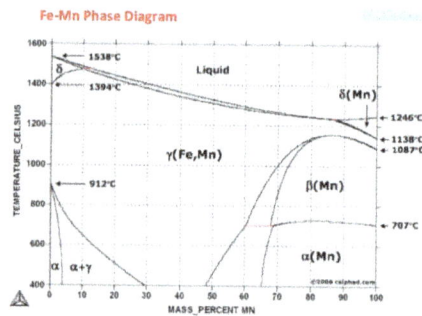

Figure 10.18 Fe-C-Ni and Fe-C-Mn equilibrium diagrams

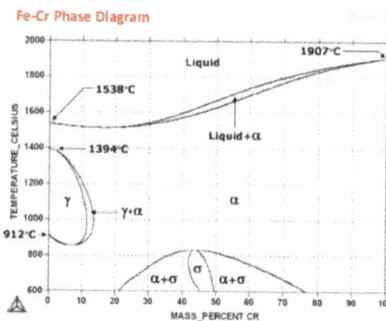

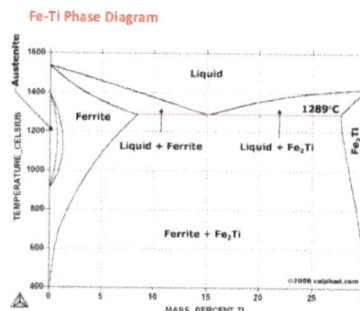

Figure 10.19 Fe-C-Cr and Fe-C-Ti equilibrium diagrams

temperature. Fe-Cr alloys in the solid state containing more than 13% Cr are ferritic at all temperatures up to incipient melting. Another example of a ferritic steel is one that is used as transformer sheet material. This is a low-carbon steel containing about 3% Si.

Multi-alloyed steels

When alloy steels contain a number of elements with different percentages, their microstructure cannot be simply represented on the simple phase diagram and it is necessary to look for other ways to assess and present the effect produced by the alloying elements

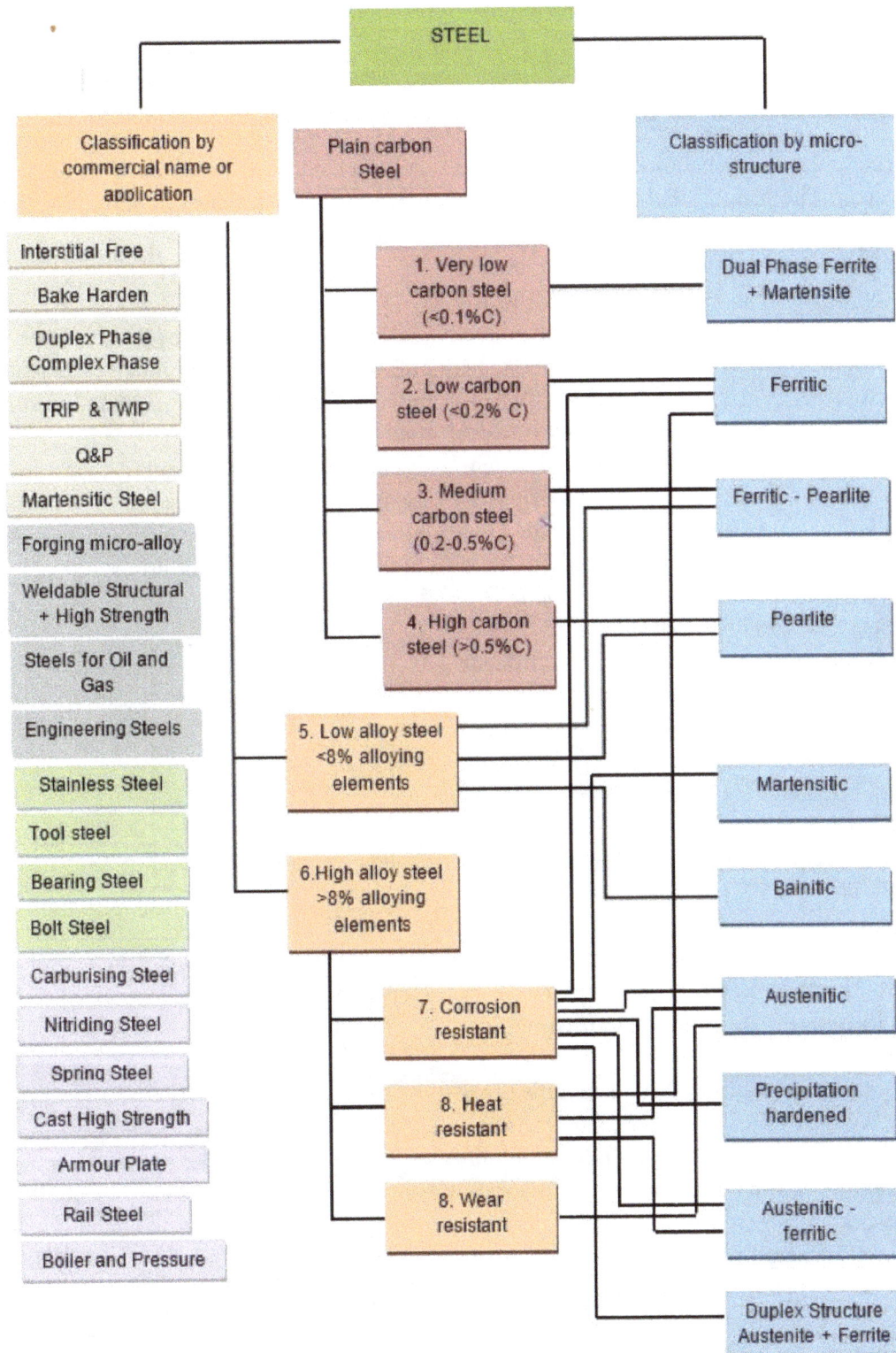

Figure 10.20 *Steels can be referenced by their microstructure or by a commercial name or application*

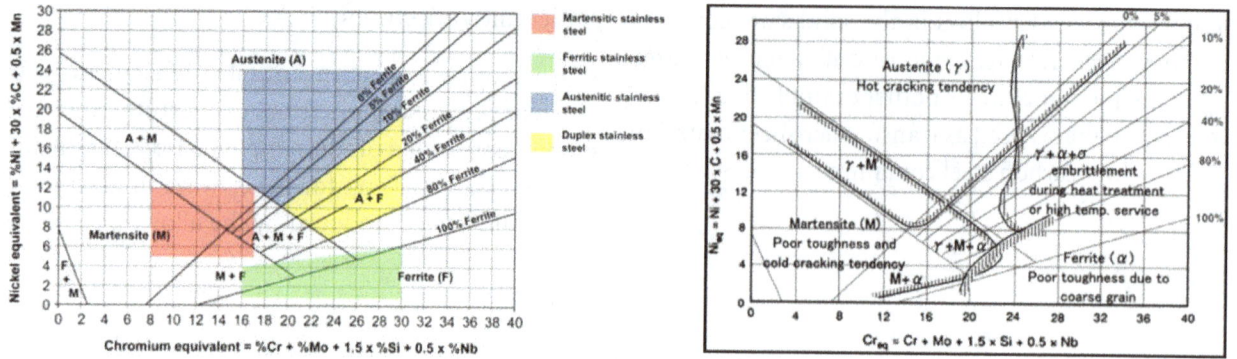

Figure 10.21 *LHS = Schaeffler diagram showing typical compositional ranges for the main stainless steels, RHS = This diagram outlines some of the known imperfections that need to be avoided*

on the structural transformations occurring during heat treatment. The method that has been used for the past 70 years was developed by Anton Schaeffler (1948).[16] The method uses the fact that some alloying elements can be classified as ferrite stabilizers, which tend to promote the formation of the bcc α-phase, or as austenite stabilizers, which tend to promote the face centred cubic (fcc) γ-phase. In predicting the room temperature microstructure of stainless steels, therefore, the balance between the ferrite and austenite formers can be considered. Based on an empirical approach the Schaeffler diagram was developed.

This was modified by H. Schneider (Foundry Trade J. 108, 562, 1960) and summarised in Fig. 10.21. This indicates the structures produced after rapid cooling from 1050°C; the axes are the chromium equivalent and the nickel equivalent. Thermo-Calc and DICTRA have recently been used to calculate the ferrite content[17]. The chromium equivalent represents the proportion of elements (expressed as weight percentage) that behave like chromium in promoting ferrite according to: Cr equivalent = Cr + 2Si + 1.5Mo + 5V + 5.5Al + 1.75Nb + 1.5Ti + 0.75W

The austenite formers give: Ni equivalent = Ni + Co + 0.5Mn + 0.3Cu + 25N + 30C

Carbide-forming elements Elements that form hard carbides in steels. These elements are chromium (Cr), tungsten (W), molybdenum (Mo), vanadium (V), titanium (Ti), niobium (Nb), tantalum (Ta), zirconium (Zr) form hard (often complex) carbides, increasing steel hardness and strength. Examples of steels containing relatively high concentration of carbides: hot work tool steels, high speed steels.

Carbide stabilizers The stability of the carbides is dependent on the presence of other elements in the steel. The stability of the carbides depend on how the element is partitioned between the cementite and the matrix. The ratio of the percentage, by weight, of the

element contained in each of the two phases is called the partition coefficient K as shown below:

Al	Cu	P	Si	Co	Ni	W	Mo	Mn	Cr	V	Ti	Nb	Ta
0	0	0	0	0·2	0·3	2	8	11·4	28	Increasing			

Note that Mn, which by itself is a very weak carbide former, is a relatively potent carbide stabilizer. In practice, Cr is the alloying element most commonly used as a carbide stabilizer.

Nitride-forming elements All carbide formers are also nitride formers. Nitrogen may be introduced into the surface of the steel by nitriding forming a surface compound layer and a diffusion zone of nitrides.

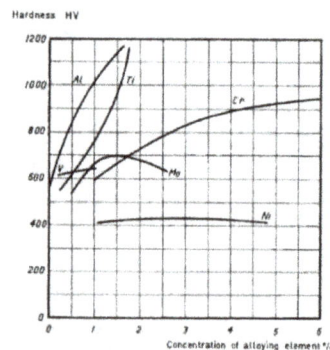

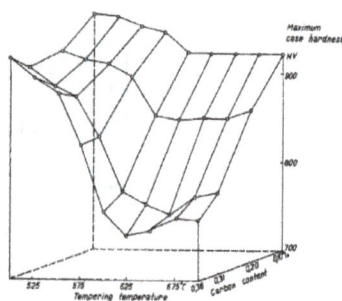

Figure 10.22 *LHS = Effect of alloying element additions on hardness after nitriding. Base composition: 0-25% C, 0-30% Si, 0-70% Mn. RHS = Relationship between carbon content, tempering temperature before nitriding and hardness after nitriding. Tempering time 100 h. Base analysis 3% Cr and 0-5% Mo[18]*

10.2.2 Introduction to Non-ferrous metals

The indigenous supply of energy and metals in the UK played a major part in the so called "Industrial Revolution". The metals antimony, arsenic, copper, gold, iron, lead, manganese, silver, tin, tungsten, zinc, have all been mined in the UK, but unfortunately, many had already become uneconomic to extract by the end of the 19th century. However, the production of metals during the "Industrial Revolution" led to the development of skills in metal refining and smelting, which continued long after the local ores were exhausted, by importing the raw materials. Figure 10.23 shows the current status of metal manufacture and potential ore deposits in the UK.[19]

Both the EU and UK governments have taken an interest in the strategic important metals and have developed an appreciation of the vital importance that an uninterrupted good quality supply of metals is for the well-being of many UK and European Companies.

- The EU report – Competitiveness of the EU Non-ferrous Metals Industries FWC Sector Competitiveness Studies Final Report KR/NZ FN97624_Final Report – 5 April, (FWC= Framework Contract)[20]
- The UK report – House of Commons Science and Technology Committee Strategically important metals Fifth Report of Session 2010–12[21]

Both reviews provided detailed analysis of industry's need for metals but positive measures to provide solutions and possible assistance, needed to keep our metals' consuming sectors competitive, appeared weak. An internet search on "A planned Economy" found the items shown in Figure 10.24. It would be rude and discourteous to think that these simple drawings could possibly suggest some of the reasons for avoidance of government intervention!

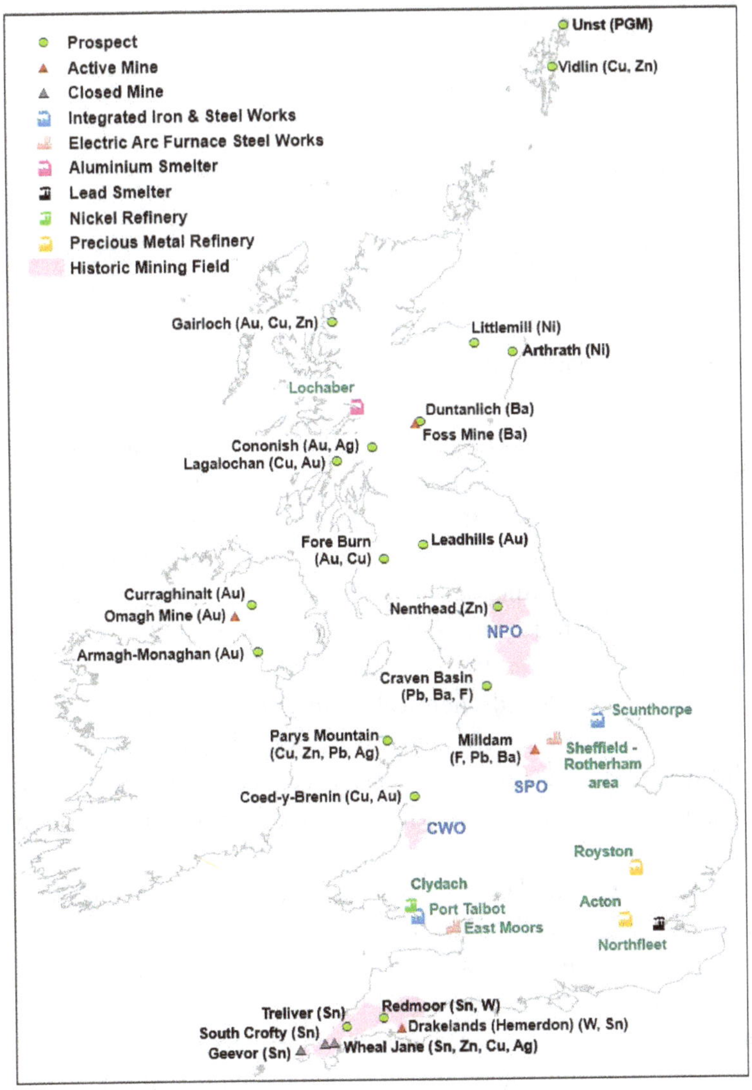

Figure 10.23 Location of prospects, historic miniming fields, metal smelters and steel mills (NPO= Northern Pennine Orefield, SPO = Southern Pennine Orefield, CWO= Central Wales Orefield, Ag, silver, Au gold, Ba barytes, Cr Chromium, Cu Copper, F Florspar, Ni Nickel, Pb lead, PGM Platinum group metals, Sn tin, W tiungsten, Zn Zinc[19]

Figure 10.24 A planned economy

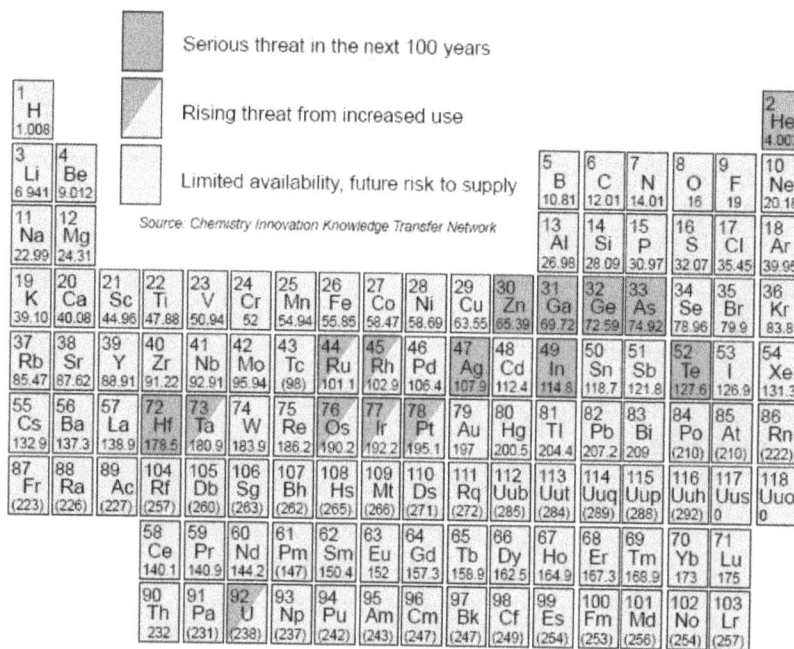

Figure 10.25 LHS = Endangered elements Ref 22. RHS = LME prices gold=per troy oz. https://www.lme.com/

It was interesting to see in the summary reference of the UK report, of possible confusion to the terms "rare earth elements" (REES) and strategic metals. However, it was instructive to see the inclusions of a periodic classification from a publication of "endangered elements".[22]

The main summary of these governmental reviews of important metals include:

- The supply of several elements, including lithium, helium, phosphorus, indium and gallium will possibly fail to meet the demand in the near future
- Energy costs were identified as a cost burden in Europe and yet in the UK they are higher
- Rare earths used in magnets are dependent on the Chinese to supply to the west. In 2010 China was responsible for 97% of REO mining, concentration and separation as well as almost all refining

of REO into REE metal. But as China's internal use grows, alternative sources will be needed. The mining and processing REES have been fraught with difficulties. First there is pollution, both chemical and radioactive. The pollution is both in the mining and the processing. There are many rare earths, and some of them have similar physical and chemical properties and therefore it is not an easy separation. Recycling and alternative sources are being explored, but they may not arrive quickly enough to cover the short-term demand gap.[23]

- The "Competitiveness of the EU Non-ferrous Metals Industries",[20] published in 2010, identified 14 raw materials and metal groups, including all of the rare earth and platinum group metals, as well as antimony, beryllium, cobalt, fluorspar, gallium, germanium, graphite, indium, magnesium, niobium, tantalum, and tungsten where there was cause for concern regarding supply.

- Sources of ores, Tungsten, Rare earth metals and Lithium, identified in SE England, and gold in Scotland but no mining plans
- Often the problems associated with a "level playing field" and macroeconomics are subjective and impossible to effectively control. The EU report quotes:

> "The sector face three intertwining challengers to reach this vision. Firstly, it must remain competitive in relation to emerging economies so that its solutions are traded globally. This is clearly a challenge, as there is currently **a lack of global level-playing field***. Smelters located in these countries usually have better access to raw materials, cheaper energy and most of all, they are often benefit from lower environmental and social standards. These differences are often exacerbated by subsidies and dual pricing of energy in competing world regions. These trade and price distortions result in the EU having a net trade deficit for most non-ferrous metals".

*Emphasis in the original report

The events regarding REE developments are a good example of the above. Below there are three historical quotes

1. "中東有石油，中國有稀土" (The Middle East has its oil, China has rare earth) Deng Xiaoping in 1992.1

2. "Without that small amount of yeast there's no pizza; without rare metals there's no high-tech world." David S. Abraham, 2015.2

3. A quote attributed to Carl Sager in 1990. "We live in a society exquisitely dependent on science and technology, in which hardly anyone knows anything about science and technology."

Many aspects of REM metallurgy/politics are found in the proceedings of the 2013 International Conference on Rare Earth Metals.[24]

10.2.2.1 FRAME – Forecasting and Assessing Europe's Strategic Raw Materials' Need, www.frame.lneg.p

The issue of a secure supply of raw materials has regained importance in the EU in recent years and part of the Horizon 2020 funds was used to support the project FRAME. A map of the Global supply of EU critical materials is shown in Figure 10.26. It is important to identify 'critical raw materials' and to take adequate measures to reduce their 'criticality'. Project FRAME (Forecasting and Assessing Europe's Strategic Raw Materials Needs) was designed to research the critical and strategic raw materials in Europe, by employing sound strategies and a knowledgeable partner base. The objective was to acquire a better knowledge about the potential primary mineral deposits, predict new target areas/deposits and recognize the potential in secondary deposits.

Figure 10.26 Top =Map of the Global supply of EU critical materials, Lower = Map of energy critical metals in Europe.

10.2.2.2 Molycorp and the USA REE experience. China's control over key metal prices

As shown in Figure 10.27 the price of rare earths started to rise in 2009, reaching a peak in the summer of 2012. Prices for some of the minerals tripled. In 2009, China, which produced 97 percent of the world's rare earths, imposed export restrictions. In 2010, Japanese sources claimed China held up rare earth shipments during a territorial dispute between the two countries over eight islands in the East China Sea. In September of 2010, when a dispute over fishing rights and the control territorial waters resulted in the arrest of a Chinese fishing boat captain by the Japanese coastguard, China retaliated by cutting off exports of rare earths to Japan, resulting in the emergence of the term "rare earth diplomacy".

The price of praseodymium oxide, an element used in aircraft engines, rose more than tenfold between 2009 and 2011 to $248 a kilogram. Based on these

prices, Molycorp invested over a billion dollars into their processing facilities at Mountain Pass, which was the largest rare earth mine in the US. The project ran into technical and process problems due to impurities and also leaky tank liners. The target production was 20,000 metric tons a year, but the mine produced 4,769 tons in 2014. Unfortunately, the price had collapsed due to China falling in line with a WTO ruling when Molycorps production costs were about $20 a kilogram compared to a market price of about $7 a kilogram. To compound the problem in March 2012, Molycorp bought Canadian rare earth processor Neo Materials Technology Inc. for $1.1 billion, which experts say was $600 million overpayment. Molycorp filed for bankruptcy protection in late June 2015. The Company is now through the bankruptcy and began production at Mountain Pass in January 2018 and hope to help USA be self-sufficient in REE in 18 months with the assistance from the Trump 20% import tariffs. The owners Mountain Pass are JHL Capital (65 percent), which does business under the name MP Materials. New York-based QVT Financial LP (25 percent share), while Chinese firm Leshan Shenghe Rare Earth Co. holds the remaining 10%.

A Rare Rise and Fall

Molycorp, the only U.S.-based rare-earths producer, benefitted from an extraordinary bubble and has struggled to turn a profit since prices collapsed.

Sources: www.metal-pages.com (metals); FactSet (share price)   THE WALL STREET JOURNAL.

10.2.2.3 Chinese "Planned" dominance[24, 25, 26]

In 15 years, China progressed from producing almost no REE to producing almost all of the world's supply. Dr. Kevin Jianjun Tu believes that it was all planned well in advance as a strategy to dominate all future high-tech business.

> "China's dominance in the RE (rare earth) supply chain is directly related to Beijing's consistent and long-term planning."

He cites the principal reasons that gave China its REE near-monopoly as its low labour cost, heavy governmental investment, and the 1981 "Let Water Flow Rapidly" policy of little to no regulations, environmental protections, or safety considerations".

Figure 10.27 *Top = Comparison Molycorp share price and REM prices. Bottom = Relative Abundance of REE*

Figure 10.28 The grim town of Baotou, in Inner Mongolia, the largest Chinese source of REE[25]

10.2.2.4 USA Mission 2016 – The Future of Strategic Natural Resources

It is interesting to contrast the governmental reviews carried out by the UK government and the EU to the "MIT" student work carried out by undergraduate students during a team and leadership skill project.[28]

Students had the task to delve deep into a global issue, to compose a comprehensive solution that is beneficial and acceptable worldwide. Students needed to work together and guide themselves through the interdisciplinary aspects of the project. Based on the output on their website it was a worthwhile use of government funding *and well worth the development of a UK version.*

10.2.2.5 Mining in the UK[29]

The UK government review raised the prospects of self-sufficiency but did not cover the detailed costs. Russia have announced a £1 billion investment in REE processing which shows the size of investment required. For a normal business this investment in the mining sector would be a risk in an industry well known since its inception thousands of years ago, *as*

a boom-bust industry. The business philosophy that created the LME (London Metal Exchange) was to partly offset price fluctuations and encourage stable supplies. (see Figure 10.25 RHS for LME traded metals)

Tungsten-In Wales Another downfall due to China's control over key metal prices. One of the contributions given to the UK enquiry was from Wolf Minerals, an Australian company. They were about to begin a project to extract tungsten in the UK. Hemerdon in Devon has the fourth biggest reserve of tungsten in the world. However, it has been abandoned for more than sixty years, following its closure in 1944. Several attempts to reopen the mine failed either because they couldn't meet safety standards or was uneconomic due to metal prices. But in 2011, the Australian-based mining company Wolf Minerals was granted planning permission and began work to recommission the Drakelands mine. Extraction of tungsten and tin began in 2015, taking advantage of the claimed total resource of 145.2 million tonnes (Mt).

The company went into administration in October 2018 after losing £100 million in three years. The problems were first identified by the Wolf Minerals in 2016 when the company was faced with underperformance and producing tungsten of a finer size than originally planned. In addition, the tungsten price, which is heavily dependent on China's economic strategy, in 2012 when the project began was rising and had reached almost $57,000 per tonne. The current price in 2018 was $30,300 per tonne.

In 2018 November there was news that the Russian billionaire Vladimer Iorich of Pala Investments, Switzerland wanted to invest $25 to re-open the plant. Stripped of its debt repayments it could be profitable. However, this did not happen. In March 2019 a Department for the Environment spokesman said: "*Some assets, including the mine, was bought from Wolf Minerals by Drakelands Restoration Ltd earlier this year. I believe the firm is coordinating the restoration of the area.*" Drakelands Restoration Ltd was incorporated in March 2019, alongside Drakelands Mine Restoration Ltd and Drakelands Holdings Ltd. All three firms have an address in West Terrace, Esh Winning, Durham, the same address as mining company Hargreaves Services Plc.a company that lost £8m when Wolf Mineral went into administration.

The British geological Survey review covers all you need to know about Tungsten in the UK.[30]

Lithium Cornish Lithium Ltd has been established and plans to extract the mineral from brines located below hot springs in the south-west of the UK. New technologies allow for an efficient extraction process, capable of yielding a purity of up to 99.9%. In 2018 the Company was still only at the exploration stage, the company has secured the funding for the first exploration holes, which will be drilled this year with hopes to begin production within the next five years. In October 2019 it was announced that Cornish Lithium had completed its £1.4 million crowdfunding round via Crowdcube.

Tin Strongbow Exploration Inc, a Canadian company, have acquired the rights to the South Crofty tin project, which is located in the Central Mining District of Cornwall, in the towns of Pool and Camborne, South West England. Tin mining in this region dates back to 2,300 BC. Large-scale production at South Crofty first started in the mid-1600s (the first documented production dated in 1592). The mine has been in operation intermittently since then, with the last closure in 1998 coming after a prolonged period of depressed tin prices. Historical production between 1700 to 1998 totalled over 450,000 tonnes of tin from the Central Mining District.

Gold and Silver Scotgold Resources Limited, incorporated in Australia. In 2016, Scotgold held the first ever auction for gold mined and processed in Scotland. The average price accepted was £4,557.9 per ounce, a premium of 378% over the current spot price of £953. Scotgold's Cononish mine is in the Highlands, and has an ore reserve of 555,000t. Following initial exploration works and reserve estimates, Scotgold must now raise £2.65m to begin full-scale operations. The expected valuation of the gold is of the order of £100 million.

Figure 10.29 First auction of Scotgold sold at four times the gold price. 2016

10.2.2.6 Mining the Ocean Floor[31 and 32]

The aspiration to mine the seabed has been around since the 1960s. However, the extraction costs could not compete with land-based mining. As land based mineral resources are depleted then seabed mining will probably become economically viable. Many governments are reviewing the potential of offshore minerals and an EU report estimates of the business being worth 10 billion euros by 2030.[33 and 34]

"By 2020, 5% of the world's minerals, including cobalt, copper and zinc, could come from the ocean floors. This could rise to 10% by 2030. Global annual turnover of marine mineral mining can be expected to grow from virtually nothing (2014) to €5 billion in the next 10 years and up to €10 billion by 2030."

However, there have been major concerns regarding the ecological downside especially from "Greenpeace".[35 and 36] In addition, there is possible conflict between fishermen and pharmaceutical companies that harvest algae for the manufacture and development of drugs.

In 2014 the EU carried out a review of the potential of seabed mining as a solution for the problems with the supply of several materials. The EU is increasingly dependent on imports for some strategically important raw materials, while exploration and extraction of these materials is facing increased competition and a strongly regulated market environment. According to the recently updated list of EU critical raw materials 49% (in terms of volume) of these materials are resourced from China. (see Figure 10.14 top)

The high import dependence of strategic (STR) and critical raw materials (CRM) has a serious impact on the sustainability of the EU manufacturing industry. This problem can only be solved by more intense and advanced exploration for new mineral deposits on land and the marine environment. Seafloor mineral resources receive growing European interest with respect to the exploration potential of REE, cobalt, selenium, tellurium and other high-tech metals.

Many seabed minerals are essential for making the equipment that generates and uses renewable "green" energy which requires high efficiency electric motors and generators, battery storage, electric and hybrid vehicles. For example, REE are used in wind turbines. Metals such as cadmium, gallium, germanium, indium, selenium, and tellurium are important metals used in current photovoltaic cell technology.[37]. In most batteries, the critical metals include lithium, graphite, cobalt and nickel.[38]

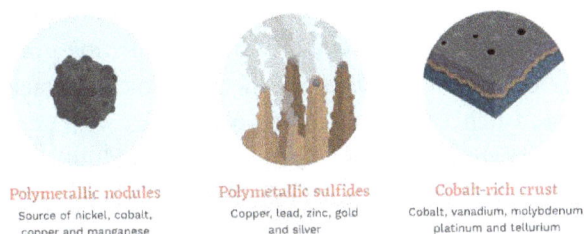

Deep-sea mineral deposits
and the metals they contain

Polymetallic nodules
Source of nickel, cobalt,
copper and manganese

Polymetallic sulfides
Copper, lead, zinc, gold
and silver

Cobalt-rich crust
Cobalt, vanadium, molybdenum
platinum and tellurium

Figure 10.30 *Seabed minerals.*

REE metals have been found on the seabed between Mexico and Hawaii. In 2017 Japan became one of the first countries to actually extract minerals from the seabed near the island group of Okinawa.[32]

This is another scientific conundrum that is probably beyond the capability of the human mind and decision-making process. Is it better to extract the "energy alloys" needed to develop clean technology or is it better to leave the ocean floor integral for future marine life?

10.2.2.7 List of non-ferrous metals

The arbitrary classification of non-ferrous metals splits the range into broad categories with the name of the category giving some insight into the use of the metals:

1. **Light metals:** Aluminium, Magnesium, Titanium, Beryllium, which includes metals with a density below 5 grams/cc. Due to their light weight, these non-ferrous metals are of special importance to the aircraft and space industries
2. **Heavy metals:** Copper, Zinc, Lead, Tin. Copper has good electrical conductivity and along with aluminium is the preferred choice for electrical conductors. Zinc, tin and lead (with low melting points) are used in special applications. Zinc for die-casting, (Zamak[39]) and tin and lead for white metal bearings[40, 41, 42]
3. **Refractory metals:** Tungsten, Nickel, Molybdenum, Chromium are used in products that must resist high temperatures. Nickel and cobalt are also suitable as heat resistant alloys
4. **Precious metals:** Gold, Silver, Platinum are not only used in jewellery, but also in many applications requiring high electrical conductivity and corrosion resistance.

The use of Aluminium is second biggest tonnage material in use next to steel. Probably greater than 60 million tonnes per annum. (Steel 1600 million tonnes.) Copper and its alloys (brass and bronze) rank second while Zinc ranks third in consumption.

The lightest metal is Lithium (with density of 0.530grms/cc) and the heaviest metal is Osmium (with density of 22.5 grams/cc). Steel has a density of 7.8 grams/cc.

1. Aluminum (Al): The most important non-ferrous metal. It has outstanding physical properties (e.g. light weight, high thermal and electrical conductivity, and good corrosion resistance). It is suitable for all machining, casting and forming operations. (See Section 10.16)
2. Titanium (Ti): is used in corrosive environments or in applications of light weight, high strength and non-magnetic properties. It has good high temperature strength as compared with other light metals. Titanium is available in many forms and alloys such as pure titanium (98.9-99.5%), Ti-Pd alloy (titanium-0.2%palladium), high strength

Ti-Al-V-Cr (alpha-beta type) alloys. Ti-6Al-4V (6%Al, 4%V, remainder Ti) and Ti-3Al-13V-11Cr (3%Al,13%V, 11%Cr, remainder Ti) is a heat treatable beta alloy titanium. This alloy is known to possess good fabricability and excellent corrosion resistance and mechanical properties. It is basically used in high strength applications 1207MPa, UTS 1276MPa, El, 8%, Rc,40, density 4.84 g/cc. (See section 10.15)

3. Beryllium (Be): is a recently developed material having several unique properties of low density (one-third lighter than aluminium), high modulus-to-density ratio (six times greater than high-strength steels), high melting point, dimensional stability, excellent thermal conductivity and transparency to X-rays. However, it has serious deficiencies of high cost, poor ductility, and toxicity. It is not especially receptive to alloying. All conventional machining operations including some non-traditional processes (e.g. EDM and ECM) are possible. However, it must be machined in specially equipped facilities due to its toxic effect. In addition, surface of beryllium is damaged after machining, and hence secondary finishing operations must be carefully conducted. It is typically used in military aircraft brake systems, missile guidance systems, satellite structures, and X-ray windows.

4. Magnesium (Mg): is the lightest "engineering" material available. The combination of low density and good mechanical strength has made it one of the most specified materials in aircraft, space, portable power tools, luggage and similar applications as competing with the aluminium alloys. Magnesium was first isolated by Sir Humphry Davy in 1808, in London (Al 1825).

   Alloys of magnesium are the easiest of all engineering metals to machine. They are amenable to diecasting, and they are easily welded. Also, magnesium parts can be joined by riveting and adhesive bonding. Other notable characteristics are high electrical and thermal conductivity as well as very high damping capacity.

   On the downside, it is highly susceptible to galvanic corrosion since it is anodic. Under certain conditions, flammability can be a problem as it is an active metal.

   Magnesium alloys are best suited for applications where lightness is of primary importance. When lightness must be combined with strength, aluminium alloys are better material alternatives.

5. Copper (Cu): One of the first engineering metal to be used (Bronze Age). Unlike other metals, it can occur in nature in the metallic form as well as an ore. It has very good heat and electrical conductivity and resists to corrosion when alloyed with other metals. However, due to lower density, aluminium has higher conductivity per unit weight.

   Copper alloys consist of the following general categories:

   1. Coppers (minimum99.3%Cu)
   2. High coppers (99.3-96%Cu)
   3. Brasses (Cu-Zn alloys with 5-40%Zn)
   4. Bronzes (mainly Cu-Sn alloy, and also alloys of Cu-P, Cu-Al, Cu-Si)
   5. Copper Nickels (Cu-Ni alloys, also known as cupro-nickels)
   6. Nickel Silvers (Cu-Ni-Zn alloys which actually do not contain silver). (See section 10.17)

6. Zinc (Zn): is an inexpensive material with moderate strength. It is chemically similar to magnesium. Mechanically, however, zinc is more ductile but not as strong. Although its metal and alloy forms are important, zinc is most commonly used to extend the life of other materials such as steel (galvanizing), rubber and plastic (as an aging inhibitor), and wood (in paint coatings).

   The zinc-base alloys have an important place as a die-casting metal as it has low melting point (419.5°C) which does not affect steel dies adversely, and hence it can be made into alloys with good strength properties and dimensional stability.

7. Tin (Sn): As with many metals, pure tin is too weak to be used alone for most mechanical applications. It is often alloyed with elements such as copper, antimony, lead, aluminium and zinc to improve mechanical or physical properties.

   It is commonly used as a coating for other metals such as tin cans, copper cooking utensils. Other applications include die-casting alloys, pewter chemicals, bronze, bearing alloys and solder.

8. Lead (Pb): is a versatile material due to special properties of high atomic weight and density, softness, ductility, low strength, low melting point, corrosion resistance and ability to lubricate. On the downside, toxicity is one of the chief disadvantages.

   There are two principal grades: chemical lead and common lead. Typical uses include chemical apparatus, batteries, and cable sheathing. For corrosion resistance and X-ray and Gamma-ray shielding, pure lead gives best performance. Lead is alloyed with tin and antimony to form a series of useful alloys employed for their low melting points. Solder is the alloy of lead and tin containing small amount of antimony and silver. The solders are mainly used in soldering electronic circuits due to their lower melting points.

Refractory metals: (with melting points above 1900°C) are characterized by high-temperature strength and corrosion resistance.

9. Niobium (Nb)&
10. Tantalum (Ta): These elements are distinguished by excellent ductility even at low temperatures. Both metals occur together in ores and they must be separated for nuclear use since tantalum has a higher neutron absorption than niobium.

    They can be fabricated by most conventional processes, usually worked at room temperature. They are usually considered together since most of their working operations are identical. They are used as anode and grid elements in medium to high-power tubes as well as in capacitors and foil rectifiers.
11. Tungsten – also known as "Wolfram" (W) & Molybdenum (Mo): Tungsten is the only refractory metal with good electrical and thermal conductivity, excellent erosion resistance, low coefficient of expansion and high strength at high temp. Although having low ductility and malleability, it can be fabricated into many forms with proper procedures. Thomas Edison's trials with tungsten during the invention of incandescent lamp was the most important application. Other uses are electrodes for inert-gas welding, electron tube filaments, anodes for X-ray and electron tubes.
12. Molybdenum is a special metal having some resistance to hydrofluoric acid. It is very similar to tungsten, but more ductile, easier to fabricate, and cheaper. Its typical applications are electron tube anodes, grids for high-power electron tubes, dies withstanding "thermal cycling", and supports for tungsten filaments in light bulbs. Both metals have brittleness at room temp. and oxidation at relatively low temp.
13. Chromium (Cr): Most refractory metals oxidize and lose their strength at high temp. whereas chromium (less subjected to oxidation) loses strength

above1090°C. It is used in chrome-plating and steel alloys to improve hardenability, strength at high temperature and corrosion resistance. Its brittleness may limit its application areas.
14. Nickel (Ni): Nickel alloys have desirable properties like ultra-high strength, high proportional limit and high modulus of elasticity. Commercial pure nickel has good electrical, magnetic and magneto strictive properties. Nickel alloys are still strong, tough and ductile at cryogenic (very low) temperatures. Most nickel alloys can be hot and cold worked, machined and welded successfully. They can be joined by shielded metal-arc, gas tungsten-arc, gas metal-arc, plasma arc, electron beam, oxy-acetylene, resistance welding, brazing and soft soldering. (See section 10.15)
15. Hafnium (Hf): Melting point of 2130°C. Exist in all zirconium ores. Not commercially available. Limited use in rectifiers and electronic applications in aerospace.
16. Rhenium (Re): High strength, good ductility and high melting point (3165°C), but strong oxidation tendency. Used in filaments in high-power vacuum tubes and high temperature thermocouples.
17. Vanadium (V): Melting point of 1900°C. Added to steels in small amounts.

Precious metals They are divided into three subgroups, gold and gold alloys, silver and silver alloys, platinum group metals (consists of six metals extracted from nickel ores: platinum, palladium, rhodium, iridium, ruthenium, osmium).

These metals are nearly completely corrosion resistant. Platinum metals withstand up to 1760°C without any evidence of erosion and corrosion.

18. Gold (Au): Extremely soft, ductile material that undergoes very little work hardening. It is often alloyed with other metals (such as copper) for greater strength. It has occasional uses as a reflecting surface, attachments to transistors and jewellery.
19. Silver (Ag): The least costly metal of this group. Very corrosion resistant. Plated on to low-voltage electrical contacts to prevent oxidation of surfaces when arcing occurs. Used in photographic emulsions due to photosensitivity of silver salts.
20. Platinum (Pt): A silver-white metal. Extremely malleable, ductile and corrosion resistant. When heated to redness, it softens and is easily worked. Since it is inert, nearly non-oxidizable and stable

even at high temperature. Used for high-temperature handling of high purity of chemicals and laboratory materials. Also used in thermocouples, spinning and wire-drawing dies, components of high-power vacuum tubes, glass working environment and electrical contacts.

21. Palladium (Pd): A silver-white metal. Very ductile, slightly harder than platinum. Principally used in telephone relay contacts and as a coating on printed circuits.

22. Rhodium (Rh): The hardest metal with the highest electrical and thermal conductivity in platinum group. Its high polish and reflectivity make it ideal as a plated coating for special mirrors and reflectors. Rhodium and its alloys are used in furnace windings and in crucibles at certain temperatures that are too high for platinum.

23. Iridium (Ir): The most corrosion resistant metallic element. Together with Cobalt-60, radio isotope Iridium-192 satisfies most of the industrial radiography requirements. It has high-temperature strength comparable to that of tungsten up to 1650°C. It is used for spark plug electrodes in aircraft and other engines where high reliability is needed.

24. Ruthenium (Ru): Very hard and brittle metal. Its tetra-oxide is quite volatile and poisonous. It is added to platinum to increase resistivity and hardness.

25. Osmium (Os): The heaviest known metal, having a density of about three times that of steel. It oxidizes readily when heated in air to form a very volatile and poisonous tetra-oxide. It has been used predominantly as a catalyst.

Other Nonferrous Metals

26. Bismuth (Bi): The common constituent of all low melting alloys (95-150°C) and ultra-low melting groups. Increases in volume on solidification, which makes its alloys excellent for casting as they pick-up every detail when expand into the mould.

27. Antimony (Sb): Also expands upon solidification. Used as a hardener in bearing alloys and semi-conductors, and as alloying element in thermo-electric materials.

28. Cadmium (Cd): Easy to deposit as a plating, excellent ductility and good resistance to salt water and alkalis. Used for plating of steel hardware and fasteners, as an alloy in bearing materials and in

special batteries. Not permitted in food industry due to toxicity.

29. Indium (In): Similar in colour to platinum. One of the softest metals. Can be deformed almost indefinitely by compression. Used in semiconductor devices, bearing metals and fusible alloys.

30. Cobalt (Co): Hard metal melting at1493°C. Used in hot-working steels, heat resisting castings, superalloys, sintered carbide cutting tools. Invar alloy (54%Co, 36%Ni) used in strain gauges. Co-Cr alloys in dental and surgical applications.

31. Zirconium (Zr): Outstanding corrosion resistance to most acids and chlorides. Increase in its mechanical strength when alloyed with hafnium. Consumed in fuel rods for nuclear reactors.

REFERENCES FROM 10.1

1. http://www.tribology-abc.com/
2. Anton Van Beek, Advanced engineering design Lifetime performance and reliability, http://www.tribology-abc.com/book/default.htm
3. M Ashby et al, Materials Engineering Science, Processing and Design, Butterworth-Heinemann, 2007, https://the-eye.eu/public/WorldTracker.org/Physics/Materials%20Engineering%20-%20Science%2C%20Processing%20and%20Design%20-%20M.%20Ashby%2C%20et%20al.%2C%20%28B-H%2C%202007%29%20WW.pdf
4. Cephas Yaw Attahu et al, INFLUENCE OF ASSEMBLY GAP AND SHIMS ON THE STRAIN AND STRESS OF BOLTED COMPOSITE-ALUMINUM STRUCTURES, ARPN Journal of Engineering and Applied Sciences, VOL. 12, NO. 5, MARCH 2017, file:///C:/Users/user/Downloads/jeas_0317_5800.pdf
5. T. Sourmail, Coatings for Turbine Blades http://www.phase-trans.msm.cam.ac.uk/2003/Superalloys/coatings/index.html
6. Metals Handbook ASM 1948
7. Metals Handbook ASM 9th Edition
8. R T Rolfe, Steel for the user, Pub Chapman Hall Ltd 1937
9. Kern, Steel Selection- A guide to improving Performance and Profits 1976
10. J A Charles Crane, Selection and use of engineering materials, second edition
11. http://etheses.whiterose.ac.uk/14433/1/237901_vol.1.pdf
12. Robert Tuttle Editor, "alloy steels" Special Issue, MDPI AG, First Edition 2018 file:///C:/Users/user/Downloads/Alloy%20Steels.pdf
13. Metallurgist, March 2016, Volume 59, Issue 11–12, pp 1053–1061
14. www.fa-fe.com/files/pdf/libri_articoli/en/2_Stainless_steels.pdf
15. Marco Boniardi e Andrea Casaroli, Stainless steels, Gruppo Lucefin Research & Development. http://www.lucefin.com/en/pubblicazioni/gli-acciai-inossidabili/
16. https://www.metallurgical-research.org/articles/metal/pdf/2017/06/metal170104.pdf

17. 1.Maria Asuncion Valiente Bermejo, Computational thermodynamics in ferrite content prediction of austenitic stainless steel weldments, Welding in the World, Published online 5th December 2018 Springer https://link.springer.com/content/pdf/10.1007%2Fs40194-018-00685-x.pdf

18. Thelning, Steel and its heat treatment, Bofors Handbook

19. Metals, Mineral Planning Factsheet, British Geological Society Crown Copyright 2015 https://www.bgs.ac.uk/mineralsuk/planning/mineralPlanningFactsheets.html

20. Competitiveness of the EU Non-ferrous Metals Industries FWC Sector Competitiveness Studies Final Report KR/NZ FN97624_Final Report – 5 April, (FWC= Framework Contract) https://ec.europa.eu/growth/tools-databases/eip-rawmaterials/en/system/files/ged/82%20fn97624_nfm_final_report_5_april_en.pdf

21. House of Commons Science and Technology Committee Strategically important metals Fifth Report of Session 2010–12 https://publications.parliament.uk/pa/cm201012/cmselect/cmsctech/726/726.pdf

22. "endangered elements" http://www.rsc.org/images/Endangered%20Elements%20-%20Critical%20Thinking_tcm18-196054.pdf

23. Hugo Royen, Uwe Fortkamp, Rare Earth Elements – Purification, Separation and Recycling, IVL Swedish Environmental Research Institute 2016. https://www.ivl.se/download/18.76c6e08e1573302315f3b85/1480415049402/C211.pdf

24. 2013 international conference on Rare Earth Metals. OCTOBER 27 to 31, 2013, MONTRÉAL, QUÉBEC, CANADA http://web.cim.org/com2014/conference/Rare_Earth_Elements_PROCEEDINGS_2013.pdf

25. https://apps.dtic.mil/dtic/tr/fulltext/u2/1032144.pdf

26. file:///F:/non%20ferrous%20metals/rem/HE-DISSERTATION-2016.pdf

27. https://books.google.co.uk/books?id=YIOqCgAAQBAJ&printsec=frontcover&source=gbs_ge_summary_r&cad=0#v=onepage&q&f=false

28. "Mission 2016 – The Future of Strategic Natural Resources." http://web.mit.edu/12.000/www/m2016/finalwebsite/problems/limitedaccess.html

29. https://www.mining-technology.com/features/mining-the-uk/

30. Tungsten in the UK, British geological Survey review, file:///C:/Users/user/Downloads/tungstenProfile%20(1).pdf

31. Kathryn A. Miller et al, An Overview of Seabed Mining Including the Current State of Development, Environmental Impacts, and Knowledge Gaps, Frontiers in Marine Science, January 2018 | Volume 4 | Article 418, https://pdfs.semanticscholar.org/2fa5/b9e4157331b82eb06042f8d4a47568db05d4.pdf?_ga=2.84276121.1016327926.1570271738-2139650799.1545429715

32. https://www.vanadiumcorp.com/news/industry/a-bus-sized-robot-will-soon-be-mining-the-ocean-floor/

33. Study to investigate state of knowledge of Deep Sea Mining, Interim report under FWC MARE/2012/06 – SC E1/2013/04, Rotterdam/Brussels, 28 March 2014, A study carried out by Ecors for the European Commission – DG Maritime Affairs and Fisheries. Plus Final report Final Report under FWC MARE/2012/06 – SC E1/2013/04 Rotterdam/Brussels, 19 November 2014

34. The EU Blue Economy Report. 2019 European Union, 2019, Project Number: 2019.2797 https://prod5.assets-cdn.io/event/3769/assets/8442090163-fc038d4d6f.pdf

35. Michelle Allsopp et al, Review of the Current State of Development and the Potential for Environmental Impacts of Seabed Mining Operations, Greenpeace Research Laboratories Technical Report (Review) 03-2013: 50pp.

36. IN DEEP WATER The emerging threat of deep sea mining, Greenpeace, protect the Oceans

37. Donald I. Bleiwas, Byproduct Mineral Commodities Used for the Production of Photovoltaic Cells, Circular 1365, U.S. Geological Survey, Reston, Virginia: 2010, https://pubs.usgs.gov/circ/1365/Circ1365.pdf

38. Battery raw materials, The British Geology Society, 2018

39. Zamak, http://www.albco.com/wp-content/uploads/2018/01/ZINC-DIE-CASTING-ALLOYS.pdf

40. White metal bearings https://digitised-collections.unimelb.edu.au/bitstream/handle/11343/24746/307385_UDS2013255-24-0013.pdf?sequence=1

41. White metal bearings http://akademikpersonel.kocaeli.edu.tr/adaletz/sci/adaletz14.03.2013_23.16.42sci.pdf

42. White metal bearings file:///C:/Users/user/Downloads/lubricants-03-00091.pdf

Books M.YASILYEV, Man and Metals, MIR Publishers Moscow, 1967. Quaint Russian style referring to the metals as 80 brothers
ASM Handbook Volume 2: Properties and Selection: Nonferrous Alloys and Special-Purpose Materials

10.3 METALLURGICAL ASPECTS OF CAST IRON

Cast iron contains silicon in addition to high carbon content. Cast irons should therefore be described as iron-carbon-silicon alloys. The presence of silicon in iron carbon alloys promotes the formation of graphite (inoculation).

The phase diagrams for each silicon level are similar but the eutectic/eutectoid points are moved to the left as the silicon is increased. The iron carbon diagrams for various silicon levels are shown in Figure 10.34. These diagrams show that the eutectic carbon moves from 4.2% carbon to 2% carbon as the silicon increases from 0 to 6.5%.

Cast irons have a long history of manufacture and use and yet there is still significant activity in detailed R&D and process control across the full range of cast irons. There is now an understanding of how events shape the graphite. The main parameters are the

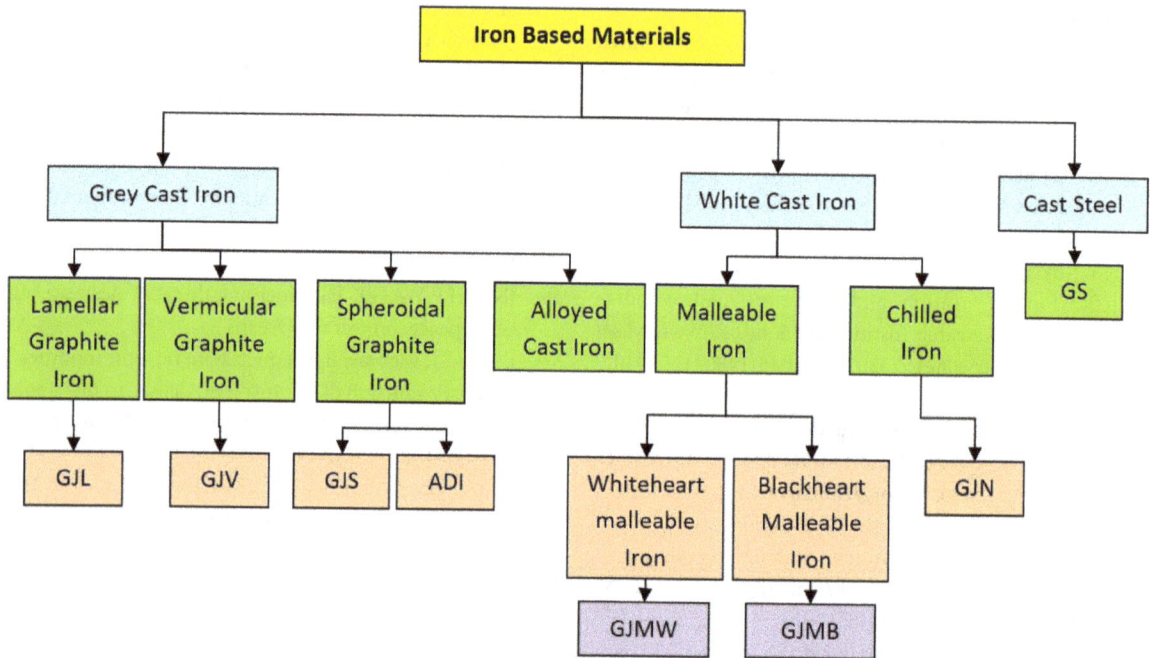

Figure 10.31 Classification of Iron based alloys

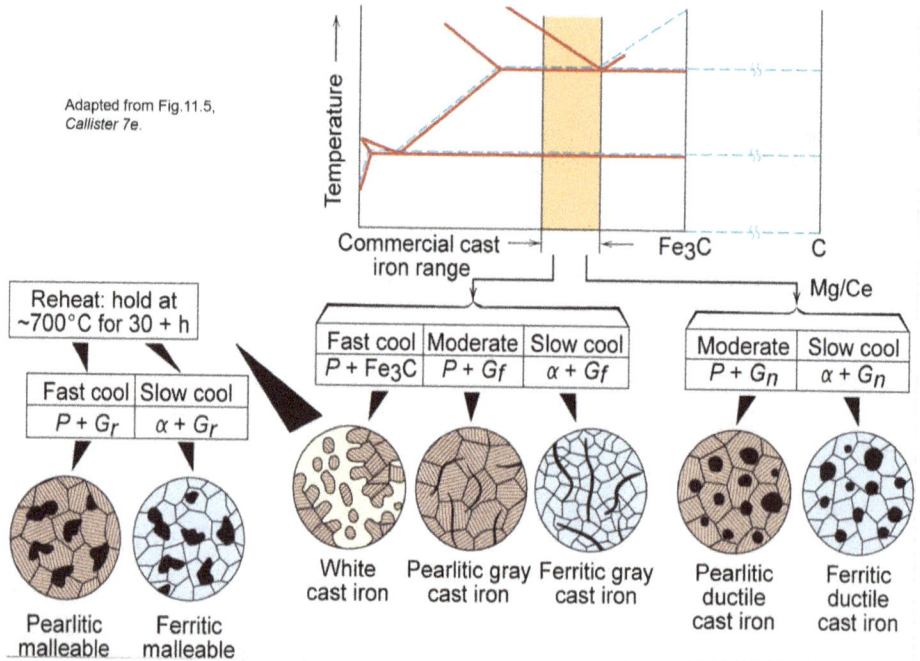

Figure 10.32
Solidification sequence of cast irons (from Callister)

- **Lamellar graphite iron (GJL)**
  - Pressure and wear-resistant
  - Corrosion-resistant
  - Good damping characteristics
  - Low tensile strength and elongation at fracture
  - Low cost
- **Vermicular graphite iron (GJV)**
  - Between GJL and GJS
  - Low thermal expansion
- **Spheroidal graphite iron (GJS)**
  - Ductile
  - High tensile strength and elongation at fracture

Un-etched photomicrograph ⊢————⊣ 100 µm

SEM image ⊢————⊣ 20 µm

Quelle: CLAAS GUSS

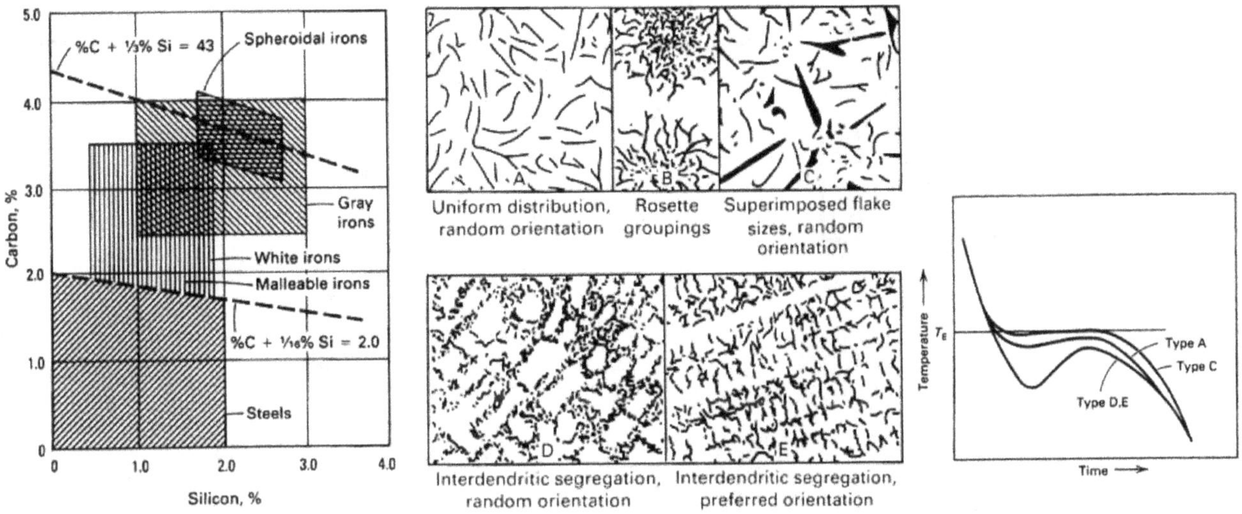

Uniform distribution, random orientation — A

Rosette groupings — B

Superimposed flake sizes, random orientation — C

Interdendritic segregation, random orientation — D

Interdendritic segregation, preferred orientation — E

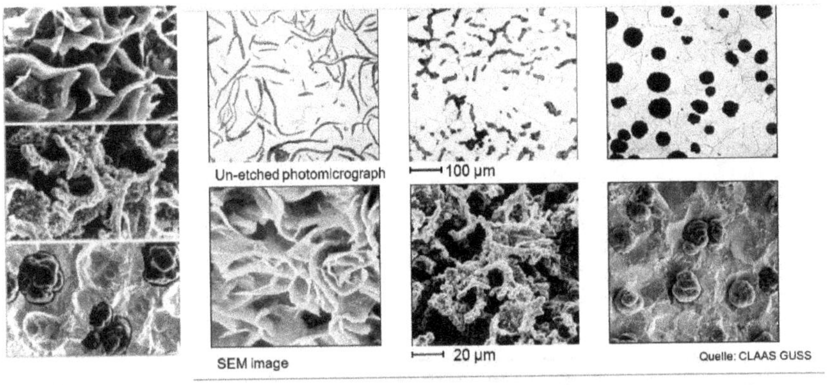

Figure 10.33 *Top photos Flake, Ductile and vermicular lower LHS- ranges of carbon and silicon for each cast iron type, Centre Cast iron flake graphite microstructure. RHS Thermal arrest curves during solidification and cooling*

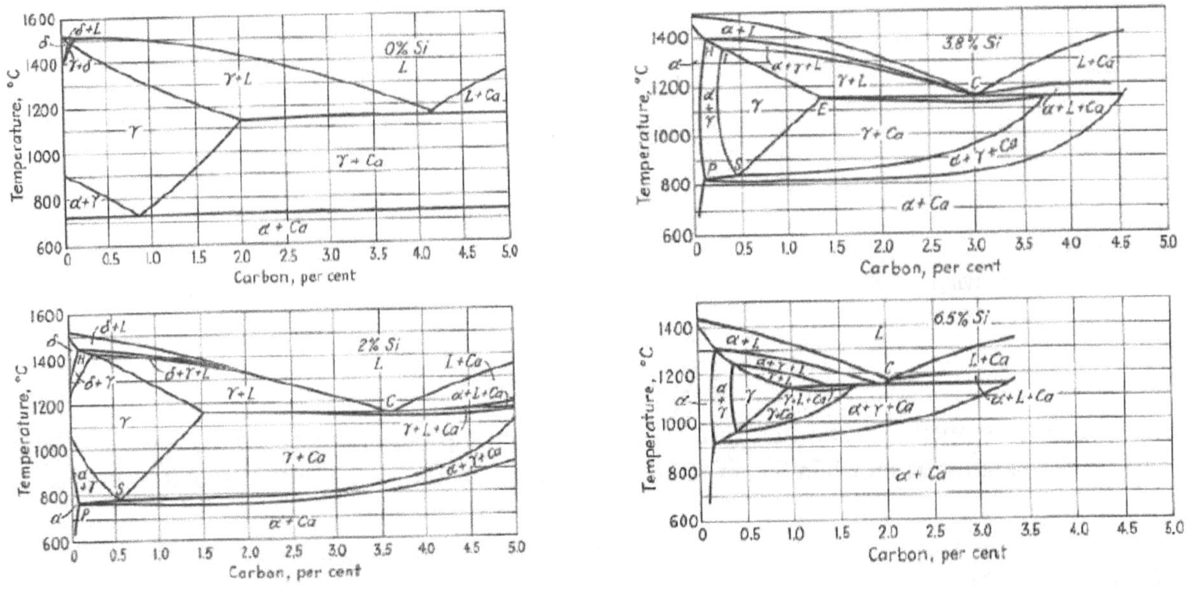

Figure 10.34 *The iron carbon diagrams for various silicon levels*

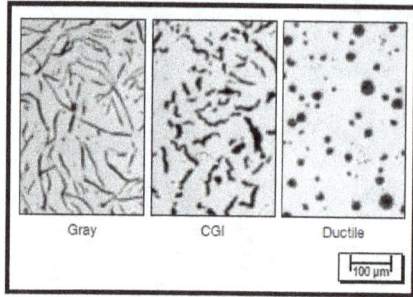

Figure 1: Gray iron, compacted graphite iron, and ductile iron are differentiated by the shape of the graphite particles

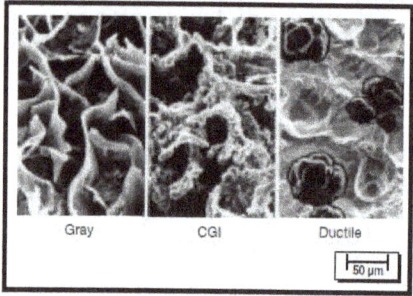

Figure 2: Deep-etched scanning electron micrographs show the three-dimensional shape of the graphite

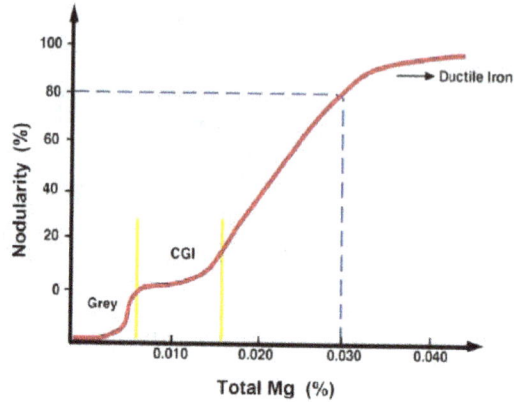

Figure 10.35 *LHS Optical and SEM grey cast iron microstructure. RHS Effect of magnesium on graphite shape and nodularity*

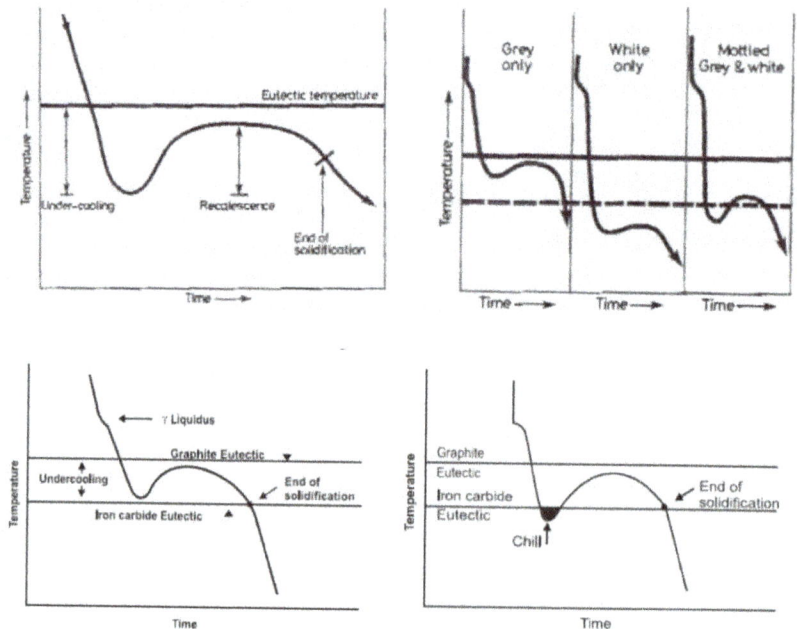

Figure 10.36 *Cooling curves of iron during solidification showing arrests at critical points*

"Graphite" eutectic temperature and the "Iron carbide" eutectic temperature as shown in Figure 10.36.

10.3.1 Inoculation

In the 1920s methods of inoculation were developed, and there were a number of competing technologies with combinations of Ba, Ca, Al. The term inoculation was not used until the 1930s but it was understood that the task was to avoid chill cast areas which formed carbides. To produce a grey-iron it was essential to have the transformation of the graphite eutectic and avoidance of the carbide eutectic. The first test used was a "wedge" test which allowed a range of cooling rates from the tip to the thickest part of the wedge.

Most commercial ferro-silicon based inoculants contain around 5% mixture of Ca, Al, Sr, Ba, Mn, Zr, Ce that are known to form micro-inclusions of complex oxides and or oxy-sulphides that have suitable crystallographic features to act as heterogeneous nuclei for the graphite. Fig 10.38[1]

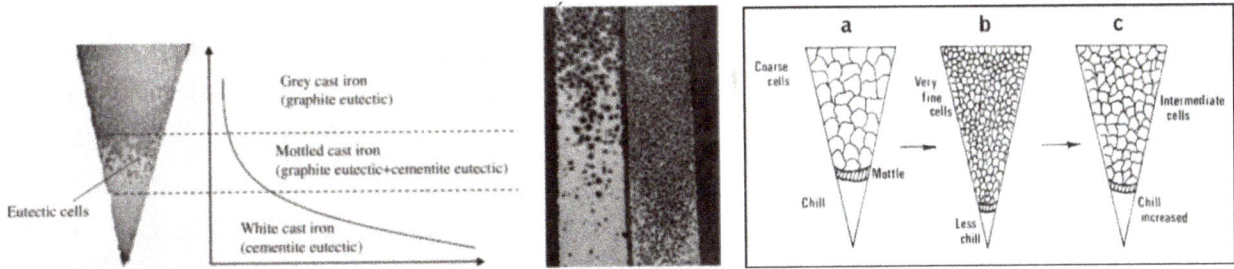

Figure 10.37 *LHS = The wedge test ASTM A 367-11. RHS = Effect of inoculation on eutectic cell size and chill depth in wedge samples (schematic). (a) before inoculation (b) immediately after inoculation (c) fading due to holding time after inoculation before pouring http://web.askewindustrial.com/ASTM2014/848aa3419a054b4b85f332dc8b87e847. pdf?tblASTMSpecsPage=17*

Inoculant Type	%Si	%Al	%Ba	%Ca	%Mn	%RE	%Sr	%Zr
Foundry Grade FeSi	75	1.2	-	1.0	-	-	-	-
FeSi - Sr	50 or 75	<0.5	-	<0.1			0.8	
FeSi - Ba1	75	1.0	1.0	1.0	-	-	-	-
FeSi - Ba2	75	1.0	2.5	1.5	-	-	-	-
FeSi - Zr	75	1.2	-	2.0	-	-	-	1.5
FeSi - Mn – Zr - Ba	65	1.2	0.8	1.2	4.5	-	-	4.5
FeSi - RE	75	1.0	-	0.8	-	2.0	-	-
FeSi - Sr - RE	75	<0.5	-	<0.1	-	2.0	0.8	-
FeSi Sr -Zr	75	<0.5	-	<0.1	-	-	0.8	1.2

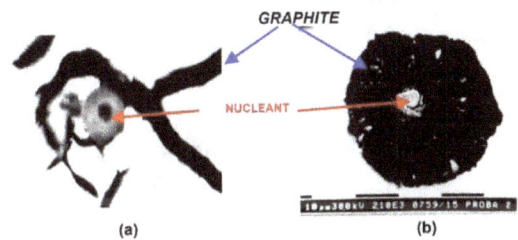

Figure 10.38 *LHS = Table showing inoculant compositions (3), RHS = Inoculant giving Nuclei for graphite (2)*

- The FeSi-RE type can also contain small controlled amounts of Oxygen and Sulphur to boost nucleation where high nodule numbers are needed in producing ferritic ductile iron.
- Magnesium ferrosilicons used as nodularising agents normally contain around 45%Si and have a range of Mg levels from 3 – 10%, some grades may contain up to 3%Ca, 1%Al, and 3%RE.
- Nodularising agents and inoculants are supplied in controlled size ranges to suit their intended modes of application e.g. 3-25mm for MgFeSi, 2-6mm for ladle inoculant, and 0.2-0.6 for late in stream inoculant.

There are always efforts to gain improvements and a review of US patents provides valuable information. 6102983 2000 Norway, Elkem, 6293988 2001 USA, 6733565 2004, USA. www.freepatentsonline.com

More recently there has been significant development of thermal analysis where samples of the molten metal are quickly cooled in a small cup which has a thermocouple to monitor the cooling rate. The data analysis and interpretation of thermal analysis of these cooling curves has been a major subject of R&D. Based on this R&D work carbon and silicon content can be estimated from the cooling curves. The iron is solidified in a tellurium-coated sand cup which causes the iron to solidify as white iron (carbide) rather than grey iron (graphite). Silicon is a major alloying constituent in cast irons. The silicon raises the graphite eutectic solidification temperature and lowers the carbide eutectic range. The percentage of carbon and silicon in the iron are obtained by measuring the arrest temperatures (TAL) and (CET).[4]

- the austenite liquidus temperature (**TAL**)
- and the carbide eutectic temperature (**CET**).

Once the eutectic temperature is determined, the silicon and carbon compositions and carbon equivalent (CE) can be obtained by the following equations

TAL = 0.556 (2962 − 212.3 C% +0.25 Si%) **CET** = 2085.4 − 22.7 Si%

Equations have also been developed to predict the values of **TEG** and **TEC**.

For example, Toshitake Kanno[5] developed the following equations for a ductile cast iron, from the chemical composition of the molten metal, as follows:

TEG = 1149.1(°C) + 4.7Si% − 4.0Mn% − 44P% + 2.7Cu% + 1.0Ni% + 1.8Co% + 13.9Al% − 17.7Mo% − 10.5Cr% − 9.3Sn% − 5.2Sb% − 6.1W% − 3.7Nb% − 14.8V% − 80.3B%

TEC = 1142.6(°C) − 11.6Si% − 0.75Mn% − 46.2P% − 1.4Cu% − 1.1Ni% − 0.7Co% − 1.8Al% − 14.5Mo% + 5.9Cr% − 6.0Sn% − 5.1Sb% − 2.8W% + 0Nb% + 3.3V% − 26.0B

10.3.2 Solidification of graphite eutectic

After undercooling below the graphite eutectic equilibrium temperature, Tr (Figure 10.30a), in the liquid alloy, graphite nuclei are created that take the form of rosettes during growth (Figure 10.30b).

On the concave surface of graphite rosettes, austenite nucleates and surrounds the central part of the rosette, rising along its branches and leading to the creation of eutectic cells. From each nucleus, a single eutectic cell is formed. Therefore, the number of nuclei also represents the number of eutectic cells.

In ductile iron, each graphite nucleus gives rise to a single graphite nodule (Figure 10.31a). As the solidification process continues, the austenite shell nucleates directly on the graphite nodule and the eutectic transformation begins. Eutectic cells may contain a lot of nodules (Figure 10.31a). Thus, in ductile iron, the number of nuclei can be identified only by the number of graphite nodules rather than the number of eutectic cells.

10.3.3 Solidification of cementite eutectic

After undercooling of cast iron below the cementite eutectic equilibrium temperature, Tc (Figure 10.30d), in the liquid alloy, cementite nuclei are created that take the form of plates during growth.

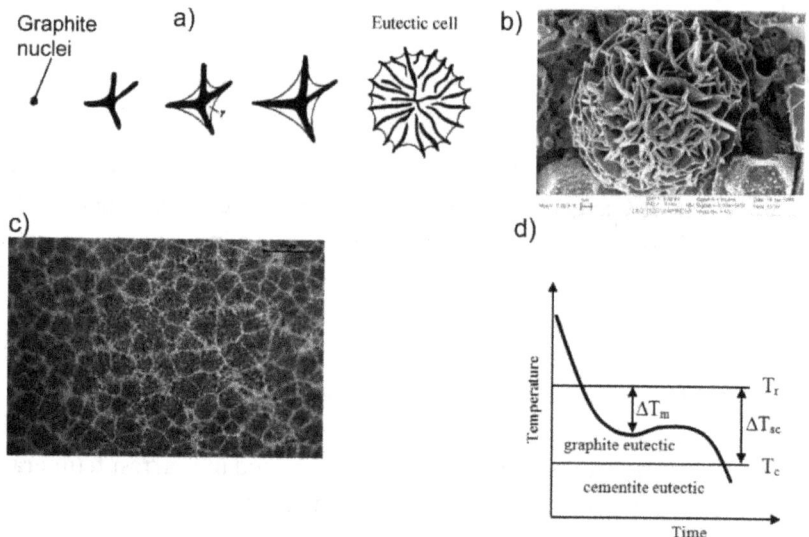

Fig 10.30 (a) Sequence of solidification of graphite eutectic cell, (b) scanning electron photography of eutectic cell, (c) graphite eutectic cell boundaries on metallographic specimen (d) cooling curve of cast iron

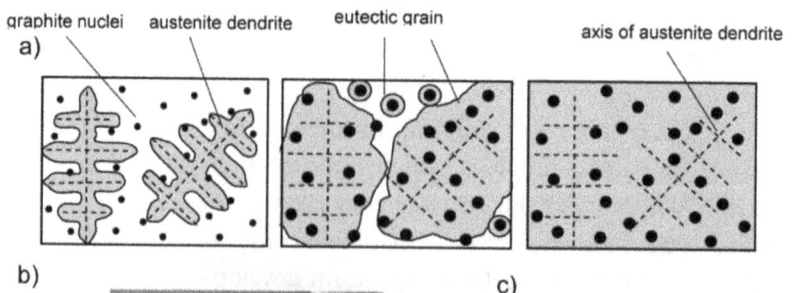

Figure 10.31 *(a) Sequence of solidification of ductile iron, (b) graphite nodule on the fracture surface of ductile iron, (c) microstructure of ductile iron*

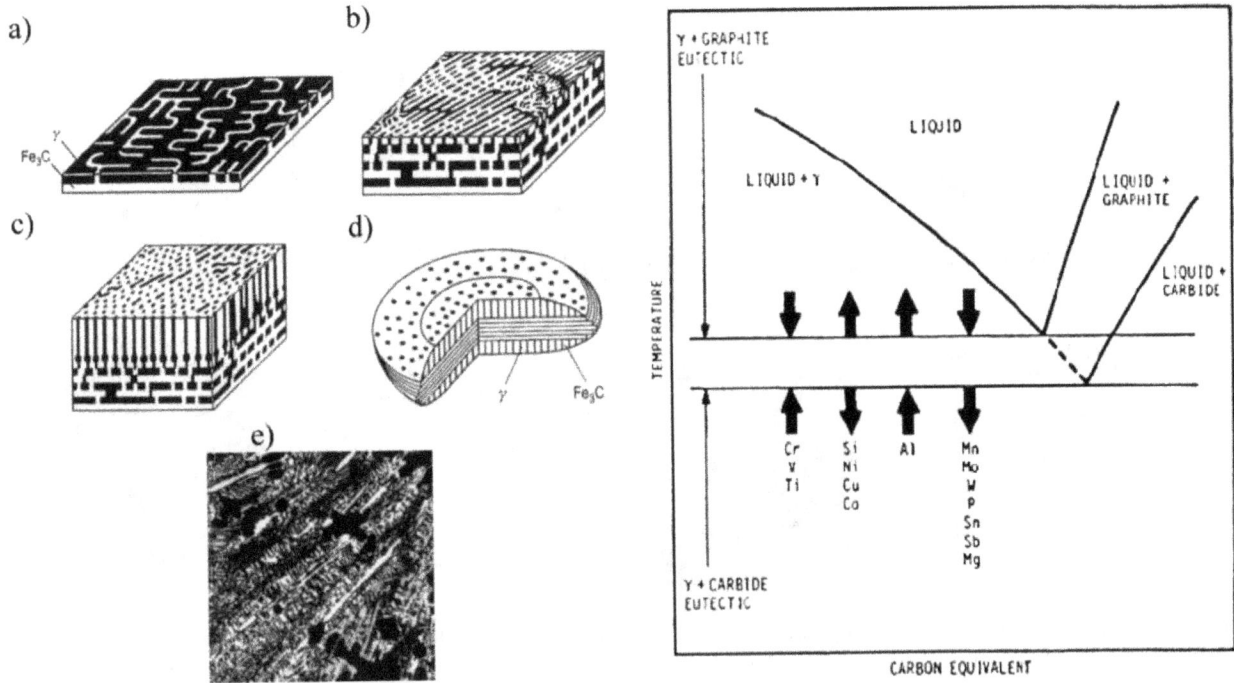

Figure 10.32 *LHS = Sequence of the development of cementite eutectic cell (a, b, c), a scheme of cell (d) and microstructure of a cross section of cementite eutectic cell (e), RHS = Effect of elements on TEG and TEC*

On this plate, the austenite nucleates and grows in a dendritic form to cover the cementite (Figure 10.32a). A common solidification front of the eutectic structure is created. During its growth, plate-to-fibre transition of cementite takes place (Figure 10.32) and a cell of cementite eutectic is formed (Figure 10.32e).

10.3.4 Morphology of eutectic cells

In a typical grey cast iron with flake graphite, there are two types of eutectic structures. The A type of eutectic cells are formed at a low degree of undercooling, while the D type is formed at a high degree of undercooling. The appearance of skeletons in the graphite eutectic cells of A and D types are shown in Figure 4a, b while microphotographs of their cross-sections are given in Figure 4c, d.

Fig 10.33 *(a) appearance of a graphite skeletons in A type eutectic cells (b) appearance of D type types of eutectic cells, (c) microstructure of cells: with A type of eutectic formed at small undercooling (with low growing rate) (d) microstructure of D type eutectic cell formed at high undercooling (with high growing rate)*

a) High undercooling
c) Low undercooling

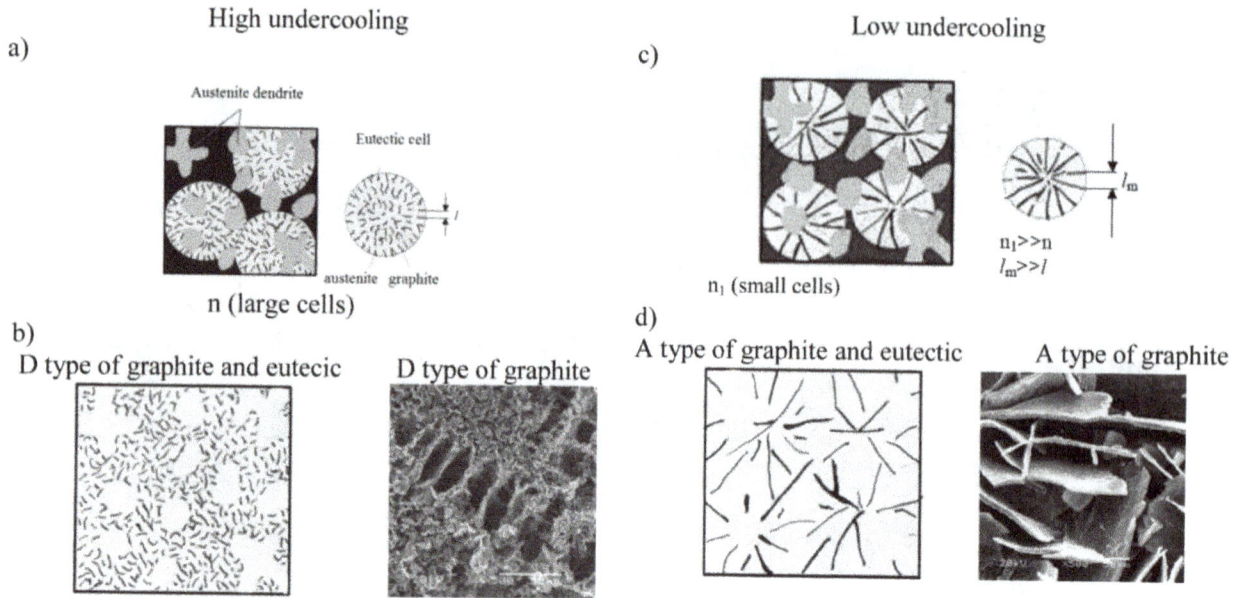

Austenite dendrite

Eutectic cell

austenite graphite

n (large cells)

n_1 (small cells)

$n_1 >> n$
$l_m >> l$

b)
D type of graphite and eutecic

D type of graphite

d)
A type of graphite and eutectic

A type of graphite

Fig 10.34 *Schemes of solidification of hypereutectic cast iron and interfacial distances between graphite precipitations (a, d), schemes of A and D type of eutectics (d, b) appearance of graphite of A and D type in scanning electron microscopy images (6)*

10.3.5 Process technology, inoculation, modification and thermal analysis

Control of the solidification characteristics of cast irons bythermal analysis has been used in the foundry industry for many years. The use has increased in recent years due to the improved systems available for temperature measurement and "expert system" software which assists the process control. Several companies have developed systems and software for thermal analysis.

• **OCC GmbH** has developed a system to measure the melt quality evaluation that uses the thermal analysis information from a sample taken from the treated melt and place into a closed thermal analysis cup named AccuVo®, which the company claims to be more accurate in the obtainment of results than a regular open cup.[7] [The operational business of

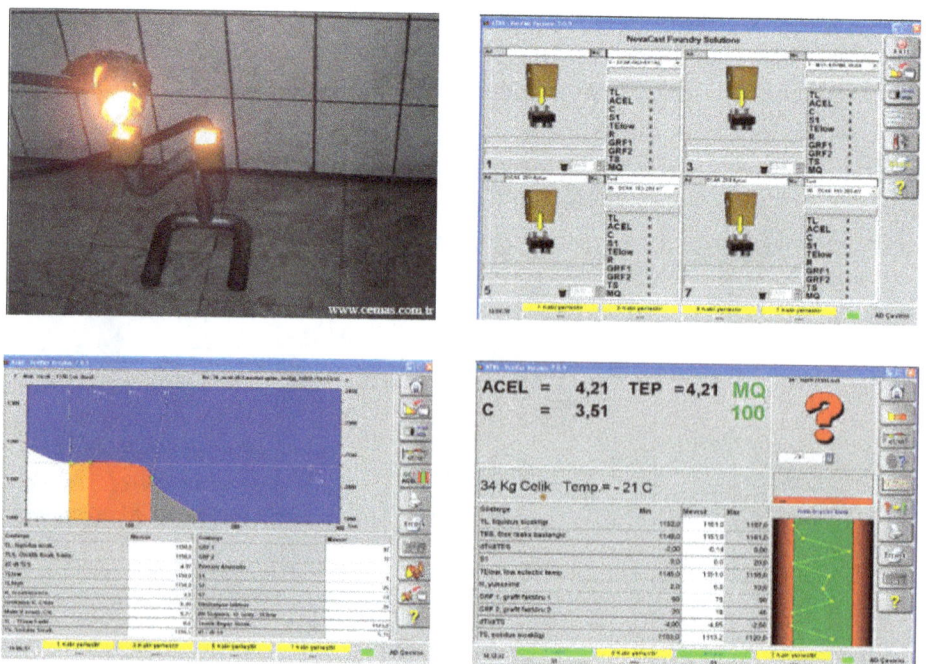

Figure 10.35 *NovaCast software*

OCC Gesellschaft für physikalische Messtechnik und kybernetische Systeme mbH („OCC ") has been transferred to Heraeus Electro-Nite GmbH & Co. KG, Hagen („Heraeus Electro-Nite ") on 1st of January 2016.]

- **SinterCast** with the System 3000[8]
- **Foseco** with the ITACA® software[9]
- **NovaCast** with the ATAS MetStartTM[10]
- **Heraeus** Quik-lab E-IBT[11]

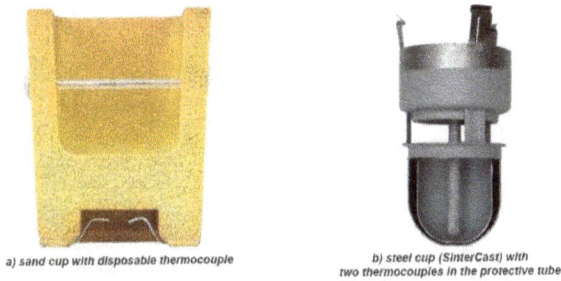

a) metal cup

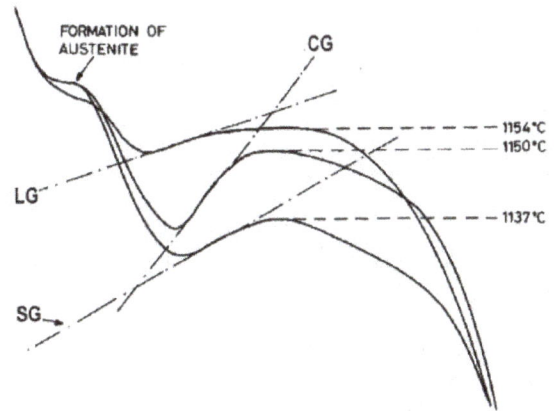

Figure 10.36 Cups used for thermal analysis

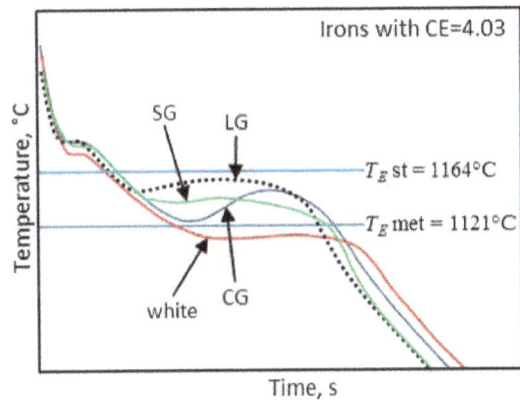

b) sand cup

Figure 10.37 Thermal analysis results

10.3.5.1 Prediction of graphite shape

Since the cooling curves are sensitive to the nucleation and growth of graphite, thermal analysis technology can be used in foundries to predict the graphite shape. The curves in Figure 10.37b show the distinctive curves relating to the different types of graphite shape.

Nodular iron melts with good nodularity, are characterized by having small recalescence (from 1°C to 4 °C). As nodularity decreases, the eutectic recalescence tends to increase, as presented in Figure 10.38.

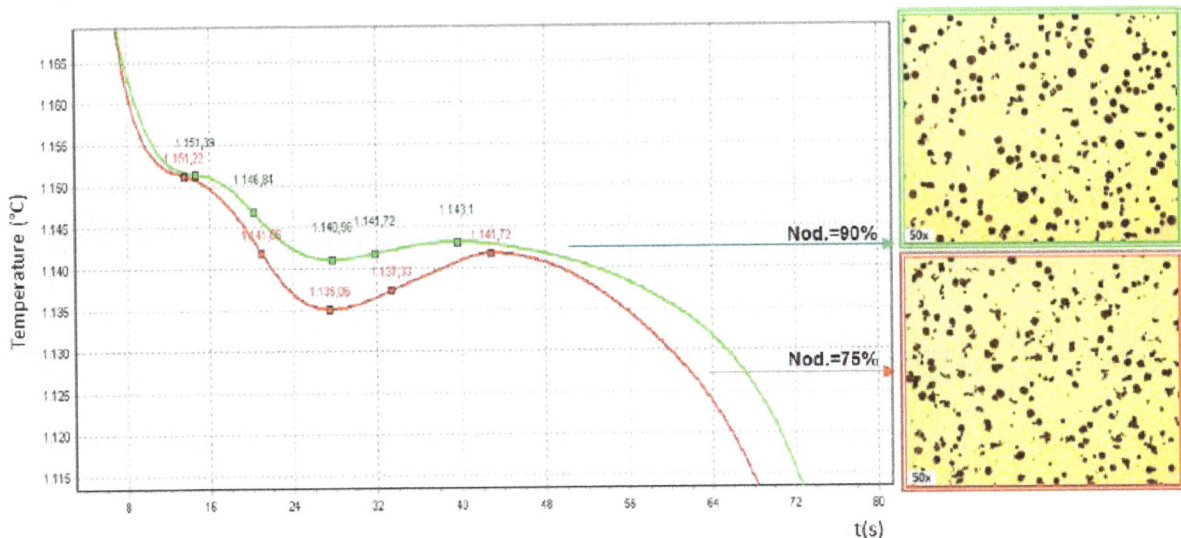

Figure 10.38 Solidification curves for nodular cast iron with high and lower nodularity

10.3.5.2 Evaluation of inoculation

Thermal analysis is often used in production of cast iron to perform the evaluation of the inoculation properties of the melt. Inoculation is used to assist the nucleation of graphite and avoid the formation of carbide phase. Through the measurement of the undercooling of the cooling curve on the eutectic plateau, it is possible to estimate if a melt is properly inoculated for the production of a given casting.

There is evidence that Rem technology should be considered an important aspect of Ductile iron manufacture.[12 and 13]

10.4 GRADES OF CAST IRONS

Pig iron, steel scrap, foundry scrap, and ferro-alloys charges are melted to give the appropriate composition. This can be carried out using a coreless induction furnace or in a small electric-arc furnace. For larger outputs a cupola, which resembles a small blast furnace, is the most common melting unit. In this process cold pig iron and scrap are charged from the top onto a bed of hot coke through which air is blown.

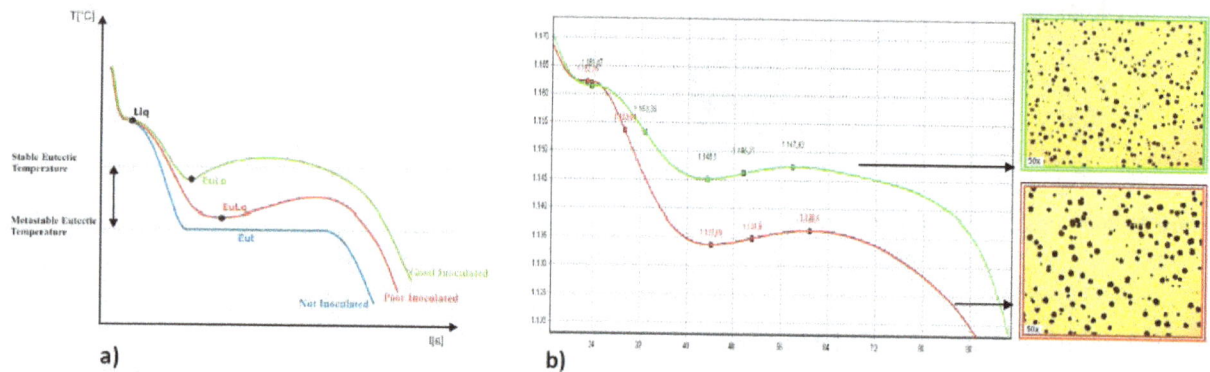

Figure 10.39 a) Illustration of the melt's inoculation influence in the cooling curves; b) real cooling curve display showing the difference between a cooling curve with low inoculation (red) and with high inoculation (green) and respective micrographs

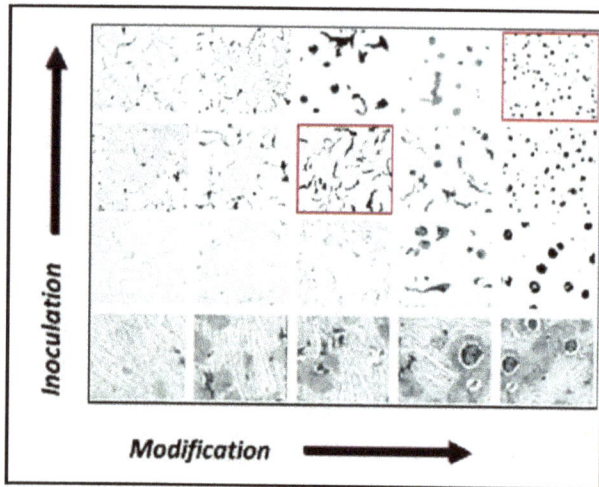

Figure 10.40 Inoculation and modification

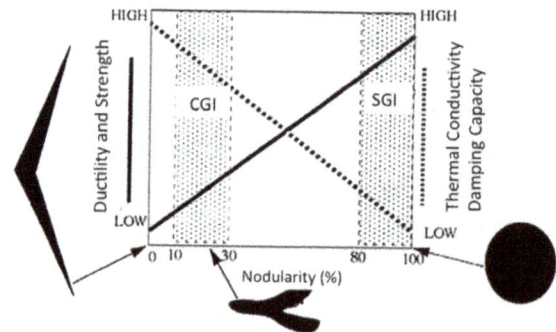

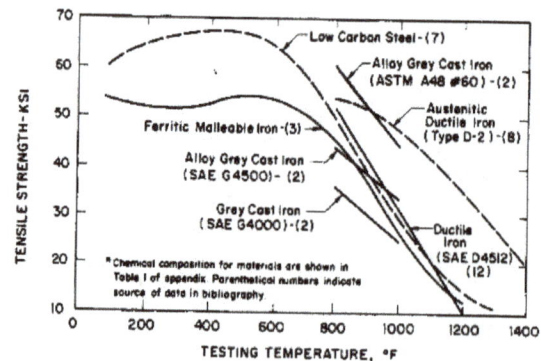

Figure 10.41 LHS Effect of graphite shape on strength and ductility and thermal conductivity. RHS Elevated temperature tensile strength

10.4.1 Grey flake cast iron

Grey iron is the most widely used, cast iron with an annual production several times greater than the total of all other cast metals. It has excellent machinability, good wear resistance, and high vibration absorption. Grey iron is valued particularly for its ability to be cast into complex shapes at relatively low cost for parts such as hydraulic control valves.

Cast irons typically contain 2-4 wt% of carbon with a high silicon content and a greater concentration of unintentionally added impurities than steels. It is convenient to combine the effect of the silicon with that of the carbon into a single factor, which is called the carbon equivalent (CE). The CE of a cast iron describes how close a given analysis is to that of the eutectic composition. The CE of the eutectic without silicon is at 4.3 %. (See Figure 10.26) As the silicon content of iron is increased, the carbon content of the eutectic is decreased. There is a linear relation and can be expressed as a simple equation

$$CE = \% \ C + 1/3 \ \% \ Si = 4.3 \ (\text{at } 0 \ \% \ \text{silicon})$$

Silicon content shifts the graphite eutectic line upwards such that the temperature interval between the graphite line and the cementite line increases from 6°C at 0% Si to 35°C at 2% Si (this increases the degree of undercooling, allowing the formation of in graphite instead of iron carbide) (See Figure 10.28). The silicon addition contains inoculant elements which favour the formation of graphite. The mechanical properties of grey cast iron are dependent on the chemical composition and the rate of cooling of the casting and therefore the section thickness. Thin sections can have reasonable tensile strength which is not maintained as the section thickness is increased. The properties of the grey cast iron are very dependent on the proportions of graphite in the matrix. If all of the carbon has separated from the molten metal (full graphitisation) then the grey cast iron resulting will

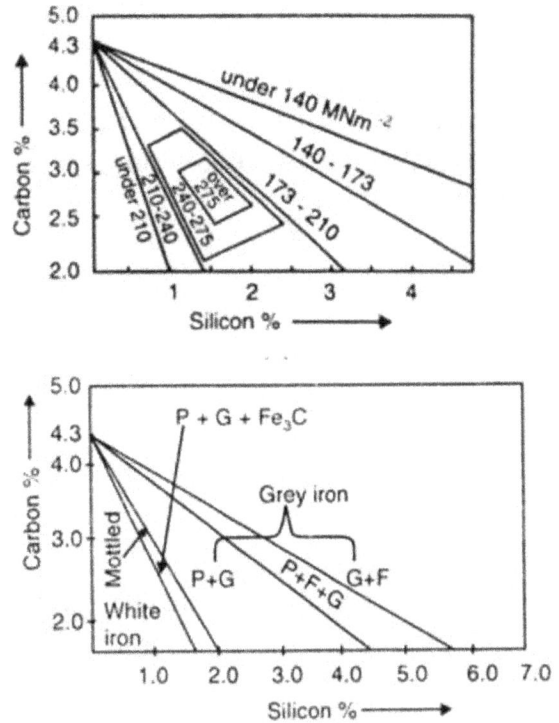

Fig. 10.42 *LHS = (a) shows that greatest structural strength is obtained when carbon is about 2.75% and silicon about 1.5%, i.e., when the matrix is completely pearlitic. RHS = Microstructure*

have graphite flakes in a ferritic matrix. If, however, between 0.5% to 0.8% carbon content remains in the form of Fe_3C the resulting grey cast iron matrix will be pearlitic and the cast iron will be stronger and harder.

The sulphur present in cast iron is kept low. Manganese is often present and combines with the sulphur to form manganese sulphide inclusion. Normally the manganese level is kept low with a level of 1.7 times the sulphur level plus around 0.12% since manganese encourages carbide formation. When tensile strengths above 350MPa are required, for thicker sections, alloying additions of chromium, nickel or molybdenum are required. A recent study recommends that the optimum level of sulphur and manganese was 0.03%S and 0.3% Mn.[14] This finding was also supported by work carried out by AFS.[15]

Ch. No.	Total Mn	Total S	Sol. [S%]	Sol. [Mn%]	[Mn] × [S]	Mn in MnS	S in MnS	MnS %	Tensile strength MPa	Hardness HB
S0.01	0.12	0.010	0.010	0.12	0.001	0	0	—	260	192
S0.025	0.24	0.025	0.025	0.24	0.006	0	0	—	290	208
S0.03	0.32	0.030	0.030	0.32	0.010	0	0	—	298	212
S0.05	0.45	0.050	0.025	0.41	0.010	0.043	0.025	0.068	290	208
S0.10	0.57	0.100	0.023	0.44	0.010	0.132	0.077	0.21	276	201
S0.30	0.95	0.300	0.021	0.47	0.010	0.477	0.279	0.76	266	197

Lower Critical temperature, °C = 730 + 28 (% Si) – 25 (% Mn)

Si Equ. Value- % Si + 3 (% C) + 0.3 (Ni% + % Cu) + 0.5 (% Al) + % P – 0.25 (% Mn) – 0.35 (% Mo) – 1.2 (% Cr)

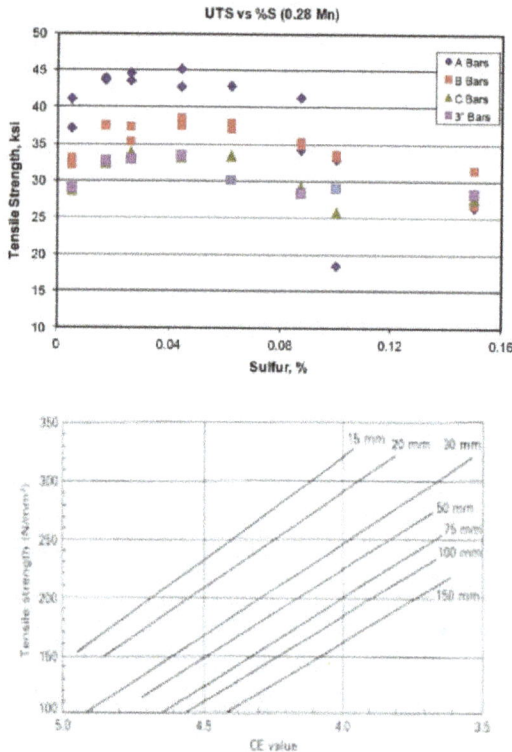

Figure 10.43 Top = Effect of CE and cooling rate (section size) on tensile strength. Bottom = Effect of sulphur

10.4.2 Nodular Cast Iron and Compacted graphite An excellent source of information = https://www.ductile.org/

Nodular cast iron also known as ductile cast iron, ductile iron, spheroidal graphite iron, spheroidal graphite cast iron and SG iron, is a type of graphite-rich cast iron discovered in 1943 by Keith Millis. During the Second World War there was uncertainty about the supply of chromium. The manufacture of Ni-hard cast iron required 1.5% chromium. Development work was carried out to find a chromium replacement. The project was carried out by Keith and he initially favoured the use of Mg to form Mg_2C_3 and MgC_2 carbides. In his own words:

"I then laid out a program in which the following additions should be made: Cr, zirconium (Zr), cerium

(Ce), bismuth (Bi), copper (Cu) and lead (Pb), tellurium (Te) – two levels since it was known to form carbides in iron-Mg and colombium (Cb). I went over the plans with my superiors and was immediately told that I could make all the additions except Mg. That was forbidden because it was dangerous.

As I recall, during Mr. Pilling's actual research days, he was experimenting with Mg additions in high-Ni iron alloys. Mg is commonly used in Ni alloys such as Monel for deoxidization purposes. He found that where the iron (Fe) content approached 25% or more, the violence of the Mg addition became intolerable. Hence, he forbade me to use Mg"

From http://www.ceesvandevelde.eu/ MillisInvention.html

Eventually on April 12, 1943 the microstructure showed that the graphite had taken a spheroidal shape . . . and ductile iron was born.

Henton Morrogh of the British Cast Iron Research Association (BCIRA) was also working on methods to

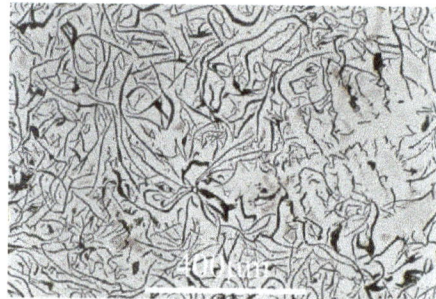

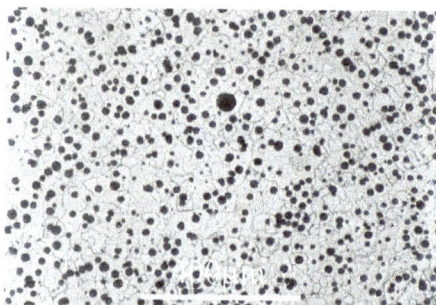

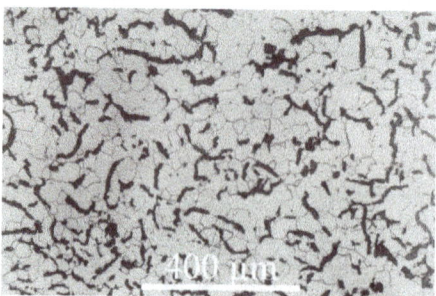

Figure 10.44 Top = Flake, Centre = Ductile Iron invented 1943, patented 1949, Bottom = Compacted graphite 1949 patent

Country/region specification	Minimum tensile strength/elongation (N/mm²/%)								
Europe EN-GS CEN 1563:1997	350-22	400-18	400-10	450-10	500-7	600-3	700-3	800-2	900-2
UK BS2789 1985	350/22	400/18	420/10	450/10	500/7	600/3	700/2	800/2	900/2
USA ASTM A356 1993[1]	60-40-18	60-42-10	65-45-12	70-50-05	80-55-06	80-60-03	100-70-03	120-90-02	
Japan JIS FCD G5502 1995	350-22	400-18	400-10	450-10	500-7	600-3	700-3	800-2	
International ISO 1083 1987	350-22	400-18	400-10	450-10	500-7	600-3	700-3	800-2	900-2
Typical hardness (HB)	<160	130–175	135–180	160–210	170–230	190–270	225–305	245–335	270–360
Typical structures[2]	F	F	F	F & P	F & P	F & P	P	P or T	TM

[1]60-40-18 refers to minimum tensile strength (lbf/in²) — minimum proof stress (lbf/in²) — % elongation.
[2]The structures are: F, ferrite; P, pearlite; T, tempered; TM, tempered martensite.

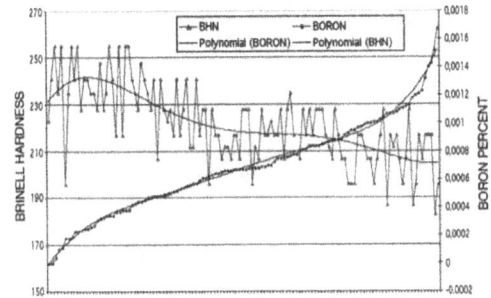

Figure 10.45 LHS = Ductile Iron Specifications. RHS = Effect of boron on pearlitic ductile iron (16)

produce spheroidal graphite. At the 1948 AFS Casting Congress in Philadelphia, Henton Morrogh presented "Production of Nodular Graphite Structures in Gray Cast Irons," which documented the work done to produce nodular (ductile) irons by the addition of rare earth elements.

During the discussion of the presentation INCO made a simple announcement that INCO had a similar process using Mg. This was done in a discussion of Morrogh's classic paper. The INCO main patent and the patent on improved cast iron was filed in Washington, D.C. on November 21, 1947 and patents 2,485,760 and 2,485,761, respectively, were granted on October 25, 1949. The patent 2,485,761 covered compacted graphite iron. Henton Morrogh filed on January 25, 1949 for a patent on his Ce process in hypereutectic irons.

The CGI was also developed in 1948 as a result of insufficient Mg or Ce treatment in melts intended to produce SGI. Its name was derived from the blunt form of graphite, thus giving properties intermediate between FGI and SGI. Their thermal conductivity is nearer flake than ductile iron and therefore it is useful for brake discs. CGI production requires close metallurgical control and addition of rare earth elements to minimize SGI formation. ASTM A 842, specifies

that the graphite shape of CGI shall be of type IV where there is no FG in the structure and for which the amount of SG is less than 20%, that is 80% of all graphite is compacted. Boron can increase the ferrite content in ductile pearlitic iron and cause a drop in hardness

Ductile iron has found wide acceptance and competes favourably with steel. A recent comparison (2013) has been made of the required Charpy value of a DI compared to steel. This has been summarised in the following table:

$C_{v\,GJS}$ [J]	$C_{v\,eq}$ [J]	$C_{v\,GJS}$ [J]	$C_{v\,eq}$ [J]	$C_{v\,GJS}$ [J]	$C_{v\,eq}$ [J]	$C_{v\,GJS}$ [J]	$C_{v\,eq}$ [J]
1	4	7	19	13	28	19	35
2	8	8	20	14	29	20	37
3	10	9	22	15	31	21	38
4	13	10	23	16	32	22	39
5	15	11	25	17	33	23	40
6	17	12	26	18	34	24	41

Figure 10.46 Comparison of the relationship between Charpy energy for Ductile iron and steel.(17)

10.4.3 Austempered ductile iron

Austempered ductile iron (ADI) has been an outstanding achievement in the metallurgy of cast iron. The application of an austempering heat treatment to a ductile iron with the correct level of hardenability

Sample Description	Nodularity (%)	Pearlite (%)	Chemical analysis (%)										
			C	Si	Mn	S	Mg	CeMM	Cu	Sn	Cr	P	Mo
Ferritic CGI	10	20	3.58	2.60	0.39	0.011	0.011		0.29	--	0.03	0.01	--
Ferritic CGI	<2	10-20	3.55	3.02	0.18	0.010	0.007		0.05	0.006	0.02		
Pearlitic CGI	0-20	70	3.6-3.8	2.1-2.5	0.2-0.4	0.005-0.022	0.006-0.014	0.01-0.03	0.3-0.6	0.03-0.05			
Pearlitic CGI	0-20	100	3.6-3.8	2.1-2.5	0.2-0.4	0.005-0.022	0.006-0.014	0.01-0.03	0.6-0.9	0.08-0.10			
Mo-CGI	5	60	3.63	2.33	0.37	0.013	0.010		0.55	0.03	0.03		0.27
Cr-Mo CGI	5	90	3.70	2.23	0.41	0.015	0.011		0.42	0.04	0.24	0.01	0.28
Hypereutectic CGI	15	95	3.85	2.26	0.40	0.010	0.008		0.30	--	0.03	0.01	--
50-Nod+50%CGI	45-55	>90	3.74	2.35	0.41	0.012	0.026		0.90	0.026	0.03		
CGI+FGI	-1	95	3.49	2.13	0.43	0.014	0.006		0.77	0.07	0.03		

Figure 10.47
Compacted Graphite Iron

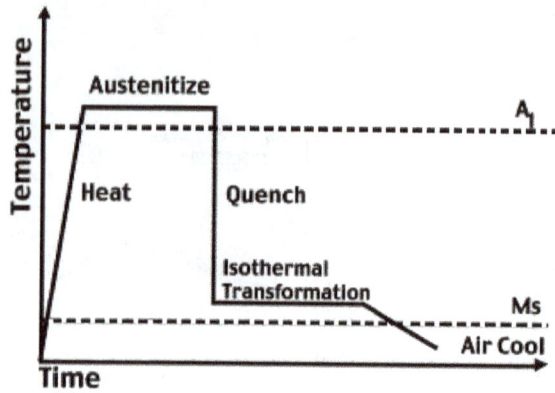

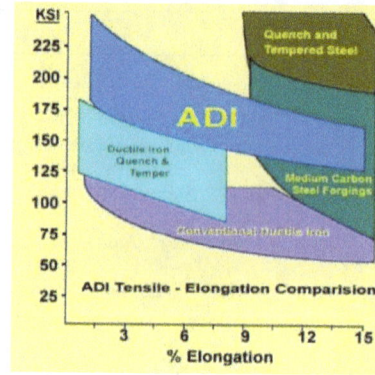

Figure 10.48 *LHS = Austempering heat treatment, RHS = mechanical properties*

results in high strength. The relatively simple heat treatment gives the cast iron a combination strength and ductility making it a very attractive material for many applications. The ADI technology involves the solution heat treatment at the temperature 850-950°C which is followed by isothermal quenching in the temperature range 250-400°C.

Depending on the heat treatment parameters, high strength ADI with Rm=1600MPa or medium strength ADI with Rm= 800MPa but 10% elongation can be obtained. These mechanical properties result from the unique microstructure of ADI, which is the mixture of carbon stabilized austenite and bainitic ferrite referred to as ausferrite. The ductility of austempered ductile iron depends mostly on relative amount of austenite. With high temperature austempering up to 40% austenite may be present. At lower treatment temperatures the high strength, low ductile ADI is obtained. This microstructure may be carbon stabilized austenite and bainitic ferrite with some martensite. The relative proportion between each component in ADI microstructure depends not only on the temperature but also time of austempering. This relationship was summarized in the form of graphs, one of which was showed in fig 10.41d. Recent work with low CE ADI found better wear resistance could be achieved.[18]

Figure 10.49 shows the potential two stages and if the castings are held too long the "stage II" reaction takes place which is undesirable since it converts the austenite to ferrite and carbide. The formation of the epsilon carbide makes the casting brittle. The aim is to allow the formation of "stage I" reaction and to then terminate the process. The time between the end of stage I and beginning of stage II is known as the "process window". Additions of Ni, Mo, and Cu are made to increase the time of the process window. These

alloy additions also increase the "hardenability" and avoid the formation of pearlite during cooling from the austenitsing temperature. Only the minimum alloy content should be added relative to the cross

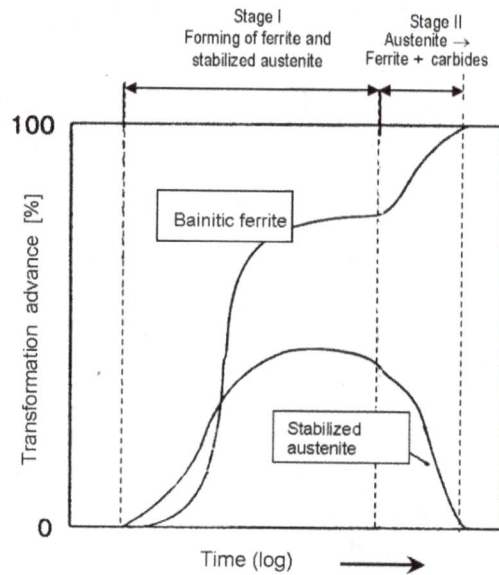

GRADE	TENSILE* STRENGTH (MPA)	YIELD* STRENGTH (MPA)	ELONGATION* (%)	TYPICAL HARDNESS (HBW)
Grade 1 900-650-09	900	650	9	269-341
Grade 2 1050-750-07	1050	750	7	302-375
Grade 3 1200-850-04	1200	850	4	341-444
Grade 4 1400-1100-02	1400	1100	2	388-477
Grade 5 1600-1300-01	1600	1300	1	402-512

*MINIMUM VALUES

Figure 10.49 *The microstructure of ADI as function of austempering time at temperature T = 300 – 350°C. RHS = Grades of ADI*

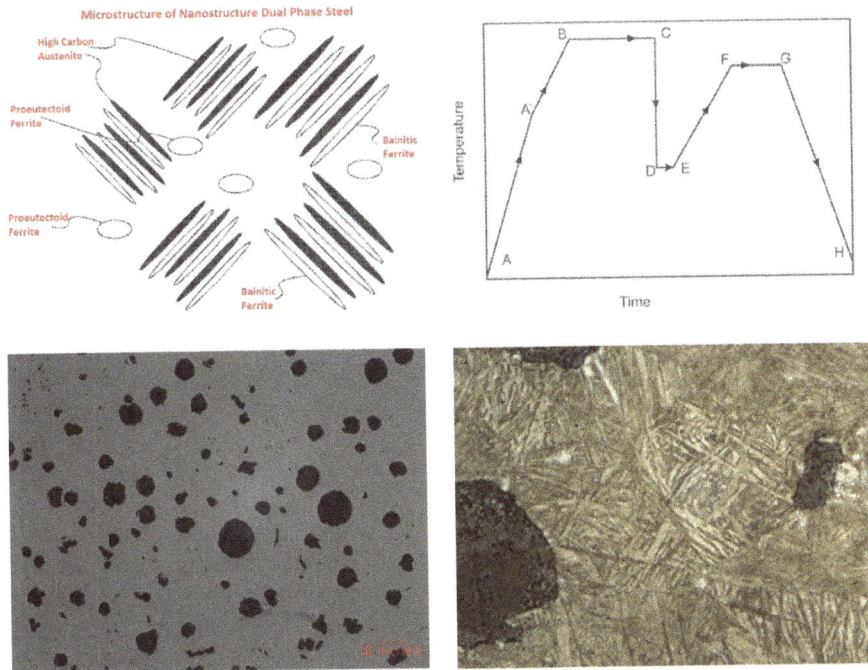

Figure 10.50
Top LHS = Microstructure of nanostructure dual phase RHS = Process time temp Lower LHS = microstructure of ADI Unetched X100 RHS = Etched in nital X1000

section for cost reasons and to ensure a good quality ductile iron. (Mo limit = 0.2% due to segregation to gbs, Cu up to 0.8% and Ni up to 2%))

Recently "nano" technology concepts have been applied to ADI in the USA By Susil Putatunda of Wayne State University. This involves the heating of an iron-silicon carbon ductile iron to near the A1, carrying out adiabatic deformation where the heat causes the casting to exceed the A1 temperature and proeutectoid ferrite and austenite then cooling to a first austempering temperature followed by heating to a second austempering temperature. This produces a fine dual phase microstructure of proeutectoid ferrite, ausferrite including bainitic ferrite and high carbon austenite. (US patent 2016 0032430[19]) A similar austempering process has been developed for low alloy steel.[20 to 22]

10.4.4 Carbidic Austempered Ductile Iron CADI

Recently a new ADI, containing carbides in the ausferrite matrix has been considered for wear applications (CADI). Carbides can provide wear resistance and are the key part of "white irons". However, the main limitation of white irons has been due to their higher brittleness. With the CADI concept areas of hard carbides are combined with the tough ausferrite matrix and the combination gives a tough wear resistant material. The main challenges relate to manufacturing

issues and the development of the compositions and process methods to allow the formation of carbides without loss of toughness. The available methods to allow carbide formation are

- Reducing the graphitizing elements such as Si
- Increasing cooling rate of the castings by chills (copper chills)
- To add carbide stablishing elements like Cr, Mo, Mn and Ti. Carbide stabilizing agents like manganese and molybdenum have a tendency to segregate in the grain boundaries and therefore the **chromium** would be preferred.

10.4.5 High Chromium White Iron

The high chromium white cast irons are a specific group of materials primarily used for abrasion-resistant applications. These irons give the best price to wear resistance for many applications. However, for some application they have insufficient toughness.

BS 4844 covers the composition ranges and hardness of the abrasion-resistant white-iron grades. The US equivalent specification is ASTM A532. High chromium white irons are readily cast into the parts needed in machinery for crushing, grinding, and handling of abrasive materials. The chromium content of high-alloy white irons also enhances their corrosion-resistant properties. The large volume fraction of primary and/or eutectic carbides in their

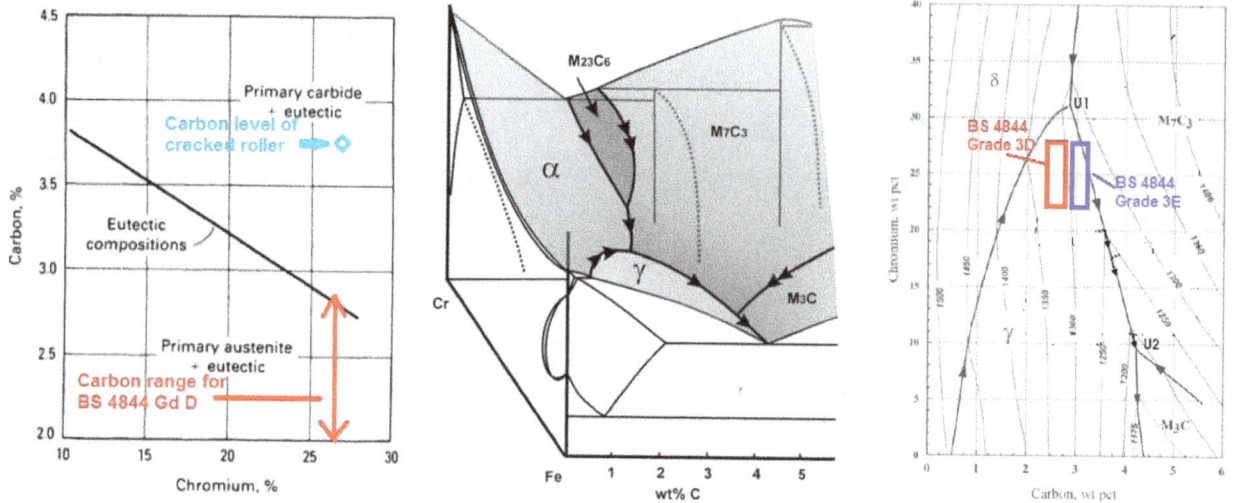

Figure 10.51 *LHS = relationship between the eutectic composition at various levels of carbon and chromium plus the carbon range for BS 4844 grade 3D and the carbon in roller that cracked during manufacture. Centre = Liquidus for the Fe-Cr-C ternary diagram where above 3% carbon pro-eutectic RHS = Carbon vs Chromium section showing the BS 4844 compositions*

microstructures provides the high hardness needed for crushing and grinding other materials. The metallic matrix supporting the carbide phase in these irons can be adjusted by alloy content and heat treatment to develop the proper balance between the resistance to abrasion and the toughness needed to withstand repeated impact.

This class of materials provides a combination of hardness and corrosion resistance. The presence of carbides in the microstructure provides the hardness needed for applications where crushing and grinding are involved, while the chromium content enhances corrosion-resistant properties. In many applications, a balance is needed between the resistance to abrasion and the toughness needed to withstand manufacturing and service stresses.

BS 4844 Class 3D white irons have 25-28% chromium, 2 to 2.8% carbon with up to 1.5% molybdenum. The Class 3E has slightly higher carbon at 0.28 to 0.32%. The diagrams in Figure 10.42 show the compositional zones against cross sections from equilibrium diagrams. These diagrams can be used to predict the microstructure. The LHS diagram shows a wear part made with higher carbon (up to 3.7%) where processing difficulties and degraded wear performance were experienced.

Figure 10.51 (LHS) shows the relationship between the eutectic composition at various levels of carbon and chromium. Wear resistant high chromium white irons are usually made as either eutectic or hypo-eutectic (*Composition below eutectic composition*) to avoid the coarse pro-eutectic carbide that would form with hyper-eutectic chemical composition (*Composition above the eutectic composition*). The presence of coarse pro-eutectic carbides (M_7C_3) can substantially reduce the toughness and wear resistance and increase the risk of cracking. Figure 10.52 from Metal Handbook Volume 14 shows a photograph of the large pro-eutectoid carbides (RHS image).

Based on published data the microstructure shown in the RHS image could potentially:

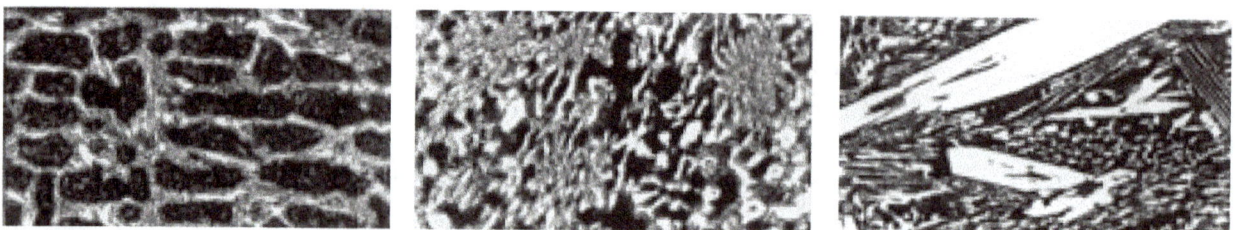

Figure 10.52 *Images of Microstructure from Metals Handbook Volume 15. LHS = hypo-eutectic, Centre = Eutectic, RHS = Hyper-eutectic*

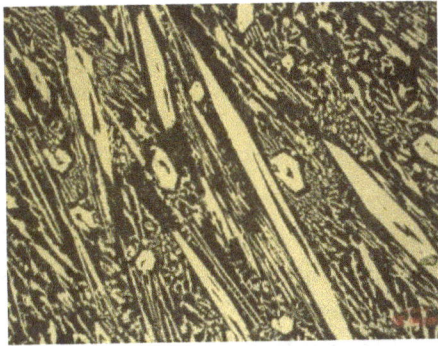

Figure 10.53 *Hollow/rod hexagonal primary carbide and eutectic carbide*

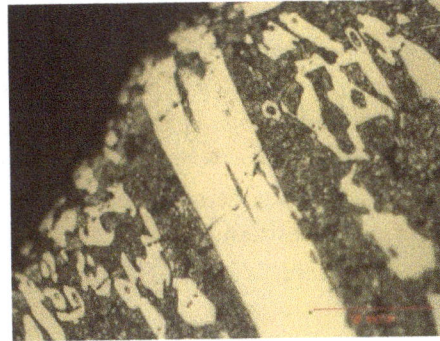

Figure 10.54 *Cracked carbide*

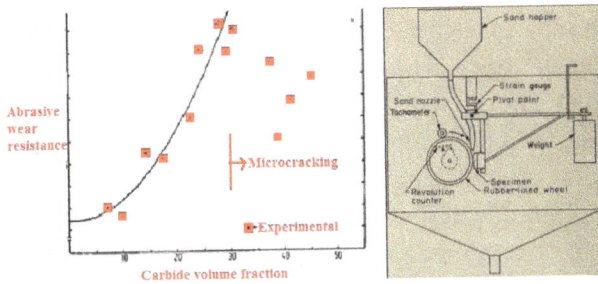

Figure 10.55 *Abrasive wear resistance of white cast irons tested in a wet rubber wheel abrasive test on 50 to 70 mesh silica sand as a function of the volume fraction of massive M_7C_3 carbides RHS = DSAW tester*

- reduce both the toughness and
- the wear resistance and
- substantially increase the risk of cracking.

The microstructure of parts made with a carbon level 1% above the upper limit of Grade are shown in Figure 10.53. The metallographic appearance of a sample showed large hollow/rod hexagonal primary pro-eutectoid carbide and eutectic carbide. The in-service wear resistance can be affected by carbide volume fraction as shown in Figure 10.55. To examine the potential wear resistance the volume fraction of carbide was measured. A eutectic composition would give a volume fraction of carbide of around 30%. With increasing carbide volume

fraction micro-cracking of the carbides occurred and a concurrent deterioration in the wear resistance.

The manufacture and wear resistance of high chromium white cast iron is a complex process which can require optimisation. The main aim in the development of improved white irons has targets high volume fraction of fine carbide.

10.4.6 Ni-hard

Nickel-chromium white irons They are low-chromium alloys containing 3 to 5% Ni and 1 to 4% Cr, with one alloy modification that contains 7 to 11% Cr. The nickel-chromium irons are also commonly identified as Ni-Hard types 1 to 4.

Ni-resist specs ASTM A439, ASTM A436, DIN 1694 and DIN EN 13835, BS 3468, ISO 2892

The NI-resist irons were developed in the 1930s and various grades were developed with nickel contents between 14% and 36% along with other elements. In addition, spheroidal graphite austenitic cast iron developed with improved mechanical properties. The compositions of the various grades and typical properties are shown in the Table 10.1.

TABLE 1 Chemical Requirements, Weight %

Class	Type	Designation	Carbon	Manganese	Silicon	Nickel	Chromium	Molybdenum	Copper	Phosphorus	Sulfur
I	A	Ni-Cr-Hc	2.8–3.6	2.0 max	0.8 max	3.3–5.0	1.4–4.0	1.0 max	. . .	0.3 max	0.15 max
I	B	Ni-Cr-Lc	2.4–3.0	2.0 max	0.8 max	3.3–5.0	1.4–4.0	1.0 max	. . .	0.3 max	0.15 max
I	C	Ni-Cr-GB	2.5–3.7	2.0 max	0.8 max	4.0 max	1.0–2.5	1.0 max	. . .	0.3 max	0.15 max
I	D	Ni-HiCr	2.5–3.6	2.0 max	2.0 max	4.5–7.0	7.0–11.0	1.5 max	. . .	0.10 max	0.15 max
II	A	12 % Cr	2.0–3.3	2.0 max	1.5 max	2.5 max	11.0–14.0	3.0 max	1.2 max	0.10 max	0.06 max
II	B	15 % Cr-Mo	2.0–3.3	2.0 max	1.5 max	2.5 max	14.0–18.0	3.0 max	1.2 max	0.10 max	0.06 max
II	D	20 % Cr-Mo	2.0–3.3	2.0 max	1.0–2.2	2.5 max	18.0–23.0	3.0 max	1.2 max	0.10 max	0.06 max
III	A	25 % Cr	2.0–3.3	2.0 max	1.5 max	2.5 max	23.0–30.0	3.0 max	1.2 max	0.10 max	0.06 max

Table 10.1 *ASTM A532 grades*

Corrosion Resistance and Wear Resistance: They have been specified to solve corrosion problems involving the handling of sour well oils, salts, salt water, acids and alkalis. They are intermediate in corrosion resistance between grey cast irons and austenitic chromium-nickel stainless steels. The excellent erosion-corrosion resistance of the Ni-Resist alloys has resulted in extensive applications for pump and valve components in sea water handling systems. Engine, pump and their parts such as pistons, wearing rings, sleeves, glands, etc. and other metal-to-metal rubbing parts have been cast in the Ni-Resist alloys.

Toughness and Low-Temperature Stability: Ni-Resist castings are considerably superior to grey iron under these service conditions.

Controlled Expansion: Expansivities from 2.2 to $10.6 \times 10\text{-}6$ in./in. per degree F are available with different types of Ni-Resist irons. This affords a conventional cast metal with low expansivity for precision machines or parts.

Heat Resistance: For unusual high-temperature service, 870°C and above, the ductile grade of Ni-Resist should be considered. The standard grades (Ni-Resist irons) exhibit a high order of heat resistance to 704°C. The alloys have a relatively low rate of oxidation, and what oxidation does occur adheres tenaciously to the base metal. This property is especially helpful in gas turbine, manifold and turbocharger applications.

Ductile Ni-Resist Irons (Austenitic Ductile Iron Castings) The ductile Ni-Resist irons are similar to standard/conventional Ni-Resist compositions but have been treated with magnesium to convert the graphite from the flake form to spheroids, which results in higher strength and ductility. Austenitic ductile iron is characterized by having its graphite substantially in a spheroidal form and substantially free of flake graphite. It contains some carbides and sufficient alloy content to produce an austenitic structure.

The ductile family of Ni-Resist is available in every type except Type 1, which because of the higher copper

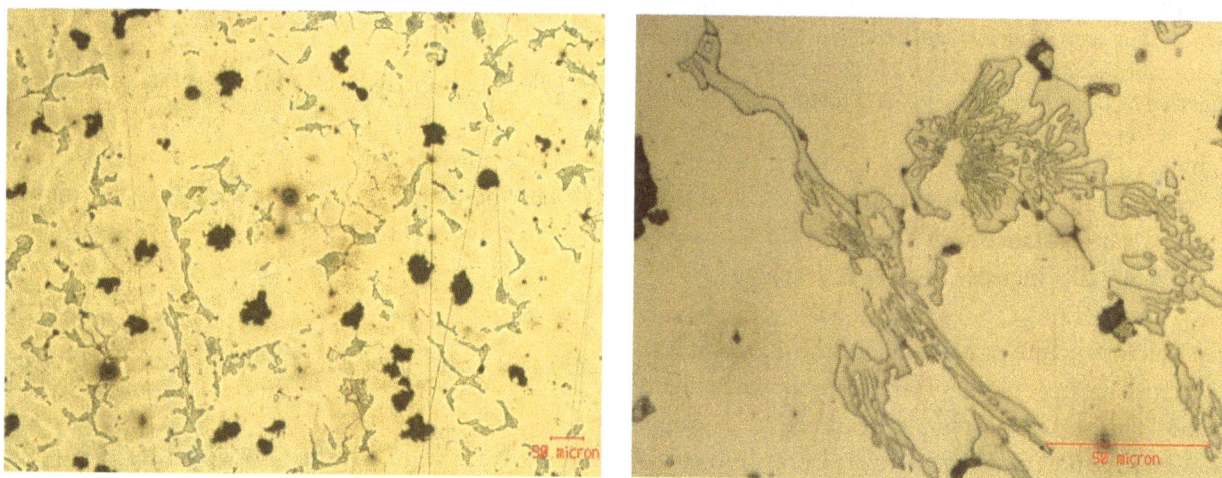

Figure 10.56 The microstructure was typical of a BS 3468 Grade S2- Niresist with spheroidal graphite and carbide areas in an austenitic matrix.

ASTM Type	UNS	TC	Si	Mn	P ≤	Cr	Ni
D-2	F43000	≤3.00	1.50-3.00	0.70-1.25	0.08	1.75-2.75	18.0-22.0
D-2B	F43001	≤3.00	1.50-3.00	0.70-1.25	0.08	2.75-4.00	18.0-22.0
D-2C	F43002	≤2.90	1.00-3.00	1.80-2.40	0.08	≤0.50	21.0-24.0
D-3	F43003	≤2.60	1.00-2.80	≤1.00	0.08	2.50-3.50	28.0-32.0
D-3A	F43004	≤2.60	1.00-2.80	≤1.00	0.08	1.00-1.50	28.0-32.0
D-4	F43005	≤2.60	5.00-6.00	≤1.00	0.08	4.50-5.50	28.0-32.0
D-5	F43006	≤2.40	1.00-2.80	≤1.00	0.08	≤0.10	34.0-36.0
D-5B	F43007	≤2.40	1.00-2.80	≤1.00	0.08	2.00-3.00	34.0-36.0
D-5S	—	≤2.30	4.90-5.50	≤1.00	0.08	1.75-2.25	34.0-37.0

Table 10.2 Chemical analysis for various types of ductile Ni-Resist irons covered by ASTM A439

ASTM Type	Tensile strength σb≥/Mpa	Yield strength σ0.2≥/Mpa	Elongation δ≥(%)	Hardness HBS
D-2	400	207	8.0	139-202
D-2B	400	207	7.0	148-211
D-2C	400	193	20	121-171
D-3	379	207	6.0	139-202
D-3A	379	207	10	131-193
D-4	414	—	—	202-273
D-5	379	207	20	131-185
D-5B	379	207	6.0	139-193
D-5S	449	207	10	131-193

Table 10.3 *Mechanical requirements for various types of ductile Ni-Resist irons covered by ASTM A439*

content does not respond properly to the magnesium treatment. There are in addition several modifications of Type 2 to 5. Various types of ductile Ni-Resist irons are covered by ASTM A439 – Standard Specification for Austenitic Ductile Iron Castings. The chemical composition of these alloys is shown in Table 10.1.

10.4.7 SiMo51

High silicon grey irons were developed in the 1930s by BCIRA and was known as Silal. These cast irons had high critical temperatures (A1), a stable ferrite and good growth and oxidation resistance. Following the development of ductile iron, the Silal was changed from flake to ductile iron. With up to 5% silicon the A1 was increased to 870°C. During the 1960s and

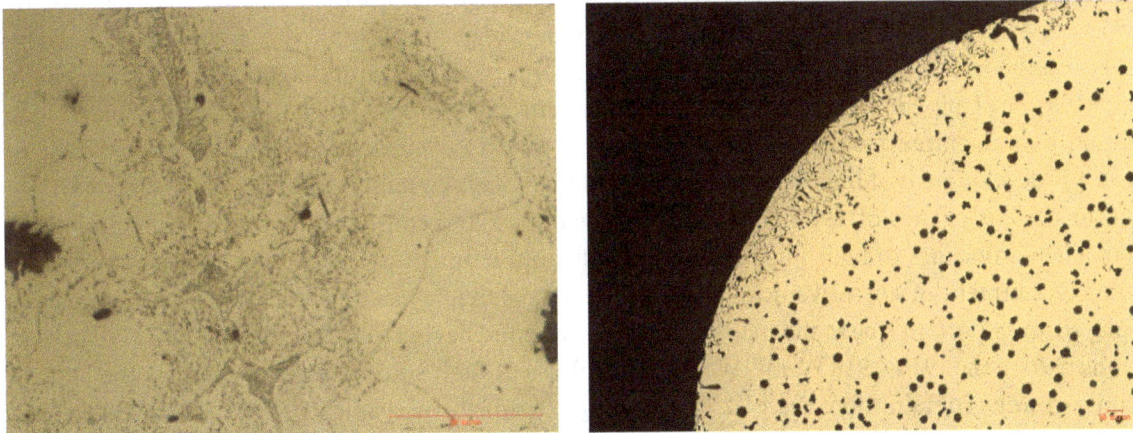

Figure 10.57 *LHS = Etched microstructure showing presence of carbides (x500 Magnification, RHS = Imperfection of Flake Graphite near cast edge due to loss of modification due to contact with mould wall (x50 Magnification)*

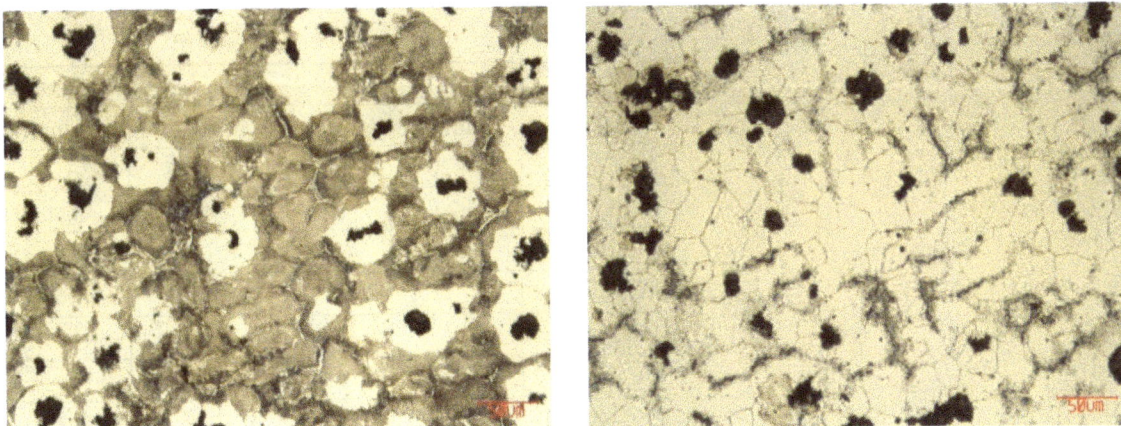

Figure 10.58 *LHS as cast x200, RHS annealed x200*

70s research was aimed at developing a material withstanding the ever-higher temperatures in a combustion engine. This work resulted in what have later been named SiMo. SiMo is a nodular cast iron typically containing 4–6 % Si and 0.5–2 % Mo and can be used in temperatures up to 850–860°C.

Molybdenum partly segregates and freezes in intercellular regions, promoting carbides while during the solid-state transformation molybdenum particles precipitate around grain boundaries. The solidified material shows a range of interesting structures and phases which will be presented There has been growing interest in the use of high silicon-molybdenum ductile irons with up to 4% Si and 1% Mo. Their good strength up to 600°C makes them a viable and cost-effective replacement for more highly alloyed irons and steels in elevated temperature applications such as turbocharger housings, engine exhaust manifolds and furnace components.

10.4.8 High silicon cast iron (15% silicon)

The anodes used for impressed current cathodic protection (ICCP) of pipelines, storage tanks and structures are manufactured from high silicon cast iron. The initial development of high silicon cast iron dates back to the early years of 1900 and pre-date World War 1. The initial use of high silicon cast iron was for the containment of corrosive acids and the material played an important role in the manufacture of munitions used in the 1914-1918 War.[23] When the high silicon cast irons contain more than 14.5% silicon the alloys are characterized by excellent corrosion resistance. However, these alloys have some known weaknesses such as low strength and toughness, high

hardness and associated brittleness. Parts made from the alloy have low thermal conductivity and low thermal expansion coefficient which results in a tendency for cast metal to have porosity and shrinkage defects.

Thus, the useful application of high silicon cast iron requires control of several metallurgical features during manufacture especially the microstructure. A study of the microstructure of commercial acid-resisting silicon-iron alloys has been carried out by J E Hurst and R V Riley.[24] The metallographic preparation of silicon-iron-alloy specimens for metallographic examination was found to be difficult. This was due to the hardness of the iron and its proneness to inter-dendritic porosity which causes the preparation media to be trapped and the specimens become vulnerable to scratching during polishing. The etching of the polished surface to reveal the underlying structure was also found to be difficult. Due to the acid-resistant qualities of the metal a very strong etchant had to be used. The choice of etchant was based on a mixture of hydrofluoric and nitric acid in water. However, for the work carried out during this project an electrolytic oxalic acid etching technique was developed. Based on the binary equilibrium diagram for the iron-silicon system[25] shown in Figure 10.59 an alloy made with less than 15.2% silicon would be a single phase of alpha. Alloys containing over 15.2% of silicon would be duplex and consist when in thermal equilibrium at room temperature, the alpha and epsilon phases. The epsilon phase is considered to be an intermetallic compound having the composition FeSi. The successful use of high silicon cast iron requires the alloy to have adequate mechanical properties. These mechanical properties are dependent to large extent on the presence of free fine graphite in the microstructure. An essential aspect of the manufacture of high silicon

Standard	ASTM A518 GR 3		BS 1591 SiCr 14 4	
Elements	Minimum	Maximum	Minimum	Maximum
Silicon	14.20	14.75	14.25	15.25
Chromium	3.25	5.00	4.00	5.00
Carbon	0.70	1.10		1.40
Manganese		1.50		1.00
Molybdenum		0.20		
Copper		0.50		
Phosphorus				0.25
Sulphur				0.10

Table 10.4 Chemical Compositions, Selected Standards

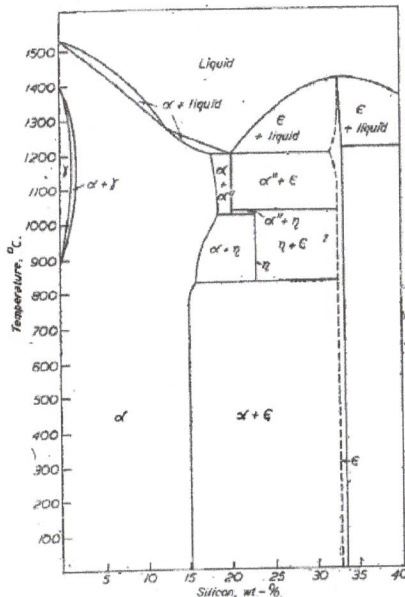

Figure 10.59 Binary equilibrium diagram

cast iron is that they are true cast irons and contain graphite and do not solidify as steel with carbides in the microstructure.

The salient factors that provides assurance that anodes will give satisfactory service have been reviewed by Levelton Engineering Ltd[26 and 27] and also discussed in early Duriron patents[28]. Levelton Engineering Ltd examined the microstructure of HSCI samples made by sand casting, chill casting and centrifugal casting and concluded that the corrosion performance was linked with:

• The shape and size and form of graphite
• The presence of secondary phases (silicides, carbides)
• The degree of segregation of alloys
• The grain size.

The Duriron patent 3129095 dated May 1963,[28] recommended a level of mechanical strength required for anodes. The patent quotes that for high silicon cast iron a minimum transverse load value of around 800 pounds and deflection of 0.025 ins should be met to avoid excessive breakage of the anodes during handling and storage. Because normal tensile testing was difficult, the patent proposed a transverse bend test. A test specimen was cast as a one-inch square bar, thirteen inches long. This was placed on twelve-inch centres in a testing machine, loaded at the centre until breakage occurred. The optimum chemical analysis used for high silicon cast iron for impressed current anodes has been established over many years

and are controlled by established National Standards. The alloy was initially used for anodes around 1950. Due to excessive corrosion in chloride environments anodes with a chromium addition were introduced in 1959. Based on this development work the silicon-chromium composition became the first material of choice for high silicon cast iron anodes. This was adopted as ASTM A518 Grade 3 which was specifically recommended for impressed current anodes. BS 1591 grade 14. 4 being essentially identical to the ASTM grade differing slightly from the ASTM A518 Gr 3 chemistry, as shown in Table 10.4.

In addition to the main elements used in the manufacture of anodes the major experienced foundries involved with anode casting are aware of other residual elements present that can have an important effect on the integrity of the product.

In addition, the importance of gas content that can arise during the manufacture such as hydrogen, nitrogen and carbon dioxide.[29] The importance of gas and the resulting gas porosity on the mechanical integrity of HSCI was outlined in a Duriron patent US 322161[30] where the benefits of vacuum degas treatments to remove the gas was demonstrated, resulting in improved mechanical integrity.

For normal castings a low hydrogen melting practice should be used (dry alloy additions and avoiding all sources of moisture) and possibly a small addition of ferro-boron to combine with the nitrogen. US patent 3129095 recommended a boron addition to act as

Figure 10.60 Microstructure of high silicon cast iron

a grain refiner.[28] Aluminium should also be kept as low as possible and therefore low aluminium ferro-silicon should be used.[31]

Cerium/rare earth metals have also been used to try convert the flake graphite to spheroidal form.[32] It has been shown that the addition of 0.02% of rare earth metals improved the corrosion resistance.[33] The trend for sulphur and phosphorous would be to target low levels. A recent European publication quoted typical levels for sulphur and phosphorous to be 0.007% and 0.1% respectively.[34]

REFERENCES

1. https://docplayer.net/21478772-Inoculation-of-cast-irons-practices-and-developments.html
2. https://ankirosfoundrycongresstr.files.wordpress.com/2014/09/manuscript4.pdf
3. https://www.foundry-planet.com/fileadmin/redakteur/pdf-dateien/1-Iulian-Riposan.pdf
4. Vasudev D. Shinde, Thermal Analysis of Ductile Iron Casting, https://cdn.intechopen.com/pdfs/57825.pdf
5. Toshitake Kanno, Prediction of Graphite Nodule Count and Shrinkage Tendency of Spheroidal Graphite Cast Iron by One Cup Thermal Analysis, 2018 Japan Foundry Engineering Society.
6. E. FRAŚ*, M. GÓRNY, AN INOCULATION PHENOMENON IN CAST IRON, ARCHIVES OF METALLURGY AND MATERIALS Volume 57 2012 Issue 3)
7. http://www.occ-web.com/index.php?id=accuvo&L=1%27%22
8. http://sintercast.com/technology/system-3000-2
9. http://www.foseco-at-gifa.com/fileadmin/GIFA2015/brochures/itaca-e.pdf
10. http://foundrymag.com/new-products/most-user-friendly-metallurgical-process-control-system
11. http://www.ibt.co.il/uploaded_files/documents/file1369645989_U1909.pdf
12. RARE EARTHS IN DUCTILE IRON PRODUCTION: A CRITICAL REVIEW, RESEARCH PROJECT NO. 27 by S. V. Subramanian, Qichuan Jiang, S. Thangirala, G. R. Purdy and D. A. R. Kay, 1996 pp83 The ductile iron society
13. Sheng Da Cast Iron containing rare earths, pp 383, 2000, Tsinghua University Press.
14. Toshitake Kanno, Quantitative Relation between Mn and S for Mechanical Properties of Flake Graphite Cast Iron, Materials Transactions, Vol. 55, No. 11 (2014) pp. 1716 to 1721)
15. Richard Gundlach, INFLUENCE OF Mn AND S ON THE PROPERTIES OF CAST IRON PART III TESTING AND ANALYSIS, International Journal of Metalcasting, Volume 9, Issue 2, 2015 file:///C:/Users/user/Downloads/Re-Examining-the-Role-on-Manganese-and-Sulfur-on-Gray-Cast-Iron%20(2).pdf
16. Richard B. Gundlach, Compensation for Boron in Pearlitic
17. https://www.trafi.fi/filebank/a/1477641190/e22f005127f9ffc5644f2701f8f98ca2/22895-WNRB_research_report_90.pdf
18. Priyanshu Bajaj et al Effect of Austenitising Temperature on Microstructure and Wear Properties of Low Carbon Equivalent Austempered Ductile Iron Indian Foundry Journal Vol 59 No. 10 October 2013 page 32
19. Susil K Putatunda, Development of nanostructure austempered ductile iron with dual phase microstructure, US2016 0032430
20. Susil K.Putatunda, Abhijit Deokar and Gowtham Bingi, MECHANICAL PROPERTIES OF A MEDIUM CARBON LOW ALLOY STEEL PROCESSED BY TWO STEP AUSTEMPERING PROCESS, Materials Science Forum Online:2010-01-12 ISSN: 1662-9752, Vols. 638-642, pp 3453-3458
21. Susil K.Putatunda, Development of a high strength high toughness bainitic steel, US 2011/0114233.
22. Susil K.Putatunda, AUSTEMPERING OF A SILICON MANGANESE CAST STEEL, Pages 743-762 | Published online: 23 Aug 2006 Download citation https://doi.org/10.1081/AMP-100108696
23. Duriron and Durichlor 51M Materials Data Sheet. Flowserve. https://www.flowserve.com/files/Files/Literature/FPD/bulletin_a2.pdf
24. J E Hurst and R V Riley: The microstructure of commercial acidresisting Silicon-Iron Alloys; Journal of the Iron and Steel Institute, 155, 172-178 (Feb 1947)
25. M. C. M. Farquhar, H. Lipson and A. R. Weill: Journal, of the Iron and Steel Institute.1945, No.II .p. 457
26. R.S. Chariton, BH Levelton & Associates, Richmond, B.C. Canada: Effect of Microstructure on the Corrosion Resistance and Mechanical Properties of High Silicon Cast Iron Anodes. http://www.farwestcorrosion.com/technical-paper-effect-of-microstructure-on-high-silicon-cast-iron
27. www.anotec.com/articles/.../Art-23-Metallic-Structure-for-HSCI-Anodes-Rev01.pdf
28. US patent 3129095 May 1963
29. ASM Specialty Handbook Cast Iron Page 151
30. US {Patent 3222161 December 1965
31. H T Angus Cast Irons Physical and Engineering properties. Butterworths 1976
32. Chinese Patent CN 103146989A. Centrifugal casting method of chromium containing high silicon cast iron anode tubes
33. Li Ju-cang et al: Corrosion properties of high silicon iron based alloy in nitric acid, China Foundry. Vol 14 No4 p276. 12. Veselinka Dordovic et al: Corrosion resistant high silicon cast iron, Proceedings of 3rd BMC-2003 Macedonia p333

Ductile Iron Project Proposal presented to the Ductile Iron Society 2017

Books

Roy Elliott, Cast Iron Technology, Butterworth & Co. (Publishers) Ltd., 1988.

Peter Beeley, Foundry Technology, pub Butterworth-Heinemann 2001

John Campbell, Castings, pub Butterworth-Heinemann, 2003

10.5 STAINLESS STEEL

Steels containing at least 10% chromium are called stainless steels because they have good corrosion resistance. These alloys also have improved resistance to high temperature. A separate system of AISI designations is used which employs a three-digit number, such as AISI 316 and AISI 409, with the first digit indicating a particular class of stainless steels. Figures 10.61 and 10.62 provide details of the important stainless steel types and Figure 10.73 shows their historical development.

10.5.1 Dual Phase 3CR12

The initial composition of 3CR12 established in the 1980s in South Africa was based on a small but significant composition change from the established AISI 409 grade. Because of its unique properties several "European" equivalent versions were developed similar to 1.4003 grade. The steel is dual phase ferrite and martensite and several ferrite and austenite forming elements were balanced in order to give a ferrite factor in the range 8-12 which is calculated from the formula given by Catechiser. KALTENHAUSER R.H. Met. Eng. Q., May 1971, pp. 41–47. This gives the steel the process capability to develop dual phase ferrite/martensite This steel has been used in the manufacture of bus and rail vehicles and bridges based on the high strength and corrosion resistance.[1 to 7] Ferrite Factor (FF) = %Cr+ 6(%Si) + 8(%Ti) + 4(%Mo) + 2(%Al) + 4(%Nb) − 2(%Mn) − 4(%Ni) − 40[%(C+ N)].

10.5.2 Martensitic Stainless

In 1912, Elwood Haynes applied for a US patent on a martensitic stainless-steel alloy. This patent was not granted until 1919. Also in 1912, Harry Brearley of the Brown–Firth research laboratory in Sheffield, England, while seeking a corrosion resistant alloy for gun barrels, discovered and subsequently industrialized a martensitic stainless-steel alloy. The discovery was announced two years later in a January 1915 newspaper article in The New York Times. Brearley applied for a US patent during 1915. The 400-series stainless steels have carbon in various percentages and also small amounts of metallic alloying elements in addition to the chromium. If the chromium content is less than about 15%, as in the types 403, 410, and 422, the steel can in most cases be heat treated using quenching and tempering to have a martensitic structure, so that it is called a martensitic stainless steel (see Figure 10.64).

The ISSF have prepared a concise summary of martensitic grades with the slogan "Rediscover the Martensitics" with the view that their range of properties are underestimated.[8] The various grades were summarised in the table shown in Figure 10.63 (LHS). Group 1 = Fe-C-Cr alloys, Group 2= Ni bearing grades, where some of the carbon is replaced by Ni. These grades offer higher toughness particularly at low temperature. The higher chromium offers higher corrosion resistance. Mo can be added to both group 1 and 2 to enhance corrosion resistance. Group 3= Precipitation hardening grades which offer the best strength and toughness. Group 4= Creep resistant grades. The chromium content is around 11% with additions of Co, Nb, V and B they have creep resistance up to 650°C.

- 3CR12
- 400 Series (martensitic)
- F6NM (Soft martensitic wrought)
- CA6MN (Soft martensitic cast)
- 409,410, 430 (Ferritic)
- 29Cr-4Mo (Super-ferritic)
- 200 series (austenitic)
- 300 series 316L (austenitic)
- 6Mo UNS S31254 W 1.4547 ASTM A403 (Super-austenitic)
- 22Cr (duplex)
- 25 Cr (Super-duplex)

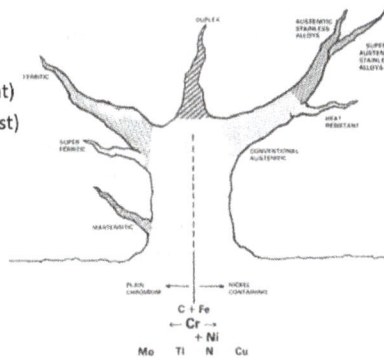

Figure 10.61 Grades of stainless

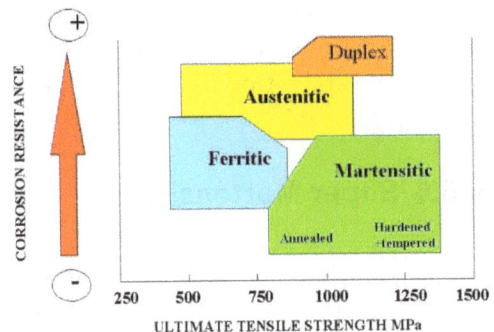

Figure 10.62 Strength and corrosion resistance

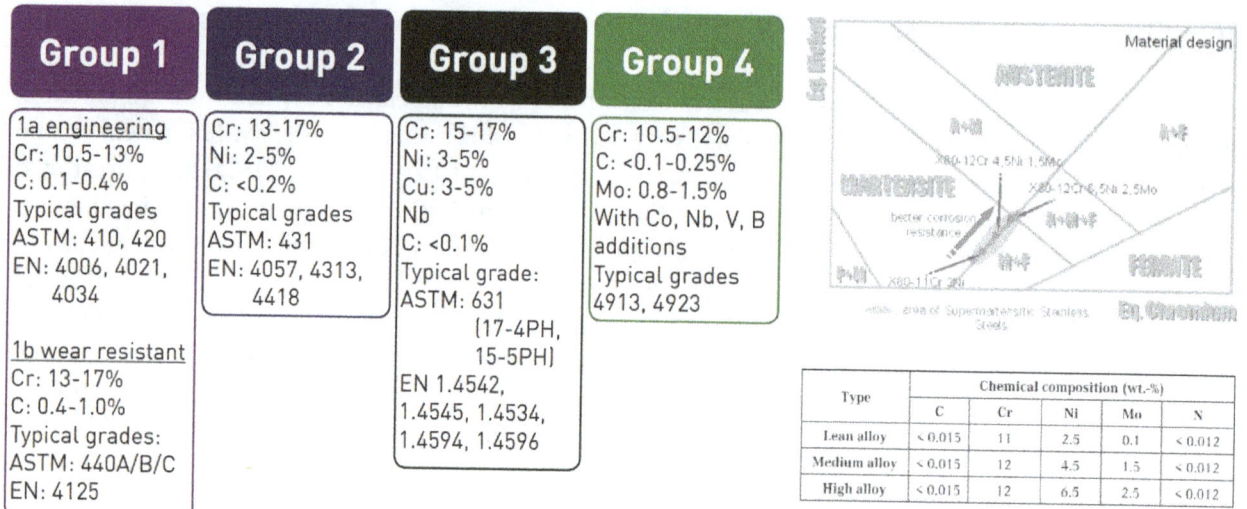

Figure 10.63 LHS = Conventional martensitic stainless steel, RHS = Super martensitic stainless steel

Figure 10.64 LHS = Martensitic, Centre= 410 stainless with delta ferrite ASTM A1021 grade B (Failed Charpy-See Figure 10.66 for SEM images of the fracture surface. RHS = Hardness of martensitic stainless grades

10.5.3 Soft Martensitics (Also known as CA6MN and F6MN)

Based on the 12% chromium with reduced carbon (below 0.06%) and 4% to 6% nickel and 0.5% to 1.5% molybdenum. This results in better quality castings, higher toughness and improved weldability with only a small change in strength properties. One concern during the processing of these grades is the heat treatment parameter to meet the NACE hardness requirement of 23Rc. A recent publication shows that the steel should have a low C and N with carbon below 0.02% and nitrogen below 100ppm.[9]

10.5.4 Super Martensitic

Super martensitic stainless steels were developed from the soft martensitic steels with a lower carbon and nitrogen level. They are also known as weldable grades of martensitic stainless steels with less than 0.015 % C, 4-6 % Ni, up to 2.5 % Mo and a small amount of Ti content. Although these grades have significantly lower resistance in H_2S and chloride environments as compared to that of duplex stainless steels in a sweet and mildly sour environment, they offer sufficient corrosion resistance and are cost effective to the use of carbon steels and process inhibitors. (See Figure 10.63 RHS for typical compositions and microstructure.) Super martensitic stainless steels have high proof strengths of 550 MPa and above.[10] Some failures of centrifugal compressor impellers made from SMSS that involved the environmentally assisted cracking working in sour environments containing CO_2 and H_2S led to a doctorate study in France at "l'Ecole des Mines de Saint-Etienne"[11]. The work was carried out for a GE owned company and demonstrates the complexity of the SMSS and the high-tech equipment needed for the in-depth study.

10.5.5 Ferritic Stainless

An ISSF publication on ferritic grades can be found with the title "The Ferritic Solution properties, advantages, applications the essential guide to ferritic stainless steels",[12] which provides a comprehensive introduction to ferritic stainless grades. The chromium content in ferritic stainless steel is typically 17 to 25% as shown in Table 10.5 from a Sandvik brochure and also Figure 10.65.

The ferritic stainless steels have a BCC (body centred cubic) lattice. Due to the presence of interstitial carbon and nitrogen the toughness can be low. Therefore, the commercialisation and use of the ferritic stainless grades had to wait until the 1970s and the development of the AOD process so that steels with low levels of carbon could be manufactured. The AOD vessels are similar to the vessels used in the Bessemer process and are also "bottom" blown but use inert gas which also shrouds the oxygen entry. The AOD process allowed the reduction of carbon and nitrogen to low levels. However, since the carbon is kept low the ferritic stainless steels have low strength. The steels are not hardenable by heat treatment and have annealed yield strengths in the range of 275 to 350 MPa. In addition, partly because of the removal

of the carbon there is no natural grain size control and the steels can have a coarse grain size which tends to increase with section size. The ferritic grades are therefore limited to a relatively thin strip.

The ferritics are usually lower in cost compared to the austenitic steels due to the absence of nickel. When nickel prices are high the interest in ferritics increases. The cost difference can be around £12 per kg. They have often been thought of as also having lower corrosion resistance but the higher grades can match the corrosion resistance of the austenitics. The stabilised grades such as 1.4509 and 1.4521 are broadly similar in corrosion resistance to 1.4301 (304) and 1.4401 (316). While austenitic stainless steels suffer from sensitivity to stress corrosion cracking ferritic stainless steels (FSS) are almost immune. The main disadvantages of the ferritics are:

- Limited toughness – Not acceptable for sub-zero temperatures
- Formability – Good for deep drawing but not stretch forming due to lower ductility
- Weldability – Rapid grain growth in thick sections (greater than about 3mm) leads to poor weld toughness in comparison to austenitic stainless steel.

Alloy	C	N	Cr	Ni	Mn	Si	Mo	P	S	Other	Applications
409	0.03	0.03	10.5-11.7	0.5	1	1	-	0.04	0.02	≤0.5 Ti	Automotive exhaust systems
410	0.04	0.1	10.5-12.5	0.6-1.1	1.0	1.0	-	0.04	0.03		
430	0.12	-	16-18	-	1	1	-	0.04	0.03		Washing drums, decorative trim, domestic appliances,
439	0.07	0.04	17-19	0.5	1	1	-	0.04	0.03	≤1.1 Ti	Washing drums, decorative trim, domestic appliances,
444	0.025	0.035	17.5-19.5	1	1	1	0.75-1.25	0.04	0.03	Ti+Nb	
446-1	0.17	0.17	23-27	-	0.8	0.5	-	0.03	0.015		Recuperators, soot blower tubes, thermocouple tubes, muffle tubes in furnaces
UNS K 91500			20-24							5-6 Al	Automotive exhaust catalysts, resistance heating in ovens and furnaces

Table 10.5 *The chemical analysis of ferritic stainless steel*

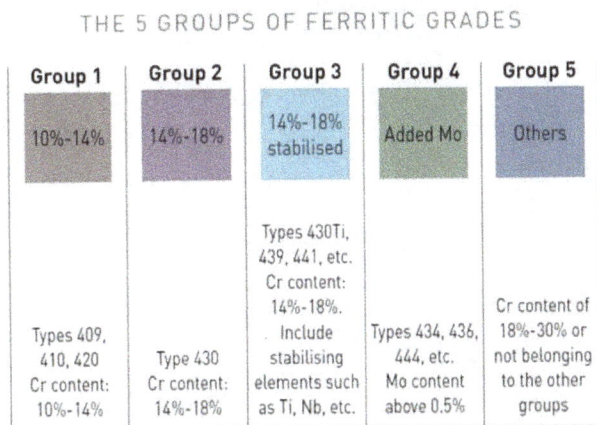

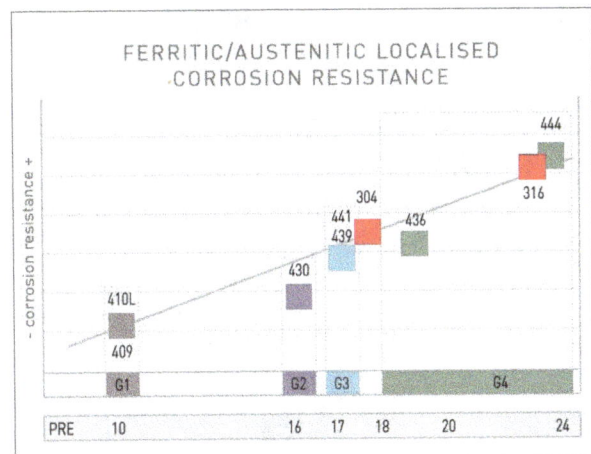

Figure 10.65 *Top= ISSF classification of ferritic grades Lower = PREN values for stainless grades*

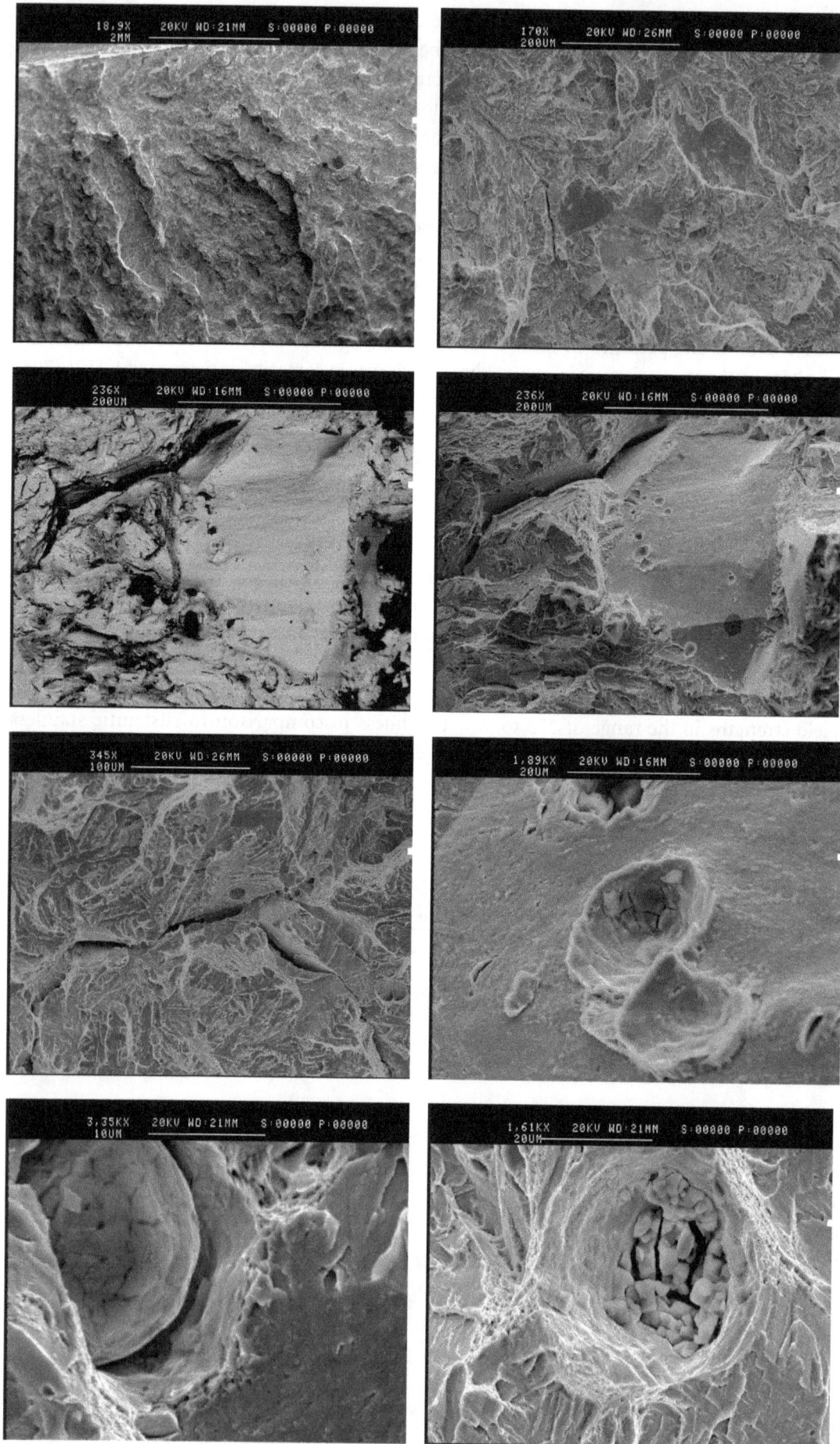

Figure10.66 SEM examination of the fracture surface of a 410 stainless low energy Charpy

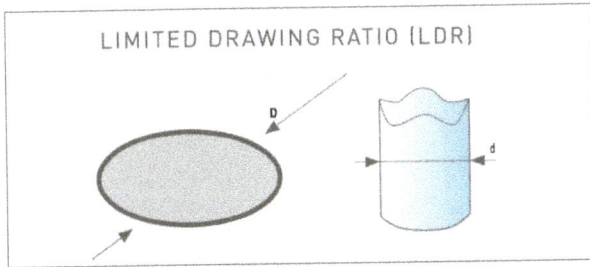

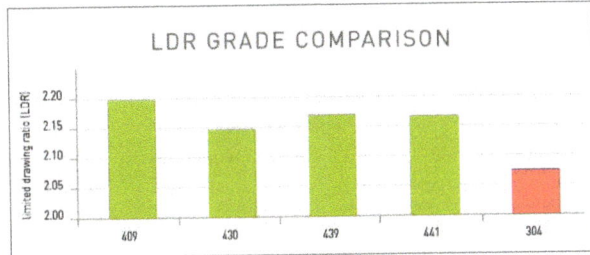

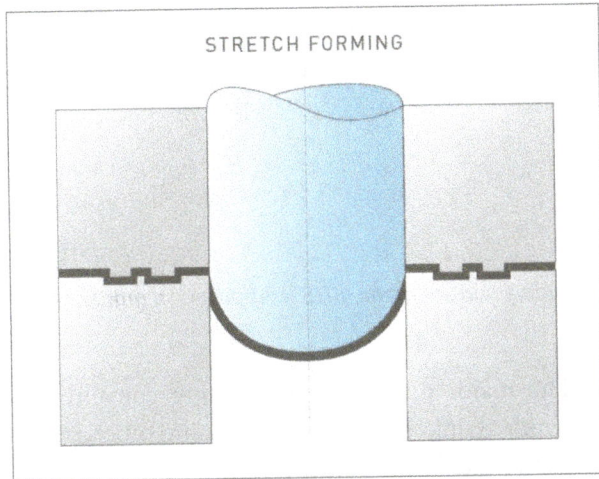

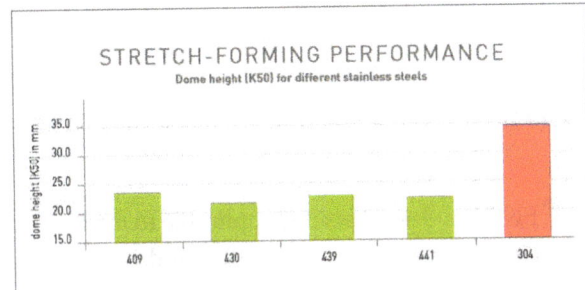

Figure 10.67 *Ferritic stainless steel formability*

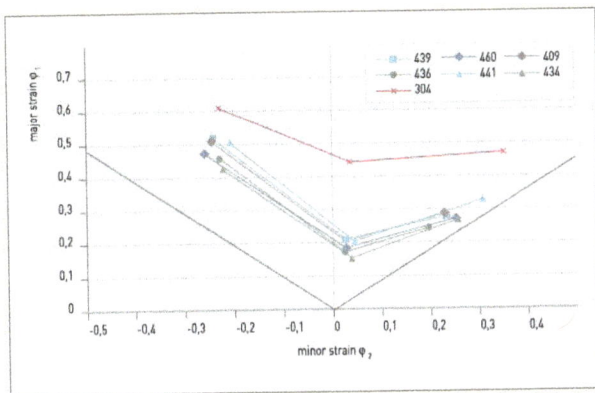

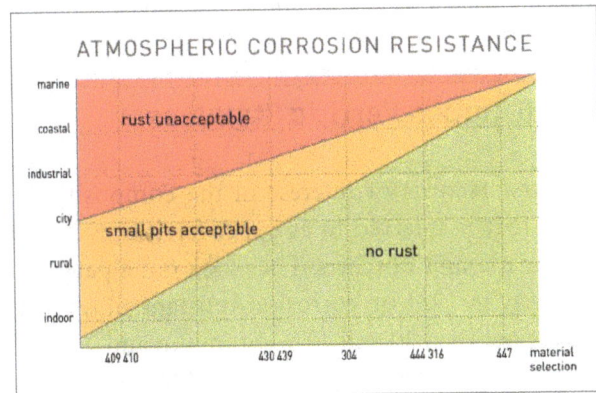

Figure 10.68 *LHS = Ferritic stainless steel forming limit diagram, RHS = Stainless atmospheric corrosion*

Ferritics have higher LDR values than austenitics, which makes them particularly suitable for drawing. However, Ferritic grades are inferior to austenitics in pure stretch-forming as shown in Figure 10.67.

In practice, industrial forming operations involve a combination of both pure drawing and pure stretch-forming deformation, in a series of "passes". Forming limit curves are a useful guide to maximum deformation before failure, in both deep drawing and stretching processes. Established for the principal stainless steel grades, they can be used to analyse a forming operation (see Figure 10.68).

A major European collaborative project began in in 2010 entitled Structural Applications of Ferritic Stainless Steels (hereafter referred to as SAFSS). The principal aim of the study was to develop the information needed for comprehensive structural design guidance to be included in relevant parts of the Eurocodes and other accompanying standards and guidance.

Despite being widely used in the automotive and domestic appliance sectors, structural applications of ferritic stainless steels are scarce owing to a lack of knowledge, performance data and design guidance. They have been specified for cladding and roofing applications as well as in the transportation sector for load-bearing members, for example for tubular bus frames. The Eurocode dealing with structural stainless

WELDING FERRITIC STEELS: REMEDIES

Stainless steel group	Special feature	Phenomenon	Cause	How to avoid
Unstabilised grades	Sensitisation	Poor corrosion resistance in welded zone	Cr-carbide precipitation in grain boundary	Annealing in temperature range 600–800°C
Stabilised grades	Grain coarsening	Poor toughness in welded zone	Excessive grain growth due to high temperature	Minimising the heat input of welding
Cr from 15%	475°C embrittlement	Embrittlement occurs in the range from 400-540°C	Decomposition of the matrix into 2 phases, one rich in iron, one rich in chromium.	Reheating at 600°C and cooling rapidly
High Cr-Mo grades	Sigma (σ) phase embrittlement	Embrittlement occurs at 550–800°C	Sigma (σ) phase formation due to decomposition of delta (δ) ferrite	Reheating above 800°C and cooling rapidly
Unstabilised grades	Martensitic phase embrittlement	Embrittlement occurs in lower Cr and higher-C types	Martensitic phase formation due to faster cooling	Removing the martensitic phase by long annealing in the 600-700°C range

RULES OF THUMB

- In the case of an aggressive environment, select a grade with a higher chromium and/or molybdenum content.

- Avoid rough surface finishes – favour a fine-polished surface with a low Ra value.

- Optimize design for "washability" (e.g. min. 15° slope on upward-facing surfaces).

- Avoid "crevice-like" geometries.

- Keep surface clean, by regular washing, to avoid staining and dust accumulation.

Figure 10.69 LHS = Ferritic stainless steel potential welding problems and solutions, RHS = Application guidance

steel, EN 1993-1-4, states it is applicable to three traditional ferritic grades (1.4003, 1.4016 and 1.4512). The SAFSS project aimed to ensure that adequate design guidance is available to allow the use of ferritic grades. One aspect of this study was the weldability which was carried out by Otokumpu[15] (See Figure 10.69).

10.5.6 Super-Ferritic Stainless

The first records of interest in the compositional range that is referred to as super ferritics was that of Monartz and Borchers. They took out a patent in Germany in 1911 on corrosion-resistant alloys containing between 30 and 40% chromium and 2 to 3% molybdenum. It was not until the 1960s using electron-beam refining the Airco Vacuum Metals, was established to manufacture a ferritic stainless steel known as E-Brite 26-1, which contained less than 100 ppm of total carbon and nitrogen29-31. E-Brite (now often called just 26-1). This alloy enjoyed a modest success on account of its exceptional corrosion resistance in aggressive chloride-containing media but was too expensive.[16]

Economy versions had to wait until the development of the AOD (argon-oxygen decarburization), which was introduced by Union Carbide in 1968, also continuous casting and, to a lesser extent but especially in Japan, the vacuum oxygen decarburization process (VOD). The AOD technique can bring the carbon-plus-nitrogen content down to 0.016%, whereas the VOD can bring it to less than 0.004%.

The major disadvantage in using conventional austenitic stainless steels in seawater and other aggressive waters has been the susceptibility of these steels to chloride pitting and crevice corrosion. Conventional austenitic stainless steels are particularly susceptible

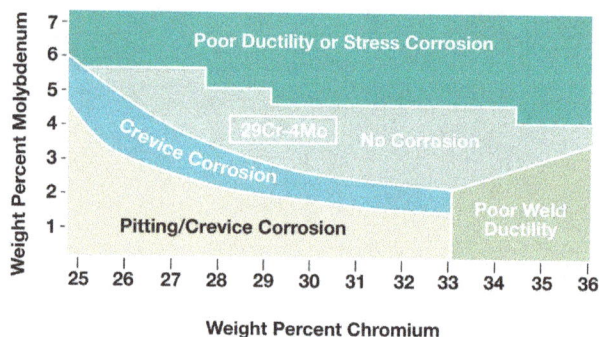

Element	Weight Percent
Chromium	29
Molybdenum	4
Nickel	0.3
Manganese	0.5
Phosphorus	0.03
Sulfur	0.01
Silicon	0.4
Cobalt	0.03
Carbon	0.02
Nitrogen	0.02
Titanium + Niobium	0.5

Figure 10.70 Effect of Chromium and Molybdenum on the Properties of Fe-Cr-Mo Alloys. Composition of AL 29-4C

to corrosion if deposits form or low flow conditions exist in high chloride waters.

Increased resistance to chloride pitting and crevice corrosion correlates with increased chromium and molybdenum content in stainless steels. The basic composition of AL 29-4C alloy (29% chromium and 4% molybdenum) was derived from the work of M. A. Streicher (Corrosion 30 (3), 1974, pp. 77-91)[17]. Streicher used two laboratory tests to establish the chromium and molybdenum content necessary for optimum resistance to chloride crevice and pitting corrosion. One test utilized a 2% $KMnO_4$ – 2% NaCl solution at 195°F (90°C) while the second test utilized a 10% $FeCl_3$ solution at 122°F (50°C). The $FeCl_3$ test was conducted in a manner similar to ASTM procedure G48B and used teflon blocks held with rubber bands to create crevice conditions. Streicher's results are shown schematically in Figure 10.70.

AL 29-4C alloy, is a superferritic stainless steel developed by ATI Allegheny Ludlum in the early 1980s specifically for power plant surface condenser tubing. Since that time over 60 million feet of superferritic condenser tube has been put into service. These limits have come to define the family of superferritic stainless steels S44735 and S44660 known as AL 29-4C and SEA-CURE respectively.

10.5.7 Austenitic Stainless

The 300-series stainless steels, such as types 304, 310, 316, and 347, contain around 10 to 20% nickel in addition to 17 to 25% chromium. The nickel further enhances corrosion resistance and results in the FCC crystal structure being stable even at low temperatures, so that these are termed austenitic stainless steels. These materials either are used in the annealed condition or are strengthened by cold work, and they have excellent ductility and toughness. Uses include nuts and bolts, and pressure vessels and piping.

Figure 10.71 *Mixed grained size austenitic stainless*

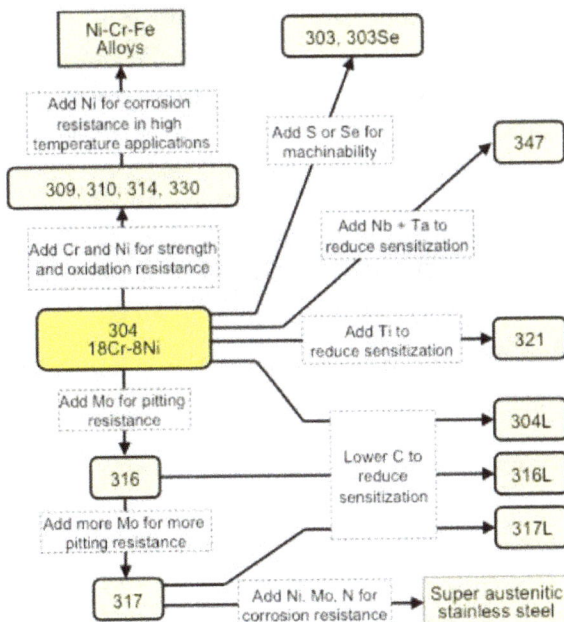

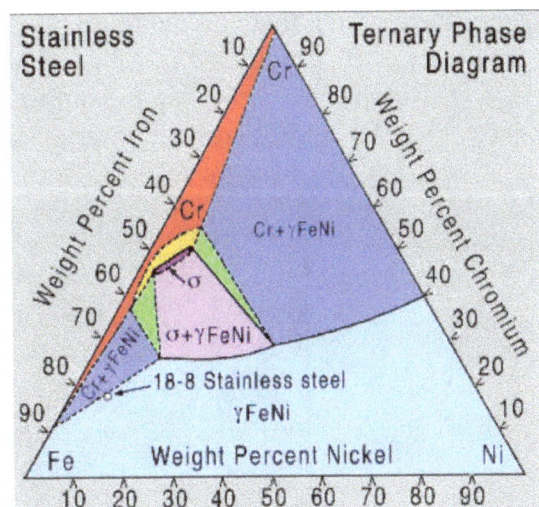

Figure 10.72 *LHS = Austenitic steel types. RHS = Austenitic and a ternary phase diagram*

10.5.8 Super Austenitic Stainless

Super austenitic stainless steels contain high levels of chromium and higher levels of nickel together with additions of molybdenum and nitrogen. The result is a series of austenites, stronger than conventional 300 series stainless and with superior resistance to pitting, crevice corrosion, and stress corrosion cracking.

UNS S31254 F44 is a 6% Mo Super Austenitic stainless steel. The material combines moderately good mechanical strength and high ductility with excellent corrosion resistance in a variety of environments. This material is generally supplied in the annealed condition giving yield strengths in excess of 44 KSI (300Mpa), this material cannot be hardened by heat treatment but stronger surface strengths can be achieved by cold working. The typical PREN of 42-44 of this alloy ensures that the resistance to crevice and pitting corrosion is high, giving this alloy particular use as an alternative to the 300 series alloys in applications where higher mechanical properties are required (see Figure 10.72).

10.5.9 Precipitation-hardening martensitic stainless steels

These chromium and nickel containing steels can be precipitation hardened to develop very high tensile strengths. Precipitation-hardening stainless steels are usually designated by a trade name rather than by their AISI 600 series designations. The most common grade in this group is '17-4 PH', also known as Grade 630, with a composition of 17% chromium, 4% nickel, 4% copper and 0.3% niobium. The main advantage of these steels is that they can be supplied in the 'solution treated' condition – in which state the steel is just machinable. Following machining, forming, etc. the steel can be hardened by a single,

fairly low temperature 'ageing' heat treatment that causes no distortion of the component. Precipitation hardening generally results in a slight increase in corrosion susceptibility and an increased susceptibility to hydrogen embrittlement.

10.5.10 Austenitic 200 Series

The reduced nickel content of the 200 series chrome-manganese grades makes them significantly cheaper. However, depending on their chemistry, they also offer good formability (ductility) and/or strength. Some grades (201, 202 and 205 series) offer about 30% higher yield strength than the classic 304-series chrome-nickel grade – allowing designers to reduce weight. There are several key publications on the development and use of the 200 series austenitic stainless steels. Jindal steel, one of the main proponents to the replacement of the 300 series stainless with 204 grade has a website www.J4stainless.com/CrMn.html which expresses the firm view that the 204 type stainless "provides a highly cost effective alternative to the costly 300 series for several applications".

http://www.aperam.com/uploads/stainlesseurope/ TechnicalPublications/Thenew200-series_EN_9p_ 988Ko_Sevilla2005.pdf This publication outlines the development of these grades in India and China due to problems of nickel supply and price pressures. This publication reviews the corrosion properties and weldability and touches on the potential recycling problems that these grades can cause. The history and development of the austenitic 200 grade steel in India can be reviewed at https://www.metalbulletin. com/events/download.ashx/document/speaker/6546/ a0ID000000X0jVtMAJ/Presentation The J4 grade matches the chemical analysis of these plates. The website shows a table of the "family" of J4 type grades as shown in Table 10.7.

Chemical Composition of J4 Family

Grades	%C	%Mn	%P	%S	%Si	%Cr	%Ni	%Cu	%N
J4	0.10 Max	8.5 to 10.0	0.08 Max	0.015 Max	0.75 Max	15.5 to 17	1.0 to 2	1.5 to 2.0	0.1 to 0.2
204Cu	0.10 Max	6.5 to 9.0	0.06 Max	0.015 Max	0.75 Max	16 to 17.5	1.5 to 3.5	2.0 to 4.0	0.1 to 0.2
JSL-Aus	0.08 Max	6.0 to 8.0	0.07 Max	0.015 Max	0.75 Max	16 to 18.0	1.0 to 2.0	1.5 to 2.0	0.1 Max
JT	0.12 Max	9.0 to 11.0	0.08 Max	0.015 Max	0.75 Max	14 to 16.0	0.2 to 0.5	0.5 to 1.0	0.1 to 0.2

Table 10.7 Composition of J4 type series 200 austenitic stainless

Table 10.8 Stainless compositions

Type of Steel	% Nickel	% Chromium	% Carbon	% Manganese	% Silicon	% Copper	% Moly	% Nitrogen
420								
201	3.5 – 5.5	16.0 – 18.0	0.15 max	5.5 – 7.5	1.0 max			0.25 max
301	6.0 – 8.0	16.0 – 18.0	0.15 max	2.0 max	1.0 max			0
302	8.0 – 10.0	17.0 – 19.0	0.15 max	2.0 max	1.0 max			0
304	8.0 – 10.5	18.0 – 20.0	0.08 max	2.0 max	1.0 max			0
309	19.0 – 22.0	24.0 – 26.0	0.2 max	2.0 max	1.0 max			0
F6MN	3.5-5.5	11-14	0.05	0.5-1.0	0.6		0.5-1.0	
F44	17.5-18.5	19.5-20.5	0.02	1.0 max	0.8 max	0.5-1.0	6.0-6.5	0.18-0.22

10.5.11 An addition important range of high alloy steels- Tool steels

Tool steels are described by their application, cold work = A and D grades, Hot work = H grades, Cutting tools=M and T grades. Table 10.9 shows some typical compositions

Compositions and Application of some Tool steels

AISI Number	UNS Number	Composition (wt.%)						Typical Allpications
		C	Cr	Ni	Mo	W	V	
M1	T11301	0.85	3.75	0.30 max	8.70	1.75	1.20	Drills, saws, lathe and planer tools
A2	T30102	1.0	5.15	0.30 max	1.15	-	0.35	Punches, embossing dies
D2	T30402	1.5	12	0.30 max	0.95	-	1.10 max	Cutlery, drawing dies
O1	T31501	0.95	0.50	0.30 max	-	0.50	0.30 max	Shear blades, cutting tools
S1	T41901	0.50	1.40	0.30 max	0.50 max	2.25	0.25	Pipe cutters, concrete drills
W1	T72301	1.10	0.15 max	0.20 max	0.10 max	0.15 max	0.10 max	Balcksmith tools

Table 10.9 Tool steels

REFERENCES

1. Utility Ferritic Stainless steel, COLUMBUS STAINLESS [Pty] https://www.columbus.co.za/downloads/Ferritics_Utility.pdf

2. Robert Douglas Knutsen, A MICROSTRUCTURAL EXAMINATION OF DUPLEX FERRITE-MARTENSITE CORROSION RESISTING STEELS, A thesis submitted to the Faculty of Engineering, University of Cape Town in fulfilment of the degree of Doctor of Philosophy, University of Cape Town August 1989

3. J. HEWITT, High-chromium controlled-hardenability steels. INFACON 6. Proceedings of the 1st International Chromium steels and alloys congress, Cape Town. Volume 2. Johannesburg, SAIMM, 1992. pp. 71-88.

4. CHARENTON et al, Stainless steel with 11% chromium and high yield strength. for welded constructions resistant to corrosion and abrasion. INFACON 6. Proceedings of the 1st International Chromium steels and alloys congress, Cape Town. Volume 2. Johannesburg, SAIMM. 1992. pp. 229-234

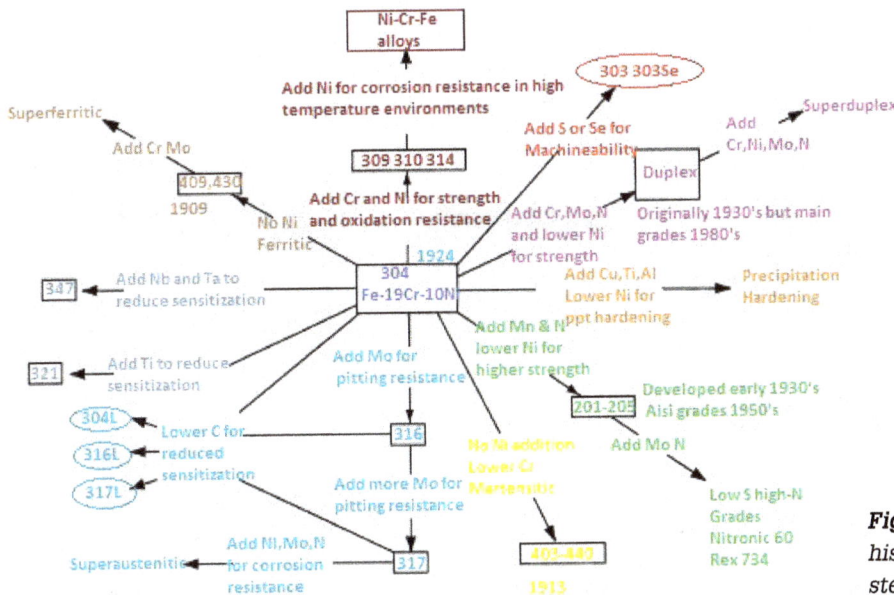

Figure 10.73 Inter-relationship and historical development of stainless steels

5. European Commission Research Fund for Coal and Steel Innovative stainless steel applications in transport vehicles 2007

6. Hormoz Seradj, WELDABILITY OF ASTM A 1010, Iowa DOT A1010 Steel Workshop Ames, Iowa March 18, 2015

7. 7. Mohamed Soliman et al, Life-Cycle Cost Evaluation of Conventional and Corrosion-Resistant Steel for Bridges, J. Bridge Eng. 2015.20. https://ascelibrary.org/doi/10.1061/% 28ASCE%29BE.1943-5592.0000647

8. ISSF Martensitic Stainless Steel. http://www.worldstainless. org/Files/issf/non-image-files/PDF/ISSF_Martensitic_ Stainless_Steels.pdf

9. Massimo De Sanctis et al, Study of 13Cr-4Ni-(Mo) (F6NM) Steel Grade Heat Treatment for Maximum Hardness Control in Industrial Heats, Metals 2017, 7, 351, file:///C:/Users/user/ Downloads/metals-07-00351-v2%20(3).pdf

10. Emel Taban et al, Properties, weldability and corrosion behavior of supermartensitic stainless steels for on- and off-shore applications, Materials Testing, (2016) 6 p58 file:///C:/ Users/user/Downloads/120.110884%20(1).pdf

11. Clément GAYTON, Embrittlement mechanisms of the X4CrNi16-4 super-martensitic stainless steel Virgo™38: Effects of heat treatments and corrosive environments containing Na_2S or H_2S, l'Ecole des Mines de Saint-Etienne, Ecole Doctorale N° 488, 2018 file:///C:/Users/user/Downloads/ Gayton%20(1).pdf

12. ISSF The Ferritic Solution properties | advantages | applications THE ESSENTIAL GUIDE TO FERRITIC STAINLESS STEELS http://www.worldstainless.org/Files/issf/non-image-files/PDF/ISSF_The_Ferritic_Solution_English.pdf

13. K.A. Cashell and N.R. Baddoo, Ferritic stainless steels in structural applications, https://bura.brunel.ac.uk/bitstream/2438/ 9834/1/Fulltext.pdf

14. N Baddoo et al, Structural applications of ferritic stainless steels, SAFSS, European Commission Research Program, Technical Group TGS8, Final report June 2013.

15. Structural applications of ferritic stainless steels, SAFSS, Study of weldability, Final Report, Otokumpu March 2014.

16. M.B. Cortie, History and development of ferritic stainless steels, J S. Afr. Inst. Min. Metall., vol. 93, no. 7. Jul. 1993. pp. 165-176. https://www.saimm.co.za/Journal/v093n07p165. pdf

17. 17. M. A. Streicher Corrosion 30 (3), 1974, pp. 77-91

10.6 STEEL FOR AUTOMOTIVE APPLICATIONS

Steel is used in many industries and in many forms, shapes, sizes and mechanical property grades. The welding and jointing methods are well established and measures to prevent corrosion, such as galvanising and paint systems, are also well established. This also applies to other steel grades such as stainless and high-temperature steels and low and high alloy steels. Carbon steel plates, sections and strip products offer many different ranges of mechanical properties to meet a wide range of applications from offshore oil rigs to ship building and automotive application. For these reasons, combined with a relatively low production cost, steel is now used in a wide range of products. Figure 10.74 and 10.75 shows the material flow through the process facilities and the output of steel shapes and products on the right-hand side. The products range from pipe, plate and strip, rolled sections and rod.

The heat treatment aspects of steel have been covered in Chapter 4. The aspects covered here will focus on specific applications/grades of steel. The developments in steel have followed the strengthening methods outlined in Chapter 2, grain refining, solid solution strengthening, precipitation hardening, work hardening, transformation hardening.

The development of the key grades often requires special processing facilities to obtain the optimum properties in the individual products. Steel

Iron and Steel Making Process

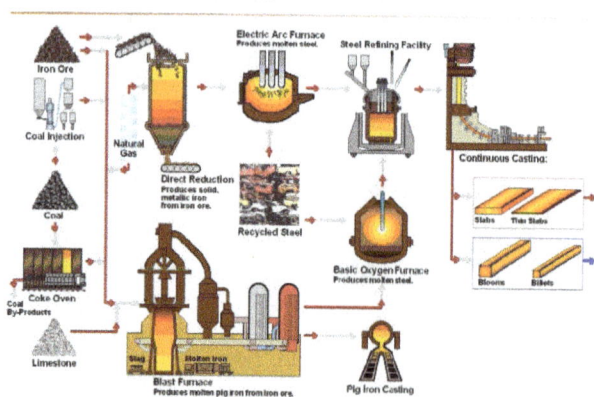

Figure 10.74 *Iron and steel making processes*

Semi Finished Products

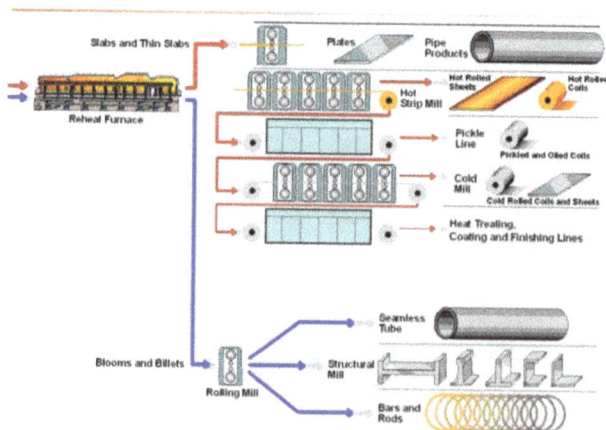

Figure 10.75 *Rolling processes for semi-finished products.*

Figure 10.76 The 4 principles of material science

development adopts the 4 principles of material science, ability to process/fabricate, microstructural development, mechanical property requirements and service performance. This combination of options is shown in Figure 10.76.

There are many products that demonstrate these principles, for example the development of high strength weldable steels. The main source of understanding of steel quantitative relationships was the work from 1963 onwards by Brian Pickering and Terry Gladman who carried out detailed work at Swinden Laboratories and published many of their findings. An excellent major review of Micro Alloyed steel was published in 2016 by TN Baker of Strathclyde University, which included reference to the many aspects of the important history and the recent innovations such as direct rolled thin slab.[1]

In the 1970s there were a few main specifications, BS4360 for weldable plate and sections, BS1501 for pressure vessel steel, BS5500 for boiler and pressure vessel, BS 1449 hot and cold rolled Carbon Manganese steel strip and BS970 for engineering steels. A basic understanding of these specifications gave insight into the use of steels. Maybe after Brexit they will all return!

For several important applications such as oil and gas that require stringent quality assurance, inspection authorities such as DVN, API, Lloyds, now certify the full steel manufacturing route from steelmaking to final products. The users of steel must then use steel from these authorised/certified steel makers and have full traceability.

The environmental pressures for reduced emissions and better fuel economy combined with increased safety requirements has forced car manufacturers to reduce the weight of the vehicles and to strengthen the passenger compartment. These requirements led to major cooperation between the

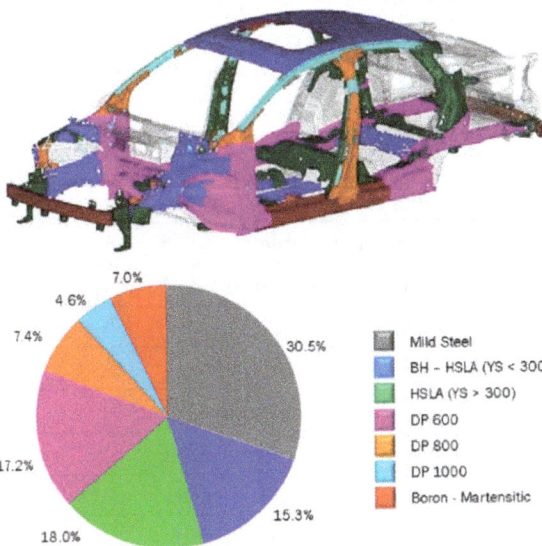

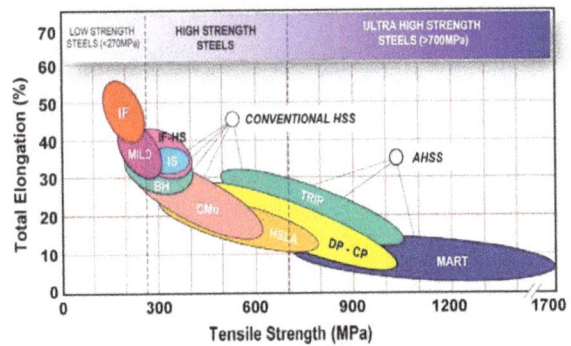

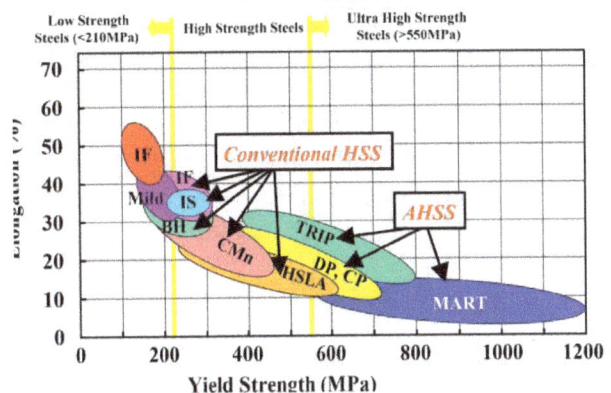

Figure 10.77 Optimised steel grade for each part of the chassis Figure 10.78 Steel types[2,3,6]

worldwide steelmakers and the car manufacturers. This has resulted in the development and optimisation of steel grades for each part of the chassis. Figure 10.77 and Figure 10.78

There has been substantial technical development in strip for automotive applications. These are well presented in "Advanced High-Strength Steels Application Guidelines" prepared by World Auto Steel, https://www.worldautosteel.org/ In 1994, a consortium of 35 sheet steel producers began cooperation on the application and development of steels for auto bodies. This organization is still in business and still targeting further developments. The development work has been directed at materials that would meet the demanding range of safety and performance targets.

These achievements in steel strip can be seen in one of the primary graphical displays (see Figure 10.78) used to show the combinations of strength vs elongation. This is often described as a banana relationship as the elongation (tensile ductility) drops as the strength increases. The work is now achieved phase 1 and phase 2 and is now targeting a steel to meet the third-generation development requirements as shown in the diagram which targets medium levels of manganese. (see Figure 10.80 which shows the individual phases of development). The TWIP steels were developments of the Hadfield's steel (austenitic manganese steel AMS). The work carried out by World Auto Steel mastered many of the problems such as:

*   Formability
*   Spot welding and associate zinc LME (liquid metal embrittlement).

The latest published review "Advanced High-Strength Steels Application Guidelines" should be consulted for a full background.

This long list of steel types shown in Figure 10.79 can appear quite disconcerting and prevent an understanding of how the selection and use of the various steel types/grade is carried out. Some perspective can be gained by the past history.

In the 1970s when I began working in the steel industry the main formability steels were "rimming steels". These were cast with a high level of oxygen retained in the steel and during teeming into the ingot moulds a "rimming" action was initiated by adding oxide mixtures. This caused the oxygen to react with the carbon to produce a 1 to 2 inch rim of pure iron due to reaction between the carbon and the oxygen in the steel. This was followed by blowholes formed from the reaction which compensated for solidification shrinkage. The blowholes were "welded" during rolling and an ingot to product yield of over 90% was achieved. Obviously, the death knell for this grade of steel was the introduction of continuously cast steel.

This technology was replaced by aluminium killed steel simply referred to as Al-killed. These steels were low carbon with low silicon and an aluminium level of 0.025 to 0.04% This was followed in the 1980s by the development of Interstitial Free Steels (IF). Therefore, steels with high formability are basically:

*   Al-killed (AK) steels. This is still based on the traditional formable steel with low carbon and silicon. The carbon forms iron carbides/pearlite in the microstructure. The aluminium addition provides deoxidation and to form aluminium nitride (AlN) which is used to control the final properties (texture and anisotropy)
*   Interstitial Free (IF) steels. This steel was developed following the introduction of steelmaking process facilities to remove the carbon and the nitrogen to very low levels in the 1980s. Additions

Abbreviation	Steel type	Abbreviation	Steel type
Mild	Mild steel	DP	Dual phase
IF	Interstitial free	SF	Stretch flangeable
BH	Bake hardenable	TRIP	Transformation induced plasticity
CMn	Carbon manganese	CP	Complex phase
HSLA	High strength low alloy	MS	Martensitic (MART)
IS	Isotropic $\Delta r = 0$	MnB	Hardenable manganese boron
FB	Ferritic Bainitic	Q&P	Quenching and Partitioning
HF	Hot formed	TPN	Three Phase Nano-Precipitation
Sandwich	Sandwich[21]	TWIP	Twinning-Induced Plasticity

Figure 10.79 *Steel types used for automotive chassis application*

of titanium and niobium were added to combine with the carbon and nitrogen to make the micro-structure free of interstitial carbon and nitrogen the imperfection that had caused Luders bands or stretcher strain marks due to discontinuous or non-uniform yielding related to potential strain ageing effects.

The automotive steels can be considered as:

1. Hot and cold rolled low carbon steels for form-ability applications
2. The conventional high strength steels HSS
3. The advanced ultra-high strength steels AHSS.

The HSS and AHSS have suitable chemistry and pro-cessing to achieve strength levels above 600MPa (See Figure 10.80).

elements in solid solution, formation of carbides, precipitation of carbides and/or nitrides, and grain refinement. Drawing Quality (DQ) and Aluminium Killed (AKDQ) steels are examples and often serve as a reference base because of their widespread applica-tion and production volume. The classification based on formability can be seen in Figure 10.82 and 10.84.

The appropriate European supply specification can be seen in Figure 10.81. This shows grades DC01 to DC07 with DC02 missing and DC07 the steel with the highest formability. Al-killed steel could be used for grades DC01 to DC04 but the higher grades would require IF steel. The highest grade DC07 would require a well optimised process route. A Turkish doc-torate thesis outlined the process optimisation carried out for new facilities in 2006.[8]

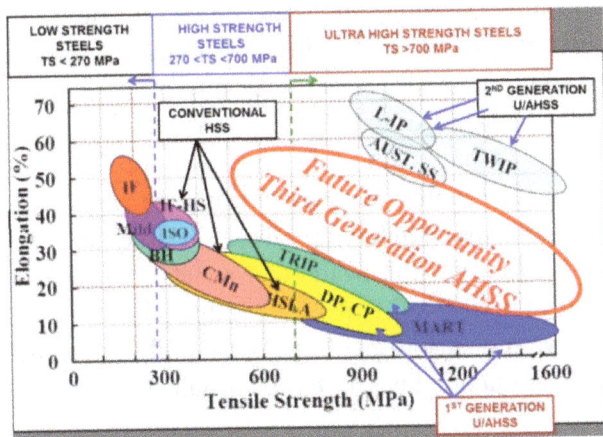

Figure 10.80 2005 – Conventional, current and prospective automotive steels and their corresponding elongations and strengths[5]

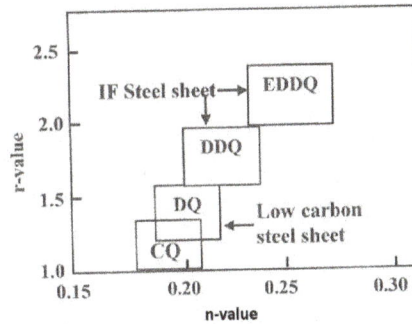

Figure 10.82 r-value vs n-value

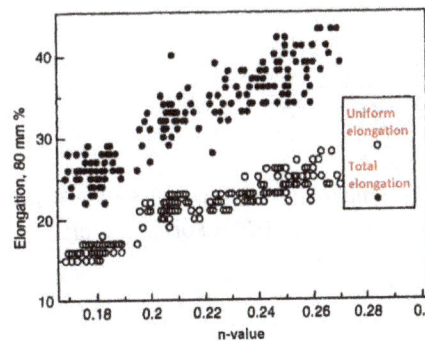

Figure 10.83 Tensile elongation vs n-value

10.6.1 Mild Steels

Mild steels have an essentially ferritic microstructure. The main strengthening is due to a combination of

Steel grade	Norms and specifica-tions	Test dir.	Yield strength $R_{p0.2}$ [MPa]	Tensile strength R_m [MPa]	Total elong. A_{80} min. [%]	r value min. [-]	n value min. [-]	BH_2 min. [MPa]	Exposed
Mild steels									
EN 10130			R_e	R_m	A_{80}	r_{90}	n_{90}	BH_2	E
DC01	EN 10130	Trans.	140 - 280	270 - 410	28	-	-	-	✓
DC03	EN 10130	Trans.	140 - 240	270 - 370	34	1.3	-	-	✓
DC04	EN 10130	Trans.	140 - 210	270 - 350	38	1.6	0.18	-	✓
DC05	EN 10130	Trans.	140 - 180	270 - 330	40	1.9	0.20	-	✓
DC06	EN 10130	Trans.	120 - 170	270 - 330	41	2.1	0.22	-	✓
DC07	EN 10130	Trans.	100 - 150	250 - 310	44	2.5	0.23	-	✓

Figure 10.81 Classification of cold draw grades according to EN 10130

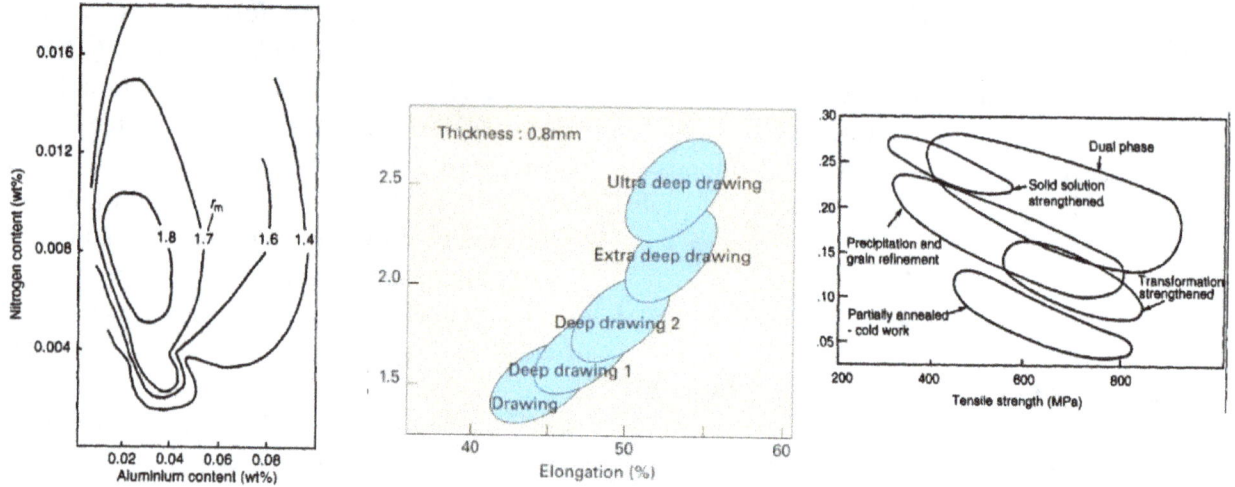

Figure 10.84 *LHS = Effect of Al and nitrogen on the r-value Centre = r-value vs total elongation (from jfe brochure) RHS = n-value vs UTS, Next= Comparison of draw ratio for EDD and UDD grade (from jfe brochure)*

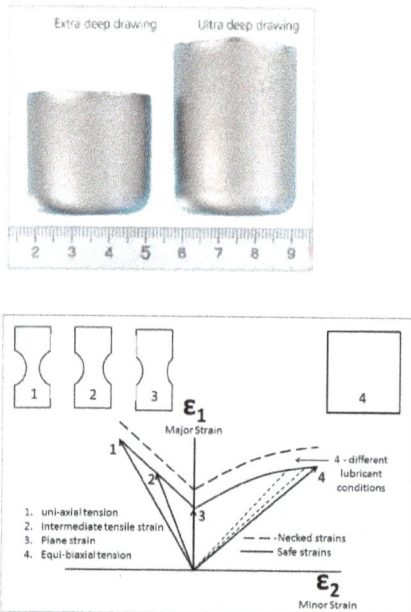

Figure 10.85 *LHS = Comparison of draw ratio for EDD and UDD grade (from jfe brochure), RHS = Forming limit diagram (FLD)*

10.6.2 Interstitial free (IF) steels

The possibility of improving the cold formability of sheet steel by using interstitial free steels has been well known for several decades. The initial development began in Japan in the early 1980s. The adoption of IF was made possible because of the steel making technology such as vacuum degassing which allowed the manufacture of high volumes of steel with levels of carbon and nitrogen below 30 ppm for each element. This level of interstitials needed relatively small additions of titanium and niobium which reduced

the alloy costs. Another development that improved the process capability and cost was the introduction of continuous annealing, which allows for time and cost savings compared to the batch annealing process. In addition, the thermal uniformity of this process reduced the variation in mechanical properties over the length and width of the sheet and also from coil to coil. The continuous annealing is also the standard processing route in hot dip galvanising lines for coating steel sheet with zinc or other metals. The usage of such corrosion resistant sheet products for the automotive industry is growing very rapidly and many new production lines have been installed worldwide.

10.6.3 Bake hardenable (BH) steels

BH steels have a basic ferritic microstructure and are strengthened by solid solution strengthening. A unique feature of these steels is the chemistry and processing designed to keep carbon in solution during manufacture and then allowing this carbon to precipitate during paint baking. This increases the yield strength of the formed part.

10.6.4 Carbon-manganese (CM) steels

High strength carbon-manganese steels are primarily strengthened by solid solution strengthening.

10.6.5 High-strength low-alloy (HSLA) steels

This group of steels are strengthened primarily by micro-alloying elements contributing to fine carbide precipitation and grain-size refining.

10.6.6 Dual phase (DP) steels

Dual phase steels consist of a ferritic matrix containing a hard martensitic second phase in the form of islands. This microstructure provides a wide range of good and mechanical properties. This gives DP steels high ultimate tensile strength (UTS) (due to the martensite) combined with low initial yielding stress

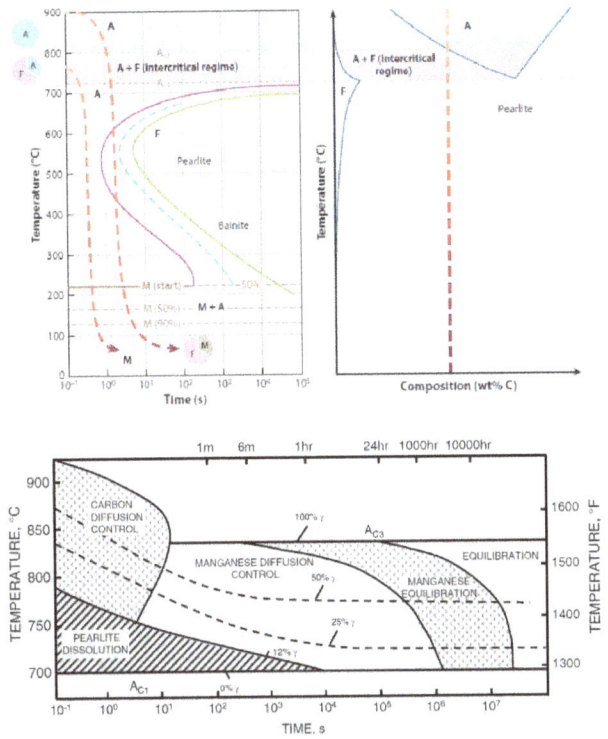

Figure 10.86 *Dual phase steels produced by inter-critical annealing. Heating rate, soaking time and cooling rate are important parameters LHS from Ref 31, RHS from Ref 30*

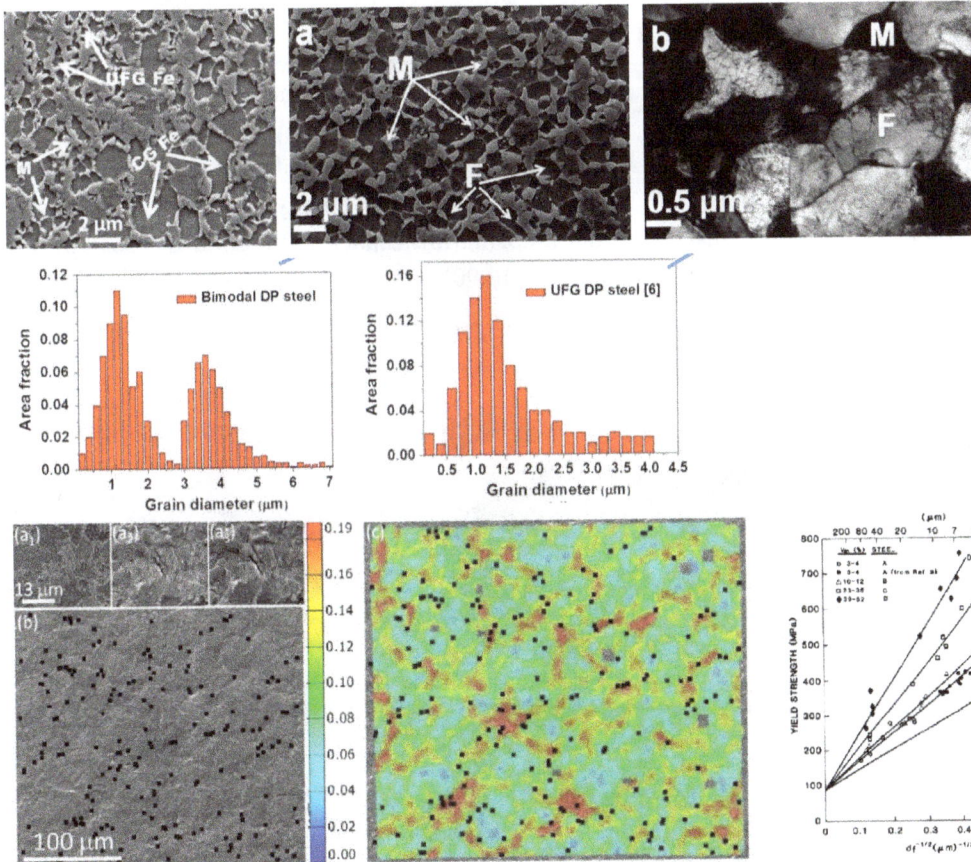

Figure 10.87 *Top SEM images showing fine grained ferrite and martensite after inter-critical annealing a cold-rolled sample at 760°C for 2 min. photo (b) bright field TEM micrograph showing dislocation structures in the ferrite and block type of martensite. M: Martensite, F: Ferrite Centre= Grain size distributions associated with the above SEM images. Lower LHS = Damage events captured during an in-situ SEM experiment. High magnification images showing micro-cracks near a martensite-ferrite phase boundary. Each damage event is shown in the low-magnification image and is represented with black dots. The damage incidents are overlaid with the local strain map. RHS = Hall-Petch slope increases with increase in volume fraction of martensite. Yield strength increases with grain refinement. Therefore, motivation for producing ultra-fine grained DP steels*

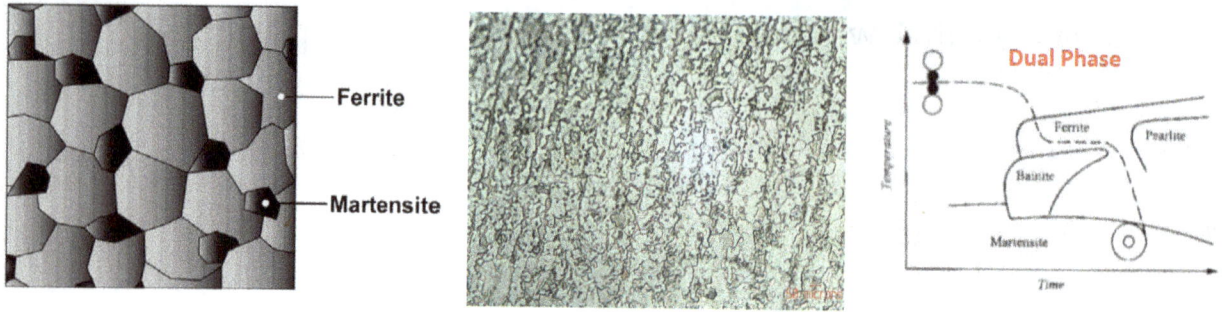

Figure 10.88 *LHS = Schematic of microstructure. Centre = Optical microstructure. RHS = TTT*

(enabled by the ferrite) DP shows high early-stage strain hardening, and macroscopically homogeneous plastic flow (enabled through the absence of Luders effects) The properties are controlled by martensite volume fraction (VM), martensite grain size (SM), martensite carbon content (CM), martensite/ferrite morphology, ferrite grain size (SF), ferrite texture, density of transformation induced geometrically necessary dislocations (GNDs), micro- and mesoscale segregation, and the chemical decoration state of the hetero-interfaces (see figures 10.86 to 10.88).[7, 13, 33-35]

10.6.7 Transformation induced plasticity (TRIP assisted) steels

The microstructure of TRIP steels is retained austenite embedded in a primary matrix of ferrite. In addition to a minimum of 5% by volume of retained austenite, hard phases such as martensite and bainite are present in varying amounts. The retained austenite progressively transforms to martensite with increasing strain, thereby increasing the work hardening rate at higher strain levels (Figure 10.89).

10.6.8 Complex phase (CP) steels

CP steels consist of a very fine microstructure of ferrite and a higher volume fraction of hard phases that are further strengthened by fine precipitates. Complex phase steels typify the transition to steel with very high ultimate tensile strengths. CP steels are characterized by high energy absorption, high residual deformation capacity and good hole expansion (Figure 10.90).

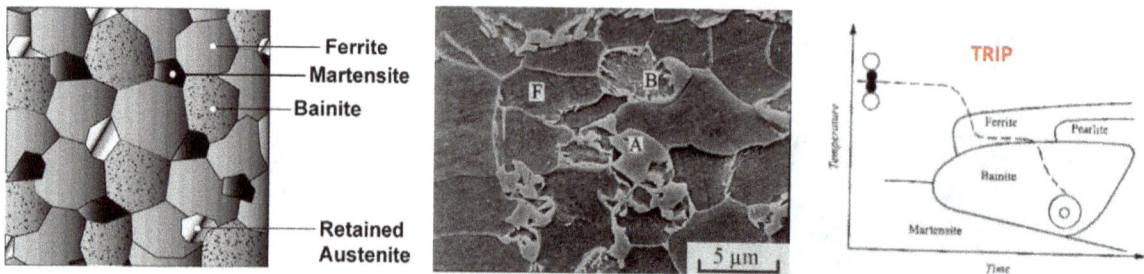

Figure 10.89 *LHS = Schematic of microstructure. Centre = SEM microstructure. RHS = TTT*

Figure 10.90 *LHS = Schematic of microstructure. Centre = Optical microstructure. RHS = TTT*

10.6.9 Twinning-Induced Plasticity (TWIP) Steel

TWIP have a high manganese content (17-24%) that causes the steel to be fully austenitic at room temperatures. A large amount of deformation results from the formation of deformation twins. This deformation mode leads to the naming of this steel class. The twinning causes a high value of the instantaneous hardening rate (n-value) as the microstructure becomes finer and finer. The resultant twin boundaries act like grain boundaries and strengthen the steel. TWIP steels combine extremely high strength with extremely high stretchability. The n-value increases to a value of 0.4 at an approximate engineering strain of 30% and then remains constant until both uniform and total elongation reach 50%. The tensile strength is higher than 1000 MPa (Figure 10.91).

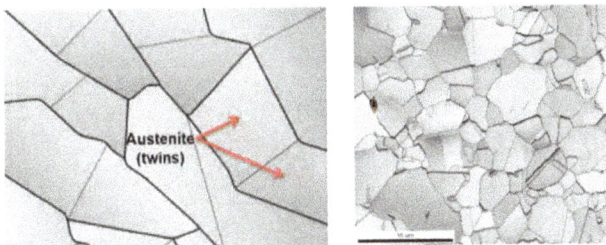

Figure 10.91 LHS = Schematic of microstructure. RHS = Optical microstructure

10.6.10 Ferritic-Bainitic (FB) Steel

FB steels sometimes are utilized to meet specific customer application requirements that require Stretch Flangeable (SF) or High Hole Expansion (HHE) capabilities for improved edge stretch capability. FB steels have a microstructure of fine ferrite and bainite. Strengthening is obtained by both grain refinement and second phase hardening with bainite. The primary advantage of FB steels over HSLA and DP steels is the improved stretchability of sheared edges as measured by the hole expansion (Figure 10.92).

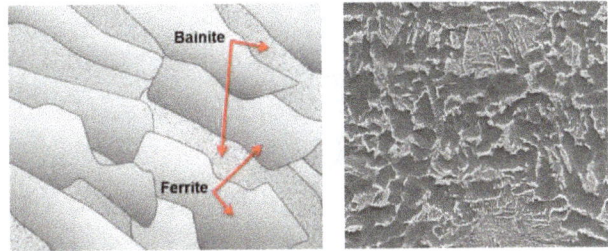

Figure 10.92 LHS = Schematic of microstructure. RHS = SEM microstructure

10.6.11 Martensitic (Mart) steels

To create martensitic steels, the austenite that exists during hot-rolling or annealing is transformed almost entirely to martensite during quenching on the run-out table or in the cooling section of the continuous annealing line. This structure can also be developed with post-forming heat treatment. Martensitic steels provide the highest strengths, up to 1700 MPa ultimate tensile strength (Figure 10.93).

10.6.12 Manganese–Boron Steels

For hot forming and hardening, the manganese–boron steels offer the highest strengths of up to 1650MPa in the hardened condition. (Also possible to coat with Usibor 1500P[27]). After having heated the steel to the austenite region, and hot forming a subsequent controlled cooling leads to a martensitic structure, and thus, to a high strength of the material. An alternative route would be to form cold and then subsequently heat the part for hardening, which is referred to as the indirect hot stamping.[28] Manganese boron steels are

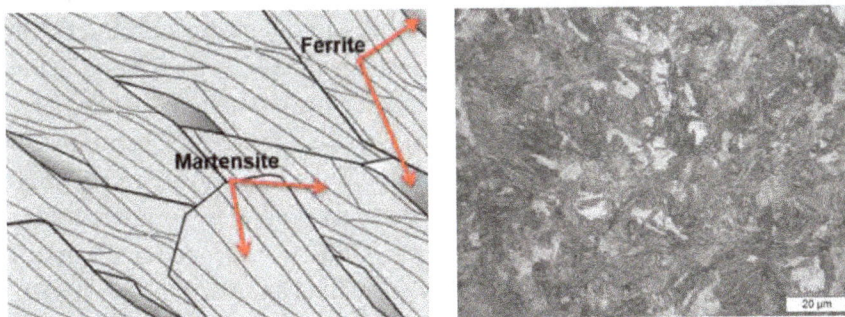

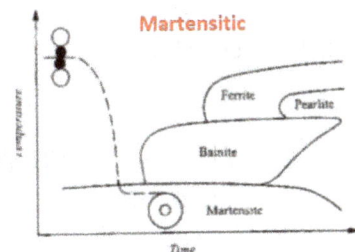

Figure 10.93 LHS = Schematic of microstructure. Centre = Optical microstructure. RHS = TTT

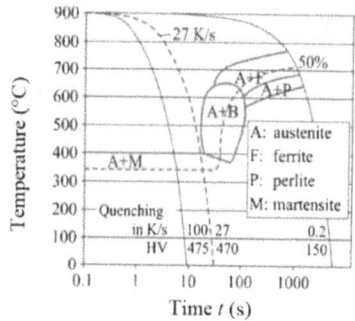

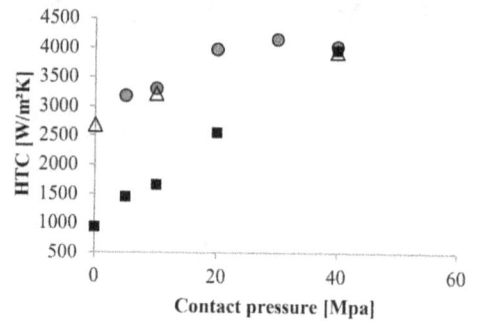

Steel	Al	B	C	Cr	Mn	N	Ni	Si	Ti
20MnB5	0.04	0.001	0.16	0.23	1.05	-	0.01	0.40	0.034
22MnB5	0.03	0.002	0.23	0.16	1.18	0.005	0.12	0.22	0.040
8MnCrB3	0.05	0.002	0.07	0.37	0.75	0.006	0.01	0.21	0.048
27MnCrB5	0.03	0.002	0.25	0.34	1.24	0.004	0.01	0.21	0.042
37MnB4	0.03	0.001	0.33	0.19	0.81	0.006	0.02	0.31	0.046

Steel	Martensite start temperature in ^{0}C	Critical cooling rate in K/s	Yeild stress in MPa		Tensile strength in MPa	
			As delivered	Hot stamped	As delivered	Hot stamped
20MnB5	450	30	505	967	637	1354
22MnB5	410	27	457	1010	608	1478
8MnCrB3	-	-	447	751	520	882
27MnCrB5	400	20	478	1097	638	1611
37MnB4	430	14	580	1378	810	2040

Figure 10.94 Top LHS = Schuler assumes responsibility for turnkey hot forming lines including automation, heating technology, press and quality assurance issues. Centre = CCT 22MNB5. RHS = HTC Lower Composition of Boron steels and Process conditions[29-31]

of special interest for parts with complex geometries, and high demands concerning strength. The mechanical properties, which can be influenced by tempering, correspond to the highest demands, and enable significant weight saving when these steels are used in the production of structural and safety parts of vehicles, such as bumper supports, side impact beams, column, and body reinforcing panels (see Figures 10.94 and 10.101 Lower RHS).

10.6.13 Evolving AHSS Types and the "Third generation" of steels

In response to automotive demands for additional AHSS capabilities, steel industry research continues to develop new types of steel. These steels are designed to reduce density, improve strength, and/or increase elongation. Examples of these developing steels are steels with nano-size particles for increasing strength and improving stretch flangeability.

Based on the above review there are some metallurgical concepts that can offer a steel product with the satisfactory strength-ductility balance or a better final part price. It is interesting to note that the large range of steels available is counter to "economies of scale" resulting in complexity in steel logistics and potential mixed material problems. Some topics are carbide free bainite (CFB) as developed by Professor Bhadeshia, TPN, quenching and partitioning (Q&P), XPF and "Medium Mn" steel (MMS).[9, 10, 11, 12]

10.6.14 TPN – Three-Phase Steel with Nano-Precipitation

A hot rolled steel that results from cooperation between ThyssenKrupp Steel Europe and the Japanese steel group JFE. A new family of high-performance cold forming steels consists of a ferritic matrix containing bainite, residual austenite and nano-precipitations. During the forming, a high percentage of retained austenite is converted to martensite. This work-induced transformation has a positive effect on the hardening behaviour of the material to create a good combination of high formability and high strength. Due to their high resistance to local thinning, these grades allow component geometries that are very difficult to achieve with other high-strength steels of the same strength class. Several nanoprecipitation grades are produced for automotive applications[9] (Figure 10.95).

10.6.15 Q&P – Quenching and Partitioning

Q&P steels are a series of C-Si-Mn, C-Si-Mn-Al steel compositions subjected to the quenching and partitioning (Q&P) heat treatment process as shown in Figure 10.96. With a final microstructure of ferrite (in the case of partial austenitization), martensite and retained austenite, Q&P steels exhibit an excellent combination of strength and ductility, which permitted their use in a new generation of advanced high strength steels (AHSS) for automobiles. Q&P steels are suitable for cold stamping for those structure and safety parts of automotive with relatively complicated shape to improve fuel economy while promoting passenger safety.[10] Leeds University used a 0.7C, 5%Mn, 1.5%Si steel)

10.6.16 Tata Steel XPF steel

A stronger, lighter more formable steel for chassis applications. A key measure of the formability of steel is hole-expansion capacity (HEC) – how far the material around a stamped hole will stretch before fracturing. XPF combines the hole-expansion performance associated with bainitic and martensitic grades, with the elongation characteristics of dual-phase (DP) and high-strength low-alloy (HSLA) steels.[11]

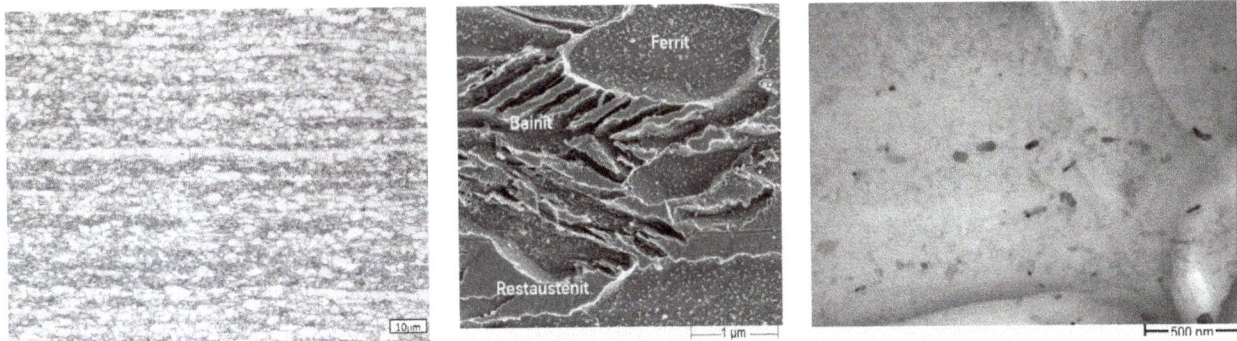

Figure 10.95 Eine neue Stahlfamilie: TPN®-W 18.11.2010 Fachpresseforum, Dr. Heller, ThyssenKrupp Steel Europe

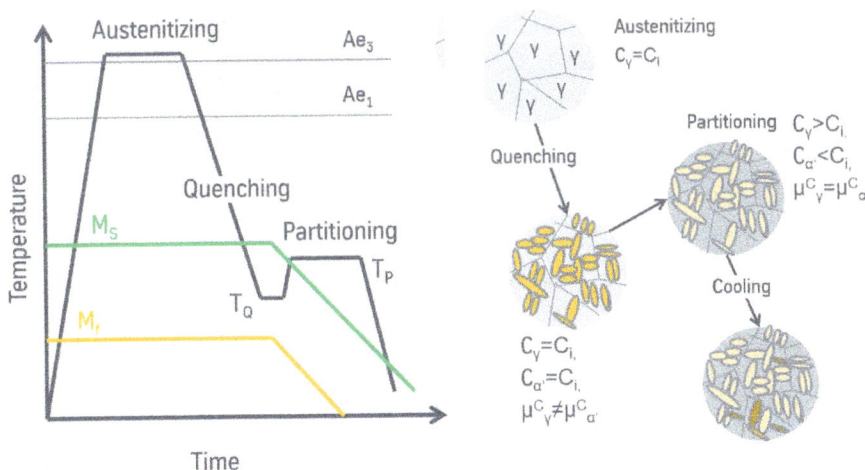

Figure 10.96 Q&P Modeling and Simulation of Q&P steels / G. Paul, 02.09.2016 (TC user meeting 2016, Aachen) thyssenkrupp Steel Europe

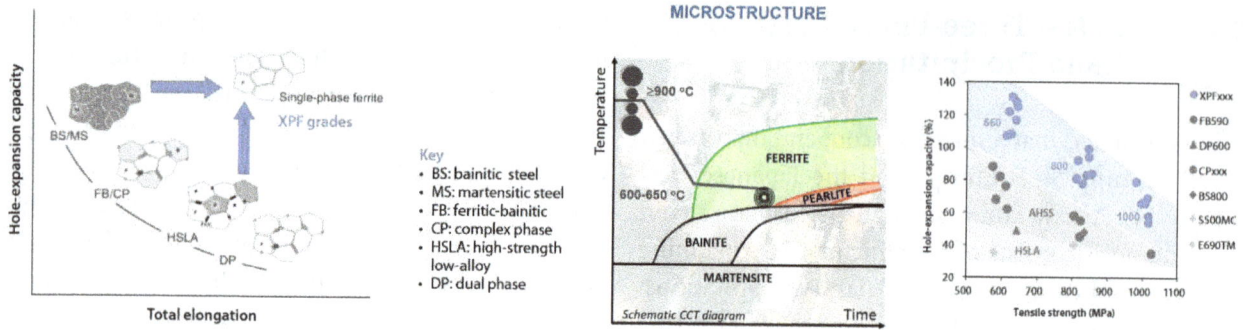

Figure 10.97 *Development ideas from "DARE" of Tata XPF steel*

10.6.17 Medium manganese steels

Medium Mn TRIP steel has been regarded as a promising solution to get high strength steels with good formability based partly on the good performance of the high manganese TWIP steels. A recent doctorate thesis has studied the development of a 5% manganese steel. Part of the work was a review of published data.[32] The results of this review are shown in Figure 10.98. The reported frequent observation of Luders bands during the testing of the steel samples would be a warning of potential stretcher strain marks on parts.

10.6.18 NanoSteel 3rd Generation AHSS

This is a novel development from a US company Nanosteel Inc https://nanosteelco.com/ The process is described in US patent 9284635. The process is based on rapid solidification based on strip continuously cast steel. The composition in the patent is 55 to 88 atomic % Iron, 0.5 to 3.8 atomic % boron, 0.5 to 12 atomic % silicon, 1 to 19 atomic % manganese, 0.1 to 9 atomic % nickel, 0.1 to 19 atomic % chromium, 0.1 to 6 atomic percent copper, 0.1 to 1 atomic % titanium, 0.1 to 4 atomic percent carbon. Claimed to be scale-able.[10]

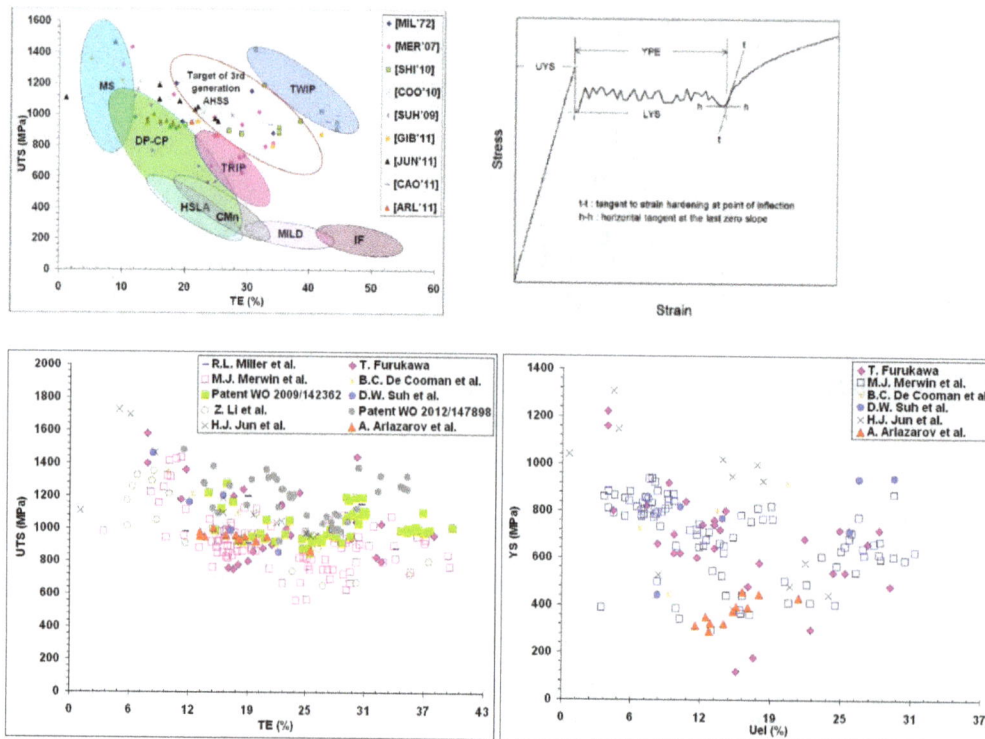

Figure 10.98 *Top RHS = Published results of mechanical properties of Medium Mn steels – ultimate tensile strength (UTS) as a function of total elongation (TE); target of third generation high strength steels is marked with red ellipse. Top LHS = Stress-strain diagram showing Yield Point Elongation (YPE) and Upper (UYS) and Lower (LYS) Yield Strengths Bottom LHS = UTS-TE. Bottom RHS = YS-U-el (when the data were available) charts using the most relevant*[32]

Figure 10.99 The novel enabling mechanisms key to development of new class of 3rd Generation AHSS (Ref10)

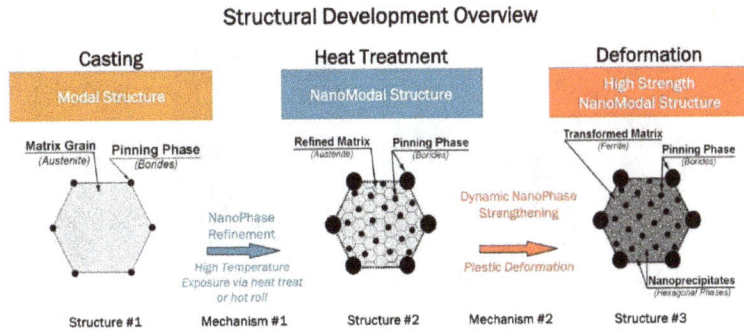

10.6.19 Commercial Vehicles

The weight reduction of commercial vehicles has also been carried out.[13-16] In Europe, trailer manufacturers use 700 strength grade.[17] In 2014 there was a Chinese conference which outlined their efforts to match European and Japanese vehicles which were

Figure 10.100 Grades of steel

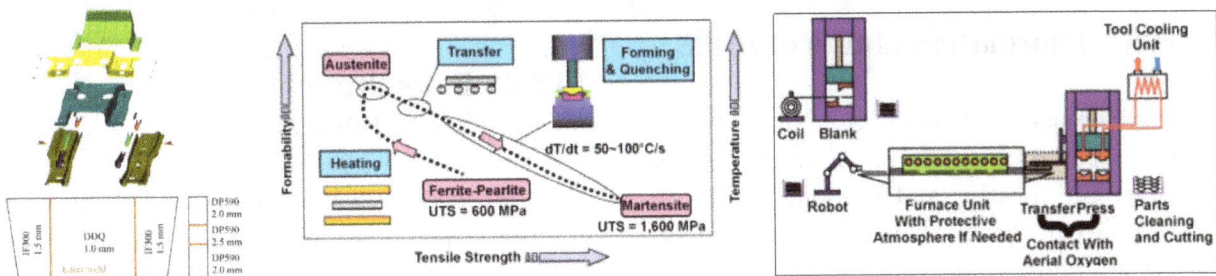

20% lighter. Similar grades of steel were used as shown in Figure 10.100 This has involved roll forming, Hot stamping and direct quenching, Tailored blanks, laser welding and hydroforming.

The improved materials used to manufacture these vehicles can be contrasted against a metallurgical failure carried out on a UK built dumper truck chassis made with low strength steels in 2013, which failed in 2015 (despite 2003 Euro project which showed the advantages of high strength steels).

Based on a review of the material certificates for the steel used to manufacture the trailer, there were several grades of structural steel used. (S275 JR +AR, S355 JR + AR, S355J2 +AR and S235 JRH cold formed hollow square tube.) The steel was imported from several countries: Russia, Holland, China, Germany, UK and Turkey using steel makers Arcelor Mittal, Bengang, Severstal, Tata, Toscelik and Wisco.

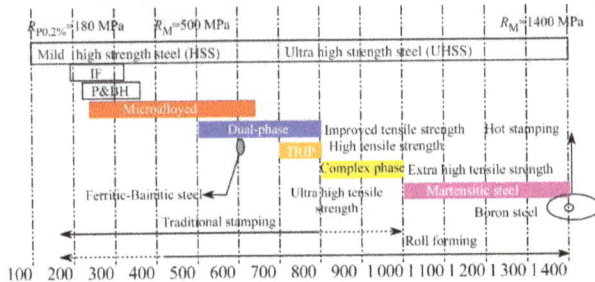

Fig.10.101 *Top= Roll forming[12] Lower LHS = Tailored blank for cabin floor with parts reduction[17]. Lower RHS = Strength and microstructure change during hot stamping. Usibor 1500P is used for the pillars and sills of the new VW Passat and the Mercedes S Class; for A-pillars in the Land Rover; and bumper beams in the Ford Mustang. The hot stamping precoated boron-alloyed steel 22MnB5 is increasingly being used in new Euro and US cars[13]*

Figure 10.102 View from the rear of the chassis showing chassis bend around the landing leg position

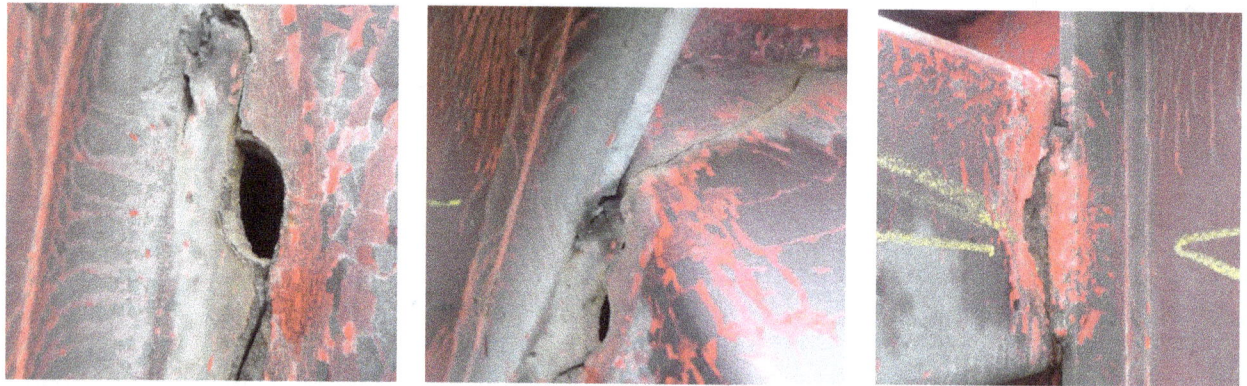

Figure 10.103 Cracks of landing legs cross member. Plus chassis crack

Based on a metallurgical examination the root cause of the chassis cracks was that the chassis had insufficient durability for the service conditions. The poor durability was caused by weld cracking of the welds associated with the cold formed square tube at the landing leg position (Figures 10.102 and 10.103).

The trailer was only two years old, and yet the paint showed an unexpected level of flaking and the steel work showed unexpected areas of corrosion damage. The main failure of the structure was the significant permanent distortion of the chassis that was associated with the loss of structural support due to the cracking of the landing leg cross member.

10.6.20 Deformation Limits of AHSS

The development and manufacture of many different steel types has made the manufacturing processes involving the shearing and stamping of the strip metal more complex. In many cases this has involved greater attention to *tool condition, lubrication and plant maintenance*. The "Advanced High strength steels application guidelines" Version 6.0 from "WorldAutoSteels" gives a comprehensive introduction to the aspects of formability and welding.

The guide presents three different types of forming limits.

- The traditional forming limit curve – applicable to stretching modes of sheet metal forming (global formability). Keeler and Backofen (RHS of forming limit diagram) and Goodwin (LHS of forming limit diagram).
- Edge stretching limit – that applies to the problem of stretching (hole expansion, stretch flanging and blank edge extension) of the cut edge of sheet metal (*local formability*).
- Shear fracture – encountered during small radii bending of DP and TRIP steels.

10.6.21 Edge Stretching Limits – Local Formability

Metal cut edges have always been a vulnerable feature in terms of potential risk of cracking during subsequent processing. Unlike the body of the sheet metal where the "Forming Limit Curve" (FLC) data can be used to give warning of failure, there is no similar method/failure criteria to control the potential risk of cracking the cut edge. In addition, there are several

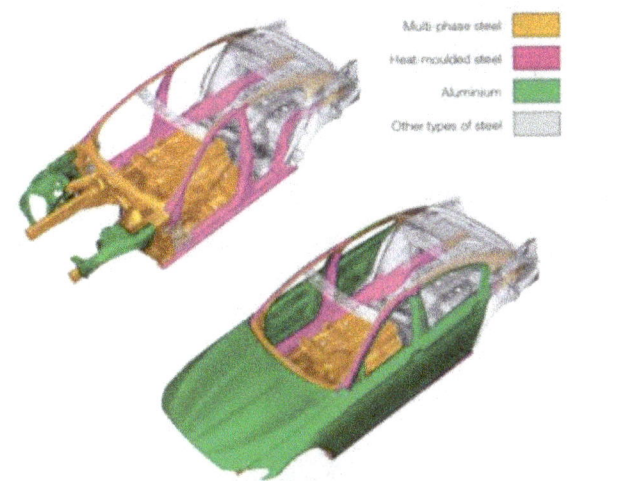

Multi-phase steel
Heat-moulded steel
Aluminium
Other types of steel

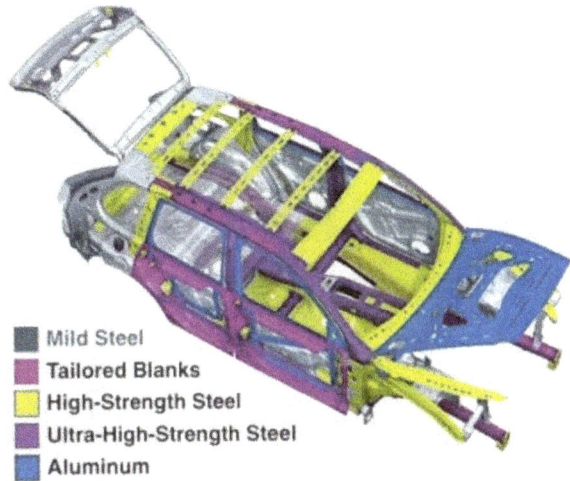

Mild Steel
Tailored Blanks
High-Strength Steel
Ultra-High-Strength Steel
Aluminum

Advanced high strength steels
Magnesium
Aluminum

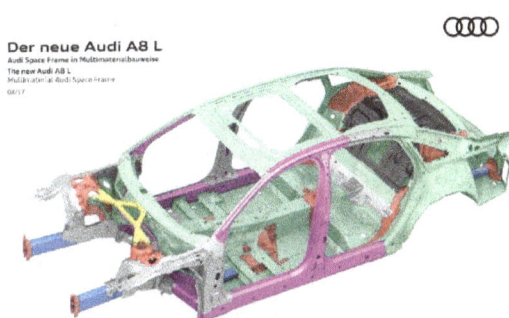

Der neue Audi A8 L
Audi Space Frame in Multimaterialbauweise
The new Audi A8 L
Multimaterial Audi Space Frame

Aluminium-Blech
Aluminium-Profil
Aluminium-Guss
Ultrahochfester Stahl (warmumgeformt)
Konventioneller Stahl
Kohlenstofffaserverstärkter Kunststoff (CFK)
Magnesium

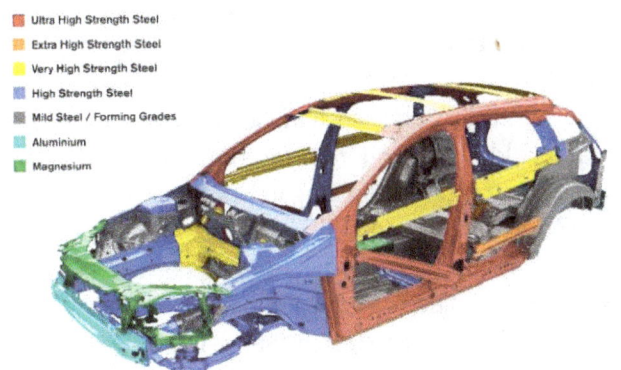

Ultra High Strength Steel
Extra High Strength Steel
Very High Strength Steel
High Strength Steel
Mild Steel / Forming Grades
Aluminium
Magnesium

Material Specification
Mild Steel
BH – HSLA (YS < 300)
HSLA (YS > 300)
DP 600
DP 800
DP 1000
Boron - Martensitic

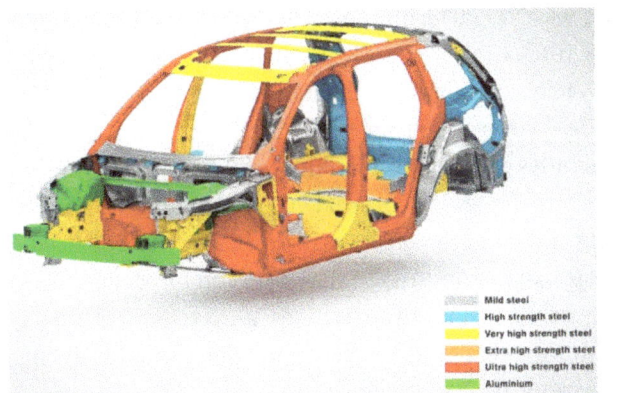

Material Breakdown by Mass
Press Hardened / Hot Stamped
Aluminum
Dual or Multiphase
Bake Hardened
High Strength Low Alloy
Martensite
Mild / Low Carbon
17%
5.70%
2.90%
29.60%
4.80%
22.60%
17.50%

Mild steel
High strength steel
Very high strength steel
Extra high strength steel
Ultra high strength steel
Aluminium

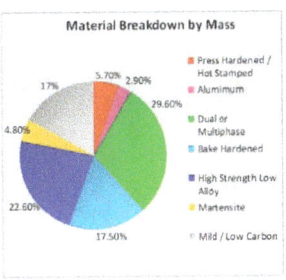

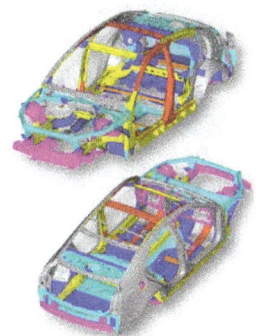

Figure 10.104 *Material selection for body structures 1. BMW Series 5. 2. Porsche 3. Volvo 4. Audi A8 5. Volvo 60 6. Ford Focus 7 Volvo 8 Cadillac Source*[19 to 22]. *Design procedure – identify part duty and specification – select steel to meet requirements, including both manufacturing capability and service performance, build proto-type, test, confirm durability and function*

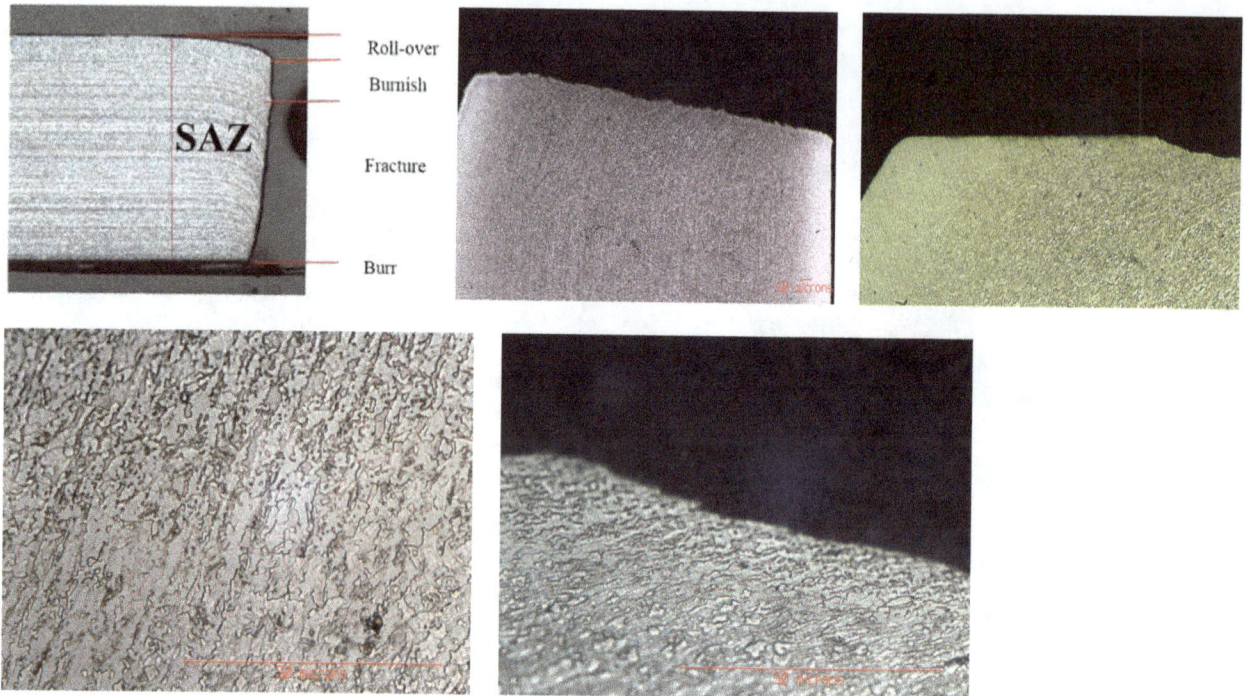

Figure 10.105 *Top LHS = Nomenclature for the sheared edge. Top Centre and RHS = DP 600 sheared edge. Lower LHS = DP 600 strip microstructure. RHS = DP 600 microstructure at the sheared edge*

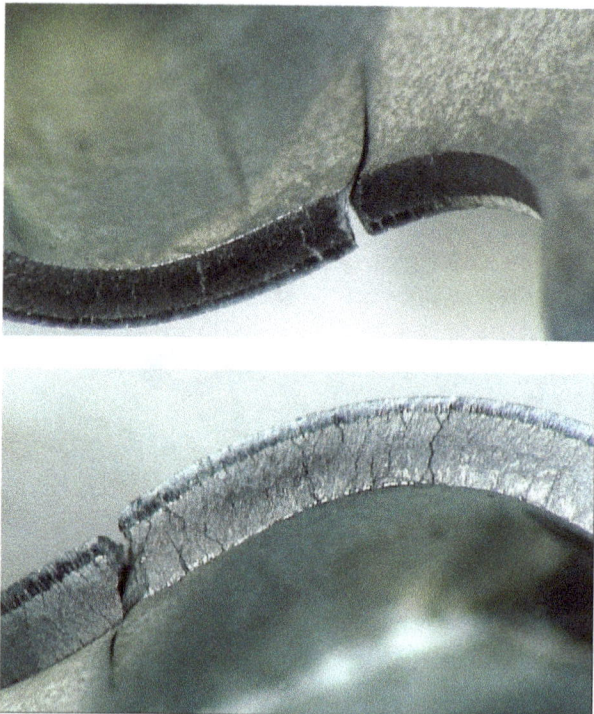

Figure 10.106 *Local edge cracking on a DP600 steel showing minimal wall thinning*

ways to cut the edge such as coil slitting, blanking, laser cutting. The metal shearing methods are based upon part shear and then the controlled progression of a crack through the remaining thickness. Figure 10.105 shows the nomenclature for describing the sheared edge. The "damage" surface of the crack zone

obviously has the ability to generate new cracks if forced to do so by subsequent high tensile manufacturing or service stresses depending upon the steel's resistance to crack formation (toughness at a micro level).

In addition, the mechanical shearing can introduce cold work and reduces the work hardening exponent (n-value), leading to less edge stretchability during future manufacturing stages. However, it is well known that shearing parameters such as clearance, shear angle and rake angle also play a large part in improving edge stretch. Since local formability failures are not preceded by significant thinning and necking of the steel, the "forming limit curve" cannot be used for determination of failure criteria (Figures 10.106).

Sheared edges are created at several stages as the steel progresses from the coil to the final part. This includes coil slitting, blanking, skeleton web trimming (external edges or internal cut-outs) and hole punching. Damaging tensile stresses can be used during hole expansion, stretch flanging, and blank extension processes. It is important to understand the variables that can affect edge stretch performance during forming. This knowledge can be gained by the use of appropriate test methods.

10.6.22 Hole Expansion Test[6, 39, 40, 41]

The Hole Expansion Test (HET) is one of the most common tests to quantify the "local formability" related edge cracking with AHSS. The HET is also useful for grading the relative performance of various "competing" AHSS grades, both in terms of different suppliers and the different grades. The results of these "trial evaluations" can then be used to modify the tailored blank for the intended application, through modifications of chemistry, rolling and thermal practices to achieve the desired edge stretching performance. *In addition, they can be used to "benchmark" the current quality level associated with an established forming process for any future problem-solving events.*

There are two main versions of the HET:

- a flat bottom punch test. The flat punch test often results in an early strain concentration during the test due to the free edge. Therefore, the hole expansion ratio (HER) generated by a conical punch was consistently higher than that created by a flat bottom punch, so when companies test results for various conventional and AHSS grades, the HET test method needs to be specified
- the other test method uses a conical or hemispherical punch. The conical punch test has become more common because it gives a better simulation of a stretch flanging operation.

Widely used standards for conducting the HET include ISO/TS 16630-2003 and JFS T 1001-1996. In both versions, a square steel blank is cut and a hole pierced in the sample. The quality of the punched hole is critical when trying to obtain consistent data. Special effort is needed to keep the tools sharp and damage free in the lab test environment to maintain the consistency in edge conditions. The quality of the hole punching and the identification of the failure height have been the causes of laboratories reporting different results on the similar material[41] (See Figure 10.107). Hard and wear resistant tools, preferably a coated PM grade are highly recommended. Clearances should be monitored, especially if thickness changes are made to the tested material as trim clearances are a critical variable when working with AHSS. If different steel thicknesses are being tested, different diameter punch dies may have to be used to maintain a consistent clearance for varying sheet thicknesses during the hole piercing process.

The flat blank with pierced hole is then clamped in the hole expansion test device, and either a flat or conical punch is pushed up into the hole. The test continues while the hole edge is examined, until a through-thickness edge crack is observed, at which time the punch is stopped. Detecting the point at which edge fracture initiates can be done visually or more commonly, with a camera magnified to get a close-up of the specimen during the test. The hole expansion ratio (HER) or percent hole expansion is then calculated. The increase in the ratio of the final/

Examples of bad cut-edge quality and hole punching conditions

(a) Variable Burnishing and local excessive burr. The sizes of sheared and fracture zones can vary locally from one point to another. The excessive burr is probably caused by a defect on punch or die punching tools or by a tooling-sample misalignment.

(b) "V-shape" on fracture zone through thickness. Can appear when the contact between tools and sheet is non-uniform (misalignment, local defects on tools).

(c) Grooves in fracture zone. Seem by the darker patches in the lighter toned fracture zone, these grooves are often found when the clearance is set too high.

(d) Parallel cracks in fracture zone. Usually this imperfection is benign as hole expansion will compress the cracks, rather than propagate them. Often due to a poor alignment or inappropriate clearance.

(e) Double shearing zones This is seen when cutting clearance is too low.

(f) Jagged boundaries between the sheared and fracture zones. This indicates non-uniform tooling contact in punching. Potential causes are the poor tool alignment, improper clearance setting, or deterioration of tool cutting edges.

(g) Smearing- This is a sign of tool contact during the punch travel. It is normally caused by the improper cutting tool clearance and alignment.

(h) Abnormal edge profiles- Abnormal edge profile can happen due to various reasons, such as the improper cutting tool clearance, poor alignment and equipment irregularities.

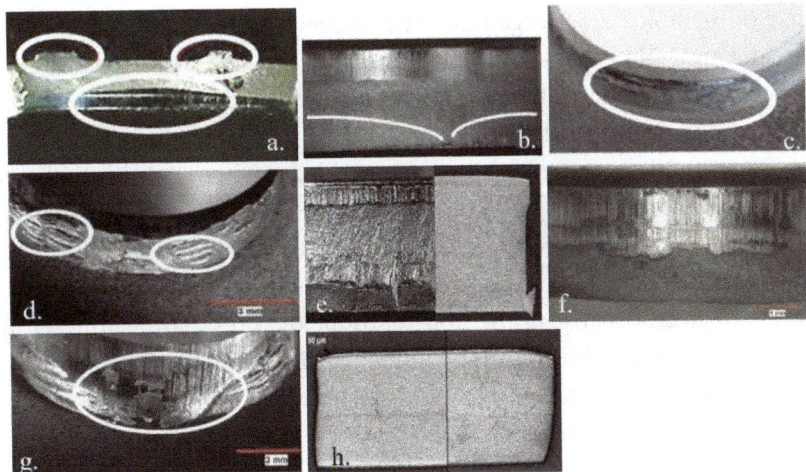

Figure 10.107 *Examples of bad cut-edge quality and hole punching conditions*[41]

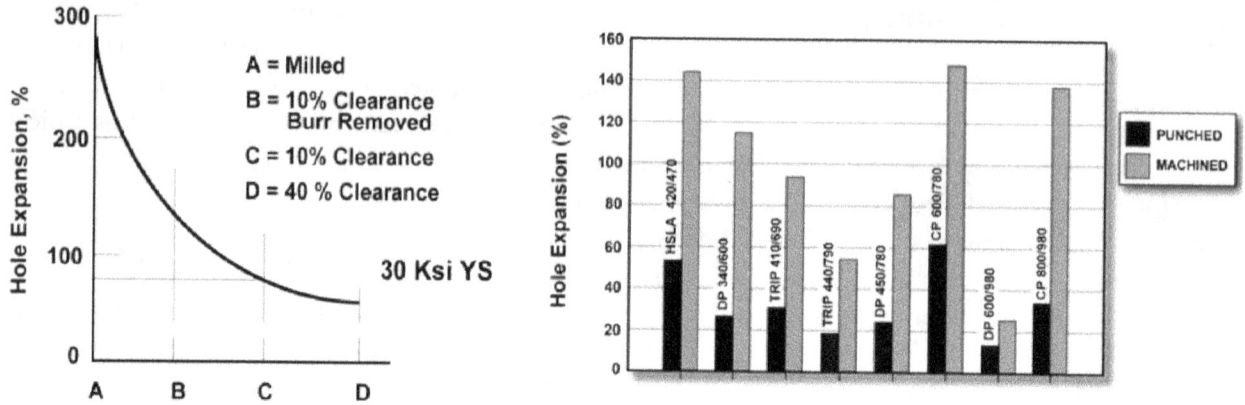

Figure 10.108 *Schematic showing hole expansion capability for a 200 MPa mild steel for various punch condition. RHS = HET results for punched and machined holes showing effect of damage to sheet metal stretchability*[6]

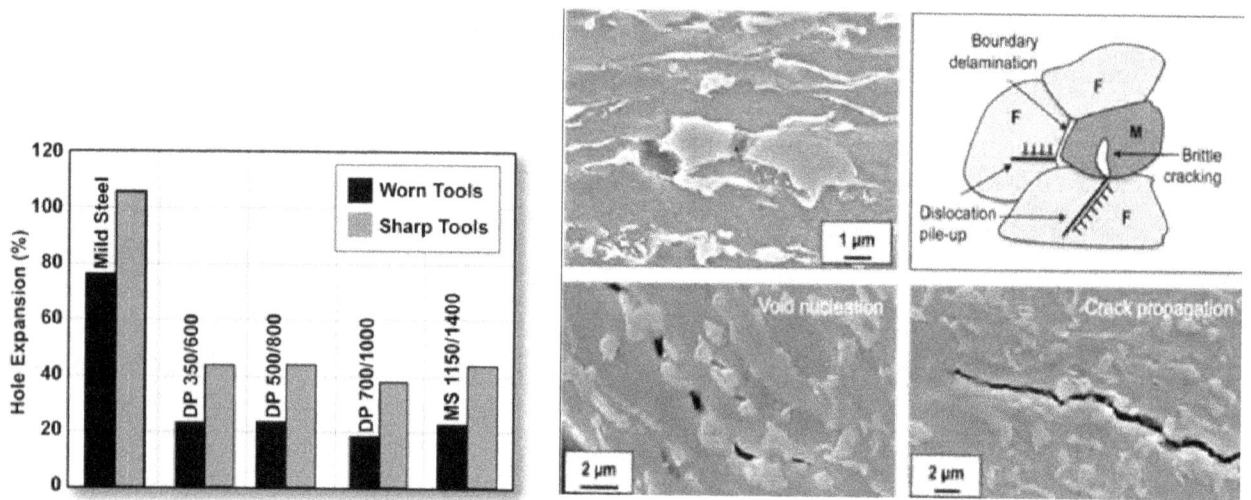

Figure 10.109 *LHS = Impact of production tooling condition on hole expansion performance (tests conducted with 50 mm diameter conical punch). RHS = Features and mechanisms of damage initiation and propagation in DP steel (From ref 6 and RHS = ref 42 Fig 3)*

initial hole diameter is typically given by the symbol lambda (λ) and is the percentage increase in the size of the hole before edge fracture is observed. The greater the hole expansion ratio of the material, the more likely that the material will have improved local formability characteristics.

An example of the results that are provided can be seen in Figure 10.108.

A European round robin test revealed high deviations among several laboratories in testing of high strength steels because of the non-reproducible pre-damaged edge condition and the influence of several operators. The HER showed a standard deviation of 14.6% with an average value of 58.8% for a martensitic steel[44]. Therefore, there were further efforts to find better test methods.

A solution for improved sheared edge stretchability would be a homogeneous microstructure, so

where this is clearly a problem, steelmakers would offer a complex phase steel. Ideally coupled with good grain refinement (reducing the size of the ferrite and martensite grains) is achieved. Such a steel would have less damage resulting from shearing, raising the critical stress for subsequent crack propagation. Additionally, reducing the difference in hardness between the soft ferrite phase and the hard martensite phase improves the hole expansion ratio. Changes in chemistry, hot rolling conditions and inter-critical annealing temperatures are some of the methods of product improvement. These changes can include a single phase of bainite or multiple phases including bainite and removal of large particles of martensite. This trend is shown in Figure 10.110.

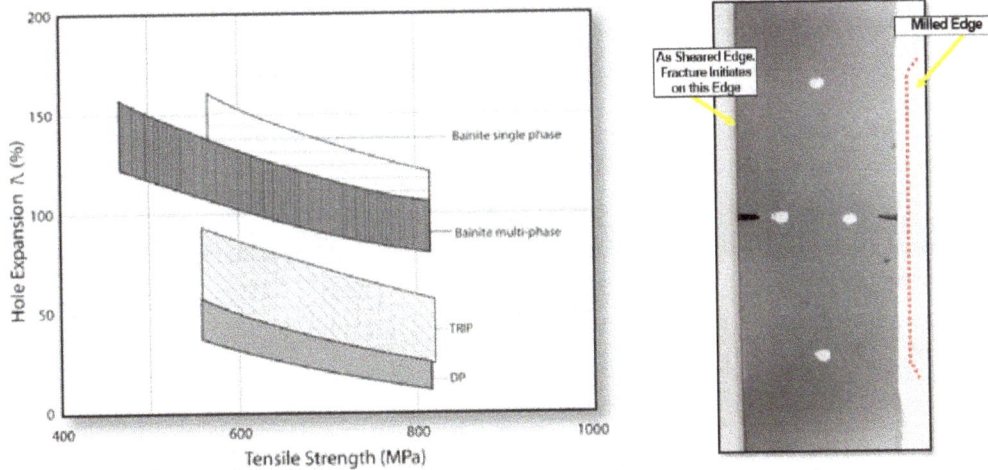

Figure 10.110 LHS = Improvements in hole expansion by modification of microstructure. RHS = 2D edge tension test specimen sometimes referred to as a "half dog bone" as only one side is milled

10.6.23 Sheared Edge Tension Test (SET) Tata version SETi[6 and 39, 40, 41]

Additional tests have been developed to further investigate strip cutting parameters to assist with the optimization of the non-steelmaking related variables. One of the tests, used to evaluate edge stretchability, has been the edge tension test. There are multiple versions of this type of test, but they all are based on the same concept.

The 2-D edge tension test is similar to a standard tensile test in that a steel specimen is pulled under tension until failure. Unlike a standard tensile test where both sides of the tensile specimen are milled into a "dog bone", the 2D tension test only mills one edge of the sample whereas the other edge is cut under a predetermined set of conditions which are being evaluated. Figure 10.109.

The cutting methods being examined could involve laser cutting, EDM, water jet cutting, milling, slitting or mechanical cutting at various trim clearances, shear angles, rake angles or with different die materials or different grades or suppliers of AHSS. The output data is typically measured as total elongation and comparisons can be made with different process parameters using the same material to better define optimum process parameters.

The Sheared Edge Tensile test can be performed on a regular tensile test machine. Part of the original procedure, the test was stopped just upon failure, by detecting a load drop. Then the area of the test piece where the fracture will occur is measured. This area is transformed to a "length strain at fracture" which is the actual test result. Alternatively, the total elongation can be used to carry out the comparison.

10.6.24 Mechanical properties and damage mechanisms[44, 45, 46, 47]

The analysis of the knowledge of the failure modes with the high strength steels has led to the consideration of problems in terms of "local formability" and "global formability". One of the findings was that some microstructures may be good for global but may not be appropriate for high local strains. This applied to the DP steels.

A recent publication provides many answers and some simplification of test requirements. Reference 37 is worthy of detailed consideration. The work begins by establishing definitions for global formability, local formability, damage tolerance, edge-crack sensitivity and fracture toughness for sheet materials. (Table 10.10)

The classical view of the microstructure of DP steels is based on the distribution of hard martensitic islands in a soft ferritic matrix. The fact that the stress heterogeneity during forming can result in cracks at the ferrite to martensite interface has been previously discussed (see Figure 109 RHS). This results in the conclusion that ductile damage evolution is strongly dependent on the local microstructural. The work examined the damage during tensile test work.

Two new test methods gained in status. The 3-point-bending test according to VDA 238-100, and the tensile test! The thickness reduction during the

Table 10.10 *Definition of terms regarding the mechanical properties related to fracture behaviour of high-strength multiphase steels*

Term	Definition
Global formability	The ability of a material to undergo plastic deformation without the formation of a localized neck respectively to distribute strains uniformly
Local formability	Ability to undergo plastic deformation in a local area without fracture
Damage	The decrease in the load-bearing capacity of materials due to the appearance and evolution of voids
Damage tolerance	The ability of a material to undergo sever damage evolution until rupture
Edge-crack sensitivity	The tendency of a material to crack initiation due to further loading at a sheared edge
Fracture toughness	The ability of a material to withstand the growth of existing cracks
Bendability	The ability of a material to undergo bending operations without crack initiation along the bending line

tensile test and the true stain to fracture were found to give information on the steel formability.

10.6.25 The newly identified importance of the tensile test and its correlation with both global and loacal formability

Throughout the history of the development of formable steels there have been many formability tests developed. In recent years the full circle has been completed and there is now growing awareness of that important information that can be obtained from a simple tensile test. The importance of the "True uniform strain" and the thickness reduction at the centre of the fracture the "True thickness strain at fracture", regarding the ranking of steel formability, have only recently been identified.

Global formability correlated with the **True uniform strain**

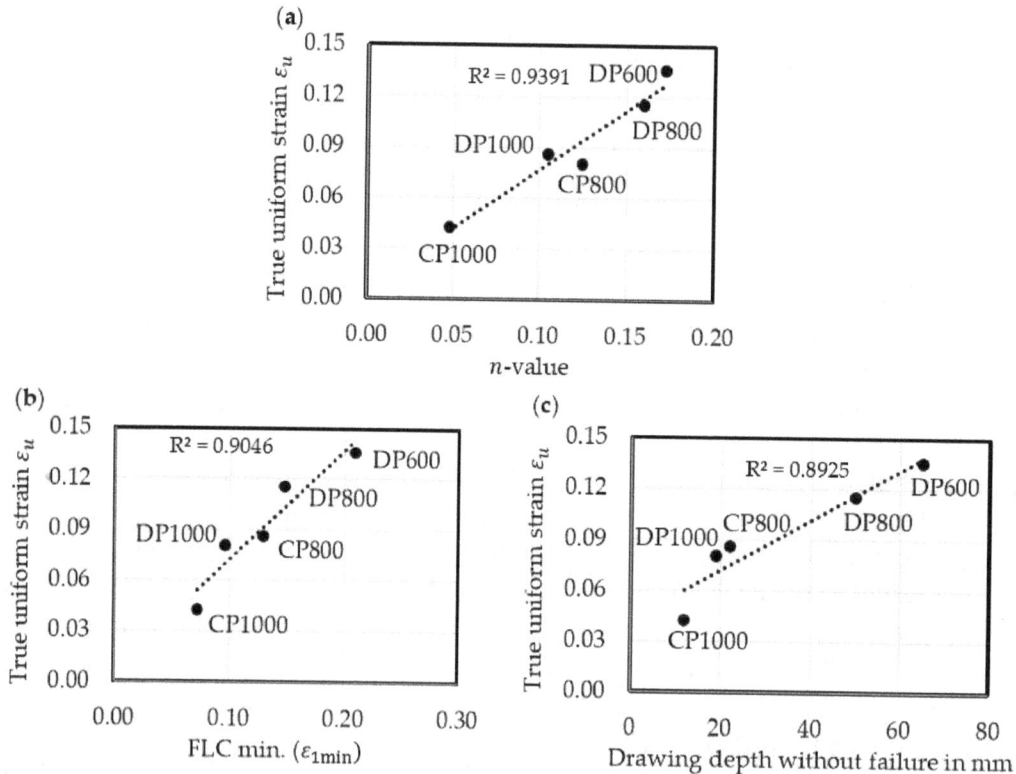

Figure 10.111a *True uniform strain ε_u as measure for global formability: (a) Correlation with n-value; (b) Correlation with forming limit curve (FLC) min.; (c) Correlation with drawing depth without failure of cross-die sample*

Local formability is correlated with the **True thickness strain at fracture**

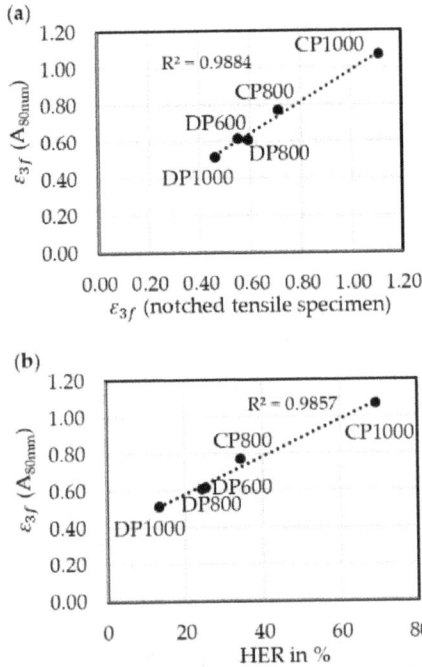

Figure 10.111b True thickness strain at fracture ε_{3f} (A80mm) as measure for global formability: (a) Correlation with ε_{3f} (notched tensile specimen); (b) Correlation with hole expansion ratio (HER)

The increased need for a more suitable description and ranking of newly developed advanced high strength sheet steels with regard to local ductility (stretch-flangeability, bendability, crash-ability) versus global ductility (deep-drawability) appear to be the tensile "Z-value".

10.6.26 Sheet Metal Forming

The formability is a measure of the capability of sheet metal to undergo plastic deformation to a given shape without defects. There are two basic method of forming:

* deep drawing and
* stretch forming.

The defects associated with the two methods need to be considered separately. The difference between these types of stamping procedures is based on the mechanics of the forming process. For deep-drawing, the usual defects of the produced parts are shown in Figure 10.112a LHS. Some of these defects are caused

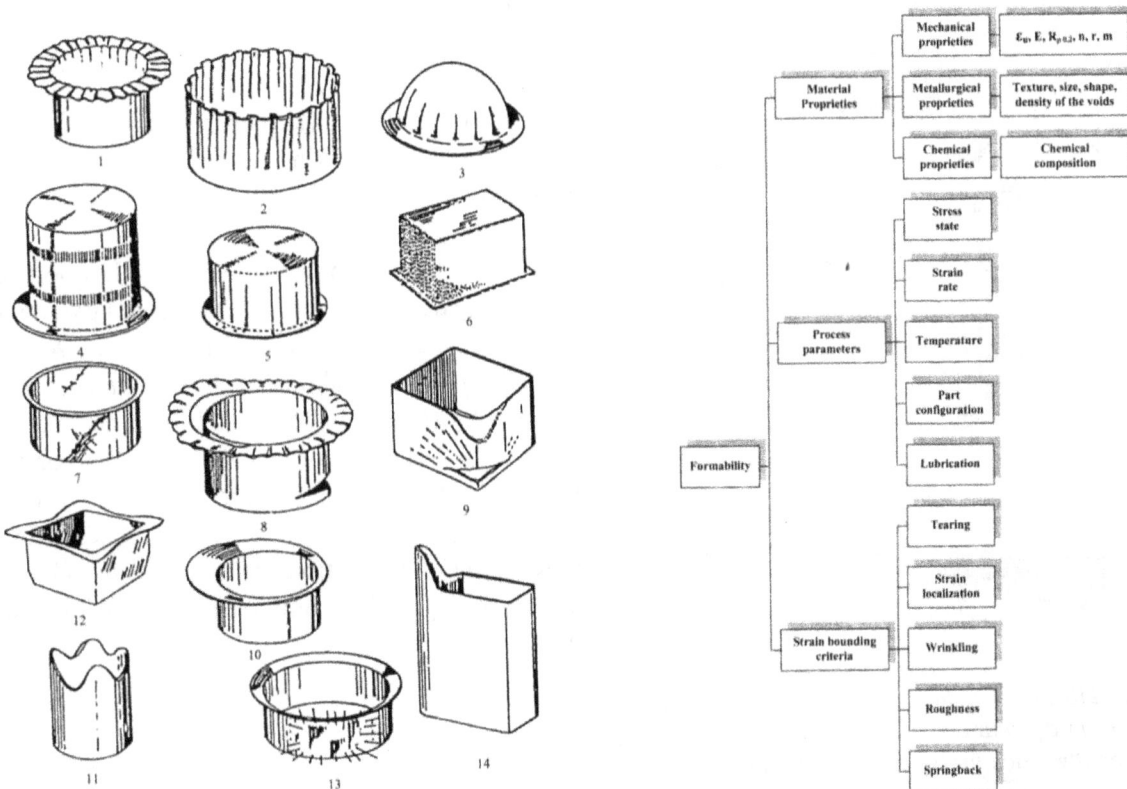

Figure 10.112a LHS = Defects in deep-drawing: 1—flange wrinkling; 2—wall wrinkling; 3—part wrinkling; 4—ring prints; 5—traces; 6—orange skin; 7—Lüders strips; 8—bottom fracture; 9—corner fracture; 10, 11, 12—folding; 13, 14—corner folding. RHS = Parameters influencing sheet metal formability which demonstrates that formability is a complex characteristic[48]

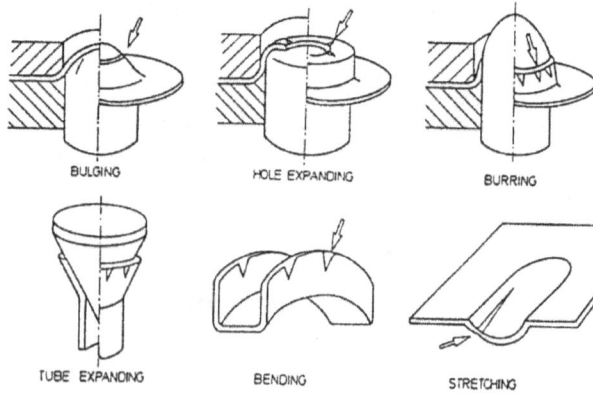

Figure 10.112b Examples of processes where deformation is limited by necking

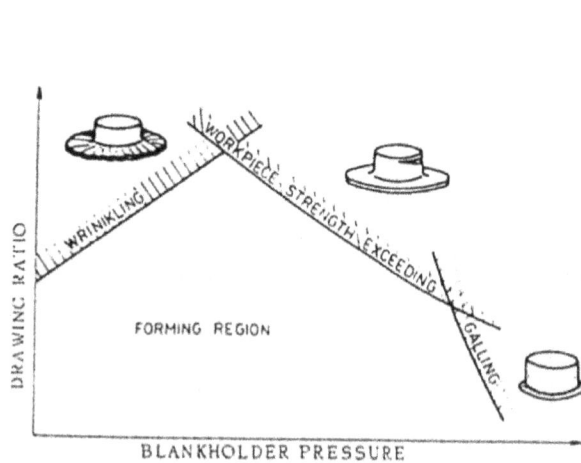

Figure 10.113 LHS = Influence of various parameters on formability in deep-drawing. RHS = Factors influencing the formability index for the most important sheet metal forming processes

by the forming tools (types 5, 9, 10, 14), by the friction regime (types 4, 13) or by the mechanical and metallurgical properties of the material as well as by geometrical parameters (types 1, 2, 3, 6, 7, 8, 11, 12).

Only the defects of type 3, 6, 8 are related to stretching processes, the others are specific to deep-drawing. Concerning the defect type 3 (e.g. in hemispherical punch stretching), the tear is oriented along the circumference and located near the pole. Tearing is usually preceded by strain localization (necking) which causes a reduction of the part's strength, worsens its appearance and is a reason for rejecting it. Necking, tearing, wrinkling, modification of the roughness or a poor appearance are factors that generally define a limit to the deformation by stretching.

Necking is a limiting criterion not only for stretching but also for other processes leading to similar strain states in the plastic zone (Fig.10.115).

10.6.27 Evaluation of the Sheet Metal Formability

Various method for evaluating sheet metal formability have been developed. One may subdivide them into four classes:

- simulating tests
- methods based on mechanical tests
- method of the limiting dome height
- methods based on forming limit diagrams.

Many aspects of formability are replaced by Software which now dominates the development of part design and tooling, see https://www.autoform.com/en The TriboForm software allows the user to quickly simulate the effects of tool coatings, lubricants, material surface characteristics or new sheet materials on friction and ultimately on product quality. Through a more realistic consideration of tribological effects, a new level of simulation accuracy can be achieved. TriboForm's products complement AutoForm's product portfolio.

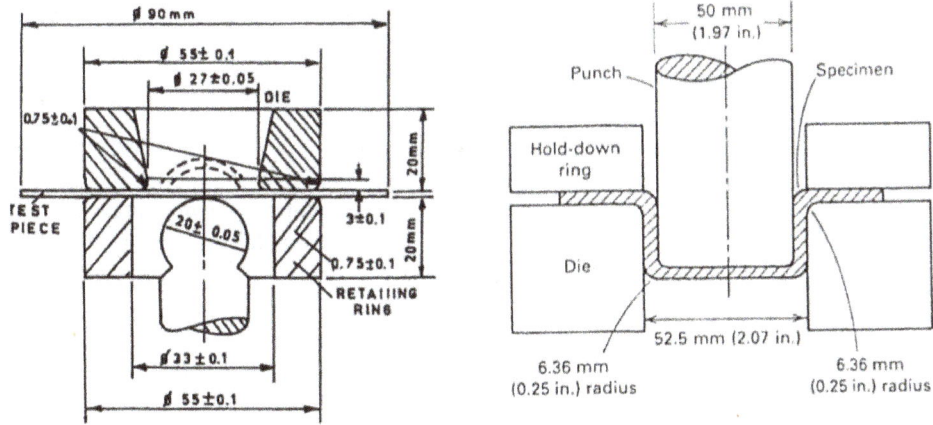

Figure 10.114a *LHS = Scheme of Erichsen test. RHS = Swift's cup-drawing test*

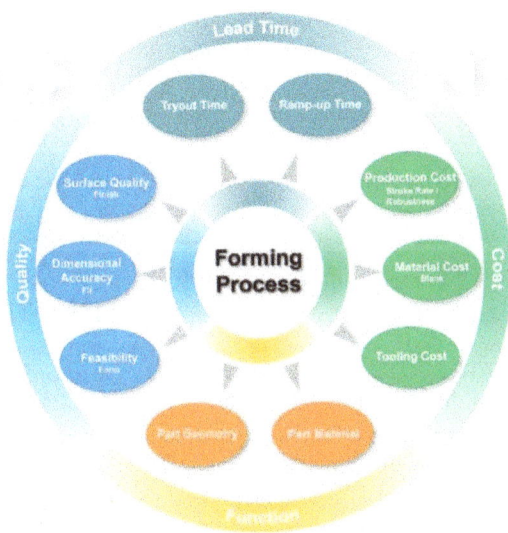

Figure 10.114b *Autoform software plus photos of the tribo module From https://www.autoform.com/en*

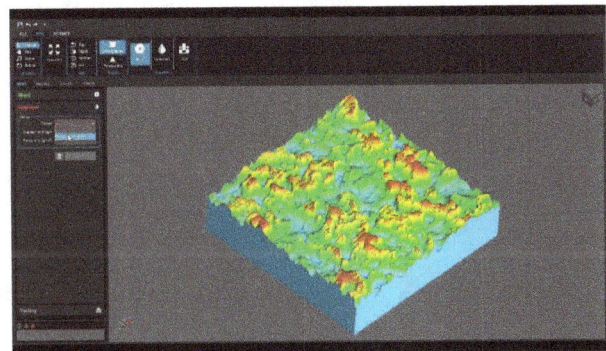

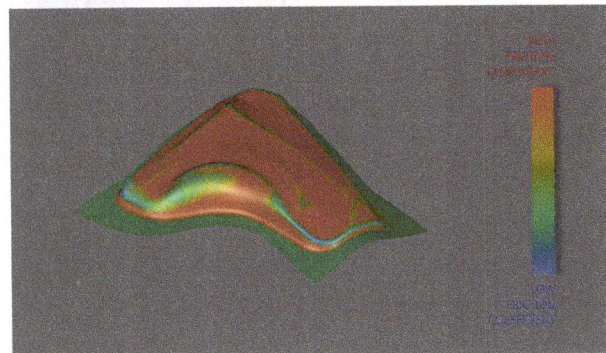

10.6.28 A thin coat of paint between success and failure! (hopefully not)

It could be a nightmare scenario regarding whether sufficient attention has been paid to the possible risk of hydrogen embrittlement due to corrosion related hydrogen charging of the high strength steel used for the "lightweighting" of vehicles for fuel efficiency. The risk would depend as to whether the parts are "external" where paint failure would lead to their aqueous-corrosion or whether the parts were "internal" and fully protected from wet corrosion even in the event of a paint failure.

The failure due to hydrogen embrittlement is not unique to the automotive industry. Grade 12,9 bolts (hardness 44Rc) have been tried and were withdrawn from use after hydrogen related failures. Helical suspension springs are another part made from high strength steel that could be embrittled by environmentally generated hydrogen. From a personal perspective I have had vehicles with spring failures (helical springs in the engine compartment) where the paint has failed and areas of rust were present together with complete fracture.

There are now several recent published papers regarding the hydrogen risk.

1. Wayne State University and Ford Motor Company Detroit Material has a range of high strength cast low alloy steels strengthened by an innovative austempering process. The marketing for the company shows that they have identified the automotive market as having strong potential for their castings. It is probable that concerns regarding hydrogen has resulted in the recently published R&D, "Fatigue Crack Growth Behavior of Austempered AISI 4140 Steel with Dissolved Hydrogen" 2017.[49] The main conclusion was that: *"austempering of 4140 steel appears to provide a processing route by which the strength, hardness, and fracture toughness of the material can be increased with little or no degradation in the ductility and fatigue crack growth behaviour".* However, further work was recommended.

2. BMW carried out work in 2011 and published the findings with the title "Hydrogen Embrittlement in HSSs Limits Use in Lightweight Body" confirming embrittlement was a potential problem.[50] Further work by Tom Depover et al was reported in 2018 which also confirmed that AHSS were susceptibility to hydrogen embrittlement concluding that HELP (H Enhanced Localized Plasticity) was probably the most probable mechanism. *"HELP considers an increase in dislocation mobility due to the presence of H, resulting into highly localized plastic deformation and accelerated failure."*[51]

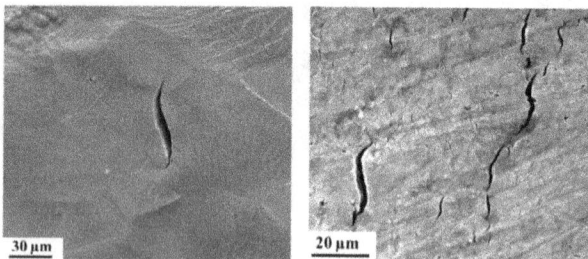

Figure 10.115 SEM images showing similar S-shape of crack for pure iron and TRIP steel[51]

In a review in 2016 Professor Bhadeshia suggested the cures were the use of coating to prevent entry or methods to safely accommodate/fix the hydrogen (carbide interfaces)[52]

Organisations

www.worldautosteel.org
www.darealloys.com Designing alloys for resource efficiency. Improving the technical competitiveness of UK metal industries
https://nanosteelco.com/ Steel alloy design and commericalization
https://www.autoform.com/en/ Process to Success: Software Solutions for Sheet Metal Forming

REFERENCES

1. T. N. Baker, REVIEW OR CRITICAL ASSESSMENT Microalloyed steels, Ironmaking and Steelmaking, September 2015. https://www.researchgate.net/publication/283182435_Microalloyed_steels

2. Porsche Engineering Services, Inc. ULSAB Program Phase 2 Final Report to the Ultra Light Steel Auto Body Consortium March 1998

3. Structural materials and processes in transportation Edited Dirk Lehmhus et al Wiley -VCH 2013

4. .Posco Automotive brochure http://www.posco.com/homepage/docs/eng5/dn/company/product/k_car_pdf_2014.pdf

5. Ömer Necati Cora, Muammer Koç, PROMISES AND PROBLEMS OF ULTRA/ADVANCED HIGH STRENGTH STEEL (U/AHSS) UTILIZATION IN AUTOMOTIVE INDUSTRY, OTEKON'14 7. Otomotiv Teknolojileri Kongresi, 26 – 27 Mayıs, 2014, BURSA

6. Stuart Keeler et al, Advanced High strength steels application guidelines Version 6.0 WorldAutoSteels April 2017 Plus earlier version Advanced High strength steels application guidelines Prepared by INTERNATIONAL IRON & STEEL INSTITUTE Committee on Automotive Applications, March 2005. www.worldautosteel.org

7. Ylva Granbom, Structure and mechanical properties of dual phase steels – An experimental and theoretical analysis, Doctoral thesis, Royal Institute of Technology School of Industrial Engineering and Management Materials Science and Engineering, 2010

8. Dursun Ali YAŞACAN, OPTIMIZATION OF ALLOYING ELEMENTS AND PROCESS PARAMETERS FOR EXTRA DEEP DRAWABILITY PROPERTY OF COLD ROLLED AUTOMOTIVE STEELS, ISTANBUL TECHNICAL UNIVERSITY, Ph.D. Thesis, Feb 2006

9. TPN, https://www.metaalmagazine.nl/wp-content/uploads/2011/01/TPN_Staehle.pdf

10. Q&P,http://thermocalc.micress.de/proceedings/proceedings2016/Presentations/Paul_20160901_TCmeeting_QP_model.pdf

11. Tata Steel XPF https://www.tatasteeleurope.com/static_files/Downloads/Automotive/Case%20Studies/Tata%20Steel%20AM%20-%20XPF%20steel%20-%20case%20study%20EN.pdf

12. Nanosteel, www.smdisteel.org/~/media/Files/Autosteel/Great%20Designs%20in%20Steel/GDIS%202014/Daniel%20Branagan%20-%20NanoSteel.pdf

13. Report EUR 21909 EN, Manufacturing guidelines when

using ultra high strength steels in automotive applications, European Commission, Contract No 7210-PR/179 1 July 1999 to 30 June 2003, Final report 2005.

14. Hardy Mohrbacher et al, Innovative manufacturing technology enabling light weighting with steel in commercial vehicles, published with open access at Springerlink.com, 2015

15. Taylan Altan, R&D Update: Hot-stamping boron-alloyed steels for automotive parts – Part II, https://www.thefabricator.com/article/stamping/hot-stamping-boron-alloyed-steels-for-automotive-parts---part-ii

16. DR. M. VENKATRAMAN, Design of high strength steel for long member of trucks and commercial vehicles, http://www.essarsteel.com/upload/pdf/design_of_high_strength_steel1.pdf

17. Peter Seyfried et al, Light weighting opportunities and material choice for commercial vehicle frame structures from a design point of view, published with open access at Springerlink.com 2015. https://link.springer.com/content/pdf/10.1007%2Fs40436-015-0103-8.pdf

18. In Truck Innovation on the road, https://www.thyssenkrupp-components-technology.com/media/publikationen/intruck_en.pdf

19. Audi 8 www.carmagazine.co.uk/car-news/tech/new-2017-audi-a8-active-suspension-explained/

20. Volvo V60 https://www.sjf.tuke.sk/transferinovacii/pages/archiv/transfer/28-2013/pdf/168-171.pdf

21. Volvo XC90 https://www.boronextrication.com/2014/08/22/2015-volvo-xc90-body-structure/

22. Ford Focus and Cadillac https://www.autosteel.org/-/media/images/autosteel/research/growth-of-ahss/ahss---cadillac-ats.jpg

23. K.Logesh et al, A REVIEW ON SHEET METAL FORMING AND ITS LATEST DEVELOPMENT OF SANDWICH MATERIALS, International Journal of Mechanical Engineering and Technology (IJMET) Volume 8, Issue 8, August 2017, pp. 369–385,

24. Hande GÜLER, Investigation of Usibor 1500 Formability in a Hot Forming Operation, MATERIALS SCIENCE (MEDŽIAGOTYRA). Vol. 19, No. 2. 2013

25. Junying Mina, Indirect Hot Stamping of Boron Steel 22MnB5 for an upper B-pillar, Advanced Materials Research Vols. 314-316 (2011) pp 703-708 Online available since 2011

26. A. Erman Tekkaya, Thermo-mechanical coupled simulation of hot stamping components for process design, Prod. Eng. Res. Devel. (2007) 1:85–89.

27. Nuraini Aziz et al, OPTIMIZATION OF QUENCHING PROCESS IN HOT PRESS FORMING OF 22MNB5 STEEL FOR HIGH STRENGTH PROPERTIES, 2nd International Conference on Mechanical Engineering Research 2013 (ICMER2013)

28. R Vollmer, C Palm, IMPROVING THE QUALITY OF HOT STAMPING PARTS WITH INNOVATIVE PRESS TECHNOLOGY AND INLINE PROCESS CONTROL, 36th IDDRG Conference – Materials Modelling and Testing for Sheet Metal Forming, Journal of Physics: Conf. Series 896 (2017)

29. ARTEM ARLAZAROV, Evolution des microstructures et lien avec les propriétés mécaniques dans les aciers ‹Médium Mn›,

'UNIVERSITÉ DE LORRAINE, Doctorate Thesis 2015 http://docnum.univ-lorraine.fr/public/DDOC_T_2015_0086_ARLAZAROV.pdf

30. Professor Subramanya Sarma Vadlamani, IIT Madras, Metallurgical and Materials Engineering https://mme.iitm.ac.in/vsarma/mm5025/TRIP-DP-TWP-Notes.pdf

31. Ylva Granbom, Structure and mechanical properties of dual phase steels –An experimental and theoretical analysis, Doctoral thesis, Royal Institute of Technology 2010, https://www.diva-portal.org/smash/get/diva2:353680/FULLTEXT10.pdf

32. G R Speich et al, Metallurgical transactions, August 1981, Volume 12, Issue 8, pp 1419–1428 http://www.academia.edu/11109516/Formation_of_Austenite_During_Intercritical_Annealing_of_Dual-Phase_Steels

33. Nina Fonstein, Advanced High Strength Sheet Steels: Physical Metallurgy, Design, Processing and Properties Springer, 2015

34. C.C. Tasan et al, An Overview of Dual-Phase Steels: Advances in Microstructure-Oriented Processing and Micromechanically Guided Design, Annu. Rev. Mater. Res. 2015. 45:391–431

35. Timothy David Bigg, Quenching and Partitioning A New Steel Heat Treatment Concept Doctorate Thesis The University of Leeds February 2011.

36. Mai Huang & Jugraj Singh, A/SP Standardization of Hole Expansion Test www.autosteel.org/~/media/Files/Autosteel/Great%20Designs%20in%20Steel/GDIS%202014/Mai%20HuangJ%20Singh.pdf

37. Mai Huang, Liwei Zhang, Standardization and Automation of Hole Expanding Test https://www.autosteel.org/~/media/Files/Autosteel/Great%20Designs%20in%20Steel/GDIS%202011/09%20-%20Mai%20Huang%20-%20Standardization%20and%20Automation%20of%20Hole%20Expanding%20Test.pdf

38. J Goncalves, Importance of hole punching conditions during Hole Expansion test, International Deep Drawing Research Group 37th Annual Conference, IOP Conf. Ser.: Mater. Sci. Eng. 418 012060 (2018) http://iopscience.iop.org/article/10.1088/1757-899X/418/1/012060/pdf

39. Hardy Mohrbacher, Advanced metallurgical concepts for DP steels with improved formability and damage resistance http://www.niobelcon.com/NiobelCon/resources/Advanced-metallurgical-concepts-for-DP-steels-with-improved-formability-and-damage-resistance.pdf

40. Eisso Atzema, SETi, Sheared Edge Tensile test improved, IDDRG 2015 May conference 31-June 03, 2015, Shanghai, China http://www.iddrg.com/mm/15/C_91_15.pdf

41. Sebastian Heibel, Thomas Dettinger, Winfried Nester, Till Clausmeyer, and A. Erman Tekkaya, Damage Mechanisms and Mechanical Properties of High-Strength Multiphase Steels, Materials 2018, 11, 76 file:///C:/Users/user/Downloads/Damage_Mechanisms_and_Mechanical_Properties_of_Hig.pdf

42. P Larour et al, Reduction of cross section area at fracture in tensile test: measurement and applications for flat sheet steels, 2017 J. Phys.: Conf. Ser. 896 http://iopscience.iop.org/article/10.1088/1742-6596/896/1/012073/pdf

43. Leopold Wagner, Erich Berger, Patrick Larour, Heinrich Pauli, FORMING FRACTURE LIMITS OF AHSS SHEETS

AS RELATED TO DIFFERENT CHARACTERIZATION TESTS, Forming Technology Forum 2018 July 2 & 3, 2018, Zurich, Switzerland file:///C:/Users/user/Downloads/14_FTF2018_Wagner%20(2).pdf

44. 11th Forming Technology Forum Zurich 2018, Experimental and numerical methods in the FEM based crack prediction https://ethz.ch/content/dam/ethz/special-interest/mavt/virtual-manufacturing/ivp-dam/News_Events/ftf2018/FTF2018_Proceedings_digital_release_2018-10-23.pdf

45. Dorel Banabic, Sheet Metal Forming Processes Constitutive Modelling and Numerical Simulation, Pub Springer 2010 file:///C:/Users/user/Downloads/DorelBanabicauth.-SheetMetalFormingProcesses_ConstitutiveModellingand NumericalSimulation-Springer-VerlagBerlinHeidelberg2010%20(2).pdf

46. Varun Ramasagara Nagarajan, Susil K. Putatunda and James Boileau, Fatigue Crack Growth Behavior of Austempered AISI 4140 Steel with Dissolved Hydrogen, Metals 2017, 7, 466 file:///C:/Users/user/Downloads/metals-07-00466%20(1).pdf

47. Matthias Loidl O. Kolk, Hydrogen Embrittlement in HSSs Limits Use in Lightweight Body, ADVANCED MATERIALS & PROCESSES • MARCH 2011, page 22.

48. https://www.asminternational.org/c/portal/pdf/download?articleId=AMP16903P22&groupId=10192

49. Tom Depover et al, Understanding the Interaction between a Steel Microstructure and Hydrogen, Materials 2018, 11, 698 https://www.ncbi.nlm.nih.gov/pmc/articles/PMC5978075/

50. H.K.D.H. Bhadeshia, Prevention of Hydrogen Embrittlement in Steels, ISIJ International, 56 (2016) 24-26 https://www.phase-trans.msm.cam.ac.uk/2016/preventing_hydrogen.pdf

10.7 NITRIDING, NITROCARBURIZING AND CARBURISING STEEL

10.7.1 Nitriding and nitrocarburizing

10.7.1.1 Introduction and background

Nitriding is the surface hardening process that diffuses nitrogen into the surface of a metal. The process can be classified as a thermochemical surface hardening process. Nitrocarburising, is a similar process which would be carried out at a similar temperature within similar facilities but with a small addition of carbon. (in gas and plasma this would be either carbon dioxide, methane or butane). These processes can be used to surface harden low-carbon, low-alloy, medium and high-carbon steels, titanium, aluminium and molybdenum.

Nitriding and nitrocarburising are two hardening processes that can be carried out with very low distortion occurring during the treatment. The temperature used for nitriding would be in the range 490 to 530°C and the chosen temperature for nitrocarburizing would be in the range 550 to 580°C. The treatments are therefore carried out with the steel below the AC1. Therefore, this avoids any distortion that could result from passing through the allotropic transformation which occurs above the AC1. The lower distortion associated with nitriding presents a major advantage compared to through hardening, induction hardening carbonitriding and carburizing processes.

The surface hardening nitriding processes were first developed in the early 1900s and their development can be traced back to the initial patents. The initial gas used for nitriding was ammonia. It dissociates at the surface of the steel, which acts as a catalyst. The nitrogen, which is then present in atomic form, diffuses into the steel surface and combines with the iron to form iron nitride ($Fe_{2-3}N$, Fe_4N). The nitrogen diffuses further into the material during the treatment, and can result in the formation and precipitation of the alloy nitrides which strengthen the diffusion zone.

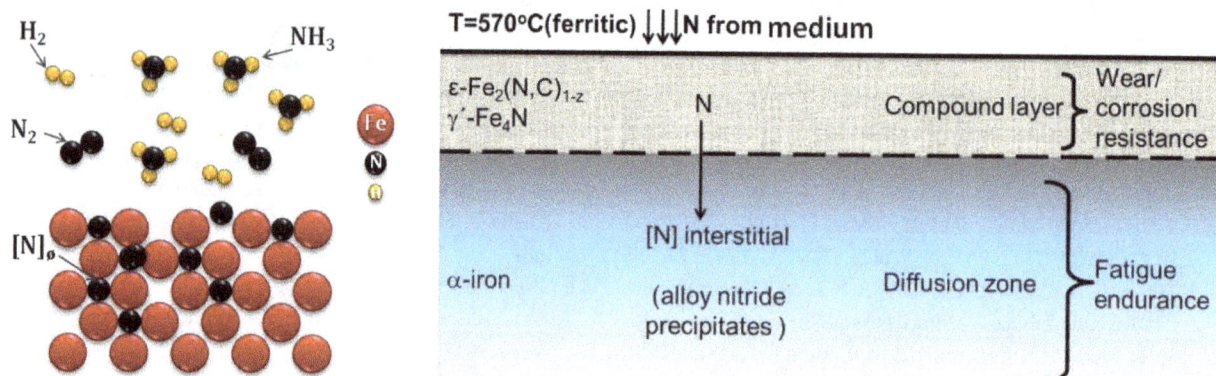

Figure 10.116a LHS = Schematic illustration of gas nitriding. RHS = Schematic compound layer and diffusion zone structure of a nitrided surface[2]

In addition, the relatively low temperatures of nitriding, allows the retention of the beneficial compressive residual stresses which enhance the fatigue endurance.[28] Therefore, two enhancements are achieved through nitriding, the compound layer can give wear resistance and corrosion resistance and the diffusion zone can give improved fatigue endurance. A schematic illustration of gas nitriding is shown in Figure 10.116a LHS. Figure 10. 116a RHS shows a schematic of the compound layer and diffusion zone structure of a nitrided surface.[2]

A form of gas nitriding was patented in 1913 in the USA which modified the process by adding hydrogen to the dissociated ammonia by Dr. Adolph Machlet. He also patented, in 1914, a process for nitrocarburizing using an ammonia and hydrocarbon gas mixture. In 1921, in Germany Adolph Fry who was regarded by many as the "father of nitriding" used dissociated ammonia and explored the effects of temperature concluding that higher temperature should be avoided since they can result in high subsurface levels of nitrogen with the possibility of forming nitride networks particularly at corners during the cooling from nitriding temperature.

10.7.1.2 Selection of steels for nitriding

Adolph Fry was also a key person in the development of steels suitable for nitriding and developed alloy steels with additions of chromium and aluminium that could form strengthening nitride precipitation in the diffusion zone. The first steel was named "Nitralloy" and was made by Krupp Steel in Germany. It was also made under licence in the UK by Firth Brown Steel which later became the BS 970 En 40 and En 41 grades of steel.[1 and 29]

The selection of a suitable nitriding steel should still be an important consideration. The achievable surface hardness is dependent not only on the nitriding temperature and the amount of nitrogen available, but particularly also on the quantity of nitride-forming elements and the strength of the components in quenched and tempered condition prior to nitriding. The higher the content of chromium and aluminium, the higher the surface hardness after nitriding. In addition, the higher the quench and tempered hardness prior to nitriding the higher the nitride hardness since the alloy element are still in solution and not precipitated as carbides. Steels used for nitriding are shown in Table 10.11

Maximum hardness of the nitride case (900-1200 HV) can be attained in aluminium alloyed steels. For steels containing chromium and molybdenum, the obtainable case hardness is somewhat lower (750-900 HV). The typical surface hardness and nitride depth obtained with nitriding are shown in Table 10.12

Steel family	Grade	Typical composition (mass %)														
		C	Si	Mn	P	S	Cr	Mo	V	Al	Ni	Co	Ti	Cu	W	Fe
Plain carbon	C45* (Ck45)	0.42-0.5	≤ 0.4	0.5-0.8	≤ 0.035	≤ 0.035	≤ 0.4	≤ 0.1			≤ 0.4					bal.
	C35* (Ck35)	0.32-0.39	≤ 0.4	0.5-0.8	≤ 0.035	≤ 0.035	≤ 0.4	≤ 0.1			≤ 0.4					bal.
Low alloyed	31CrMoV9** (En40A, En40B, En40C)	0.27-0.34	≤ 0.4	0.4-0.7	≤ 0.025	≤ 0.035	2.3-2.7	0.15-0.25	0.1-0.2							bal.
	32CrMoV13 (AMS6481)	0.29-0.36	0.1-0.4	0.4-0.7	≤ 0.025	≤ 0.020	2.8-3.3	0.7-1.2	0.15-0.35		≤ 0.3					bal.
	34CrAlMo5** (En41A, En41B)	0.30-0.37	≤ 0.4	0.4-0.7	≤ 0.025	≤ 0.035	1.0-1.3	0.15-0.25		0.8-1.2						bal.
	42CrMo4* (AISI4140)	0.38-0.45	≤ 0.4	0.6-0.9	≤ 0.025	≤ 0.035	0.9-1.2	0.15-0.3								bal.
	35CrNiMo6 (AISI4340)	0.36-0.44	0.1-0.35	0.45-0.7	≤ 0.035	≤ 0.04	1.0-1.4	0.2-0.35			1.3-1.7					bal.
	20NiCrMo2*** (AISI8620)	0.14-0.19	≤ 0.4	1-1.3	≤ 0.035	≤ 0.035	0.8-1.1	0.15-0.25			0.4-0.7					bal.
	60Si7 (AISI9260)	0.57-0.65	1.5-2	0.6-0.9	≤ 0.035	≤ 0.035	≤ 0.3				≤ 0.25			≤ 0.2		bal.
High alloyed	M50NiL	0.11-0.15	0.1-0.25	0.15-0.35			4-4.25	4-4.5	1.13-1.33		3.2-3.6					bal.
	XT87W6Mo5Cr4V2 (AISI M2)	0.78-1.05	0.2-0.45	0.14-0.4	≤ 0.035	≤ 0.035	3.75-4.5	4.5-5.5	1.75-2.2		≤ 0.3			≤ 0.25	5.5-6.75	bal.
	80MoCrV42-12 (AISI M50, ASM 6490)	0.77-0.85	≤ 0.25	≤ 0.35	≤ 0.015	≤ 0.015	3.75-4.25	4-4.5	0.9-1.1		≤ 0.15	≤ 0.25		≤ 0.1	≤ 0.25	bal.
Maraging	350 maraging	≤ 0.03	≤ 0.1	≤ 0.1	≤ 0.01	≤ 0.01	≤ 0.5	4.6-5.2		0.05-0.15	18-19	11.5-12.5	1.3-1.6	≤ 0.5		bal.

Table 10.11 Composition of Nitriding steels (*EN10083, **EN10085, ***EN10084 standards)

Table 10.12 *Typical surface hardness and nitride depth.*

Steel	Surface Hardness Hv	Case Depth mm
EN3A, 070M20, 080M40, Mild Steels	350 - 500	0.100 – 1.00
EN19, 708M40, 4140, 4340, P20	550 – 850	0.025 – 1.00
EN40, 722M24, 31CrMoV9	750 – 1000	0.025 – 0.70
EN41, 905M39, Nitralloy	750 – 1100	0.025 – 0.70
H13, H11, D2, Hot Work Tool Steel	800 - 1200	0.025 – 0.50

10.7.1.3 Current Nitriding processes

Nitriding has now well established as a major thermochemical treatment which along with ferritic nitrocarburizing represents the dominant volume of industrial surface hardening technologies.[27]

The nitriding and nitrocarburizing can be carried out by several competing technologies each with certain advantages and disadvantages.

- Classical Gas nitriding
- Technically Refined "Controlled gas nitriding" using control probes for hydrogen and oxygen and microprocessor control, Nitrex, Seco Warwick, United Process Controls
- Plasma or Ion nitriding[32]
- Salt bath the main process being Tufftride[44]
- Fluidised bed[46,47 and 48]
- Ion implantation[22, 23]
- High temperature solution treatment[38]

Nitriding, has achieved proven performance in application of parts that operate under severe loading and temperature conditions. Examples include:

- Gears[3]
- Shafts and bearings[4]
- Crank shafts[5]
- Valve springs[6]
- Tools and dies for die casting and hot forging[7]
- Hydraulic radial pumps[8]
- High temperature solution treatment of austenitic stainless steel[19, 20 and 21]

10.7.1.4 White layer problems and benefits

The early use of nitriding involved single-stage nitriding processes that were carried out at maximum nitrogen potential and resulted in brittle cases with high porosity as well as networks of iron carbonitrides at the grain boundaries in the diffusion zone. (see Figure 10.116b) For most component application,

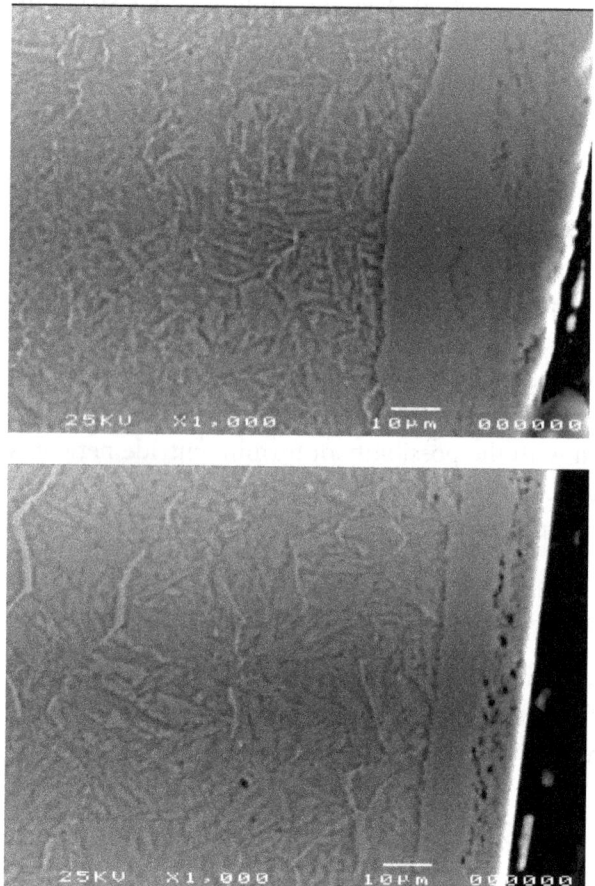

Figure 10.116b *Top = SEM image of single stage compound layer. Lower = SEM of controlled layer improved but shows coarse nitrides in the diffusion zone*

this poor microstructure required removal of the "white layer", typically by grinding

For many years, gas nitriding was used for gear applications in combination with a final grinding operation to remove the white layer. This grinding was necessary because the white layer that formed during gas nitriding had a brittle and hard layer which were frequently too thick and porous. If the white layer was excessively porous and brittle it could spall from the gear surface and cause scuffing. For applications such as gears, the compound layer could be detrimental to gear life. For aerospace applications the compound zone should be restricted depending upon nitriding class specified.[9] In some applications its presence should be avoided. In other applications it was found that the compound layer thickness and its integrity are of primary importance for component service performance.[37] It has been shown that nitrided steels with ε-phase in the compound layer acted more brittle than those containing Υ'.[38]

The depth of porosity in the outer part of the compound layer on unalloyed and low-alloy steels is of the order of 30–40% of the total compound layer depth. The degree of porosity increases with increased process time and increased nitrogen potential. For high-nitrogen activities, the porosity depth may exceed 50% of the total compound layer depth. High-alloy steels less prone to porosity formation, probably due to the lowering effect of alloying elements on nitrogen activity.[33] Pores are formed at discontinuities (grain boundaries, slag inclusions) because of the de-nitriding step:

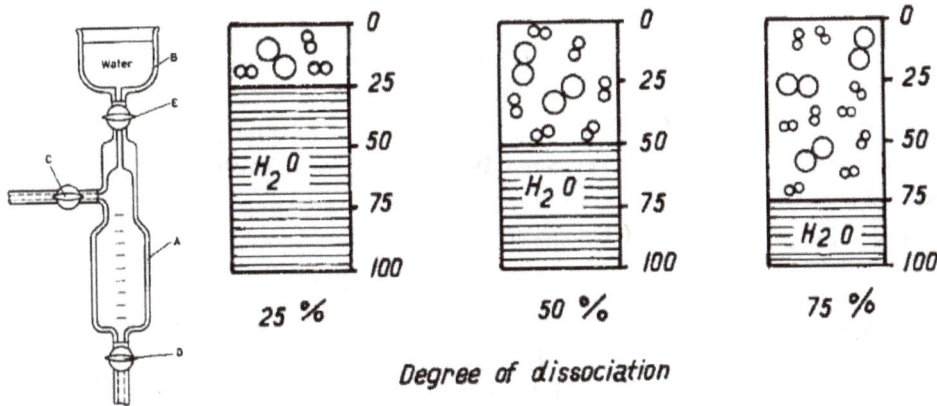

$$2\,N \rightarrow N_2$$

The equilibrium nitrogen gas pressure is high enough to create pores in the compound layer. The probability of pore formation increases quickly at high nitrogen activities.[34]

When plasma nitriding was introduced in the 1970s this process gave greater control of the compound layer and the integrity and surface roughness. Plasma nitriding became a popular method of nitriding. The plasma method was developed further to carry out plasma carburising.

The first nitriding that I carried out I used Thelning's "Steel and its heat treatment" for guidance. This book shows the early control of nitriding was to check that the ammonia gas in the nitriding furnace had decomposed (dissociated) according to the requirements of the process. The composition of the exit gas was determined by means of a dissociation pipette as shown in Figure 10.116c. The exhaust gas was allowed to enter into the pipette and then the pipette was filled with water. The ammonia then dissolved in the water and as shown in the RHS of the

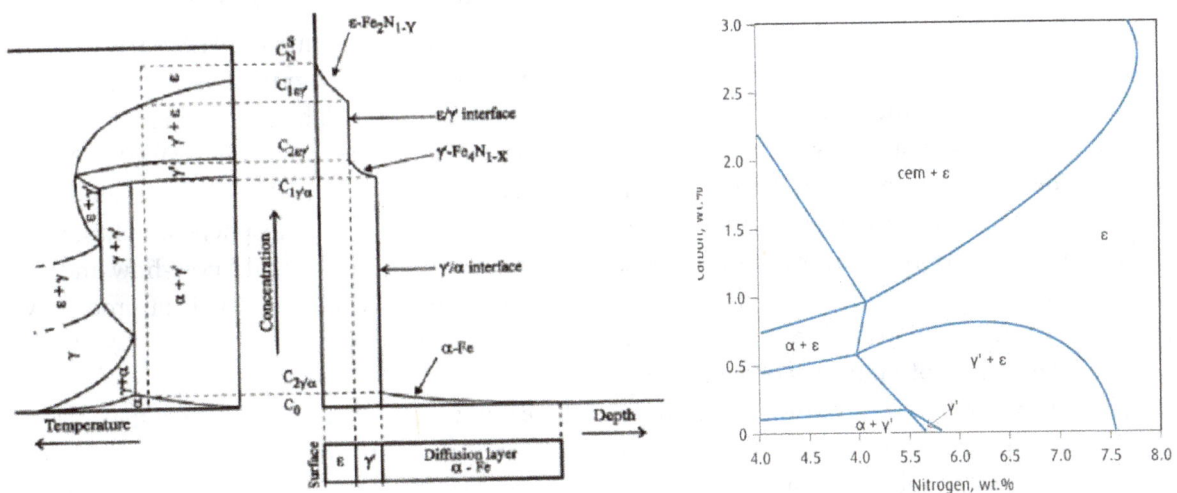

Figure 10.116c Top LHS = Dissociation pipette for nitriding. Top RHS = Height of water column in meter for different degrees of dissociation[10] Lower Relation between Fe-N phase diagram and concentration/depth for growth of a dual layer of ε-Fe$_2$N$_{1-x}$ and γ'-Fe$_4$N into a substrate α-F[2]. Lower RHS = Isothermal section at 580 °C of the ternary Fe-N-C phase diagram[33]

diagram the remaining gas, i.e. the dissociated ammonia, which consists of nitrogen and hydrogen, does not dissolve in the water. The dissociation rate can be converted to nitriding potential using appropriate equations.[2]

In the 1930s, E. Lehrer established for the iron-nitrogen (Fe-N) system, a diagram relating temperature and nitrogen potential. This allowed the accurate determination the phase boundaries in the iron-nitrogen (temperature-composition) system. The Lehrer diagram shows the solubility of nitrogen in ferrite as a function of the nitriding potential. This allowed the nitriding process and resultant case microstructure to be better understood. Unfortunately, the Lehrer only applies to pure iron and has limitations when applied to alloy steels.

Based on this knowledge methods to reduce the nitriding potential and the problematic white layer were developed and in 1943 a process was patented by Carl Floe using a two-stage nitriding process. The process created a nitrogen-rich layer with ammonia dissociation of 15 to 30%, and then the dissociation of ammonia was increased to 75 to 85% to allow the diffuse of the compound layer. This allowed the production of better white layer microstructures but it was not a universal solution.

The use of the Fe-N equilibrium diagram can be used to show the development of the white compound layer and diffusion layer as shown in Figure 10.116c

10.7.1.5 The introduction of nitriding potential probes and microprocessor control

Unfortunately, single-stage and two-stage gas nitriding processes, as originally developed, are still in use today and can produce erratic results.

The success of the plasma process stimulated more research and development in the gas nitriding field. In the 1990s, Leszek Małdziński modified the Lehrer diagram to incorporate iso-concentration curves. (Figure 10.117a).

This major contribution to the science of nitriding allowed the development of a nitriding potential-controlled process, which allowed control of the white layer structure, integrity and thickness during gas nitriding. This overcame a shortcoming in the Lehrer diagram, which predicted the phase structure of the nitrided layer but does not provide information about the concentration of nitrogen (in α, γ' and ε phases) as a function of temperature and nitriding potential.

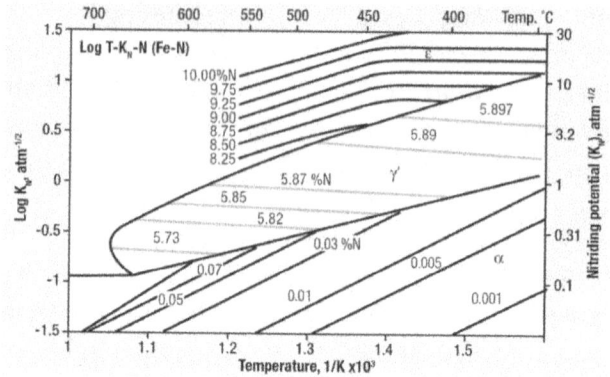

Figure 10. 117a Representation of gas nitriding parameters on the Lehrer diagram.

This diagram allowed the implementation of superior process control methods and reliable atmosphere measuring devices such as hydrogen and oxygen probes combined with microprocessor controls. These developments resulted in several systems:

- Nitrex[8]. The controlled gas nitriding Nitreg®, uses a mixed-gas atmosphere, consisting of ammonia and an additive gas. The process is not controlled by the dissociation rate but by the nitriding potential of the furnace atmosphere.[11]
- Seco Warwick, ZEROFLOW®GAS NITRIDING ZeroFlow is a fully automatic and single-component process (only ammonia), adjusted by the flow rate of NH_3 through the retort (including temporary complete flow cut-off), and at the same time the "equilibrium" process (in terms of metal/gas interaction), gives full control of the phase development of the nitride layer.[12 and 13]
- United Process Controls, H_2SmartTM hydrogen analyzer, an integrated system that measures dissociation with high accuracy in nitriding/nitrocarburizing process atmospheres.[14, 15 and 16]
- Linde and BOC, NITROFLEX trademark. The package covers the atmosphere supply solutions, flow control units, related know-how and specifications of the complete details of process cycles and of safety instructions.[33]

There is now a better understanding of the differences in plasma and gas nitriding. Plasma nitriding can be considered as a low nitriding-potential method, unable to produce a thick compound zone. Plasma nitriding has difficulty in entering small holes. In some applications this can be an advantage since the plasma process can be used for hardening sintered

Figure 10.117b LHS = Compound layer after controlled gas nitriding, porosity in the surface layer (Nitreg process). RHS = reduced compound layer after plasma nitriding, reduced porosity[17]

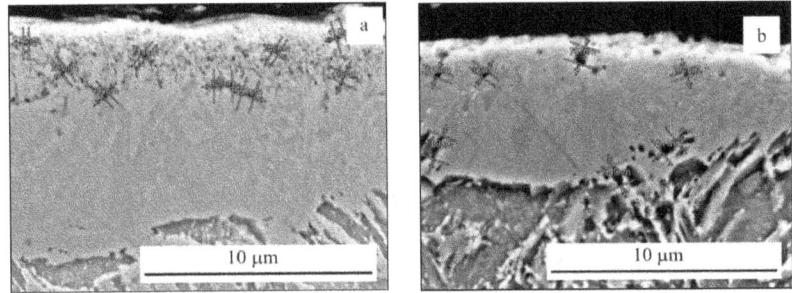

Phase	Nature	Content N	Properties
δFe	Solid solution of N in phase δFe	0 to ~ 0,9	
γFe	Solid solution of N in phase γFe	0 to 2,8	
αFe	Solid solution of N in phase αFe	0 to 0,1	Hard
γ'Fe₄N	alternating phase Fe₄N	5,7 to 5,9	
ε	alternating phase Fe₂.₃N	~ 4 to ~ 11	Very hard

Figure 10.117c Surface of grey cast iron samples after gas nitrocarburizing at 566°C (a and b), and plasma nitrocarburizing at 566°C (e and f). RHS = Content of Nitrogen in phases[30]

low-density iron parts. This feature of preventing active plasma coverage can also be used to prevent nitriding by using a simple mechanical cover.

Gas nitriding can produce a compound zone (white layer) of potentially any thickness by controlling the nitriding potential using hydrogen or oxygen sensors and modern automatic closed loop process control. Gas nitriding can be used for hardening parts having small holes, such as in diesel injector plungers.

A comparison of the roughness of cast iron after a plasma nitrocarburizing treatment and gas nitriding treatment found that the gas nitride parts had a significantly worse surface roughness. The explanation was ammonia molecules entering the space between the graphite and the ferrite matrix and then converting to nitrogen and the growth causing bulging of the surface. With plasma this cannot occur.[18]

10.7.1.6 Plasma nitriding history and process development[24]

Plasma nitriding method was invented and patented by B.Berghaus in 1932, (DPR 668639 1932 Process for surface treatment of metal surfaces). In the 1930s Berghaus was the first person to use the glow discharge phenomenon which he used to develop the plasma nitriding technology creating new opportunities for plasma surface alloying. During world war II the plasma nitride process was used for tank turret slewing gears and long gun barrels. Many physicists involved with plasma technology regarded the work of Berghaus and his invention as key event in the foundation of modern plasma surface engineering. By 1972 the technology was commercialised by Klockner and Klockner began to export plasma nitriding equipment around the world.

10.7.1.7 Plasma processing is based on the glow discharge

Plasma nitriding is a surface alloying technique based on the glow discharge. The glow discharge was first discovered applying a DC voltage across two electrodes in a diode. The observed characteristics can be seen in Figure 10.118a. The glow discharge can be separated into several regions. Plasma nitriding processes operate in the region of the abnormal glow discharge. In addition, there is a need to operate as near as possible to the peak to achieve the required current density on the parts. Unfortunately, this raises the possibility of an arc discharge which causes arc

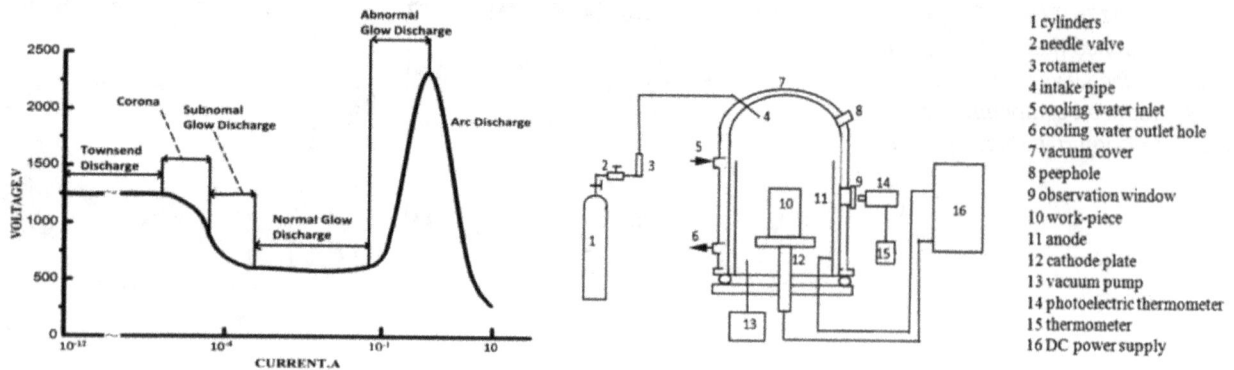

Figure 10.118a *LHS = Voltage-current characteristics of different types of high voltage discharge in argon gas. RHS = Schematic of plasma chamber and equipment*

damage to the workpiece. The plasma nitride equipment must be equipped with "Arc suppression" which senses the event and turns off the power to prevent damage to the parts

The plasma nitride process uses a vacuum chamber which is connected to the anode, and the cathode is connected to an insulated hearth on which the parts are place. A schematic of the plasma chamber and associated equipment is shown in Figure 10.118a RHS. A nitriding atmosphere consists of a mixture of nitrogen and hydrogen gases is used at a pressure that can range from 1 to 10 Mbar. The voltage is applied between the anode and the cathode in the range of 300 to 800 volts. The applied potential ionizes the gas mixture. The plasma generated by ionization surrounds the surface of the workpiece with a blue-violet colour. This can be seen in Figure 10.118b. The charged positive ions (nitrogen)accelerate towards and hit the negatively charged workpiece at a relatively high level of kinetic energy. Up to 90% of the kinetic energy is transferred as heat, which heats up the work piece to nitriding temperature. The electrical attraction between the nitrogen ions and the parts allows the nitrogen to reach the workpiece which penetrates by diffusion leading to the formation of a white layer followed by a diffusion layer.

In the plasma-nitriding process, nitrogen gas (N_2) can be used instead of ammonia because the gas is dissociated to form elemental nitrogen under the influence of the glow discharge. Therefore, the nitriding potential can be precisely controlled by the regulation of the N_2 content in the process gas. This control allows precise determination of the composition of the entire nitrided case, selection of a mono-phase layer of $Fe_{2-3}N$ or Fe_4N or total prevention of white-layer formation

Plasma nitriding exhibits a number of the following additional advantages.

- Metallurgical control of the process is much simpler than with conventional gas processes. Case formation can be either single-phase, dual-phase or diffusion zone only.
- Improved control of compound layer and diffusion depth
- The process is conducted at a lower temperature due to plasma activation.
- Plasma nitriding typically exhibits lower distortion.
- Reduced treatment times for plasma nitriding. There is no environmental hazard since no ammonia is involved – the process gas is a mixture of H_2 and N_2.

10.7.1.8 Research and development and production equipment

One of the great privileges that I had in the 1990s was to work with Dr Cristian Ruset in the building of plasma facilities to carry out plasma boriding R&D project at Metaltech. This resulted in a publication. (C Ruset, G Forster, R Scott, Plasma boronising with boron trifluoride gas, Heat Treatment of Metals 28(4):91-96 · January 2001). The facilities are shown in Figure 10.118b.

Plasma nitriding equipment consists of a vacuum vessel (retort), a pumping system to control the furnace pressure, a plasma generator and a cooling system. In cold-wall furnaces the plasma energy absorbed by the parts is the only form of heating, whereas hot-wall furnaces have external resistive heating elements which heat the parts to nitriding temperature. Therefore, since cold-wall furnaces require energy supplied by the plasma for heating and

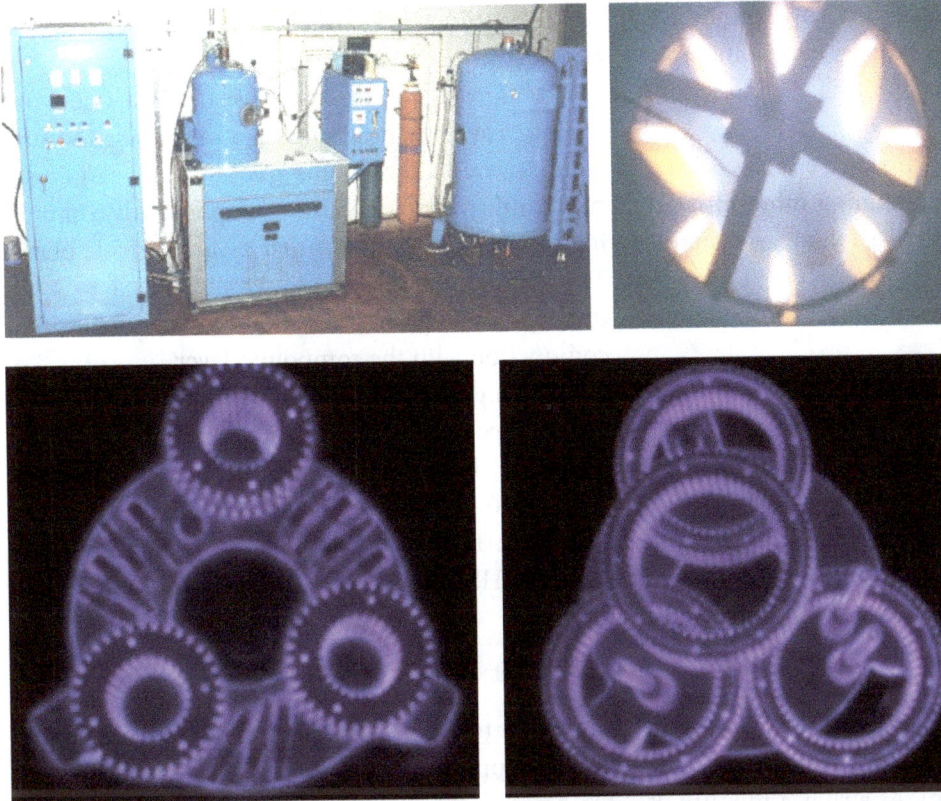

◀ *Figure 10.118b*
*Top LHS = R&D
plasma facilities, Top
RHS = Experimental
hollow cathode
heater, Lower Parts
surrounded by blue-
violet plasma. Visual
observation confirms
good distribution of the
plasma*

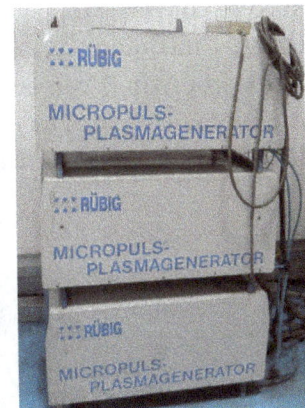

Figure 10.118c *Plasma nitriding production facilities (Rubig)*

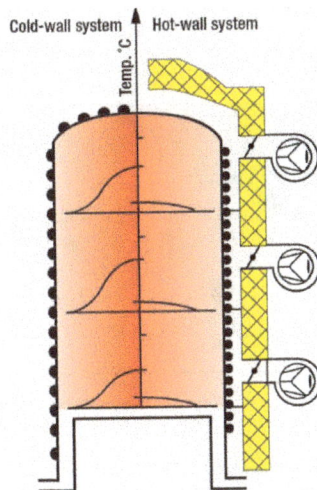

Figure 10.118d
*Problems and
differences between
a hot wall and a cold
wall furnace*[45]

maintaining temperature, the restrictions on the voltage and current control are limited, which gives less control of the nitriding process. Figure 10.118d shows an additional problem associated with cold wall systems, which is a temperature gradient from the wall to the centre. There is better control of temperature uniformity in a hot wall system

10.7.1.9 Process Capability

A key factor in the wear performance of nitride parts is known to be a good quality white layer

The compound layer which consists of iron nitrides and/or carbonitrides determines the hardness and toughness of the surface. The compound layer may consist of a variety of mixtures depending upon concentration distributions of nitrogen and carbon:

- a γ'-compound layer (γ'-nitride: Fe_4N)
- an ε-compound layer containing more nitrogen and/or carbon (ε-nitride: $Fe_{2-3}N$, ε-carbonitride: $Fe_{2-3}NC$)
- a mixed-phase compound layer (γ'-nitride and ε-nitride).

γ'-compound layers are tougher than ε-compound layers, but they grow more slowly; at 2 to 6 μm thickness they are significantly thinner than typical ε-compound layers (10 to 20 μm thick). The composition of the compound layer may also be modified by the presence of special nitrides and a more or less pronounced porous zone. (Figure 10.118f LHS)

The considerable range of control options for the plasma-based nitriding process enables the growth of the compound layer to be optimized for a specific application (Figure 10.118e). As a rule, a mixture of both nitride phases is obtained. Depending on the process control parameters, nearly single-phase γ' (plasma nitriding) or ε-compound layers (plasma nitrocarburizing) can be produced. Both types of compound layer are characterized by high resistance to wear. As the nitrogen content increases, hardness,

corrosion resistance and ceramic character increase and ductility decreases.

To improve the corrosion resistance, friction and sliding properties or aesthetics, nitrocarburized surfaces can be oxidized subsequently. This requires a sufficiently thick compound layer preferably comprising the more densely diffused ε-carbonitride phase. The magnetite layer (Fe_3O_4), would only 1 to 2 μm thick, provides good protection against corrosion in conjunction with the compound layer.

Alloying elements in ferrous materials such as chromium, vanadium, molybdenum and aluminum form special nitrides which affect the surface hardness (Figure 10.118f LHS) and the depth of the diffusion layer. Typical nitriding hardness depths (Figure 10.118f RHS) are between 10 μm and 0.8 mm

10.7.1.10 Process problems[45]

There are two power-system types used for plasma power supplies: continuous DC power and pulsed DC power combined with further innovations such as "active screen" systems where the plasma is developed on a screen surrounding the parts.[24] Most current systems use pulsed power supplies. Also, there are two types of hardware systems, which are cold-wall and hot-wall (Figure 10.118d). The potential problems include:

1. Part Overheating

Apart from obvious errors such as setting the furnace at an incorrect temperature or badly place thermocouples there are other potential sources of overheat. Overheating can occur as a result of the parts being too close together forming what is known as hollow-cathode effect. Figure 10.118b Top RHS shows an extreme version of a "hollow cathode" effect used as a

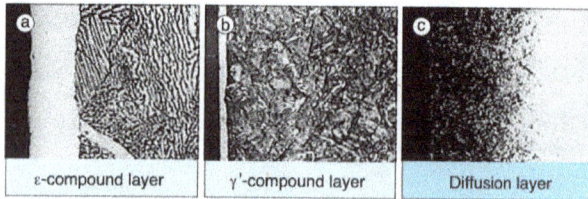

ε-compound layer   γ'-compound layer   Diffusion layer

Figure 10.118e *Surface layers after plasma nitriding; a)10 hours at 550°C b)16 hours at 530°C c)20 hours 510°C*

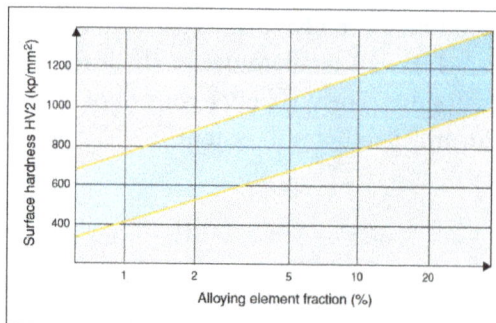

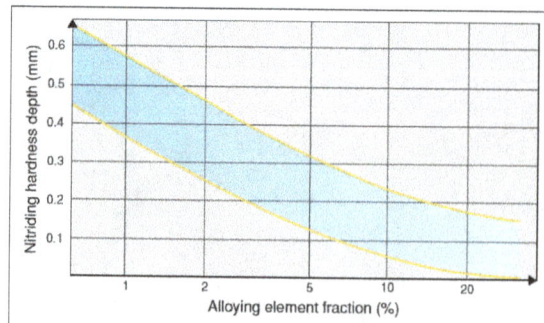

Figure 10.118f *LHS = Influence of the alloying elements chromium (Cr) and molybdenum (Mo) on surface hardness. RHS = Influence of the alloying elements chromium (Cr) and molybdenum (Mo) on the nitriding depth*

heater. The hollow-cathode effect can usually be seen during the process by looking through the process sight glass. After the process is completed overheating can be identified by a dark area on the part. The part will also have a lower surface hardness.

2. Lack of nitriding and soft areas

An important control feature on nitriding furnaces is the presence of visible ports. The photographs shown in Figure 10.118b were taken through a port. As shown in these photographs the uniformity of the plasma glow can be examined. Adjustments can be made by adjusting the chamber operating pressure.

3. Arc Discharge

The plasma nitride equipment must be equipped with "Arc suppression" which senses an arc the event and turns off the power to prevent damage to the parts. The cause can be dirty parts or too high a process voltage. A cure would be to reduce the plasma voltage until the arcing no longer occurs.

4. Part Chipping

This can occur at sharp corners on the parts. The most probable cause is "nitride networking." This means that the corner that has chipped has been oversaturated with nitrogen. This condition can apply also to other nitriding processes. The cause is that too much nitrogen is present in the corner due to the "corner effect." When oversaturation occurs, the nitrogen will precipitate out of solution during the cooling stage of the process and cause iron nitride compounds to form at the grain-boundary locations especially in the corners of a component. To prevent this problem, would require a reduction in the nitrogen potential or the provision of a larger radius on the corners of the part.

5. Low Surface Hardness

Assuming there are no major issues such as parts with insufficient metal removal and decarb being present or that the incorrect steel gradehas been used, there are some process issues. Low surface hardness can be caused by low nitrogen availability with inadequate nitrogen in solution with the steel to form sufficient nitrides at the surface. Another condition that can cause low surface hardness is that the steel itself is too low in nitride-forming elements. The remedy is to change either the steel that the component is manufactured from or increase the nitrogen and thus the nitride potential of the process gas.

6. Surface Flaking

The cause of this condition can be related to surface contaminant being carried into the process on the part surface. Check the manufacturing method for the type of coolant or cutting fluid used during the machining operation and then verify the method of precleaning prior to the nitride procedure.

Some surface contaminants can be removed by sputter cleaning at the commencement of the ion-nitride process using hydrogen as the sputter-clean gas. If hydrogen is not aggressive enough, a blended mixture of hydrogen/argon can be used. However, be cautious with the use of argon, because this gas has a high atomic weight and can cause surface etching. The maximum suggested volume of argon would be 10% with 90% hydrogen. Generally, the mixture ratio is 5% argon and 95% hydrogen.

10.7.1.11 Controlled gas nitriding

Controlled nitriding is accomplished by the use of a control system combined with sensors that determines the actual nitriding potential inside the furnace. The atmosphere composition can then be adjusted to a chosen set point value. This could be achieved either by manual analysis and flow adjustments or by automatic closed loop control, with automatic control being the best option. Section 10.6.1.5 listed several companies that provide automatic comtrol of the gas nitriding process.

The process control could involve altering the flow of ammonia to change the dissociation rate or the addition of either hydrogen or nitrogen to alter the nitriding potential. The addition of hydrogen would give the greater control. Typical sensors and control schematic are shown in Figure 10.118g

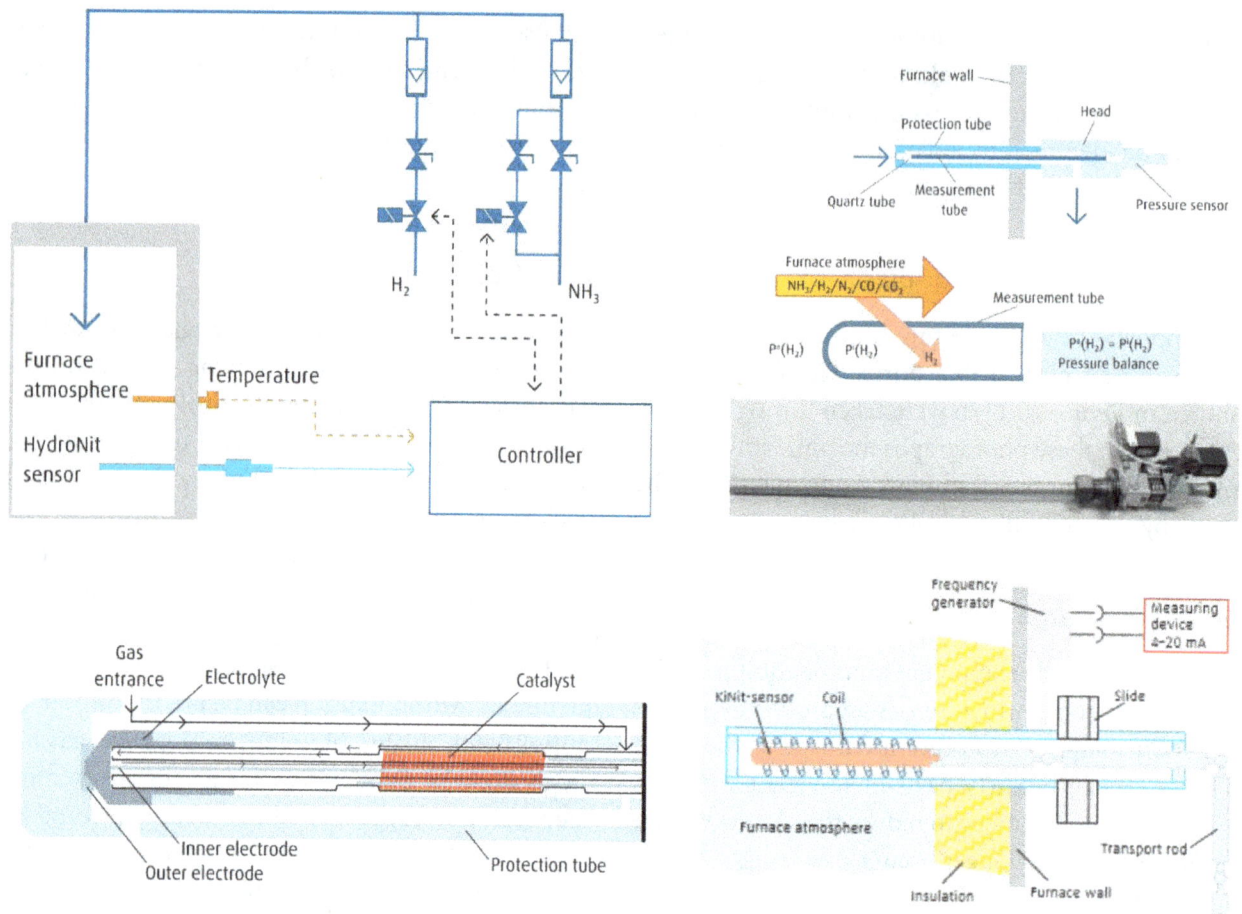

Figure 10.118g *Top LHS = Nitriding potential control (Courtesy Ipsen International GmbH). Top RHS = a) Principle and b) appearance of the hydrogen sensor [Courtesy of Ipsen International GmbH]. Lower LHS = Oxygen probe as a nitrogen sensor. Lower RHS = Schematic principle of the installation of an electromagnetic sensor for registration of compound layer growth*[33]

10.7.1.12 Salt bath nitrocarburizing

In the late 1950s, Imperial Chemical Industries (ICI), developed one of the first of salt bath nitrocarburizing processes. The process was successful with but there were issues with cleaning the solution off because it was not very water soluble. Since the salts were cyanide based this was a disincentive to use the process. In the late 1970s, Degussa acquired the ICI patents and subsequently development of a more environmentally friendly salt bath process now widely known as the Tufftride process. Other trade names for this process are Melonite and Tenifer.

The salt melt mainly consists of alkali cyanate and alkali carbonate. It is operated in a pot made from titanium, and the pot is fitted with an aeration device. The active constituent in the TF 1 bath is the alkali cyanate. During the nitrocarburizing process a reaction takes place between the surface of the components and the alkali cyanate, resulting in the formation of alkali carbonate. By adding specific amounts of the

non-toxic regenerator REG 1, the nitriding active constituents are again produced in the salt melt and the activity of the TF 1 bath is kept within very strict tolerances by appropriaye test procedures

Nitrocarburizing process operates in temperature range 460°C-560°. The tufftride process forms asimilar white layer and a similar diffusion zone to other nitriding processes. The compound layer consists of (gamma) γ- $Fe_4(C.N)$ and ε –$Fe_{2-3}(C,N)$ and is 2–30 microns thick depending upon the steel and process time. The surface layer of the sample after post oxidation consist of the γ- $Fe_4(C.N)$ and ε–$Fe_{2-3}(C,N)$ but contain additionally Fe_2O_4(ferric oxide). Corrosion resistance and tribological properties (friction and wear) are mainly determined by the compound layer. Below the compound layer the "diffusion zone", is typically 0.1–0.5 mm depending upon the treatment time and the steel type. The diffusion layer contains precipitated nitrides and carbides that increase the hardness of the diffusion zone. The diffusion zone mainly determines the load bearing capacity fatigue

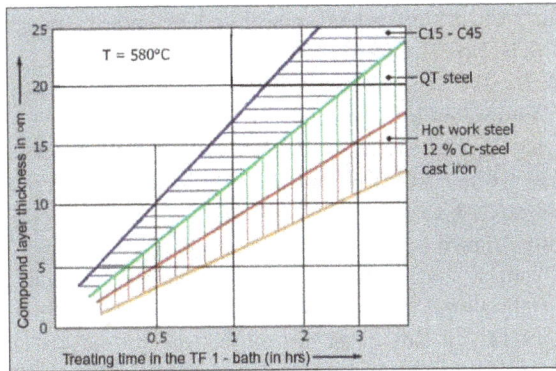

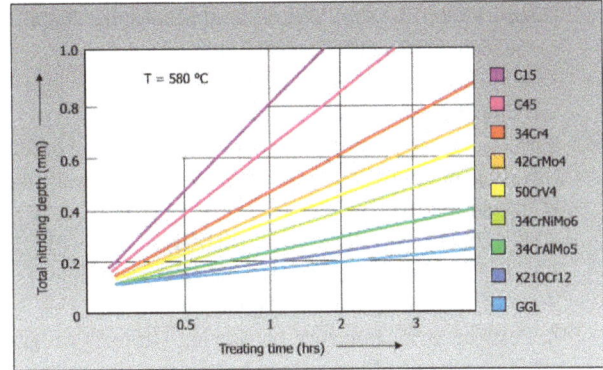

Figure 10.119a LHS = Compound layer thickness after Tufftride treatment RHS = Total nitriding depth after Tufftride treatment[44]

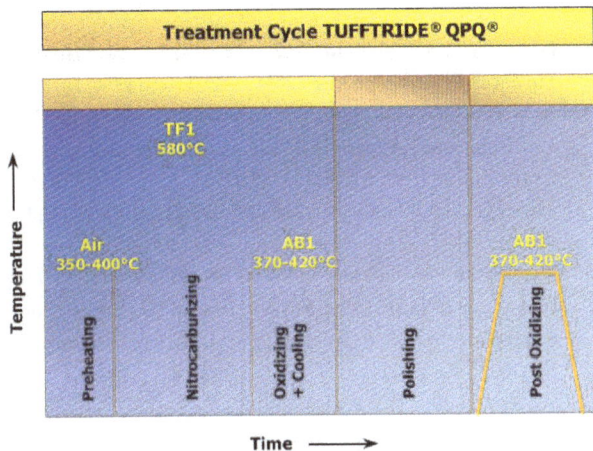

Figure 10.119b Tufftride QPQ[44]

strength. Tufftriding is widely used to increase the fatigue strength, hardness, and corrosion resistance and wear resistance of engineering parts such as dies, tools, automobile parts, and machine parts.

From the 1980s, a two-stage salt bath heat treatment was introduced which is a combination of Tuffftride and post-oxidation. This was partly driven by the success of oxidation after nitrocarburizing in the Lucas, Nitrotec process. Post-oxidation in nitrate/nitrite salt bath at 350°C for 10 min and then cooled in water to room temperature gave an improvement in corrosion resistance. About 1–2 microns of Fe_3O_4 layer is formed on top of the compound layer. The oxidation treatment gives the processed parts a black colour. The process can be combined with a polish treatment which results in a corrosion resistant black polished finish.

REFERENCES

1. David Pye, ASM International, 1 Jan 2003 – Technology & Engineering
2. Mei Yang, Nitriding – fundamentals, modelling and process optimization, Worcester Polytechnic institute, Doctorate thesis In Material Science and Engineering, April 2012, https://web.wpi.edu/Pubs/ETD/Available/etd-041912-093451/unrestricted/Gaseous_Nitriding.pdf
3. Hideki IMATAKA et al, Development of High-strength Nitriding Steel for Gear, NIPPON STEEL & SUMITOMO METAL TECHNICAL REPORT No. 116 September 2017, page 9. https://www.nipponsteel.com/tech/reports/nssmc/pdf/116-03.pdf
4. Daniel GIRODIN, Deep Nitrided 32CrMoV13 Steel for Aerospace Bearings Applications, NTN TECHNICAL REVIEW No.76 (2008), https://pdfs.semanticscholar.org/db8e/b5df17ddf4250691e7afcdb4bf48eb6438b6.pdf
5. https://www.nitrexheattreat.com/nitriding-applications-materials/applications/
6. Sumie SUDA et al, The Past and Future of High-strength Steel for Valve Springs, KOBELCO TECHNOLOGY REVIEW NO. 26 DEC. 2005, https://www.kobelco.co.jp/english/ktr/pdf/ktr_26/021-025.pdf
7. S.S. Akhta et al, Nitriding of Aluminum Extrusion Die: Effect of Die Geometry, Journal of Materials Engineering and Performance Volume 19(3) April 2010 page 401, https://www.researchgate.net/publication/225630472_Nitriding_of_Aluminum_Extrusion_Die_Effect_of_Die_Geometry
8. https://www.wallworkht.co.uk/content/salt_bath_nitrocarburise_tufftride/
9. Aerospace Materials Specification 2759-10A, Automated Gaseous Nitriding Controlled by Nitriding Potential, 2006
10. K. E. Thelning, "Steel and its heat treatment", Butterworths, 1984
11. Nitreg controlled nitriding, https://www.nitrex.com/news-information/downloads/
12. Gas Nitriding using a ZeroFlow® method, https://www.secowarwick.com/wp-content/uploads/2017/03/ATM-Zero-Flow6.pdf
13. Leszek Maldzinski et al, CONTROLLED NITRIDING USING A ZEROFLOW PROCESS, Technical contribution to the 70º Congresso Anual August 17th-21st, 2015, Rio de

Janeiro, RJ, Brazil, https://pdfs.semanticscholar.org/7ff8/a74 9445d7666ed1560f3b9b5818b20d3782a.pdf

14. Karl-Michael Winter, Gaseous nitriding in theory and in real life, United process controls white paper, 2009

15. Karl-Micheal Winter et al, State-of-the-art Controlled Nitriding and Nitrocarburizing, Umited process controls, White paper 2010

16. NITRIDING & NITROCABURIZING CONTROL SOLUTIONS https://group-upc.com/wp-content/uploads/Broc702_Nitriding_Solution_R4_EN.pdf

17. Z. Pokorny et al, Influence of alloying elements on gas nitriding process of high-stressed machine parts of weapons, Kovove Mater. 56 2018 97–103. file:///C:/Users/user/Downloads/56_2_POKORNY%20(1).pdf

18. Edward Rolinski et al, Influence of nitriding mechanisms on surface roughness of plasma and gas nitrided/nitrocarburized grey cast iron. HEAT TREATING PROGRESS • MARCH/APRIL 2007 page 39, https://www.asminternational.org/documents/10192/1910327/htp00702p039.pdf/d83390ae-3af0-450d-b405-ee38a2b7c483

19. Bernd Edenhofer et al, Solution Nitriding A Cost-Effective Case-Hardening Process for Stainless Steels, Ionic Technologies Inc 2008 https://www.industrialheating.com/articles/88049-a-cost-effective-case-hardening-process-for-stainless-steels

20. Heat Treatment of Stainless Steels Using the SolNit Process Ipsen USA, https://www.ipsenusa.com/processes-and-resources/processes/heat-treating-stainless-steels

21. C. M. GARZON et al, New high temperature gas nitriding cycle that enhances the wear resistance of duplex stainless steels, JOURNAL OF MATERIALS SCIENCE 3 9 (2004) 7101 – 7105 http://www.pmt.usp.br/academic/antschip/JMSL.pdf

22. Ortac ONMUS, MICROSTRUCTURAL AND MECHANICAL CHARACTERIZATION OF NITROGEN ION IMPLANTED AND PLASMA ION NITRIDED PLASTIC INJECTION MOULD STEEL, MSc Thesis, İzmir Institute of Technology İzmir, Turkey August, 2003

23. Orhan Oztqrk et al, Microstructural, mechanical, and corrosion characterization of nitrogen-implanted plastic injection mould steel, Surface & Coatings Technology 196 (2005) 333 – 340, http://openaccess.iyte.edu.tr:8080/xmlui/bitstream/handle/11147/1946/1946.pdf?sequence=1&isAllowed=y

24. J.J. JASINSKI et al, EFFECTS OF DIFFERENT NITRIDING METHODS ON NITRIDED LAYER STRUCTURE AND MORPHOLOGY, Arch. Metall. Mater. 63 (2018), 1, 337-345, http://www.imim.pl/files/archiwum/Vol1_2018/46.pdf

25. D Kovacs et al, The Effects of Screen Sizes on the Surface Properties of Tepered Steel Treated by Active Screen Plasma Nitriding, 7th International Conference on Advanced Materials and Structures, Materials Science and Engineering 416 (2018) page 1

26. D Kovacs et al, Effects of plasma nitriding on tempered steel, 11th Hungarian Conference on Materials Science, Materials Science and Engineering 426 (2018) page 1

27. Frank Czerwinski, Thermochemical Treatment of Metals, https://www.intechopen.com/books/heat-treatment-conventional-and-novel-applications/thermochemical-treatment-of-metals

28. H. C. Child, "Surface hardening of steel, Engineering, design Guides No. 37", Oxford Univ. Press., 1980

29. A. Fry, The theory and practice of nitrogen casehardening, J. Iron and Steel Inst., 125,1932,191-212

30. Nguyen Duong Nam et al, CONTROL GAS NITRIDING PROCESS: A REVIEW, Journal of Mechanical Engineering Research & Developments (JMERD) 42(1) (2019) 17-25, https://jmerd.org.my/Paper/Vol.42,No.1(2019)/17-25.pdf

31. Wendi Liu, The Effects of Contaminants on the Gas Nitriding of Nitralloy-135, Thesis WORCESTER POLYTECHNIC INSTITUTE Nov 2008. https://web.wpi.edu/Pubs/ETD/Available/etd-112408-113232/unrestricted/WendiThesis.pdf

32. Thomas auf dem Brink at al, Plasma-Assisted Surface Treatment Nitriding, nitrocarburizing and oxidation of steel, cast iron and sintered materials, verlag moderne industrie, ISBN-13: 978-3-937889-39-9, 2006

33. Furnace atmospheres no. 3. Gas nitriding and nitrocarburising. Linde 2018, https://www.linde-gaz.pl/pl/images/Furnace%20Atmospheres%20No.%203_tcm48-461074.pdf

34. B. SCHWARZ et al, Pore Formation Upon Nitriding Iron and Iron-Based Alloys: The Role of Alloying Elements and Grain Boundaries, METALLURGICAL AND MATERIALS TRANSACTIONS A VOLUME 45A, DECEMBER 2014 page 6173

35. Zinchenko V.M at al, NITROGEN POTENTIAL: current status of knowledge, 9th Seminar of the International Federation for Heat Treatment and Surface Engineering NITRIDING TECHNOLOGY. Theory and practice» 21-23 September 2003 Warsaw, Poland

36. I. Altinsoya et al, Gas Nitriding Behaviour of 34CrAlNi7 Nitriding Steel, Proceedings of the 3rd International Congress APMAS2013, April 2428, 2013, Antalya, Turkey, ACTA PHYSICA POLONICA A, Vol. 125 (2014) No 2 page 414 http://przyrbwn.icm.edu.pl/APP/PDF/125/a125z2p079.pdf

37. Hirotaka Kato, SLIDING WEAR OF NITRIDED STEELS, Thesis, Doctorate, Department of Materials Technology, Brunel University August 1993

38. Viktoria Westlund, Fundamental friction phenomena and applied studies on tribological surfaces, Licentiate Thesis Department of Engineering Sciences Uppsala University, September 2017 https://uu.diva-portal.org/smash/get/diva2:1192061/FULLTEXT01.pdf

39. ISABELLA FLODSTRÖM, Nitrocarburizing and High Temperature Nitriding of Steels in Bearing Applications, Master of Science Thesis, CHALMERS UNIVERSITY OF TECHNOLOGY Göteborg, Sweden, 2012 Report No. 76/2012

40. Yasushi Hiraoka et al, Effect of Compound Layer Thickness Composed of γ'-Fe4N on Rotated-Bending Fatigue Strength in Gas-Nitrided JIS-SCM435 Steel Materials Transactions, Vol. 58, No. 7 (2017) pp. 993 to 999

41. C. X. Li, J. Georges, Active Screen Plasma Nitriding of Austenitic Stainless Steel, University of Birmingham,

42. Proceedings of the Nitriding Symposium 3, November 14-15 Las Vagas, USA www.nitriding.info

43. Proceedings of the Nitriding Symposium 4 November 17-18 2016 Las Vagas USA

44. Dr. Joachim Boßlet, TUFFTRIDE®-/QPQ®-PROCESS, Durferrit GmbH.

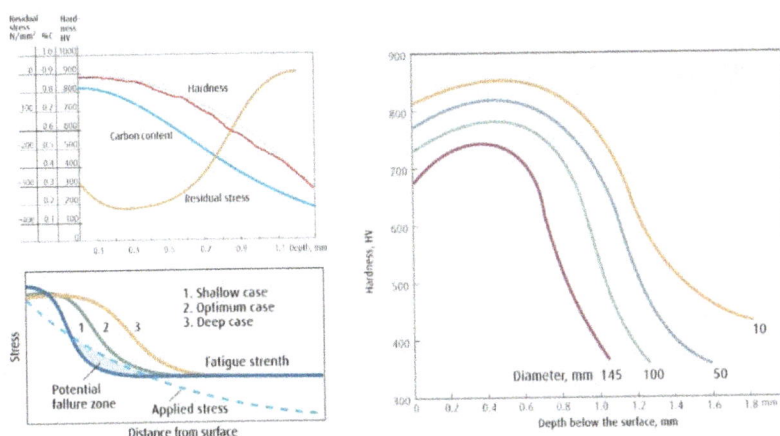

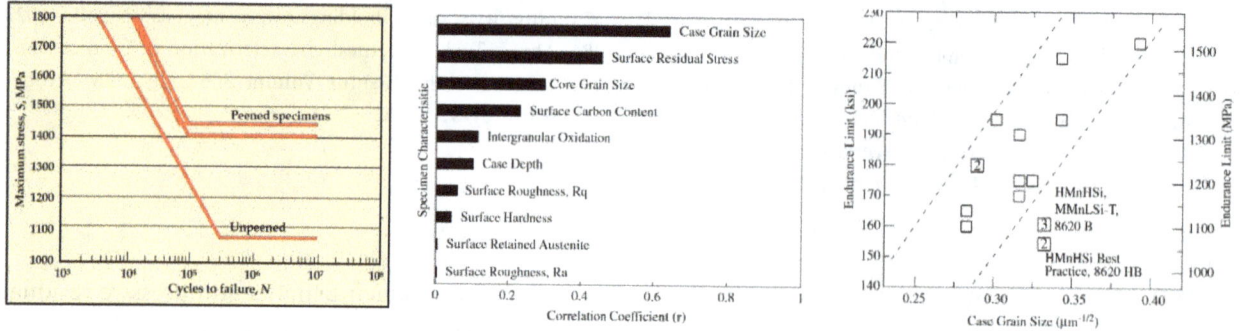

Figure 10.121 *LHS Effect of shot peening on endurance limit. RHS = Factors that affect the bending fatigue. The grain size of the steel is important. Steels with Ti and Nb grain refiners have been developed*[2, 3 and 4]

10.7.2.3 Carburising process control

An additional consideration is that the carburising process also needs to be optimised for each grade of steel. There is an optimum case carbon for each alloy content to minimise the retained austenite. No details were provided of the actual carburising process and it is assumed that all the samples have been given the same treatment. Table 10.13 shows some guidance on the actual case carbon target. More details of the aspects of process control can be found in several publications such as "Production Gas Carburising Control".[5]

The past use and testing of carburised parts has resulted in an understanding of the important aspects of the microstructure that result in "good" case quality in terms of delivering integrity and durability. The main imperfections are retained austenite, intergranular oxidation (see Figure 10.122), surface bainite/pearlite, subsurface bainite, potential carbide networks, ferrite in the case, and microcracks. Details of these imperfections can be found in a book by Parish.[6] One of the best standards for judging the acceptability of these imperfections has been prepared by Caterpillar Inc. This standard shows microstructure images of the imperfections and levels of severity that are satisfactory.[7]

Table 10.13 *Typical case carbon levels and expected surface hardness for various steel types*

Major alloy elements	Carbon concentration, %C	Surface hardness, HV
Ni (1-4%)	0.60-0.75	620-670
1.5%Cr, 2%Ni, 0.2%Mo	0.65-0.70	840
1.5%Mn, 0.004%B	0.85	815
Mn, Cr	0.70	840
Mo, Cr	1.0	940

In 1992 Ray Sieber in an SAE publication (922533)[8] noted that grinding with 0.15 mm metal removal out of the root resulting in 6 times the average bending fatigue life compared to no grinding. Though the ground surface finishes were the same as the carburized finishes, the influence of grinding appeared to be the removal of the oxide films and the intergranular oxidation of alloy elements in the carburizing furnace atmospheres.

The specified case depth requirements are determined by the surface load, wear conditions, and static and dynamic bending fatigue and torsional service stresses. These are established for gears and shafts. An additional problem after carburizing and quenching can be distortion. This can result in the part dimensions not meeting the specified tolerances and can require subsequent grinding or hard turning. When

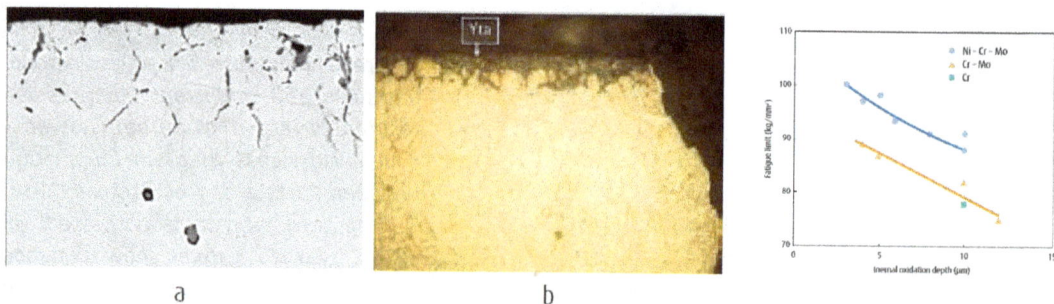

Figure 10.122 *Typical intergranular oxidation (IGO) imperfection which results in alloy depletion near the surface and the formation of sub-surface pearlite/bainite (centre photo). RHS = Effect of IGO on endurance.*

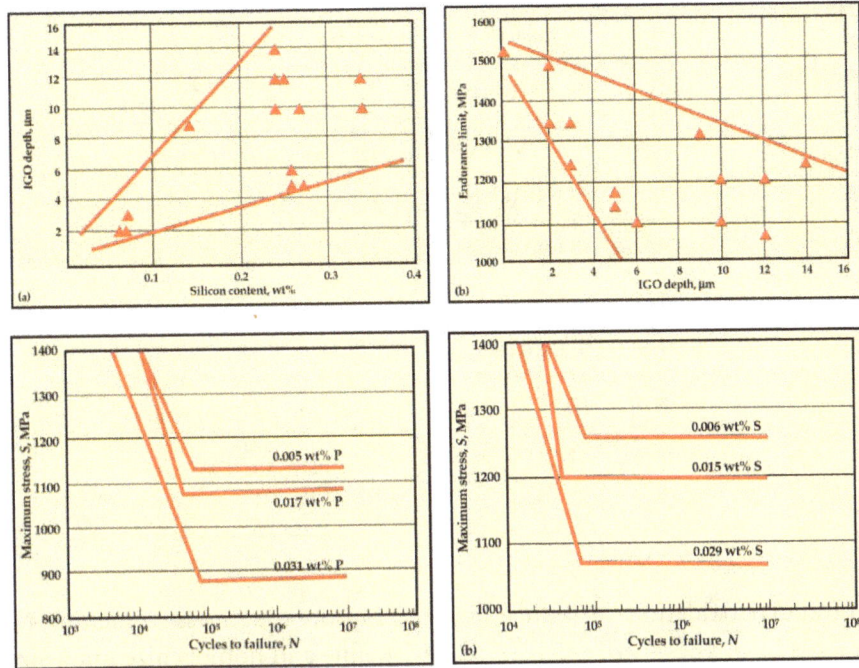

Figure 10.123 *Lower RHS = top – MnS inclusions improve machinability, but also act as stress concentrators to facilitate fatigue crack nucleation. Increasing the sulphur content of the steel from 0.006 wt% to 0.029 wt% caused a drop in endurance limit from 1260 MPa to 1070 MPa. Lower LHS – Effect of phosphorus content on the S-N curve of a carburized SAE 4320 steel. High phosphorus levels enhance segregation of the element to austenite grain boundaries and considerably lower bending fatigue endurance limits. Upper – Control of alloy chemistry can virtually eliminate intergranular oxidation (IGO) and subsequently increase fatigue performance. (a) Effect of silicon content on IGO depth in a carburized, modified SAE 4320 steel. Silicon levels were systematically reduced, resulting in corresponding reductions in IGO depth and stress concentration. (b) Influence of IGO depth on the steel's bending fatigue endurance limit[2]*

this is expected carburizing depth must therefore be high enough to attain the final specified case depth after grinding. Grinding allowance is typically of the order of 0.1-0.2 mm.

Figure 10.123 shows the effects of sulphur, phosphorous and IGO (intergranular oxidation) on the endurance.

In areas where products are to be made more compact with increased loads and there is a need for continuous improvement then consideration should be given to the steel development work carried out by FZG gear research centre at the university of Munich to improve gears[9]. Although the work was linked with gear steels the ideas and improvements apply to other case hardened parts.

10.7.2.4 Steel selection

This work gives a concise summary of the main requirements for gear steels:

- Chemical composition/hardenability
- Homogeneity/microscopic and macroscopic cleanness

- Mechanical properties (tensile strength, fatigue strength and toughness)
- Wear resistance, contact fatigue strength, bending strength and vibration resistance
- High and uniform dimensional stability.

The main carburising steels used for medium to large gears are shown in Figure 10.124.

The work examines the requirements for gears and then concludes the needed developments.

Improvements needed for medium to large gears:

- Improving hardenability
- Increasing core tensile strength and toughness
- Increasing fatigue strength in both the case and core
- Reducing quench distortion and thus negative acting tensile residual stresses
- Improving microstructural stability to withstand elevated temperatures during manufacturing and service.

The methods recommended to solve these issues were:

Figure 10.124 Choice of steels for various geographical locations

Steel Grade	Standard		Alloy Addition in wt %								Region
			C	Si	Mn	P	S	Cr	Mo	Ni	
20MnCr5	EN 10084 (1.7147)	min.	0.17	-	1.10	-	-	1.00	-	-	Western Europe
		max.	0.22	0.40	1.40	0.035	0.035	1.30			
18CrNiMo7-6	EN 10084 (1.6587)	min.	0.15	-	0.50	-	-	1.50	0.25	1.40	
		max.	0.21	0.40	0.90	0.025	0.035	1.80	0.35	1.70	
15CrNi6	EN 10084 (1.5919)	min.	0.14	-	0.40	-	-	1.40	-	1.40	France, Germany
		max.	0.19	0.40	0.60	0.035	0.035	1.70		1.70	
17NiCrMo6-5	EN 10084 (1.6566)	min.	0.14	-	0.60	-	-	0.80	0.15	1.20	Italy, France
		max.	0.20	0.40	0.90	0.025	0.035	1.10	0.25	1.50	
SAE 8620	SAE J1249	min.	0.18	0.15	0.70	-	-	0.40	0.15	0.40	North America
		max.	0.23	0.35	0.90	0.030	0.040	0.60	0.25	0.70	
SAE 9310	SAE J1249	min.	0.08	0.15	0.45	-	-	1.00	0.08	3.00	
		max.	0.13	0.35	0.65	0.025	0.040	1.40	0.15	3.50	
20CrMnTi	GB T 3077-1999	min.	0.17	0.17	0.80	-	-	1.00	0.00	-	China
		max.	0.23	0.37	1.10	0.035	0.035	1.30	0.15	0.30	
20CrMnMo	GB T 3077-1999	min.	0.17	0.17	0.90	-	-	1.10	0.20	-	
		max.	0.23	0.37	1.20	0.025	0.035	1.40	0.30	0.30	
SCM420	JIS	min.	0.18	0.15	0.60	-	-	0.90	0.15	-	Japan
		max.	0.23	0.35	0.85	0.030	0.030	1.20	0.30		

- Minimize intergranular oxidation → reduce Si, Mn and Cr
- Prevent MnS inclusions → reduce S, limit Mn
- Prevent TiN inclusions → control Ti/N wt % ratio close to three
- Improve hardenability → increase Mo

- Improve toughness → increase Ni and Mo
- Refine and homogenize grain size → balance Nb, Ti, Al and N microalloying addition
- Strengthen grain boundaries → reduce P and S, add Mo and Nb.

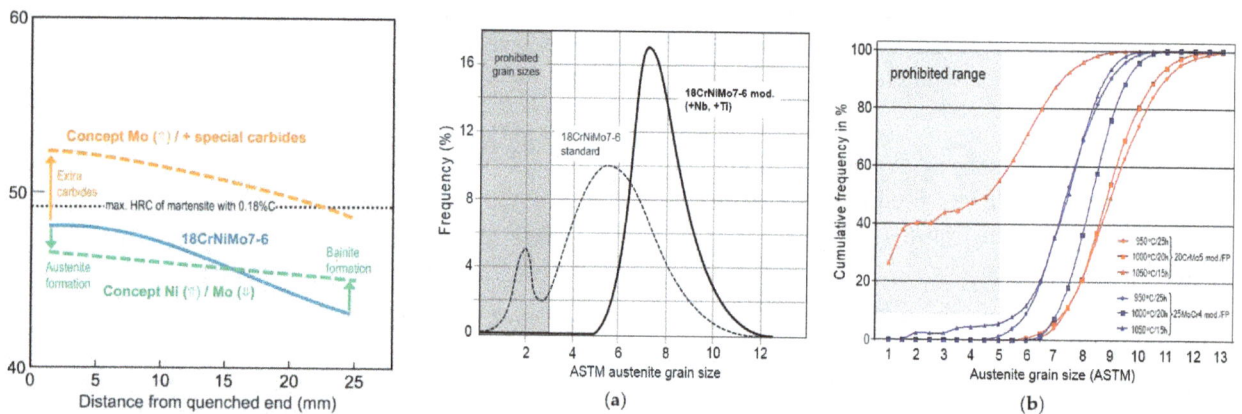

Figure 10.125 *LHS = Effect of alloy modifications on hardenability as compared to standard 18CrNiMo7-6 steel. RHS = Controlling grain size in carburizing steel: (a) conventional 18CrNiMo7-6 compared with modified steel (b) Nb treated steel grain size under various carburizing conditions*

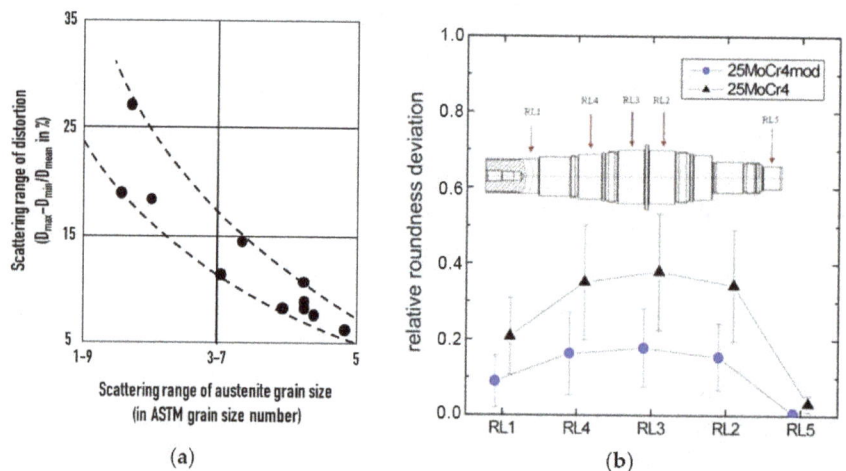

Figure 10.126 Reduced quench distortion (a) influence of mean prior austenite grain size scattering in steel 16MnCr5 (b) roundness deviation of a heat treated transmission shaft measured at five positions

The preferred steel development was to increase the molybdenum and grain refine with Titanium and niobium as shown in Figure 125. Figure 126 shows the reduced distortion achieved with grain size control.

During the selection of new grades of steels consideration must be given to a holistic approach. Frequently the established steel choice has given several years of satisfactory performance. The technical knowledge, trial work and test work needed to validate a replacement steel should not underestimated.

REFERENCES

1. KARL-ERIK THELNING, Steel and its Heat Treatment Bofors Handbook, Butterworths, ISBN 0 408 70934 0 1975.
2. John P. Wise, David K. Matlock, and George Krauss, Bending Fatigue of Carburised Steels, HEAT TREATING PROGRESS • AUGUST/SEPTEMBER 2001 Page 33.
3. David K. Matlock, Khaled A. Alogab, Mark D. Richards, John G. Speer, Surface Processing to Improve the Fatigue Resistance of Advanced Bar Steels for Automotive Applications, Materials Research, Vol. 8, No. 4, 453-459, 2005
4. J.P. Wise, G. Krauss, D.K. Matlock, Microstructure and Fatigue Resistance of Carburized Steels, 20th ASM Heat Treating Society Conference Proceedings, 9-12 October 2000, St. Louis, MO, ASM International, 2000.
5. C Dawes and D F Tranter, Production gas carburising control, Heat Treatment of Metals, 2004, 4, p99-108.
6. G. PARRISH and G. S. HARPER, PRODUCTION GAS CARBURISING, PERGAMON PRESS, 1985
7. Caterpillar Inc Microstructure Standards- Carburize hardening heat treatments 1E2532 2004
8. Ray Sieber, Bending Fatigue Performance of Carburized Gear Steels, SAE paper 920533 1992
9. Thomas Tobie, Frank Hippenstiel, and Hardy Mohrbacher, Optimizing Gear Performance by Alloy Modification of Carburizing Steels, Metals 2017, 7, 415 file:///C:/Users/user/Downloads/metals-07-00415-v3%20(2).pdf

10.8 STEEL CASTINGS

Steel Castings can be produced in a wide range of sizes and weights, from a few grams to hundreds of tonnes, with the limits being controlled by:

- The casting process used
- The required mechanical properties
- The required surface finish.

As discussed previously the UK Foundry Industry has gone through hard times and some of the surviving Foundries have diversified in terms of steel grades and size of cast product. For example, William Cook Foundry Group https://www.william-cook.co.uk/gb/ is one of the unique foundries that covers the full range of cast metal sizes having the capability ranging from investment casting to large mould sand casting.

With the selection of the correct composition a casting can have both high strength and high ductility and give satisfactory performance in the most onerous conditions. Castings can be produced with good surface finish and good weldability. Cast steel grades can be selected to operate at high temperatures and low temperatures, under high pressures and a range of environmental conditions. In addition, castings are free of the "directionality" effect or "mechanical fibre" that can influence the performance of wrought. One of the key selling points often quoted in favour of cast part is the isotropic mechanical properties that can be achieved in a cast metal.

Steel is perhaps the most versatile of the structural engineering materials, and therefore, there are more grades of steel available for casting than any other alloy. In simplistic terms any steel composition used for a wrought or forged steel can be duplicated in a cast metal. The large range of steel compositions and grades available has advantages in terms of customer choice but can cause technical constraints and therefore a degree of rationalisation of the many competing alloy grades is essential. This is the area where the experienced foundry can give suitable guidance on steel selection and cast shape. Therefore, it is good practice to involve a foundry at early stages of product development.

In addition, there are still on-going research and development projects worldwide working on higher strength better toughness and lower cost of ownership especially in the fields of automotive, oil and gas, wear resistance, corrosion resistance and military applications. These include:

- Elgin steel
- M-Steel
- Detroit Material
- AMS (austenitic manganese steel)
- Fully austenitic cast steels Fe-30%Mn-9%Al-0.9%C.

Figure 10.127 *Magma software output*

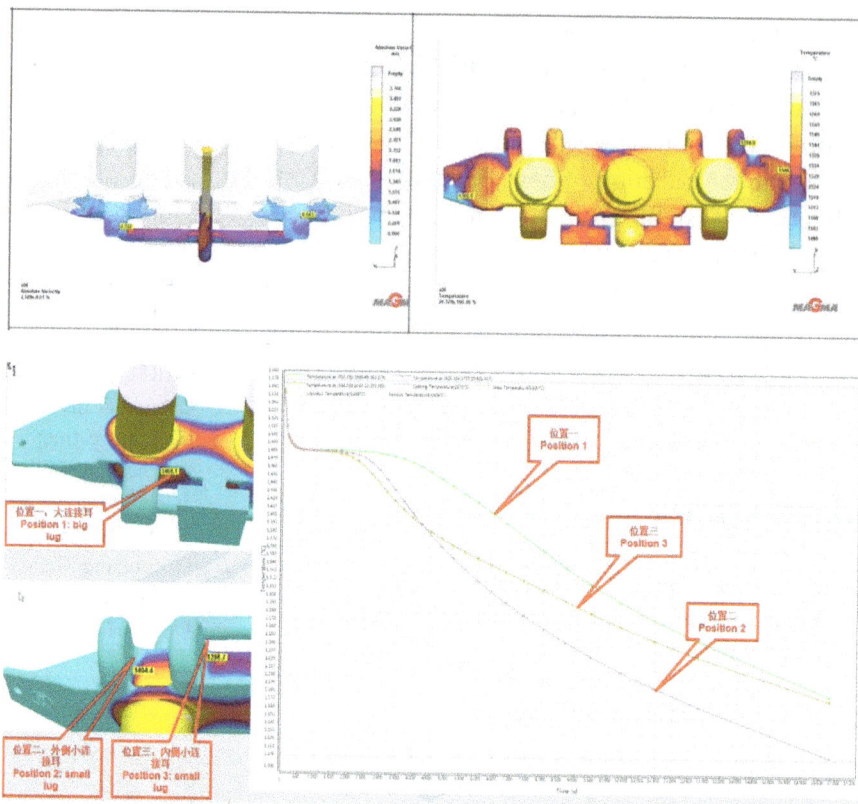

Figure 10.128 World Class Foundry, China Coal

The casting of steel presents two main technical problems:

- the high casting temperature, typically 1550–1600°C, which requires the selection of the best available refractories for liquid metal containment during processing
- and the volume shrinkage (6.0–10.0%) that occurs on freezing, which requires appropriate micro and macro feeding of liquid metal to compensate for the shrinkage.

Therefore, validated procedures have to be followed since steel castings are often used in critical situations where achieving the optimum mechanical properties to provide the durability is important and freedom from cast metal imperfections is essential to guarantee the cast metal integrity.

As in other industries the application of software has been a major assistance with the manufacture of quality castings. Figure x shows the solidification simulation of a large track shoe using Magma software. https://www.magmasoft.com/en/company/about-magma/

Specifications for steel castings UK Industry has a long history with BS 3100 but it is now withdrawn and BS EN 10293 is the current document.

There are a number of new European Standards that have been issued over the years giving physical and mechanical properties of grades of steel castings. Although BS 3100 Specification has been superseded and withdrawn, it is still used by many in the UK partly because it was so simple to understand. The numbering system for carbon and alloy steel consisted of one or two numbers followed by a number.

The first letter
A = carbon and carbon-manganese steel
B= low alloy steel

The second letter if appropriate
No letter= for general purpose use
L = for low temperature toughness
W = for wear resistance
M = for specified magnetic properties.

The number
This is arbitrary

Two numbers frequently remembered are A4 (G20Mn5 (1.6220) in BS EN 10293), the weldable carbon manganese steel used by the construction industry and BT2 (GS25NiCrMo8-5-4.), the high tensile low alloy steel.

It is convenient to classify steels into three groups:

- Carbon steels
- Low and medium alloy steels
- High alloy steels.

10.8.1 Carbon steels

These are steels in which carbon is the main alloying element; other elements such as manganese and

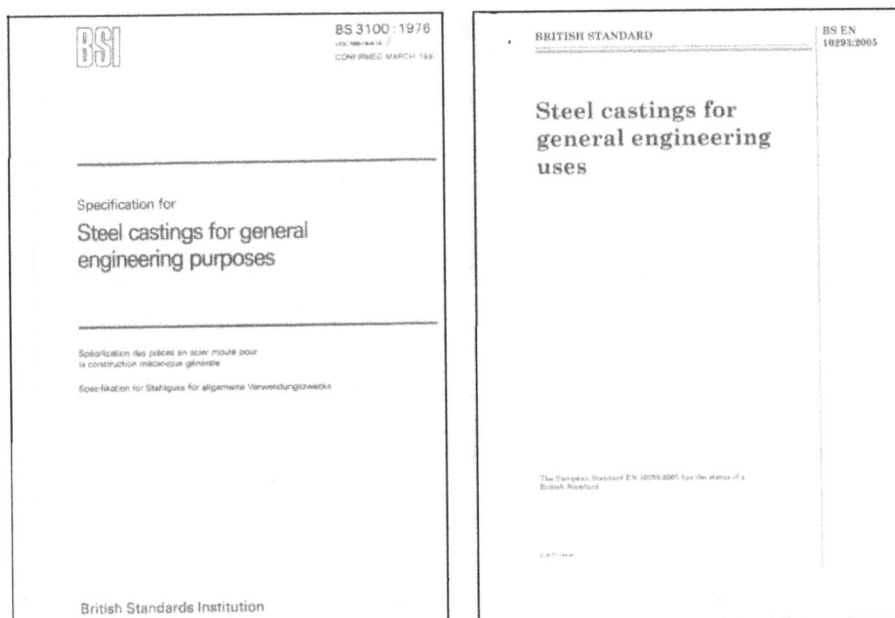

Figure 10.129 LHS = Now withdrawn. RHS = Current specification

Figure 10.130 *Carbon and carbon manganese steels*

BS 3100:1976 Steel Castings for General Engineering Purposes
Carbon and Carbon–Manganese Steel Castings

Chemical composition (%)	Carbon Steel Castings for General Purposes						1½% Manganese Steel Castings for General Purposes					
	A1* min	max	A2* min	max	A3* min	max	A4 min	max	A5 min	max	A6 min	max
C	–	0.25	–	0.35	–	0.45	0.18	0.25	0.25	0.33	0.25	0.33
Si	–	0.60	–	0.60	–	0.60	–	0.60	–	0.60	–	0.60
Mn	–	0.90(1)	–	1.0 (1)	–	1.0 (1)	1.2	1.6	1.2	1.6	1.2	1.6
P	–	0.06	–	0.06	–	0.06	–	0.05	–	0.05	–	0.05
S	–	0.06	–	0.06	–	0.06	–	0.05	–	0.05	–	0.05
Cr	–	0.25 (2)	–	–	–	–	–	–	–	–	–	–
Mo	–	0.15 (2)	–	–	–	–	–	–	–	–	–	–
Ni	–	0.40 (2)	–	–	–	–	–	–	–	–	–	–
Cu	–	0.30 (2)	–	–	–	–	–	–	–	–	–	–
Mechanical properties												
TS (N/mm²)	430	–	490	–	540	–	540	690	620(5)	770(5)	690(6)	850(6)
0.2%PS (N/mm²)	230	–	260	–	295	–	320	–	370(5)	–	495(6)	–
Elongation (%)	22	–	18	–	14	–	16	–	13(5)	–	13(6)	–
Angle of bend (°)	120 (3)	–	90 (3)	–	–	–	–	–	–	–	–	–
Radius of bend (t)#	1.5 (3)	–	1.5 (3)	–	–	–	–	–	–	–	–	–
Charpy impact (J)	25 (3)	–	20 (3)	–	18 (3)	–	30	–	25(5)	–	25(6)	–
Hardness (HB)	–	–	–	–	–	–	152	207	179(5)	229(5)	201(6)	255(6)

Notes: *If attached test samples are used, mechanical properties expected only where max. section thickness of casting is < 500 mm.
t = thickness of the test piece.
(1) For each 0.01%C below the max., an increase of 0.04%Mn will be permitted up to a max. of 1.10.
(2) Residual elements: the total shall not exceed 0.80%.
(3) Either a bend test or an impact test may be specified.
(4) Impact test mandatory only if specified by purchaser.
(5) Limiting section thickness, 100 mm.
(6) Limiting section thickness, 53 mm.
This table is intended only as a guide, refer to the British Standard for details.

silicon are present for the purposes of deoxidation. The effectiveness of heat treatment in increasing the strength of the steel depends on carbon content and carbon steels may be further subdivided into:

Low carbon steel <0.20%C, Medium carbon steel 0.20–0.50%C, High carbon steel >0.50%C.

Applications of carbon steel castings include Ship's structural castings, Railway rolling stock, Automotive castings, Mine and quarry equipment, Oil and petroleum equipment.

Carbon steels are most frequently used in the annealed, normalised or annealed and normalised conditions.

10.8.2 Low alloy and medium alloy steels

Castings in this class normally contain amounts of manganese or silicon greater than is needed for deoxidation (e.g. >1% Mn) or to which other elements have been specially added. Alloy steels will contain Ni, Cr or Mo as main alloying constituents and smaller amounts of V, Cu and possibly B. These elements are added primarily to provide the required hardenability so that the required properties can be obtained in different section thicknesses. Frequently used alloy steels are 1.5% Mn-Mo, 1.25% Cr-Mo, 0.75% Ni-Cr-Mo,

BS 3100:1976 Steel Castings for General Engineering Purposes
Low Alloy Steel Castings

Chemical composition (%)	C-½Mo B1 min	max	1¼Cr-Mo B2 min	max	2¼Cr-Mo B3 min	max	3%Cr-Mo B4 min	max	1¼%Cr-Mo BW4 min	max
C	–	0.20	–	0.20	–	0.18	–	0.25	0.55	0.65
Si	0.20	0.60	–	0.60	–	0.60	–	0.75	–	0.75
Mn	0.50	1.00	0.50	0.80	0.40	0.70	0.30	0.70	0.50	1.00
P	–	0.05	–	0.05	–	0.05	–	0.04	–	0.06
S	–	0.05	–	0.05	–	0.05	–	0.04	–	0.06
Cr	–	0.25	1.00	1.50	2.00	2.75	2.50	3.50	0.80	1.50
Mo	0.45	0.65	0.45	0.65	0.90	1.20	0.35	0.60	0.20	0.40
Ni	–	0.40	–	0.40	–	0.40	–	0.40	–	–
Cu	–	0.30	–	0.30	–	0.30	–	0.30	–	–
Mechanical properties										
TS (N/mm²)	460	–	480	–	540	–	620	–	–	–
0.2%PS (N/mm²)	260	–	280	–	325	–	370	–	–	–
Elongation (%)	18	–	17	–	17	–	13	–	–	–
Angle of bend	120	–	120	–	120	–	120	–	–	–
Radius of bend (t)	1.5t	–	1.5t	–	3t	–	3t	–	–	–
Charpy Impact (J)	20	–	30	–	25	–	26	–	–	–
Hardness (HB)	–	–	140	212	156	235	179	255	341	–

Figure 10.131 *Low alloy steel*

Notes:
t = thickness of the test piece
This table is intended only as a guide, refer to the British Standard for details.

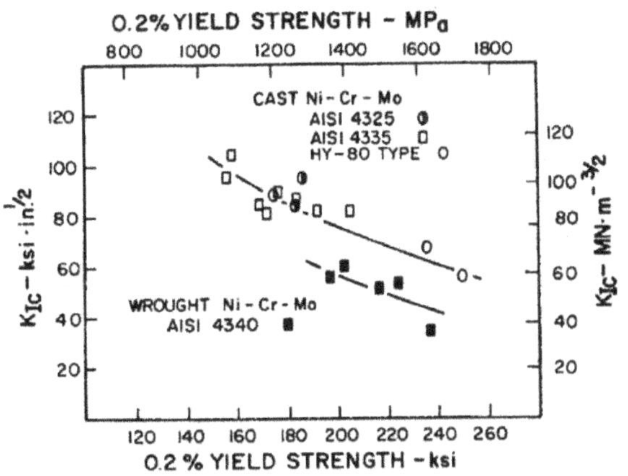

Figure 10.132 *Plain strain fracture toughness to 0.2 proof strength for Ni-Cr-Mo cast steels (ref1)*

1.25% Ni-Cr-Mo and 3% Cr-Mo (Fig 130). Such higher strength steels find use for gears, shaft pinions, impellers, rolls, rollers, crusher heads, mill liners, crankshafts, cement processing machinery, fittings and components in the oil industry. Steel castings can be surface hardened by flame or induction hardening or carburising or nitriding.

10.8.3 High alloy steels

These are used where corrosion resistance or heat resistance is required. The alloys may be subdivided into three main groups:

Fe-Cr, Fe-Cr-Ni and Fe-Ni-Cr.

The iron-chromium steels contain little or no nickel and 10–30%Cr. The Cr content influences their resistance to scaling, the 13%Cr steels having reasonable oxidation resistance up to 650°C while 28%Cr steel can be used up to around 1100°C. The 13%Cr steels are used for pumps and steam turbine equipment. 20–30%Cr steels are used in the chemical industry for pump parts and valves, also for furnace parts, grates, kiln parts etc. The iron-chromium-nickel alloys include the well-known 18/8 stainless steels to which additions of Nb or Ti may be used to stabilise the austenite on heating. Mo may also be added and Ni raised to 10% to allow the steel to be used at higher temperatures (up to 1100°C) and improve the corrosion resistance.

10.8.4 Duplex steels

The severe conditions experienced in off-shore drilling platforms and other marine applications require high strength corrosion resistant steels. Alloys having a duplex ferrite-austenite microstructure have been developed for this purpose. High hardness is produced by the high copper content (3%). The steel is heat treated by austenitising at around 1100°C when

Duplex steel GX2CrNiMoCN25-6-3-3

C	*Si*	*Mn*	*P*	*S*	*Cr*	*Mo*	*Ni*	*Cu*	*N*
0.03	1.0	1.50	0.035	0.025	24.5–26.5	2.5–3.5	5.0–7.0	2.75–3.5	0.12–0.22

Mechanical properties:	
UTS (N/mm²)	650–850
0.1%PS (N/mm²)	480
Elongation (%)	22
Impact (J)	50

Figure 10.133 *Duplex stainless*

Designation		C		Si	Mn		P	S	Cr		Mo		Ni		V		W
Name	Number	min.	max.	max.	min.	max.	max.	max.	min.	max.	min.	max.	min.	max.	min.	max.	max
GE200	1.0420	-	-	-	-	-	0,035	0,030	-	-	-	-	-	-	-	-	-
GS200	1.0449	-	0,18	0,60	-	1,20	0,030	0,025	-	-	-	-	-	-	-	-	-
GE240	1.0446	-	-	-	-	-	0,035	0,030	-	-	-	-	-	-	-	-	-
GS240	1.0455	-	0,23	0,60	-	1,20	0,030	0,025	-	-	-	-	-	-	-	-	-
GE300	1.0558	-	-	-	-	-	0,035	0,030	-	-	-	-	-	-	-	-	-
G17Mn5	1.1131	0,15	0,20	0,60	1,00	1,60	0,020	0,020 [a]	-	-	-	-	-	-	-	-	-
G20Mn5	1.6220	0,17	0,23	0,60	1,00	1,60	0,020	0,020 [a]	-	-	-	-	-	0,80	-	-	-
G24Mn6	1.1118	0,20	0,25	0,60	1,50	1,80	0,020	0,015	-	-	-	-	-	-	-	-	-
G28Mn6	1.1165	0,25	0,32	0,60	1,20	1,80	0,035	0,030	-	-	-	-	-	-	-	-	-
G20Mo5	1.5419	0,15	0,23	0,60	0,50	1,00	0,025	0,020 [a]	-	-	0,40	0,60	-	-	-	-	-
G10MnMoV6-3	1.5410	-	0,12	0,60	1,20	1,80	0,025	0,020	-	-	0,20	0,40	-	-	0,05	0,10	-
G15CrMoV6-9	1.7710	0,12	0,18	0,60	0,60	1,00	0,025	0,020 [a]	1,30	1,80	0,80	1,00	-	-	0,15	0,25	-
G17CrMo5-5	1.7357	0,15	0,20	0,60	0,50	1,00	0,025	0,020 [a]	1,00	1,50	0,45	0,65	-	-	-	-	-
G17CrMo9-10	1.7379	0,13	0,20	0,60	0,50	0,90	0,025	0,020 [a]	2,00	2,50	0,90	1,20	-	-	-	-	-
G26CrMo4	1.7221	0,22	0,29	0,60	0,50	0,80	0,025	0,020 [a]	0,80	1,20	0,15	0,30	-	-	-	-	-
G34CrMo4	1.7230	0,30	0,37	0,60	0,50	0,80	0,025	0,020 [a]	0,80	1,20	0,15	0,30	-	-	-	-	-
G42CrMo4	1.7231	0,38	0,45	0,60	0,60	1,00	0,025	0,020 [a]	0,80	1,20	0,15	0,30	-	-	-	-	-
G30CrMoV6-4	1.7725	0,27	0,34	0,60	0,60	1,00	0,025	0,020 [a]	1,30	1,70	0,30	0,60	-	-	0,05	0,15	-
G35CrNiMo6-6	1.6579	0,32	0,38	0,60	0,60	1,00	0,025	0,020 [a]	1,40	1,70	0,15	0,35	1,40	1,70	-	-	-
G9Ni14	1.5638	0,06	0,12	0,60	0,50	0,80	0,020	0,015	-	-	-	-	3,00	4,00	-	-	-
GX9Ni5	1.5681	0,06	0,12	0,60	0,50	0,80	0,020	0,020	-	-	-	-	4,50	5,50	-	-	-
G20NiMoCr4	1.6750	0,17	0,23	0,60	0,80	1,20	0,025	0,015 [a]	0,30	0,50	0,40	0,80	0,80	1,20	-	-	-
G32NiCrMo8-5-4	1.6570	0,28	0,35	0,60	0,60	1,00	0,020	0,015	1,00	1,40	0,30	0,50	1,80	2,10	-	-	-
G17NiCrMo13-6	1.6781	0,15	0,19	0,50	0,55	0,80	0,015	0,015	1,30	1,80	0,45	0,60	3,00	3,50	-	-	-
G30NiCrMo14	1.6771	0,27	0,33	0,60	0,60	1,00	0,030	0,020	0,80	1,20	0,30	0,60	3,00	4,00	-	-	-
GX3CrNi13-4	1.6982	-	0,05	1,00	-	1,00	0,035	0,015	12,00	13,50	-	0,70	3,50	5,00	-	-	-
GX4CrNi13-4	1.4317	-	0,06	1,00	-	1,00	0,035	0,025	12,00	13,50	-	0,70	3,50	5,00	-	-	-
GX4CrNi16-4	1.4421	-	0,06	0,80	-	1,00	0,035	0,020	15,50	17,50	-	0,70	4,00	5,50			-
GX4CrNiMo16-5-1	1.4405	-	0,06	0,80	-	1,00	0,035	0,025	15,00	17,00	0,70	1,50	4,00	6,00	-	-	-
GX23CrMoV12-1	1.4931	0,20	0,26	0,40	0,50	0,80	0,030	0,020	11,30	12,20	1,00	1,20	-	1,00	0,25	0,35	0,50

[a] For castings of ruling thickness < 28 mm, S ≤ 0,030 % (by mass) is permitted.

Figure 10.134 *Steel grades in BS EN 10293*

the copper dissolves. Oil quenching retains the copper in solution; subsequent ageing at around 480°C produces a ferrite-austenite structure with precipitated copper increasing strength and hardness.

The steel grades specified in BS EN 10293 are shown in Figure 10.134.

10.8.5 Eglin Steel

The steel was developed by the US Air Force and Ellwood National Forge as a lower cost ultrahigh strength steel for use in armaments, aerospace and commercial applications. The steel is referred to as ES1 in the wrought condition and CES in the as cast condition.

Figure 10.135 shows the composition of similar steels taken from a doctorate thesis[3] carried out to determine the continuous cooling transformation diagram developed for Eglin steel to be used as a guideline during processing. Dilatometry techniques performed on a Gleeble thermo mechanical simulator were combined with microhardness results and microstructural characterization to develop the diagram.

The steel was patented in 2009[4]. The following chemical analysis range and typical analysis were quoted. The patent document for Eglin steel quoted that the silicon is introduced to enhance toughness by making cementite precipitation sluggish at the tempering temperatures used for Eglin steel, vanadium and nickel are included to increase toughness,

Alloy	Composition (weight percent)								
	C	**Mn**	**Si**	**Ni**	**Cr**	**Mo**	**V**	**Co**	**W**
Eglin Steel	0.26	0.65	1.00	1.00	2.70	0.42	0.10	----	1.00
AerMet® 100	0.21-0.25	0.10 max	0.10 max	10.5-12.5	2.85-3.35	----	----	12.5-14.5	----
HP 9-4-20	0.17-0.23	0.20-0.40	0.20 max	8.50-9.50	0.65-0.85	0.90-1.10	0.06-0.12	4.25-4.75	----
HP 9-4-30	0.28-0.34	0.10-0.35	0.10 max	7.00-8.50	0.90-1.10	0.90-1.10	0.06-0.12	4.00-5.00	----
AF1410	0.13-0.17	0.10 max	0.10 max	9.50-10.50	1.80-2.20	0.90-1.10	----	13.5-14.5	----
4340	0.38-0.43	0.60-0.80	0.15-0.30	1.65-2.00	0.70-0.90	0.20-0.30	----	----	----
300M	0.38-0.46	0.60-0.90	1.45-1.80	1.65-2.00	0.70-0.95	0.30-0.65	0.05 min	----	----

Element	Weight %	
Carbon (C)	0.16-0.35%	
Manganese (Mn)	0.85% Maximum	1
Silicon (Si)	1.25% Maximum	
Chromium (Cr)	1.50-3.25%	
Nickel (Ni)	5.00% Maximum	
Molybdenum (Mo)	0.55% Maximum	
Tungsten (W)	0.70-3.25%	1
Vanadium (V)	0.05-0.30%	
Copper (Cu)	0.50% Maximum	
Phosphorous (P)	0.015% Maximum	
Sulfur (S)	0.012% Maximum	2
Calcium (Ca)	0.02% Maximum	
Nitrogen (N)	0.14% Maximum	
Aluminum (Al)	0.05% Maximum	
Iron (Fe)	Balance	2

Element	C	Mn	P	S	Ni	Cr	Al	W	Si	Mo	N	V	Cu	Ca
Weight %	.28	.74	.012	.003	1.03	2.75	.011	1.17	1.00	.36	.0073	.06	.10	.02

Figure 10.135 *Chemical composition of some high strength steels and Eglin steel*

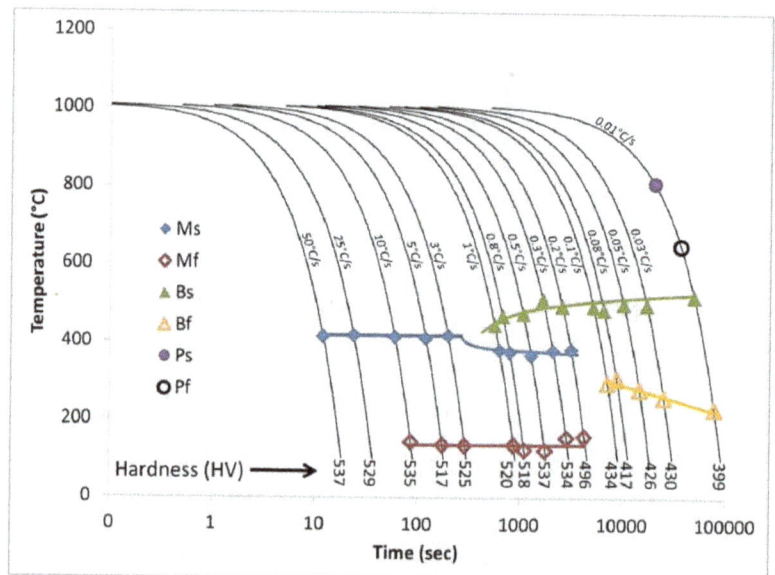

Figure 10.136 *CCT diagram from reference 3*

chromium is included to enhance strength and hardenability, molybdenum is included to enhance hardenability, and tungsten is included to enhance strength and wear resistance.

The thermal processing of the steel and the mechanical properties achieved were outlined in an AFS (American Foundry Society) published paper[5]. The publication criticised the cast metal quality.

The continuous cooling transformation diagram shown in Figure 10.136 provides guidance on the heat treatment, fusion welding and casting, which is useful for optimizing microstructure and properties. Understanding the variation in weld properties in the fusion zone and the heat affected zone will allow proper selection and manipulation of welding parameters.

10.8.6 M-Steel

In 2013 another high strength steel for military applications was patented (U.S. Patent No 8,414,713)[6]

that had been designed as a lower cost substitute to the Eglin steel. New high strength steel ("M-steel") has higher strength to Eglin steel at the same level of ductility and toughness. In addition, alloy cost was reduced compared with Eglin steel, while the cost of melting, hot forging, and heat treatment were comparable.

The table below shows the ASTM standard room temperature quasi-static tensile test results of the air melted quenched and low tempered M-steel and the air melted quenched and low tempered Eglin steel.

The reasons for the change was quoted as:
M-steel is superior to Eglin steel due to:

- Higher strength and the same ductility and toughness
- Reduction in total raw material cost by 50% or more by significantly lowering concentrations of W, Mo, and Cr
- Better formability at hot working
- Better machinability
- Better weldability.

	Density, lbs/in³	HRC	YS, ksi	UTS, ksi	El, %	RA, %	CVN, ft-lb
M-steel	0.285	52-53	220-235	275-285	11-14	44-48	26-32 at r.t. (12-16 at -40°F)
Eglin steel	0.285	48-49	195-205	250-255	11-14	44-48	26-32 at r.t. (11.5-15.5 at -40°F)

Figure 10.137 Comparison of the mechanical properties of Eglin and M-steel

Sheet Alloying Element % weight	New Steel	Eglin Steel*
C	0.3 to 0.45	0.16 to 0.35
Cr	1.0 to 3.0	1.50 to 3.25
Mo	0.1 to 0.55	0.55 max
W	0.1 to 2.0	0.7 to 3.25
Ni	0.1 to less than 3.0	5.00 max
Mn	0.1 to 1.0	0.95 max
Si	more than 0.3 to 1.0	1.25 max
Cu	0.1 to 0.6	0.50 max
V	more than 0.1 to 0.55	0.05 to 0.30
Ti or Nb	0.02 to 0.2	N.A.
Ca	N.A.	0.02 max
N	N.A.	0.14 max
Al	N.A.	0.05 max
Fe	remainder	remainder

Mechanical Properties	Eglin Steel*	New Steel Quenched Low Tempered	New Steel Quenched Refrigerated Low Tempered	New Steel Quenched High Tempered
Rockwell Hardness Scale C	46.6	52 - 54	54 - 56	48 - 50
Ultimate Tensile Strength, (ksi)	244.4	285 - 295	290 - 305	240 - 250
Yield Strength, (ksi)	201.9	215 - 220	225 - 235	225 - 235
Elongation, (%)	17.5	13 - 14	13 - 14	10 - 11
Reduction of Area, (%)	N.A.	48 - 50	47- 50	48 - 50
Charpy V-notch Impact Toughness Energy (ft-lb)	27.3	26 - 30	26-28	20 - 22

Figure 10.138 The chemical composition and mechanical comparison of M-steel and Eglin

Alloying Element, weight %	1st embodiment	2nd embodiment	3rd embodiment
C	0.18% to 0.55%	0.18% to 0.55%	0.18% to 0.55%
N	0.001% to 0.05%	0.001% to 0.05%	0.001% to 0.05%
Mn	2.0% max (excludes 0%)	2.0% max (excludes 0%)	2.0% max (excludes 0%)
Cu	1.5% max	1.5% max	1.5% max
Ni	1.0% max (excludes 0%)	-	-
Cr	3.0% max (excludes 0%)	3.0% max (excludes 0%)	3.0% max (excludes 0%)
Mo+W	0.20% max (excludes 0%)	-	-
V	0.30% max (excludes 0%)	0.30% max (excludes 0%)	-
Ti+Nb	0.1% max (excludes 0%)	0.1% max (excludes 0%)	0.1% max (excludes 0%)
Si	2.0% max (excludes 0%)	2.0% max (excludes 0%)	2.0% max (excludes 0%)
Al	0.0% to 0.2%	0.0% to 0.2%	0.0% to 0.2%
Ca	0.001% to 0.05%	0.001% to 0.05%	0.001% to 0.05%
P	0.035 max	0.035 max	0.035 max
S	0.04 max	0.04 max	0.04 max
Fe	remainder	remainder	remainder

	C, wt.%	HRC	UTS, ksi	YS, ksi	El, %	RA, %	CVN, ft-lb
Embodiment #1	0.20	46	220	175	15	50	32
	0.30	50	255	210	13	44	26
	0.40	53	282	238	11	38	20
	0.50	57	320	260	8	30	12
	0.55	59	340	270	7	22	8
Embodiment #2	0.20	46	220	170	14	44	26
	0.30	50	252	206	12	40	23
	0.40	53	280	230	10	36	18
	0.50	57	320	255	8	24	10
	0.55	59	340	265	7	20	6
Embodiment #3	0.20	45	215	160	10	38	22
	0.30	49	250	200	9	34	20
	0.40	52	270	220	8	30	14
	0.50	56	310	240	6	22	10
	0.55	58	330	260	5	14	4

Figure 10.139 *Inventor Gregory Vartanov (US 9869009) "New steel"*

The chemistry and mechanical properties of "M" steel shown on the patent document.

A recent patent (September 2018) by the same inventor Gregory Vartanov (US 9869009) describes a range of similar steels as "high strength low alloy steels" and the method of manufacture. The chemical analysis and mechanical properties are shown in Figure 10.139.

10.8.7 Detroit Materials

Detroit Materials has developed extremely strong, castable low-alloy steel. The DM steel offers the performance advantages of exotic-alloy steels (1300 MPa UTS, 16% elongation) with the ability to cast thin wall sections (3mm wall) and complex geometries at comparable cost per performance to ADI (Austempered Ductile Irons) and GJS ductile irons.

DM steels					
	YS (MPa)	UTS (MPa)	El (%)	Modulus (MPa)	HRC
DM 800-15	469	809	15	228,190	12
DM 1300-16	1049	1302	16	229,246	35
DM 1400-11	1124	1440	11	251,628	39
DM 1700-18	1270	1738	8	204,934	42

Figure 10.140 *The range of cast steel developed by Detroit Materials for "light weighting" trucks*

The company was recognized at a SAE World Congress where it was announced as winner of the 7th Annual Global Automotive Innovation Challenge for its ultra-high strength castable low-alloy steel. The market interest for affordable light-weighting of vehicle structures is high, due to both light-duty fuel economy mandates and forthcoming freight efficiency rules for medium- and heavy-duty trucks.

The background to the steel was identified as

Wayne Steel Tech, Detroit, MI, United States Wayne State University
A team of researchers led by Susil Putatunda, Ph.D., professor of chemical engineering in Wayne State University's College of Engineering, have been working to create advanced materials with high-yield strength, fracture toughness and ductility. Their efforts have led to the development of a portfolio of bainitic steels and austempered ductile irons exhibiting an excellent combination of mechanical properties currently only available in the form of highly alloyed and costly exotic steels. The information given in the announcement of a research award was:

The objective of this research award is to advance fundamental understanding of the synthesis of a high strength, high toughness bainitic steel with extremely fine scale microstructure consisting of ferrite and carbon stabilized austenite. This novel steel will be synthesized using the concept of austempered ductile cast iron technology and by applying a novel concept of adiabatic deformation together with a two-step austempering process in a medium carbon low alloy steel with high silicon content. This novel process will result in extremely fine bainitic ferrite and carbon stabilized austenite which will in turn provide simultaneous high yield strength, fracture toughness, impact strength and fatigue strength. Currently such combination of unique properties cannot be obtained in conventional low alloy steels.

One of the initial publications from Professor Putatunda was in 2002 regarding the austempering of a silicon manganese cast steel.[8] This development was similar to the nano-bainitic steels developed by Professor Bhadeshia in the late 1990s.

In 2010 there was a further publication which outlined the dual austempering principle and an initial low temperature followed by a second higher temperature[9].

C-0.42	Mn-0.40
Si-2.03	Ni-0.97
S-0.005	Mo-0.29
P-0.005	Cr-0.82
Cu-0.50	

Figure 10.141 *Steel composition chosen by Professor Susil Putatunda for austempering*

The chosen steel is similar to 300M, the high silicon type SAE 4140.

There are also several patents US 8657972,[10] 10066278.

REFERENCES

1. Steel casting handbook Supplement 5, General properties of steel castings, SFSA

2. John R. Brown ed, Foseco Ferrous Foundryman's Handbook, pub Butterworth-Heinemann 2000

3. Brett M. Leister, Mechanical Properties and Microstructural Evolution of Welded Eglin Steel, Doctorate Thesis, Lehigh University September 2014

4. US 7,537,727 B2 EGLIN STEEL—A LOW ALLOY HIGH STRENGTH COMPOSITION

5. T O Webb, D C Van Aken, S N Lekakh, Evaluating the Homogeneity in the Performance of Eglin Steel, A F S Proceedings Schaumbury IL USA, Paper 14-017, 2014

6. US 8,414,713 High Strength Military Steel.

7. US 9869009 High strength low alloy steels and the method of manufacture, September 2018

8. Susil K. Putatunda, AUSTEMPERING OF A SILICON MANGANESE CAST STEEL, Microsc. Microanal. 8 (Suppl. 2), 2002, https://www.cambridge.org/core/services/aop-cambridge-core/content/view/3EDA804143BAA442B8D65292001D963B/S1431927602105009a.pdf/mechanical_properties_of_an_austempered_high_carbon_high_silicon_and_high_manganese_steel.pdf

9. Susil K.Putatunda, Abhijit Deokar and Gowtham Bingi, MECHANICAL PROPERTIES OF A MEDIUM CARBON LOW ALLOY STEEL PROCESSED BY TWO STEP AUSTEMPERING PROCESS, Materials Science Forum, Vols. 638-642, pp 3453-3458 2010 https://www.researchgate.net/publication/250353104_Mechanical_Properties_of_a_Medium_Carbon_Low_Alloy_Steel_Processed_by_Two_Step_Austempering_Process

10. US 8657972, Development of a high strength high toughness steel, Susil K. Putatunda

LESCALLOY® 300M VAC-ARC®
HIGH STRENGTH ALLOY STEEL

Typical Composition	C	Mn	Si	Ni	Cr	Mo	V
	0.42	0.75	1.65	1.80	0.80	0.40	0.07

Austempering Temperature °C (°F)	Yield Strength (MPa)	Ultimate Tensile Strength (MPa)	% Elongation	Hardness (HRC)	Strain hardening exponent (n)	Fracture Toughness (MPa√m)
260 °C-288°C (500°F)- (550°F)	1645.8*	1961*	2.4*	52	---	87.4
316°C - 344°C (600°F)- (650°F)	1457.6	1655.0	6.1	47	0.05	98.9
344°C -372°C (650°F)- (700°F)	1363.5	1502.8	8.0	43	0.05	92.7
371°C -399°C (700°F)- (750°F)	1236.5	1381.5	10.6	40	0.05	73.9

*Estimated values

Case Study – Thin Wall Steel Casting
Detroit Materials delivers AHSS wrought steel capable of 3 mm wall with tensile strength over 1300MPa/190ksi and 16% elongation

Figure 10.142 *LHS = Mechanical properties achieved by two stage austempering medium carbon high silicon steel[8]. RHS = From Detroit Material brochure*

10.9 AUSTENITIC MANGANESE STEEL

When service conditions require a combination of very high toughness and wear resistance then the best choice of material would be Austenitic Manganese Steel (AMS) castings. For applications that require only wear resistance and high toughness is **not** needed then AMS would not be the best choice.

The high chromium white irons, Ni-hard type steels would provide better wear resistance. Published wear tests have shown that the high chromium white iron can give three times the life compared to AMS (Figure 10.143 RHS). However, the procedure used was a pin on disc with a specimen 10mm X10mm being pressed against a 120 grit SiC abrasive paper a force of 1kg which would probably not allow full work hardening of the AMS.[1] This reference shows the difference in toughness between well manufactured white iron and chromium containing AMS (2% chromium), based on un-notched Charpy values, the difference was surprisingly small. Note the sulphur and phosphorus levels for the AMS are low (see Figure 10.143).

Invented by Robert Hadfield in 1882, the austenitic manganese steel (AMS also known as Hadfield's steel) has been established as a key alloy steel for wear applications where there is a requirement for high toughness. AMS is used for railroad components such as crossings and rock-handling equipment because the austenitic matrix shows high toughness (High Charpy energy or Fracture Toughness) together with wear resistance. The high toughness is due to the austenitic microstructure which also gives the ability to work harden at the surface contact points up to 500 Brinell, if the working environment has sufficient energy of impact between the material being moved and the AMS cast surface.

10.9.1 Importance of contact impact energy

A recent evaluation of the energy required to harden AMS has been published. This also summarised an approximate calculation for the energy input into the casting for hammer mills and ball mills.[30]

The main conclusion of this work was that the effective hardening of the AMS requires an impact of 80J cm^{-2}. Based on the data shown in Figure10.144 AMS would be only suitable for the larger ball mills.

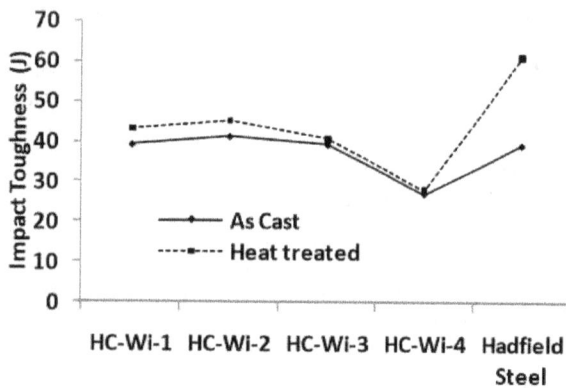

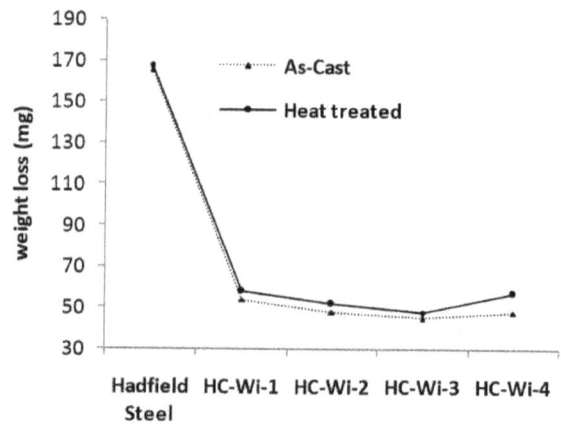

Material	C	Mn	Si	Cr	P	S	Mo	Ni	Al	Fe
HC-Wi-1	1.55	0.63	0.63	24.87	0.03	0.03	0.02	0.01	0.01	Balance
HC-Wi-2	2.22	0.63	0.62	24.60	0.03	0.03	0.01	0.01	0.01	Balance
HC-Wi-3	2.73	0.59	0.60	24.20	0.03	0.03	0.01	0.015	0.02	Balance
HC-Wi-4	3.26	0.56	0.60	23.60	0.03	0.03	0.01	0.01	0.02	Balance
Hadfield Steel	1.30	14.52	0.61	1.85	0.01	0.00	-	-	0.15	Balance

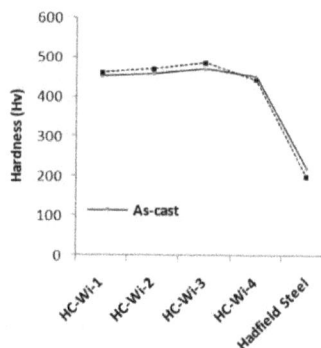

Figure 10.143 Top LHS = Toughness comparison. RHS = Wear test results. Bottom LHS = Chemical compositions. RHS = Hardness.

To calculate the impact energy, some assumptions were made, for example, assuming the hammer to be a rigid body. Table 1 lists the related parameters and the calculated impact energies of three types of hammers. Take the PC-S 0808 hammer mill as an example, the calculation process is as follows:

Angular velocity of the hammer:

$$\omega = 2\pi \cdot n = 2\pi \cdot \frac{980}{60} = 102.6 \text{ rad} \cdot \text{s}^{-1}$$

where n is rotational speed.

Line speed of the hammer:

$$v = \omega \cdot r = 102.6 \times \frac{0.8}{2} = 41 \text{ m} \cdot \text{s}^{-1}$$

where r is gyration radius.

The mass of the limestone:

$$m = \rho \cdot \frac{4}{3}\pi \cdot r^3 = 2 \times 10^3 \times \frac{4}{3} \times 3.14 \times (\frac{120}{2} \times 10^{-3})^3 = 1.8 \text{ kg}$$

where ρ and r are density and granularity of limestone, assuming the limestone is spherical, and the maximal diameter is 120 mm.

Impact energy endured by the hammer:

$$E' = \frac{E}{s} = \frac{0.5 \ mv^2}{s} = \frac{0.5 \times 1.8 \times 41^2}{10 \,(\text{to } 20)} = \frac{1514}{10 \,(\text{to } 20)} = 75 \text{ to } 151 \text{ J} \cdot \text{cm}^{-2}$$

where E is the maximal impact energy, s is the impact contact area. According to the shape and size of the hammer and the limestone, we assumed it was approximately 10 to 20 cm^2 (Table 1).

Figure 10.144 *Calculation of the impact energy for hammer mills*[30]

Model	Rotation speed (r·min⁻¹)	Rotation diameter (m)	Calculated Max. impact energy (J)	Calculated impact energy (J·cm⁻²) [Impact area (cm²)]
PC-S 0808	980	0.8	⩽1514	⩽75–151 [10–20]
PCK-M 1430	735	1.43	⩽1449	⩽120–241 [6–12]
MB70/90	209	2.7		>>200

Model	Diameter of grinding ball (mm)	Height of ball sinking (m)	Calculated impact energy (J·cm⁻²)
Φ3.2	120	2.13	147
Φ2.5	100	1.67	67
Φ2.0	100	1.33	52
Φ1.5	80	1.0	21

Figure 10.145 *Calculated energy of impact available to harden the AMS for hammer and ball mills*[30]

	C	Si	Mn	Cr	Mo	Fe
Sample A	0.8	0.28	12.8	1.78	0.48	Bal.
Sample B	1.23	0.505	14.33	2.25	0.863	Bal.

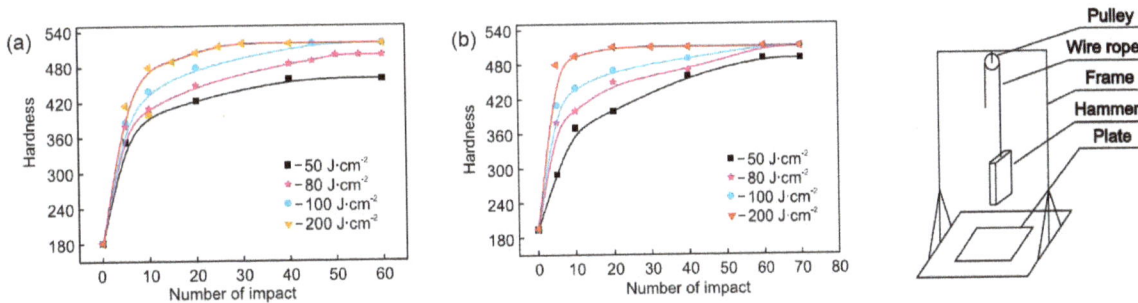

Figure 10.146 *Hardness increase due to impact for two AMS grades. RHS = test rig*[30]

This is in line with the early experience of Climax Molydbendum where they developed lean AMS and developed white irons for ball mills.

10.9.2 Specifications

Commercial alloys usually contain between 1.0 and 1.4% carbon and 10 to 14% manganese as listed in ASTM A128 and other national specifications. The compositions of the grades listed in ASTM A128 are shown in Table 10.13.

Table 10.13 *The compositions of the grades listed in ASTM A128*

ASTM A128 Grade	Composition %						
	C	Mn	Cr	Mo	Ni	Si (max)	P (max)
A	1.05-1.35	11.0min	-	-	-	1.00	0.07
B-1	0.9-1.05	11.5-14	-	-	-	1.00	0.07
B-2	1.05-1.2	11.5-14	-	-	-	1.00	0.07
B-3	1.12-1.28	11.5-14	-	-	-	1.00	0.07
B-4	1.2-1.35	11.5-14	-	-	-	1.00	0.07
C	1.05-1.35	11.5-14	1.5-2.5	-	-	1.00	0.07
D	0.7-1.3	11.5-14	-	-	3.0-4.0	1.00	0.07
E-1	0.7-1.3	11.5-14	-	0.9-1.2	-	1.00	0.07
E-2	1.05-1.45	11.5-14	-	1.8-2.1	-	1.00	0.07
F	1.05-1.35	11.5-14	-	0.9-1.2	-	1.00	0.07

10.9.3 Control of quench severity and avoidance of carbide on grain boundaries

The compositions in Table 10.1 do not allow any austenite transformation when the steels are water quenched from above the Acm and results in a fully austenitic microstructure. However, if the quenching rate is not fast enough carbides can form along grain boundaries. This has a known effect on toughness and has to be avoided. Control of quench severity is important.

After the appropriate heat treatment which consists of solution treatment and water-quenching AMS consists of an austenitic microstructure. The austenitic microstructure developed has low hardness (200 to 230BHN) and allows AMS to achieve good wear resistance, but it must be used where the loads are sufficient to produce work-hardening.

10.9.4 Small Grain Size is essential to developed work hardening and wear resistance

The development of the appropriate work hardening requires a fine grain size. Some early work from a Russian publication established the effect of grain size.

Coarse grained steels had low strength, tensile ductility and toughness. As the grain size was reduced all

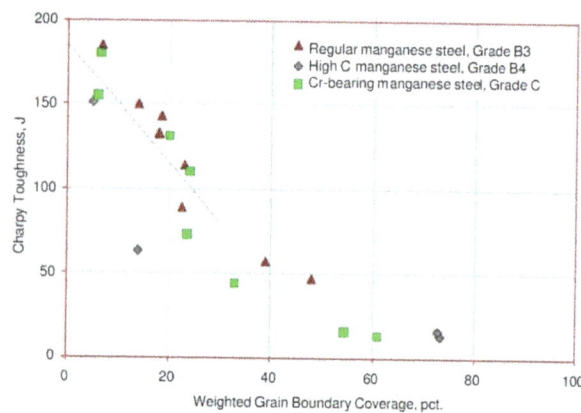

Weighted g.b. Coverage	Toughness	CVN, J
< 20 %	Excellent	>100
20 - 40%	Good	>50
> 50%	Poor	< 20

Impact toughness decreased at a rate of 3.5 J per percent weighted grain boundary coverage, from a nominal value of 180-200 J.

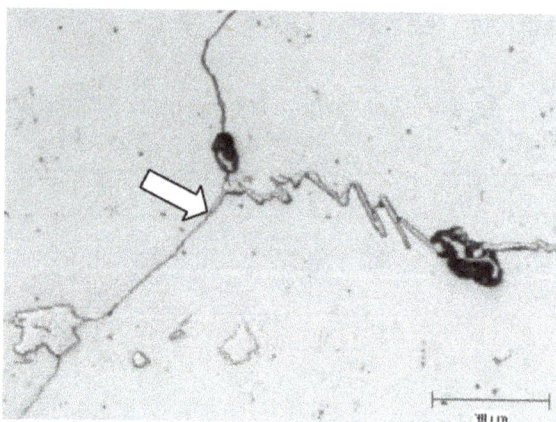

Figure 10.147 *Top = Effect of grain boundary carbide coverage on the toughness[26] Bottom = Thick carbides have a double phase boundary that distinguishes them from the thin carbides. Arrow shows transition from a thin to thick carbide*

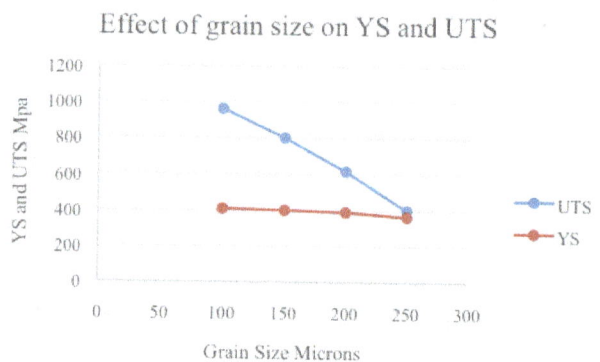

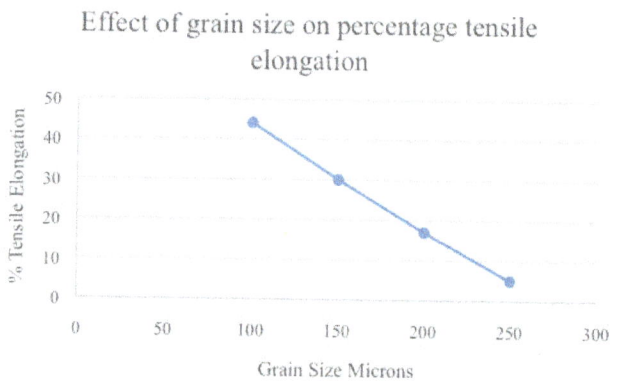

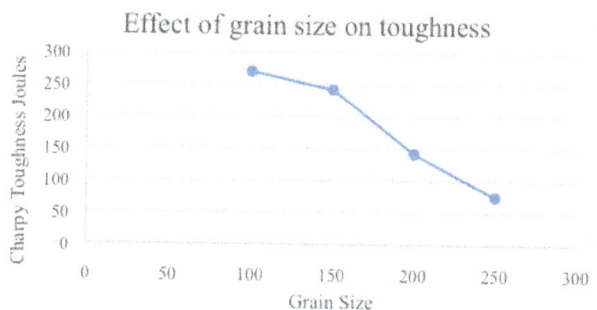

Figure 10.148 *Effect of grain size on the mechanical properties of AMS (1.28%C, 12.2%Mn, 0.77%Si, 0.05%P, 0.008%S)[19]*

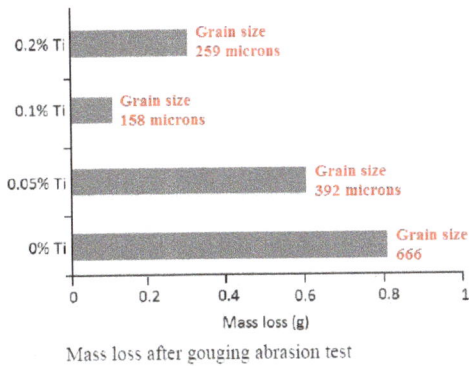

Mass loss after gouging abrasion test

Figure 10.149 *Effect of grain size on the metal loss during a gouging abrasion test[25]*

these properties improved. It was also found that for coarse grained steels the work hardening was confined to the grain boundaries but with fine grained the shear deformation work hardened the full grain which increased the degree of work hardening.

In coarse-grain steels subjected to impact abrasion the plastic deformation was localized on grain boundaries and the grain bodies participated in the deformation weakly, which caused early fracture and elevated wear. As a result, the coarse-grain steel 110GI3L had a low abrasion strength under impact loading. In fine-grain steels the grain bodies participated in the deformation, and the load was distributed over an entire grain, which increased the impact abrasion strength.[19] Figure 10.149 shows data from reference 25 on the gouging wear metal loss and grain size.

Variations of solution heat treatment can be used to enhance specific desired properties such as yield strength and abrasion resistance. Usually, a fully austenitic structure, essentially free of carbides and reasonably homogeneous with respect to carbon and manganese, is desired in the as-quenched condition, although this is not always attainable in heavy sections or in steels containing carbide-forming elements such as chromium, molybdenum, vanadium, and titanium. If carbides exist in the as-quenched structure, it is desirable for them to be present as relatively innocuous particles or nodules within the austenite grains rather than as continuous envelopes at grain boundaries.

10.9.5 The service environment

Unfortunately, there are working conditions where AMS grade does not give the best wear resistance.

- In corrosive condition where the overall material wear is related to corrosive-wear. The main improvement under these conditions are AMS with up to 2.5% chromium[20]
- In conditions where there is insufficient impact loading to create the needed work hardening. The main methods to improve the properties of the AMS involve the use of second phase particles (titanium, vanadium, chromium). Steel 2 in the Table shown in Figure 10.156 shows a heat treatment to produce a dual phase steel with carbides and ferrite in austenite with a hardness of 300 to 400 BHN.

Plastic deformation is essential for work hardening of manganese steel and therefore, the performance of manganese steel will depend on whether the wear mechanism is associated with heavy impact or moderate impact or no impact. The typical types of wear found in service are classified as:

- Gouging abrasion – Impact is always involved in applications such as digger teeth cutting into rock, or gyratory crusher or an impact crusher. In such cases manganese steel would be able to absorb energy of impact and undergo extensive plastic deformation without cracking. Even if crack is eventually developed, manganese steel has such high toughness that early detection would be possible before equipment damage occurs
- Grinding or high stress abrasion – In this type of application high stresses results from a crushing action such as in a ball mill. In the past the choice has been AMS, mainly to avoid premature failures. However, as the quality of new white irons has been achieved, hard martensitic white cast irons have been used.
- Scratching or low stress abrasion. With this mode of wear there is no impact loading. The scratching results from loose particles lightly abrading the steels. With this mode of low stress abrasion AMS will not work harden sufficiently to prevent surface wear.

AMS would be the main choice of cast metal where there needs to be resistance to both wear and impact loads such as sizing of minerals, crusher jaws and rail junctions. AMS grades have a large market share for these critical applications and due to the importance of high integrity there is significant R&D work aiming at improved performance.

10.9.6 Introduction to AMS manufacture

There are two good guides to AMS

- Austenitic Manganese Steel-A Complete Overview by David Havel, P.E. Columbia Steel Casting Co. Inc https://www.sfsa.org/doc/2017-4.1%20 Columbia%20-%20Havel.pdf[2]
- Cast Austenitic Manganese Steels-Some Practical Notes by Subrata Chakrabartti http://foundrygate. com/upload/artigos/5EOqaqG66tpALbISY4sYL5c dLii3.pdf[3]

The typical composition was 1.2% carbon and 12% manganese although the initial patent gave broader ranges of chemical analysis up to 20% manganese. AMS remains austenitic when cooled to ambient temperature. To achieve fully stable austenitic

microstructure the following limits on composition apply: Wt%Mn + 13Wt%C >17

In most cases the steel is put into service in an austenitic condition at a hardness of around 200HB and under service loads undergoes strain hardening and converts to strain induced martensite with a hardness up to 500HB. The depth of the work hardening is quoted as 0.5mm from shot peening, 2.5mm from crushing hard rock and 38mm from explosive hardening.

For some applications where there are insufficient loads to induce the hardening a low temperature heat treatment can be carried out to pre-harden the steel to a hardness of around 350 BHN. Steel 2 in the Table shown in Figure 10.156 shows a heat treatment to produce a dual phase steel with carbides and ferrite in austenite with a hardness of 300 to 400 BHN. In addition, for some applications "explosive hardening" has been used prior to use.[4 and 5]

Figure 10.150 *Turnout, most frequent in train operation, Crossings, more frequent in tram operation[5]. RHS = RBM frog casting removed from mould viewed from underside with knock–off risers still in place. Many feeder heads needed to produce sound frogs[6]*

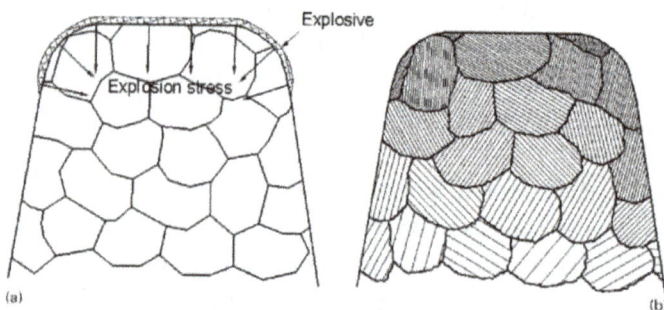

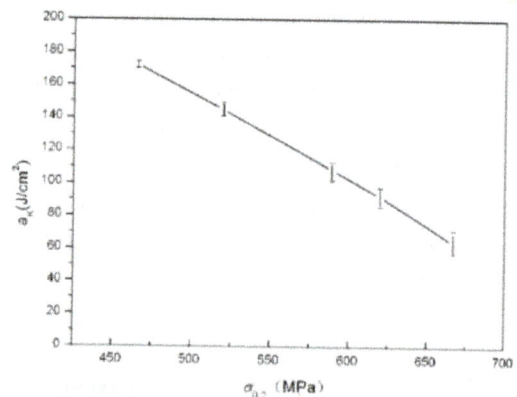

Figure 10.151 *LHS = Sketches of in situ deformation of high manganese steel crossing a before and b after explosion. RHS = Relationship between yield strength and impact toughness for high manganese steel after explosion hardening[4]*

There is unanimous agreement that the key factors that affect the quality of the parts particularly for thick components are:

- The pouring temperature and the use of grain refining elements (fine grain essential)
- The heat treatment
- The chemical composition.

10.9.6.1 The pouring temperature and the use of grain refining elements

During the melting and pouring of all metals it is well known the high superheat can result in coarse grain size and a high degree of alloy segregation. The freezing range of manganese steel is about 1371°C to 1260°C. In practice the optimum superheat should be around 50°C. At superheats above 100°C teeming cracks and coarse grain size are produced. Too low a pouring temperature would result in cold laps and excessive ladle skulls.

To produce good quality casting there is a need to produce a fine grain size. This requires control of the tapping temperature. **To achieve this there is a need to use a bottom pour ladle and high alumina ladle refractories with synthetic slag cover on the ladle to prevent heat loss.** In addition, this is a topic of current R&D and several recent publications describe work that has added **grain refining elements** to assist with nucleation to allow the production of a fine grain size.[2, 3 and 11] The use of grain refining element has been known for many years and expired patent US 6572713B2 shows a detained metal treatment procedure for appropriate addition of grain refining elements. A more recent patent US8636857B2 (2014) shows a similar procedure for addition to ensure a fine grain size. In addition a 2019 publication on grain size

and precipitation control used a heating procedure that was called "segmented heat preservation process (SHPP)" which had holds at 450 and 650°C.

10.9.6.2 Heat treatment

Another important part of the process is the heat treatment which requires appropriate soak and rapid quench to remove carbides and produce an austenitic microstructure.

The phase diagram for a steel with 13% manganese is shown in Fig 10.152 RHS. With a manganese at 13%, it can be seen as the eutectoid temperature changes from 723°C for a plain carbon steel to 588°C and the carbon content of the eutectoid for a plain carbon steel from 0.77% to 0.3%. In addition, a metastable zone is created between 698 – 500°C and 0.08 – 1.6% C, which presents a combination of ferrite, austenite and cementite which, as the manganese content increases and displaces to the right the pearlitic zone ($\alpha + Fe_3C$).

10.9.6.3 Method for refining the grain size of AMS

The metastable zone between 698 – 500°C and 0.08 – 1.6% C, allows a potential method of grain refinement. If the AMS is held in this temperature range for 12 hours or longer some of the austenitic steel transforms to ferrite and cementite and forms pearlitic areas ($\alpha + Fe3C$) which will consist of areas of new grain structure. If this is followed by a solution heat treatment around 1010°C the austenite produced will have a finer grain size. This process is mentioned in patents US4531974 (1985) and EP0136433B1 (1983) both expired. Steels 4 and 8 in the Table in Figure 10.156 show the application of this procedure.

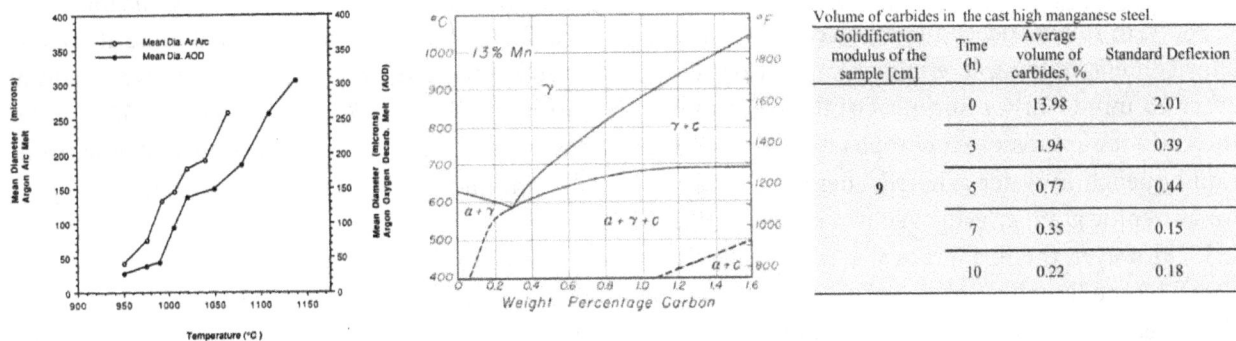

Solidification modulus of the sample [cm]	Time (h)	Average volume of carbides, %	Standard Deflexion
9	0	13.98	2.01
	3	1.94	0.39
	5	0.77	0.44
	7	0.35	0.15
	10	0.22	0.18

Figure 10.152 *Mean Grain Size vs Solution Treatment Temperature, One Hour at Temperature and Water Quench for 2% Mo AMS.(AOD Fe-12.5Mn-2.01Mo-1.15C)[7], Centre = Phase diagram for a steel containing 13% manganese[6], Right = Time to reduce carbides at 1050°C[9]*

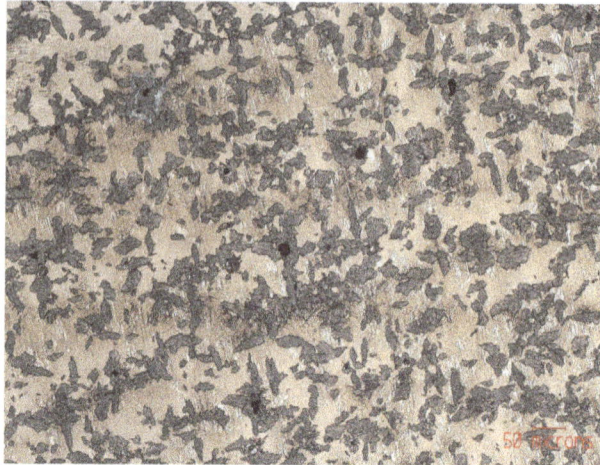

Figure 10.153 *Microstructures of specimens held at 600°C for 40 hrs.*

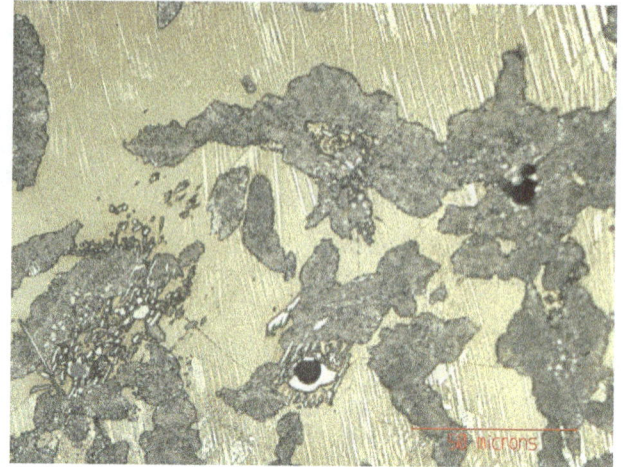

Figure 10.154 *The grey areas are ferrite/cementite formed in the austenite*

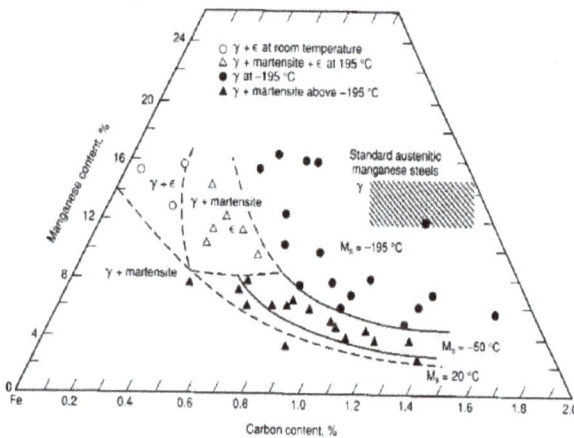

Figure 10.155 *Austenite stability and Ms temperatures for various compositions.*

Sl. No.	COMPOSITION						Heat Treatment	BHN	U.T.S. Kg/mm2	Y.S. Kg/mm2	%E	Impact data
	%C	%Mn	%Si	%S	%P	%Others						
1	1.0-1.25	11.0-14.0					1050°C. W.Q.	185-200	60-100	34-42	15-40	Izod: 20°c 16ft-lb
2	1.0-1.4	10.0-14.0	0.10-0.30	0.01-0.03	0.1		1040°C.w.q 540-700°C a.c.48hr, 730-820°C, 2hr, w.q.	300-400	94-116	84-98	10-16	
3	1.34	12.0				Mo-0.5	1090°C.w.q	207	74	42	26	
4	1.22	11.9	0.51		0.019	Mo-2.0	1040°C.w.q pearlitized 590°C.a.c. austenized 980°C.w.q.	217	82	45	29	
5	Regular Manganese Steel(Keel Block)						1040°C.w.q.	175	80	38	40	
6	1.34	12.0				Mo-0.5 Cr-3.0	1090°C.w.q	228	75	47	18	
7	1.07	10.6	0.5	0.014	0.028	Ni-2.78	As forged condition.		122		33	
8	1.23	13.6	0.57				1050°C.1hr, w.q. 600°C, 8hr, a.c. 980°C, 1hr, w.c.	207	92	42	68	Izod:- at+20°C 115 ft-lb

Figure 10.156 *Some unusual heat treatments for AMS[3]*

These phases formed in the metastable area of the phase diagram are in general undesirable in the steel but can form during the cooling of the casting following solidification or can be created by holding at a temperature and thus it is necessary to heat treat to develop an austenitic microstructure. Heat treatment of austenitic steel manganese involves slow heating up to 980°C to 1090°C, depending upon carbon level and alloy content, holding at temperature for 1 or 2h, for every 25 mm (1 in) of thickness of the piece so that the carbides are dissolved completely followed by a rapid quench in water agitated There is a tendency for austenitic grain growth as shown in Figure 10.152 LHS. However, the final grain size is strongly influenced by the liquid metal pouring temperature and the speed of solidification. Consequently, to determine representative grain size and distribution, it is often required to prepare and etch an entire cross section. The relatively high temperature combined with the furnace atmosphere can cause decarburization of the surface due to the oxidation and also a loss of manganese. Figure 10.155 shows that the composition change can cause the rise in the Ms temperature and the formation of martensite Therefore the areas of decarb and loss of alloy martensite can form. This can be identified since this layer will be magnetic.

A. Chojeckia[9] examined the soak time at 1050°C to remove the large carbides. Samples were held at temperature for 2, 5, 7 or 10 hours at 1050°C. After water quenching the samples were metallographically examined. The results are presented in Figure 10.152. This work also included hot tensile testing and SEM examination of the fracture surfaces. The main conclusion was that AMS tend to experience more micro segregation and this imperfection would be more likely to cause failure rather than retained carbides which are much smaller. Failure investigation that have identified carbide[15] have possibly

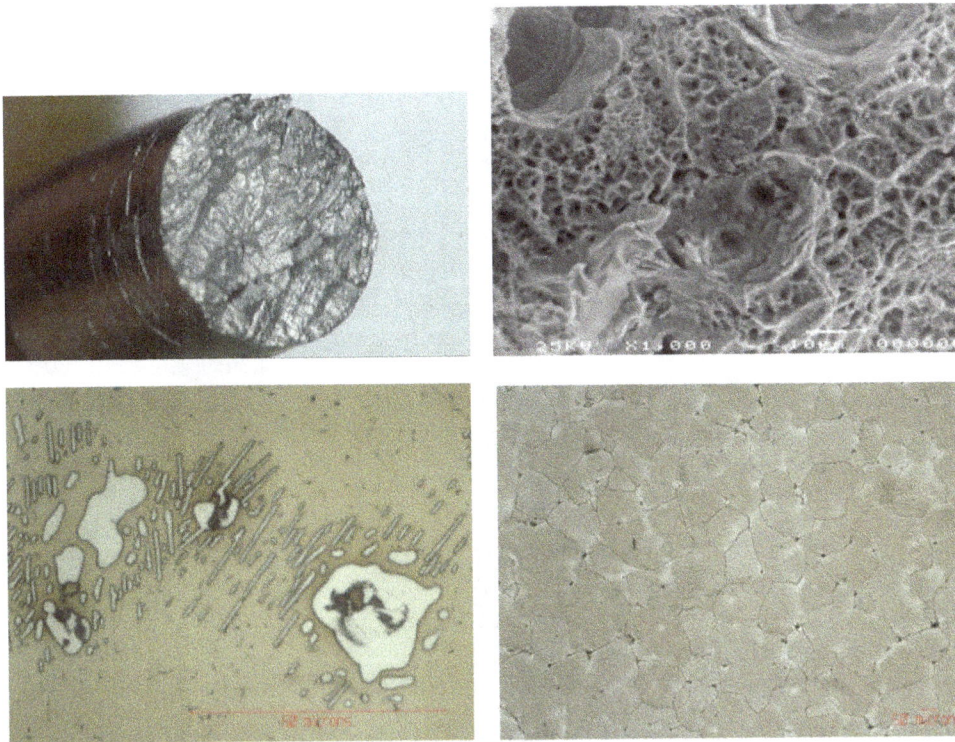

Figure 10.157
Top LHS = Evidence
of coarse grain size on
AMS tensile specimen.
Top RHS = SEM image
of a tensile fracture,
Lower LHS = Carbide
in 2.5Cr AMS after HT
and RHS = Equiaxed
grain structure in AMS

been incorrect and the porosity imperfections were of greater importance. The complete removal of chromium carbide from AMS can be difficult.[17] Recent work (2015) has looked at elimination of the need to be heat treated by treating the liquid steel with magnesium.[16]

10.9.7 Chemical composition

Phosphorous

The preferred practice is to aim for the phosphorus content to be below 0.04% even though 0.07% is permitted by ASTM A 128. With levels above 0.06%, which formerly were prevalent, this contributes to hot shortness and low elongation at very high temperatures and frequently is the cause of hot tears in castings and underbead cracking in weldments. It is particularly advantageous to keep phosphorus at the lowest possible level in the grades that are welded, and

in manganese steel welding electrodes, and in heavy section castings. The effect on the mechanical properties is available in several published documents. The tensile data on 25mm test bars shows little change in properties up to 0.10% (Figure 10.158).

Performance data for crusher applications indicate a significant effect of phosphorous on the toughness. In the "real" service condition the final toughness can be influenced by other factors such as effective section thickness, carbon, silicon levels and other alloy additions. However, based on service performance there is a main conclusion that the phosphorus level should be held as low as practically and economically possible.

Phosphorus above 0.02% progressively promotes inter-granular cracking in manganese steels. Above 0.06%, the high temperature plasticity of manganese steel is severely reduced and the steel becomes extremely susceptible to hot tearing. At such a high phosphorus level, microstructural evidence of grain boundary films of phosphide eutectic can be observed. Below 0.06% phosphorus, no microstructural evidence can be observed but phosphorus still has an effect on the incidence of hot tearing. The maximum tolerable phosphorus content is dependent upon the severity of the stress system which is related to casting design, size and riser location. For large, complex castings it is advisable to hold the phosphorus below 0.04%. Phosphorus also has an effect on possible cracking during the removal

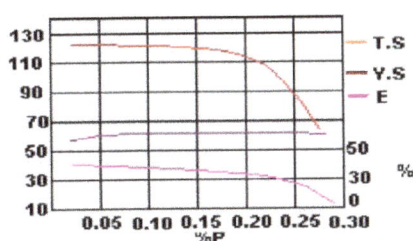

Figure 10.158 Effect of phosphorous on AMS

of large risers and in welding operations which can have higher levels of phosphorous due to segregation during solidification. The same cracking mechanisms that are associated with hot tearing also apply in the case of weld integrity. A general guideline is that the phosphorus content of the weld metal (combination of base metal and weld rod due to dilution effects) should not exceed 0.03% if cracking in the weld repair areas is to be prevented.

Sulphur level

There is a general statement that has been made since the early days of Hadfield's discovery of AMS that sulphur is seldom a quality factor in 13% manganese steel. This was mainly attributed to the effect of manganese combining with the sulphur to form manganese sulphide inclusions avoiding the formation of iron sulphide which causes grain boundary embrittlement. The content of round shaped manganese sulphide inclusions is now recognised as a potential imperfection and the content should be minimised. A check of published chemical analysis showed that the sulphur was frequently as low as 0.002%.

Alloy additions

Titanium addition (0.5 and 0.1% Ti) has been found to refine the grain size[13, 25]. The added chromium at 2.2% and titanium at 1.5% both improved the wear.[14]

10.9.8 Determination of Work-Hardening Rate

The work-hardening rate is usually determined from ordinary tensile or compression tests as the slope of the true stress-true strain curve, which is usually linear in the plastic region.

Another measure of the tendency of austenitic manganese steels to work harden is based on a determination of the Meyer Index or exponent. The technique uses a 10 mm diameter Brinell ball indentor and a series of loads. The test loads are plotted against the diameters of the corresponding indentations on logarithmic scales. The result is a straight line that fit the equation:

$$P = A \cdot d^n$$

Figure 10.159 *LHS = As received, Hardness = 214BHN, n= 2.21, m=0.21. RHS = Additional heat treatment, Hardness = 200BHN, n=2.40, m=0.40*

where P represents the applied load; d, the diameter of the indentation; A, a constant; and n, a measure of the tendency of the metal to strain harden (called the Meyer Index or exponent).

The relationship between the Meyer Index (n) and the tensile work hardening index (m) was derived by Tabor[33]. Tabor was able to relate the experimentally derived quantity, n, with the work-hardening index, m, of the test materials by calculating the mean strain surrounding spherical indentations and correlating this with the results from uniaxial tensile tests. This relationship of m = n — 2 was further verified by Hill et al. but it applies only in the case of indentation by spherical indentors.

The measurement of the work hardening rate has been measured in the two samples of coarse grained steel in Figure 10.159 (the target for n should be 2.4 to 2.9 and m 0.4 to 0.9).

REFERENCES

1. Johnson O. Agunsoye, On the Comparison of Microstructure Characteristics and Mechanical Properties of High Chromium White Iron with the Hadfield Austenitic Manganese Steel Journal of Minerals and Materials Characterization and Engineering, 2013, 1, 24-28 file:///C:/Users/user/Downloads/4.onthecomparison.pdf

2. David Havel, P.E. Columbia Steel Casting Co., Inc, Austenitic Manganese Steel- A Complete Overview, September 2017, https://www.sfsa.org/doc/2017-4.1%20Columbia%20-%20Havel.pdf

3. Subrata Chakrabartti, Cast Austenitic Manganese Steels- Some Practical Notes, http://foundrygate.com/upload/artigos/5EOqaqG66tpALbISY4sYL5cdLii3.pdf

4. F. C. Zhang et al, Explosion hardening of Hadfield steel crossing, Materials Science and Technology 2010 VOL 26 NO 2 p 223 file:///C:/Users/user/Downloads/MaterSciTechnolExplosionhardeningofHadfieldsteelcrossing.pdf

5. LINDA M. E. NORBERG, Fatigue Properties of austenitic Mn-steel in explosion depth hardened condition Material used in highly stressed railway components CHALMERS UNIVERSITY OF TECHNOLOGY, Diploma work No. 33/2010 http://publications.lib.chalmers.se/records/fulltext/138643.pdf

6. Nigel W. Peters, THE PERFORMANCE OF HADFIELD'S MANGANESE STEEL AS IT RELATES TO MANUFACTURE, www.arema.org/files/library/2005_Conference_Proceedings/00040.pdf

7. John F. Chinella, PROCESSING AND CHARACTERIZATION OF HIGH STRENGTH, HIGH DUCTILITY HADFIELD STEEL, MTL TR 90-21, April 1990, a221769, https://apps.dtic.mil/dtic/tr/fulltext/u2/a221769.pdf

8. Keyur Panchal, Life Improvement of Hadfield manganese steel castings, May 2016 IJSDR I Volume 1, Issue 5 file:///C:/Users/user/Downloads/LifeImprovementofHadfieldmanganesesteelcastings.pdf

9. A. Chojeckia I. Telejko, Cracks in high-manganese cast steel, ARCHIVES of FOUNDRY ENGINEERING, Volume 9 Issue 4/2009 17 – 22 https://pdfs.semanticscholar.org/74fe/9309583057c2421b8b32e785f6cd9f36ae4a.pdf

10. Dimitrios Siafakas, On deoxidation practice and grain size of austenitic manganese steel, Licentiate Thesis, Jönköping University School of Engineering Dissertation Series No. 029 • 2017, https://www.diva-portal.org/smash/get/diva2:1153020/FULLTEXT01.pdf deox ngs neg

11. S.A. Balogun, Effect of Melting Temperature on the Wear Characteristics Austenitic Manganese Steel, Journal of Minerals & Materials Characterization & Engineering, Vol. 7, No.3, pp 277-289, 2008 http://file.scirp.org/pdf/JMMCE20080300007_78392480.pdf

12. Eva Schmidova, Ivo Hlavaty, Petr Hanus, THE WELDABILITY OF THE STEEL WITH HIGH MANGANESE, Tehnički vjesnik 23, 3(2016), 749-752, file:///C:/Users/user/Downloads/tv_23_2016_3_749_752.pdf

13. Mohammad Bagher Limooei, Shabnam Hosseini, OPTIMIZATION OF PROPERTIES AND STRUCTURE WITH ADDITION OF TITANIUM IN HADFIELD STEELS, Metal 2012 23. – 25. 5. 2012, Brno, Czech Republic, EU http://metal2012.tanger.cz/files/proceedings/02/reports/349.pdf

14. B. Kalandyk, Cast High-Manganese Steel – the Effect of Microstructure on Abrasive Wear Behaviour in Miller Test, Archives of Foundry Engineering, Volume 15 Issue 2/2015 35 – 38 https://www.degruyter.com/downloadpdf/j/afe.2015.15.issue-2/afe-2015-0033/afe-2015-0033.pdf

15. Olawale J. O et al, Workhardening Behaviour and Microstructural Analysis of Failed Austenitic Manganese Steel Crusher Jaws Materials Research. 2013; 16(6): 1274-1281 http://www.scielo.br/pdf/mr/v16n6/aop_1627.pdf

16. Mohamed K. El-FawkhryAyman et al, Eliminating Heat Treatment of Hadfield Steel in Stress Abrasion Wear Applications, International Journal of Metal Casting, January 2014, Volume 8, Issue 1, pp 29–36

17. G. Tęcza et al, Effect of Heat Treatment on Change Microstructure of Cast High-manganese Hadfield Steel with Elevated Chromium Content, ARCHIVES of FOUNDRY ENGINEERING, Volume 14 Special Issue 3/2014 67

18. Vuki Lazi, THEORETICAL AND EXPERIMENTAL ESTIMATION OF THE WORKING LIFE OF MACHINE PARTS HARD FACED WITH AUSTENITE-MANGANESE ELECTRODES, Materials and technology 46 (2012) 5, 547–554 file:///C:/Users/user/Downloads/vlazic-materintehnolog.pdf weld repair

19. A A Astafev, Effect of grain size on the properties of manganese austenite steel 110G13L, Material Science and Heat Treatment May 1997, Volume 39, Issue 5, pp 198-201

20. Eduardo R. Magdaluyo et al, Gouging Abrasion Resistance of Austenitic Manganese Steel with Varying Titanium, Proceedings of the World Congress on Engineering 2015 Vol II WCE 2015, July 1 – 3, 2015, London, U.K.

21. US4531974 1985

22. EP0136433B1 1983

23. I El-Mahallawi et al, Evaluation of effect od chromium on wear performance of high managanese steel, Materials Science and Technology August 2001 Vol 17 page 1

24. Zaifeng Zhou et al, Influence of Heat-Treatment on Enhancement of Yield Strength and Hardness by Ti-V-Nb Alloying in High-Manganese Austenitic Steel, Metals 2019, 9, 299,

25. Eduardo R. Magdaluyo et al, Gouging Abrasion Resistance of Austenitic Manganese Steel with Varying Titanium, Proceedings of the World Congress on Engineering 2015 Vol II WCE 2015, July 1 – 3, 2015, London, U.K.

26. Selçuk Kuyucak et al, Heat-treatment Processing of Austenitic Manganese Steels, Conference Paper · September 2004, Conference: 66th World Foundry Congress, At Istanbul, Turkey, Volume: Steel Castings, https://www.researchgate.net/publication/273458792_Heat-treatment_Processing_of_Austenitic_Manganese_Steels

27. Fredrik Haakonsen et al, Grain Refinement of Austenitic Manganese Steels, AISTech 2011 proceedings volume II p763, https://www.researchgate.net/publication/286333017_Grain_refinement_of_austenitic_manganese_steels

28. Dimitrios Siafakas et al, The Influence of Deoxidation Practice on the As-Cast Grain Size of Austenitic Manganese Steels, Metals 2017, 7, 186,

29. Bianka Nani Venturellia et al, The effect of the austenite

grain refinement on the tensile and impact properties of cast Hadfield steel, Materials Research. 2018; 21(5)

30. Li Xiaoyun1 et al, Influence of impact energy on work hardening ability of austenitic manganese steel and its mechanism, CHINA FOUNDRY Vol.9 No.3 page 248, August 2012. http://www.foundryworld.com/uploadfile/2012082747991625.pdf

31. Hui CHEN et al, Effects of impact energy on the wear resistance and work hardening mechanism of medium manganese austenitic steel, Friction 5(4): 447–454 (2017), https://link.springer.com/content/pdf/10.1007%2Fs40544-017-0158-6.pdf

32. O Bouaziz et al, High manganese austenitic twinning induced plasticity steels : A review of the microstructure properties relationships, Solid State and material Science 15 (2011) p 141-168,

33. Philip M Sargent, Indentation size effect and strain-hardening, Journal of material science letters 8 (1989) p 1139-1140.

10.10 BACKGROUND METALLURGICAL ASPECTS OF RAIL STEEL

1. Early development
2. Reasons for rail steel improvements
3. Current status – Product Benchmark
4. Software developments
5. Process improvements

10.10.1 Early development

The first steel rail tracks were laid in the UK in 1857 following Robert Mushet's invention of treating the Bessemer steel with manganese ore. The manganese allowed the deoxidation of the steel and also converted the sulphur to manganese sulphide. This treatment allowed the manufacture of rail steel with around 0.2% carbon.

During the following 100 years user experience led to the development of the current standard specifications for rails which are typically based on 0.80% carbon steel with a manganese level of around 0.4- 0.9%. In a recent review of the selection guidelines for rail the following tables show that the UK uses the R260 grade (Figure 10.160 RHS).[2]

In the 1970s there were some important technical conferences which summarised the status, knowledge and experience associated with rail steel. For example, "Rail steels – development, processing and use. 1976" the proceedings of symposium were published as ASTM 644 in1978 Edited by Stone and Knupp.[3] The second paper (The effect of mechanical properties upon performance of rail road steel. D H Stone and R K Steele) outlines the composition of the early steels and the important relationships between the pearlite microstructure and hardness Fig 10.160. Rail steels have been developed using a pearlitic microstructure. From Chapter 1 we should remember that pearlite consists of alternating lamellae of soft iron and very hard cementite obtained by relatively slow cooling. As shown in Figure 10.161 the hardness increases as distance between the lamellae

Steel grade		Hardness range (HBW)	Description	Branding lines
Steel name	Steel number			
R200	1.0521	200 to 240	Non-alloy (C-Mn) Non heat treated	No branding lines
R220	1.0524	220 to 260	Non-alloy (C-Mn) Non heat treated	———
R260	1.0623	260 to 300	Non-alloy (C-Mn) Non heat treated	— ———
R260Mn	1.0624	260 to 300	Non-alloy (C-Mn) Non heat treated	—————— —
R320Cr	1.0915	320 to 360	Alloy (1 % Cr) Non heat treated	— ———
R350HT	1.0631	350 to 390 b	Non-alloy (C-Mn) Heat treated	—— —— ——
R350LHT	1.0632	350 to 390 b	Non-alloy (C-Mn) Heat treated	— ——— —
R370CrHT	t.b.a.	370 to 410	Alloy (C-Mn) Heat treated	—
R400HT	t.b.a.	400 to 440	Non-alloy (C-Mn) Heat treated	——

R [m]	≤ 300	≤ 400	≤ 500	≤ 600	≤ 700	≤ 800	≤ 1 500	≤ 3 000	> 3 000
UIC	R350HT			R350HT/R260			R260		
DB	R350HT (≥ 30 000 t/d)					R260			
DB new	R350HT (≥ 50 000 t/d)							R260	
CH	R350LHT			R350LHT/R320Cr		R320Cr R350LHT	R260		
CH (pro-posal)	R370CrHT			R350LHT		Bainite up to 1 200 m		R260	
AT	R350HT		R260						
SWE	R350HT		R260						
SWE (HH)	R350HT								R260
NOR	R350HT			R260					
UK	R260								
IT	R260								
BE LUX	R350HT					R260			
NL	R350HT R370CrHT	R370CrHT					R370CrHT		R260
DK	R350HT		R260						
PL	R350HT		R260						
H	R350HT		R260						
RO	R350HT		R260						

Figure 10.160 LHS = Rail grades. RHS = Overview of National Guidelines[2]

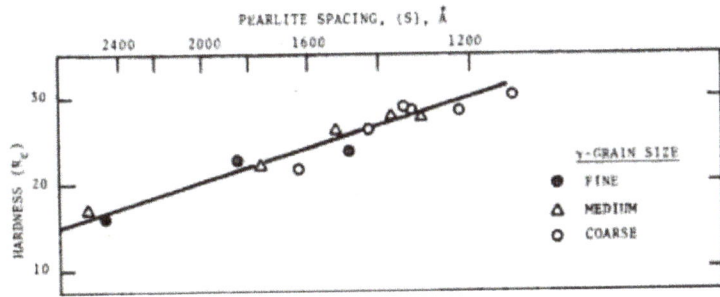

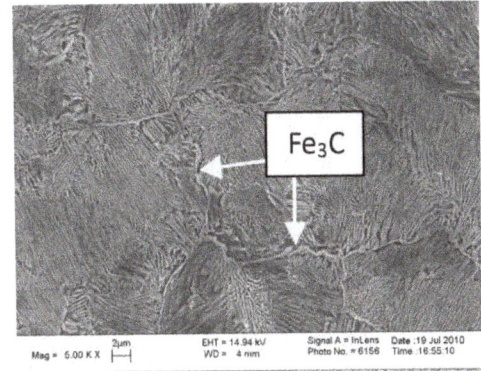

Figure 10.161 *Relationship between interlamellar spacing of pearlite and hardness. This relationship has been subsequently revalidated at various times in following years[3]. RHS = SEM image of one of the premium rail steels indicating Fe3C observed at the grain boundaries of the pearlite grains in the head region[6] The presence of Fe3C may be linked to the occurrence of rolling contact fatigue (RCF)*

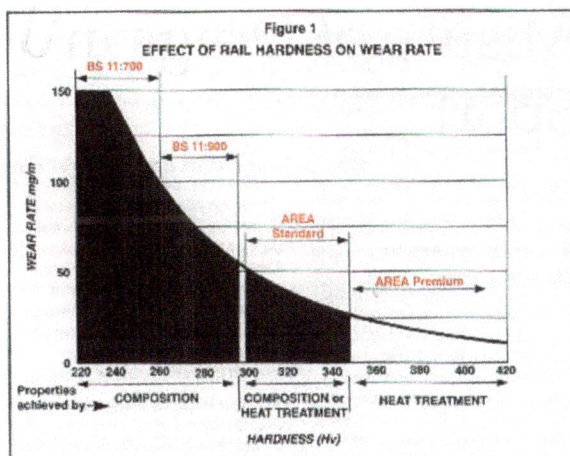

Figure 10.162 *LHS = Plot of the wear rate (mg m-1) versus hardness (HV) of rail steel for different grades of rail steel. RHS = Charpy transition curves for new bainitic steel.*

decreases. In a similar way the tensile strength will increase.

The quality, durability and wear resistance of rail steels made from high carbon steel requires the control of the pearlitic microstructure, especially the pearlite morphology in terms of colony size and interlamellar spacing. Thermal treatment that can refine the colony size and reduce the inter-lamellar spacing of the cementite such as control of the rolling process (thermo-mechanical treatment) and accelerated cooling to reduce the pearlite transformation temperature are used.

However, the pearlitic rail widely used in the construction of railways, has reached its maximum strength of around 1300 MPa. At strength levels beyond this value the toughness of pearlite rail significantly decreases and brings a risk of brittle fracture. In 1996, Yates[1] outlined the situation and gave the details regarding a new bainite steel developed by

Professor Bhadeshia and his co-workers. The new "bainitic" steel was used for the channel tunnel. Yates used Figure 10.162 to show the range of hardness vs wear of rail steels. There are other development programs developing grades of bainitic rail.[4]

10.10.2 Reasons for rail steel improvements

There has always been active interest in the product and process development of rail steel to achieve improvements in the integrity and durability of rail steel. This has been driven by the trend to reduce infrastructure costs and minimise the risk of future rail related accidents.

One of the most recent rail accidents in the UK that has been fully reported was the Hatfield accident.[5] An extract from the summary of the report:

"On 17 October 2000 the 12.10 train travelling from London Kings Cross to Leeds derailed south of Hatfield station. The train was an intercity 225 Mark 4 express train operated by Great North Eastern Railway (GNER). The location of the derailment was between Welham Green and Hatfield, approximately 16.7 miles (27 km) from Kings Cross. The left hand rail fractured on the down fast line1 (i.e. going North). At the time the train was travelling between 115 and 117 mph (185 and 188kph). There were 170 passengers and 12 GNER staff on the train. As a result of the derailment, four passengers were killed, over seventy people suffered injuries including four seriously injured; two of the seriously injured were GNER staff."

Following the conclusion of the investigation, the CPS prosecuted six men and two companies (BBRML and Network Rail (as the successor to Railtrack)) with manslaughter due to gross negligence and offences under the HSWA. This was partly because the poor condition of the rail was known and replacement rail to carry out the repair was on site but never installed. In addition, measures to slow down the trains due to the known poor condition of the track were not carried out.

At the court proceeding no one was found guilty.

The rail was significantly damaged and it was clear almost immediately that poor maintenance and neglect had allowed cracks to develop in much of the country's track.

An explanation of the severity of the cracking was given by M Ashby et al, in the book "Materials Engineering Science, Processing and Design", Butterworth-Heinemann, 2007. The cracks develop because of the high contact stresses (up to 1 GPa) at the point where the wheels contact the rail. This deforms the rail surface, and because the driving wheels also exert a shear traction, the steel surface is deformed in a direction opposite to that of the direction of travel.

Cracks nucleate in this layer after about 60 000 load cycles, and lie nearly parallel to the surface. It was noted that the cracks propagate much faster when the rails are wet than when they are dry. The explanation was that as a wheel approaches a surface crack it can be forced open, and water enters and is trapped, which is compressed to a high pressure which propagates the crack (See Figure 10.164).

The poor microstructure and the decarb found on the rail was a cause for concern (Fig 10.163 RHS). Newcastle University was commissioned by the HSE to check if this poor microstructure had assisted the fatigue damage.[6] The input data on decarburisation comes from both the experimental data collected in project JR31.086, and alternative data available from published literature. The report found that for the conditions modelled, the presence of a decarburised

Figure 10.163 Top Multiple rail fractures re-assembled. Lower LHS = General rail head condition. Centre = The most southerly fracture. RHS = Intergranular ferrite network

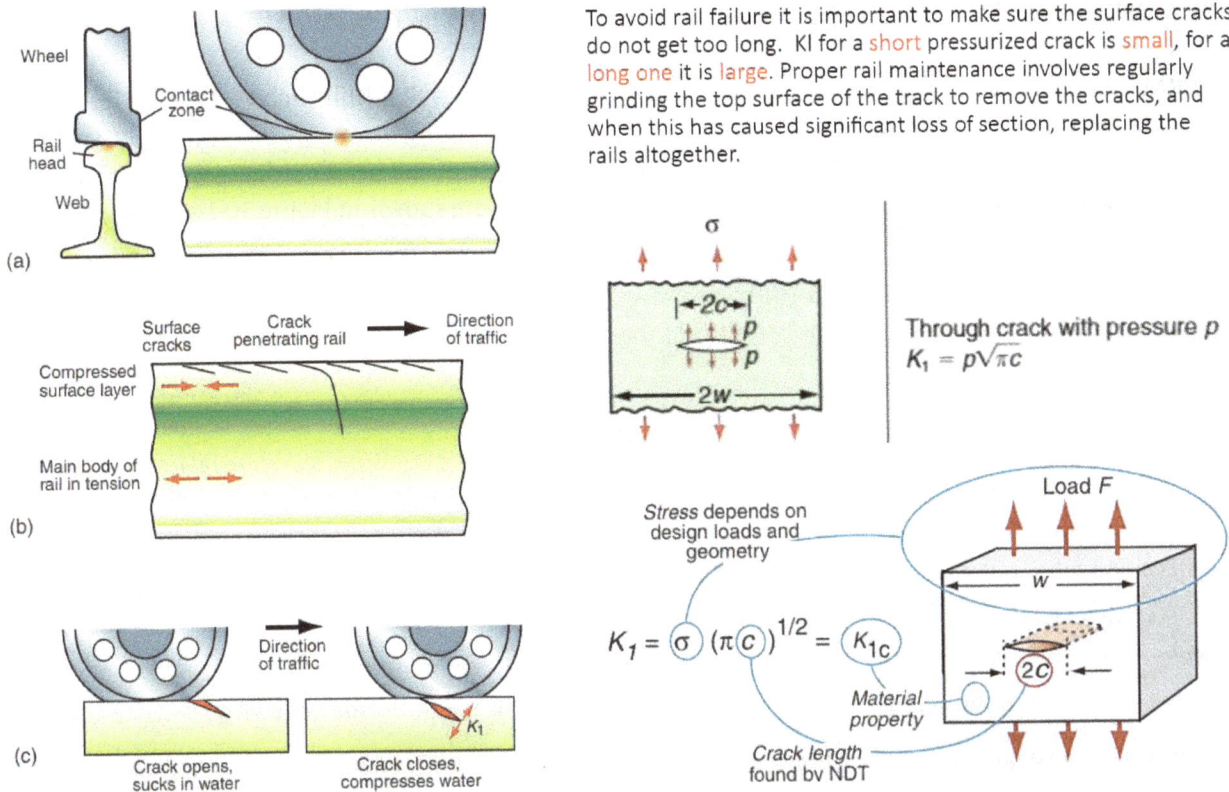

To avoid rail failure it is important to make sure the surface cracks do not get too long. KI for a short pressurized crack is small, for a long one it is large. Proper rail maintenance involves regularly grinding the top surface of the track to remove the cracks, and when this has caused significant loss of section, replacing the rails altogether.

Through crack with pressure p
$$K_1 = p\sqrt{\pi c}$$

Stress depends on design loads and geometry

$$K_1 = \sigma\,(\pi c)^{1/2} = K_{1c}$$

Material property

Crack length found by NDT

Figure 10.164 *LHS = Rail cracking. (a) The rolling contact. (b) The surface cracks, one of which has penetrated the part of the rail that is in tension. (c) The mechanism by which the crack advances. RHS = Preventing failure, keep cracks short by rail grinding. From M Ashby et al, Materials Engineering Science, Processing and Design, Butterworth-Heinemann, 2007.*

layer is predicted to produce a peak depth to which cracks may initiate of up to 340–380µm. The comparable depth for a non-decarburised steel was found to be 5–10µm, indicating that the decarburisation has led to around a thirty-times increase in the size of cracks.

This incident and several others demonstrate the importance of rail steel quality.

Rail wear and rolling contact fatigue from train operations are inevitable. In the period 2000 to 2004 500,000 tons of rail tracks have been replaced in the USA with 10% being curved products. A European study has also suggested that the cost of rolling contact fatigue to the European railway network, including inspection, train delay, rail replacements and weld repair, rail grinding and derailments, is about 300 million euros per year.

10.10.3 Current status – Product Benchmark[7]

An important document which outlines the current status and benchmarks the main market leader products in terms of microstructure and wear resistance is "Development and Evaluation of High Performance Rail Steels for Heavy Haul Operations" by Daniel Szablewski, SemihKalay, Joseph LoPresti of the Transportation Technology Center, Inc. (TTCI) Pueblo, Colorado, USA. The report covers the current premium rail performance of seven participating manufacturers and 10 different rail grades. The manufacturers, countries of origin and number and type of rail grades in test are as follows:

- ERMS Rail Mill (USA) – 1 grade: OCP
- Corus Rail Mill (France) – 1 grade: MHH HE (head hardened)
- Nippon Rail Mill (Japan) – 1 grade: NSC-HEX (control rail)
- JFE Rail Mill (Japan) – 2 grades: JFE-A (SP2), JFE-B (SP3)
- Mittal Rail Mill (USA) – 1 grade: HC
- Panzhihua Rail Mill (China) – 1 grade: PG4 (head hardened)
- Voestalpine Rail Mill (Austria) – 3 grades: VAS-1, VAS-2, 400NEXT.

For the IH (Intermediate Hardness) rail wear test, there are six participating manufacturers with the following 8 rail grades:

- Corus Rail Mill (France) – 1 grade: MHH HE (as rolled)
- ERMS Rail Mill (USA) – 3 grades: ERMS-1 (IH), ERMS-2 (IH HS), SS (control rail)
- Panzhihua Rail Mill (China) – 1 grade: PG4 (as rolled)
- Mittal Rail Mill (Spain) – 1 grade: ML

- TrineckéZelezárny Rail Mill (Czech Republic) – 1 grade: TZ
- Lucchini RS Rail Mill (Italy) – 1 grade: IH.

An example of the head hardness values taken from the report are shown in Figure 10.165.

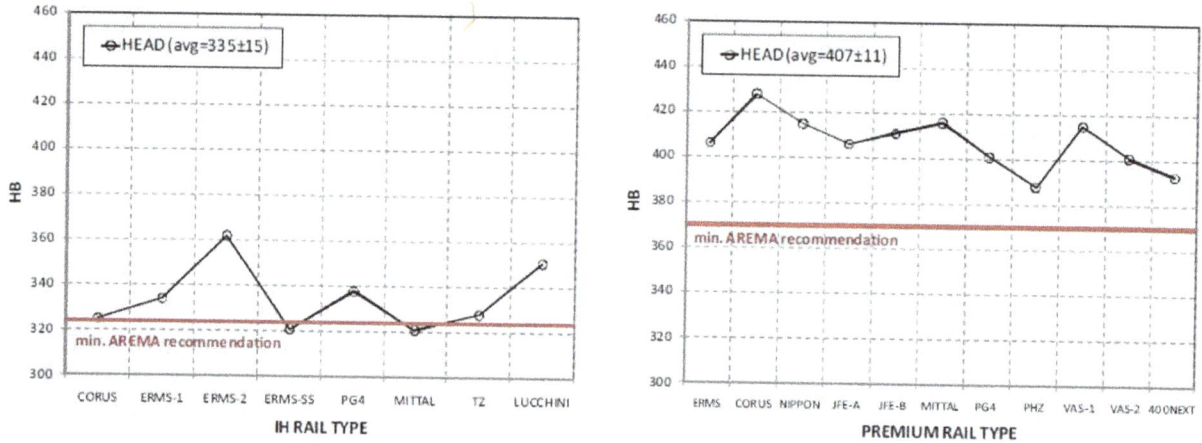

Figure 10.165 Top of the rail hardness measurements for the IH and premium grades' Head hardness values

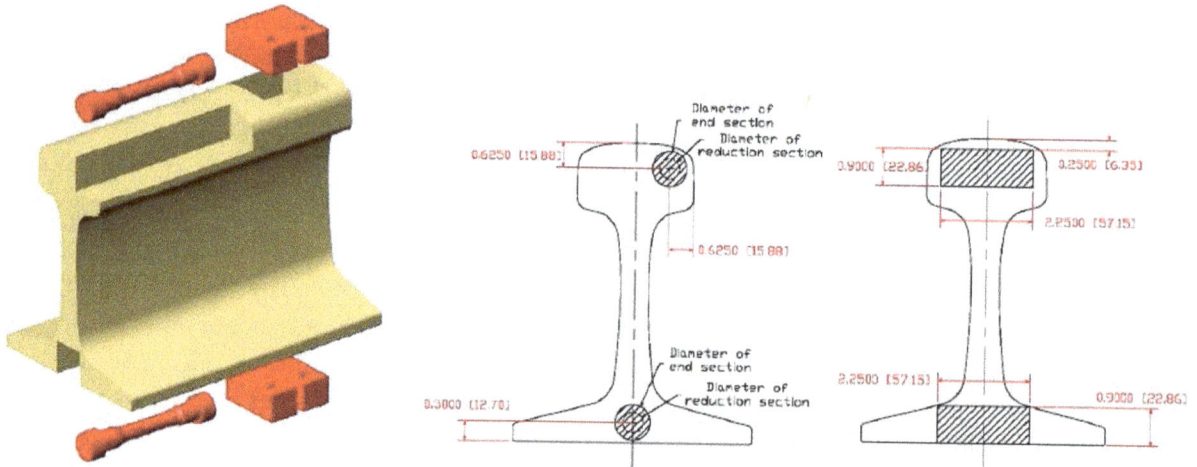

Figure 10.166 The sampling plan used in the report.

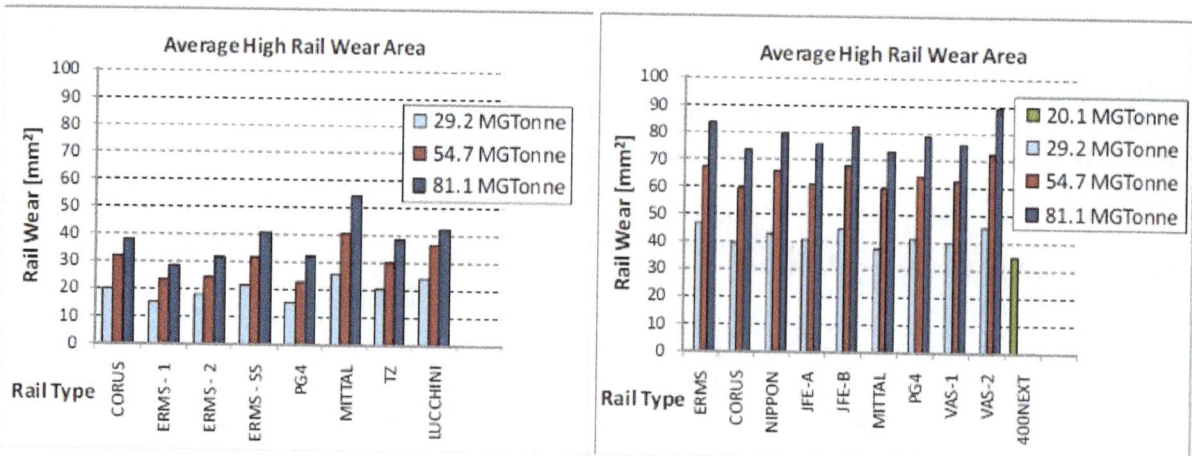

Figure 10.167 Average wear rates

The sampling plan for mechanical and metallurgical samples (Figure 10.166) were taken and evaluated.

The report is detailed and covered a number of processes of manufacture and a range of suppliers. The above benchmarking gives an indication of the quality of rail steels manufactured in China, Japan, Europe and the United States.

10.10.4 Software developments

There are several publications that provided data to develop appropriated software models to predict the mechanical properties.[8]

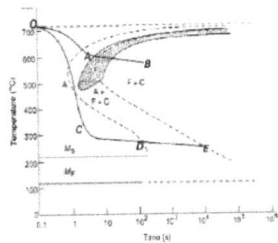

Fig. 1 CCT Diagram for AISI-1080 steel [3]

Path OAB represents the type of schedule used to produce fine-grained pearlite in the hardened layer of the rail head. Segments OA and AB correspond to cooling rates of 20 °K/s and 1 °K/s, the first easily achievable by means of forced air, water spray, or immersion quenching, and the second easily achievable in passive air cooling.

Path OCDE represents the type of schedule that would be required to produce a bainitic microstructure

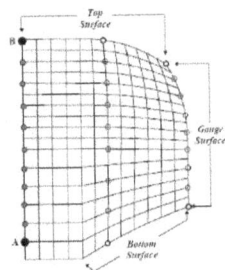

Fig. 3 Detail of finite element model grid in rail head

In the validation study, a 30 s quench with a heat transfer coefficient of 2200 W/m²°K was applied to the top surface and gauge face, but neither a simultaneously applied milder quench on the web and bottom of the base nor brief interruptions for surface re-heating were simulated. These latter details reflect current rail making practice, but the simplified model was

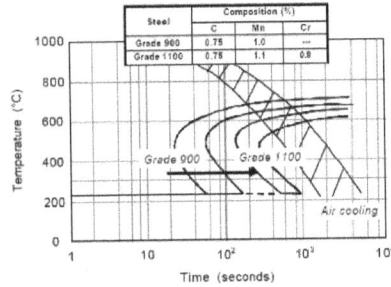

Overview of Pearlitic Rail Steel: Accelerated Cooling, Quenching, Microstructure, and Mechanical Properties[9]

Satyam S. Sahay, Goutam Mohapatra, and George E. Totten

Journal of ASTM International, Vol. 6, No. 7 Paper ID JAI102021

TABLE 3—Influence of cooling rate on the microstructure of rail steel [3].

Cooling rate	Phases
<240°C/min	Pearlite
250°C/min	Pearlite + bainite
400°C/min	Pearlite + bainite + martensite
>643°C/min	Martensite
Interrupted cooling	Fine pearlite

NIOBIUM IN RAIL STEEL Harald de Boer and Hiroki Masumoto[10 and 11]

Figure 4. CCT – diagram showing the effect of alloying to achieve pearlite refinement.

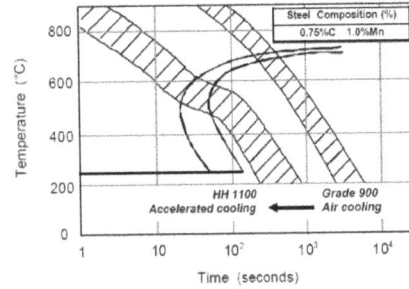

Figure 5: CCT-diagram showing the effect of cooling rate to achieve pearlite refinement.

10.10.5 Process improvements

An innovative process for the manufacture of the blooms for rail steel manufacture was outlined in 2017 which was a combination of concast and ESR (Electro-Slag-Refining).[12]

REFERENCES

1. Yates, J.K. Innovation in rail steel. Sci. Parliam. 1996, 53, 2–3 https://www.phase-trans.msm.cam.ac.uk/parliament.html

2. M. TOMIČIĆ-TORLAKOVIĆ, GUIDELINES FOR THE RAIL GRADE SELECTION, METALURGIJA 53 (2014) 4, 717-720 file:///C:/Users/user/Downloads/MET_53_4_717_720_Tomicic_Torlakovic.pdf

3. Rail steels- development, processing and use. 1976 ASTM special technical publication 644

4. Min Zhu et al, Effects of Tempering on the Microstructure and Properties of a High-Strength Bainite Rail Steel with Good Toughness, Metals 2018, 8, 484 file:///C:/Users/user/Downloads/metals-08-00484.pdf

5. Train Derailment at Hatfield: A Final Report by the Independent Investigation Board https://webarchive.nationalarchives.gov.uk/20130903223925/http://www.rail-reg.gov.uk/upload/pdf/297.pdf

6. Comparison of the Hatfield and alternative UK rails using models to assess the effect of residual stress on crack growth from rolling contact fatigue. Prepared by University of NewcastleUponTyne for the Health and Safety Executive 2006 RESEARCH REPORT 461 2006 http://www.hse.gov.uk/research/rrpdf/rr461.pdf

7. Daniel Szablewski et al, Development and Evaluation of

High Performance Rail Steels for Heavy Haul Operations, Challenge C: Increasing Freight capacity and services http://www.railway-research.org/IMG/pdf/c1_lopresti_joseph.pdf

8. TAILORING HEAT TREATMENT AND COMPOSITION FOR PRODUCTION OF ONLINE HEAD-HARDENED BAINITIC RAIL J.A. Jones, A.B. Perlman, and O. Orringer Mechanical Engineering Department Tufts University, Medford, MA

9. Satyam S. Sahay, Goutam Mohapatra, and George E. Totten, Overview of Pearlitic Rail Steel: Accelerated Cooling, Quenching, Microstructure, and Mechanical Properties, Journal of ASTM International, Vol. 6, No. 7 Paper ID JAI102021

10. Harald de Boer and Hiroki Masumoto, NIOBIUM IN RAIL STEEL International symposium, Niobium; science &

technology; 2001; Orlando, FL in Niobium; science & technology; 821-844

11. A. Ray and H. K. D. H. Bhadeshia, NIOBIUM IN MICROALLOYED RAIL STEELS, Conference Paper · November 2015 file:///C:/Users/user/Downloads/railchina-wordtopdf.pdf

12. Ganna Polishko, ESR FOR RAILROAD WHEEL AND RAILS, Proceedings of the Liquid Metal Processing & Casting Conference 2017 file:///C:/Users/user/Downloads/ESRforRailroadWheelandRails.pdf

13. P.KAYDA, ESR POSSIBILITIES TO IMPROVE RAILROAD RAIL STEEL PERFORMANCE, Conference Paper · September 2015 file:///C:/Users/user/Downloads/04.pdf

10.11 SPRING STEEL

There are many mechanical mechanisms that use springs to provide force. These range from automotive accelerator pedals to sub-sea actuators. When springs are used in safety critical applications the strength and endurance performance must always match or exceed the product design requirements. Subsea actuators are held open by electrical power and in the event of power failure or an emergency command to close, the closure must be accomplished by the spring energy. Leaf and large helical springs are used on rail and automotive suspension applications. Smaller springs are critical items on engine valve springs. In all applications the spring quality begins with the selection and quality control of spring steel. "Materials for Springs"[1] provides an excellent guide on all aspects of spring steel technology. The document was prepared by the Japan Society of Spring Engineers and translated into English in 2007.

10.11.1 Spring Material Selection

Some of the simple but key guidelines are:

1. The selected material and manufacturing process should ensure that the quality of finished springs satisfies customers' quality requirements
2. Availability of selected material
3. The whole material cost and method of manufacture should meet the required cost
4. The manufacturing processes must not adversely affect the material quality
5. Due care and attention should be paid to environmental issues and the responsibility of "cradle to grave" concept and disposal issues.

Spring steels Composition

SUP9, Mn–Cr steel, shows good hot deformability and good hardenability and can be applied for the relatively large sized stabilizers, torsion bars, and coil springs. SUP9A, which is equivalent to SAE5160 steel, has basically the same chemical composition as the SUP9 but with a slightly higher C and higher range of Mn and Cr to improve its hardenability. SUP10 (ISO 13) is a Cr–V steel having good hardenability with high toughness, and is used for higher hardness application. SUP11A is the same material as boron treated SUP9A of Mn–Cr steel, and its hardenability is excellent, generally the mechanical properties are similar to SUP9A. The Si–Cr steel, SUP12, is mostly used for oil tempered wire material of cold-formed springs. SUP13 is Cr–Mo steel with higher hardenability than the boron treated SUP11A. The SUP13 (ISO 12) can be applicable for extremely large coil spring with over 100mm diameter.

10.11.2 Spring Steel Developments

Several published papers provide guidance of hardenability predictions[3 and 5] and estimation of fracture toughness from the Charpy energy and the Rockwell hardness.[2 and 7] In addition, there are studies on the spring critical defect size,[4] fatigue,[6 and 8] heat treatment[9] and hydrogen[13]. Warm shot peening at 220°C gave a 5 times improved life on a vehicle suspension helical spring[16]. This information can provide excellent assistance for the spring designer or for any failure investigations.

The work carried out determined that the relationship between surface hardness, Charpy toughness and fracture toughness was YS and hardness.[7]

(a) JIS (JIS G 4801-1984) Spring steel

Grade	Chemical compositions %								
	C	Si	Mn	P	S	Cr	Mo	V	B
SUP3	0.75–0.90	0.15–0.35	0.30–0.60	≤ 0.035	≤ 0.035	—	—	—	—
SUP6	0.56–0.64	1.50–1.80	0.70–1.00	≤ 0.035	≤ 0.035	—	—	—	—
SUP7	0.56–0.64	1.80–2.20	0.70–1.00	≤ 0.035	≤ 0.035	—	—	—	—
SUP9	0.52–0.60	0.15–0.35	0.65–0.95	≤ 0.035	≤ 0.035	0.65–0.95	—	—	—
SUP9A	0.56–0.64	0.15–0.35	0.70–1.00	≤ 0.035	≤ 0.035	0.70–1.00	—	—	—
SUP10	0.47–0.55	0.15–0.35	0.65–0.95	≤ 0.035	≤ 0.035	0.80–1.10	—	0.15 0.25	—
SUP11A	0.56–0.64	0.15–0.35	0.70–1.00	≤ 0.035	≤ 0.035	0.70–1.00	—	—	≥ 0.0005
SUP12	0.51–0.59	1.20–1.60	0.60–0.90	≤ 0.035	≤ 0.035	0.60–0.90	—	—	—
SUP13	0.56–0.64	0.15–0.35	0.70–1.00	≤ 0.035	≤ 0.035	0.70–0.90	0.25–0.35	—	—

(b) ISO 683-14 (1992-08-15) Spring steel (%)

No.	Steel Grade	ISO 683-14:1973 Grade	C	Si	Mn	P_max	S_max	Cr	Mo
1	59 Si 7	5	0.55–0.63	1.60–2.00	0.60–1.00	0.030	0.030	—	
2	56 SiCr 7	—	0.52–0.59	1.60–2.00	0.70–1.00	0.030	0.030	0.20–0.40	
3	61 SiCr 7	7	0.57–0.65	1.60–2.00	0.70–1.00	0.030	0.030	0.20–0.40	
4	55 SiCr 63	—	0.51–0.59	1.20–1.60	0.50–0.80	0.030	0.030	0.55–0.85	
5	55 Cr 3	8	0.52–0.59	0.15–0.40	0.70–1.00	0.030	0.030	0.70–1.00	
6	60 CrMo 31	—	0.56–0.64	0.15–0.40	0.70–1.00	0.030	0.030	0.70–1.00	0.08–0.15
7	60 CrB–3	10	0.56–0.64	0.15–0.40	0.70–1.00	0.030	0.030	0.60–0.90	— B : 0.0008min
8	60 CrMo 33	12	0.56–0.64	0.15–0.40	0.70–1.00	0.030	0.030	0.70–1.00	0.25–0.35
9	51 CrV 4	13	0.47–0.55	0.10–0.40	0.60–1.00	0.030	0.030	0.80–1.10	— V : 0.10–0.25
10	52 CrMoV 4	14	0.48–0.56	0.15–0.40	0.70–1.00	0.030	0.030	0.90–1.70	0.15–0.25 V : 0.07–0.15

(c) BS 970 : Part 2 (1988) Spring steel (%)

	Gr	C	Si	Mn	P_max	S_max	Cr	Mo	V	Ni	Cu	Sn
Si–Mn	251 A 58	0.55–0.60	1.80–2.10	0.80–1.00	0.035	0.035	0.15–0.30	0.10max	—	≤0.40	≤0.35	≤0.035
	251 A 60	0.57–0.62	1.80–2.10	0.80–1.00	0.035	0.035	0.25–0.40	0.12max	—	≤0.40	≤0.35	≤0.035
Alloy steel	525 A 58	0.55–0.60	0.20–0.35	0.80–0.95	0.035	0.035	0.70–0.85	0.10max	—	≤0.40	≤0.35	≤0.035
	525 A 60	0.57–0.62	0.20–0.35	0.85–1.00	0.035	0.035	0.80–0.95	0.06min	—	≤0.40	≤0.35	≤0.035
	525 A 61	0.57–0.63	0.20–0.35	0.85–1.00	0.035	0.035	0.85–1.00	0.08–0.15	—	≤0.40	≤0.35	≤0.035
	685 A 57	0.55–0.60	1.20–1.60	0.70–0.90	0.035	0.035	0.60–0.85	—	—	≤0.40	≤0.35	≤0.035
	704 A 60	0.57–0.62	0.20–0.35	0.85–1.0	0.035	0.035	0.80–0.95	0.15–0.25	—	≤0.40	≤0.35	≤0.035
	705 A 60	0.57–0.62	0.20–0.35	0.85–1.0	0.035	0.035	0.85–1.00	0.25–0.35	—	≤0.40	≤0.35	≤0.035
	735 A 51	0.48–0.54	0.20–0.35	0.70–1.00	0.035	0.035	0.90–1.20	—	0.10–0.20	≤0.40	≤0.35	≤0.035
	735 A 54	0.52–0.57	0.20–0.35	0.90–1.15	0.035	0.035	1.05–1.20	—	0.12–0.20	≤0.40	≤0.35	≤0.035
	925 A 60	0.55–0.65	1.70–2.10	0.70–1.00	0.035	0.035	0.20–0.40	0.20–0.30	—	≤0.40	≤0.35	≤0.035

Figure 10.168 JIS, ISO and BS grades of spring steel

Steel grade	Heat treatment		Mechanical properties				Hardness HB
	Quench (C°)	Temper (C°)	Yield strength (MPa)	Tensile strength (MPa)	El. % JIS No. 4 or No. 7	RA. % JIS No. 4	
SUP3	830–860	450–500	≥ 834	≥ 1079	≥ 8	—	341–401
SUP6	830–860	480–540	≥ 1070	≥ 1226	≥ 9	≥ 20	363–429
SUP7	830–860	480–540	≥ 1079	≥ 1226	≥ 9	≥ 20	363–429
SUP9	830–860	460–510	≥ 1079	≥ 1226	≥ 9	≥ 20	363–429
SUP9A	830–860	460–520	≥ 1079	≥ 1226	≥ 9	≥ 20	363–429
SUP10	840–870	470–640	≥ 1079	≥ 1226	≥ 10	≥ 30	363–429
SUP11A	830–860	460–520	≥ 1079	≥ 1226	≥ 9	≥ 20	363–429
SUP12	830–860	510–570	≥ 1079	≥ 1226	≥ 9	≥ 20	363–429
SUP13	830–860	510–570	≥ 1079	≥ 1226	≥ 10	≥ 30	363–429

Materials		Size obtained the 80% of martensite at the center, by oil quenching mm	
ISO Steel type No.	JIS steel type	thickness	diameter
2	SUP 3	8	12
5	SUP 6	8	12
6	SUP 7	14	20
8	SUP 9	18	28
9	SUP 9A	22	33
13	SUP 10	27	40
10	SUP 11A	24	35
12	SUP 13	47	70

Figure 10.169 Spring mechanical properties and maximum cross section with 80% martensite at the centre

$$K_{IC} = 3.965 CVN^{0.9537} HRC^{0.042}$$

and

$$\sigma_{ys} \approx 47.65 HRC^{0.9102}$$

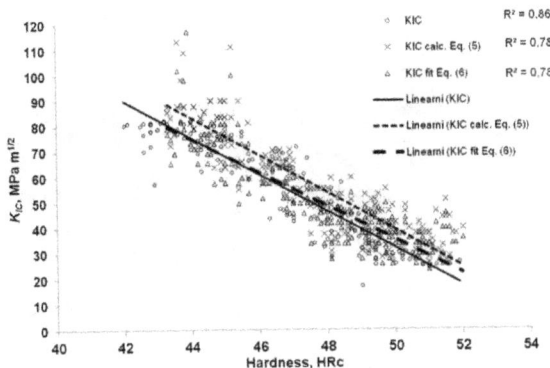

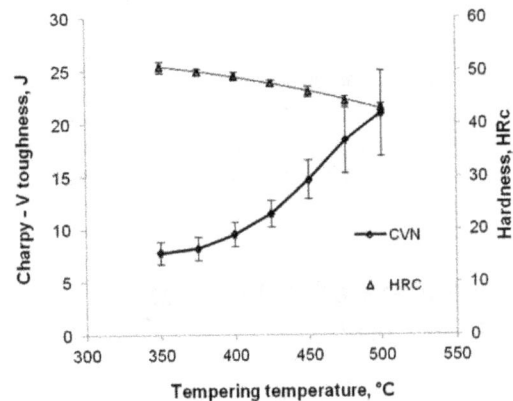

Figure 10.170 LHS = Comparison between experimentally obtained values of KIc and related hardness, and values calculated from CVN test results and respective hardness for 51CrV4 spring steel. RHS = Combined 'KIc–HRC–tempering temperature' diagram for continuously cast, hot rolled, flat spring steel 51CrV4 for austenitising temperature of 870°C

For hardenability of 51CrV4 from Reference 3

The linear regression model is:

$$63.462 - 0.284 \cdot D + 3.875 \cdot C - 0.981 \cdot Si + 1.845 \cdot Mn + 37.209 \cdot P + 1.445 \cdot S$$
$$- 2.365 \cdot Cr + 12.553 \cdot Mo - 17.477 \cdot Ni - 33.779 \cdot Al + 0.263 \cdot Cu + 18.319 \cdot Ti$$
$$- 10.336 \cdot V - 47.154 \cdot Sn + 93.117 \cdot N$$

$W = w(W)$; W = C, Si, Mn, P, S, Cr, Mo, Ni, Al, Cu, Ti, V, Sn, N
with average deviation for training data (74 batches) 4.25 %.

10.11.3 Other Topics

Comparison ERS spring 51CrV4 steel (0.003) S with a CCC (0.007) S Reference 6

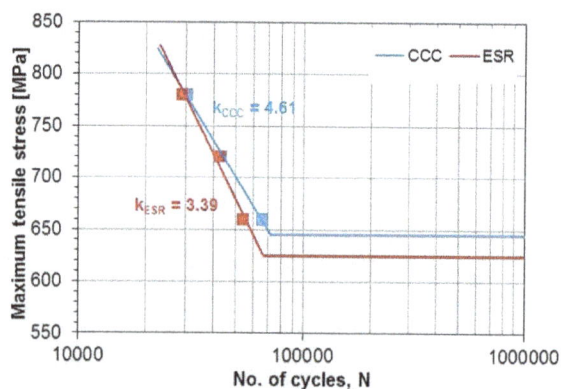

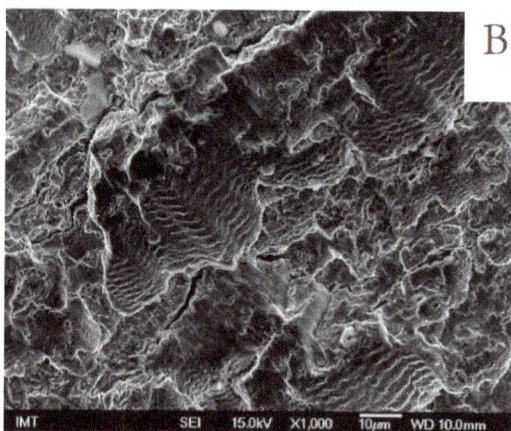

Figure 10.171 *LHS = Tensile-compression fatigue properties of CCC and ESR spring steel, tempered at 475°C. RHS = The ESR samples led to formation of staircase-like fracture regions as seen on the SEM image, with reduced crack propagation resistance and consequently to reduced fatigue*

10.11.4 Accelerated Spheroidizing Treatment

(AST) 2017 Reference 9 An induction based spheroidization treatment gave finer carbide distribution and allowed full dissolution of carbides at lower austenitising temperatures.

Austeniti-zation temperature [°C]	Hardness HV10 after treatment			
	SA, quenching	SA, quenching and tempering	ASR, quenching	ASR, quenching and tempering
800	624 ±2	406 ±5	677 ±4	433 ±7
820	650 ±2	419 ±3	705 ±7	443 ±8
840	660 ±9	424 ±5	692 ±7	446 ±9
860	679 ±7	434 ±8	706 ±9	447 ±2

Figure 10.172 *Induction heat treatment – ASR RHS = Hardness HV10 after quenching and tempering (SA – Conventional soft annealing; ASR – Accelerated Spheroidisation)[9]*

10.11.5 Hydrogen

The potential damage from hydrogen has been examined.[13] A critical level of above 0.6ppm was found to cause a loss of tensile ductility and to increase the risk of failure.

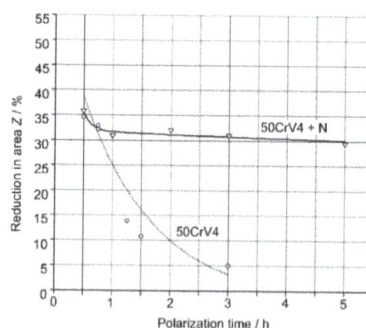

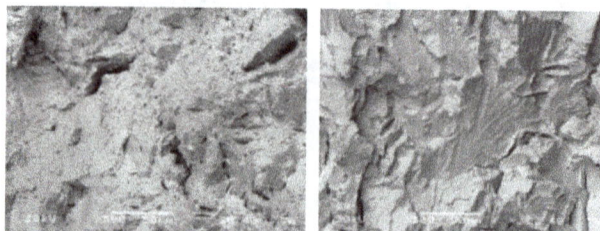

Figure 10.173 *LHS = Effect of hydrogen on tensile ductility maximum value of hydrogen for the 51CrV4 was recommended as H_{crit} = <0.60ppm. Centre = 0.5 ppm still ductile. RHS = 0.7 ppm and the fracture is brittle[13]*

10.11.6 Large helical springs

Large helical springs are manufactured from rods which are hot coiled in the form of a helix. The design parameters of a coil spring are the rod diameter, spring diameter and the number of coil turns per unit length. Some manufacturers have adopted a hot coil procedure followed by a direct oil quench practice which requires greater control compared to a cool followed by a conventional quench. Two important tables are shown in Figure 10.174. The LHS shows the type of loading pattern and the important factors to be considered. The table on the RHS shows the expected failure.

It is rare that a spring fails in service due to faulty design. The causes of failures are mainly related to poor microstructure and/or presence of stress concentration raisers such as inclusions and poor surface condition caused by the method of manufacture or the service environmental condition.

The adverse effect of inclusions on fatigue behaviour is well known. Their effect is more pronounced at high stress amplitudes. Residual stress on the surface is another well-known factor for influencing fatigue behaviour. (Spring relaxation can be reduced to acceptable levels by a process known as 'hot prestressing' or 'hot scragging' and cold scragging can be used to make the spring more stable.)[15] Tensile stress at the surface promotes fatigue failure, and compressive stress improves the fatigue behaviour. The effect of adverse residual stresses on the surface can be reduced either by proper stress relief treatment, cold scragging by compressing the spring solid and then releasing or by giving a shot peening operation, which imparts compressive stress on the surface. Recent published work suggests that warm shot peening at 220°C can give a 5 times improved life on a vehicle suspension helical spring[16]. Double shot peening and super surface finish can also improve the fatigue life[17]. This document has some SEM images of shot peened surfaces.

Raw material quality has an effect on coil springs' performance. The main requirements are:

- Good steel cleanliness
- Appropriate microstructure –normally tempered martensitic microstructure produced by an appropriate quench and temper treatment
- The avoidance of decarburization during manufacture especially during rod manufacture and heat treatment
- Since spring rod is primarily subjected to torsional stresses, maximum stress levels occur at the rod surface. Therefore, material surface defects (i.e. seams, laps, pits, corrosion pits etc.) can dramatically reduce a spring's fatigue life.

Corrosion is a more common cause of spring breakage and usually occurs when there is insufficient corrosion protection applied or the corrosion protection fails.

The stresses that act on the spring are also important (see Figure 175 and 178).

Visual examination of the spring shown in Figure 10.179 confirmed that this had resulted in heavy corrosion of the lower seat. Due to the relatively high hardness of the spring steel there would be a serious risk of "cathodic" charging with hydrogen and the occurrence of environmentally assisted cracking when the corrosive environment broke down the protective paint layer.

The fracture face showed evidence of a heavily oxidised initial crack which subsequently propagated probably when the stress levels increased when the valve was actuated. The scanning electron microscope examination of the fracture confirmed the presence of areas of inter-granular fracture and grain boundary

	Types of load	Figures to be acquainted	Properties required for material
Static load	A constant and invariable load (permitted insignificant variation of load)	Load and deflection	High elastic limit
Repeated load	Constant loads repeatedly applied	Mean load Load amplitude Deflection Number of cycles	High fatigue strength
Impact load	A load applied abruptly at high speed	Impact force Deflection Number of cycles	High elastic limit High impact value
Load for measuring load	Accurate load being ensured for a wide range of deflection, like a spring balance	Spring constant Maximum load	High elastic limit High dimensional accuracy

Fracture	Fracture with repeated stresses • Fatigue with no corrosion • Corrosion fatigue • Fatigue from fretting corrosion or wear Fracture with impact stress • Brittle fracture (Low temperature brittle fracture) • Ductile fracture Fracture with static stress • Stress corrosion cracking • Delayed fracture (Hydrogen embrittlement fracture)
Deformation (Permanent set)	• Yielding, plastic deformation (due to over stressing) • Static creep • Dynamic creep • Stress relaxation
Decrease of cross-sectional dimensions	• Wear • Fretting • General corrosion • Local corrosion • Errosion

Figure 10.174 *Top = Types of load of a spring, Bottom = Principal types of spring failures*[1]

widening, which are both features of a hydrogen embrittled microstructure. Corrosion in the presence of water especially saline water results in the type of corrosion known as aqueous corrosion. In one part of the steel surface the steel forms oxide/rust called the "anodic reaction" and in another area normally nearby called the "cathodic reaction", "nascent" hydrogen can be produced which can be absorbed into the spring steel. Due to the high hardness of the spring steel the hydrogen pick-up results in hydrogen embrittlement and reduced toughness. In this condition cracks can initiate and the service stresses eventually result in fracture and failure of the spring.

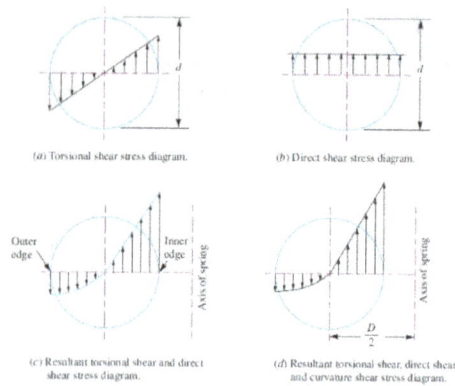

(a) Torsional shear stress diagram.

(b) Direct shear stress diagram.

(c) Resultant torsional shear and direct shear stress diagram.

(d) Resultant torsional shear, direct shear and curvature shear stress diagram.

Figure 10.175 *Summation of stresses acting on a compressive helical spring. Max stress on inside surface*[14]

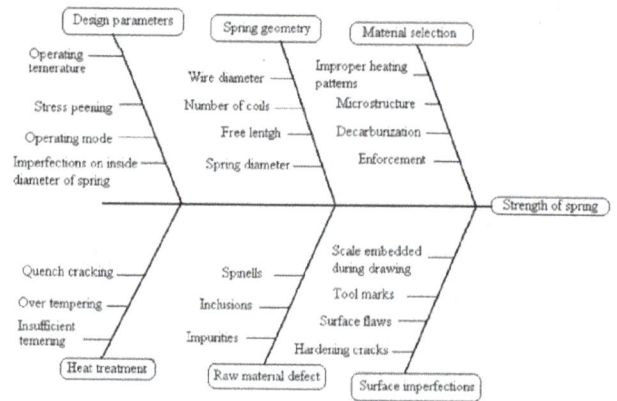

Figure 10.176 *Shows a "fishbone" diagram of the factors affecting the strength of a spring*

Figure 10.177 *Typical FEM mesh for a spring*

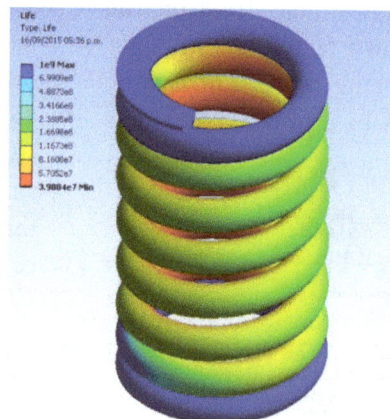

Figure 10.178
The stresses on a spring showing maximum stress on the inside surface[10]

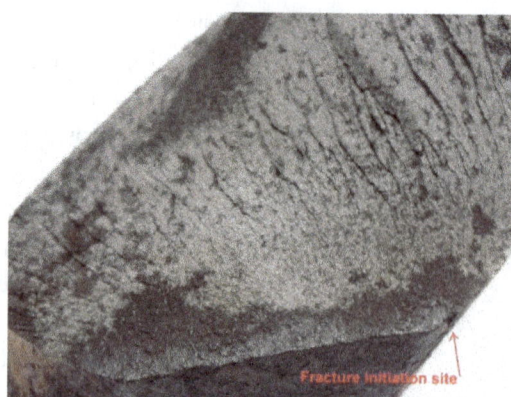

Figure 10.179 *LHS = Spring failure View of the spring from the direction of the upper seat. RHS = Shows the fracture surface after 3 hours' ultrasonic cleaning. The oxide at the spring seat remains intact and difficult to remove. This heavy oxide indicates an initial pre-crack had formed prior to the final fracture. The chevron marks suggest that the final fracture initiated at the inside corner of the spring which would be the position of maximum stress during operation*

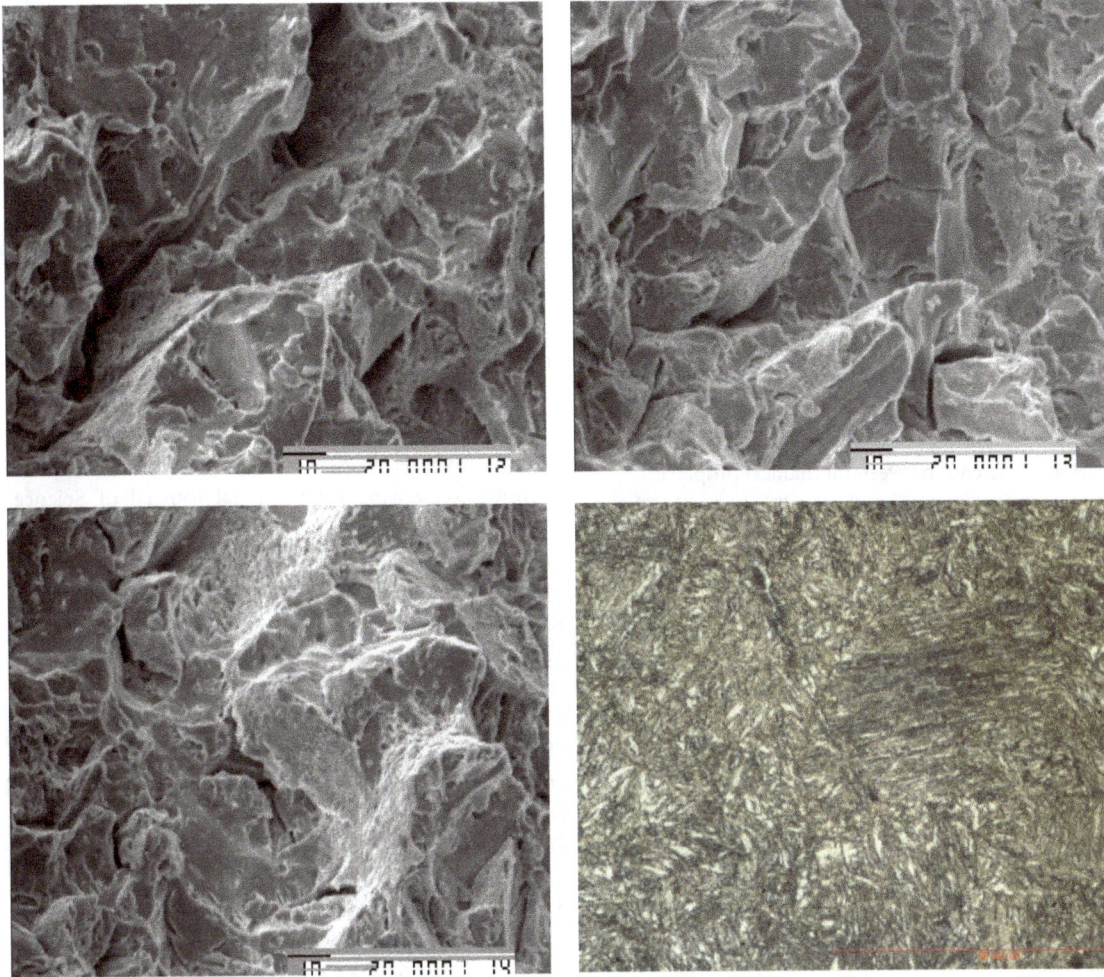

Figure 10.180 *SEM images of the fracture surface near the heavy rust showing indications of inter-granular fracture and grain boundary widening Lower RHS shows the quench and tempered microstructure*

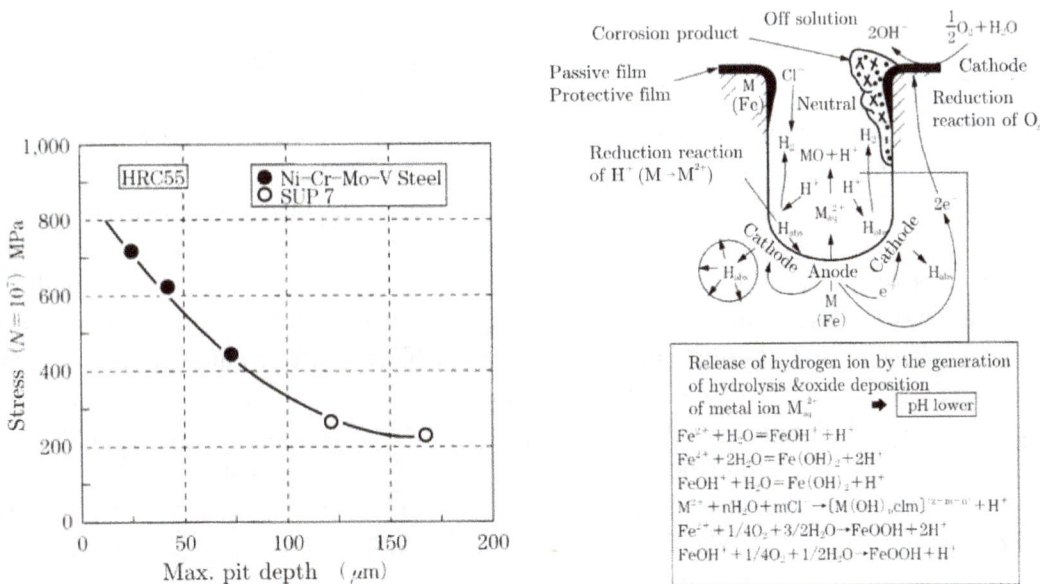

Figure 10.181 *LHS = Relationship between fatigue strength and corrosion pit depth of high strength steel. RHS = schematic diagram of the hydrogen pick-up caused by the hydrolysis in corrosive pit where the pH fall (acidic) due to rust the products of corrosion*

REFERENCES

1. Y. Yamada (chief Ed.), Materials for Springs, Japan Society of Spring Engineers, Springer-Verlag Berlin Heidelberg 2007

2. Bojan Sencic, FRACTURE TOUGHNESS OF THE VACUUM-HEAT-TREATED SPRING STEEL 51CrV4 / Materials and technology 45 (2011) 1, 67–73 http://mit.imt.si/Revija/izvodi/mit111/sencic.pdf

3. Miha Kovačič, Hardenability modelling, RMZ – Materials and Geoenvironment, Vol. 57, No. 2, pp. 173–180, 2010 http://www.rmz-mg.com/letniki/rmz57/RMZ57_0173-0180.pdf

4. Miha Kovačič, Critical inclusion size in spring steel and genetic programming, RMZ – Materials and Geoenvironment, Vol. 57, No. 1, pp. 17–23, 2010

5. Sanja Šolić et al, Difference between mechanical properties of 51CrV4 high strength spring steel modelled by hardenability software and obtained properties by heat treatment INFOTEH-JAHORINA Vol. 12, March 2013

6. B. Žužek et al, Effect of segregations on mechanical properties and crack propagation in spring steel, Frattura ed Integrità Strutturale, 34 (2015) 160-168

7. B. Sencic et al, Fracture toughness–Charpy impact test–Rockwell hardness regression based model for 51CrV4 spring steel Materials Science and Technology 2013

8. Borivoj Šuštaršič, FATIGUE STRENGTH AND MICROSTRUCTURAL FEATURES OF SPRING STEEL, STRUCTURAL INTEGRITY AND LIFE Vol. 11, No. 1 (2011), pp. 27–34 http://divk.inovacionicentar.rs/ivk/ivk11/027-034-IVK1-2011-BS-PB-WE-GG-AJ-BS.pdf

9. D. HAUSEROVA et al, STRUCTURE REFINEMENT OF SPRING STEEL 51CrV4 AFTER ACCELERATED SPHEROIDISATION, Arch. Metall. Mater. 62 (2017), 3, 1473-1477 file:///C:/Users/user/Downloads/[23001909%20-%20Archives%20of%20Metallurgy%20and%20Materials]%20Structure%20Refinement%20of%20Spring%20Steel%2051Crv4%20after%20Accelerated%20Spheroidisation.pdf

10. Rosendo Franco Rodríguez, Fatigue failure analysis of vibrating screen spring by means of finite element simulation: a case study, Presentations to the XIV International Conference on Computational Plasticity (COMPLAS 2017) file:///C:/Users/user/Downloads/download.pdf

11. Lauralice C.F. Canale et al, Overview of factors contributing to steel spring performance and failure Int. J. Microstructure and Materials Properties, Vol. 2, Nos. 3/4, 2007

12. V. Anilkumar et al, Heat Treatment Studies on 50CrV4 Spring Steel, Materials Science Forum Vols. 830-831 (2015) pp 139-142

13. U Rotnic et al, The influence of hydrogen in solid solution on spring steel, Metabk, 43 (2) 77-82 2004 file:///C:/Users/user/Downloads/MET_43_2_077_082_Rotnik%20(1).pdf

14. Dhiraj V. Shevale, Review on Failure Analysis of Helical Compression Spring International Journal of Science, Engineering and Technology Research (IJSETR) Volume 5, Issue 4, April 2016 http://ijsetr.org/wp-content/uploads/2016/04/IJSETR-VOL-5-ISSUE-4-892-898.pdf

15. R. G. Slingsby, A new heat-treatment process overcomes temperature relaxation problems for spring users https://www.shotpeener.com/library/pdf/1979030.pdf

16. M. R. Isa et al, IMPROVING THE FATIGUE LIFE OF COIL SPRING USING WARM SHOT PEENING METHOD, J Fundam Appl Sci. 2018, 10(7S), 30-39, file:///C:/Users/user/Downloads/5158-10351-1-PB.pdf

17. Doug Hombach, OPTIMIZATION OF SPRING PERFORMANCE THROUGH UNDERSTANDING AND APPLICATION OF RESIDUAL STRESS Spring Industry Technical Symposium 1999 111 https://www.lambdatechs.com/wp-content/uploads/275.pdf

18. Youli Zhu, Failure analysis of a helical compression spring for a heavy vehicle's suspension system, Case Studies in Engineering Failure Analysis 2 (2014) 169-173

10.12 OIL AND GAS

The design, manufacture and maintenance of equipment used to extract and process oil and gas has challenged several generations of engineers and metallurgists.

The current status is that the challenge has broadened and become more difficult since service conditions have become more demanding, with higher temperatures and higher pressures combined with corrosive conditions associated with sour fields (high hydrogen sulphide content) with high carbon dioxide levels. The need to cope with these conditions has to be set against a backdrop of the Macondo (Mexican Gulf) and Kashagan (Kazakhstan, the Caspian Sea) failures, which led to major loss of confidence within the Industry and the need for a major review of corporate policy, improved technical specifications and guidance to allow the rebuilding of the design, development and test activities.

There have been several major US governmental and industry discussions, debates and reviews to plan the way forward. The main conclusions are that the future challenges require proven technology with improved equipment reliability combined with the best available materials and methods of manufacture to ensure reliable and safe production of oil and gas. Low alloy steels (LAS) have already been replaced with a range of Corrosion Resistant Alloys (CRAs) that are better suited to meet increasingly demanding requirements in exploration and production and the demand for special alloy steels, stainless steels and nickel alloys for upstream oil and gas applications has grown substantially and this trend is expected to continue.

10.12.1 The range of CRAs and the selection procedure in the oil and gas industry

CRAs are typically defined as various stainless-steel grades and nickel alloys. Alloy steels and stainless steels account for around 95% of the tonnage used but the 5% of nickel alloy tonnage contributes 25% of the total value. The demand and selection of CRAs is driven by the production/reservoir conditions and include:

1. Corrosiveness (hydrogen sulphide, carbon dioxide, chloride)
2. Temperature
3. Pressure.

Corrosion engineers calculate the expected corrosion rate per year, which is then multiplied by the design life. When corrosion is expected to exceed a certain level, either CRAs are selected, or carbon steel with a higher wall thickness. Instead of solid CRA material, cladding or weld overlay can also be applied on carbon or alloy steels. Other options to prevent corrosion include the use of chemical inhibitors or cathodic protection.

The main guidance has been provided by NACE MR-0175 / ISO 15156-1 – "Petroleum and natural gas industries – Materials for use in H_2S-containing environments in oil and gas production". This document is a standard that all corrosion and material specialists refer to for guidance on H_2S corrosion related phenomena and material selection for the oil and gas industry. NACE MR-0175 / ISO 15156-1 is a detailed document divided into 3 parts. Part 1 is "General principles for selection of cracking-resistant materials". Part 2 is "Cracking-resistant carbon and low alloy steels, and the use of cast iron". Part 3 is "Cracking-resistant CRAs (corrosion-resistant alloys) and other alloys". Of particular interest are the annexures of parts 2 and 3 which provide actual corrosion data on the various grades of carbon steel and CRAs as well as give material selection guidelines for the various grades in a specific oilfield environment.

10.12.2 Balance between CAPEX and OPEX (capital expenditure and operational expenditure)

Safety is now paramount and the number one criterion. The selection of the material also depends on the calculated life cycle costs. Although CRAs require higher capital investment, they may well be the cheaper option in the entire life cycle as they have to be replaced less frequently, require less maintenance and do not need chemical inhibitors. Figure 10.182 LHS shows how the steel selection changes to more corrosion resistant grades as the service condition becomes more demanding. Carbon and alloy steels (LAS) are used whenever possible as these are the lowest cost materials.

The main material grades in the oil and gas industry:

- Alloy Steel: 4145H, 4130, 4140,4330, 8630 mod, F22
- Martensitic: 13Cr, Super 13Cr,410, 420, F6NM
- Austenitic: 316, 304, 321, 317L, Nitronic 50/60, 904L, 254SMO (6Mo)
- Duplex: 2205, 2507, LDX2101
- PH Grades: 17-4, 15-5, 13-8

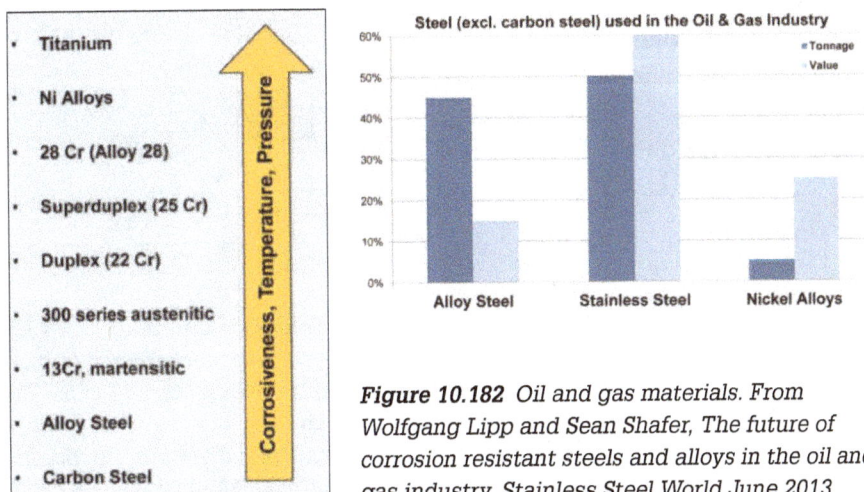

Figure 10.182 Oil and gas materials. From *Wolfgang Lipp and Sean Shafer, The future of corrosion resistant steels and alloys in the oil and gas industry, Stainless Steel World June 2013*

- Ni Alloys: 825, 625, 718, 925, Alloy 28
- Non-Magnetic: special chromium manganese austenitic grades

CRAs are required in all product forms including forgings, rolled rings, forged and rolled bars, seamless and welded tube and pipe, clad pipe, plates, sheet and strip.

Nickel-based superalloys are increasingly finding applications in the oil and gas sector. Nickel-base alloys 718, 725, and 925 are commonly used in oil and natural gas production. These alloys contain chrome and molybdenum which aid in resisting corrosion. Alloy 718 was initially developed for use in aerospace and gas turbines, but has become the preferred material for the manufacture of wellhead components, auxiliary and down-hole tools, and subsurface safety valves.

10.12.3 Applications for CRAs

Special alloy steels and CRAs are required for a wide range of applications, especially on offshore production. They are used for drill string components, tubing and casing, downhole completion equipment, wellheads, blowout preventers, subsea trees, manifolds, riser systems, flowlines, umbilical, valves, pumps, topside processing equipment. Here are some examples listed:

- Seamless tubes are used for tubing and casing (OCTG). Depending on the fluid composition either 13Cr, Super duplex or nickel alloy grades are selected. Cold working is required to achieve the required strength
- Welded clad pipes are often liner (e.g. 316L, 825, 625). Two different technologies are available: metallurgical and mechanical bonding
- Instead of rigid flowlines and risers, flexible pipes can also be used. The carcass of these pipes is made of cold rolled coils. The main grades are 316L, lean duplex, duplex and super duplex
- Large size forgings are required, e.g. for blow-out preventers and subsea trees. They are usually made of alloy steel (e.g. F22, 8630, 4130) and a nickel alloy weld overlay (grade 625) is applied on all process wetted areas to prevent corrosion. Smaller forgings, e.g. for valve bodies, are often made of duplex grades
- Large size, thin wall, welded pipes are used for piping for LNG projects (mainly 304L and 316L). Cryogenic transfer lines for LNG consist of Invar

(Alloy 36) or 9% nickel steel. Invar is also applied for membranes of LNG tankers.

10.12.4 Carbon structural steels

The main supply conditions:

- As Rolled
- Normalise Rolled
- Thermo-mechanically Control Rolled (TMCR)
- Normalised
- Quenched and Tempered
- Shot-blast and Primed.

In the past, S355J2 type material was supplied Normalised.

Development of HSLA means supplied NR to ~50mm

Typical applications:

- AR – low strength – mild steel / structural applications (275 yield)
- NR and N – high strength e.g. bridges (355 yield)
- TMCR / TM&AC – very high strength and excellent toughness e.g. pipelines (460 yield)
- Q&T – highest strength and toughness (690 yield and beyond).

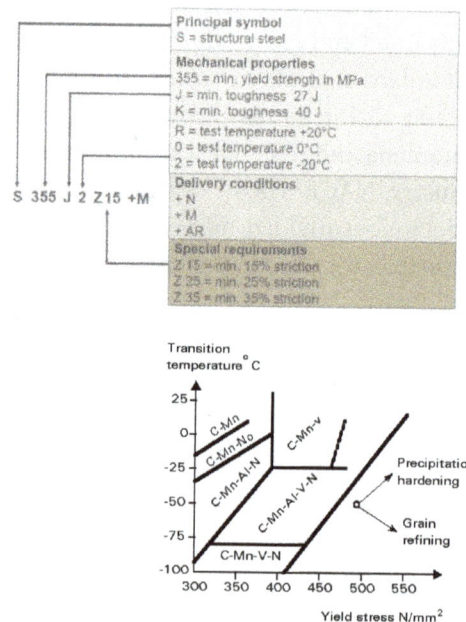

Figure 10.183 *LHS = Designations according to BS EN 10025 2004[1]. RHS = strength toughness ranges for normalised ferrite /pearlite steels.[1] Oliver Hechler et al, The right choice of steel according to the Eurocode, http:// sections.arcelormittal.com/uploads/tx_abdownloads/ files/25_The_right_choice_of_steel_Ver2.pdf*

Product Standards in the Steel Industry

Hot Rolled Products of Non-Alloy Structural Steels

BS EN 10025-1:2004 General Requirements

BS EN 10025-2:2004 Non-Alloyed

BS EN 10025-3:2004 Normalised / Normalise Rolled Fine Grained

BS EN 10025-4:2004 Thermomechanical Controlled Rolled Fine Grained

BS EN 10025-5:2004 Improved Atmospheric Corrosion Resistance

BS EN 10025-6:2004 High Yield Quench & Tempered

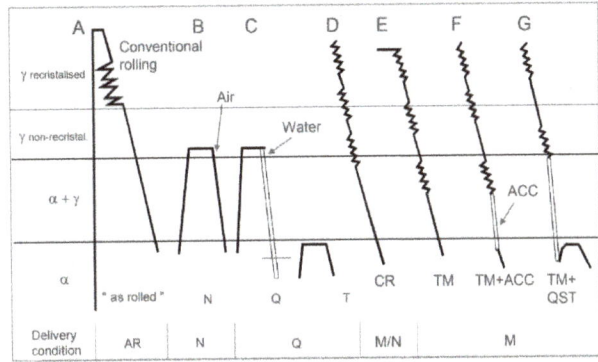

Figure 10.184 *Explanation of delivery conditions*[1]

Product Standards in the Steel Industry

Hot Rolled Products of Non-Alloy Structural Steels

Normative References

BS EN 10021:1993 General Technical Delivery Requirements of Steel Products

BS EN 10029:1991 Tolerances on Hot Rolled Plate

BS EN 10034:1993 Tolerances on I & H Sections

BS EN 10204:2004 Metallic Products – Types of Inspection Documents

BS EN 10164:2004 Steel Products with Improved Perpendicular Properties

BS EN 10160:1999 Ultrasonic testing of Flat Steel Products

BS EN 10052:1994 Heat Treatment vocabulary

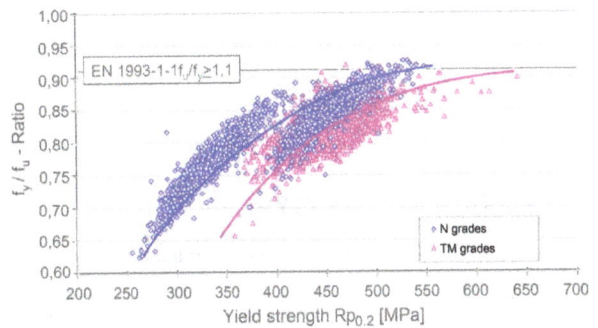

Figure 10.185 *YS/UTS ratio*[1]

Failure mode	Deformation of crystal lattice	Fractography
Ductile failure – shear – slipping – toughness – dull		
Brittle fracture – cleavage – decohesion – brittleness – shiny		

Figure 10.171 *Toughness*[1]

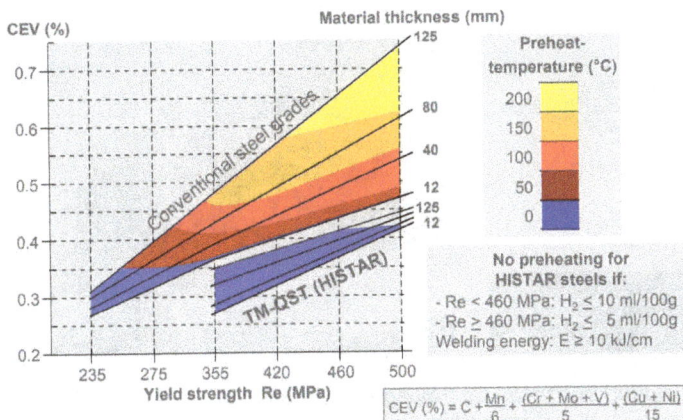

$$CEV (\%) = C + \frac{Mn}{6} + \frac{(Cr + Mo + V)}{5} + \frac{(Cu + Ni)}{15}$$

Figure 10.186 *Weld preheat in accordance with EN 1011-2*[1]

Proposed new modified grades of low alloy steels (LAS) After Macondo ASTM A1099/ A1099M 2017 For AISI 4130, 4140,8230, ASTM A182 F22 (2.25Cr 1Mo)

Table 2 – Grade Chemical Composition																						
Grade	C	Mn	P	S	Si	Cr	Mo	Ni	Cu	V	Ti	Al	Sn	As	Sb	Pb	Bi	N	H	O	Cb	B
F22OF	0.10-0.15	0.30-0.60	0.015	0.010	0.15-0.50	2.0-2.50	0.87-1.13	0.50	0.25	0.04	0.025	0.055	0.015	0.020	0.020	0.02	0.010	0.0120	2 PPM	25 PPM	0.02	0.0005
F22OFA	0.15-0.20	0.40-0.80	0.015	0.010	0.50	2.0-2.50	.90-1.10	0.50	0.25	0.04	0.025	0.055	0.015	0.020	0.020	0.02	0.010	0.0120	2 PPM	25 PPM	0.01	0.0005
4130OF	0.25-0.33	0.60-0.90	0.015	0.010	0.20-0.35	1.20-1.50	0.65-0.75	0.25	0.25	0.04	0.025	0.055	0.015	0.020	0.020	0.02	0.010	0.0120	2 PPM	25 PPM	0.02	0.0005
8630OF	0.27-0.33	0.80-0.95	0.015	0.010	0.20-0.35	0.85-1.00	0.35-0.45	0.80-0.90	0.25	0.04	0.025	0.055	0.015	0.020	0.020	0.02	0.010	0.0120	2 PPM	25 PPM	0.02	0.0005

Notes:
1) Chemical composition by weight %.
2) Specified values are considered maximum unless otherwise specified.
3) Calcium may be added for inclusion shape control. Amount of calcium shall not exceed .002%.

10.13 NICKEL ALLOYS

(Reference 1, dated 1968, provides a comprehensive review of nickel alloys.)

There is evidence that nickel alloys have been used since the Bronze Age. However, at that time the existence of nickel was not known and the only possible source of the alloy was meteorites that had landed. It is now known that these were collected and forged into armaments! Recent XRF chemical analysis checks have confirmed the presence of nickel, iron and cobalt in some of the early artefacts such as the dagger found inside the sarcophagus of Tutankhamun. The composition was nickel (10 wt%) and cobalt (0.6 wt%), which was typical of the concentrations found in iron meteorites.

A major driving force for the development of nickel alloys was the development of "Whittle's Jet Engine". His initial patent was in 1930 and the first engines were built in the early 1940s. These engines needed improved high temperature alloys since the first engines lasted 25 hours before a rebuild. Whittle had difficulty with funding the development of his engine due to the negative advice given to the Air Ministry in 1929, by senior RAE scientist Dr Arnold Griffith (the same A Griffith that gave us fracture toughness).

"Encouraged by his commanding officer, in late 1929 Whittle sent his concept to the Air Ministry to see if it would be of any interest to them. With little knowledge of the topic they turned to the only other person who had written on the subject and passed the paper on to Griffith. Griffith appears to have been convinced that Whittle's "simple" design could never achieve the sort of efficiencies needed for a practical engine. After pointing out an error in one of Whittle's calculations, he went on to comment that the centrifugal design would be too large for aircraft use and that using the jet directly for power would be rather inefficient. The RAF returned his comment to Whittle, referring to the design as being "impracticable."[2]

© Wiley
This dagger and gold sheath were one of the many artefacts found in King Tutankhamun's tomb

Figure 10.187 *2014 researchers revealed that a dagger found inside the sarcophagus of Tutankhamun has a blade made from meteoric iron. It had levels of cobalt, nickel and phosphorus[2]. RHS = Whittle's Jet Engine*

The evolution of modern nickel-base superalloys was closely linked with the development of Whittle's gas-turbine engine. This provided significant motivation for the development of superalloys for high-strength, high-temperature applications. The Mond Nickel Company, in Wales, developed the first precipitation hardenable nickel-base superalloy. The alloy, developed in 1941, was called Nimonic 80 and used titanium to achieve precipitation hardening.

Pure nickel was first extracted by Baron Axel Fredrik Cronstedt, a Swedish mineralogist and chemist in 1751. It was possibly known to exist much earlier since Chinese documents from around 1500BC make reference to 'white copper' (baitong), which was very likely an alloy of nickel and silver.

The steel industry identified nickel as a useful alloy element in the late 1890s and in the following years the steel industry used significant tonnage of nickel. With the development of stainless steels in the 1930s, nickel was found to improve the stability of the protective oxide film that provides corrosion resistance. Its major contribution was in conjunction with chromium in austenitic stainless steels, in which nickel enables the austenitic structure to be retained at room temperature.

Nickel has a relatively high melting point of 1,453°C and a face-centred cubic crystal structure, which gives the metal good ductility. Nickel alloys exhibit a high resistance to corrosion in a wide variety of media and have the ability to withstand a range of high and low temperatures.

As an electroplated coating, nickel is widely used in several industries for corrosion and wear resistance (electroless nickel).

10.13.1 Monel

Nickel–copper alloys possess excellent corrosion resistance, notably in seawater. Monel was developed by metallurgist Robert Crooks Stanley and patented in 1905 by the International Nickel Company. The

Designation	Cu %	Al %	Ti %	Fe %	Mn %	Si %	Ni %
Monel 400	28-34	-	-	2.5 max.	2.0 max.	-	63 min.
Monel 405	28-34	-	-	2.5 max.	2.0 max.	0.5 max.	63 min.
Monel K-500	27-33	2.3-3.15	0.35-0.85	2.0 max.	1.5 max.	-	63 min.

Figure 10.188 Monel alloys

metal was given the name of a director called Monel of International Nickel Company. By 1908, Monel was being used as a roofing material for Pennsylvania Station in New York.

During the 1920s and later, Monel was used for countertops, sinks, appliances, and roof flashing. While Monel was among the most popular metals on the market through the 1940s, it was largely replaced by the more versatile stainless steels from the 1950s onward. The Monel (~30% Cu) series of alloys is used for turbine blading, valve parts and for marine propeller shafts, because of their high fatigue strength in seawater.

10.13.2 Nickel–chromium alloys and superalloys

These form the basic alloys for jet engine development – the nickel-based superalloys, e.g. the Nimonic alloys. The earliest of these, Nimonic 80A, was essentially 'Nichrome' (80/20 Ni/Cr) precipitation hardened by the γ' phase (Ni_3Ti,Al).

There are three groups of superalloys, which are named on the predominant metal present in the alloy. They are:

- Nickel-based superalloys
- Iron-based superalloys. The AISI 600 series
- Cobalt-based superalloys (basically developed by Haynes based on Stellite alloys) used for components in the combustion chamber, for temperatures up to1100°.

Alloy	Percent										
	C	Si	Cu	Fe	Mn	Cr	Ti	Al	Co	Mo	B	Ni
Nimonic alloy 75	0.08/0.15	1.0 max	0.5 max	5.0 max	1.0 max	18 /21	0.2/0.6	----------	2.0 max	----------	0.008 max	bal
Nimonic alloy 80A	.1 max	1.0 max	.2 max	3.0 max	1.0 max	18 /21	1.8/2.7	1.0/1.8	2.0 max	----------		bal
Nimonic alloy 90	.13 max	1.5 max	----------	3.0 max	1.0 max	18 /21	1.8/3.0	.8/2.0	18 /21	4.5/5.5		bal
Nimonic alloy 105	.2 max	1.0 max	.5 max	2.0 max	1.0 max	13.5/15.75	0.9/1.5	4.5/4.9	18 /22	3 /5		bal
Nimonic alloy 115	.20 max	1.0 max	.2 max	1.0 max	1.0 max	14 /16	3.5/4.5	4.5/5.5	13.5/16.5			bal

Figure 10.189 Chemical composition of the Nimonic alloys[1]

Nickel-base superalloys currently constitute over 50% of the weight of advanced aircraft engines. The nickel base superalloys are modifications of the heat-resistant, corrosion-resistant alloys and many of the nickel alloys can be referred to as superalloys. The high-nickel superalloys are typically of the Al-Ti age-hardenable type. In these alloys, chromium is present to provide oxidation resistance along solid solution strengthening elements, niobium, molybdenum, tungsten, and tantalum. The major part of the strengthening at high temperatures, however, is due to the precipitation of the Ni_3(Al, Ti) compound, generally referred to as the gamma prime phase. Fig. 10.190 shows the strengthening effect of additions of Al+Ti.[2]

Subsequent alloy development involved composition changes to increase the volume fraction of the γ' phase. This made the alloys difficult to forge into turbine blades and eventually the higher γ' volume fraction alloys were manufactured by casting. (See Figure 10.176.) Unfortunately, the initial cast superalloys had a lack of creep ductility due to cavitation at the grain boundaries lying perpendicular to the maximum tensile stress. Consequently, the casting process was changed to directionally solidified (DS) and then to single crystal (SC) casting methods.

By incorporating channels into the turbine blades through which cooling air was passed, which was originally suggested by Whittle, it was possible to use the alloys at significantly higher engine temperatures than was possible with uncooled blades. Their maximum operating temperature is limited by the tendency of the γ' phase to return into solid solution and, thus, by choosing an insoluble phase, powder metallurgically produced oxide dispersion-strengthened (ODS) superalloys have been developed by a technique known as mechanical alloying.

More recent increases in operating temperatures have been achieved by deposition of thermal barrier coatings (TBCs) on high-temperature gas turbine components. TBCs are complex films (typically 100 μm to 2 mm in thickness) of a refractory material that protect the metal part from the extreme temperatures. One important TBC material is yttria-stabilized zirconia, or YSZ. In spite of their high creep strength, the low ductility of present ceramics limits their use in bulk for the turbine blades themselves.

Figure 10.190 RHS illustrates how, over the years, the gas turbine entry temperature has been progressively increased as the properties of the superalloy blades have been enhanced by the methods referred to

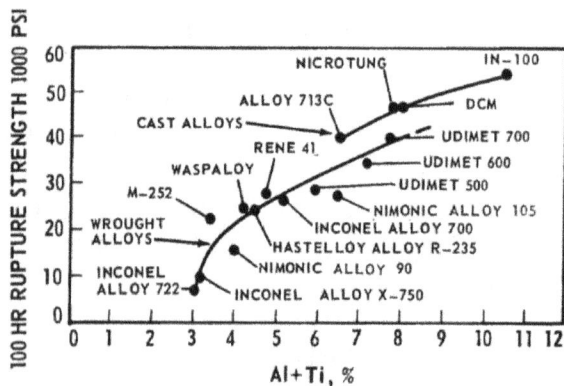

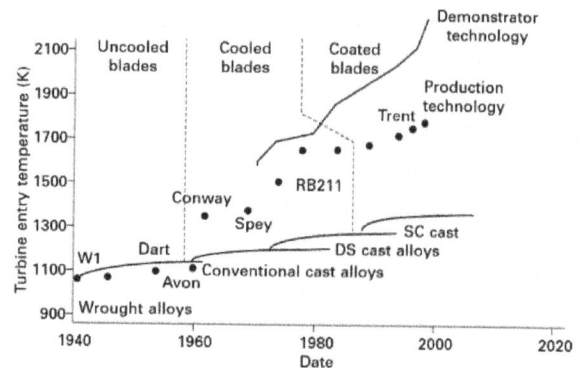

Figure 10.190 LHS = The effect of aluminium plus titanium content on the 100-hr rupture life at 1600 "F of several high-temperature nickel alloy[1]. RHS = Development of turbine blades

Alloy	Percent							
	Ni	Mo	Fe	Cr	W	Si	Cu	Others
Hastelloy alloy B.	62	28	5	---		---	---	
Hastelloy alloy C.	54	17	5	15	4	---	---	
Hastelloy alloy D.	85	---				10	3	
Hastelloy alloy F.	47	7	17	22		---	---	
Hastelloy alloy G.	44	6.5	20	22.2	1.0 max	---	2.0	Co 2.5 max. 2.1 Cb+Ta
Hastelloy alloy N.	70	17	5	7	---	---	---	
Hastelloy alloy W.	62	24.5	5.5	5	---	---	---	
Hastelloy alloy X.	47	9	18	22	---	---	---	

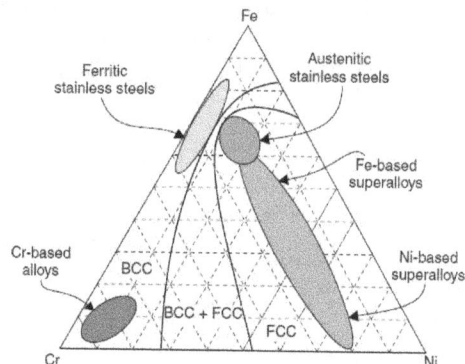

Figure 10.191 LHS = Hastelloy compositions. RHS = Constitution of the ternary system nickel-chromium-iron 650°C

above. The lower series of curves illustrate the behaviour of uncooled, uncoated blades of the wrought or cast alloys. The series of dotted lines are data for production materials in specific Rolls Royce aeroengines (the names of which are shown). This illustrates the magnitude of the additional benefits obtained with cooled blades and with cooled and coated blades. Even greater high temperature capability has been demonstrated on a research (rather than production) basis and this is indicated by the upper line of the diagram.

10.13.3 Hastelloy

The first of the high nickel corrosion-resistant alloys known as Hastelloy alloys were nickel-molybdenum-iron alloys, but in subsequent alloys the composition was considerably changed.

Hastelloy alloy B is notable for its unusually high resistance to all concentrations of hydrochloric acid at temperatures up to the boiling point.

Hastelloy alloy C possesses an unusual degree of resistance to oxidizing solutions, especially those containing chlorides, and to hypochlorite solutions and moist chlorine.

10.13.4 Inconel (tm)

Inconel (tm) is a specialty alloy that uses higher percentages of Nickel and Chrome than Stainless steel, as well as many other elements in small quantities. It is

Type	% Ni	% Cr	% C	% Mn	% Si	% Fe	% S	% Cu	% Al	% Ti	% P	% Co	% Nb	% B	% Mo
600	72.0 min	14.0 - 17.0	0.15 max	1.0 max	0.5 max	6.0 - 10.0	.015 max	0.5 max	0	0	0	0	0	0	0
601	58.0 - 63.0	21.0 - 25.0	0.1 max	1.0 max	0.5 max	bal	.015 max	1.0 max	1.0 - 1.7	0	0	0	0	0	0
625	58.0 min	20.0 - 23.0	0.1 max	0.5 max	0.5 max	5.0 max	0.015 max	0	0.4 max	0.4 max	.015 max	1.0 max	3.15 - 4.15	0	8.0 - 10.0
718	50.0 - 55.0	17.0 - 21.0	0.08 max	0.35 max	0.35 max	bal	.015 max	0.3 max	0.2 - 0.8	0.65 - 1.15	.015	1.0 max	4.75 - 5.5	.006 max	2.8 - 3.3
800	32.5	21.0	0.1 max	0.8 max	.008 max	46.0	0	0.4	0.4	0.4	0	0	0	0	0

Figure10.192 Composition of chemical analysis

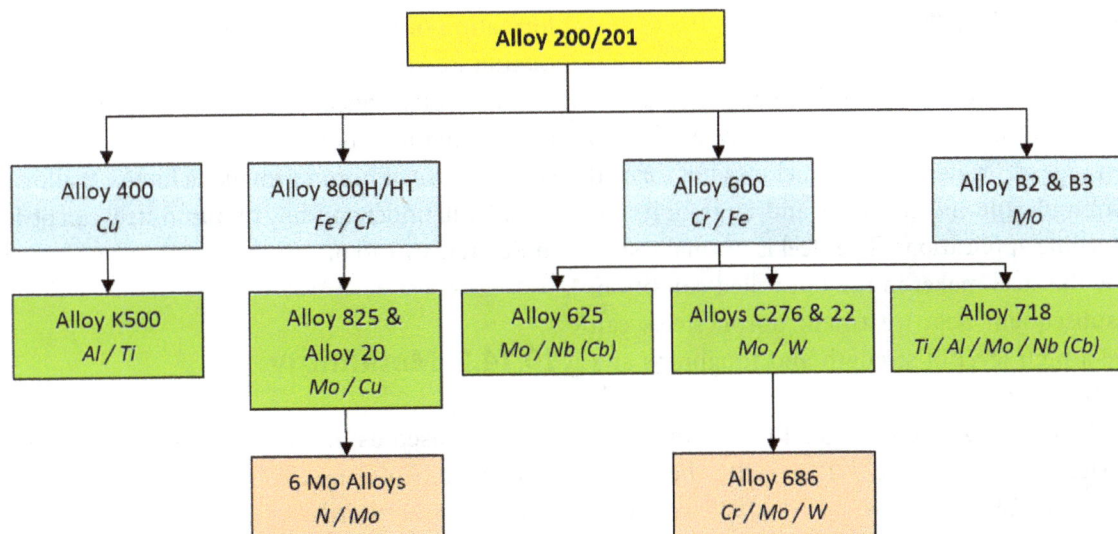

Figure 10.193 The Inconel Family Tree

actually a trademark name of Inco Alloys International and is in a group of metals known as the Nickel based super alloys.

The 80 Ni-14 Cr-6 Fe alloy, made by adding ferrochromium to nickel and originally designated as Inconel, was first offered in 1932 to the dairy industry for its resistance to corrosion by milk. It is now designated as Inconel alloy 600 with a slightly different composition.

Inconel alloy 625 is a high strength corrosion resistant material in which the nickel-chromium matrix is solid-solution strengthened by additions of molybdenum and columbium.

Inconel alloy 718 is a nickel-chromium-iron-molybdenum alloy made age-hardenable by the addition of niobium. The microstructure of this alloy is shown in (ref 7). It has a number of unique characteristics which distinguish it from the family of nickel-chromium-iron alloys hardened by aluminium and titanium.

Incoloy alloy 800 is an austenitic solid solution alloy and was developed to provide a material of good strength and resistance to oxidation and carburization at elevated temperatures. Some of its more important uses are in the industrial heating field for furnace equipment, baskets, trays, muffles, radiant tube.

These small additions of other elements results in solid-solution hardening. It is quite expensive and therefore usually reserved for applications when some type of stainless steel won't suffice. Figure 10.68 compares the chemical analysis of Inconel alloys.

Corrosion resistance

One outstanding characteristic of high-nickel alloys, such as Inconel (tm), is their good resistance to a wide variety of corrosives. In general terms, high-nickel alloys perform better than martensitic, ferritic, and austenitic stainless steels in corrosive environments.

REFERENCES

1. Samuel J. Rosenberg, Nickel and Its Alloys, National Bureau of Standards Monograph 106 Issued May, 1968, https://nvlpubs.nist.gov/nistpubs/Legacy/MONO/nbsmonograph106.pdf
2. https://www.chemistryworld.com/news/tutankhamuns-burial-dagger-is-extra-terrestrial-in-origin/1010362.article
3. Chester T. Sims, A HISTORY OF SUPERALLOY METALLURGY FOR SUPERALLOY METALLURGISTS https://www.tms.org/superalloys/10.7449/1984/Superalloys_1984_399_419.pdf
4. C. H. Lund, PHYSICAL METALLURGY OF NICKEL-BASE SUPERALLOY, DEFENSE METALS INFORMATION CENTER Battelle Memorial Institute 1961 https://apps.dtic.mil/dtic/tr/fulltext/u2/258041.pdf
5. N. Clement et al, LOCAL ORDER AND MECHANICAL PROPERTIES OF THE y MATRIX OF NICKEL-BASE SUPERALLOYS, https://www.imeche.org/news/news-article/thesis-on-the-development-of-the-jet-engine 3rd April 2012)
6. Enes Akca, A Review on Superalloys and IN718 Nickel-Based INCONEL Superalloy, PERIODICALS OF ENGINEERING AND NATURAL SCIENCES Vol. 3 No. 1 (2015) http://pen.ius.edu.ba/index.php/pen/article/viewFile/43/47
7. API STANDARD 6A718 THIRD EDITION, Age-Hardened Nickel-Based Alloys for Oil and Gas Drilling and Production Equipment

10.14 BORON STEEL

Boron steels are now accepted by many users as an alternative to alloy steel and they are now applied to a wide range of products. These includes hot formed high strength automotive parts and large bolts for wind turbine applications. The steel is normally supplied by the steel maker in the hot rolled condition. With subsequent heat treatment boron steels can achieve a high level of strength and toughness at a reasonable price.

The first studies of the original development of boron steel began in the mid-1920s with the initial US patent by Walters US1509624.[1] In 1948 work carried out by the National Bureau of Standards was the first to suggest that the effect of boron was to delay the

formation of ferrite during the cooling and transformation from austenite.[2]

The metallurgical principles and practice regarding the manufacture of boron steel are important since the use of boron provides a low-cost alloy addition. A full understanding of boron steel can be found in References 1 to 6.

10.14.1 Metallurgy

Boron is used as an alloy addition to increase the hardenability of a steel. The addition of 0.001% boron can have a hardenability effect equivalent to 0.5% chromium. Therefore, the use of boron is a very cost-effective alloy addition. The quench and tempering of an alloy steel to give a high proportion of the preferred microstructure of tempered martensite requires an appropriate level of hardenability relative to the section size and quenchant used. The heat treatment of an alloy steel needs to provide two properties: the strength, but also the degree of toughness to prevent brittle failure. Obviously, the toughness aspect of steel selection is of great practical importance since attainment of high strength is of little value unless sufficient toughness exists to meet service requirements without risk of premature fracture.

As discussed in Chapter 3 "Hardenability" refers to the depth of hardening or to the size of the piece of steel which can be fully hardened with a minimum of 50% martensite at the centre of the part. The hardness level achieved in a steel is almost entirely dependent on the carbon content, while hardenability is primarily dependent on the alloy content. (The carbon level and austenitic grain size also effect hardenability.) To achieve the best combination of strength and toughness in alloy steels it is necessary to transform

the microstructure from austenite to the lower temperature transformation product martensite (lower bainite may also be satisfactory for certain applications). Plain carbon steels have low hardenability, and the lower the carbon content, the lower the hardenability. In engineering applications there is a need to increase the alloy content, to achieve the required hardenability, to allow the quench and temper treatment to produce a high percentage of martensite.

Early research established that all engineering steels, having sufficient hardenability to quench to 90% to 100% martensite, will all have similar mechanical properties at any given hardness level within the range of 200 to 400 Brinell hardness after tempering. The tensile data shown in Table 2 (Figure 10.194) illustrates that similar mechanical properties are obtained.

These steels have different chemical analysis ranges and therefore different levels of hardenability, and in heavier sections some grades of steel would not provide equivalent properties since they would probably

Table 2. PROPERTIES OF COMMERCIAL STEELS AT 150,000 PSI YIELD STRENGTH

Grade	Quench	Grain Size	Temper (°F)	C	Mn	Si	Ni	Cr	Mo	UTS (ksi)	Elon. (%)	R.A. (%)	Brinell Hardness Number
6130	H_2O	6-8	1025	0.33	0.61	0.18	-	1.03	-	160	18	58	341
2330	H_2O	6-8	840	0.31	0.70	0.26	3.45	-	-	163	15	61	331
4140	H_2O	6-8	925	0.31	0.53	0.28	-	1.04	0.20	165	15	57	331
8630	H_2O	6-8	950	0.30	0.80	0.27	0.65	0.48	0.18	160	16	64	331
86B30	Oil	6-8	840	0.33	0.62	0.24	0.31	0.28	0.13	162	17	60	331
1340	Oil	6-8	925	0.43	1.70	0.23	-	-	-	160	15	55	331
3140	Oil	6-8	925	0.39	0.76	0.25	1.20	0.65	-	167	16	61	331
4140	Oil	6-8	1020	0.41	0.85	0.20	-	1.01	0.24	165	16	55	331
4340	Oil	6-8	1050	0.41	0.67	0.26	1.77	0.78	0.26	163	17	58	341
4640	Oil	6-8	975	0.41	0.70	0.24	1.83	-	0.28	163	17	56	341
8740	Oil	6-8	1100	0.39	1.00	0.25	0.53	0.52	0.28	160	16	57	331
9440	Oil	6-8	925	0.39	1.06	0.28	0.39	0.32	0.11	165	16	59	331
4150	Oil	7-8	1160	0.50	0.76	0.21	-	0.95	0.21	165	15	54	341
5150	Oil	7-8	1000	0.49	0.75	0.25	-	0.80	-	160	15	53	331
6152	Oil	6-8	1125	0.49	0.78	0.29	-	1.00	-	160	16	51	331
8750	Oil	6-8	1040	0.51	0.80	0.24	0.53	0.52	0.25	166	14	50	341

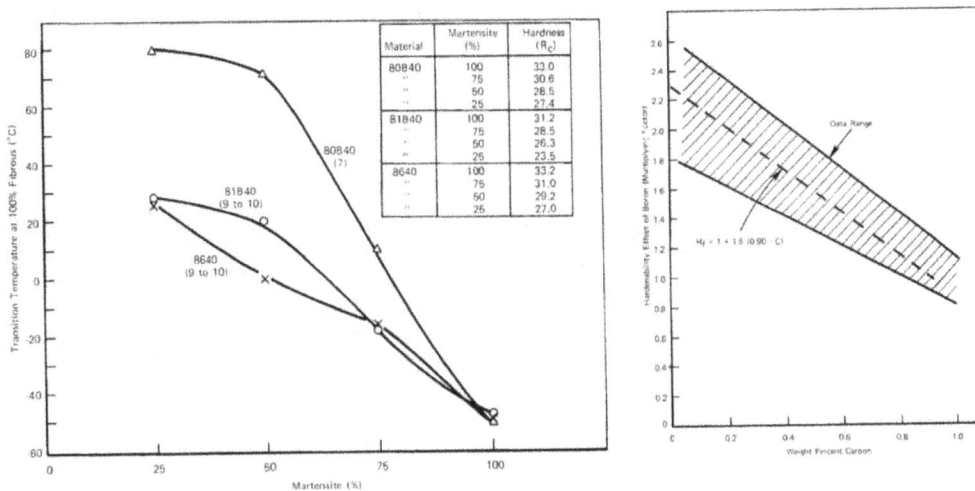

Figure 10.194 Comparison of impact transition temperatures for two boron and one non-boron steel as a function of martensite (ASTM grain size marked in brackets). With 10% martensite low ITT. RHS = Relationship between boron factor and carbon level

have some higher temperature austenite transformation products in their microstructures (ferrite, pearlite and upper bainite). As non-martensitic transformation products increase in the microstructure, reduced toughness can occur, or in ductile-brittle transition temperatures. Figure 10.194 LHS shows the effect of percentage martensite on the impact transition temperature (ITT). This is a US version of the German Figure 3.43 in Chapter 3.

10.14.2 Factors Controlling effective boron

An important aspect of steel selection would be to achieve the appropriate hardenability at the lowest possible cost and therefore boron provides an economic advantage due to the small addition required and the availability of ferro boron (Turkey, China, India). Figure 10.195 shows that the hardenability effect of boron typically occurs at a range of 10–15ppm. To be effective, boron must be in a solid solution with austenite and protected from any nitrogen or oxygen. Boron increases hardenability at concentrations as low as 0.0010% of boron, by delaying the formation of ferrite and pearlite, although it has only a minimal effect on the bainite transformation rate. As the grain size decreases, the hardenability effect increases. Increasing hardenability by boron addition does not decrease the Ms temperature and with low carbon steel there is tendency for "self" tempering of the martensite and a reduced risk of quench cracking.

It is well established that boron increases hardenability of steels by retarding the heterogeneous nucleation of ferrite at the austenite grain surfaces. It is probable that this effect is due to the reduction in interfacial energy as the boron segregates to the grain boundaries. This in turn makes grain boundary less effective as heterogeneous sites. Excessive boron produces a reduction in hardenability effects and can cause embrittlement, hot shortness and reduced impact properties.

Steelmaking control is needed to ensure that soluble boron is present in the finished product to ensure the hardenability response of boron. This requires good deoxidation practice and the protection of boron from combining with oxygen or nitrogen. This can be achieved by limiting the nitrogen content and by the addition of elements that have a greater affinity for nitrogen (Ti, Zr, Al). These additions (especially

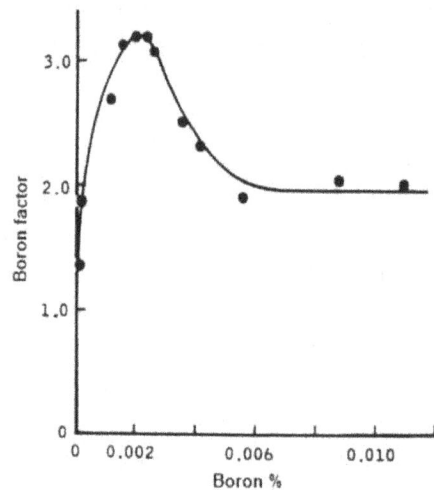

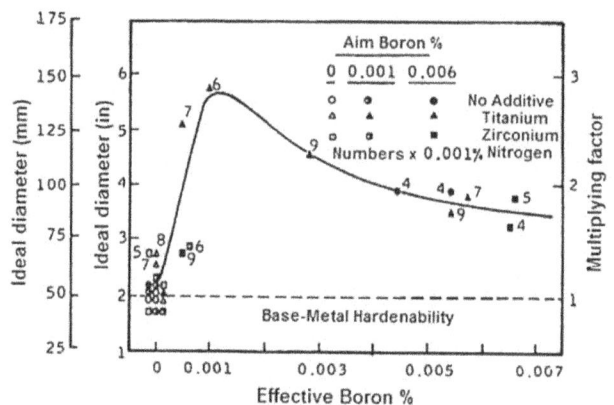

Figure 10.195 Top = Forged steel with 0-.02% boron. Centre = Correlation of the boron factor with the boron content in the steel alloy. From Report CDL/ESOP Report; Technical Universität Graz, Institut für Werkstoff kunde, Schweiß technik und Spanlose Formgebungs verfahren: Graz, Austria, Oct 2007; pp 1–36. Bottom = The correlation of the effective boron content of construction steels containing nitrogen, titanium, and zirconium with the ideal diameter (hardenability). Reproduced from Kapadia, B. M.; Brown, R. M.; Murphy, W. J. The Influence of Nitrogen, Titanium and Zirconium on the Boron Hardenability Effect in Constructional Alloy Steel. Trans. Metall. Soc. AIME 1968, 242, 1689–1694.

Ti) protect the boron and ensure the hardenability effect.

K-E Thelning established the following equation to calculate the required titanium content in boron steels.

$$Ti = 5 (N - 0.003) \text{ Eq 1.}$$

The hardenability effect of boron (the boron factor Bf) can be calculated from the following equation:

$$Bf = DI \text{ obtained from Jominy/DI calculated from Chemistry (excluding B) Grossman Eq 2}$$

For boron steels containing 0.25 to 0.3C the Bf value is usually about 2. When Bf values of below 1.5 are calculated this is associated with insufficient titanium and ineffective protection of boron from nitrogen. When unusually high boron factors are found, a large excess of titanium can be calculated. This has been attributed by Thelning to the added hardenability effect from the excess titanium (Figure 10.196 RHS).

In the early 1970s Uqine Steel[7] and later MaitreDierre[8] demonstrated that the titanium protective addition caused reduced touqhness and fatigue properties. This was associated with coarse TiN particles. If the nitrogen levels are kept low this does not occur. An alternative method of protection used by Uqine in France and in Japan is to protect with an addition of 0.06 to 0.08 Al. However, some published results[9] show that the Bf factors with this practice are low and variable.

The use of protected boron additives is essential in achieving consistent results. Complex ferro-boron alloys (20Ti, 13Al 4Zr, 8Mn, 5Si and 0.5B sold with the trade name "Batts Alloy") have been shown to give better performance than a "ferro-boron" addition (Figure 10.196)[11]. The determination of "effective-boron" by chemical analysis is thought to be time consuming, costly, and of questionable accuracy[13]. In practice the method usually used to establish the quality of boron-steel is the Jominy or hardenability test.

10.14.3 Effect of carbon and alloy content on the boron hardenability

Boron has the greatest effect on hardenability in low carbon steels and becomes less effective as the carbon content of the steel increases. (See Figure 10.194 RHS) Several investigations have expressed this mathematically:

$$Bf = 1 + 2.7 (0.85 - \%.C) \text{ Eq 3 (Lewellyn \& Cook)}$$
$$Bf = 1 - 1.76 (0.74 - \%.C) \text{ Eq 4 (Brown \& Walters)}$$

The effect of alloy content follows a similar trend and therefore boron is used to its best advantage in low carbon or low carbon alloy steels where the hardenability is controlled by the proeutectoid reaction. Work carried out by Caterpillar Tractor made corrections to the original Grossman method for the effect of carbon. The relationships they established for carbon and alloy steel between boron factor and carbon level are shown in Figure 10.197.

Recent work has established the importance of avoiding boro-carbide precipitates which can result in low toughness.[21] This can be avoided by slightly

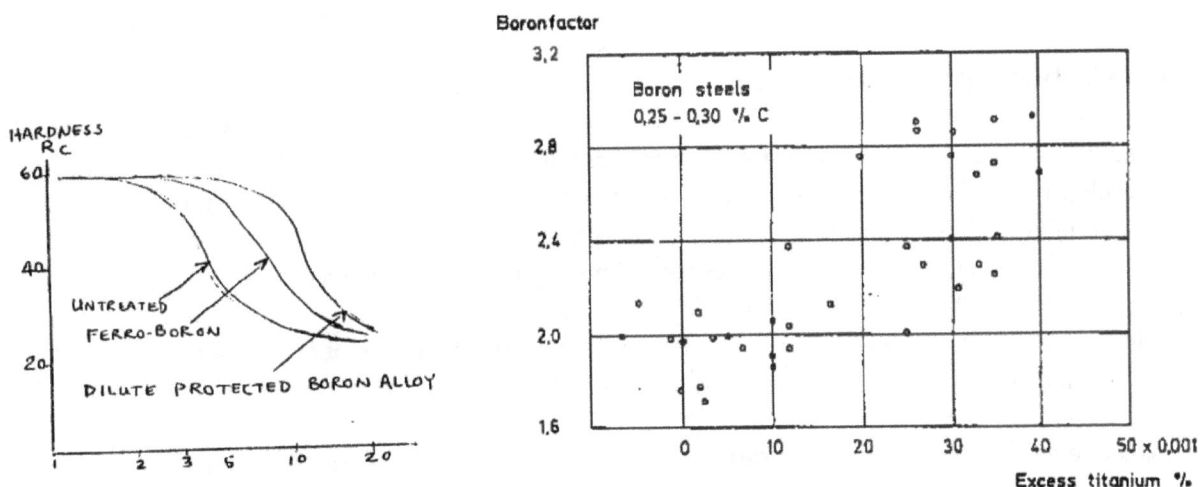

Figure 10.196 LHS = Effect of method of boron addition on boron factor, RHS = Effect of excess titanium

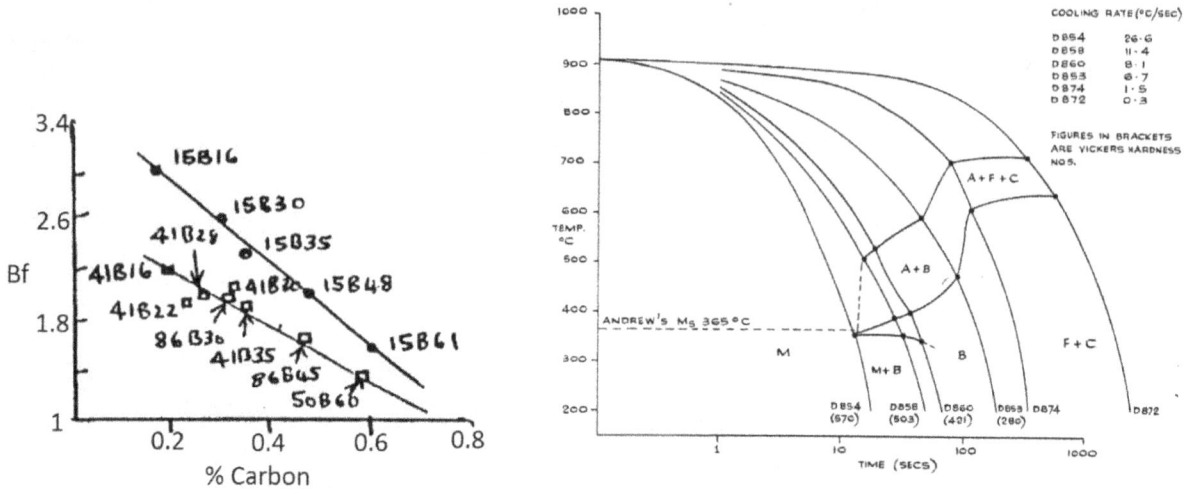

Figure 10.197 *Effect of carbon on the boron factor for various SAE steel grades Boron CCT diagram. Boron steel made at Consett steelworks in the mid1970s*

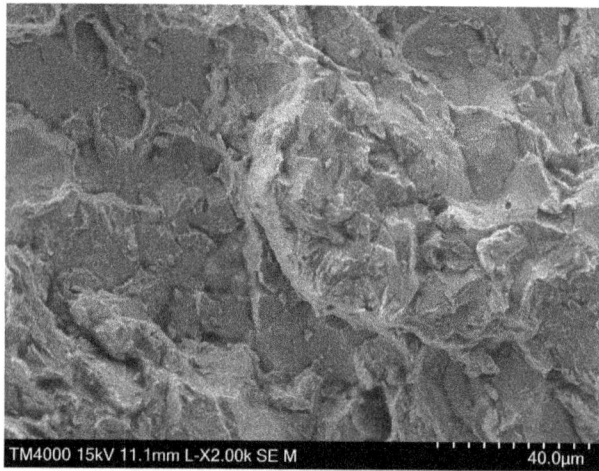

Figure 10.198 *SEM/BSE image of fracture surface from CVN specimen with Indication of the large boro-carbide particles*

higher austenitising temperature (930°C). Figure 10.198 show SEM images with boro-carbides visible on the BSE image (black dots).

10.14.4 Carburising of boron steel

In the mid-1960s considerable attention was given to the development of boron treated carburizing steels containing up to 1.5% manganese and controlling the level of residual elements. The grades were designated CM60, CM70 and CM80 and were developed as lower cost Ni-Cr and NI-Cr-Mo carburizing grades. Although these steels were reasonably successful, some problems were encountered in trials on large gears due to distortion. During the early 1970s W T Cook[14] carried out extensive research work on heat treatment distortion in carburized steels. It was

shown that for boron with a carbon level of 0.3% the distortion was reduced compared to that of a low alloy steel.

An additional problem with carburized boron steel, which was outlined in a previous section, is that as the carbon increases the effect of boron on the hardenability decreases. Consequently, boron carburizing steels will have lower "case" hardenability than a low alloy steel. However, Cook examined this aspect in detail and concluded that the effect was not significant on the section sizes used for small to medium gears. In 2012 A Vermin described the development of boron steel for transmission gears[20]. It was also found[1] that the surface regions after carburizing became rich in BN due to the nitrogen in the carburizing atmosphere. The authors noted that they were unaware of any adverse effects in terms of brittleness or general service performance. It should be noted that in low alloy carburized steel oxidation of silicon and chromium occurs that is likely to be more damaging than the formation of BN.

REFERENCES

1. JOSEPH W. SPRETNAK et al, A CRITICAL EVALUATION OF THE BORON HARDENABILITY EFFECT IN STEEL, JUNE 1952, WADC TECHNICAL REPORT 52-140, AD857775,

2. Thomas G. Digges et al, Effect of Boron on the Hardenability of High-Purity Alloys and Commercial Steels, National Bureau of Standards, Research Paper RP1938, Volume 41, December 1948, https://nvlpubs.nist.gov/nistpubs/jres/041/6/V41.N06.A01.pdf

3. W T Cook: The Metallurgy of boron-treated low alloy steels Metals Technology 1(1):517-529, December 1974

4. K-E Thelning Heat treatment of steel, second edition Ch 6.3 Hardening and Tempering of Boron-Alloyed Steels

5. EUR 22446 Final Report 2007 Optimization of the influence of boron on the properties of steel

6. A063645, Assessment of boron steel for army use, Army Materials and Mechanics Research Centre, Watertown, Massachusetts

7. C Siebert et al, The Hardenability of steel ASM 1977

8. R T Rincot et al, Reve de Met Nov 1972 (11) p721-735

9. Ph. Mailrepierre et al, Boron in Steel, AIME Conference 1980 p1

10. K E Thelning, Boron in Steel, AIME Conference 1980 p127

11. R Habu et al, Trans ISIJ Vol 18 1978

12. R L Szuch, Blast Furnace and Steel, October 1967 p930

13. F L Porter, Boron in Steel AIME Conference 1980 p199

14. W T Cook, Metals Technology, May 1977 p265

15. Melvin R. Meyerson et al, Impact Properties of Slack-Quenched Alloy Steels, Journal of Research of the National Bureau of Standards Vol. 59, No.4, October 1957 Research Paper 2799

16. Saeed N. Ghali et al, Influence of Boron Additions on Mechanical Properties of Carbon Steel, Journal of Minerals and Materials Characterization and Engineering, 2012, 11, 995-999,

17. Hande Güler et al, EFFECT OF HEAT TREATMENT ON THE MICROSTRUCTURE AND MECHANICAL PROPERTIES OF 30MnB5 BORON STEEL, Materials and technology 48 (2014) 6, 971–976

18. D. E. Lescano et al, Study of microstructure and tempered martensite embrittlement in AISI 15B41 steel, 11th International Congress on Metallurgy & Materials SAM/CONAMET 2011.

19. Dr Eng. Frydman S., PROPERTIES OF BORON STEEL AFTER DIFFERENT HEAT TREATMENTS, SCIENTIFIC PROCEEDINGS IX INTERNATIONAL CONGRESS "MACHINES, TECHNOLOGIES, MATERIALS" 2012, VOLUME 3, P.P. 72-74 (2012

20. A. Verma et al, Boron Steel: An Alternative for Costlier Nickel and Molybdenum Alloyed Steel for Transmission Gears, The Journal of Engineering Research Vol. 8 No. 1 (2011) 12-18

21. João Paulo Gomes Antunesa et al, Characterization of Impact Toughness Properties of DIN39MnCrB6-2 Steel Grade, Materials Research 2017

10.15 TITANIUM AND TITANIUM ALLOYS

(an excellent information source http://www.dierk-raabe.com/titanium-alloys/mechanical-properties-of-titanium/)

Titanium was discovered in 1790 by Reverend William Gregor during the chemical analysis of Cornwall rocks. However, the commercial extraction of titanium metal was not available until the 1950s. Figure 10.199 LHS shows the main reasons for the use of titanium alloys and the main metallographic alloy grades, commercially pure, α alloys, α+β alloys and β alloys.

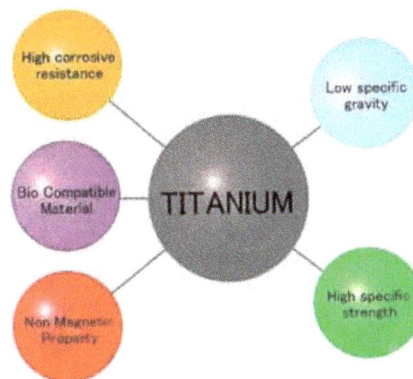

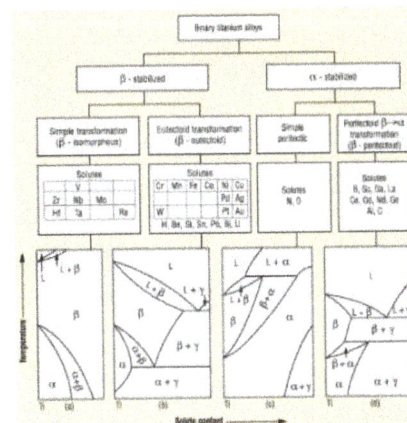

Figure 10.199 Top = Attributes of titanium and titanium alloys from http://www.supraalloys.com/titanium-grades.php, Bottom = Illustration of influence of alloying elements on the extent of α and β fields. (3)

10.15.1 Extraction of titanium

During the 1930s, Kroll began experiments on the extraction of titanium that resulted in a patent in 1940 for alkaline earth reduction of $TiCl_4$. In 1938, the U.S. Bureau of Mines began a research program to find a method to produce commercial quantities of titanium. By 1941, the U.S. Bureau of Mines was using a small Kroll reactor to produce 100-gram quantities of titanium. By 1948, the U.S. Bureau of Mines was making 91-kilogram (200-pound) batches at a small pilot plant. During that same year, (DuPont) began commercial production of titanium.[1] Figure 10.200 shows the titanium alloys available together with the cross-reference of various national specifications. Figure 10.202 shows the alloys used for aerospace applications and Figure 10.203 the titanium alloys used for automotive application.[2]

The extraction of this reactive metal was expensive which kept the price of the metal high and limited the applications to aerospace and golf clubs where the performance gains made the choice cost effective. Many different extraction processes have been developed for the production of titanium metal but only the Kroll process has prevailed. However, there are several methods that are being developed/ claim to be developed using electrolysis in molten salt. The most famous being the FFS (Fray Farthing Chen) process that was accidentally discovered at the University of Cambridge in the 1990s.[4-10, 12]

10.15.2 Applications and key properties

Titanium alloys were developed in the early 1950s for defence and aeronautical applications because of their high strength-to-weight ratio. In recent years the application of the alloys in several diverse markets, such as automotive, desalination plants and consumer goods has resulted in the increased production of titanium which has increased the availability of titanium and its alloys. This increase in production and lower cost has allowed more industries to utilize its unique combination of strength, weight, and corrosion resistance.

Titanium is a light metal (4.5 g/cm³), strong and highly resistant to corrosion. Titanium is highly resistant to heat with a melting temperature as high as 1668°C; its melting point is higher than that of steel. Although heat conductivity of titanium is almost the same as that of stainless steel, its weight is almost half of stainless steel. Titanium is also non-toxic and non-allergenic, often used in surgical implants and piercing jewellery.

ASM	ASTM (UNS)	DIN	GOST	BS	UNI 10221	JIS	Rm, Mpa Rp0,2, Mpa A, %
				unalloyed titanium			
Gr -1	1 (R50250)	3.7025	BT1-00	1	Ti1-Type 1	Class 1	240/170/24
Gr -2	2 (R50400)	3.7035	BT1-0	2,3,4,5	Ti2-Type 2	Class 2	340/280/20
Gr -3	3 (R50500)	3.7055			Ti3-Type 3	Class 3	450/380/18
Gr -4	4 (R50700)	3.7065		6,7,8,9	Ti4-Type 4	Class 4	550/480/15
Gr -7	7 (R52400)	3.7235			Ti2Pd-Type7	Class 13	340/280/20
Gr -11	(R52250)	3.7225			Ti1Pd-Type11	Class 12	240/170/24
				α and near α alloys			
Gr -12	12 (R53400)	3.7105			TiNiMo-Type 12		480/350/18
Ti-6Al-2Sn-4Zr-2Mo	(R54620)	3.7145	BT-25 BT-18y				900/830/10
Ti-5Al-2,5Sn (Gr-6)	6 -(R54520)	3.7115	BT5-1		TiAl5Sn2,5-Type6		830/790/10
Ti-5Al-2,5Sn ELI	(R54521)		BT5-1кт				720/690/10
Ti-8Al-1Mo-1V	(R54810)		BT-14				930/830/10
Ti-6Al-2Sn-4Zr-2Mo (+Si)	(R54620)	3.7145					1000/830/10
				α – β alloys			
Gr -5	5 (R56400)	3.7165	BT-6	10,11,12, 28,56,59	TiAl6V4-Type5	Class 60	900/830/10
Gr -5ELI	(R56401)		BT-6C		TiAl6V4ELI-Type 5.1		830/760/10
Ti-4Al-4Mo-2,5Sn	-	3.7185		45,57,57			1100/960/9
Ti-3Al-2,5V (Gr-9)	9 (R56320)	3.7195	ПТ-3B		TiAl3V2,5-Type9	Class 61	620/520/15
Ti-3Al-2,5V with Ruthenium (Gr-28)	28 (R56323)						620/480/15
Ti-6Al-4V ELI with Ruthenium (Gr-23)	23 (R56407)					Class 60 E	860/800/10
Ti-6Al-6V-2Sn	(R56620)	3.7175					1030/970/10
Ti-6Al-2Sn-4Zr-6Mo	(R56260)						1170/1100/10
Ti-5Al-2Zr-4Mo-4Cr	(R58650)						1165/1110/10
Ti-7Al-4Mo	(R56740)		BT-8				1030/970/10
				β alloys			
Ti-10V-2Fe-3Al	(R54610)						1190/1100/9
Ti-3Al-8V-6Cr-4Mo-4Zr (Gr-19)	(R58640)						790/760/15
Ti-13V-11Cr-3Al			TC6				1170/1100

Figure 10.200 Designation of titanium alloys From (2)

ASM – Aerospace Specifacation Metals;
ASTM – American Society for Testing and Materials;
DIN - German Institute for Standardization;
UNI – Italian Organization for Standardization;

UNS - Unified Numbering System
BS - British Standards
GOST -Russian Interstate standard
JIS – Japanese Industrial Standards

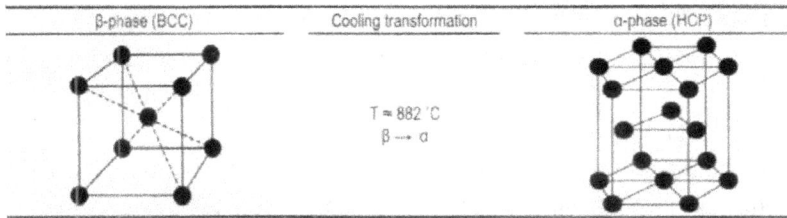

Figure 10.201 *The crystallographic cell and allotropic transformation of pure titanium*

Figure 10.202 *Titanium and alloys used for aerospace parts. From (2)*

Alloy	Application
Ti-3Al-2.5V (Gr-9)	Used for hydraulic high pressure lines, replacing the stainless steel pipe and thus reducing the weight by 40%. It is used for the production of cell structures.
Ti-5Al-2.5Sn (Gr-6)	Used in tempered state in cryogenic technique because it keeps good strength and ductility in low temperatures. Used in the turbo-pumps high-pressure space shuttles.
Ti-8Al-1Mo-1V	It is used for the blades of military engines.
Ti-6Al-2Sn-4Zr-2Mo (+Si)	Used mainly in the parts of gas turbine engines, including disks and rotors at temperatures up to about 540 ° C, in the high pressure compressors.
Ti-6Al-4V	It is used in gas turbine engines for both static and rotating components, including all parts of the aircraft - fuselage, nacelles, landing gear, wing and tail surfaces, as well as the structure for the support on the floor.
Ti-6Al-2Sn-2Zr-2Mo-2Cr + Si	Used for F22 program for Lockheed / Boeing.
Ti-6Al-2Sn-4Zr-6Mo	It is used at temperatures up to about 315 ° C, primarily for military engines, such as F-100 and F-119, with yield strength of 1035 MPa.
Ti-5Al-2Sn-2Zr-4Mo-4Cr	It is used at temperatures below 400 ° C for fans and compressor disks.
Ti-13V-11Cr-3Al	Widely used in aircraft SR-71 for the wings and body, frames, partitions and ribs.
Ti-10V-2Fe-3Al	Almost the whole main landing gear of Boeing 77 is produced from this alloy which leads to a weight saving of about 270 kg per airplane.

Systems and parts	Materials		Manufacturer	Application	Introduction
	Frame structures		Mitsubishi	Ti-22V-4Al in the AMG engine retainers of the Gallant 1	989
Suspension springs	Ti-6.8Mo-4.5Fe-1.5Al;		Honda Motors	Ti-3Al-2.5V + REM in the connecting rods of the sport cars NSX	1990
	Ti-6Al-4V		Toyota	Sintered titanium alloys Ti-6Al-4V/TiB and Ti-Al-Zr-Sn-Nb-	1998
Armor	Ti-6Al-4V			Mo-Si/TiB in the intake and exhaust engine valves, respectively,	
Body	CP-Ti (Grade 4);			in the Altezza	
	Ti-6Al-4V		Nissan Motor	Ti-6Al-4V and Ti-6Al-2Sn-4Zr-2Mo-Si in engine inlet and	2000
	Engines			exhaust valves, respectively, for the CIMA	
Outlet valves	(TiAl); Grade 2;		Volkswagen	Ti-4.5Fe-6.8Mo-1.5Al in suspension spring of Lupo FS	2001
	Ti-6Al-4V; Ti-6Al-		Kawasaki	Titanium alloys in the muffler of the large sports-type motorcycle ZX-9	1998
	2Sn-4Zr-2Mo-0.1Si		General Motors	Titanium alloys in dual mufflers of the Corvette Z06	2001
Intake valves	Ti-6Al-4V				
Turbocharger rotors	(TiAl)				
Connecting rods	Ti-6Al-4V				
Exhaust system	Grade 2				

Figure 10.203 *Titanium used for automotive parts From (3)*

Due to its high strength, lightness, and corrosion resistance, titanium has emerged as the metal of choice for aerospace, industry and medical, leisure and consumer products, notably golf clubs and bicycle frames. Furthermore, due to its strength and lightness, titanium is currently being tested in the automobile industry where the use of titanium for connecting rods and moving parts has resulted in fuel efficiency.

Ti-6Al-4V is the most widely used of the titanium alloys as it can be heat-treated to different strength levels, and is readily weldable and has good machineability. The many uses of Ti-6Al-4V include blades and discs for aircraft turbines and compressors, rocket motor cases, marine components, steam turbine blades, structural forgings and fasteners. To enhance durability a range of thermal, mechanical, chemical and other treatments have been developed to modify the surface characteristics and a considerable body of data on its use and properties is available.

At low temperatures, pure titanium has a hexagonal close-packed structure (hcp), called α-titanium. But at high temperatures the stable structure is body-centred cubic (bcc), which is referred to as β-titanium. The atomic unit cells of the referred structure are illustrated in Fig.10.201. The α-transus temperature for pure titanium is around 882°C depending upon the impurities present. The existence of the two different crystal structures is the basis for the large variety of titanium alloys' properties.

Commercially Pure (CP) Titanium, Ti 6Al-4V and Ti 6Al-4V ELI (extra low interstitials) are the most common titanium alloys used in industry. The additional grades usually have more limited availability for more specialized circumstances.

10.15.3 China's output of Titanium and their new method of extraction

China is the largest maker of titanium sponge and has over 27% of the world's known reserves. The titanium is first produced as sponge which then requires subsequent re-melts and refining to produce ingot grade titanium. The main source of sponge titanium:

China has reported to have developed an improved method for the extraction of titanium. The USTB process has several claimed advantages with the cost being 60% reduction compared to the Kroll process together with lower energy consumption and an environmentally better process. The process is a molten salt electroless method and was developed by Professor Zhu at the university of science and technology at Beijing.[11 and 13]

However, it appears to be partly based on the Fray Farthing and Chen process (FFS) developed at Cambridge University in the 1990s currently used by Metalysis. Metalysis entered administration during June 2019 after experiencing financial difficulties due to an extended investment round. Metalysis were acquired during July 2019 by PRG, a mining company, and now specialise in additive manufacture.

Rank	Country/Region	2010	2011	2012	2013	2014	2015	2016
	World	137,000	186,000	200,000	209,000	194,000	160,000	170,000
1	China	57,800	60,000	80,000	105,000	110,000	62,000	60,000
2	Russia	25,800	40,000	44,000	44,000	42,000	40,000	38,000
3	Japan	31,600	56,000	40,000	42,000	25,000	42,000	54,000
4	Kazakhstan	14,500	20,700	25,000	12,000	9,000	9,000	9,000
5	Ukraine	7,400	9,000	10,000	6,300	7,200	7,700	7,500
6	India	-	-	-	-	-	500	500

Figure 10.204 Titanium sponge output

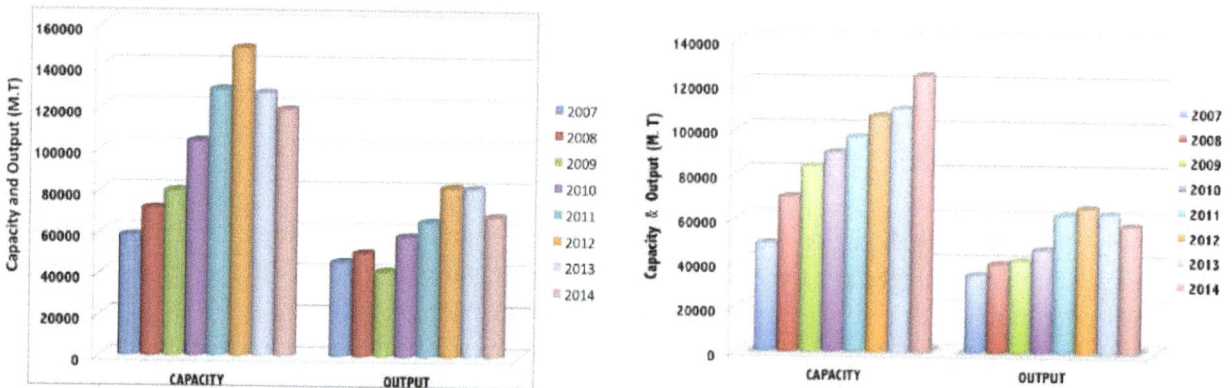

Figure 10.205 LHS = China titanium sponge capacity and output RHS = China titanium ingot capacity and output (11)

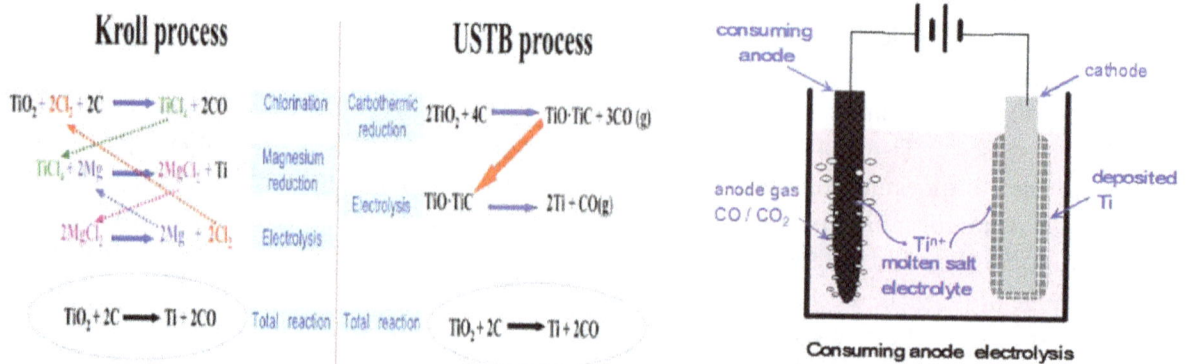

Figure 10.206 The Chinese USTB process (11)

Figure 10.207 Metalysis the holder of the FFS patents

10.15.4 The three structural types of titanium alloys[3]

1. **Alpha alloys** are non-heat treatable and are generally very weldable. They have low to medium strength, good notch toughness, reasonably good ductility and possess excellent mechanical properties at cryogenic temperatures. The more highly alloyed alpha and near-alpha alloys offer optimum high temperature creep strength and oxidation resistance as well.

2. **Alpha-Beta alloys** are heat treatable and most are weldable. Their strength levels are medium to high. Their hot-forming qualities are good, but the high temperature creep strength is not as good as in most alpha alloys.

3. **Beta or near-beta alloys** are readily heat treatable, generally weldable, and capable of high strengths and good creep resistance to intermediate temperatures. Excellent formability can be expected of the beta alloys in the solution treated condition. Beta-type alloys have good combinations of properties in sheet, heavy sections, fasteners and spring applications.

Alpha alloys

Alpha alloys (α-alloys) are easily welded and are relatively tough even at cryogenic temperatures. Aluminium is the main alloying element apart from Zr and Sn. The combined effect is expressed as:

aluminium equivalent, wt% = Al + (1/3) Sn + (1/6) Zr + 10 (O + C + 2N)

If this exceeds about 9 wt% then there may be detrimental precipitation reactions.

The presence of a small amount of the more ductile β-phase in nearly α alloys is advantageous for heat treatment and the ability to forge. The alloys may therefore contain some 1wt% of Mo, e.g.

Ti-6Al-2Sn-4Zr-2Mo where the Zr and Sn give solid solution strengthening.

Ti-5Al-2.5Sn is an α alloy which is available commercially in many forms. Because it is stable in the α condition, it cannot be hardened by heat treatment. It is therefore not particularly strong, but can easily be welded. The toughness at cryogenic temperatures increases when the oxygen, carbon and nitrogen concentrations are reduced to produce a variant designated ELI, standing for extra low interstitials. The fact that the strength increases at low temperatures, without any deterioration in toughness, makes the alloy particularly suitable for the manufacture of cryogenic storage vessels, for example to contain liquid hydrogen.

Near-α alloys

A near-α alloy has been developed, with good elevated temperature properties (T<590°C):

Ti-6Al-4Sn-3.5Zr-0.5Mo-0.35Si-0.7Nb-0.06C

The niobium is added for oxidation resistance and the carbon to allow a greater temperature range over which the alloy is a mixture of $\alpha+\beta$, in order to facilitate thermomechanical processing. This particular alloy is used in the manufacture of aero engine discs and has replaced discs made from much heavier nickel base super alloys. The final microstructure of the alloy consists of equiaxed primary-α grains, Widmanstätten α plates separated by the β-phase.

Alpha-Beta Alloys ($\alpha+\beta$ Alloys)

Most $\alpha+\beta$ alloys have high-strength and formability, and contain 4-6 wt% of β-stabilizers which allow substantial amounts of β to be retained on quenching from the $\beta \rightarrow \alpha+\beta$ phase fields, e.g. Ti-6Al-4V.

Al reduces density, stabilizes and strengthens α while vanadium provides a greater amount of the more ductile β phase for hot-working. This alloy, which accounts for about half of all the titanium that is produced, is popular because of its strength (1100 MPa), creep resistance at 300°C, fatigue resistance and castability. One difficulty with the β phase, which has a body-centred cubic crystal structure, is that like ferritic iron, it has a ductile-brittle transition temperature. The transition temperature tends to be above room temperature, with cleavage fracture dominating at ambient temperatures.

A powder metallurgical variant of Ti-6Al-4V, containing small concentrations of boron and carbon, has been developed with an approximately 25% higher strength and modulus, but significantly lower ductility. The alloy contains stable TiB precipitates which prevent grain growth during the hot-processing operations.

REFERENCES

1. Paul C. Turner et al, LOW COST TITANIUM – MYTH or REALITY, DOE/ARC-2001-086 https://www.osti.gov/servlets/purl/899609

2. Danail Gospodinov et al, Classification, properties and application of titanium and its alloys, PROCEEDINGS OF UNIVERSITY OF RUSE – 2016, volume 55, book 2 https://www.researchgate.net/publication/324152759_Classification_properties_and_application_of_titanium_and_its_alloys/download

3. C. Veiga et al, PROPERTIES AND APPLICATIONS OF TITANIUM ALLOYS: A BRIEF REVIEW, Rev. Adv. Material Science 32 (2012) 14-34.

4. .DI HU et al, Development of the Fray-Farthing-Chen Cambridge Process: Towards the Sustainable Production of Titanium and Its Alloys JOM 2017, file:///C:/Users/user/Desktop/Titanium/extraction/10.1007_s11837-017-2664-4.pdf

5. S.J. Oosthuizen, In search of low cost titanium: the Fray Farthing Chen (FFC) Cambridge process, The Journal of The Southern African Institute of Mining and Metallurgy VOLUME 111 MARCH 2011,

6. Sarah Lubik et al, Commercialising advanced material processing technology for additive manufacturing: the case of Metalysis, R&D Management Conference 2016 "From Science to Society: Innovation and Value Creation" 3-6 July 2016, Cambridge, UK file:///C:/Users/user/Downloads/Commercialisingadvancedmaterialprocessing technologyforadditivemanufacturing.pdf

7. file:///C:/Users/user/Desktop/Titanium/extraction/v111n03p199.pdf

8. Bernd Friedrich et al, Molten Salt Electrolysis – Latest Developments, EMC 2005 September 18th – 21th, Dresden, file:///C:/Users/user/Downloads/Titan-VortragEMC2005 Dresden.pdf

9. Claudia A. Möller et al, Molten salt electrolysis of Titanium using a TiO2 -C composite anode in halide electrolytes – proof of concept – Titanium 2009, September 13th-16th, Waikoloa, Hawaii, file:///C:/Users/user/Downloads/Titanium2009-Hawaii.pdf

10. Naomi A. Fried and Donald R. Sadoway, Titanium Extraction by Molten Oxide Electrolysis, Department of Materials Science & Engineering Massachusetts Institute of Technology Cambridge, Massachusetts, Sadoway, MIT TMS Meeting, Charlotte, NC March 15, 2004, http://web.mit.edu/dsadoway/www/MOE_Ti.pdf

11. Hui Chang et al, Progress of titanium industry, technologies and research in China, Proceedings of the 13th World Conference on Titanium, 2016.

12. 6th International round table on titanium production in molten salts Reykjavik University, June 10-13 2018 The journal Materials Transactions (The Japan Institute of Metals and Materials) will publish a special issue on Titanium Production by Molten Salt Electrochemical Process as of March 2019 based on contributions for the 6th International Round Table on Titanium Production in Molten Salts.

13. Qiuyu Wang et al, A new consumable anode material of titanium oxycarbonitride for the USTB titanium process, Physical Chemistry Chemical Physics · March 2014, file:///C:/Users/user/Downloads/-PCCP.pdf

10.16 ALUMINIUM

10.16.1 Historical aspects of aluminium alloy development

A major contribution to the world of metallurgy came about 113 years ago by the accidental discovery of "age hardening". In 1901, Alfred Wilm was appointed metallurgist at the Neubabelsberg Scientific and Technical Analysis Centre near to Berlin and in 1903, the Centre was commissioned by the German War Munitions factory of Berlin to find an aluminium alloy with the characteristics of brass that could be used for the manufacture of ammunition. The work was to replace the brass cartridge shell with a lightweight metal.

Alfred was working on the development of aluminium-copper-mangnesium (Al-Cu-Mg) alloys which were water quenched. During the trail work

Primary Aluminium Production (Million t/yr)

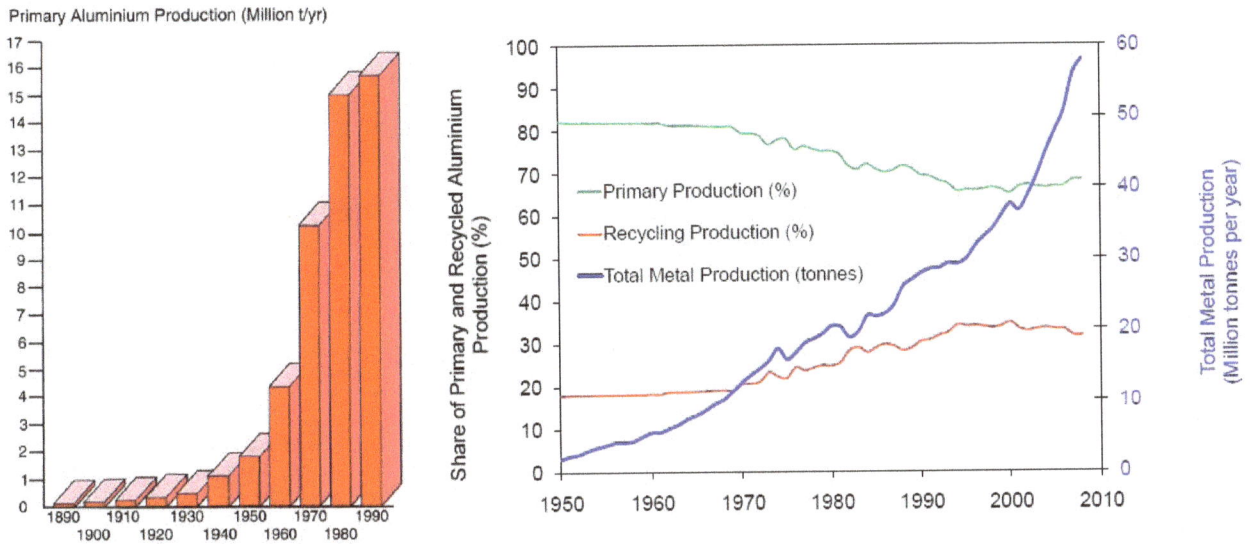

Figure 10.208 *LHS = Growth in primary aluminium output. RHS = Ratio of primary and secondary*

with a 0.5% Mg alloy following water quenching he requested Fritz Jablouski to carry out hardness testing. It was a Saturday afternoon and Fritz wanted to leave early. They came to an agreement to do "one" hardness indent before he left, which gave a disappointing result. On the following Monday morning the hardness test work was completed and gave results that were considerably higher. This was the discovery of the concept of "age hardening".[1] This event occurred around1906 and the alloy was patented in1909. It was originally made only at the company Dürener Metallwerke at Düren, Germany and the town's name was used to name the alloy "Duralumin".

The discovery led to subsequent R&D work at the then United States Bureau of Standards around 1919 resulting in the development of several age hardening alloys. Around 1932, fourteen base metals had been discovered to harden by precipitation in a total of more than one hundred different alloy combinations. The initial list turned out to be an under-estimation. Most of today's high strength commercial aluminium and nickel-based alloys are precipitation hardened, as are many titanium, copper and iron-based alloys.[2]

Initially several theories were developed to explain the hardening. In 1938 A. Guinier and G. D. Preston independently identified the precipitation hardening mechanism. The initial stage was the movement of atoms to sites in the lattice in preparation for forming the final precipitate. These sites were rich in copper atoms in an aluminum-copper alloy, and are called G-P zones in honour of Guinier and Preston.[3]

Supersaturated Solid Solution (SSS)→GP zones→ θ''→θ'→θ GP zones, θ'' and θ', are metastable phases,

but θ is the equilibrium phase in the binary system. One of the early uses of Duralumin was aircraft. There was over 13 miles (21kms) of Duralumin girders used to build the Hindenburg airship (manufactured 1932 to 1936). These airships were the largest man-made objects ever to fly.[4]

Aluminium is a lightweight metal and it is one of the most abundant metallic elements on earth, about 8% of earth's crust, making it second after silicon (28%). The aluminium production is now split between primary or secondary aluminium production routes. Primary production involves extraction from bauxite. The secondary aluminium industry is basically the recycling of scrap aluminium. The secondary manufacturing route now represents about 30% of the total output. There has been an increasing interest towards the recycling of aluminium, because the energy required to re-melt aluminium scrap is only 5% of that required to produce the primary aluminium.

10.16.2 The importance of aluminium alloys

In the late 1800s and early 1900s, three important industrial developments allowed aluminium to become a very important essential metal and the aluminium industry entered an era of exponential growth. The three main uses were:

1. The development of the first internal combustion engine

2. The start of electrification
3. The invention of the aeroplane by the Wright brothers.

Aluminium offers a wide range of properties that make it a suitable metal for many applications in different industrial fields such as: the automotive industry, the aeronautical and aerospace area, the electrical and electronic industries.

Some of the principal properties of this metal are:

1. A density $\rho = 2{,}7$ g/cm3, that is one-third that of the steel
2. A high strength to weight ratio
3. A high corrosion resistance under the majority of service conditions
4. The excellent thermal and electrical conductivity
5. The high reflectivity
6. It is a non-ferromagnetic metal, which is important especially in electrical and electronic industries
7. It is nontoxic, so for this reason is used routinely for food and beverages packages
8. It is recyclable.

10.16.3 Aluminium alloys are classified in two categories

1. cast aluminium alloys
2. wrought aluminium alloys.

Figure 10.209 shows the aluminium alloys designation system, used by the Aluminium Association of the United States, for both cast and wrought aluminium alloys. This designation system uses a four-digit numerical system to identify the different aluminium alloys. The nomenclature for wrought alloys has been accepted by most countries and is now called the International Alloy Designation System (IADS).[5] The first digit indicates the alloy group and the last two digits identify the aluminium alloy or indicate the aluminium purity. The second digit indicates modifications of the original alloy or impurity limits. The designation of cast alloys, the first digit is essentially the same as for wrought alloys while the second two digits serve to identify a particular composition.

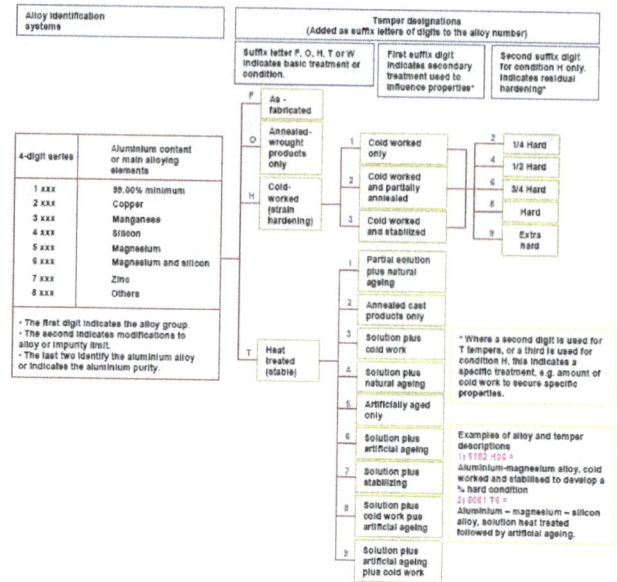

Figure 10.209 Designation system of aluminium wrought alloys and Temper designations

10.16.4 Cast aluminium alloys

Cast aluminium alloys are finding new applications in many industry fields. About 80% of all aluminium casting parts are made from aluminium scrap, a percentage that is significantly higher than for wrought products. In the last 10 years casting technologies have improved allowing the production of higher quality alloys. (See John Campbell[6])

Grade	Major alloy elements
1XX.X	Al, 99.00% or greater
2XX.X	Cu
3XX.X	Si with added Cu and or Mg
4XX.X	Si
5XX.X	Mg
7XX.X	Zn
8XX.X	Sn
9XX.X	Other elements
6XX.X	Unused series

Code	Condition
M	As Cast
TB	Solution treated and naturally aged
TB7	Solution treated and stabilised
TE	Artificially aged after casting
TF	Solution treated and artificially aged
TF7	Solution treated and artificially aged and stabilised
TS	Thermally stress relieved

Figure 10.210 Cast aluminium alloy

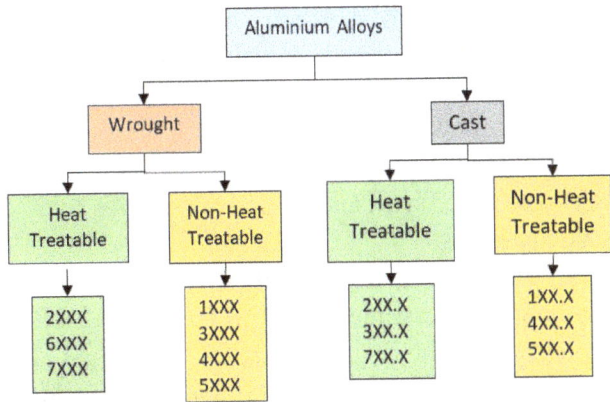

Figure 10.211a Heat treatable and non-heat treatable gardes

Due to the low-density, cast aluminium parts have found increased application in the automotive and aerospace industries. The properties such as good castability, high mechanical properties, ductility and good corrosion resistance, have allowed aluminium parts to substitute steel and cast iron for the production of critical components. Generally, these alloys contain a high percentage of alloying elements; the most important alloying elements are:

- **Silicon**: is one of the most important alloying elements used for cast aluminium alloys. The typical level of silicon represents between 5-12 wt%. This alloying element allows for good castability, and reduced thermal expansion coefficient. The low density (2.34g/cm^3) allows a reduction of cast components' weight and finally its low solubility in aluminium gives a microstructure with precipitation of pure, hard Si particles which improve the abrasion resistance of the alloy.

- **Copper**: increases both the mechanical strength and the machinability of alloys; reduces the coefficient of thermal expansion but has a negative effect on the corrosion resistance of alloys.

- **Magnesium**: allows an increase in the mechanical properties through the precipitation of Mg$_2$Si hardening precipitates, enhancing the corrosion resistance and the weldability of alloys.

- **Manganese**: improves the tensile properties as well as increases the low cycle fatigue resistance. Addition of manganese improves the corrosion resistance of the alloy.

- **Iron**: is the most common impurity in Al-Si foundry alloys. Iron can form different types of intermetallic compounds which are brittle and have a deleterious effect on the mechanical strength of components. Several types of Fe-rich phase exist, such as β-Al5FeSi, α-Al15Fe3Si2

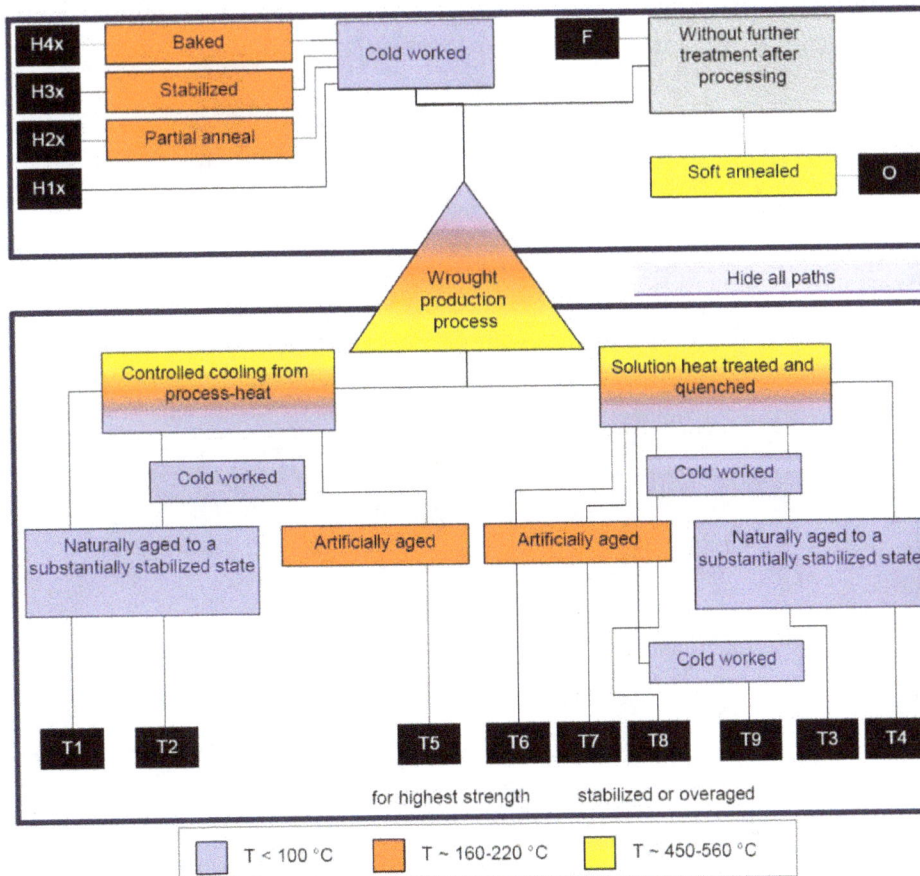

Figure 10.211b Thermal treatment, F= Without treatment, O= Solution annealed, T= Heat treated, H= Strain hardened (for none age hardenable alloys)

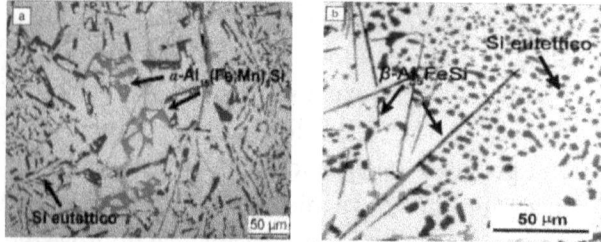

Figure 10.212 *Microstructures showing: a) the α-Al15(Fe, Mn)3Si2 phase with Chinese script morphology and b) the β-Al5FeSi platelets*

and α'-Al8Fe2Si. When the Mn concentration in the alloy is increased the total volume fraction of intermetallic phase linearly increases, but the β-Al5FeSi phase is converted into α-Al15-(Fe, Mn)3Si2 phase, that is stable at low cooling rate. The α-Al15-(Fe, Mn)3Si2 (Figure10.44a particles are more compact than the β ones (Figure 10.44b) and they show a Chinese script, star-like or polyhedral morphology which are less damaging.

The alloys used for cast components typically contain larger proportions of alloying elements such as silicon and copper than wrought alloys. This can cause a heterogeneous cast structure that consists of primary aluminium solid solution, Si-Al eutectic and intermetallic second phases.

The second phases that are known to form can have an effect on the mechanical properties particularly the tensile ductility. The eutectic and intermetallic phases can consist of large, sometimes needle like and brittle constituent, which can create harmful internal notches and nucleate cracks when loads are applied to the cast metal. The fatigue properties are very sensitive to the size, shape and volume fraction of second phase particles.

In the past 20 years a clarification of the factors that affect the cast metal quality, the as cast microstructure and the origins of imperfections (gas and shrinkage porosity [3 to 6% shrinkage]), inclusions, oxide bi-films[6], and unwanted inter-metallics volume fraction, size and shape has allowed a better understanding of the casting process.[6-9]

The recent second edition of the "Complete Casting Handbook" by John Campbell[6] stresses the importance of avoiding oxide bi-films. The message of the book, summarised in the section "Bifilm-free Properties", is that "a quality improvement of astonishing scale is possible now. When I first started to experiment with novel filling systems for castings, there were naturally many disappointments. However, those days are long gone. The concepts of entrainment and bifilm creation laid out in the book are now proven. Some foundries are already being designed to take advantage of a unique and easily affordable quality revolution and scrap reduction. More needs to follow. The risks are minimal and the rewards are great".

Association number	BS 1490 LM number	Casting process	Si	Fe	Cu	Mn	Mg	Cr	Ni	Zn	Ti	Other
150.1	LM 1	Ingot	†	†	0.10					0.05		99.5 Al min
201.0		S	0.10	0.15	4.0-5.2	0.20-0.50	0.15-0.55				0.15-0.35	Ag 0.40-1.0
208.0		S	2.5-3.5	1.2	3.5-4.5	0.50	0.10		0.35	1.0	0.25	
213.0		PM	1.0-3.0	1.2	6.0-8.0	0.6	0.10		0.35	2.5	0.25	
	LM 4	S and PM	4.0-6.0	0.8	2.0-4.0	0.20-0.6	0.15		0.30	0.50	0.20	
238.0		PM	3.5-4.5	1.5	9.0-11.0	0.6	0.15-3.5		1.0	1.5	0.25	
242.0	LM 14	S and PM	0.7	1.0	3.5-4.5	0.35	0.15-0.35	0.25	1.7-2.3	0.35	0.35	
295.0		S	0.7-1.5	1.0	4.0-5.0	0.35	0.03			0.35	0.25	
308.0		PM	5.0-6.0	1.0	4.0-5.0	0.50	0.10			1.0	0.25	
319.0	LM 21	S and PM	5.5-6.5	1.0	3.0-4.0	0.50	0.10		0.35	1.0	0.25	
328.0		S	7.5-8.5	1.0	1.0-2.0	0.20-0.6	0.20-0.6	0.35	0.25	1.5	0.25	
A332.0	LM 13	PM	11.0-13.0	1.2	0.50-1.5	0.35	0.7-1.3		2.0-3.0	0.35	0.25	
355.0	LM 16	S and PM	4.5-5.5	0.6‡	1.0-1.5	0.50‡	0.40-0.6	0.25		0.35	0.25	
356.0	LM 29	S and PM	6.5-7.5	0.6	0.25	0.35	0.20-0.40			0.35	0.25	
A356.0	LM 25	S and PM	6.5-7.5	0.20	0.20	0.10	0.20-0.40			0.10	0.20	
357.0		S and PM	6.5-7.5	0.15	0.05	0.03	0.45-0.60			0.05	0.20	Be 0.04-0.07
360.0	LM 9	D	9.0-10.0	2.0	0.6	0.35	0.40-0.6		0.50	0.50		
380.0	LM 24	D	7.5-9.5	2.0	3.0-4.0		0.50	0.10	0.50	3.0		
A380.0	LM 24	D	7.5-9.5	1.3	3.0-4.0	0.50	0.10		0.50	3.0		
390.0	LM 30	D	16.0-18.0	1.3	4.0-5.0	0.10	0.45-0.65			0.10		
	LM 6	S, PM and D	10.0-13.0	0.6	0.10	0.50	0.10		0.10	0.10	0.20	
413.0	LM 20	D	11.0-13.0	2.0	1.0	0.35	0.10		0.50	0.50		
	LM 2	D	9.0-11.5	1.0	0.7-2.5	0.50	0.30		0.50	2.0	0.20	
443.0	LM 18	S	4.5-6.5	0.8	0.6	0.50	0.05	0.25		0.50	0.25	
514.0	LM 5	S	0.35	0.50	0.15	0.35	3.5-4.5			0.15	0.25	
518.0		D	0.35	1.8	0.25	0.35	7.5-8.5		0.15	0.15		
520.0	LM 10	S	0.25	0.30	0.25	0.15	9.5-10.6			0.15	0.25	
535.0		S	0.15	0.15	0.05	0.1-0.25	6.2-7.5				0.10-0.35	
705.0		S and PM	0.20	0.8	0.20	0.40-0.6	1.4-1.8	0.2-0.4		2.7-3.3	0.25	
707.0		S and PM	0.20	0.6	0.20	0.40-0.6	1.4-1.8	0.2-0.4		4.0-4.5	0.25	
712.0		PM	0.15	0.50	0.25	0.10	0.50-0.65	0.4-0.6		5.0-6.5	0.10-0.25	
713.0		S and PM	0.25	1.18	0.40-1.0	0.6	0.20-0.50	0.35	0.15	7.0-8.0	0.25	
850.0		S and P	0.7	0.7	0.7-1.3	0.10	0.10		0.7-1.3		0.20	Sn 5.5-7.0

Notes: Compositions are in % maximum by weight unless shown as a range

S = sand casting; PM = permanent mould (gravity die) casting; D = pressure diecasting

†Ratio Fe: Si minimum of 2:1

‡If iron exceeds 0.45% manganese content must be less than one-half the iron content

Figure 10.213

Cast aluminium alloys

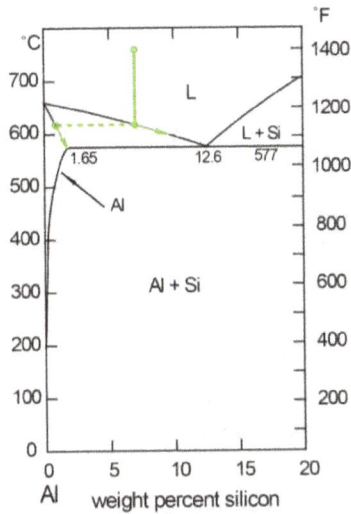

element	k	m	%max
Ni	0.007	-3.3	6
Fe	0.02	-3	1.8
Si	0.13	-6.6	12.6
Cu	0.17	-3.4	33.2
Zn	0.4	-1.6	50
Mg	0.51	-6.2	34
Mn	0.94	-1.6	1.9
Nb	1.5	13.3	≈0.15
Cr	2	3.5	0.4
Hf	2.4	8	≈0.5
Ta	2.5	70	0.10
Mo	2.5	5	≈0.1
Zr	2.5	4.5	0.11
V	4	10	≈0.1
Ti	≈9	30.7	0.15

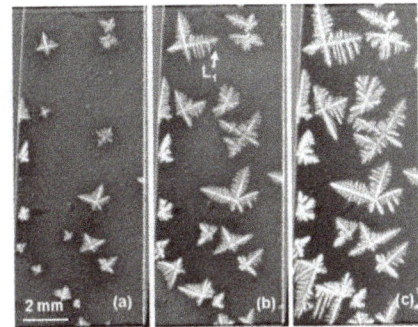

$$k_{Si} = \frac{\%Si_{solid}}{\%Si_{liquid}} \cong \frac{1.65}{12.6} \cong 0.13 \qquad m_{Si} = \left[\frac{dT_{liquidus}}{d\%Si}\right]_{\%Si \to 0}$$

Figure 10.214 *Top LHS = Detail from the Al-Si phase diagram for 0-20% Si, Top centre= Alloy Constants for Several Elements Calculated from Phase Diagrams, the lower equations show k and m. Top RHS = Real time confirmation of dendrites These are dendrites found in Al-20% Cu Liquid. (Pictures were taken (a) 110, (b) 139 and (c) 360 seconds after the first grains appeared.) (Photo courtesy of Henri Nguyen-Thi, Aix-Marseille University, Marseille, France) Lower LHS = Equation for distribution coefficient. RHS = Depression of the melting point of aluminum (8)*

The publications by Geoffrey K. Sigworth[7,8 and 9] gives a detailed introduction to the use of equilibrium diagrams and the solidification aspects of aluminium alloys.

Geoffrey K. Sigworth's paper then covered the development of dendritic solidification, micro-segregation and macro-segregation. In addition, it outlined the measurement of the secondary dendrite arm spacing (SDAS) and that it can be used to determine the local solidification time at any point in a casting.

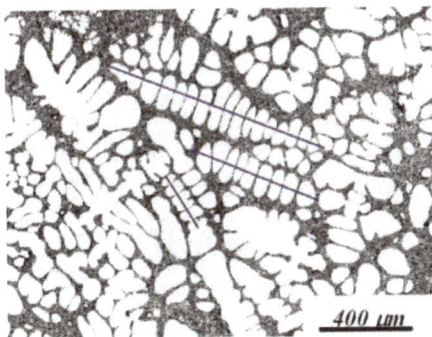

Figure 10.215 *LHS = Measuring SDAS by linear intercept method. RHS = SDAS versus solidification time in aluminium casting alloys (8)*

10.16.5 Microstructure control in aluminium foundry alloys

The microstructure of as-cast aluminium alloys depends on the alloy composition, but typically shows primary aluminium grains surrounded by a eutectic mixture (as would be predicted from the equilibrium diagram shown in Figure 10.187 LHS). For a hyper-eutectic Al-Si alloys, primary silicon phase may form before the eutectic. The processing of the cast alloys requires methods to produce a fine grain size and methods to refine the eutectic structure.

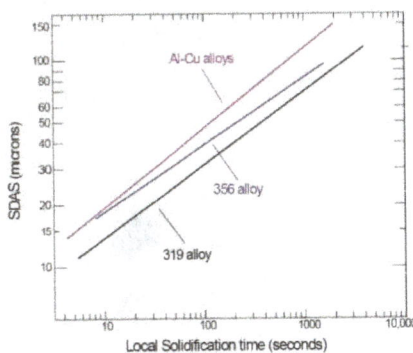

10.16.6 Grain refinement

All wrought aluminium alloys have grain refiner additions however, it is not always required for aluminium castings. For premium castings where mechanical properties are required, grain refinement is needed.

The grain refiners used in foundry alloys are the same as used in wrought alloys. The most common grain refiners would be Al-5wt%Ti-1wt%B. Additions are usually made in rod or waffle form, and addition levels can be around 1 kg/tonne and the main particles produced (hexagonal TiB_2) are theoretically expected to nucleate aluminium solidification.

10.16.7 Eutectic modification

For Al-Si alloys it is essential to carry out a treatment called eutectic modification to change the eutectic silicon from coarse plate-like into fine fibrous morphology. This is commonly done by trace additions (<400 ppm) of certain elements. Aladar Pacz is credited with this discovery in his patent in 1920 (US1387900). His statement "before casting this metal into the finished forms I treat it at temperature with an alkaline fluoride, or combination of flourides – I prefer to employ sodium fluoride as a base and use 1%." Strontium has become the most popular modifier because it is more durable and longer lasting than the other common modifier – Na. Modification treatments improve strength, ductility, pressure tightness and machinability.[21, 22, 23]

The control of cast metal defects and imperfections requires good metallurgical and foundry practice; however, despite the best controls, the elongation and strength and endurance properties of most cast products are lower than those of wrought products. John Campbell stresses the importance of contact casting and avoiding contact with air: "I have always been aware of the potential benefit of contact pouring, but had completely underestimated its effects. It achieves miraculous improvements to castings by eliminating the 50% air mixing step. Contact pouring is strongly recommended in this volume as a major but low-cost step forward".

Since the solid solubility of iron in aluminium is less than 0.05% at equilibrium, almost all iron forms compounds and appears as second phases in the aluminium. The binary Al-Fe and ternary Al-Fe-Si phases are the main Fe-rich phases in aluminium alloys.

The most important Fe-rich phases in aluminium alloys containing silicon are β-phase and α-phase. The α-phase is identified most commonly as α-Al8Fe2Si. The α-phase has a compact morphology such as Chinese script, star like and polygon, which was thought to be much less harmful than the platelet β-Al5FeSi to the mechanical properties of Al alloys.

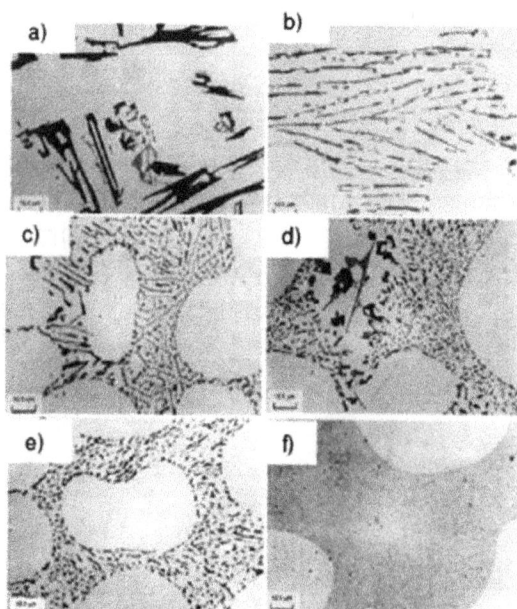

Type		Formation	Morphology	Size
Porosity	Hydrogen-induced	Hydrogen dissolution in the liquid metal		Micrometer-scale in diameter
	Shrinkage-induced	Lack of liquid feeding when directional solidification is lost		Can be as large as in millimeter-scale in thickness
Inclusions	Oxide film	Young oxide film		Nanometer-scale in thickness
		Old oxide film		Micrometer-scale in thickness
	Other exogenous inclusions	Entrained particles arising from furnace linings, die components and coatings, and upstream processes.	—	Micrometer-scale in thickness
Cracks	Hot tearing	Due to presence of thermal stresses in the semi-solid phase		Micrometer to millimeter in scale
	Cold cracking	Due to contraction-induced stress below the solidus temperature		Micrometer to millimeter in scale

Figure 10.216 LHS = Microstructures showing the effect of modification of the eutectic a) not modified, b) lamellar, c) partly modified, d) non lamellar, e) modified, f) over modified. RHS = Known imperfections

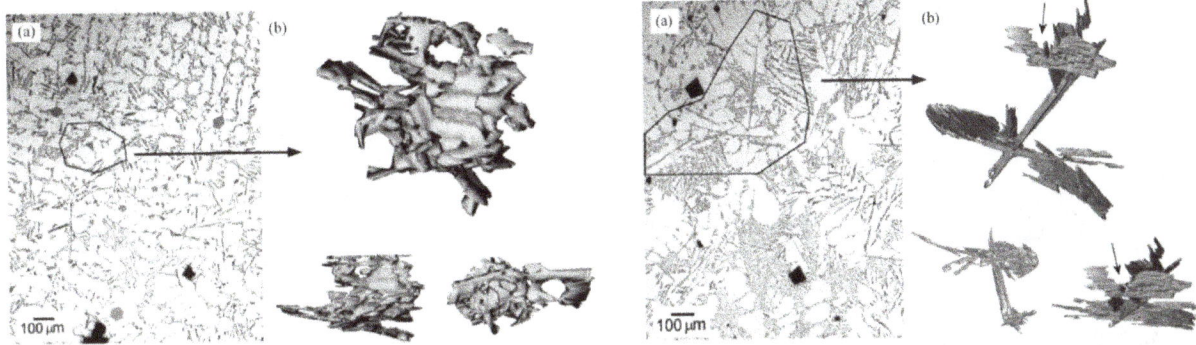

Figure 10.217 *LHS = Three-dimensional reconstruction of α-phase: (a) original two-dimensional photo; (b) three dimensional α-phase with high convoluted arms observed. RHS = Three-dimensional reconstruction of β-phase, (a) original two-dimensional phases; (b) three-dimensional β-phase (19)*

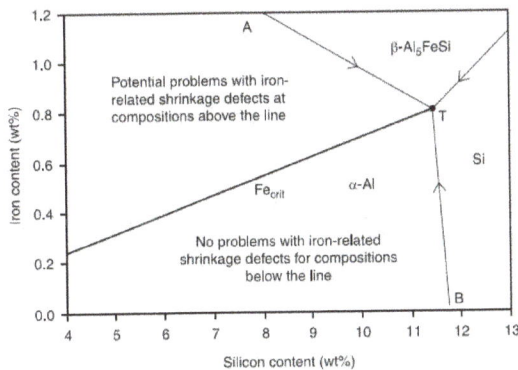

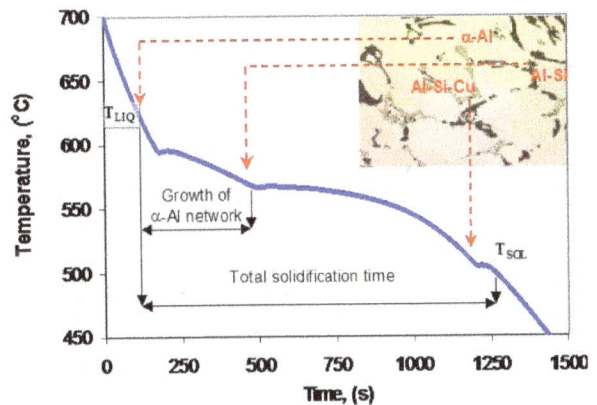

Figure 10.218 *LHS = Level of Fe to give problems[6]. RHS = Thermal analysis of cast alloy solidification*

10.16.8 Stacast

There has been a major European project called StaCast,[12 and 13] carried out by various institutions from European countries, which was constituted to carry out:

1. A survey of European aluminium alloy foundaries. The survey has been based on an on-line questionnaire

   i) Company profile
   ii) Data on Production: annual production, processes, alloys, application of castings
   iii) CEN Standards currently used
   iv) Needs for new Standards and guidelines
   v) Techniques usually employed for Defects Analysis
   vi) Defects more frequently detected.

2. To develop a method and tooling to provide mechanical test specimens for the standard CEN/

TR 16748:2014.[10 and 11] This was based on the idea that the real mechanical potential of Al-based cast alloys needs to be accurately defined. StaCast Project provided information for a new CEN Technical Standard. The intention was that this standard should show the mechanical potential (in terms of Ultimate Tensile Strength, Yield Strength and Elongation) which can be achieved by Al-Si alloys, cast by high pressure, low pressure and gravity (permanent mould) processes is introduced and described. Mechanical properties are measured on test specimens produced by using optimized test specimens.

3. The project is aimed at developing a new classification of the casting's structural defects and a statement of the limits of acceptability for the final application.[13]

Hot Tear An important cast aluminium alloy used in the automotive industry is the grade AlSi$_8$Cu$_3$Fe. A study was carried out of the cracking of an engine block associated with the withdrawal of a core pin

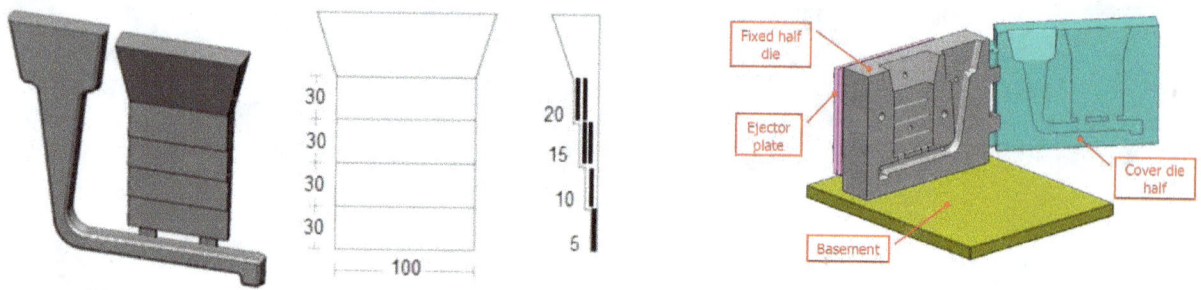

Figure 10.219 *LHS = Reference casting for gravity die casting (11). RHS = Layout of the die (11)*

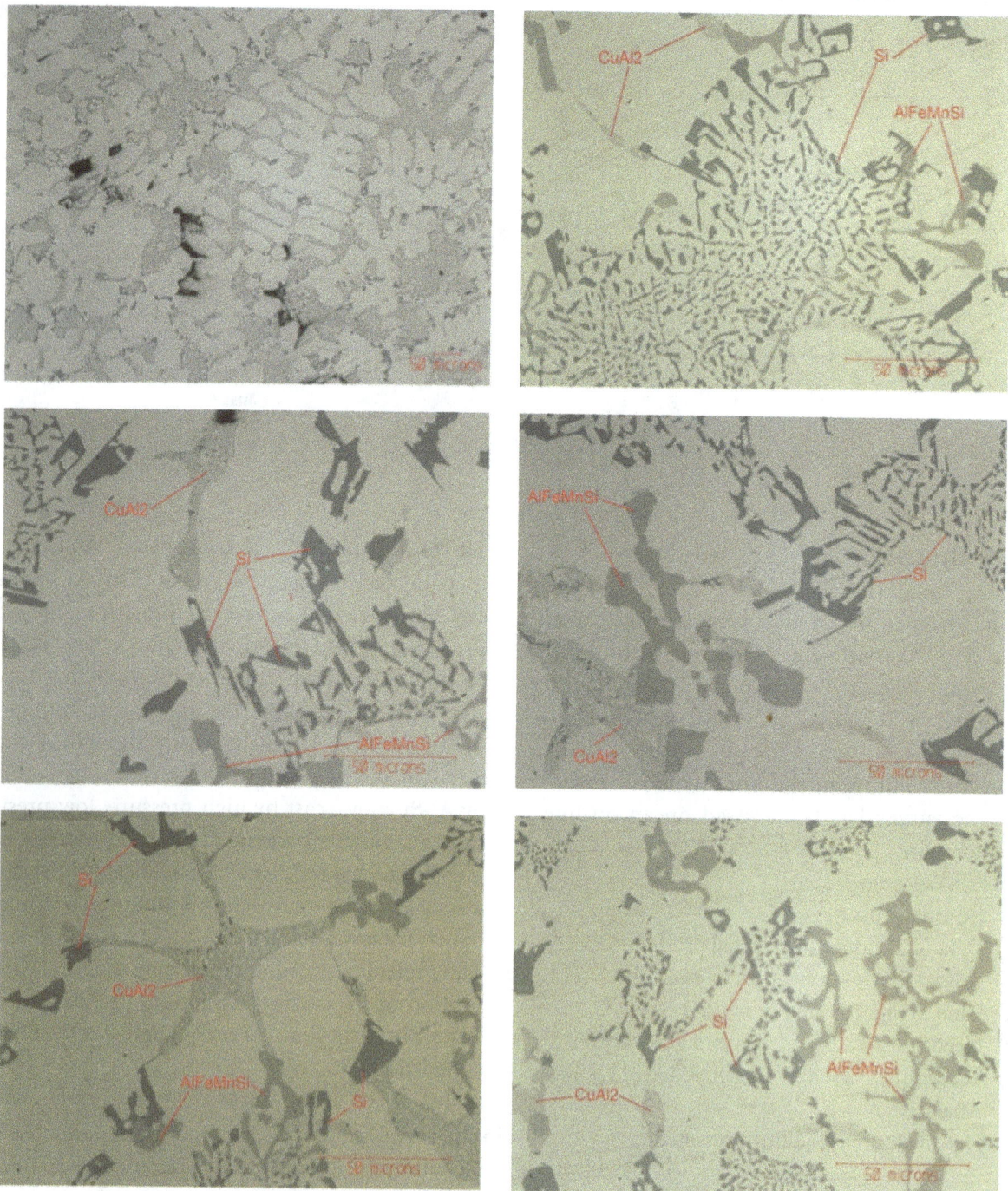

Figure 10.220 *Cast aluminium microstructures*

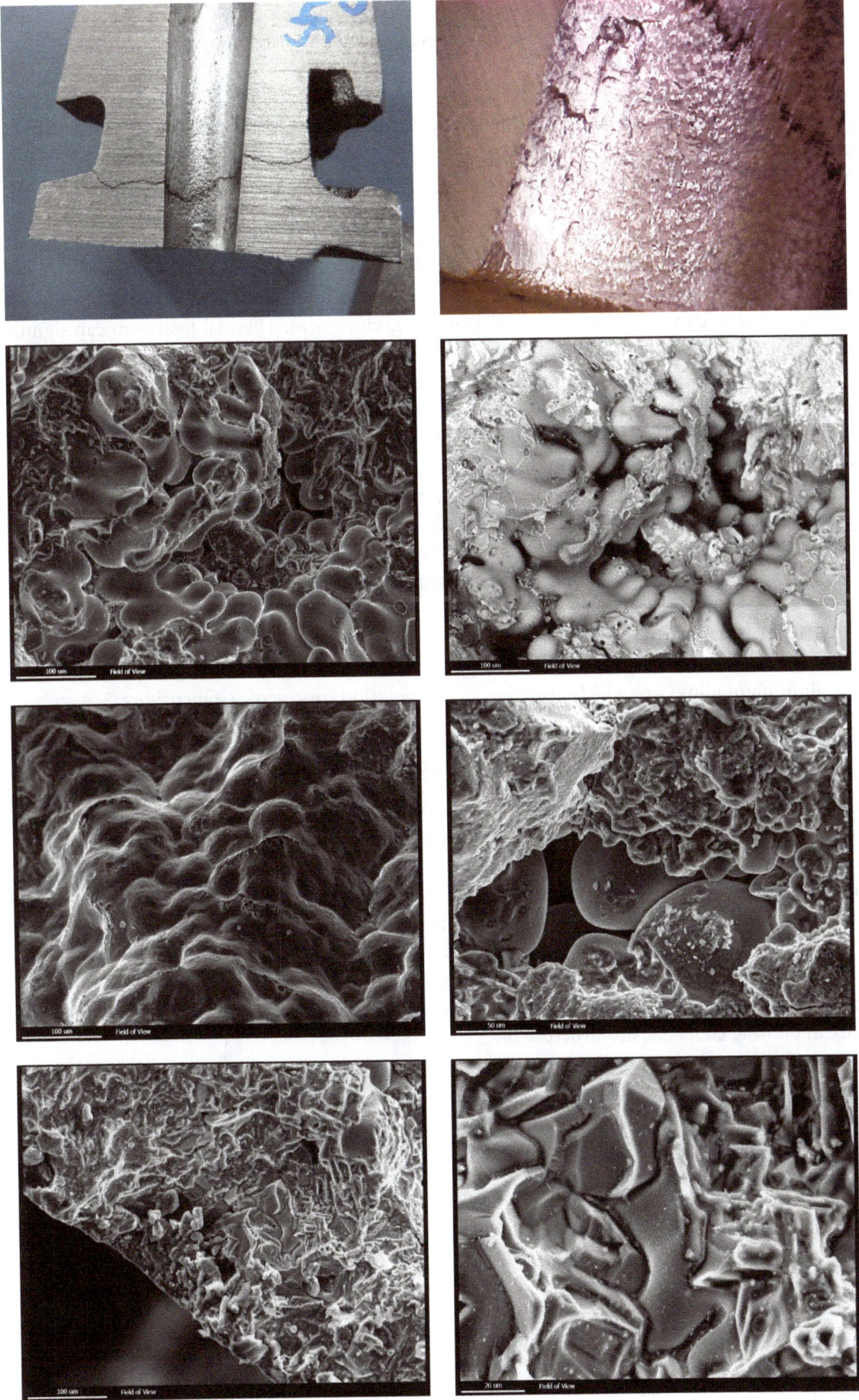

Figure 10.221 *Cast Aluminium cracks and SEM images*

from one side of the casting. The casting process was a bottom fed low pressure die cast process with sand cores.

The removal of the steel core pin appears to have "bonded" and caused surface damage as the pin was removed as shown in Figure 10.221 top. The forces associated with the pin removal caused hot tearing of the surface and another part of the casting which had a tight radius. The preparation of metallographic examination of specimens near the fracture surfaces can allow information to be gained on solidification rate, the presence of inter-dendritic shrinkage and the volume fraction and refinement of the eutectic and other precipitates. In addition, SEM examination of the fracture surfaces can show how the surfaces separated with partially solidified metal present and additional oxidation of the initial cracks.

The fracture surface showed areas of dendritic appearance, some with a covering of eutectic and oxide as shown in Figure 10.221 suggesting a hot tear.

10.16.9 Wrought aluminium alloys

The wrought aluminium alloys are widely used in automotive and aerospace industry due to their level of mechanical properties, which are higher than those obtained for cast aluminium alloys. Wrought aluminium alloys represent about 85% of aluminium applications. They are initially cast as ingots or billets and subsequently hot and/or cold worked mechanically into the desired form. The crystal structure of Al, face centered cubic system (fcc) offers a good cold formability. For wrought applications, the addition of alloying elements improves most of the mechanical properties; even if they have a comparatively small quantity of alloying elements, the structure of wrought alloys offers better mechanical properties than cast alloys.

Plastic deformation has increased the degree of grain refinement and allowed the alloy to homogenize, giving better uniformity of the microstructure. There are four main processes applied to wrought alloys to obtain different products:

1. Rolled products: plates, flat sheets, coiled sheets, and foils.
2. Extruded products: extruded rods, solid and hollow shapes, profiles, or tubes.
3. Forming products: rolled or extruded products are formed to achieve complex shapes.

4. Forged products: they have complex shapes with superior mechanical properties.

10.16.10 Thermal treatment used for aluminium cast alloys

Generally, heat treatments are widely used in Al foundries to increase the mechanical strength of aluminium alloys. There is a trend to minimize the energy consumption in order to have a minor impact on the environment. Thermal treatment can significantly influence properties such as strength, ductility, fracture toughness, thermal stability, residual stresses, dimensional stability, resistance to corrosion and stress corrosion cracking. The most important heat treatment procedures are homogenization, annealing and precipitation hardening which involves solution heat treatment, quenching and ageing.

The Aluminium Association has developed the classification of temper-designation system shown in Table 1.

The heat treatment is divided into three steps:

1. Solution treatment 2. Quenching 3. Ageing.

Solution treatment

The solution treatment is carried out at a temperature as close as possible to the eutectic temperature in order to obtain:

- the dissolution of soluble phases, containing Cu and Mg formed during the solidification
- the homogenization of the alloying elements
- the spheroidization of eutectic silicon particle.

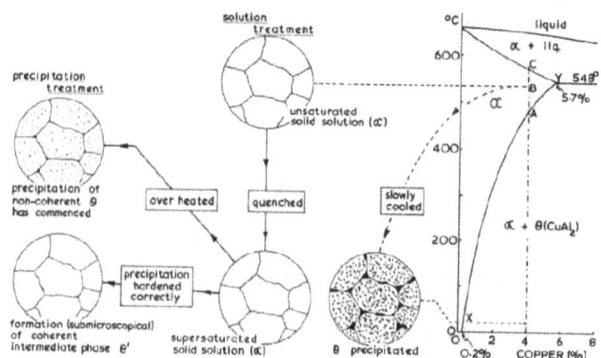

Figure 10.222 Representation of the mechanisms associated with solution, quench and ageing in the case of Al-Cu alloys

The rate of these three processes increases as the solution treatment temperature increases. The solution treatment temperature is situated between the solvus temperature and the solid temperature. For example, in the case of Al-Cu alloys, the solution temperature has to be lower than 485°C in order to avoid the incipient melting of the Cu-rich phases, because localized melting results in distortion and can significantly reduce the mechanical properties. Components have to be maintained at this temperature for a period of time necessary to reach a homogeneous solid solution.

Quenching

The role of the quenching/rapid cooling is to form a supersaturated solution of alloying elements in the aluminium matrix. To achieve this, it is necessary to choose accurately the quenching rate. If the cooling is too slow particles will be precipitated at grains' boundaries and a reduction in super-saturation of solute and a lower maximum yield strength, after ageing, will be obtained. However, a rapid quench can induce residual stresses into the heat-treated parts. Quenched parts with large thickness variations can develop large thermal gradients that can cause plastic flow, which in turn can cause distortion and residual stresses. The most common quenchant is water. To get better cooling characteristics of the quenching medium, some aqueous solutions of poly-alkylene glycol (PAG) can be used. Commonly the adopted PAG concentration is included between 4 and 30% and usually is determined by the type of product being processed.

Ageing

The purpose of ageing is to allow the precipitation of small non-coherent hardening precipitates, uniformly distributed into the α-Al matrix, to give an improvement in the mechanical strength of the alloy. The ageing can be performed at room temperature (natural ageing) or at high temperature in the range of 100–210 °C (artificial ageing). The size and the distribution of the precipitates, together with the coherency of the precipitates with the matrix, are fundamental to determine the high strength obtained during the ageing process.

The hardness-tensile strength-time curves obtained at different ageing temperatures is shown in Figure 10.223. At high temperatures the process is more rapid and the hardness's peak is achieved in a short time, due to the higher diffusion rate. However,

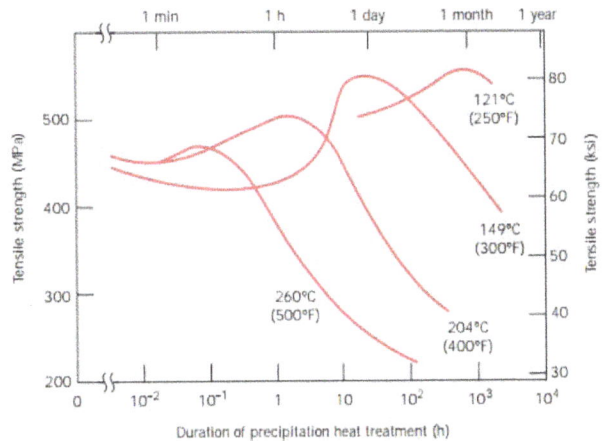

Figure 10.223 *Effect of time and temperature on hardness*

the maximum value of hardness decreases when the temperature increases.

REFERENCES

1. Charles R. Simcoe, The Discovery of Strong Aluminum, ADVANCED MATERIALS & PROCESSES • AUGUST 2011 page 35

2. Precipitation Hardening of Metal Alloys https://nvlpubs.nist.gov/nistpubs/sp958-1ide/014-015.pdf

3. O.B.M. Hardouin Duparc, The Preston of the Guinier-Preston Zones. Guinier, METALLURGICAL AND MATERIALS TRANSACTIONS, VOLUME 41A, AUGUST 2010—1873.

4. I.J. Polmear, Aluminium Alloys – A Century of Age Hardening, Proceedings of the 9th International Conference on Aluminium Alloys (2004), http://www.icaa-conference.net/ICAA9/data/papers/INV%201.pdf

5. International Alloy Designations and Chemical Composition Limits for Wrought Aluminum and Wrought Aluminum Alloys, January, 2015, The Aluminum Association, Inc. https://www.aluminum.org/sites/default/files/Teal%20Sheets.pdf

6. John Campbell, Complete Casting Handbook Metal Casting Processes, Metallurgy, Techniques and Design, Second Edition Butterworth-Heinemann 2015

7. Roger Lumley, Editor, Fundamentals of aluminium metallurgy Production, processing and applications Woodhead Publishing Limited 2011, Chapter 6 Casting of aluminium alloys, Chapter 7 Quality issues in aluminum castings, and Chapter 8 Case studies in aluminium casting alloys

8. Geoffrey K. Sigworth, FUNDAMENTALS OF SOLIDIFICATION IN ALUMINUM CASTINGS, International Journal of Metalcasting/Volume 8, Issue 1, 2014,

9. G. K. Sigworth, Understanding Quality in Aluminum Castings, Copyright 2011 American Foundry Society, https://pdfs.semanticscholar.org/34ff/f86f06bb0c3f7dcf08719e25b-41c0072b15a.pdf

10. Anilchandra R. Adamane et al, Reference Dies for the Evaluation of Tensile Properties of Gravity Cast Al-Si Alloys: An Overview, Materials Science Forum Vols. 794-796 (2014) pp 71-76, file:///C:/Users/user/Downloads/MSF.794-796.71.pdf

11. A.R. Anilchandra et al, Evaluating the Tensile Properties of Aluminum Foundry Alloys through Reference Castings— A Review, Materials 2017, 10, 1011, file:///C:/Users/user/Downloads/materials-10-01011.pdf

12. F. Bonollo et al. StaCast Project: from a Survey of European Aluminium Alloys Foundries to New Standards on Defect Classification and on Mechanical Potential of Casting Alloys 71st World foundry congress Bilbao, 2014, file:///C:/Users/user/Downloads/71STWFC_Bonollo.pdf

13. Elena Fiorese et al, NEW CLASSIFICATION OF DEFECTS AND IMPERFECTIONS FOR ALUMINUM ALLOY CASTINGs, International Journal of Metalcasting/Volume 9, Issue 1, 2015 page 55, file:///C:/Users/user/Downloads/INTJMetalcasting2015DEFECTCLASSIFICATION.pdf

14. E. Tan et al, Reproducibility of Reduced Pressure Test Results in Testing of Liquid Aluminum Gas Levels, 6th International Advanced Technologies Symposium (IATS'11), 16-18 May 2011, Elazığ, Turkey http://web.firat.edu.tr/iats/cd/subjects/Metallurgy&Material/MSM-63.pdf

15. D. Dispinar and J. Campbell, Critical assessment of reduced pressure test. Part 2: Quantification, International Journal of Cast Metals Research 2004 Vol. 17 No 5, file:///C:/Users/user/Downloads/ijcmr526e.pdf

16. Yu-bo Zuo et al, A new high shear degassing technology and mechanism for 7032 alloy, China Foundry Research & Development, Vol.12 No.4 July 2015

17. Alan Kaye and Arthur Street, Die Casting Metallurgy, Butterworth Scientific 1982

18. Xixi Dong and Shouxun Ji, Si poisoning and promotion on the microstructure and mechanical properties of Al–Si–Mg cast alloys J Mater Sci (2018) 53:7778–7792, https://link.springer.com/content/pdf/10.1007%2Fs10853-018-2022-0.pdf

19. Lifeng Zhang et al, REMOVAL OF IRON FROM ALUMINUM: A REVIEW, Mineral Processing & Extractive Metall. Rev., 33: 99–157, 2012 file:///C:/Users/user/Downloads/6RemovalofIronFromAluminum_Review.pdf

20. M. B. Djurdjević and M. A. Grzinčić, The effect of major alloying elements on the size of the secondary dendrite arm spacing in the as-cast Al-Si-Cu alloys, ARCHIVES of FOUNDRY ENGINEERING Volume 12, Issue 1/2012, 19 – 2 4. http://www.afe.polsl.pl/index.php/en/3499/the-effect-of-major-alloying-elements-on-the-size-of-the-secondary-dendrite-arm-spacing-in-the-as-cast-al-si-cu-alloys.pdf

21. MÓNIKA TOKÁR et al, THE EFFECT OF STRONTIUM AND ANTIMONY ON THE MECHANICAL PROPERTIES OF Al-Si ALLOYS, Materials Science and Engineering, Volume 39, No. 1 (2014), pp. 69–79. https://matarka.hu/koz/ISSN_2063-6792/vol_39_1_2014_eng/ISSN_2063-6792_vol_39_1_2014_eng_069-079.pdf

22. Cameron M. Dinnis, Arne K. Dahle, John A. Taylor, Three-dimensional analysis of eutectic grains in hypoeutectic Al–Si alloys, Materials Science and Engineering A 392 (2005) 440–448, https://pdfs.semanticscholar.org/77c3/43d46b1728ce52d223f37648a34c42270500.pdf

23. Chikezie W. Onyia et al, Structural Modification of Sand Cast Eutectic Al-Si Alloys with Sulfur/Sodium and Its Effect on Mechanical Properties, World Journal of Engineering and Technology, 2013, 1, 9-16, https://file.scirp.org/pdf/WJET_2013081215543445.pdf

24. R. Gitter, Design of Aluminium Structures: Selection of Structural Alloys, Structural Design according to Eurocode 9: Essential Properties of Materials and Background Information, Selection of structural alloys; Brussels 2008, https://eurocodes.jrc.ec.europa.eu/doc/WS2008/EN1999_4_Gitter.pdf

Books

Fundamentals of aluminium metallurgy Production, processing and applications, Edited by Roger Lumley, Woodhead Publishing Limited, 2011

Aluminium alloys, theory and applications, Edited by Tibor Kvačkaj and Róbert Bidulský, 2011 InTech. A free online edition of this book is available at www.intechopen.com

Complete Casting Handbook Metal Casting Processes, Metallurgy, Techniques and Design, John Campbell Second Edition Butterworth-Heinemann 2015

Foseco Non-Ferrous Foundryman's Handbook Eleventh edition 1999, Revised and edited by John R. Brow, Butterworth-Heinemann

10.17 COPPER AND COPPER ALLOYS

10.17.1 History of copper alloys

Humans may have started smelting copper as early as 6,000 B.C. in the Fertile Crescent, a region often called "the cradle of civilization" (Mesopotamia, the area between the Tigris and Euphrates Rivers in modern day Iraq), where agriculture and the world's first cities emerged.

The beginning of the Bronze Age marked the first-time humans started to work with metal. Bronze tools and weapons soon replaced earlier stone versions.

Ancient Sumerians in the Middle East may have been the first people to enter the Bronze Age. Humans made many technological advances during the Bronze Age, including the first writing systems and the invention of the wheel. In the Middle East and parts of Asia, the era lasted from roughly 3300 to 1200 B.C., ending abruptly with the near-simultaneous collapse of several prominent Bronze Age civilizations.

Archaeological evidence suggests the transition from copper to bronze took place around 3300 B.C. The invention of bronze brought an end to the Stone Age. Different human societies entered the Bronze Age at different times. Civilizations in Greece

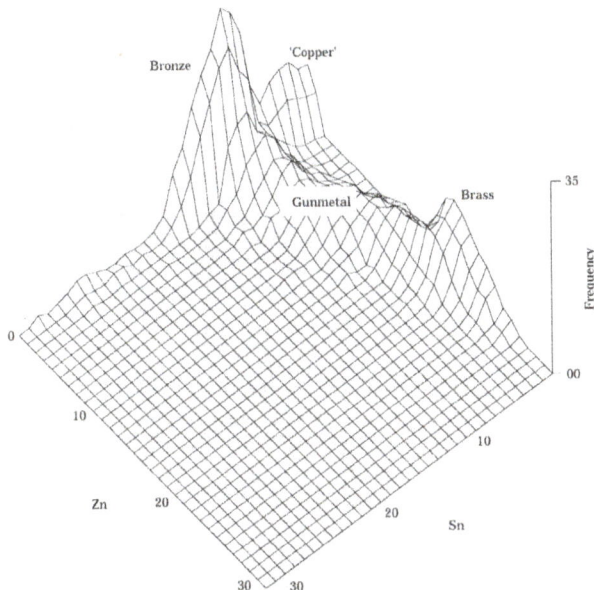

Figure 10.224 *Three dimensional plot (smoothed) of the Sn and Zn content of 1163 Roman copper alloys ref1 1.David Dungworth, Roman Copper Alloys: Analysis of Artefacts from Northern Britain, Journal of Archaeological Science (1997) 24, 901–910*

Figure 10.225 *Photographs of the "ORICHALCUM" ingots. Total weight was 40kg Eugenio Caponetti, FIRST DISCOVERY OF ORICHALCUM INGOTS FROM THE REMAINS OF A 6TH CENTURY BC SHIPWRECK NEAR GELA (SICILY) SEABED, Mediterranean Archaeology and Archaeometry, Vol. 17, No 2, (2017), pp. 11-18*

began working with bronze before 3000 B.C., while the British Isles and China entered the Bronze Age much later around 2000 B.C. and 1700 B.C., respectively. The Bronze Age ended around 1200 B.C. when humans began to forge an even stronger metal: iron. The archaeological remains provide insight into the melting and casting practices. As well as bronze there was a lot of brass artefacts. In 1997 the analysis of over 1063 copper alloy items found in Northern Britain were examined[1]. The range of composition was fascinating, ranging from brass to bronze and to gun-metal compositions (Figure 10.224).

10.17.2 Orichalcum

In addition, samples of "Platos" mythical metal "ORICHALCUM" of ancient "Atlantis" fame have been found by divers.

> "Plato describes the ports and forts of Atlantis as covered by a rare metal called ὀρείχαλκος (Balouglou, 2010). The Greek word ὀρείχαλκος (ὄρος, upstream and χαλκός, Copper) is attested in several literary sources like the pseudo-Homeric Hymn to Aphrodite, where the birth of Aphrodite is described."

The divers recovered 40 metal ingots of orichalcum which were carried by a ship that wrecked near the Gela coast. Pieces of pottery were found nearby, which have been dated around the end of the sixth century BC, and come from the Aegean Sea and the eastern Mediterranean. In ancient times, the legendary orichalcum was a very precious alloy. The only method of production of orichalcum would be a long and laborious cementation process with the copper and the source of zinc held in a sealed pot below the melting point of copper (probably around 900°C).

The earliest method of making brass was possibly the cementation process in which finely divided copper fragments were intimately mixed with roasted zinc ore (oxide) and reducing agent, such as charcoal, and heated to 1000°C in a sealed crucible. Zinc vapour formed, dissolved into the copper. It has been demonstrated experimentally that brass produced by the cementation process could not contain more than 28% zinc.

The mechanical properties and corrosion performance of copper can be optimised to meet the requirements for many industrial applications. This

	% w/w			[Cu]/[Zn]
	Cu	**Zn**	**Pb**	
S01	76	20	2.9	3.7
S02	79	14	4.3	5.7
S03	70	26	1.6	2.7
S04	75	18	5.7	4.1
S05	75	21	2.0	3.6
S06	77	20	3.3	3.9
S07	78	19	2.9	4.1
S08	76	21	3.2	3.7
S09	78	16	4.6	4.8
S10	70	26	2.7	2.7
S11	78	17	3.7	4.5
S12	75	17	6.8	4.4
S13	76	18	2.6	4.2
S14	70	26	1.8	2.7
S15	79	17	4.0	4.7

Figure 10.226 *Chemical analysis of 15 of the ingots*

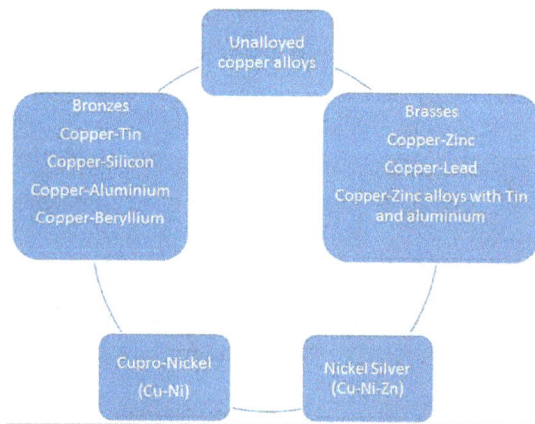

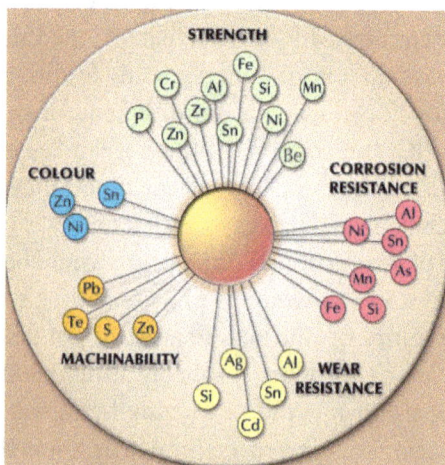

Table 9.10 Classification of copper alloys (Copper Development Association System)

	Wrought alloys
C1xxxx	Coppers* and high-copper alloys†
C2xxxx	Copper-zinc alloys (brasses)
C3xxxx	Copper-zinc-lead alloys (leaded brasses)
C4xxxx	Copper-zinc-tin alloys (tin brasses)
C5xxxx	Copper-tin alloys (phosphor bronzes)
C6xxxx	Copper-aluminum alloys (aluminum bronzes), copper-silicon alloys (silicon bronzes) and miscellaneous copper-zinc alloys
C7xxxx	Copper-nickel and copper-nickel-zinc alloys (nickel silvers)
	Cast alloys
C8xxxx	Cast coppers, cast high-copper alloys, cast brasses of various types, cast manganese-bronze alloys, and cast copper-zinc-silicon alloys
C9xxxx	Cast copper-tin alloys, copper-tin-lead alloys, copper-tin-nickel alloys, copper-aluminum-iron alloys, and copper-nickel-iron and copper-nickel-zinc alloys.

*"Coppers" have a minimum copper content of 99.3 percent or higher.
†High-copper alloys have less than 99.3 percent Cu but more than 96 percent and do not fit into the other copper alloy groups.

Foundations of Materials Science and Engineering, 5th Edn. Smith and Hashemi

Figure 10.227 *Main alloy groups*

is accomplished by the diverse range of copper alloys that have been developed. The combination of copper with other metals such as zinc, aluminium, silicon, nickel and iron means the created alloy can be manufactured to meet almost any application. There are over 475 copper alloys, each with a different combination of properties which can meet the requirement

for many applications, manufacturing processes and environments. Including the surprising ability to meet antimicrobial requirements. The main groups of alloys are shown in Figure 10.227.

10.17.3 High Purity Copper

Pure copper has the highest electrical and thermal conductivity of any commercial metal. Around 60% of the copper produced is used in electrical and electronic applications and this leads to a convenient classification of the types of copper into electrical (high conductivity) and non-electrical (engineering).

The high purity copper alloys have a copper level greater than 99.3%. They are classified into three groups:

- Electrolytic tough pitch which contains 99.9% copper with 0.045% oxygen
- Oxygen free, Oxygen free copper can be produced from electro refined cathode copper which is melted and cast in a reducing atmosphere of carbon monoxide and nitrogen to prevent oxygen pick-up
- Phosphorous deoxidized, Phosphorous is added which produces phosphorous pentoxide P_2O_3 which removes the oxygen.

The electrolytic tough pitch copper represents about 50% of the total consumption. The main areas are rail overhead electrical systems. For example, light rail metro systems and high-speed trains.

Figure 10.228 shows a photo of the Tyne and Wear Metro and a schematic showing the specific

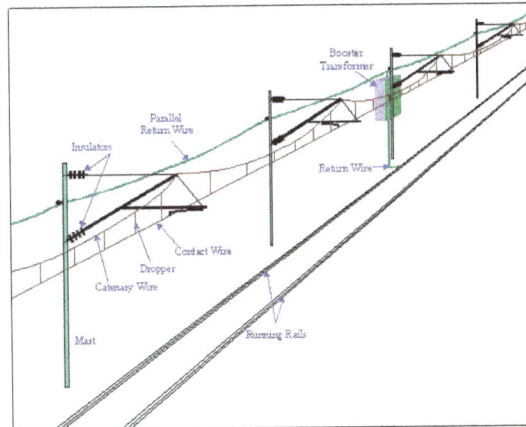

Figure 10.228 The Metro and a schematic showing the specific terms used to describe the individual components

terms used to describe the individual components of the overhead wire system. The "contact wire" can be seen below the "catenary wire" supported by "droppers". The power in the catenary wire is transferred to the contact wire via "dropper" lines. The pantograph on the train presses against the contact wire with an appropriate force to prevent separation and collects the power to run the train drive motors. The return path is through the rails which is connected to the "parallel return wire" at appropriate intervals along its length.

The failure modes expected on the contact wire are fatigue, wear damage due to the moving contact with the pantograph, and potential arc damage in the event of separation of the pantograph from the contact wire.

The contact wire for Metro systems are manufactured from Electrolytic Tough Pitch Copper (C11000) (Cu-ETP) which may have from 100 to 650 ppm oxygen present as oxide second phase particles (the nominal value is often quoted as 400ppm and expected level is from 200 to 400ppm). The oxygen is present as oxide inclusions and may have an effect on the endurance properties. The oxides are present to combine any residual elements left in the copper. If they are left in "solid solution" these residual elements can affect the wire conductivity. When these elements are present as oxides they do not affect the conductivity. Modern process technology should be effectively reducing the residual elements present as oxides in the wire. The Cu-ETP quality supplied by one company (NTK) is quoted in the NTK brochure as suitable for up to 160 km/hr speeds. Their specified minimum tensile strength is 330MPa.

For high speed trains the wire has a higher strength and is made from a Cu-Mg alloy.

Figure 10.229 shows fatigue cracks developing in a contact wire, the Upper is on a light rail and the

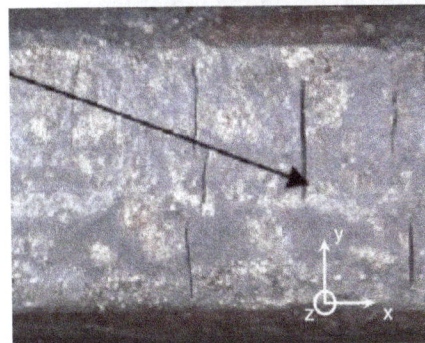

Figure 10.229 Upper = Fatigue cracks on the side of a failed wire. Lower = Published image of fatigue cracks on a contact wire from a high-speed train[1]

Figures 10.230 Outgoing part of fracture (the bottom of the contact wire shows evidence of fresh contact with the pantograph and some small arcs

Lower is on a French high speed train. Typical fracture faces and contact zone on the wire are shown in Figure 10.230.

Several studies on the interaction between the pantograph and the contact wire point out that damping of the contact wire is low and that each loading cycle is composed of several bending waves preceding and following the pantograph passage[1].

To ensure uninterrupted collection of power, while the train is in motion, the pantograph exerts a force on the contact wire to maintain contact. The force induced between the pantograph and the contact wire causes a shear wave to travel forward in the contact wire (often referred to as a bow wave). The resultant cyclical loading of components can lead to fatigue failures, particularly at features in the overhead line where there is a change of stiffness. Japanese and French R&D have measured the stresses on the contact wire to try to determine methods to avoid contact wire fatigue failure[1 and 2]. The Japanese method is based on allowing a maximum bending strain on the top of the wire of 500 micro-strain. It is interesting to note the work by N. Tanabe et al on the Fatigue of High Purity Copper Wire found an endurance limit at a bending stress of around 80MPa for Cu-ETP, which equated to a local surface strain of 500 micro-strain[3].

In addition, the contact wire is under tension to try to avoid wire movement and vibration. In accordance with the effect of mean stress (Goodman, Gerber or Soderberg for stress life endurance and Smith, Watson Topper for strain life endurance) the tension in the contact wire can amplify the effect of small stress levels on the fatigue life. The higher the tension in the wire the shorter the fatigue life. Therefore, as the wire wears in service and the tensile stress in the wire increases (assuming constant tension load in the wire), this can further reduce the fatigue life[4 and 5].

The contact stresses between the pantograph and the contact wire is currently undergoing significant worldwide R&D. Most of this work has been associated with high speed high voltage main line conditions but similar principles apply to lower voltage Metro systems. This includes test rigs to measure simulated endurance life of contact wire[4], the development of higher strength contact wire, software simulation of stresses[6], the condition monitoring of the pantograph at Birmingham University[7] and the fatigue evaluation of other parts of the overhead electrical lines[8].

The initial work carried out to evaluate the endurance properties of the contact wire used test rigs to simulate the load conditions. However, more recent methods have used conventional tensile test methods and resonant tensile-tensile fatigue tests[9]. These methods allow the determination of the effect of wire preload and to check the effect of higher stress as the wire wears. The quality assurance of the contact wires requires details specification and mechanical checks on the wire. An example of an engineering specification is shown in[10]. In addition, there are new improved grades that claim improved performance[11].

REFERENCES

1. J.P. Massat et al, Fatigue analysis of catenary contact wires for high speed trains, 9th World congress on railway research at Lille May 22 to 26, 2011. http://www.railway-research.org/IMG/pdf/e3_massat_jean_pierre.pdf

2. Atsushi Sugahara, Preventing Fatigue Breakage of Contact Wires, Railway Technology Avalanche No 24 September 19 2008. https://www.rtri.or.jp/eng/publish/newsletter/pdf/24/RTA-24-140.pdf

3. N. Tanabe et al, Fatigue of High Purity Copper Wire, JOURNAL DE PHYSIQUE IV Colloque C7, suppl6ment au Journal de Physique 111, Volume 5, November 1995

4. Yongseok Kim et al, Fatigue Safety Evaluation of Newly Developed Contact Wire for Eco-Friendly High Speed Electric Railway System Considering Wear, INTERNATIONAL JOURNAL OF PRECISION ENGINEERING AND MANUFACTURING-GREEN TECHNOLOGY Vol. 3, No. 4, pp. 353-358 OCTOBER 2016. file:///C:/Users/user/Downloads/KCI_FI002151990.pdf

5. Chikara Yamashita et al, Influence of mean stress on contact wire fatigue, QR of RTRI Vol 47 No 1 Feb 2006, https://www.jstage.jst.go.jp/article/rtriqr/47/1/47_1_46/_pdf/-char/en

6. Beagles, A et al, Validation of a new model for railway overhead line dynamics, Proceedings of the ICE – Transport, 169 (5). pp. 339-349. http://eprints.whiterose.ac.uk/103249/1/Overheadline_paper_ICE_accepted_version.pdf

7. Tingyu Xin et al, Condition monitoring of railway pantographs to achieve fault detection and fault diagnosis, Proc IMechE Part F: J Rail and Rapid Transit 0(0) 1–12

8. Liming Chen et al, Fatigue life analysis of dropper used in pantograph-catenary system of high-speed railway, Advances in Mechanical Engineering 2018, Vol. 10(5) 1–10, https://journals.sagepub.com/doi/pdf/10.1177/1687814018776135

9. Zhen Guo et al, Fatigue Life Estimation of Cold Drawn Contact Wire, INTERNATIONAL JOURNAL OF PRECISION ENGINEERING AND MANUFACTURING Vol. 15, No. 11, pp. 2291-2299 NOVEMBER 2014 / 2291

10. MESP 130500-01, TECHNICAL SPECIFICATION FOR CONTACT WIRE HARD-DRAWN COPPER 161mm2. https://documentportal.metrotrains.com.au/engineering-docs/Specifications/L1-CHE-SPE-001.pdf

11. NKTRailway_Contact_Line_Catalogue_September_2017.pdf https://www.nkt.com/fileadmin/user_upload/NKTRailway_Contact_Line_Catalogue_September_2017.pdf

10.17.4 Brass

Brass is the generic term for a range of copper-zinc alloys with differing combinations of properties, including strength, machinability, ductility, wear-resistance, hardness, colour, antimicrobial, electrical and thermal conductivity, and corrosion-resistance.

There are two classes of brass:

- The alpha brass alloys with less than 37% zinc. These alloys are ductile and can be cold worked
- The alpha plus beta or duplex alloys which have 37 to 45% zinc. These alloys have a limited cold ductility and are typically harder and stronger.

The brass can also be considered as cast and wrought. Cast alloy brasses

- Copper-Tin-Zinc alloys (red or semi-red and yellow brasses)

- Cast copper-bismuth and copper – Bismuth-Selenium alloys
- Copper-Zinc-Silicon alloys (Silicon brasses and bronzes)
- Wrought alloy brasses
- Copper-Zinc alloys
- Copper-Zinc-Tin alloys (Tin brasses)
- Copper-Zinc-Lead alloys (leaded brasses)
- Copper-Zinc-Aluminium alloys (aluminium brasses)

The examination of a heat exchanger tube stack

Copper alloys develop corrosion resistance by developing an oxide layer which separates the operating environment from the copper alloy and prevents further corrosion. In service if the operating conditions, associated with the use of the copper alloy, can disrupt the oxide scale there is a risk of pitting corrosion. If

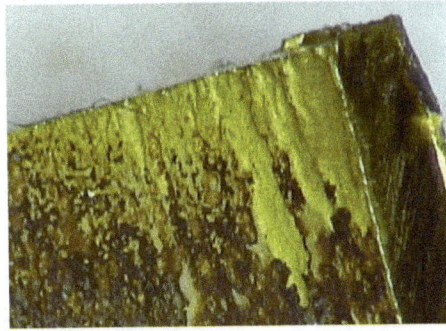

Figure 10.231 Tube stack and tube plate, Erosion-corrosion damage of the internal surface

Figure 10.232 single phase microstructure

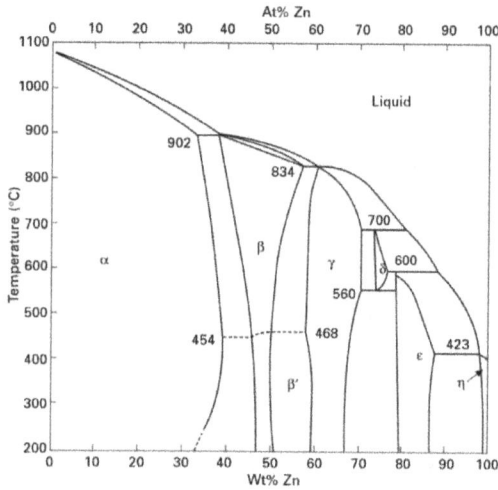

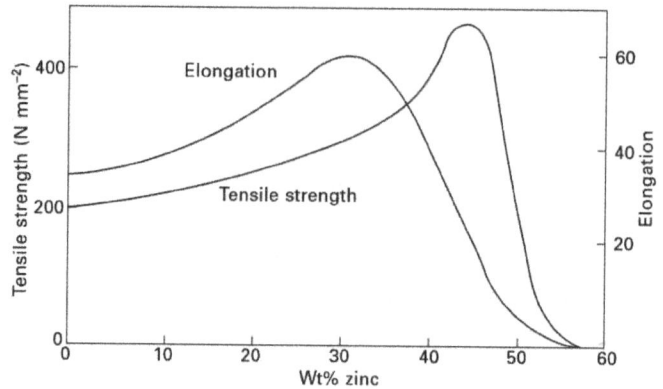

Figure 10.233 Cu-Zn equilibrium diagram

these conditions persist then complete perforation of the metal thickness can result.

The metallurgical examination of the samples provided confirmed that the composition of the aluminium brass tubes were in accordance with ASTM B111 C68700.

The metallographic examination confirmed that the grain size was uniform and showed complete recrystallization in accordance with clause 9.1.1 of ASTM B111. The grain size was measured at 32 microns and within the range of 11 to 45 microns specified in ASTH B111 clause 9.1.2.

In addition, the hardness of the tube was 115HV10 which would indicate that the strength was in accordance with ASTM B111. (Greater than 50ksi.)

Defects in brasses

*   Intergranular corrosion (Seasonal cracking)
*   Dezincification.

10.17.5 Bronzes

Cast bronze alloys

*   Copper-tin alloys (Tin Bronzes)
*   Copper-Tin-Lead alloys (Leaded and high leaded Tin Bronzes)
*   Copper-Tin-Nickel alloys (Nickel-Tin bronzes)
*   Copper aluminium alloys (Aluminium Bronzes).

Wrought alloys

*   Copper-Tin-Phosphorous alloys (Phosphorous Bronzes)
*   Copper-Tin-Lead-Phosphorous alloys (Leaded Phosphorous Bronzes)
*   Copper-Aluminium alloys (Aluminium Bronzes).

Tin Bronzes

*   Alloys of copper and tin
*   Other alloying elements may be present with the exception of zinc
*   Also known as Phosphor bronze
*   Phos content 0.01 to 0.5%
*   Tin 1 to 11%
*   High corrosion resistance, toughness, Low coefficient of friction and free from seasonal cracking.

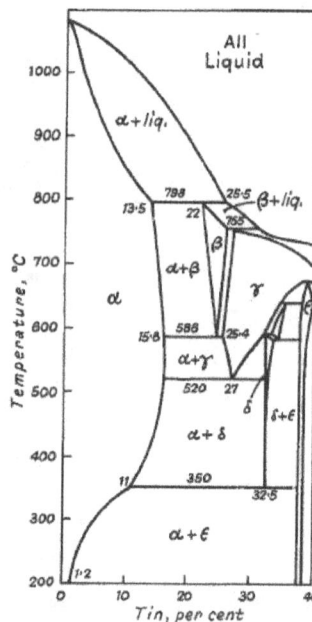

Figure 10.234 Copper tin equilibrium diagram

Silicon Bronzes

- Alloys of copper and silicon
- Maximum solubility in alpha phase is 5.3% and decreases with temperature
- Alloys contains less than 5% silicon and are single phase alloys
- These represent the strongest of the work hardened copper alloys and can achieve strength levels similar to low carbon steel and corrosion resistance similar to copper
- Silicon bronzes are used for Tanks, Pressure vessels, Marine construction, hydraulic pressure lines.

Aluminium Bronzes

- Alloys of Al and Cu
- Maximum solubility of Al in alpha solid solution is 9.5%
- Commercial Al bronzes contain 4 to 11% Al
- Alloys containing up to 7.5% Al are single phase, others are double phase
- Other elements are also added intentionally e.g. Fe, Mn, Si. Ni
- Iron increases strength, silicon improves machinability, Mn improves casting soundness
- Single phase bronze shows good cold working properties along with high corrosion resistance to water and atmospheres
- These alloys are used for water condenser tubes, bolts, corrosion resistance vessels and in marine applications.

Beryllium Bronzes

- Alloys of copper and beryllium
- Maximum solubility in alpha solid solution is 2.1% and decreases to 0.25% to room temperature
- The alloys can therefore be precipitation hardened
- They have excellent formability high tensile strength, creep resistance, high electrical conductivity
- These are used in diaphragms, surgical instruments, bolts and screws, firing pins and dies.

10.17.6 Cupro-Nickel alloys

- Alloys of copper and nickel contain up to 30% nickel
- Cupro-nickels are single phase alloys
- No heat treatment required

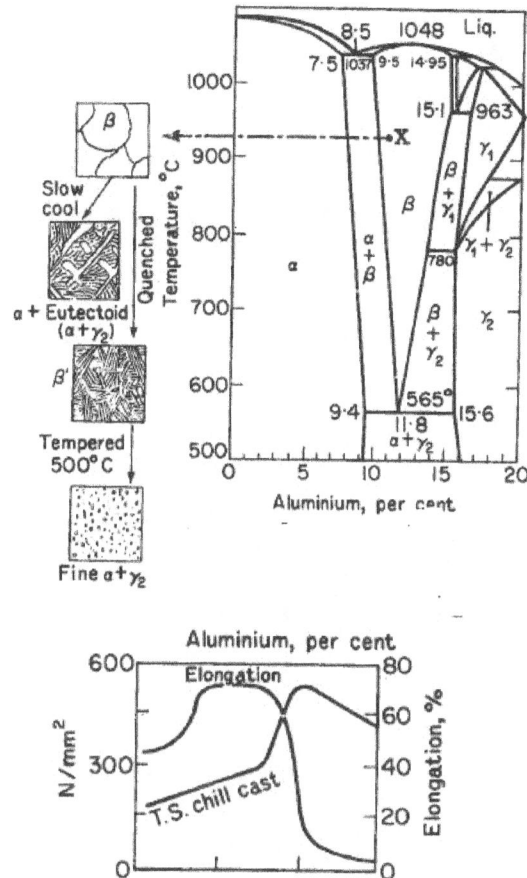

Figure 10.235 *Copper aluminium equilibrium diagram and mechanical properties*

- Properties are improved only by cold working
- They have high fatigue resistance, high corrosion resistance and erosion resistance towards sea water
- These alloys are widely used in condensers, heat exchanger tubes, coastal power plants

10.17.7 Nickel Silver

- Alloys of Cu-Ni-Zn
- Commercially alloys contain Cu (50 to 70%) Ni (5 to 10%) Zn (5 to 40%)
- If copper is more than 60% these are single phase ductile and easily workable at room temperature
- Additions of Zn imparts silver-blue-white colour good corrosion resistance
- These are excellent base metals for plating Cr, Ni, Ag
- These are used for rivets, screws, costume jewellery, name plates

CHAPTER 11

Prediction of Micro-Structural Evolution in Steel

11.1 HOW DO WE PRODUCE THE CORRECT OR "BEST" MICROSTRUCTURE?

The history of steel metallurgy contains reference to many different kinds of microstructures that have been identified from the decomposition/transformation of austenite. The main microstructure for the best combination of strength and toughness would be fine carbides in a ferritic matrix. The best way to achieve this would be a quench to produce the maximum amount of martensite followed by tempering.

For any particular steel composition, the microstructure that forms depend upon the steel type and composition, section size and cooling rate. For example, the development of low carbon bainitic steels the agreed nomenclature is shown in Table 11.1

Tables 11.1 and 11.2 show the complexity of the transformation kinetics, since the austenite phase can transform to many different phases, depending on cooling rates and cooling conditions and chemical composition. During industrial heat treatment processes where high cooling rates are used, three groups of microstructures can be differentiated as shown in Figure 11.2 LHS:

- Pearlite microstructures whose growth is primarily diffusion controlled
- Bainite microstructures whose growth is partly diffusion controlled
- Martensite microstructures whose growth is primarily diffusion less.

The austenite decomposition can be characterized as either:

- a diffusion mechanism (reconstructive)
- a diffusion-less one (displacive) see Figure 11.2 RHS.

Figure 11.2 shows that there are several phase transformations that can occur and therefore, there is often a need to optimize the composition of the steel and cooling rate to achieve the best mechanical properties. There is often a need to have some guidance and this is where software tools, spreadsheets and the methods

Table 11.1 Symbols and nomenclature

Symbol	Nomenclature
I_0 Major matrix-phase	
α_p	Polygonal ferrite
α_q	Quasi-polygnal α
α_w	Widmanstätten α
α_B	(Granula bainitic) α
α°_B	Bainitic ferrite
α'_m	Dislocated cubic martensite
II_0 Minor secondary phases	
γ_r	Retained austenite
MA	Martensite-austenite constituent
α_M	Martensite
aTM	Auto-tempered martensite
B	BII, B2: upper bainite
	Bu: upper bainite
	B_L: lower bainite
P'	Degenerated pearlite
P	Pearlite
θ	Cementite particle

Table 11.2 Description of the ferrite phases

- WF, Widmanstätten ferrite, elongated crystals of ferrite with a dislocation substructure (minimal dislocation substructure)

- QF, quasi-polygonal ferrite, grains with undulating boundaries, which may cross prior austenite boundaries containing a dislocation sub-structure and occasional martensite-austenite (MA) microconstituents. This is also referred to as massive ferrite.

- GB, granular ferrite, sheaves of elongated ferrite crystals (granular or equiaxed shapes) with low misorientations and a high dislocation density, containing roughly equiaxed islands of MA constituents

- BF, bainitic ferrite, packets of parallel ferrite laths (or plates) separated by low-angle boundaries and contains very high dislocation densities. MA constituents retained between the ferrite crystals have an acicular morphology. This is sometimes termed as acicular ferrite.

- Dislocated cubic martensite, highly dislocated lath like morphology, conserving prior austenite boundaries.

Figure 11.1 *Photographs of typical microstructures a=0.05%C, b=0.2%C, c=0.4%C, d=0.8%C, e=1.2%C f=0.4%C OQ, (Oil Quenched) g=0.4%C OQ+T650°C (Oil Quenched and Tempered), h=Copper*

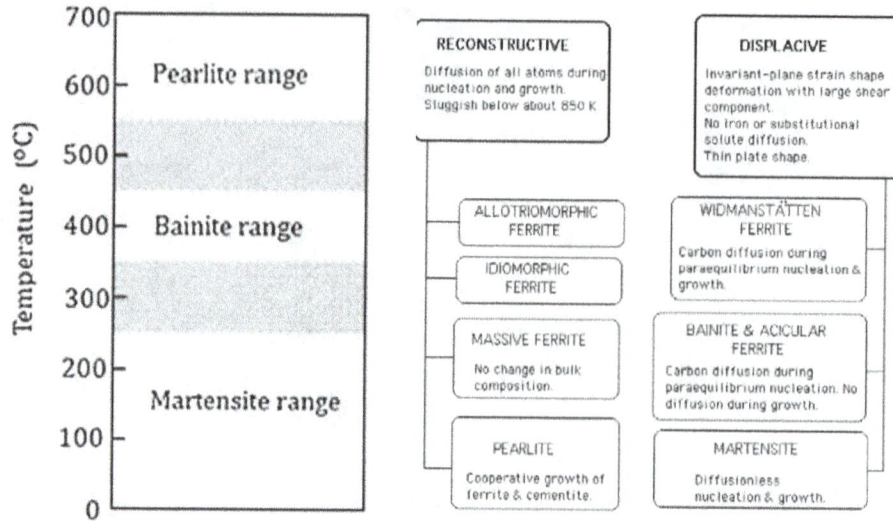

Figure 11.2 *LHS = Temperature range of the formation of microstructures in unalloyed steel, RHS = Summary of the variety of phases generated by the decomposition of austenite. The term 'Para equilibrium' refers to the case where carbon partitions but the substitutional atoms do not diffuse. The substitutional solute to iron atom ratio is therefore unchanged by transformation.*

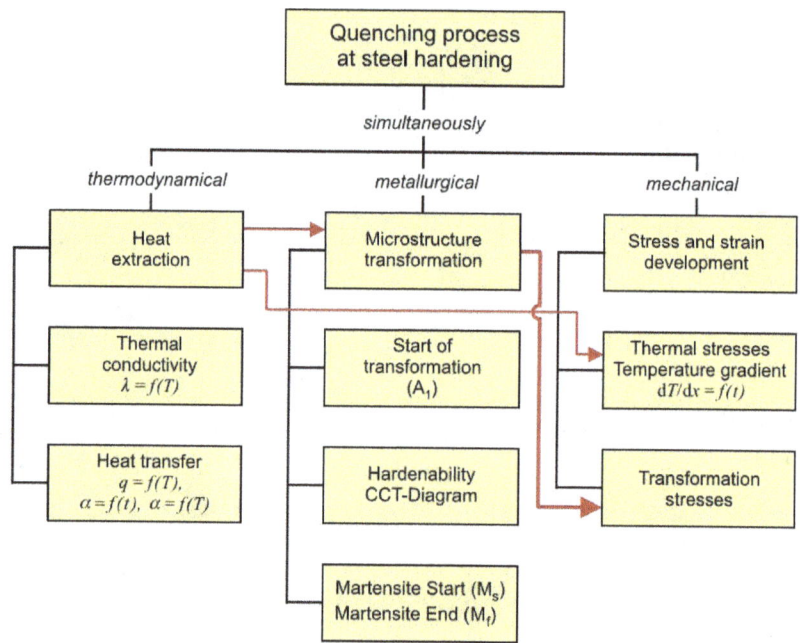

Figure 11.3 *Simultaneous mutually dependent processes during quenching. From Liscic, B. Heat Transfer Control during Quenching. Mater. Manuf. Process. 2009, 24 (7–8), 879–883.*

Legend:

T = Temperature (K)
t = Time (s)
λ = Thermal conductivity (W/mK)
q = Heat flux (W m^{-2})
α = Heat transfer coefficient (W m^{-2} K^{-1})
dT/dx = Temperature gradient (K m^{-1})
→ *= mutual influence*

devised by Grossman or Jominy can help to give a quantitative evaluation.

There are also some problems associated with the heat treatment of steel. When phase transformations occur during a heat treatment cycle, they are associated with three changes as shown in the flow diagram in Figure 11.3. These changes include thermodynamic, metallurgical and mechanical. The simultaneous modelling of these changes is difficult but essential to gain a realistic simulation of actual events. The mechanical changes included the amount of volume change which depends on the type of transformation and the amount of transformed phase. In addition, there are thermal strains due to differences in thermal expansion. Distortion and cracking may occur as the result of volume changes and thermal stress. It is therefore important for certain steels to control the rate of cooling not only for controlling the microstructure, but also to reduce distortions of the part, avoiding crack initiation, and reducing residual stresses.

Simulation of phase transformations by mathematical models can provide an understanding of the development of microstructure, and the prediction of mechanical properties. Due to the importance of the relation between steel mechanical properties and phase transformations, a short description of the transformations in steels and how they can be mathematically modelled is important. The relation between the occurrence of different phases and the cooling rate can be described with transformation diagrams. It is now possible to predict the CCT diagram by the use of established algorithms. Finite element models can then be used to predict cooling rates which can then be linked to the CCT diagram to allow the prediction of the microstructure and mechanical properties.

11.1.2 The allotropy of iron – the theoretical background

Iron and steel research and development carried out over the past 100 years eventually concluded that the hardening mechanism in steel resulted from an allotropic transformation.

The hardening of steel is possible because iron can exist in two crystallographic forms. The alpha phase (α = BCC) which is stable between ambient temperature and 910°C (in pure iron) and gamma (γ = FCC) stable above that temperature. The carbon is soluble in gamma iron up to about 2%. But in alpha iron carbon is practically insoluble. Above 910°C the carbon in the steel enters the gamma lattice. If it is then cooled quickly it can be trapped inside the gamma producing the high hardness BCT lattice known as martensite. These simple statements outline the main events, but the real situation can be more complex. A recent review of martensite formation with the title "Revisiting the Structure of Martensite in Iron-Carbon Steels"[47] suggested a revision to the current understanding. The concept of primary and secondary martensite was introduced in order to indicate that two different, sequential, martensite structures will form during quenching of Fe-C steels above 0.6 mass% C. Below 0.6 mass% C, only primary

martensite is created through the two sequential steps FCC–HCP followed by HCP–BCC. Primary martensite has a lath structure and is described as BCC iron containing a C-rich phase that precipitates during quenching.

It must be appreciated that the theory is a simplistic guide and in practice there many other details that must be considered especially since real steels are not homogeneous and contain imperfections. It is interesting that the FCC structure, although more closely packed (see Figure 11.4 which shows that the volume is smaller as the structure changes from BCC to FCC), the atomic FCC structure has larger "holes" than the BCC structure. These holes are at the centres of the cube edges, and are surrounded by six atoms in the form of an octagon, so they are referred to as octahedral holes (Figure 11.4). There are also smaller tetrahedral interstices. The largest sizes of spheres which will enter these interstices are given in Figure 11.5.

The atomic sizes of carbon and nitrogen (Figure 11.5) are sufficiently small relative to that of iron to allow these elements to enter the α- and γ-iron lattices as interstitial solute atoms. In contrast, alloying elements such as manganese, nickel and chromium

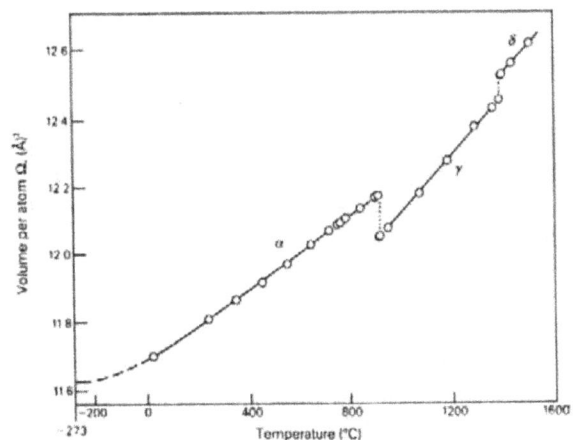

Figure 11.4 Volume change as structure changes from BCC to FCC[49]

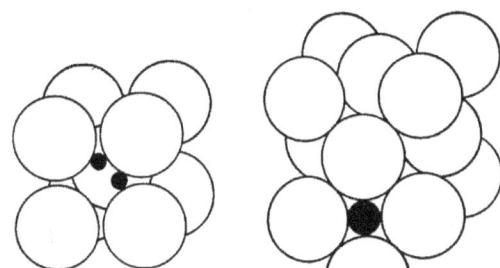

Figure 11.5a Interstices in the BCC and the FCC lattice where a carbon atom will fit

Table 1 Size of largest spheres fitting interstices in bcc and fcc iron

		Radius	Radius in iron (Å)
bcc	Tetrahedral	0.29r	0.37
	Octahedral	0.15r	0.19
fcc	Tetrahedral	0.23r	0.28
	Octahedral	0.41r	0.51

r = atomic radius of iron.

Table 2 Atomic sizes of non-metallic elements in iron

Element	Atomic radius, r (Å)	r/r_{Fe}
α-Fe	1.28	1.00
B	0.94	0.73
C	0.77	0.60
N	0.72	0.57
O	0.60	0.47
H	0.46	0.36

Table 3 Solubilities of carbon and nitrogen in γ- and α- iron

	Temperature (°C)	Solubility wt%	at%
C in γ-iron	1150	2.04	8.8
	723	0.80	3.6
C in α-iron	723	0.02	0.095
	20	<0.00005	<0.00012
N in γ-iron	650	2.8	10.3
	590	2.35	8.75
N in α-iron	590	0.10	0.40
	20	<0.0001	<0.0004

Figure 11.5b *Tables of atomic sizes and solubilities*

have much larger atoms, i.e. nearer in size to those of iron, and consequently they enter into substitutional solid solution.

Comparison of the atomic sizes of carbon and nitrogen with the sizes of the available interstices indicates that some lattice distortion must take place when these atoms enter the iron lattice. When carbon and nitrogen enter the α-iron they occupy the octahedral interstices rather than the larger tetrahedral holes since the octahedral sites are more favourably placed for the relief of strain. This can occur by movement of two nearest-neighbour iron atoms. In the case of tetrahedral interstices, four iron atoms are of nearest-neighbour status and the displacement of these would require more strain energy. Consequently, these interstices are not preferred sites for carbon and nitrogen atoms. The solubility of both C and N in austenite should be greater than in ferrite, because of the larger interstices available. Figure 11.5 shows that this is so for both elements, the solubility in γ-iron rising as high as 9–10 at%, in contrast to the maximum solubility of

C in α-iron of 0.1 at% and of N in α-iron of 0.4 at%. This difference of the solubilities of the main interstitial solutes in γ and in α enables the "transformation" hardening in the heat treatment of steels.

11.2 PREDICTION OF MICROSTRUCTURAL EVOLUTION

11.2.1 Using the Fe-Fe₃C Phase Diagram

Figure 11.6 shows a simplified the equilibrium phase diagram for plain carbon steels (Fe-Fe$_3$C).

Steels are categorized as either:

- hypoeutectoid (carbon below 0.8%)
- eutectoid (carbon = 0.8%)
- hypereutectoid (carbon greater than 0.8%).

At room temperature, the eutectoid composition of steel is formed as pearlite as seen in Figure 11.6. During slow cooling, austenite in hypoeutectoid steels will begin to transform below line A$_3$ since undercooling is necessary to provide a driving force for the solid-state nucleation process. Between A$_3$ and A$_1$ (723°C), ferrite is formed from austenite and austenite is becoming richer in carbon up to 0.8% C at 723°C, also known as the eutectoid temperature. Just below this temperature, the remaining austenite transforms to coarse pearlite. For a eutectoid composition, undercooling is necessary to provide a driving force for the solid-state nucleation process so that at room temperature the microstructure of hypoeutectoid steel will be composed of ferrite and pearlite.

The relative amounts of ferrite and pearlite are dependent on the carbon content and alloying elements in the steel. The photos in Figure 11.1 show how the pearlite fraction increases as the carbon in the steel increases from 0.05% to 0.2% to 0.4% to 0.8% (a, b, c, d). (At 0%C the pearlite = Zero, At 0.8%C the pearlite = 100%)

During slow cooling of hypereutectoid steels, austenite transformation will begin to transform below the A$_{CM}$ line. The driving force of this reaction is also undercooling. Austenite begins to transform to cementite around the grain boundary and the remaining austenite decreases in carbon proportional to the degree of transformation. At 723°C, the remaining austenite contains 0.8%C and below the eutectoid temperature transformation to pearlite

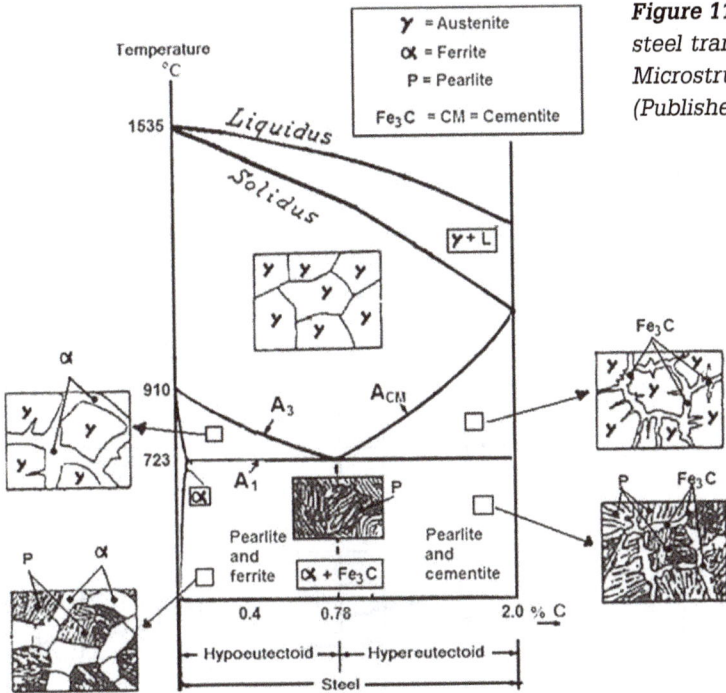

Figure 11.6 *Simplified Fe–C phase diagram related to the steel transformations. From Honeycombe, R. W. K. Steel Microstructure and Properties, 2nd ed.; Edward Arnold (Publishers) Ltd, 1995, p 324.*

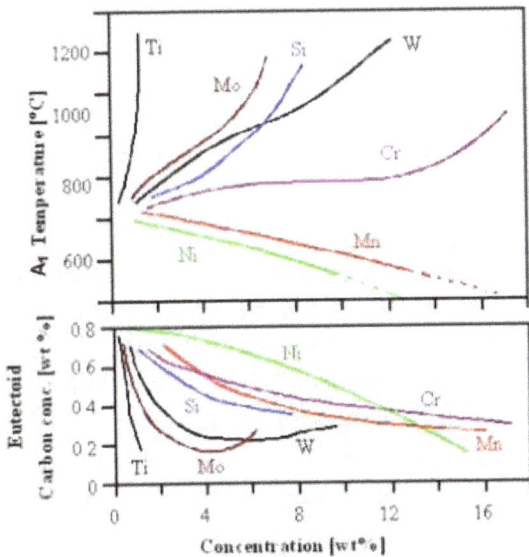

$$A_{c1} = 723 - 20.7Mn - 16.9Ni + 29.1Si + 16.9Cr + 290As + 6.38\,W$$

$$A_{c3} = 910 - 203\sqrt{C} - 15.2Ni + 44.7Si + 104V + 31.5Mo + 13.1W$$

$$M_s = 539 - 423C - 30.4Mn - 17.7Ni - 12.1Cr - 11.0Si - 7.0Mo$$

$$B_s = 630 - 45Mn - 40V - 35Si - 30Cr - 25Mo - 20Ni - 15W.$$

$$CE = C + \frac{Mn}{6} + \frac{Cr + Mo + V}{5} + \frac{Cu + Ni}{15}$$

Figure 11.7 *Effect of alloy content on AC_1, AC_3, Ms, Bs and CEV*

occurs. At room temperature, hypereutectoid steels contain cementite as a network around pearlite. (Figure 11.1 photo "e".) The relative phase composition of cementite and pearlite will vary according to the carbon content and the presence of alloying elements in the steel.

The eutectoid temperature 723°C is one of the most important values to consider; during the cooling of steel, the austenite (y-Fe) will start to decompose at temperatures lower than the eutectoid temperature. As shown above, the various categories are basically based on the amount of carbon content. Other elements such as Si, Mn, Cr, Ni, etc. do have an effect as shown in Figure 11.7.

Other important temperatures to be mentioned here are AC_1, AC_3 and Acm. For heat treatments involving phase transformations, the part must be heated to the austenite region. This means temperatures over AC_3 for hypoeutectoid alloys and over Acm for hypereutectoid alloys.

Grain growth of austenite can occur at elevated temperatures, which generally has a negative effect on the final mechanical properties. Therefore, it is commonly accepted not to heat the part more than 50°C above the AC_3 or Acm.

The austenitizing conditions of steels are shown on TTA diagrams as shown in Figure 11.8.

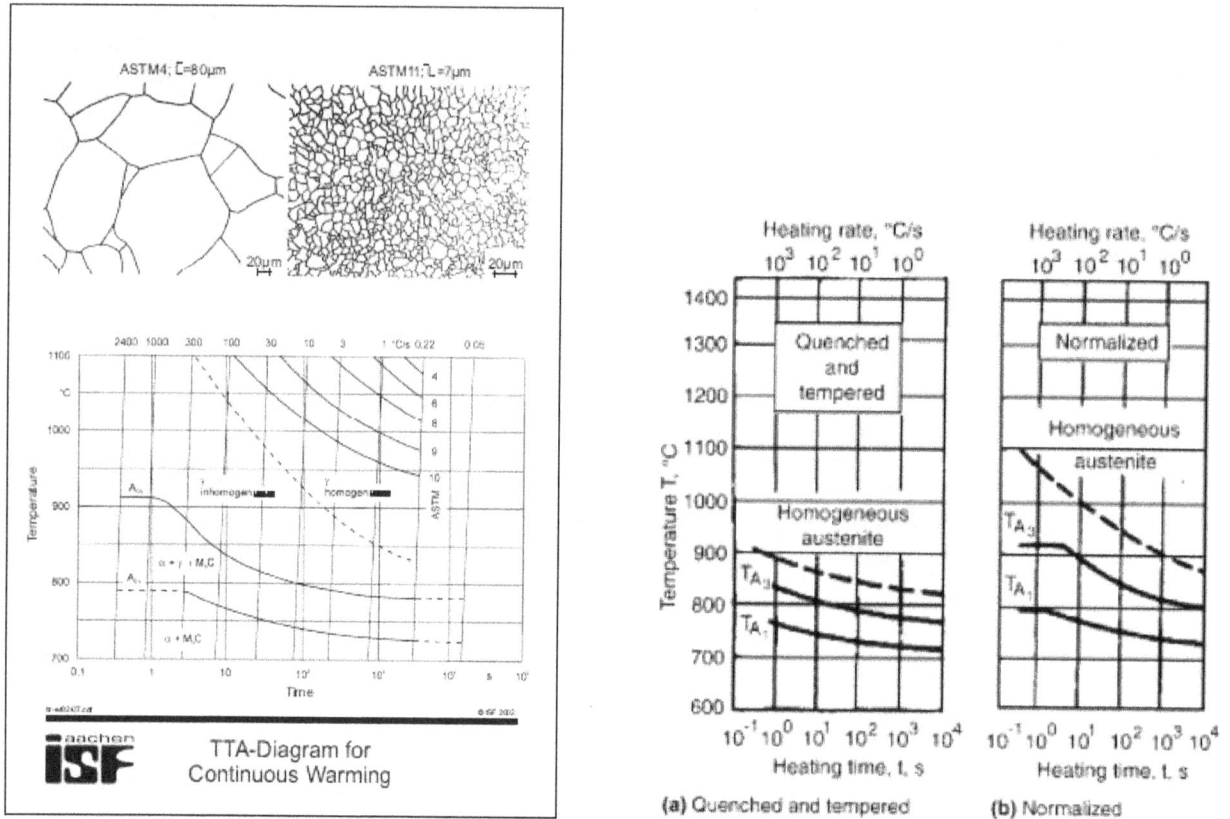

Figure 11.8 *TTA diagrams austenitising conditions have to allow the formation of austenite, allow dissolution of any compounds such as carbides into the austenite and allow time for diffusion to achieve a homogeneous austenite. But must minimise grain coarsening. Any carbides that remain undissolved will not contribute to the hardenability.*

11.2.2 Ferrite and Pearlite

Pearlite is a lamellar aggregate of ferrite and cementite and forms at relatively slow cooling rates in steel and cast iron. It results from the transformation of austenite and consists of alternating layers of alpha-ferrite and cementite, see Figure 11.9, as the following reaction during a slow cooling such as air cooling.

11.2.3 Bainite

There is a wide range of intermediate temperatures which neither pearlite nor martensite forms. Instead, fine aggregates of ferrite plates (or laths) and cementite particles are formed. The generic term for these intermediate structures is bainite. Bainite occurs during treatments at cooling rates too fast for pearlite to form, and not rapid enough to produce martensite. The nature of bainite changes as the transformation temperature is lowered. Two main forms of bainite can be identified, upper and lower bainite.

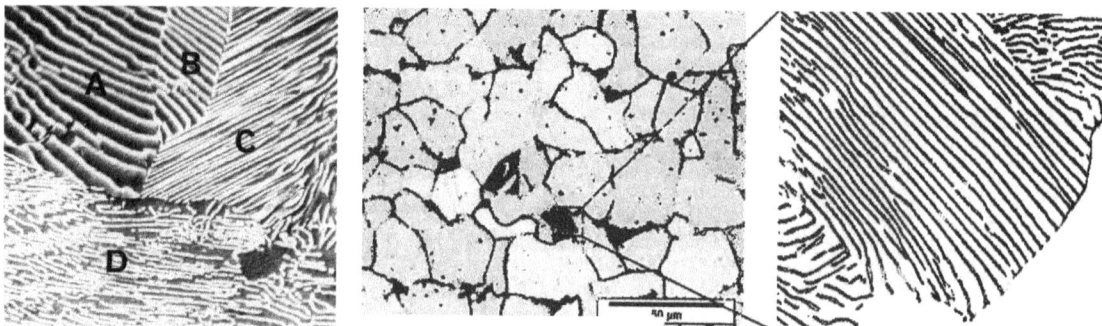

Figure 11.9 *LHS = SEM image of pearlite X6500. Shows the sectioning effect on three lamellar pearlite colonies A, B, and C. Also shows degenerate pearlite in area D. RHS = Ferrite and Pearlite*

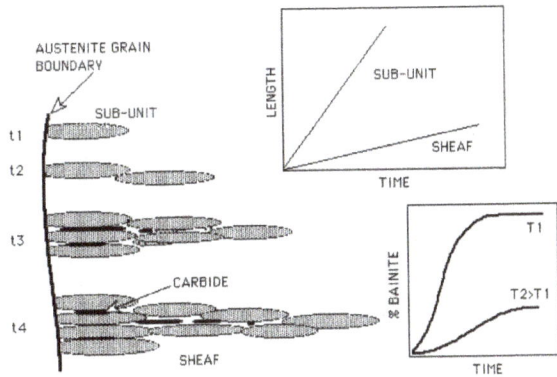

Figure 11.10 Bainite formation

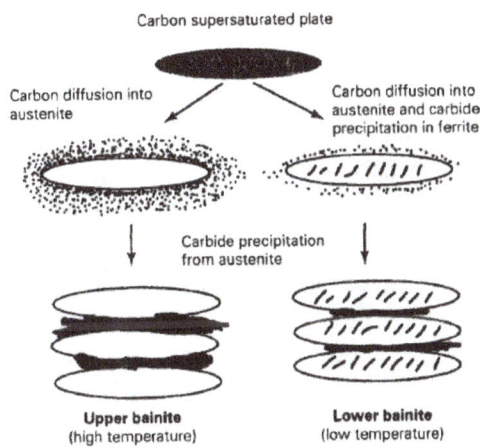

Figure 11.11 Difference between upper and lower bainite

The microstructure of upper bainite consists of fine plates of ferrite, which is about 0.2 microns thick and about 10 micron long. The plates grow in clusters called sheaves. Within each sheaf the plates are parallel and of identical crystallographic orientation. The individual plates in a sheaf are often called the 'sub-units' of bainite, see Figure 11.10.

Lower bainite has a microstructure and crystallographic features which are very similar to those of upper bainite. The major distinction is that cementite particles also precipitate inside the plates of ferrite, see Figure 11.11. There are, therefore, two kinds of cementite precipitates: those which grow from the carbon-enriched austenite which separates the platelets of bainitic ferrite, and others which appear to precipitate from supersaturated ferrite. These latter particles exhibit the 'tempering' orientation relationship which is found when carbides precipitate during the heat treatment of martensite. Carbon has a large effect on the range of temperature over which upper and lower bainite occur.

The Bs temperature, the temperature for the start of transformation into bainite, is affected by the alloying elements but carbon has the greatest influence, as indicated by the following empirical equation (Stevens).

$$Bs\,(^{\circ}C) = 830 - 270C - 90Mn - 37Ni - 70Cr - 83Mo$$

11.2.4 Martensite

The third important microstructure that forms during the cooling of iron-carbon alloys was named martensite, after the German Professor Adolf Martens (1850-1914) who carried out work on the martensite formed in high carbon steels. Martensite is formed during rapid cooling of austenitic microstructure to the Ms martensite start temperature which depends upon the carbon and alloy content. The rate of cooling must be sufficiently high so that diffusion transformation products do not occur. (Critical cooling rate CCR.) The formation of martensite was believed to be coordinated movement of iron atoms that resulted in a distortion of the BCC structure that should form from the austenite to a body-centered-tetragonal structure. For this reason, the formation of martensite is known as a "diffusion less transformation".

The microstructure of martensite in iron-carbon alloys appears as two main forms: lath (or massive) and lenticular (or plate). Lath martensite consists of blocks of parallel blades of the martensite phase

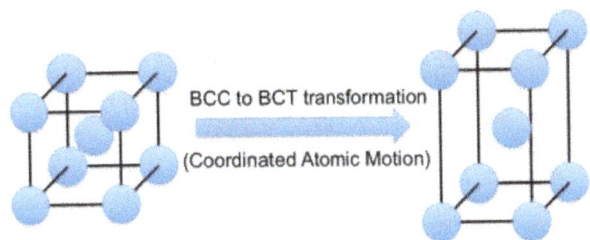

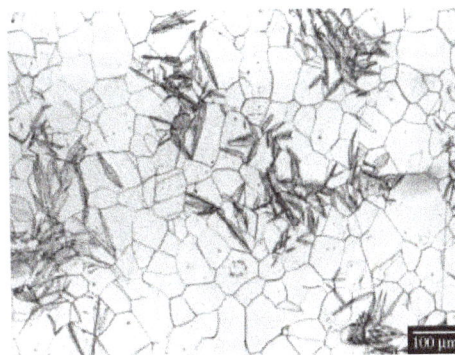

Figure 11.12 Martensite (From Formation of Lath Martensite, Sarah Löwy, INSTITUT FÜR MATERIALWISSENSCHAFT DER UNIVERSITÄT STUTTGART Stuttgart 2015)

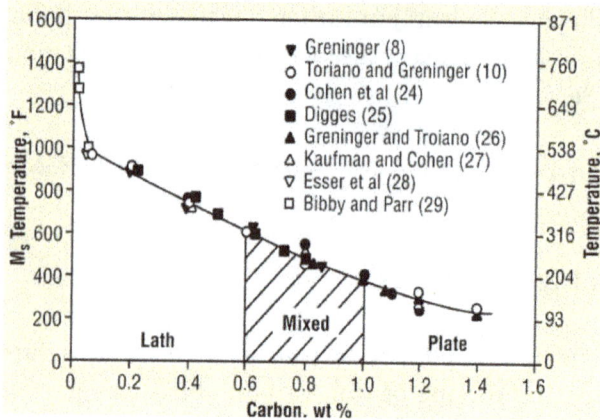

Figure 11.13 *Lath, mixed and plate martensite*

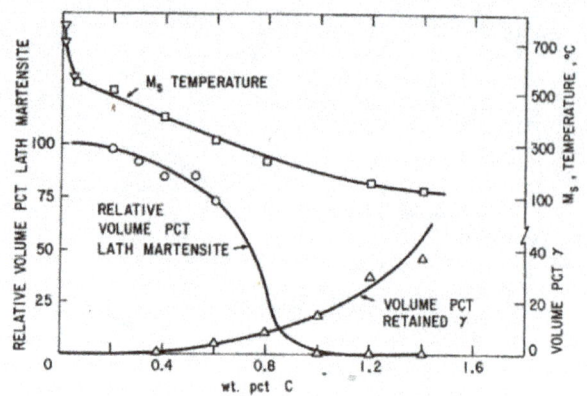

separated by the untransformed austenite (or other phases). This type of martensite is common for alloys containing less than about 0.6 wt% C. For higher carbon alloys, lenticular martensite forms, where the

Figure 11.14 *Lath martensite From Multiple mechanisms of lath martensite plasticity L. Morsdorf et al, Acta Materialia 121 (2016) 202-214*

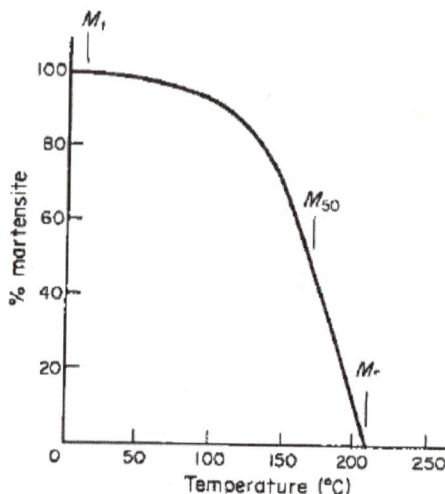

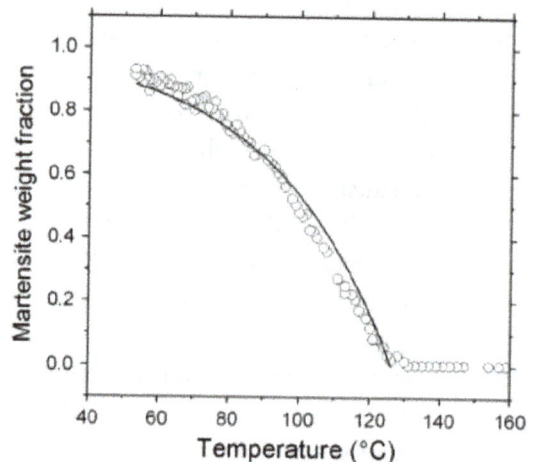

martensite takes on a lens or lenticular shape within the untransformed matrix. (Figure 11.13)

Unlike the ferrite/pearlite and bainite microstructures, martensite forms only when the temperature decreases and is known as an athermal process. The transformation is nearly instantaneous, and begins when the temperature reaches the Ms (martensite start temperature) and is complete when the Mf temperature is reached.

The carbon content of the parent austenite phase determines whether lath (low-carbon) or plate (high-carbon) martensite, or mixtures of the two will be produced, assuming the quench rate and steel hardenability are adequate for full hardening. Lath martensite produces higher toughness and ductility, but lower strengths, while plate martensite produces much higher strength, but may be rather brittle and non-ductile. For a given alloy content, as the carbon content of the austenite increases, the martensite start, Ms, temperature and the martensite finish, Mf, temperature will be depressed which results in

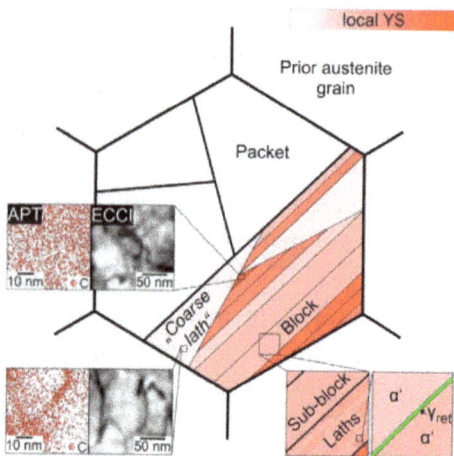

Figure 11.15 *A-thermal formation of martensite. RHS = Actual results for 17.4ph.*

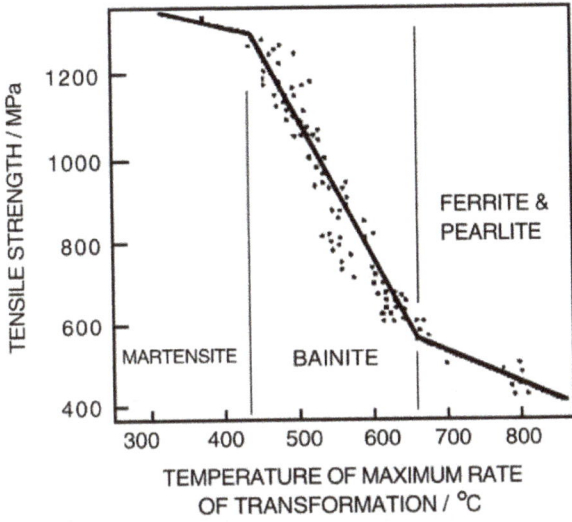

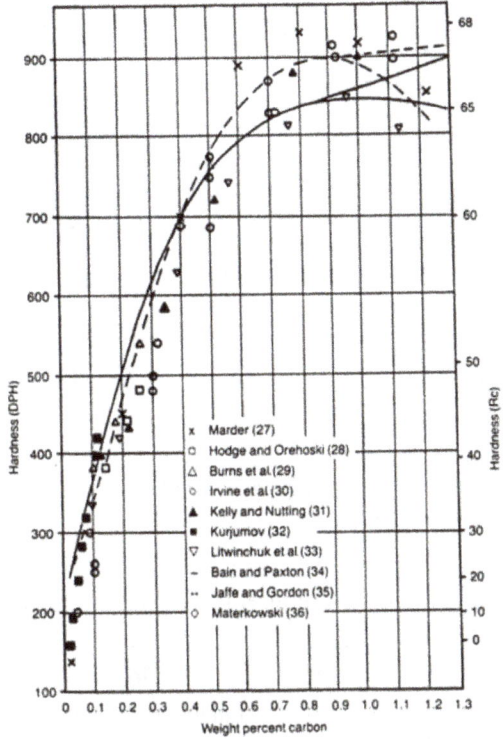

Figure 11.16 *Summary showing the effect of transformation temperature on tensile strength for structural steels (see Figure 1 photo f for martensite microstructures). RHS = Martensite Hardness*

incomplete conversion of austenite to martensite. When this happens retained austenite, which may be either extremely detrimental or desirable under certain conditions, is observed. The amount of retained austenite present depends upon the amount of carbon that can be dissolved in the parent austenite phase and the magnitude of the suppression of the Ms and Mf temperatures. Low carbon 9% nickel steels have small areas of retained austenite which allows toughness at -196°C and the use of the steel as weldable plates for the containment of liquified natural gas.

11.2.4 Continuous Cooling Transformation (CCT) Diagrams (roadmap to allow us to predict a steel's response to heat treating)

The continuous cooling transformation kinetics can be described by the so-called continuous cooling transformation (CCT) diagram. The CCT diagram is constructed by plotting a series of cooling curves (temperature against cooling time) and then connecting

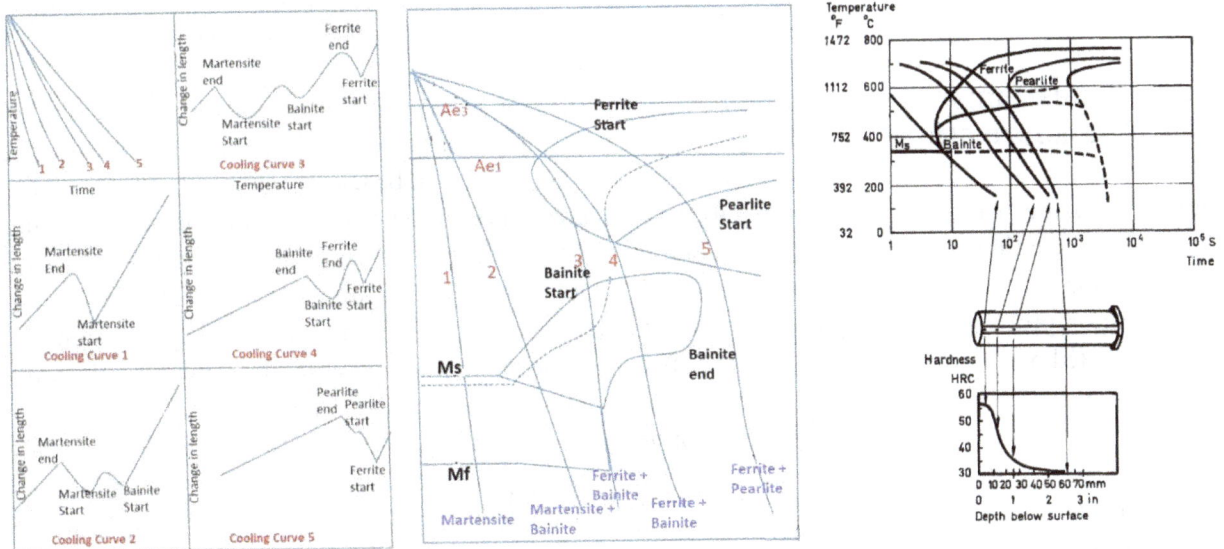

Figure 11.17 *Determination of CCT diagrams LHS = Dilatometry. RHS = Data from Jominy*

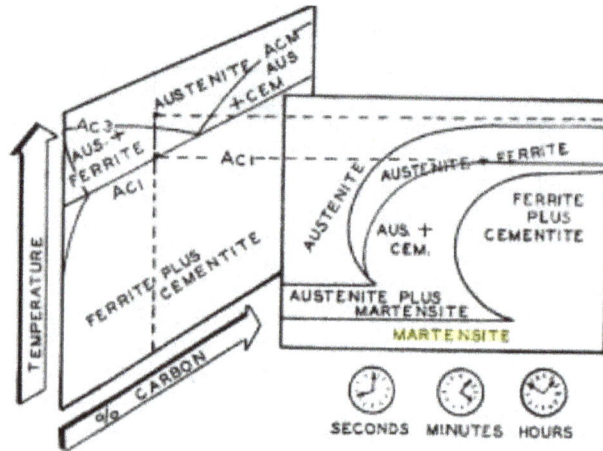

Figure 11.18 Relationship between equilibrium and CCT (USA origin – not quite correct orientation)

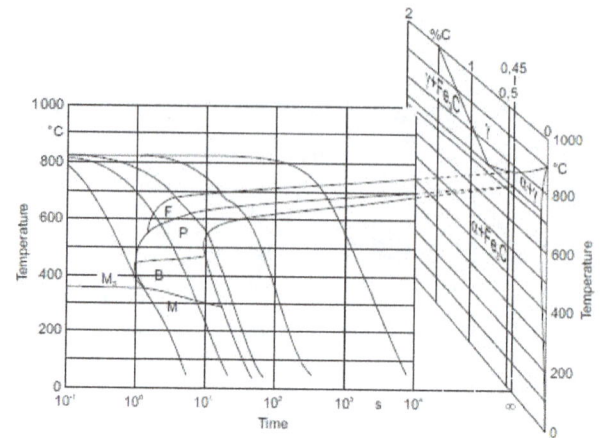

Figure 11.19 Relationship between equilibrium and CCT (German origin – shows the correct orientation)

the transformation start temperatures (Ts) and transformation-finish temperatures (Tf) with separate lines. The start temperature for diffusion independent transformations such as martensite is generally regarded as independent of cooling rate. This is clearly shown in Figure 11.17 in which the Ms appears as a horizontal line in CCT diagram. For very slow cooling rates, the austenite will decompose into ferrite and pearlite. With faster cooling rate, the bainitic structures are formed. For some cooling rates (see cooling curves 3 and 4 in Figure 11.17), the entire austenite can be transformed into bainite before reaching the martensite curve, meaning no martensite will be produced due to completion of the bainitic transformation.

The relationship between the CCT and the equilibrium diagram are shown in Figures 11.18 and Figure 11.19 with the diagram from the German origin being the most accurate showing the equilibrium diagram at "long time".

11.2.5 Time-Temperature-Transformation (TTT) Diagrams

Isothermal transformation kinetics can be portrayed graphically by the time temperature-transformation (TTT) diagram. The TTT diagram shows how the transformation proceeds depending upon the degree of under-cooled of the austenite below the AC_1. The steel is first austenitised and then quickly cooled to each transformation temperature, and then kept at that temperature constant, until the transformations are completed.

The TTT diagram for a hypo-eutectoid 0.4%C steel is shown in Figure 11.20 (top LHS). At transformation

temperatures between A_3 and A_1, only proeutectoid ferrite is formed, and the transformation of the austenite does not go to completion. Below the A_1 temperature and down to a transformation temperature approaching the nose of the TTT diagram, the formation of proeutectoid ferrite forms before the pearlite, but the amount of proeutectoid ferrite decreases with decreasing temperature, and consequently the pearlite becomes not only finer, but also more dilute. At the nose of the diagram, only dilute pearlite is formed. The structures below the nose of the diagram are bainites.

Both proeutectoid ferrite and pearlite nucleate mainly at the austenite grain boundaries, so that refinement of the austenite grain size produces more nucleation sites, and hence transformation is accelerated. This effect becomes less pronounced in the bainite region, particularly for lower bainite, which is not mainly nucleated at austenite grain boundaries, and austenite grain size has little effect on the Ms temperature.

While TTT diagrams cannot strictly be applied to transformations occurring during continuous cooling, they can be used qualitatively to explain the significance of the critical cooling velocity (cooling rate to give only martensite). If cooling curves are schematically superimposed on the TTT diagram, it can be seen that with increasing rate of cooling the transformation temperature decreases and the pearlite (and ferrite) becomes finer.

The TTT diagram depicts three isothermal reactions, pro-eutectoid ferrite, pearlite and bainite and one athermal reaction, martensite. Each of the isothermal reactions is itself depicted by a C curve, and the whole TTT diagram is the envelope of these three

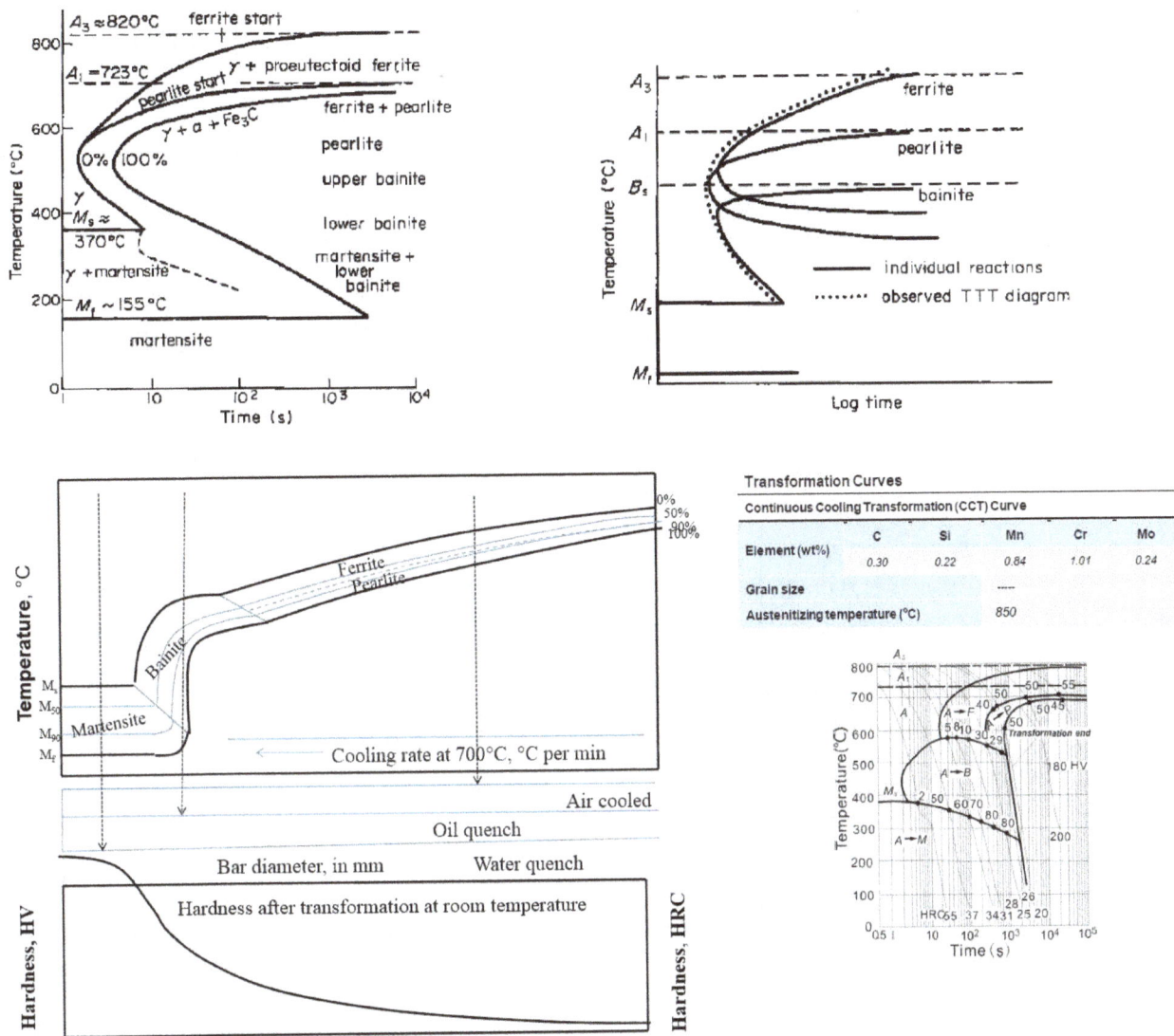

Figure 11.20 *TTT and CCT diagrams. Lower LHS = Atkins type CCT diagram*

C curves. In plain carbon steels, these individual reaction C curves overlap considerably (Figure 11.20 top RHS diagram). The bainite C curve has a very flat top, which is asymptotic to a fixed temperature Bs above which bainite does not form. In this respect, bainite has similarities with martensite, but its formation is a diffusion-controlled reaction rather than a diffusion less reaction.

11.3 EMPIRICAL MODELS

One of the first references to the development of a model to predict the microstructure and mechanical properties can be found in the Crafts and Lamond's Hardenability and steel selection 1949. (see Figure 11.21)

▶ *Figure 11.21 From Crafts and Lamond-1949*

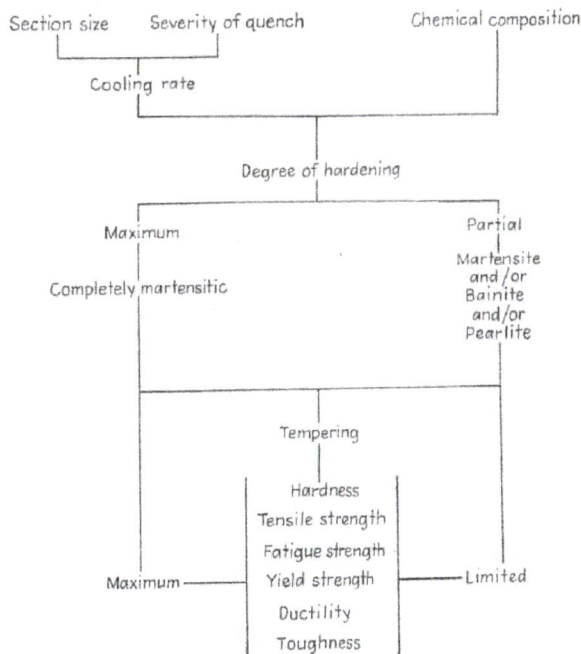

The first successful prediction of CCT diagrams and then linking to cooling rate was done by Creusot Loire with the flow diagram in Figure 11.22 as shown in Figure 11.23 and Table 11.3.

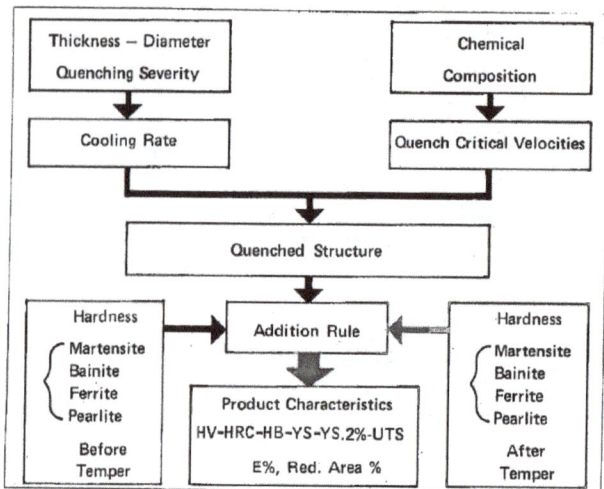

Figure 11.23 *CCT diagram showing Creusot Loire method*

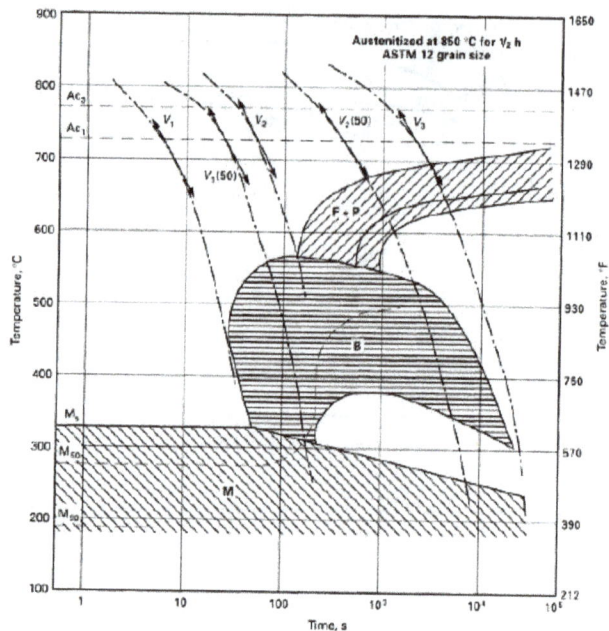

Figure 11.22 *Flow diagram for Creusot Loire method*

Table 11.3 *Creusot Loire equations*

$\log V_1 = 9.81\text{-}4.62\text{C}\text{-}1.05\text{Mn}\text{-}0.54\text{Ni}\text{-}0.5\text{Cr}\text{-}0.66\text{Mo}\text{-}0.00183P_a$

$\log V_1(10) = 8.76\text{-}4.04\text{C}\text{-}0.96\text{Mn}\text{-}0.49\text{Ni}\text{-}0.58\text{Cr}\text{-}0.97\text{Mo}\text{-}0.0010P_a$

$\log V_1(50) = 8.50\text{-}4.13\text{C}\text{-}0.86\text{Mn}\text{-}0.57\text{Ni}\text{-}0.41\text{Cr}\text{-}0.94\text{Mo}\text{-}0.0012P_a$

$\log V_2 = 10.17\text{-}3.80\text{C}\text{-}1.07\text{Mn}\text{-}0.70\text{Ni}\text{-}0.57\text{Cr}\text{-}1.58\text{Mo}\text{-}0.0032P_a$

$\log V_2(90) = 10.55\text{-}3.65\text{C}\text{-}1.08\text{Mn}\text{-}0.77\text{Ni}\text{-}0.61\text{Cr}\text{-}1.49\text{Mo}\text{-}0.0040P_a$

$\log V_2(50) = 8.74\text{-}2.23\text{C}\text{-}0.86\text{Mn}\text{-}0.56\text{Ni}\text{-}0.59\text{Cr}\text{-}1.60\text{Mo}\text{-}0.0032P_a$

$\log V_3(90) = 7.51\text{-}1.38\text{C}\text{-}0.35\text{Mn}\text{-}0.93\text{Ni}\text{-}0.11\text{Cr}\text{-}2.31\text{Mo}\text{-}0.0033P_a$

$\log V_3 = 6.36\text{-}0.43\text{C}\text{-}0.49\text{Mn}\text{-}0.78\text{Ni}\text{-}0.26\text{Cr}\text{-}0.38\text{Mo}\text{-}0.0019P_a\text{-}2\sqrt{Mo}$

11.4 NEURAL NETWORKS MODELS

Due to complex relations between the chemical compositions and metallurgical reactions, methods that could possibly give improved predictions were examined. There are several published results on the application of neural networks. Trzaska et.al. propose the following relationships after application of neural networks for over 300 samples.

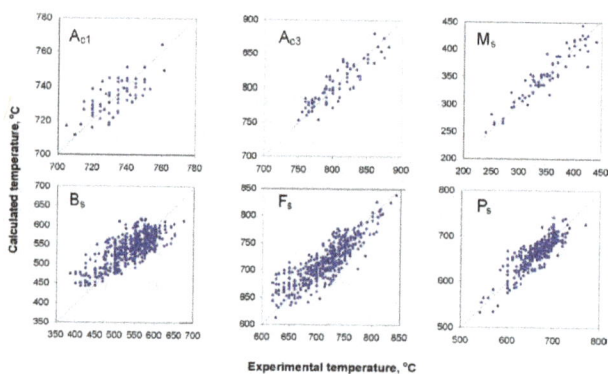

Figure 11.24 *Calculated vs actual*

$$A_{c1} = 739.3 - 22.8 \cdot C - 6.8 \cdot Mn + 18.2 \cdot Si + 11.7 \cdot Cr$$
$$-15 \cdot Ni - 6.4 \cdot Mo - 5 \cdot V - 28 \cdot Cu \qquad (4)$$

$$A_{c3} = 937.3 - 224.5 \cdot C^{0.5} - 17 \cdot Mn + 34 \cdot Si - 14 \cdot Ni$$
$$+21.6 \cdot Mo + 41.8 \cdot V - 20 \cdot Cu \qquad (5)$$

$$B_{smax} = 752 - 223.5 \cdot C - 55 \cdot Mn - 21.6 \cdot Si - 46.8 \cdot Cr$$
$$-36.9 \cdot Ni - 47.4 \cdot Mo - 70 \cdot V - 11 \cdot Cu \qquad (6)$$

$$M_s = 532.6 - 396.7 \cdot C - 33 \cdot Mn - 1.4 \cdot Si - 14 \cdot Cr$$
$$-18 \cdot Ni - 11 \cdot Mo + 49.7 \cdot V + 31 \cdot Cu \qquad (7)$$

$$\log t_B = -1.52 + 2.93 \cdot C + 0.68 \cdot Mn + 0.25 \cdot Si + 0.73 \cdot Cr$$
$$+0.31 \cdot Ni + 0.69 \cdot Mo - 0.23 \cdot V \qquad (8)$$

$$\log t_F = -1.56 + 2.67 \cdot C + 0.95 \cdot Mn - 0.1 \cdot Si + 1.27 \cdot Cr$$
$$+0.22 \cdot Ni + 2.27 \cdot Mo - 1.06 \cdot V - 0.47 \cdot Cu \qquad (9)$$

$$\log t_P = 0.19 - 0.3 \cdot C + 0.64 \cdot Mn + 0.06 \cdot Si + 1.02 \cdot Cr$$
$$+0.4 \cdot Ni + 4.9 \cdot Mo + 0.38 \cdot V - 0.42 \cdot Cu \qquad (10)$$

$$F_s = 968.7 - 254 \cdot C - 71 \cdot Mn + 27.6 \cdot Si - 30 \cdot Cr$$
$$-44 \cdot Ni - 54 \cdot Mo + 95.8 \cdot V - 0.02 \cdot T_A - 62.8 \cdot v_r^{0.25} \qquad (11)$$

$$P_s = 789.8 - 12.7 \cdot C - 61 \cdot Mn + 13.7 \cdot Si - 5 \cdot Cr$$
$$-30.4 \cdot Ni - 70.7 \cdot Mo - 1.4 \cdot V - 0.016 \cdot T_A$$
$$-47.3 \cdot v_r^{0.25} \qquad (12)$$

$$B_s = 678.9 - 239.6 \cdot C - 35.2 \cdot Mn - 1.6 \cdot Si - 19.8 \cdot Cr$$
$$-27.9 \cdot Ni - 18 \cdot Mo - 171 \cdot V - 0.03 \cdot T_A$$
$$-15.5 \cdot v_r^{0.25} \qquad (13)$$

Table 11.4 *Neural network equations*

11.5 APPLICATION TO MICROSTRUCTURALLY ENGINEERED FORMABLE STEELS

As outlined in the introduction of Chapter 5, Antonio Gorni has documented a summary of empirical and theoretical equations developed over the years for microstructure development and is a major source of data for microstructural studies. This document can be found on the Internet "Steel Forming and Heat-Treating Handbook" 2018 by Antonio Gorni. The document has been updated over the years. (http://www.gorni.eng.br/e/Gorni_SFHTHandbook.pdf)

The recent review of the application mathematical models to the design and development of formable strip steels (outlined in section 10.4) has been reviewed. The models show the progression of the key microstructures (ferrite, martensite, bainite, retained austenite) into potential steels by using the composition of the steel, the cooling rate, and the application of thermal processing. The equipment available (for example rolling mills and continuous annealing lines) have certain process capabilities and the selected steel has to achieve the outcome in the process time available. In addition, the process often includes galvanizing for corrosion protection. Therefore, methods to predict the microstructure and mechanical properties without expensive development facilities have been examined. (ref 1 G. Anand et al, Deterministic Approach for Microstructurally Engineered Formable Steels, International Journal of Metallurgical Engineering 2013, 2(1): 69-78, file:///C:/Users/user/Downloads/10.5923.j.ijmee.20130201.11%20(1).pdf)

Figure 11.25 shows microstructural development and the changes made during the AHSS development phases.

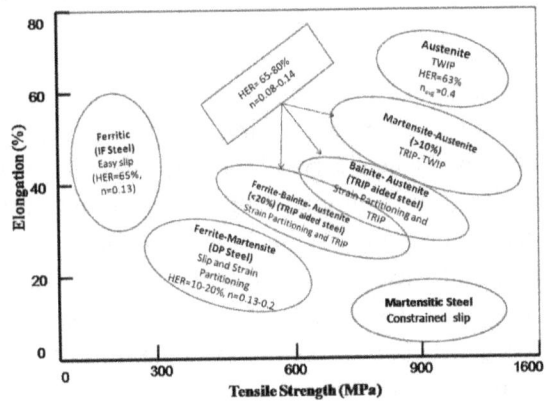

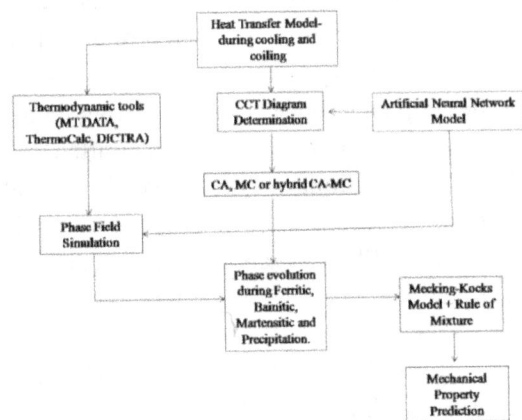

Figure 11.25 *LHS = Microstructures of formable steels. RHS = Model for phase transformation during cooling and coiling in hot-strip mill (ref1)*

The steels developed for high strength formable steels are displayed on the 'banana' type diagram shown in Chapter 10 on automotive steels. The name was given because of the typical fall in tensile ductility as strength increases. The strain hardening and formability parameters have been included. The diagram depicts the properties of three single phase microstructures namely, ferritic, martensitic and austenitic, and the evolution of mixed microstructures obtained by combining two or three of the phases at different proportions. The multiphase varieties, the strength level of the DP and TRIP steels are constrained by the presence of inter-critical ferrite (>60%) The single-phase austenitic steels require addition of large amounts of alloying elements (15-25%) (based on Hadfield's steel).

The austenite-bainite/martensitic verities of steel (Complex Phase CP) are possibly a leading contender for the strength-tensile ductility/ formability balance.

11.6 ENGINEERING STEELS AND NANO-GRAIN SIZED BAINITIC STEELS.

https://www.phase-trans.msm.cam.ac.uk/map/map.html

This web site contains many material related algorithms suitable for engineering steels. The Materials Algorithms Project serves as a centre for the "validation" and distribution of algorithms of use in the modelling of materials, in the context of materials science and metallurgy. Validation in this context means that effort has been expended to check that the program reproduces the example output from the example inputs, that there is a reasonable level of documentation, and that appropriate references are provided. The MAP Library consists of a perpetual library of elementary subroutines that enables the user to develop new concepts using existing methods as a foundation. The software may be written in any standard programming language. MAP originated from a joint project of the University of Cambridge and the National Physical Laboratory. It is a non-profitmaking venture which will distribute the library at cost. The project was sponsored for four years by the Engineering and Physical Sciences Research Council (EPSRC) of the United Kingdom. It is now run without explicit funding, for the good of the subject.

11.7 IMPLEMENTATION INTO FINITE ELEMENT METHOD

11.7.1 Background

Theoretical modelling has been an integral part of many aspects of physical metallurgy applied to thermal or thermo-mechanical processing. This includes both the evolution of microstructure with time or strain, and the dependence of properties on microstructure. For a thermally controlled process, the evolution of microstructure can be described by a simultaneous set of differential equations for each of the independent state variables. An integrated process model generally consists of three components, i.e., a numerical heat flow model, a microstructure model and a mechanical model that are coupled implemented into a FE package.

11.8 QT STEEL BASIC FUNCTIONALITY

Basic functions of the QTSteel are available from folders in its main window. These folders guide the user to specify requested inputs, to compute temperature curves, to make metallurgical predictions and to postprocess their results.

11.8.1 Specification of steel properties

Heat treated steel is specified by its chemical composition and by properties of initial austenite as grain size, austenitizing temperature and soak time. Corresponding CCT diagram is calculated from this

Figure 11.26 *Specification of steel properties with an example of predicted CCT diagram*

information and offered for the next metallurgical processing. Modification of the predicted CCT diagram is enabled.

11.8.2 FEM temperature calculations and specification of cooling curves

QTSteel contains built-in 2D FEM module that enables calculation of temperatures for special plane and axi-symmetric bodies (rectangular and rounded bars, tubes, cylinders and rings) with cooling conditions changing in time and in position on the surface of selected body. Special knowledge of FEM theory and meshing is not requested, the FEM preprocessing is fully automatic. Temperature calculation takes about a few minutes.

The cooling process consists of time sequence of cooling conditions. Each cooling condition is specified by its duration in seconds, by the number of time steps and by the type of cooling media. A built-in database containing heat transfer coefficients for basic types of cooling media (water, oils, polymers etc.) is available.

There are two possibilities available in the QTSteel how to specify cooling curves:

- cooling curves are computed by the FEM temperature module in specified points of cross-section of special bodies
- cooling curves are entered by keyboard.

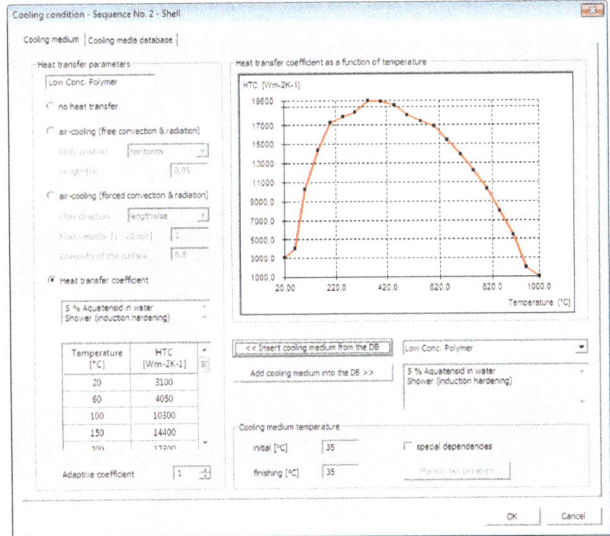

Fig.11.28 *Specification of cooling conditions*

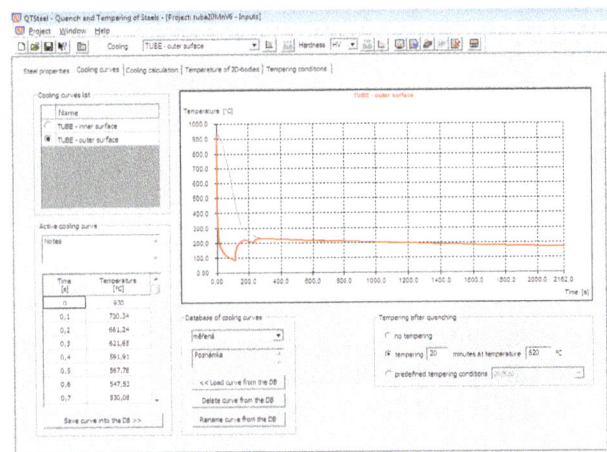

Fig.11.29 *Specification of single cooling curves*

The cooling curve describes the process of quenching primarily but it can consist of reheating sequences as well. It enables modelling of the process of self-tempering that is characteristic for heat treatment of large bodies. Separate tempering after quenching can be added by the coupling of cooling curve with specified tempering parameters as tempering time and temperature or selected more complex tempering regime.

11.8.3 Prediction of steel properties after heat treatment

QTSteel enables us to predict the CCT diagram of specified steel, microstructure shares after finishing of cooling and final mechanical properties (hardness, yield stress and ultimate tensile strength) after quenching and tempering.

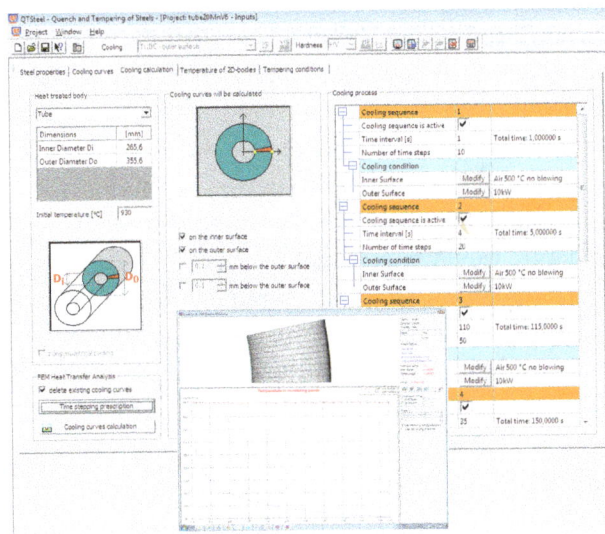

Figure 11.27 *Specification of cooling process for heat treatment of a tube with running 2D FEM temperature module in separate window*

Fig.11.30 Metallurgical prediction for the cooling curve 33.5mm below surface of a rounded bar

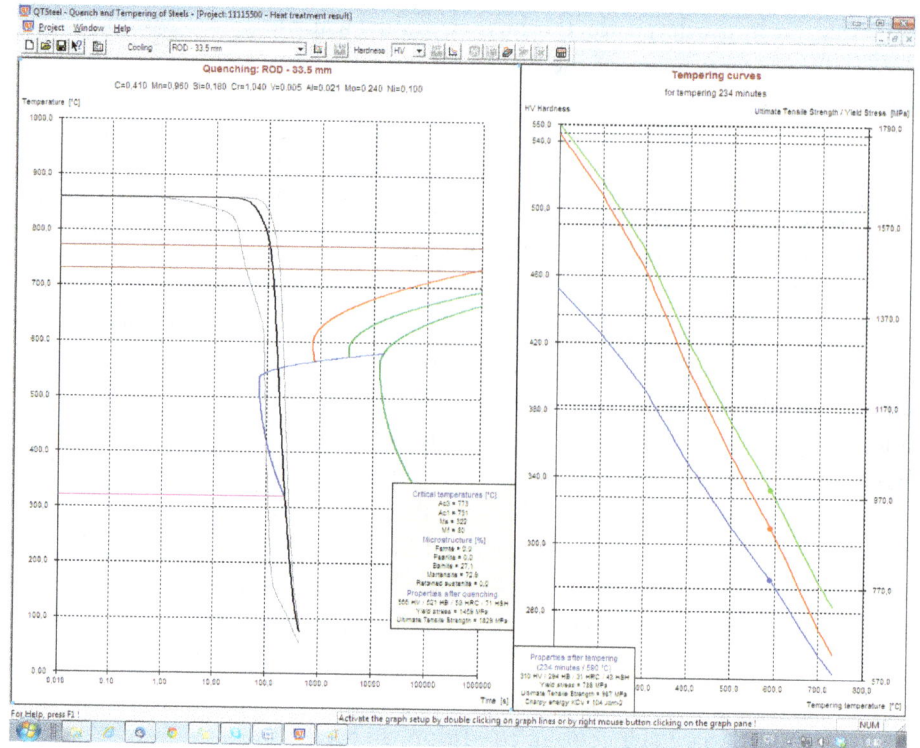

There are two types of metallurgical predictions available in the QT Steel:

- metallurgical predictions for one selected single cooling curve
- metallurgical predictions across complete 2D-body or in specified direction below surface.

Metallurgical calculation for one cooling curve takes about a few seconds.

11.8.4 Functionality available in the QTSteel 3.2

QTSteel version 3.2 (released in October 2012). The version enables the following functionality:

- Non-symmetrical cooling of cylinders and rings is allowed
- New models of heat transfer due to free and forced convection were implemented
- FEM meshes of all 2D bodies are finer to refine calculation of temperatures
- Cooling medium temperature can be specified as dependent on surface temperature or time dependent when medium is heated or cooled in dependence of heat energy transferred through surface of the body

- New reporting in HTML and PDF file formats is available
- New thermal balance of heat treatment process is displayed.

11.9 THE IMPORTANCE OF MICROSTRUCTURE

The influence of the microstructure on the mechanical properties of steel has been recognized for many years. In addition, it has been generally accepted through the work of Hall and Petch that the best combinations of strength and toughness require a small grain size. This trend is observed whether the microstructure is ferrite/pearlite or bainitic or martensitic. In addition, it is frequently concluded that for maximum strength and toughness a tempered martensitic structure would be essential.

To achieve these objectives requires application of the principles of heat treatment. The successful application of heat treatment requires process control and the assessment of the process capability. This requires detailed understanding of three major metallurgical topics:

- Quench speed
- Hardenability
- Tempering response.

The interplay between these factors generates the microstructural features that control the strength and toughness of heat-treated steel.

In an attempt to produce a quantitative tool to evaluate the response of steel to a specified heat treatment "QT Steel" software has been developed.

Some of the current application for the software has been the provision of assistance during the checking the test certification of delivered parts.

11.9.1 Examples of application of QT Steel

Developments in many areas of engineering such as oil and gas have necessitated an improved quality of steel. The mutual satisfaction of steel users and suppliers requires first that the steel user to have detailed realistic standards and secondly that the steel producer established reliable control of the processing conditions and the resultant mechanical properties.

Some examples are considered where the control of microstructure and mechanical properties are involved. Despite some criticism of the oil and gas sector there are several examples of where oil companies have published accounts of poor experience or failures in an attempt to provide warnings to others.

- The first involves the re-examination of a cracked Outer housing on subsea Wellhead based upon a published account of the failure investigation.
- The second examines a published report on an API sponsored project on hot tensile test work. Oil companies contributed samples of AISI 4130 steel bar which had valid test certification to 75ksi strength grade. The re-tests carried out on these samples found that 3 out of 5 samples failed to meet the 0.2% proof strength with one sample 17% below the required minimum.

11.9.2 Oil and gas – re-examination of the published details of the Mensa failure

The failure of Deepwater Horizon in the Mexican Gulf demonstrated that the environmental and financial consequences of an in-service Wellhead Component failure can be substantial. Therefore, the task of selection and approval of potential suppliers of Wellhead Components needs to include a full metallurgical evaluation to validate that the components are "fit for service" and manufactured in accordance with the API 6A specification requirements.

In addition, prior to approval suppliers need to demonstrate that they possess the required technical knowledge base in the metallurgical aspects of the manufacture and design, manufacturing process capability and an appreciation of the importance of working in accordance with established and validated quality control and quality assurance procedures that are based on the mandatory requirements of API 6A. Documented evidence that appropriate records have been kept during the manufacture of the components are an essential requirement. In 2001 a publication by Shell outlined a manufacturing route that failed to demonstrate the required technical control and resulted in the failure of a wellhead that had been operational for 6 months. All safety equipment functioned correctly and the well was shut down and there was minimum level of contamination. Figure 11.31. shows the abstract.

The test certification showed that the steel grade was AISI 4140. The composition of the steel was disclosed and it is shown in Figure 11.32. Figure 11.33 shows the microstructure. Figure 11.34 shows the elevated temperature Charpy test results carried out to determine the Impact Transition Temperature. Figure 11.34 shows that the 27J ITT found to be around +140Deg C and the 34J ITT was found to be around +160DegC.

ABSTRACT

Mensa is a gas field with three subsea wells that are located in 5300 feet (1600 m) of water in the Gulf of Mexico. On January 2, 1998, the outer housing (OH) of the upper tree connector of the Mensa A-1 wellhead failed in service in a completely brittle manner and directly caused the immediate shut-in of the A-1 well. All safety systems worked as designed; therefore, a minimal amount of hydrocarbons were released to the environment.

The fracture initiated at the root of two diametrically opposed keyways in the OH and propagated rapidly, rendering the OH into three large pieces.

The cause of the failure was the extremely poor fracture toughness of the forged, quenched and tempered, AISI 4140 OH. The Charpy impact toughness of the material at room temperature was determined to be 2.5 to 4.0 ft-lbs (3.4 to 5.4 J); the fracture mode was 100% brittle. Therefore, the flaw tolerance of the OH was extremely poor.

Contrary to the actual properties of the A-1 OH, information in the material certificates indicated very good impact properties as measured using quality test coupons (QTC) per API 6A.

The results of the failure analysis of the OH, and the results from the evaluation of other OH's purchased for Mensa and in service at the time of the A-1 failure are presented in this paper.

The results from this study strongly suggest that "small" QTC samples are inadequate as a QA/QC tool to assure the fitness-for-service of large, low alloy steel forgings. Therefore, it is proposed

Figure 11.31 Abstract from the publication "Brittle fracture in an upper tree connector system at Mensa". Robert Mack. Shell E&P Technology Applications and Research

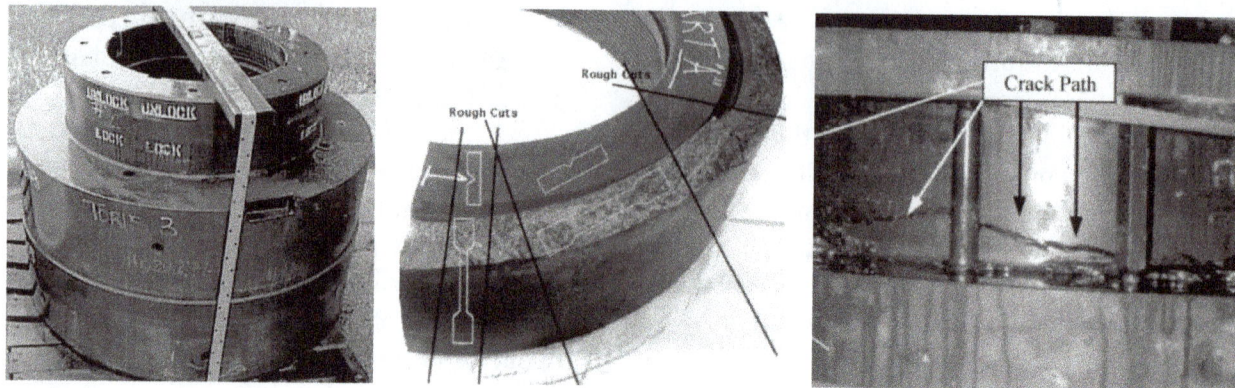

Table 1: Composition of A-1 OH - Actual and Specification Limits

Material	C	Cr	Ni	Mo	Mn	Si	P	S	Cu	Al	Ti	V	Sn
(I) A-1 OH Compositions													
Internal Analysis	0.45	1.04	0.11	0.23	0.92	0.27	0.010	0.019	0.15	0.01	0.03	0.006	0.012
Mill Report	0.41	0.96	0.11	0.20	0.85	0.24	0.009	0.013	0.15	---	---	0.006	---
(II) Specifications per UNS (UNS G41400)													
AISI 4140	0.38-	0.80-		0.15-	0.75-	0.15-	0.00 -	0.00 -					
	0.43	1.10		0.25	1.00	0.35	0.035	0.040	---	---	---	---	---

Notes: (a) Internal analysis of A-1 OH also indicated: Co = < 0.010; B = < 0.0005; Nb < 0.005; As = 0.0007; Sb = 0.003; Bi = 0.0002; W = < 0.020

Table 3: Charpy Impact Data for Mensa A-2, A-3, & Other OH's
Full Size Charpy Impact Test Data

Test Temperature (°C)	(°F)	Absorbed Energy (ft-lbs)	(J)	Shear (%)	Lateral Expansion (mils)	Hardness (HRC)
(I) A-2 OH (boat coupon from OH intended for service)						
21	70	13.2	17.9	0	5	32
0	32	9.5	12.9	0	2	27-28
(II) An OH Intended for Service at A-2 (data from Vendor on 2/2/98)						
-18	0	16-18	22-24	0	---	---
-18	0	12-13	16-18	0	---	---
(III) A-3 OH Removed from Service (boat coupon; see Figures 10 and 11b)						
23	73	14.0	19.0	5-10	9	29-31
0	32	10.5	14.2	0	6	29-31
-16	4	7.5	10.2	0	7	29-31

Figure 11.32 *Upper tree connector, chemical composition and Charpy properties*

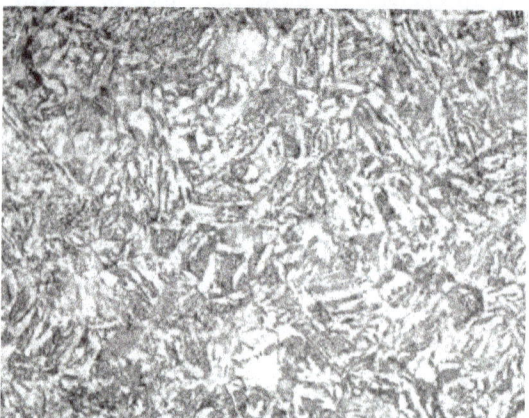

Figure 11.33 *A1 failure x300 Very coarse grain size A3 X700*

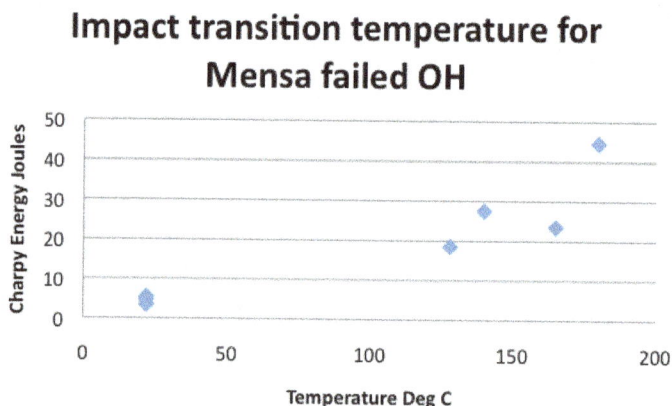

Figure 11.34 *Elevated temperature Charpy tests taken to establish the impact transition temperature*

What went wrong?

Some background to steel key quality factors.

There are many steel making factors that affect the quality of steel (see chapter 3) and process control requires good technical co-operation and communication with the steel supplier.

Hydrogen

There is a minimum hydrogen level required to avoid hydrogen cracking and any effect on the mechanical test specimens. Hydrogen levels should be lower for low sulphur steel.

Steel cleanliness

For high quality forgings there should be a low content of non-metallic inclusions in the steel and therefore good steel cleanliness should be a high priority. Normally this would need a vacuum degas treatment and steel made to a fine-grained practice made with an appropriate aluminium addition. The aluminium addition would also ensure a fine grain size and a significant reduction in the free nitrogen content that can result in a high Charpy ITT (impact transition temperature). This also requires both low sulphur levels and oxygen levels which will ensure a high Charpy shelf energy and a tough ductile steel. This will also ensure minimum risk of cracking during the thermal processing such as forging and heat treatment such as quenching.

Segregation

There is a need for minimum levels of segregation which will ensure uniformity of mechanical properties. This requires appropriate ingot mould design with the required H/D ratio (height to diameter) and wall taper. For concast supply it would involve appropriate EMS (electro-magnetic stir) capability.

Appropriate forge/rolling ratio.

The intention of the forge ratio is to consolidate any "as cast" porosity and help make the segregation pattern more uniform through the thermo-mechanical treatment which allows some diffusion of elements in the steel and grain refinement.

Control of chemical analysis/hardenability

Appropriate hardenability (DI) for the section size (ED) and the quench severity (H) to give a micro-structure that will guarantee the required YS/UTS ratio and low impact transition temperature (ITT)

Heat treatment

For water quenching the maximum cooling rate is required and the quench tank should operate with maximum agitation and at as low a temperature as possible and additional cooling may be needed in warm climates and high throughputs.

The effective control of these factors requires a strongly technically based audit system with the understanding "of what can go wrong". It also requires a standard of comparison that relates to "best world class standard" to allow a quantitative comparison with other manufacturers.

Process procedures and control specifications should be used to ensure that the standard of operation is to world class standards.

QT Steel simulations carried out

1. 1200°C and air cool
2. 870°C and oil quench plus temper 510°C

Some possible output data from QT Steel simulations are shown in Figure 11.35 and 11.36. The software also provides data showing the distribution of the yield strength, UTS and the microstructure. Using the YS/UTS ratios for the actual and predicted values and the microstructure and the graphs shown in Chapter 4 Figure 4.45 and Figure 4.46 a better understanding

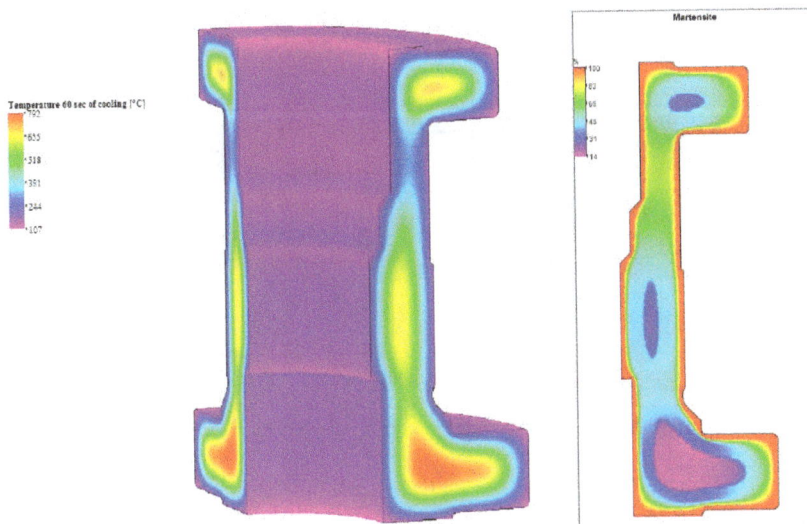

Figure 11.35 Cooling data output from QT Steel Figure 11.36 Simulation of martensite distribution (red)

of the mechanical properties can be gained, including the identification of any deficiency and remedial measures

Based on the simulations the main observations are different from the published conclusions. The review of this failure indicates that the failed OH had probably not been quenched and tempered but had been supplied in the as-forged condition. The other two had been quenched and tempered but had inadequate hardenability for the section size.

11.9.3 AWHEM High temperature tensile project

A published report on an API sponsored project on hot tensile test work describes how oil companies contributed samples of AISI 4130 steel bar which had valid test certification to 75ksi strength grade. The re-tests carried out on these samples found that 3 out of 5 samples failed to meet the 0.2% proof strength with one sample 17% below the required minimum. This represented a major cause for concern.

Appendix A. AISI 4130

Heat Code	Cylinder	Barstock or Forging	Test Agency	Test Temp. (°F)	VENDOR REPORTE D RT YIELD	Tensile	Yield 2%
1	A	B	A	75	83,000	101,100	74,300
1	B	B	B	75	83,000	103,100	74,900
1	C	B	C	75	83,000	98,300	71,400
2	A	F	A	75	79,600	102,800	82,100
2	B	F	B	75	79,600	104,600	81,700
2	C	F	C	75	79,600	100,200	78,700
3	A	B	A	75	76,000	95,100	67,500
3	B	B	B	75	76,000	96,600	66,900
3	C	B	C	75	76,000	92,400	64,300
4	A	F	A	75	80,500	101,700	81,000
4	B	F	B	75	80,500	104,000	81,700
4	C	F	C	75	80,500	101,900	81,500
5	A	B	A	75	87,159	94,500	62,900
5	B	B	B	75	87,159	96,700	63,000
5	C	B	C	75	87,159	92,400	60,700

Figure 11.37 *Extract from the report. Heat codes 1, 3, and 5 have re-test yield values below specification (75ksi). (Compare results in column 6 and column 8)*

11.9.4 Examination of the process capability and test position on a sample of SAE 4130

In view of the above non-compliance for grade SAE 4130 steel some European steel was examined that had test certification showing 90 KSI 0.2% proof strength on 150 mm diameter bar that had been water quenched. The steel was manufactured and heat treated by Gerdau Sidor at their Reinosa plant as 6-metre-long bars.

A metallurgical examination was carried out on a sample of the 150mm diameter bar. The test certification stated 90.3 KSI (0.2 Proof strength) 110.6 KSI UTS (YS/UTS ratio = 82%) at 32mm below the surface.

The 0.2 proof strength appeared to be high for a 150mm diameter cross section tested 32mm below the surface. The test certificate showed an ideal diameter of 3.524ins (calculated in accordance with ASTM A255). Predictions of the mechanical properties were made with QT Steel software. A sample was sectioned and tested at:

- The sub-surface position,
- 32mm below the surface
- At the centre-line

The actual test values compared with QT Steel software predictions. In addition, a comparison was made with the predictions from a Jominy predicted curve combined with the Lamont method to predict hardness values at various depths of round samples. The test results are shown in the following Tables

Table 11.5 *Mechanical test results compared with test certification*

Property	Subsurface	32mm Below surface	Test Cert Result (32mm)	Centre Line
UTS MPa	762	743	**760**	718
0.2 Proof strength MPa	601	578	**621**	529
Hardness HB	241	229	**232**	223
YS/UTS %	79	78	**82**	74

Table 11.6 Based on the predicted Jominy curve and the Lamont method the following hardness values were determined. (Assumed the severity of Quench was Grossman H =2)

Position	As Quenched Rc	As Quenched HB	As Quenched Martensite	Temper 650°C	Actual HB
15mm from Surface	42	390	84%	247	241
37mm from surface	33	311	<30%	231	229
Centre line	27	264	<30%	222	223

Table 11.7 QT Steel Prediction (in blue) compared with test certificate value (in red) and the MSL product tensile test (in green)

Property	Surface	10mm below surface	32mm below surface	Test Cert 32mm	MSL Check 32mm	Centre line
UTS	796	750	742	**760**	**743**	725
0.2 Proof strength	716	611	592	**621**	**578**	558
Hardness	241	226	223	**232**	**229**	218
YS/UTS%	90	81	80	**82**	**78**	77

The check tensile test carried out in a similar position indicated that the 0.2 Proof strength was 43MPa lower than the test cert. A possible explanation for the test certification result was that the actual test position was nearer to the surface. Both the conventional Grossman approach and the QT Steel finite element/ CCT predictions gave values similar to the re-test value. Table 11.7 shows very good agreement between the QT Steel results and the mechanical test results. The software was able to demonstrate that lower levels of DI = 3ins would not achieve the requirements of 75ksi for an ED= 5ins at the mid-radial position.

SELF-ASSESSMENT QUESTIONS

1. The Fe-C equilibrium phase diagram allows the design of different steels with different properties. Using a hypo-eutectoid steel composition explain the reasons for doing the following heat treatment processes and the microstructures that would result.

   a) Normalising
   b) Sub-critical anneal
   c) Full anneal

2. Explain the formation of martensite and bainite and explain why these microstructures are not shown in the equilibrium diagram. What type of diagram would be used to show the formation of these microstructures?

3. If a piece of steel having 0.2% carbon is quenched after soaking at a temperature just above AC1 what type of structure will you get? Estimate approximate constituents of phases present and their compositions.

4. Which of the two would require more severe cooling rate to get fully hardened structure? (a) 0.2 carbon steel (b) SAE 4130 steel.

5. A piece of steel which was quenched after prolonged holding at 710°C was found to have ferrite martensite structure. Explain when you would expect this to happen.

6. List the factors that determine hardenability of steel. Which of these are preferred? Give reasons.

7. What is meant by severity of quench? How could the severity of quench of a tank of water be increased?

8. Hardness of quenched and tempered steel is reported to be Rc 35. What additional test will you recommend to know that it has indeed been given this heat treatment?

9. A service failure has suggested that hardened and tempered steel is brittle. Suggest a suitable test to establish if the steel has poor toughness.

10. Austenite phase in Iron-Carbon equilibrium diagram:

   a) is face centred cubic structure
   b) has magnetic phase
   c) exists below 727°C
   d) all of the above.

11. What is the crystal structure of δ-ferrite?

   a) Body centred cubic structure
   b) Face centred cubic structure
   c) Orthorhombic crystal structure
   d) None of the above.

12. Based on the transformation diagrams for eutectoid steel shown below, what microstructure would result from the following cooling histories? Assume the steel starts above the eutectoid temperature. Distinguish between coarse and fine pearlite when applicable.

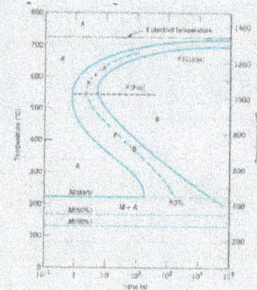

Isothermal TTT Diagram

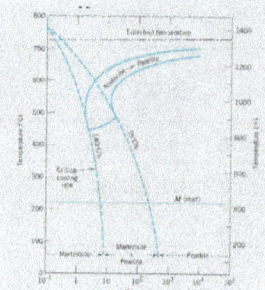

Continuous Cooling TTT Diagram

   a) Rapidly cooled to 600°C, held for 30 seconds. Then rapidly cooled to 20°C. Quenched to room temperature.
   b) Rapidly cooled to 600°C, held for 1 minute, then rapidly cooled to 450°C and held for 10 seconds. Quenched to room temperature.
   c) Rapidly cooled to 600°C, held for 7 seconds, then rapidly cooled to 170°C and held for 1 hour. Quenched to room temperature.
   d) Cooled at a rate of 10°C/s.
   e) Cooled at a rate 165°C/s.

APPENDIX 1

Steel production at Ovako Steelworks in Hofors

The Hofors Steelworks is located about 2 hours' drive north of Stockholm Images from www.ovako.com

Figure A1.1 Top Location of Hofors, Centre Plant layout, Bottom, ID of buildings www.ovako.com

The steel is manufactured using the electric arc furnace process using steel scrap sourced from the Baltic geographical region. The main processes are:

- Production of steel from steel scrap and casting into WEU ingots.
- Rolling of ingots into bloom, followed by scarfing and rolling to billets.
- Hot piercing of billets to form tubes.

The manufacturing route consists of several individual processes as shown in Figure A1.2.

Production of steel ingots at the Steel Mill

Iron scrap and slag forming elements are charged into an electric arc furnace that has a charge weight of around 110 tons. The process using a "hot heal" practice of about 10 tonnes that remains in the furnace after tap. The furnace has been upgraded with a spray cooled roof in 2018 which is a safety system that avoids high pressure water leaks. (US company Systems Spray-Cooled information release). The furnace was also upgraded with an AMI-GE electrode regulation system, (DigiARC) which is an industry accepted control system to give reliable good steelmaking control. The installation of the DigiARC control system and the Spray-Cooled™ roof allows for more energy input to the furnace, increasing steel production with less maintenance disruptions. These events show that substantial investment is taking place to maintain productivity and quality.

In cold climates a cause for concern is that if the scrap is not under cover and is not dried prior to use there can be a risk of snow and ice to entering the arc furnace. An observation from a Swedish student review states:

Figure A1.2 *Steel Making at Hofors www.ovako.com*

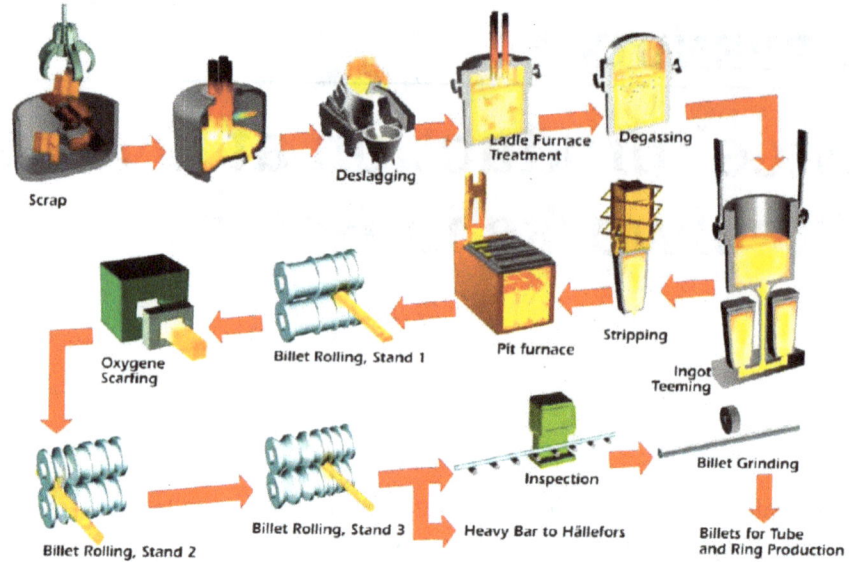

"In some cases the scrap is dried before charging (preheating to approximately 200 degrees C). Scrap drying can have a large effect on energy consumption and the safety at charging if the outside temperature is below zero degrees (high risk of ice and snow explosions)".[1]

Three graphite electrodes are lowered into the furnace and when the scrap starts to melt slag forming elements are added limestone, anthracite (hard coal) are added. The furnace is equipped for oxygen injection into the furnace. Following full melt down which takes about 70 mins the steel is refined and alloyed in a ladle furnace, where it is also degassed. The steel is then teemed uphill into ingot moulds. Each heat is teemed into 24 ingots. The moulds are removed and the ingots are heated in a soaking pit furnace to the proper rolling temperature.

Steel processing at the pouring stage requires appropriate measures to prevent surface imperfections. There are several known sources of seam imperfections formed on steel surfaces:

Mould surface cleanliness – with oxide debris from previous ingots. The method used to clean moulds after stripping uses a 500 bar water spray that is inserted into the top and moves progressively down the ingot mould.

Mould surface condition and mould temperature – It is common practice to control the mould surface such as the number of uses prior to removal from service and scrapping. Old ingot moulds need to be removed from service as the inner surface deteriorates.

Teeming flux – The correct type of teeming flux is important and avoidance of flux detachment and entering the sub-surface of the steel. A Master's Thesis,

Figure A1.3 *LHS = Spray cooled roof, RHS = Scrap charging www.ovako.com*

Royal Institute of Technology, in 2012 identified the teeming flux as a major contributor to sub-surface inclusions at Hofors and recommended further work to be carried out[2]. In the UK, during the 70s trial work was carried out with water simulating metal flow, this was found to be associated with air entrained and carried through the runner system. Lots of work carried out on this topic including, modelling with CFD, turbo-swirl mould entry and restrictors in the trumpet and runners. [3-5]

◀ *Figure A1.4* *Steel Plant Layout www.ovako.com*

A plan of the facility shows the steel plant's different areas and stations. Drawing no. 25819. Year 1991. Ovako Hofors drawing archive. Revised.

1. Scrap baskets.
2. Alloy addition station with silos for alloying materials.
3. Control room with operations lab.
4. Furnace 15, electric arc furnace.
5. Slag removal station
6. Furnace 16, ladle furnace with reheating station and vacuum station.
7. Control room for Furnace 16.
8. Casting shop.
9. The Norberg Transporter
10. Stripping bay.
11. Mould cooling with cooling tunnels.
12. Mould preparation hall.

◀ *Figure A1.5*
Top LHS = Taking slag off ladle, Top RHS = Ingot teeming, Bottom LHS = Ingot stripping, Bottom RHS = WEUIngots www.ovako.com

Figure A1.6 *LHS = Charging an ingot into the soaking pit. RHS = Ingots in soaking pit www.ovako.com*

Total gas content – When solidification takes place the gas in the liquid steel falls out of solution and forms porosity.

Rolling of ingots into billets at the Rolling Mill

In the soaking pit furnace the ingots are heated to the rolling temperature and then the ingots are rolled into bloom and then into billets. The soaking pits are fired with oxy-propane flameless technology. Hofors has used this technology since 1994[6] and have carried out a number of recent studies using CFD to predict heat transfer and heating rates and temperature distribution.[7 and 8] The flameless oxy-propane gave significant cost savings and reduce NOx Figure 3.13.

The purpose of the soaking pits is to reheat the cast ingots to the correct ingot rolling temperature to allow bloom and billet production. The soaking pit also acts as a buffer between the steel making and rolling capacity. Therefore, the demands for soaking pits

can be described as requiring maximum flexibility to cope with fast heating and high productivity periods of manufacture and also holding for stand-by periods of time in the event of slow operation or a mill outage. These working conditions have to be met with low energy consumption together with a low scale formation rate in order to reach a good surface quality, which avoids unnecessary excessive grinding.

Small surface cracks can result from thermal stress in the ingot due to non-uniform temperature cooling gradients or a large mismatch in temperatures between the ingot surface temperature and the soaking pit temperature. Procedures are often based on the "time in mould" and time in air" and for sensitive grades a "let down" soaking pit temperature is adopted.

Bloom rolling begins at stand 1. After completion of rolling, discards are taken from the positions that represent the top and bottom of the ingot. Then the surface imperfections are removed by passing the bloom through the oxygen-scarfing machine.

The billet rolling is continued in rolling stands 2 and 3, and then the billets are placed on a cooling

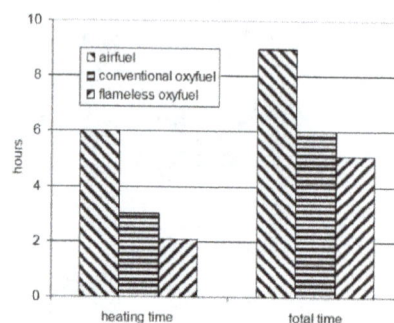

With the installation of oxyfuel in the pit furnaces at Ovako, the heating time was reduced from six to three hours   The more uniform heating of flameless oxyfuel has further reduced the heating time down to 2.1 hours. Total time, which includes soaking, has been reduced from nine to 5.1 hours with flameless oxyfuel.

Figurea A1.7 *Reheating times for different process parameters*[8]

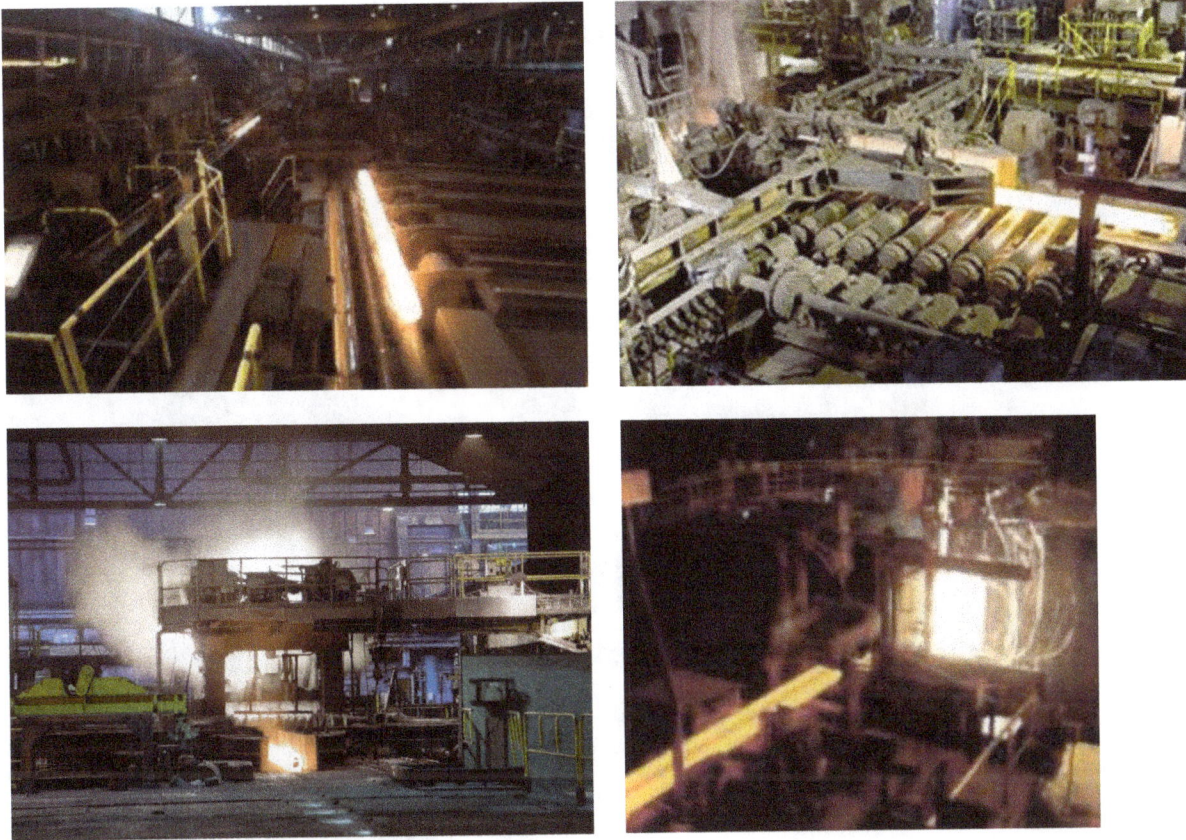

Figure A1.8 Bloom and billet rolling www.ovako.com

bed. After rolling stand 2 and 3, some billets are delivered to customers. The other billets are sand blasted, inspected by automatic MPI and surface defects, if any, are removed by grinding. These billets are delivered to either external customers or to the hot rolling mill at Ovako. The different billet dimensions produced at the rolling mill are 150 mm square and 120, 90 or 80 mm round.

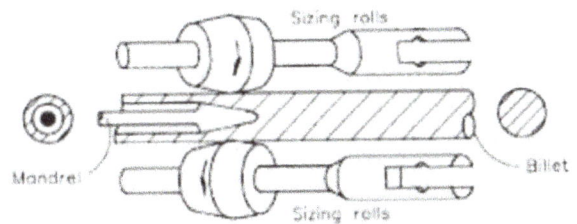

Production of hot rolled tubes at Tube Mill 5

The billets are cut to 3-metre lengths and then heated to rolling temperature, about 1,200°C in a rotary furnace. The centre is marked in one of the end surfaces of the billet, the billet is forced over a plug and the hole is pierced. The wall thickness of the tube is decided by rolling over a mandrel in the Assel mill. In the reducing mill, the outer diameter of the tube is determined. The next step is the calibrating and straightening mill, where the dimensions of the tube are finely adjusted. After that the tube is placed on a cooling bed. The hot rolled tubes are then directly transported to the customer or passed on for further processing.

Figure 1.9 LHS = A schematic illustration of seamless tube manufacture RHS = Actual piercing

Figure 1.10 *Seamless tube manufacture www. ovako.com*

REFERENCES

1. Erik Sandberg, Energy and Scrap Optimisation of Electric Arc Furnaces by Statistical Analysis of Process Data, Licentiate Thesis, page 16, Luleå University of Technology 2005, http://citeseerx.ist.psu.edu/viewdoc/download?doi=1 0.1.1.399.4857&rep=rep1&type=pdf

2. Asumadu Tabiri Kwayie, Macro Inclusion Research: Detection and evaluation of macro inclusions in special steels, Master's Thesis, Royal Institute of Technology, August, 2012.

3. Zhe Tan et al, Mathematical Modeling of Initial Filling Moment of Uphill Teeming Process Considering a Trumpet, ISIJ International, Vol. 51 (2011), No. 9, pp. 1461–1467,

4. Haitong BAI et al, Structure effect of turboswirl in an uphill teeming ingot casting process, 10th International Conference on CFD in Oil & Gas, Metallurgical and Process Industries SINTEF, Trondheim, Norway 17-19 June 2014

5. ZheTan et al, Effect of TurboSwirl on Inclusions during Ingot Casting of Steels, Hindawi Publishing Corporation Mathematical Problems in Engineering Volume 2015, Article ID 805734, 10 pages,

6. Patrik Fredriksson et al, Ovako, Hofors Works- 13 years of experience of using oxyfuel for steel reheating, Iron and steel technology, May 2008 page 323 also Steels for bearing production from Ovako

7. Axel Scherello et al, State-of-the-art oxyfuel solutions for reheating and annealing furnaces in steel industry,

8. Mersedeh Ghadamgahi et al, Design optimization of Flameless-Oxyfuel soaking pit furnace using CFD technique, The 6th International Conference on Applied Energy – ICAE2014

APPENDIX 2

Metallography

The methods used for the metallographic examination of metals and alloys have been under continuous development especially since the development of digital imaging technology, which has allowed better systems of image capture, analysis and storage. In addition, as materials have evolved, metallography has expanded to cover materials ranging from metal matrix composites to ceramics.

Metallography can provide important information regarding the quality, integrity, reliability, properties and heat-treated condition of a manufactured or processed component or part. An excellent guide to the background, history and use of metallographic examination can be found in a book prepared by George F Vander Voort, (Metallography Principles and Practice, ASM, 1999[1]. A list of sources of information regarding metalloraphy can be found at the end of this Appendix

Metallography can be regarded as the practical side of metallurgy and involves part technical knowledge and part a practical skill or workmanship that enables metal sampling, cutting, grinding, polishing and etching to a sufficient standard of competency to allow a clear image of the actual microstructure. The following page shows the stages of metallographic preparation.

The specific areas where metallography is carried out are typically for failure investigations, alloy development or for quality and integrity assessments and can be summarises as:

1. Fracture Surface Examination which can be

   a) Macro examination either with the naked eye or a low magnification
   b) Micro-examination at high magnifications X20 to x20,000 using SEM

2. Macro-examination of cross sections cut through the parts or components by:

   a) Sulphur printing
   b) Macro-etching

3. Micro-examination

Metallography or microstructural analysis can be used for the following types of analysis:

- Grain size
- Porosity and voids
- Phase analysis
- Dendritic growth
- Cracks and other defects
- Corrosion analysis
- Intergranular attack (IGA)
- Coating thickness and integrity
- Inclusion size, shape and distribution
- Weld and heat-affected zones (HAZ)
- Solder joints
- Distribution and orientation of composite fillers
- Graphite nodularity
- Recast
- Carburizing thickness
- Decarburization
- Nitriding thickness
- Intergranular fracturing
- HAZ Sensitization
- Flow-line Stress

			Specimen Preparation			
1. Cutting	**2. Mounting**	**3. Grinding**	**4. Polishing**	**5. Etching**	**6. Examination**	**7. Micro-photography**
Coolant; Cutting Disc; Sample; Specimens	Specimen; Hot mounting compound; Specimen mounting press	Coolant; Specimen; Carborundum paper	Diamond paste; Specimen; Polishing cloth; Polishing wheel	Etchant; Specimen; Keep specimen moving during etching	Camera; Specimen; Metallurgical microscope	PC and Software
1. Equipment	**1. Equipment**	**1. Equipment**	**1. Equipment**	**1. Equipment**	**1. Equipment**	**1. Equipment**
Lab cut-off saw	Hot mounting press	Grinding wheel	Polishing wheel	Etchant container	Metallurgical microscope, CCD camera and specimen leveller	Computer and printer
2. Cut-off Wheel	**2. Resin Type**	**2. Grinding Medium**	**2. Polishing Medium**	**2. Etchant**		
Resin bonded, both soft type (for hard specimen) and hard type (for soft specimen)	Phenolic resin (Bakelite) or Acrylic resin (Transparent type)	Waterproof carbon in paper grit No 240, 400, 800, 1200 and 2400	Polishing cloth and water suspended Alumina powder, 0.3 micron	2.1 Reagent 7a; 2.2 Nital 3% to 5%		
	3. Temperature	**3. Coolant**	**3. Coolant**	**3. Etching Time**	**2. Magnification**	
3. Coolant	170°C to 180°C	Water	Water	15 to 30 seconds	x50, x100, x200, x500 and x1000	
Water mixed with cooling medium and anti-bacteria solution	**4. Time**	**Note:**	**Note:**	**4. Cleaning**		
Note:	8 to 12 minutes	Grinding must start from the coarsest paper through to a fine paper. Before changing to the finer grit paper, the grinding lines from the previous paper must have disappeared and new grinding lines must be in the same direction. Before starting the next paper, the specimen must be clean.	During the polishing, the specimen must be kept clockwise and counter-clockwise to prevent comet tails. The water must be used together with the polishing paste. After polishing, the specimen must be rinsed with water and alcohol and dried with hot air.	After etching, the specimen must be rinsed with flowing water, then alcohol and immediately dried with hot air		
The specimen must be well selected and may be round, square or other irregular shape. The cutting must be carried out carefully with enough coolant to prevent overheating during cutting, particularly surface hardened parts.	**5. Coolant** / Water / **Note:** / The mounted specimen must be marked carefully to avoid any confusion or mix-up.			**Note:** 1. Nital 3% Nitric Acid (HNO$_3$): 3cc Alcohol: 97cc 2. Reagent 7a (Colour Etch) 3g potassium metabisulphite 1g sulphuric acid 100ml distilled water		

Sources of Information on Metallography

1. GEORGE F. VANDER VOORT, Metallography Principles and Practice, ASM International, 1999

2. D. San Martin et al, A New Etching Route for Revealing the Austenite Grain Boundaries in an 11.4% Cr Precipitation Hardening Semi-Austenitic Stainless Steel,

3. Donald C. Zipperian, METALLOGRAPHIC HANDBOOK, Copyright 2011 by PACE Technologies, USA,

4. George F. Vander Voort, Metallographic Techniques in Failure Analysis, 2002 https://www.georgevandervoort.com/images/Failure-Analysis/Met_Techniques_for_FA.pdf

5. BUEHLER® SUM-MET™ The Science Behind Materials Preparation A Guide to Materials Preparation & Analysis, ISBN Number: 0-9752898-0-2, 2004, ISBN Number: 0-9752898-0-2 2007, http://www.trimid.rs/upload/documents/buhler/Buehler%20Summet.pdf

6. G.F. Vander Voort, Study of Selective Etching of Carbides in Steel, https://www.researchgate.net/profile/Joseph_Michael/publication/243371454_A_Study_of_Selective_Etching_of_Carbides_in_Steel/links/02e7e529660ebd6f10000000/A-Study-of-Selective-Etching-of-Carbides-in-Steel.pdf

7. Perrin Walker, Ed, HANDBOOK of METAL ETCHANTS, CRC Press 1991, ISBN 0-8493-3623-6

8. GOUTAM DAS, Image analysis in quantitative metallography, Materials Characterization Techniques-Principles and Applications, Eds : G. Sridhar. S. Ghosh Chowdhunv & N. G. Goswanri N) NML. Jamshedpur-831007 (1999) pp. 135-150 https://pdfs.semanticscholar.org/6ef3/96e4bfe03350a21e160eec68446c6ef7e9f0.pdf

9. Leica Metallurgy Application Briefing, Application Solutions for Metallurgy, Leica Microsystems Imaging Solutions Ltd. Cambridge, UK. 2000, https://www.scribd.com/document/54219747/Metallurgy-App-Briefing-1

10. Metallography, Handbook for Sintered Components, Copyright Höganäs AB (publ.), 2015 / 0886HOG. https://www.hoganas.com/globalassets/download-media/sharepoint/handbooks---all-documents/metallography_may_2015_0886hog_interactive.pdf

11. David A. Scott, Metallography and microstructure of ancient and historic metals, The Getty Conservation Institute, The J Paul Getty Museum In association with arch Books, 1991, https://www.getty.edu/conservation/publications_resources/pdf_publications/pdf/metallography.pdf

12. Paresh U. Haribhakti, Use of In-Situ Metallography for Plant Health Assessment Studies and Failure Investigations, 13th Middle East Corrosion Conference Paper 09012, https://www.ndt.net/article/mendt2007/papers/bafna.pdf

13. C. García de Andrés, Revealing austenite grain boundaries by thermal etching: advantages and disadvantages, https://core.ac.uk/download/pdf/36044422.pdf

14. F. HAIRER, ETCHING TECHNIQUES FOR THE MICROSTRUCTURAL CHARACTERIZATION OF COMPLEX PHASE STEELS BY LIGHT MICROSCOPY, https://www.mtf.stuba.sk/buxus/docs/internetovy_casopis/2008/4mimorc/hairer.pdf

15. H. K. D. H. Bhadeshia, Interpretation of the Microstructure of Steels, https://www.phase-trans.msm.cam.ac.uk/2008/Steel_Microstructure/SM.html

16. Patrick Echlin, Handbook of Sample Preparation for Scanning Electron Microscopy and X-Ray Microanalysis, Springer, 2009, ISBN: 978-0-387-85730-5

17. C. D. Lundin et al, Literature Review, Ferrite Measurement in Austenitic andDuplex Stainless Steel Castings Submitted to: SFSA/CMC~OE August 1999, The University of Tennessee, Knoxvil, https://www.osti.gov/servlets/purl/14580

18. Metallography and Microstructures was published in 2004 as Volume 9 of the ASM Handbook

19. Albert Sauveur, THE METALLOGRAPHY OF IRON AND STEEL, FIRST EDITION FIRST THOUSAND, McGRAW-HILL BOOK COMPANY

APPENDIX 3

Greek alphabet

GREEK LETTERS

Upper Case	Lower Case	Name
A	α	Alpha
B	β	Beta
Γ	γ	Gamma
Δ	δ	Delta
E	ϵ	Epsilon
Z	ζ	Zeta
H	η	Eta
	θ	Theta
I	ι	Iota
K	κ	Kappa
Λ	λ	Lambda
M	μ	Mu

Upper Case	Lower Case	Name
N	ν	Nu
Ξ	ξ	Xi
O	o	Omicron
Π	π	Pi
P	ρ	Rho
Σ	σ	Sigma
T	τ	Tau
Y	υ	Upsilon
Φ	ϕ	Phi
X	χ	Chi
Ψ	ψ	Psi
Ω	ω	Omega

UNIT CONVERSIONS

From	Multiply by	To
A	10^{-10}	m
bar	1.019716	kg/cm²
BTU	1058.201058	J
BTU	251.9958	cal
cal	4.184	J
F	5/9 (°F-32)	°C
ft	12	inch
ft	0.30485126	m
ft.lb	1.356	J ou N.m
ft.lb	0.324	cal
ft.lb	1.355748373	J
ft.lb/s	1.355380862	W
ft.lbf	1.355818	J or N.m
ft.lbf	0.1382	kg
ft²	92.90×10^{-3}	m²
ft³	0.02831685	m³
gallon	3.78541178	liters
HP	0.7456999	kW
HP	745.7121551	W
in	25.4	mm
in²	645.2	mm²
in³	16387.064	mm³
in-lb/in²	0.000175127	J/mm²
J	9.45×10^{-4}	BTU
J	0.2390	cal
J	0.7376	ft.lb

From	Multiply by	To
J	2.389×10^{-7}	th
J	1	W.s
J	2.777×10^{-9}	kWh
Kcal/m² h °C	1,163	W/m² °C
Kg	2.205	lb
kgf	9.80665	N
kgf.m	9.80665	J
kgf/mm²	9.80665	MPa
Kip	1000	lbf
kN	224.8	lbf
kN	0.102040816	t
kN/mm	5.71×10^3	lbf/ft
ksi	6.894757	MPa
ksi	1000	psi
ksi.√in	1.098901099	MPa.√m
kW	1.341022	HP
kW	0.860	th/h
kW.h	3.6×10^6	J
kW.h	3.412×10^3	BTU
kW.h	8.6×10^5	cal
lb	0.4535924	kg
lb.in	0.1129815	J ou N.m
lb/ft³	0.016020506	g/cm³
lb/in³	27.67783006	g/cm³
lbf	4.448222	N
lbf/in²	1	psi

APPENDIX 4

Hardness conversion charts

Table 2-2 Hardness Conversion Chart according to B.S. 860/1939 [6]

Rockwell Scale C	A	Diamond Pyramid Scale HV10 HV30	Brinell Dia. Imp. for 10mm Ball	Carbide Ball	Standard Ball	Tons/in2	1000lb/in2	kg/mm2	Mpa (N/mm2)	Scleroscope Hardness Number
67.7	85.6	900								96
67	85	880								95
66.3	84.7	860								93
65.5	84.2	840								92
64.8	83.8	820								90
64	83.4	800								88
63.3	83	780								87
62.5	82.6	760								86
61.7	82.2	740								84
61	81.8	725	2.44	630	-	-	-	-	-	82
60.5	81.5	710	2.45	627	-	-	-	-	-	-
60	81.2	698	2.5	601	-	132	295	208	2039	81
58.9	80.6	670	2.55	578	-	127	284	200	1961	78
57.1	79.6	630	2.6	555	-	122	273	192	1884	75
56.1	79	609	2.65	534	-	117	262	184	1807	73
54.4	78.2	572	2.7	514	-	112	250	176	1729	71
51.9	76.9	532	2.75	495	495	108	241	170	1668	68
50.7	76.3	517	2.8	477	477	105	235	165	1621	66
49.5	75.5	497	2.85	461	461	101	226	160	1559	64
47.5	74.2	470	2.9	444	444	98	219	155	1513	62
46	73.5	452	2.95	429	429	95	212	150	1467	60
44.8	73	437	3	415	415	92	206	145	1420	58
43.7	72.5	422	3.05	401	401	88	197	139	1359	56
42.4	71.5	408	3.1	388	388	85	190	134	1312	54
41.3	71	395	3.15	375	375	82	183	129	1266	52
39.9	70.3	381	3.2	363	363	80	179	126	1235	51
38.8	69.8	370	3.25	352	352	77	172	121	1189	49
37.7	69.2	359	3.3	341	341	75	168	118	1158	48
36.7	68.8	349	3.35	331	331	73	163	114	1127	46
35	68	337	3.4	321	321	71	159	111	1096	45
34	67.5	327	3.45	311	311	68	152	107	1050	43
33	66.8	318	3.5	302	302	66	147	104	1019	42
32	66.2	308	3.55	293	293	64	143	101	988	41
30.9	65.7	300	3.6	285	285	63	141	99	973	40
29.8	65.2	292	3.65	277	277	61	136	96	942	38
29	64.6	284	3.7	269	269	59	132	93	911	37
27.5	64	275	3.75	262	262	58	130	91	895	36
26.6	63.6	269	3.8	255	255	56	125	89	865	35
25.2	62.9	261	3.85	248	248	55	123	87	849	34
24.3	62.6	255	3.9	241	241	53	118	84	818	33
23	62	247	3.95	235	235	51	114	81	787	32
22	61.6	241	4	229	229	50	112	79	772	31
20.8	60.7	234	4.05	223	223	49	110	77	756	30
	-	228	4.1	217	217	48	107	76	741	-
Rockwell 'B' Scale										
98		222	4.15	212	212	46	103	73	710	29
97		218	4.2	207	207	45	101	71	695	28
96		212	4.3	197	197	43	97	68	664	27
93		196	4.4	187	187	41	92	65	632	25
91		188	4.5	179	179	39	88	62	602	-
88.5		178	4.6	170	170	36	81	57	556	24
86		171	4.7	163	163	35	78	55	540	-
84.2		163	4.8	156	156	34	76	54	525	23
82		156	4.9	149	149	32	72	51	494	-
80		150	5	143	143	3t	69	49	479	22
77		143	5.1	137	137	30	67	48	463	21
75		137	5.2	131	131	29.5	66	47	455	20.5
72.5		132	5.3	126	126	29	65	46	448	20
70		127	5.4	121	121	28	63	44	432	-
67		122	5.5	116	116	26	58	42	401	15

Comparison of hardness scales approx. & tensile stress equivalents approx. (maximum value) in imperial and metric units.

Table 2-3 ASTM Hardness Conversion Chart[6]

HRC150 kgf	HRA60 kgf	HRD100 kgf	HR15N 15kgf	HR30N 30kgf	HR45N45 kgf	Vickers Hardness	Knoop Hardness	Brinell Hardness 3000kgf 10mm	Tensile Strength
68	85.6	76.9	93.2	84.4	75.4	940	920	..	..
67	85	76.1	92.9	83.6	74.2	900	895	..	..
66	84.5	75.4	92.5	82.8	73.3	865	870	..	
65	83.9	74.5	92.2	81.9	72	832	846	..	..
64	83.4	73.8	91.8	81.1	71	800	822	-739	
63	82.8	73	91.4	80.1	69.9	772	799	-722	
62	82.3	72.2	91.1	79.3	68.8	745	776	-688	
61	81.8	71.5	90.7	78.4	67.7	720	754	-670	
60	81.2	70.7	90.2	77.5	66.6	697	732	-654	
59	80.7	69.9	89.8	76.6	65.5	674	710	-634	351
58	80.1	69.2	89.3	75.7	64.3	653	690	615	338
57	79.6	68.5	88.9	74.8	63.2	633	670	595	325
56	79	67.7	88.3	73.9	62	613	650	577	313
55	78.5	66.9	87.9	73	60.9	595	630	560	301
54	78	66.1	87.4	72	59.8	577	612	543	292
53	77.4	65.4	86.9	71.2	58.6	560	594	525	283
52	76.8	64.6	84.4	70.2	57.4	544	576	512	273
51	76.3	63.8	85.9	4	56.1	528	558	496	264
50	75.9	63.1	85.5	68.5	55	513	542	481	255
49	75.2	62.1	85	67.6	53.8	498	526	469	246
48	74.7	61.4	84.6	66.7	52.5	484	510	455	237
47	74.1	60.8	83.9	65.8	51.4	471	495	443	229
46	73.6	60	83.5	64.8	50.3	458	480	432	221
45	73.1	59.2	83	64	49	446	466	421	215
44	72.5	58.5	82.5	63.1	47.8	434	452	409	208
43	72	57.7	82	62.2	46.7	423	438	400	201
42	71.5	56.9	81.5	61.3	45.5	412	426	390	195
41	70.9	56.2	80.9	60.4	44.3	402	414	381	188
40	70.4	55.4	80.4	59.5	43.1	392	402	371	182
39	69.9	54.6	79.9	58.6	41.9	382	391	362	177
38	69.4	53.8	79.4	57.7	40.8	372	380	353	171
37	58.9	53.1	78.8	56.8	39.6	363	370	344	166
36	68.4	52.3	78.3	55.9	38.4	354	360	336	161
35	67.9	51.5	77.7	55	37.2	345	351	327	156
34	67.4	50.8	77.2	54.2	36.1	336	342	319	152
33	66.8	50	76.6	53.3	34.9	327	334	311	149
32	66.3	49.2	76.1	52.1	33.7	318	326	301	146
31	65.8	48.4	75.6	51.3	32.5	310	318	294	141
30	65.3	47.7	75	50.4	31.3	302	311	286	138
29	64.8	47	74.5	49.5	30.1	294	304	279	135
28	64.3	46.1	73.9	48.6	28.9	286	297	271	131
27	63.8	45.2	73.3	47.7	27.8	279	290	264	128
26	63.3	44.6	72.8	46.8	26.7	272	284	258	125
25	62.8	43.8	72.2	45.9	25.5	266	278	253	123
24	62.4	43.1	71.6	45	24.3	260	272	247	119
23	62	42.1	71	44	23.1	254	266	243	117
22	61.5	41.6	70.5	43.2	22	248	261	237	115
21	61	40.9	69.9	42.3	20.7	243	256	231	112
20	60.5	40.1	69.4	41.5	19.6	238	251	226	110

Approximate hardness numbers for Non-Austenitic Steels .according to ASTM E-140.
The conversion values contained herein should be considered approximate only may be inaccurate for specific applications.

Table 2-4 Hardness conversion chart according to DIN 50 150 [9

Vickers hardness (F≥98N)	Ball indentation diameter¹⁾	Brinell hardness²⁾	Rockwell hardness		Tensile strength
HV	mm	HB	HRB	HRC	N/mm²
63	7.32	60			200
65	7.22	62			210
69	7.04	66			220
70	6.99	67			225
72	6.95	68			230
75	6.82	71			240
79	6.67	75			250
80	6.63	76			255
82	6.56	78			260
85	6.45	81	41		270
88	6.35	84	45		280
90	6.28	86	48		285
91	6.25	87	49		290
94	6.19	89	51		300
95	6.16	90	52		305
97	6.10	92	54		310
100	6.01	95	56		320
103	5.93	98	58		330
105	5.87	100	59		335
107	5.83	102	60		340
110	5.75	105	62		350
113	5.70	107	63.5		360
115	5.66	109	64.5		370
119	5.57	113	66		380
120	5.54	114	67		385
122	5.50	116	67.5		390
125	5.44	119	69		400
128	5.38	122	70		410
130	5.33	124	71		415
132	5.32	125	72		420
135	5.26	128	73		430
138	5.20	131	74		440
140	5.17	133	75		450
143	5.11	136	76.5		460
145	5.08	138	77		465
147	5.05	140	77.5		470
150	5.00	143	78.5		480
153	4.96	145	79.5		490
155	4.93	147	80		495
157	4.90	149	81		500
160	4.86	152	81.5		510
163	4.81	155	82.5		520
165	4.78	157	83		530
168	4.74	160	84.5		540
170	4.71	162	85		545
172	4.70	163	85.5		550
175	4.66	166	86		560
178	4.62	169	86.5		570
180	4.59	171	87		575
181	4.58	172			580
184	4.54	175	88		590
185	4.53	176			595
187	4.51	178	89		600
190	4.47	181	89.5		610
193	4.44	184	90		620
195	4.43	185			625
197	4.40	187	91		630
200	4.37	190	91.5		640
203	4.34	193	92		650
205	4.32	195	92.5		660
208	4.29	198	93		670
210	4.27	199	93.5		675
212	4.25	201			680

Vickers hardness (F≥98N)	Ball indentation diameter¹⁾	Brinell hardness²⁾	Rockwell hardness		Tensile strength
HV	mm	HB	HRB	HRC	N/mm²
215	4.22	204	94		690
219	4.19	208			700
220	4.18	209	95		705
222	4.16	211	95.5		710
225	4.13	214	96		720
228	4.11	216			730
230	4.08	219	96.5		740
233	4.07	221	97		750
235	4.05	223			755
237	4.03	225	97.5		760
240	4.01	228	98		770
243	3.98	231		21	780
245	3.97	233			785
247	3.95	235	99		790
250	3.93	238	99.5	22	800
253	3.91	240			810
255	3.89	242		23	820
258	3.87	245			830
260	3.85	247		24	835
262	3.84	249			840
265	3.82	252			850
268	3.80	255		25	860
270	3.78	257			865
272	3.77	258		26	870
275	3.76	261			880
278	3.74	264			890
280	3.72	266		27	900
283	3.70	269			910
285	3.69	271			915
287	3.68	273		28	920
290	3.66	276			930
293	3.64	278		29	940
295	3.63	280			950
299	3.61	284			960
300	3.60	285			965
302	3.59	287		30	970
305	3.57	290			980
308	3.55	293			990
310	3.54	295		31	995
311	3.53	296			1000
314	3.52	299			1010
317	3.50	301		32	1020
320	3.49	304			1030
323	3.47	307			1040
327	3.45	311		33	1050
330	3.44	314			1060
333	3.43	316			1070
336	3.41	319		34	1080
339	3.40	322			1090
340	3.39	323			1095
342	3.38	325			1100
345	3.36	328		35	1110
349	3.35	332			1120
350	3.34	333			1125
352	3.33	334			1130
355	3.32	337		36	1140
358	3.31	340			1150
360	3.30	342			1155
361	3.29	343			1160
364	3.28	346		37	1170
367	3.26	349			1180
370	3.25	352			1190
373	3.24	354		38	1200

¹⁾ steel ball with 10 mm diameter; ²⁾ calculated from: HB = 0.95 HV

Vickers hardness (F≧98N)	Ball indentation diameter[1]	Brinell hardness[2]	Rockwell hardness		Tensile strength
HV	mm	HB	HRB	HRC	N/mm²
376	3.23	357			1210
380	3.21	361			**1220**
382	3.20	363		39	1230
385	3.19	366			1240
388	3.18	369			1250
390	3.17	371			**1255**
392		372		40	1260
394	3.16	374			1270
397	3.14	377			1280
400	3.13	380			**1290**
403	3.12	383		41	1300
407	3.10	387			1310
410	3.09	390			**1320**
413	3.08	393		42	1330
417	3.07	396			1340
420	3.06	399			**1350**
423	3.05	402		43	1360
426	3.04	405			1370
429		408			1380
430	3.02	409			**1385**
431		410			1390
434	3.01	413		44	1400
437	3.00	415			1410
440	**2.99**	**418**			**1420**
443	2.98	421			1430
446	2.97	424		45	1440
449	2.96	427			1450
450		**428**			**1455**
452	2.95	429			1460
455	2.94	432			1470
458	2.93	435		46	1480
460		**437**			**1485**
461	2.92	438			1490
464	2.91	441			1500
467	2.90	444			1510
470	**2.89**	**447**			**1520**
473		449		47	1530
476	2.88	452			1540
479	2.87	455			1550
480		(456)			**1555**
481	2.86	(457)			1560
484	2.85	(460)		48	1570
486		(462)			1580
489	2.84	(465)			1590
490	**2.83**	(466)			**1595**
491		(467)			1600
494	2.82	(470)			1610
497		(472)		49	1620
500		(475)			**1630**
503	2.80	(478)			1640
506	2.79	(481)			1650
509		(483)			1660
510	**2.78**	(485)			**1665**
511		(486)			1670
514	2.77	(488)		50	1680
517	2.76	(491)			1690
520	**2.75**	(494)			**1700**
522		(496)			1710
525	2.74	(499)			1720
527		(501)		51	1730
530	**2.73**	(504)			**1740**
533	2.72	(506)			1750
536	2.71	(509)			1760

Vickers hardness (F≧98N)	Ball indentation diameter[1]	Brinell hardness[2]	Rockwell hardness		Tensile strength
HV	mm	HB	HRB	HRC	N/mm²
539		(512)			1770
540	2.70	(513)			**1775**
541		(514)			1780
544	2.69	(517)		52	1790
547		(520)			1800
550	2.68	(523)			**1810**
553	2.67	(525)			1820
556		(528)			1830
559	2.66	(531)			1840
560		(532)		53	**1845**
561	2.65	(533)			1850
564		(536)			1860
567	2.64	(539)			1870
570		(542)			**1880**
572	2.63	(543)			1890
575	2.62	(546)			1900
578		(549)		54	1910
580	2.61	(551)			**1920**
583	2.60	(554)			1930
586		(557)			1940
589	2.59	(560)			1950
590		(561)			**1955**
591		(562)			1960
594	2.58	(564)			1970
596		(567)			1980
599	2.57	(569)			1990
600		(570)			**1995**
602	2.56	(572)			2000
605		(575)			2010
607	2.55	(577)			2020
610		(580)			**2030**
613	2.54	(582)			2040
615		(584)		56	2050
618	2.53	(587)			2060
620		(589)			**2070**
623	2.52	(592)			2080
626		(595)			2090
629	2.51	(598)			2100
630		(599)			**2105**
631		(600)			2110
634	2.50	(602)			2120
636		(604)			2130
639	2.49	(607)		57	2140
640		(608)			**2145**
641		(609)			2150
644	2.48	(612)			2160
647	2.47	(615)			2170
650		(618)			**2180**
653		(620)			2190
655	2.46	(622)		58	2200
675				59	
698				60	
720				**61**	
745				62	
773				63	
800				**64**	
829				65	
864				66	
900				**67**	
940				**68**	

[1]) steel ball with 10 mm diameter; [2]) calculated from: HB = 0.95 HV

Values in bold face correspond exactly to DIN values. The other values are interpolated. Value

APPENDIX 5

Steel grades for various section size and strength requirements

Heat Treatment Condition	Tensile Strength Range	Hardness Range BHN	Ruling Section						
			<=13mm	>13 <=19mm	>19 <=29mm	>29 <=63mm	>63 <=100mm	>100 <=150mm	>150 <=250mm
Q	625-775 n/mm2 40-50 TSI	179-229	080M40 150M19	080M40 150M19	080M40 150M19	080M40 150M19			708M40
R	700-850 n/mm2 45-55 TSI	201-255	080M40 150M19 606M36	080M40 070M55 150M19 606M36	070M55 150M19 606M36	070M55 605M36	070M55 605M36	605M36 708M40 709M40	605M36 708M40 709M40
S	775-925 n/mm2 50-60 TSI	223-277	070M55 606M36	070M55 605M36 606M36	070M55 605M36 606M36	070M55 605M36 606M36 708M40	605M36 708M40 709M40	709M40	709M40
T	850-1000 n/mm2 55-65 TSI	248-302	070M55 605M36 606M36 708M40	070M55 605M36 606M36 708M40	605M36 606M36 708M40	605M36 708M40 709M40	605M36 708M40 709M40 817M40	817M40	817M40
U	925-1075 n/mm2 60-70 TSI	269-331	605M36 708M40	605M36 708M40 709M40	605M36 708M40 709M40	709M40 817M40	709M40 817M40	826M40	826M40
V	1000-1150 n/mm2 75-80 TSI	293-352	605M36 708M40 709M40	605M36 708M40 709M40 817M40	709M40 817M40	817M40	826M40	826M40	826M40
W	1075-1225 n/mm2 80-85 TSI	311-375	708M40 709M40 817M40	709M40 817M40	817M40	826M40	826M40	826M40	826M40

Condition	Tensile N/mm²	Yield N/mm²	Elongation %	Izod KCV J	Hardness Brinell
P	550/700	340	20	28	152/207
Q	625/775	400	18	28	179/227
R	700/850	480	16	28	201/255
S	775/925	525	14	16	223/277
T	850/1000	650	13	35	248/302
U	925/1000	755	12	42	269/331
V	1000/1150	850	12	42	293/352
W	1075/1225	940	11	35	311/375
X	1150/1300	1020	10	28	341/401
Y	1225/1375	1095	10	21	363/429
Z	1550	1235	5	9	444

APPENDIX 6

Crafts Lamont diagrams

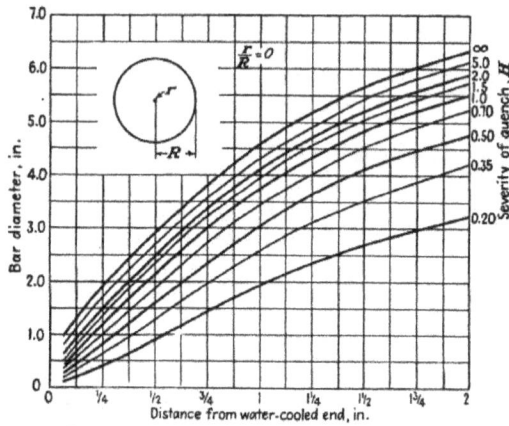

FIG. 4.14. Location on end-quenched Jominy hardenability specimen corresponding to the center of round bars. (*Lamont.*)[96]

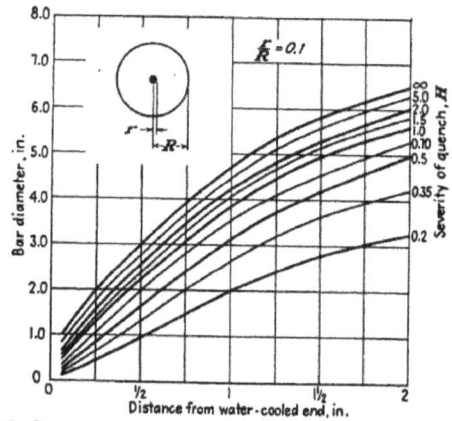

FIG. 4.15. Location on end-quenched Jominy hardenability specimen corresponding to 10 per cent from the center of round bars. (*Lamont.*)[96]

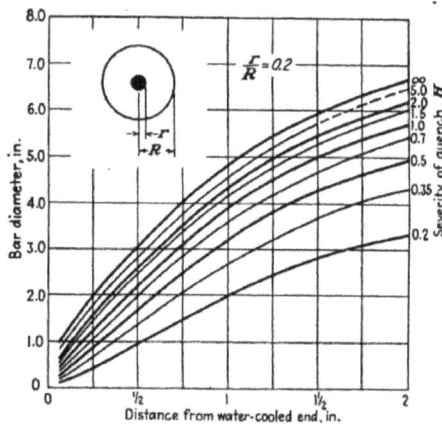

FIG. 4.16. Location on end-quenched Jominy hardenability specimen corresponding to 20 per cent from the center of round bars. (*Lamont.*)[96]

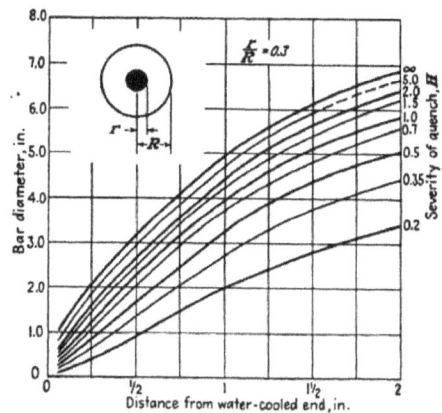

FIG. 4.17. Location on end-quenched Jominy hardenability specimen corresponding to 30 per cent from the center of round bars. (*Lamont.*)[96]

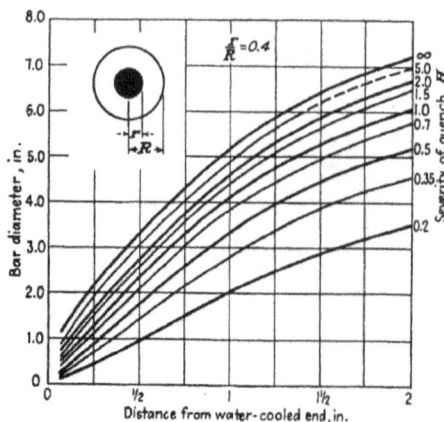

FIG. 4.18. Location on end-quenched Jominy hardenability specimen corresponding to 40 per cent from the center of round bars. (*Lamont.*)[96]

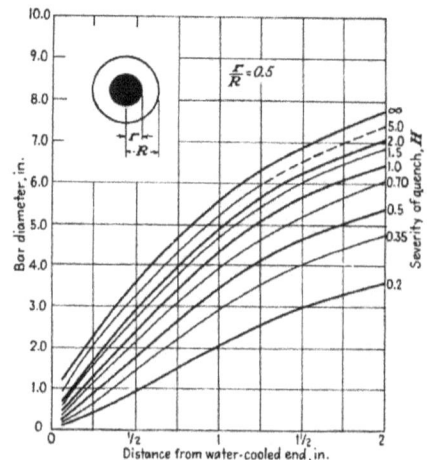

FIG. 4.19. Location on end-quenched Jominy hardenability specimen corresponding to 50 per cent from the center of round bars. (*Lamont.*)[96]

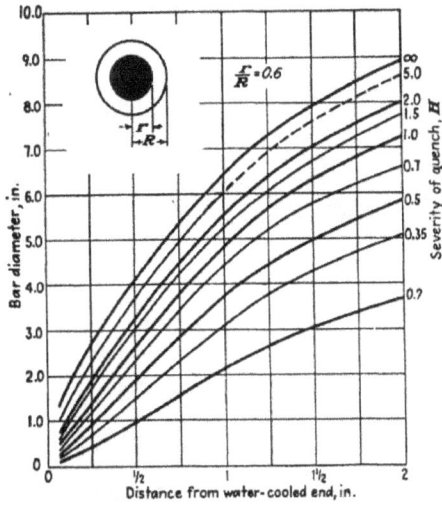

FIG. 4.20. Location on end-quenched Jominy hardenability specimen corresponding to 60 per cent from the center of round bars. (*Lamont.*)[96]

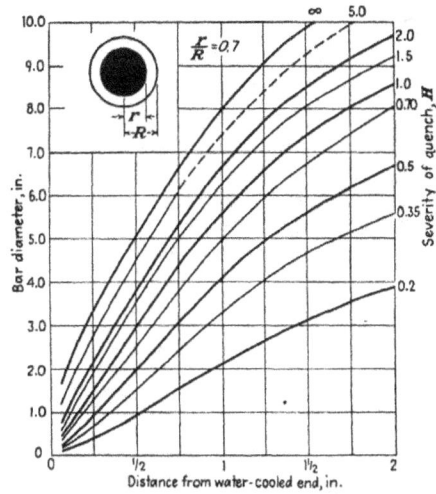

FIG. 4.21. Location on end-quenched Jominy hardenability specimen corresponding to 70 per cent from the center of round bars. (*Lamont.*)[96]

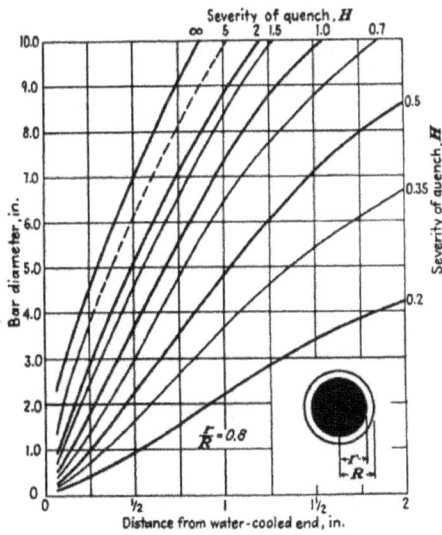

FIG. 4.22. Location on end-quenched Jominy hardenability specimen corresponding to 80 per cent from the center of round bars. (*Lamont.*)[96]

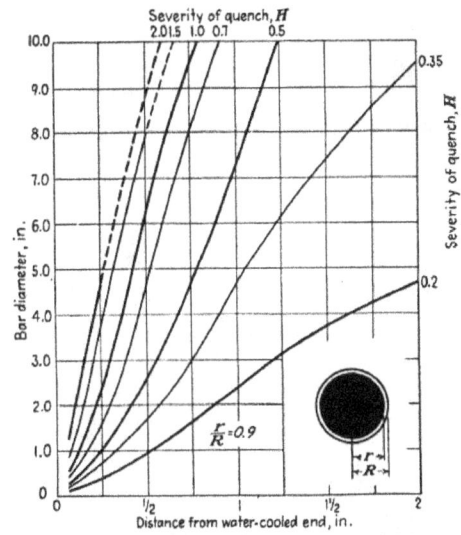

FIG. 4.23. Location on end-quenched Jominy hardenability specimen corresponding to 90 per cent from the center of round bars. (*Lamont.*)[96]

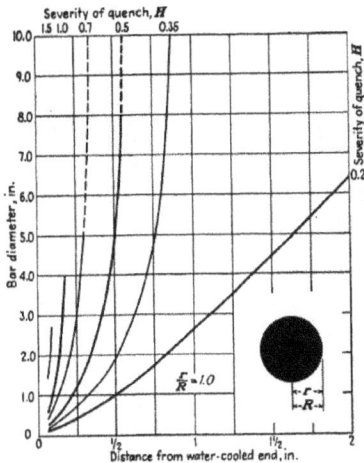

FIG. 4.24. Location on end-quenched Jominy hardenability specimen corresponding to the surface of round bars. (*Lamont.*)[96]

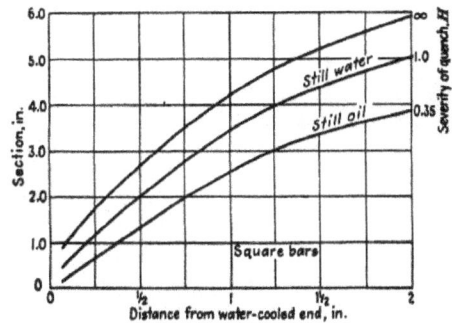

FIG. 4.25. Location on end-quenched Jominy hardenability specimen corresponding to the center of square bars. (*Lamont.*)[96]

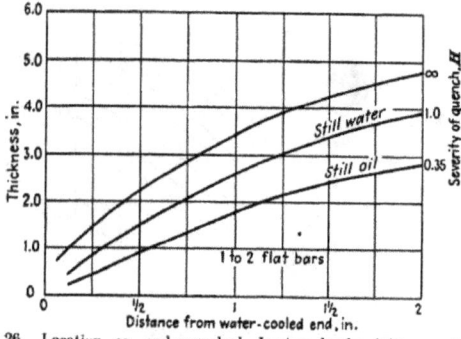

FIG. 4.26. Location on end-quenched Jominy hardenability specimen corresponding to the center of 1 to 2 flat bars. (*Lamont.*)

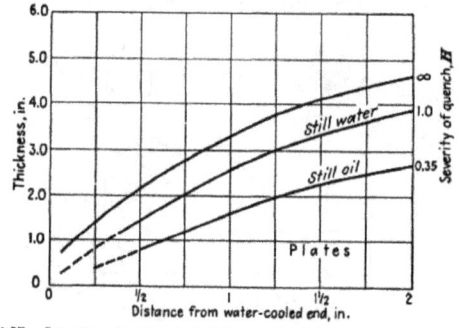

FIG. 4.27. Location of end-quenched Jominy hardenability specimen corresponding to the center of plates. (*Lamont.*)

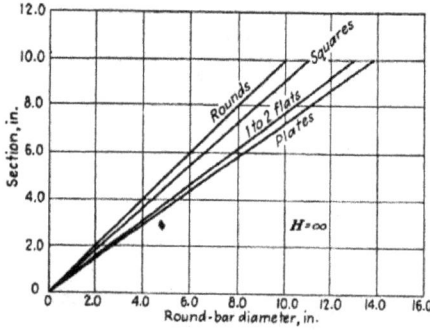

FIG. 4.28. Relation of round-bar diameter to the bar or plate thickness for an ideal quench (H = ∞). (*Lamont.*)

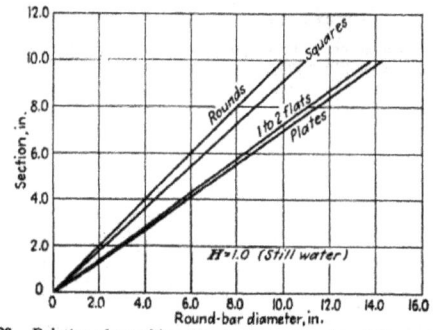

FIG. 4.29. Relation of round-bar diameter to bar or plate thickness for quenching in still water (H = 1.0). (*Lamont.*)

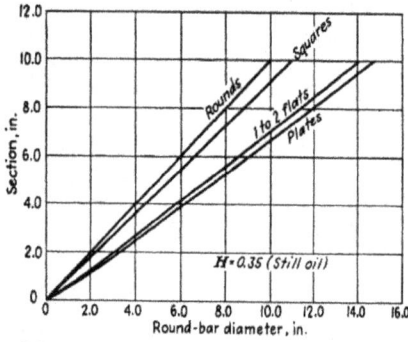

FIG 4.30. Relation of round-bar diameter to bar or plate thickness for quenching in still oil (H = 0.35). (*Lamont.*)

APPENDIX 7

Grossman method to determine the cooling power of a quench

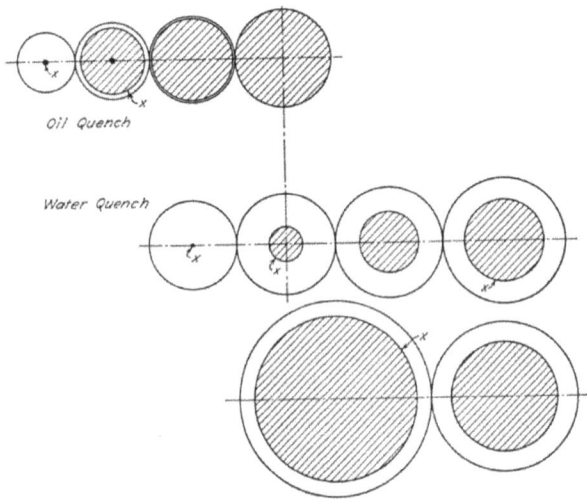

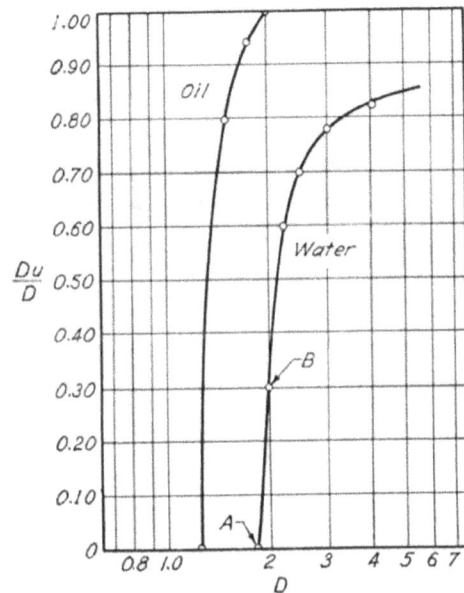

Cooling Power of Quenching Baths

Curves whereby severity of quench (heat transfer equivalent H) can be estimated. Copyright, 1939, Carnegie-Illinois Steel Corp.

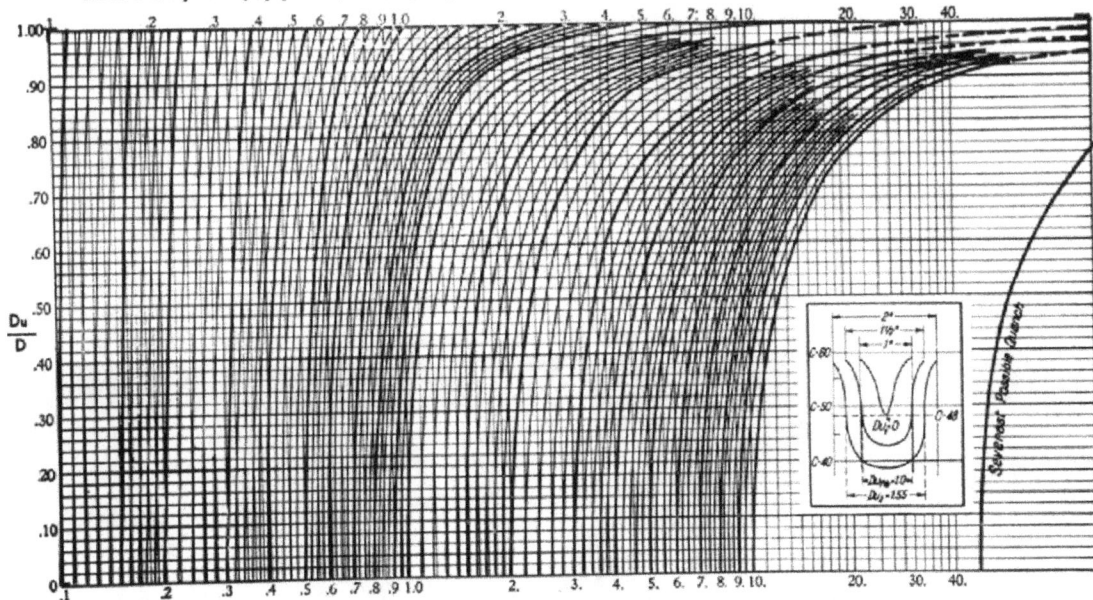

Figures *Top LHS = SAE 3140 steel quenched in oil and in water. Top RHS = unhardened cores of bars quenched in oil and water Lower=Chart for estimating the H value of a quench from EARL J. ECKEL et al, AN EVALUATION OF THE HARDENINGPOWER OF QUENCHING MEDIA FOR STEEL, UNIVERSITY OF ILLINOIS BULLETIN Vol. 48 June, 1951 No.73 https://core.ac.uk/download/pdf/4814276.pdf*

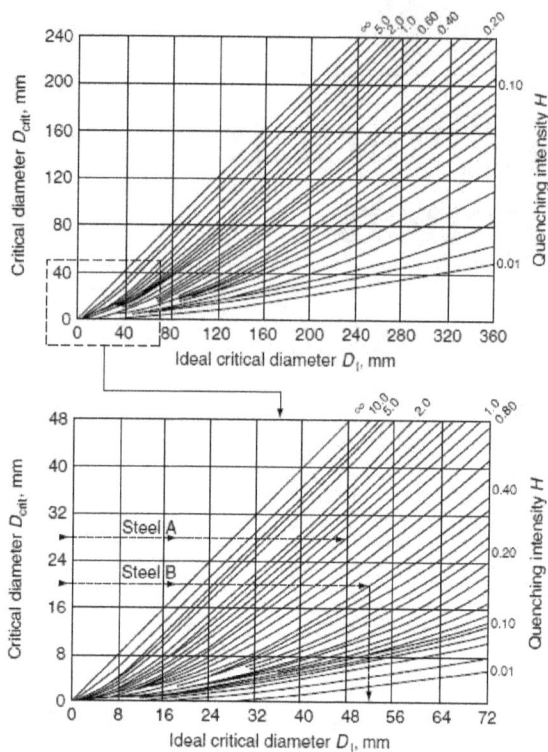

Relationship between DI, H and Dc

APPENDIX 8

Corrosion Definitions

Anion Negatively charged ions in the electrolyte. Anions are attracted to and move toward the anode under influence of a potential gradient. Some may react at the anode.

Anode The electrode of an electrochemical cell at which oxidation occurs. The metal that corrodes in a galvanic cell or the positive terminal of an electrolytic cell.

Cathode The electrode of an electrochemical cell at which reduction is the principle reaction. The metal which is protected by an anode of a galvanic cell; the negative terminal of an electrolytic cell.

Cathodic protection A technique to reduce the corrosion rate of a metal surface by making that surface the cathode of an electrochemical cell.

Cation Positivity charged ion in an electrolyte. Cations are attracted to and move toward the cathode under influence of a potential gradient. Some may react at the cathode.

Coating A liquid, liquefiable, or mastic composition that, after application to a surface, is converted into a solid protective, decorative, or functional adherent film.

Concentration cell A corrosion cell whose voltage gradient is the result of inhomogeneities or differential chemical conditions at sites on the structure within the electrolyte.

Corrosion The deterioration of a material, usually a metal, that results from a chemical or electrochemical reaction with its environment.

Corrosion cell Consists of an anode and a cathode which are electrically connected for electron flow and immersed for ion flow. Dry and wet cell batteries are common examples (when shorted across the terminals).

Corrosion monitoring system Consists of bonded or welded joints for structure electrical continuity, insulating fittings at required locations where electrical isolation of a structure is desired, and tests stations for electronic access to a structure to determine potentials

Electrical resistivity the resistance offered to the passage of current between the opposite faces of a unit cube of the material. Units are resistance times distance, such as ohm centimetres, ohm-meters, ohm-feet, or the like.

Electrolyte The medium (such as an aqueous solution, moist soil, or solution of chemicals) through which the current (positive charge) of a corrosion cell flows (i.e., from the anode to the cathode by migration of anions and cations).

Electron flow Flow of electrons in the external circuit; in the opposite direction to "conventional" current flow.

External circuit The part of a corrosion cell circuit in which electrons flow through the metal of the anode, cathode, and metallic conductor between them (the metallic part of the circuit).

Galvanic cell A corrosion cell in which there is an anode made of a different material than the cathode.

Galvanic series A list of conductive materials, especially metals and alloys, arranged according to their corrosion potentials in a given environment.

Galvanic-type corrosion Corrosion similar to that produced by a galvanic cell.

Internal circuit The part of a corrosion cell circuit in which the current flows through the electrolyte via ions or radicals (the solution part of the circuit).

Ion An electrically charged atom (e.g., Na+, Cl-, etc.); sometimes used when speaking of radicals as well.

Local cell corrosion An electrochemical cell created on a metallic surface because of a difference in potential between adjacent areas on that surface.

Long-line corrosion Current flowing through the earth between an anodic and a cathodic area that returns along an underground metallic structure. (Usually used only where the areas are separated by considerable distance and where the current flow results from concentration cell action).

Mill scale The oxide layer formed during hot fabrication or heat treatment of metals.

Mixed potential A potential resulting from two or more electrochemical reactions occurring simultaneously on one metal surface.

Noble metal A metal with a standard electrode potential more positive than that of hydrogen. Of two metals in a corrosion cell, the one with a potential more in the noble direction will be the cathode.

Polarization The change from the corrosion potential as a result of current flow across the electrode/electrolyte interface.

Potential In cathodic protection work, the voltage difference between a structure and a reference electrode, all in a continuous electrolyte.

Radical An electrically charged group of atoms (e.g., OH-, SO4^{-2}, etc.); sometimes loosely referred to as ions.

Reference electrode An electrode having a stable and reproducible potential, which is used in the measurement of other potentials.

Shunt resistor A calibrated resistor placed within a circuit to determine the current flow; calibration is typically expressed in ohms or amperage/millivolt.

Static potential The potential of a metal before any polarization and with no current flowing through the electrolyte where the potential is measured; sometimes also called native potential.

Stray current corrosion Corrosion resulting from current flowing through paths other than the intended circuit. Corrosion results when this current enters the electrolyte (e.g., ground return to a foreign cathodic protection system, streetcar line, railway system, etc.).

Structure In cathodic protection work, an item that could be monitored and/or cathodically protected (e.g., buried pipeline, submerged pump column, etc.) or an item foreign to such an article (i.e. a foreign structure).

Test station A location with electronic connection to a structure for cathodic protection testing.

Tuberculation The formation of localized corrosion products scattered over the surface in the form of knob-like mounds called tubercles.

APPENDIX 9

Effect of Chemical Elements in Steel

Carbon – is generally considered to be the most important alloying element in steel and can be present up to 2%. The higher levels being found in tool steels, spring steels and high carbon rod. Where weldability is required carbon, levels are kept low since increased amounts of carbon will allow the formation of hard HAZ microstructure and introduce a risk of cracking and therefore reduce the weldability of the steel. For the best weldability carbon levels should be below 0.1%. The use of the established CEV formula can be used to quantify the weldability.

$$CE = C + (Mn+Si)/6 + (Cr+Mo+V)/5 + (Ni+Cu)/15$$

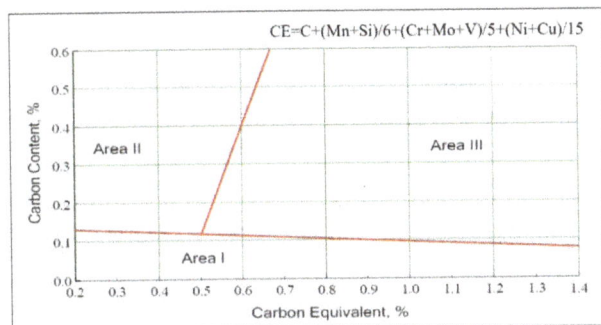

Graville diagram showing regions of weld difficulty Area I= Easy to weld, Area II=Weldable, Area III= Difficult to weld

Engineering steels were typically based on medium carbon levels of 0.3 to 0.45%. Increased amounts of carbon increase hardness and tensile strength, as well as response to heat treatment (hardenability).

Sulphur – is usually an undesirable impurity in steel and originates from the coal used to make the coke used in the blast furnace used to convert the ore to pig iron. There are two main non-metallic inclusions in steel, one being the sulphide inclusions – normally manganese sulphide – but when the liquid steel is treated with calcium then calcium sulphide inclusions are present. Typical commercial steel levels of sulphur are 0.02 to 0.04% which can assist with machineability.

However, the awareness that clean steel free from non-metallic inclusions gives improved toughness and ductility, many steels have very low sulphur levels down to less than 0.001%. Since the non-metallic inclusions play a major role in HIC steels resistant to HIC and SOHIC are made with very low sulphur levels. In amounts exceeding 0.05% it tends to cause brittleness and reduce weldability. Alloying additions of sulphur around 0.10% to 0.30% improves the machinability of a steel. Such types may be referred to as "resulphurized" or "free-machining". Free-machining alloys are not intended for use where welding is required or high levels of stress are applied in service.

Phosphorus – is generally considered to be an undesirable impurity in steels. It is normally found in amounts up to 0.04% in most carbon steels. In hardened steels, it may tend to cause embrittlement. In low-alloy high-strength steels, phosphorus is often kept below 0.010% to avoid temper embrittlement.

Silicon – Usually only small amounts (0.20%) are present in rolled steel when it is used as a deoxidizer. However, in steel castings, 0.35 to 1.00% is commonly present. Silicon dissolves in iron and tends to strengthen it. Weld metal usually contains approximately 0.50% silicon as a deoxidizer. Some filler metals may contain up to 1% to provide enhanced cleaning and deoxidation for welding on contaminated surfaces. When these filler metals are used for welding on clean surfaces, the resulting weld metal strength will be markedly increased. High silicon levels in alloy steels assist with the prevention of large carbides during the hardening treatment. Levels up to 1.5% can be used.

Manganese – Steels usually contain at least 0.30% manganese because it assists in the deoxidation of the steel, prevents the formation of iron sulphide and inclusions, and promotes greater strength by increasing the hardenability of the steel. Amounts of up to 1.5% can be found in carbon steels.

Chromium – is a powerful alloying element in steel. It strongly increases the hardenability of steel, and markedly improves the corrosion resistance of alloys in oxidizing media. Its presence in some steels could cause excessive hardness and cracking in and adjacent to welds. Stainless steels may contain in excess of 12% chromium.

Molybdenum – This element is a strong carbide former and is usually present in alloy steels in amounts less than 1%. It increases hardenability and elevated temperature strength. In austenitic stainless steels it improves pitting corrosion resistance.

Nickel – is added to steels to increase hardenability. It often improves the toughness and ductility of the steel, even with the increased strength and hardness it brings. It is frequently used to improve toughness at low temperature.

Aluminium – is added to steel in small amounts as a deoxidizer. It also is a grain refiner and also reduces the "free" nitrogen in solution and ensures good toughness. The typical level of aluminium would be 0.02 to 0.04%, which would give a steel made to a "fine grain practice".

Vanadium – The addition of vanadium will result in an increase in the hardenability of a steel. It is very effective, so it is added in small amounts typically up to 0.1%. At greater than that there may be a tendency for the steel to become embrittled during thermal stress relief treatments.